Klaus Daniels

Advanced Building Systems

Klaus Daniels

Advanced Building Systems

A Technical Guide for Architects and Engineers

English translation by
Elizabeth Schwaiger

Birkhäuser –
Publishers for Architecture
Basel · Boston · Berlin

Foreword

Building systems play a central role in today's frequently complex building projects and the smooth running of these systems ensures unlimited use of the building. Depending on the degree and standard of finishing, building systems can account for between 25 and 50 % of the entire construction costs. They are therefore an important factor in the overall project development. Most building systems fulfil service functions, making them integral components in the construction work. However, they should be developed with a view to the requirements of both current and future users, with particular focus on conceiving key parameters such as facade, structure etc. in a manner to keep the need for building systems to a minimum.

Human beings, as users of buildings, and certain work processes are important factors in conjunction with building systems and for this reason an entire chapter has been devoted to this topic. "Advanced Building Systems" is not only an instructive manual for students of architecture, structural engineering and mechanical engineering specializing in building technology, it is also a valuable reference work for all practitioners in the field, complete with sizing and scale guidelines that are helpful for preliminary sizing calculations and planning. An abundance of detailed illustrations on both construction and design has been included to facilitate the work of architects and engineers.

The book provides an overview of all essential building systems relevant to current standards, and care has been taken to include even the latest technologies, which are still in their infancy. Particular emphasis has been placed on the operation of building systems with environmental resources (sun, wind, soil, rain, etc.). Standardized regulations and industry standards have not been included, since these are subject to frequent changes in different countries and require constant adjustment.

Several members of HL-Technik AG have made considerable contributions to this work and I would like to extend a special word of thanks to:

Renate Sprenger (Elevators and hoisting systems)
Manfred Woiwode (Sanitation and fire protection)
Hans-Joachim Kast (Refrigerating systems)
Dr. Andreas Colli (senior editor and collaboration in image design)

I am especially indebted to Dr. Ralph Hammann, who provided the impetus for this English edition and accompanied the translation throughout, to Elizabeth Schwaiger for her dedicated translation of this work into English and Riemer Design for design and layout.

Munich/Zurich, June 2003

Klaus Daniels

Contents

1.
Building performance
From linear to integrated planning

At the dawn of the new millennium we are faced with a post-industrial revolution, triggered by new variables in the areas of work, communication, high-tech development and the environment. These variables unleash worldwide competition for work and wages, and radically alter our living habits and requirements. The following factors drive these global trends:
- World population growth
- Economic globalization
- Global information networks
- Increasing gap between rich and poor
- Emergence of new markets
- Environmental pollution
- Resource exploitation

Economic globalization, knowledge transfer through worldwide networks, environmental pollution and resource exploitation are the principal themes related to building development. In the future, local or national markets in industrial countries will provide insufficient opportunities and tasks for full employment. It is all the more important that architects and engineers, in particular look beyond national boundaries and moreover investigate how they might increase the range of services and the performance palette they offer to define their task in the international or global marketplace.

It is widely known that our sector – German architects, engineers, and building system engineers – has for some time been playing "second fiddle" on the international scene. British and American colleagues have been showing us how to be a "global player" on the global market.

Economists predict that only 20 % of the total workforce will enjoy the privilege of stimulating and lucrative employment, while 80 % of all employees, especially in the production and service sectors, will lose out.

As well as the financial backers, the winners will ultimately be highly qualified concept developers, coordinators, systems analysts and marketing experts. The Internet has made knowledge accessible around the globe at all times, thus becoming the determining factor in production. Knowledge-based sectors will set the tone and no longer be tied to a specific location. The growth sectors of the future are independent of mineral resources.

Environmental pollution and resource exploitation are vital themes in many countries – especially the threshold countries – for these nations are particularly dependent on foreign energy and resources and are just as much in need of a healthy environment as we are.

The development of buildings and building systems has been characterized for many years by linear thinking and planning. In the past, single products and systems were targeted at specific building components and although this approach has been proven erroneous for future sustainability it is still practised to some degree.

It is high time, therefore, to remember what we do know, or rather what we must know in the future, so that we may find solutions for the problems that face us, some of which are clearly of our own doing:
- Environmental protection
- Sustainable resource management
- Re-usable instead of single-use buildings
- Compactness instead of sprawl
- Utilization of natural resources
- Flexible real estate use

These are the themes of the future to which we must pay heed and which may assist us in establishing a foothold in the global market.

Integrated planning – the new buzzword – is a challenge and an opportunity for architects and engineers. But integrated planning goes beyond the long-established and at least partially fullfilled mandate of integrating building systems into architectural concepts. Here integrated planning means full and unified collaboration between different disciplines in the pursuit of truly optimum total building concepts.

Integrated planning concepts favour comprehensive performance capabilities over individual aspects, or aim for the appropriate use of such individual concepts within an overall context.

According to the building, the best solution may be a low-energy-house; or, it may be just as correct to equip a building with sophisticated information and communication tools to prepare it for a specific use. This by no means makes either of these building types "intelligent"; they still place too much emphasis on individual demands, giving them priority and perhaps even basing the entire architectural concept on specific requirements.

"Building with mass" may be as appropriate as "building without weight" – the answer lies in a clear definition of the task and not necessarily in the options offered by the products available on the market.

In other words, integrated planning translates into a real opportunity to save material costs and especially operating costs. Planning buildings systems, for example, is therefore not limited to the building itself; it must begin by taking the exterior space into account, or rather preparing the immediate surroundings of a building with a view to optimizing the microclimate and diminishing any negative factors that might influence the structure. In terms of planning the building proper, individual components such as the structure, the building skin, and the sequence of internal spaces can not only save resource and energy consumption but may also satisfy a multitude of demands.

In moderate climates, good planning means that a building can be naturally ventilated for most of the year, eliminating the need for mechanical ventilation or cooling. When necessary, subsidiary cooling systems are employed during peak temperatures in summer for temporary cooling; these should be water-based systems to make use of natural resources and eliminate the need for constructional adaptation (engineering rooms, shafts, ceiling cavities), as well as reducing energy consumption.

Generally speaking, building systems fulfil service functions aimed at improving the building performance. In the broadest sense the term implies an increased capacity for performance, since building systems are conceived, among other things, to deliver a high degree of thermal, hygienic and acoustic comfort, to enhance visual comfort and to prevent an occurrence of the Sick Building Syndrome (SBS). Factors such as electromagnetic compatibility, ecological building materials, surface structures and paint finishes also come into play.

Building in harmony with what the environment has to offer – that is, incorporating planted areas, building with wind, sun and water – is a fascinating challenge that may lead to innovative architectural solutions. We are all familiar with relevant examples, although some projects still fall short of the intended ideal, while others succeed in implementing the integrated approach with full rigour. Ideally, integrating planning concepts should improve the life cycle of a building – a goal that can also be achieved by creating real estate that is increasingly flexible, i.e. buildings that can be reassigned to new uses and easily converted.

Step 1

Integrated planning tends to be a step-by-step process. The first step – that of determining the demands made on the building (programming) – is essential if one wishes to create a building distinguished by the best possible cost-benefit ratio.

Step 2

The second step in integrated planning begins with finding the most cost-efficient and convenient approach to integrating systems into a building, but at the point of developing the urban concept in the first place.

The reasons are evident in the ecological circle that illustrates the integration necessary for cooling a building. This task alone is influenced by many variables, with possible solutions both in the exterior space and in the building fabric itself. The building system (here with a focus on a gas-motor-driven cooling system) comes into play only during the final stage of the planning process **(Figure 1.1)**.

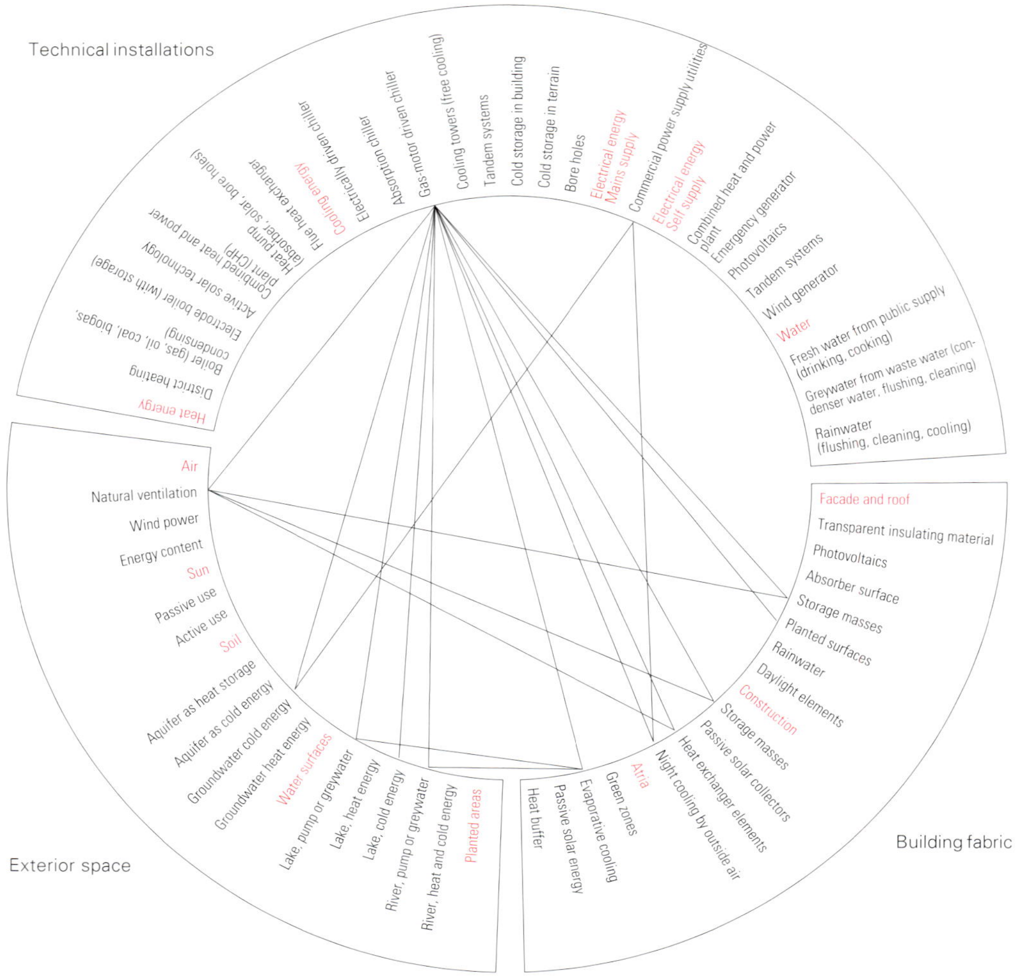

Figure 1.1
"Ecological Circle" of buildings and building technology (interactions relevant to cooling)

The ecological circle illustrates the interactions between environment, building and building systems – the factors that influence building design. The connections marked in this figure are those pertinent to the aspect of cooling.

Exterior space

Air
- Free air
  - Natural ventilation
  - Wind force
  - Energy content
  - Thermic tower

Sun
- Solar energy, diffuse radiation
- Solar energy, active radiation
- Passive use
- Active use

Soil
- Aquifer
  - Heat storage
  - Cold storage
- Groundwater
  - Cold energy
  - Heat energy

Water surfaces
- Lake
  - Pump water or greywater
  - Heat energy
  - Cold energy
- River
  - Pump water or greywater
  - Heat energy
  - Cold energy

Planted areas
- Microclimate
- Shading, foliage
- Dust absorption
- View
- Oxygen supply

Technical installations

Heat energy
- Direct
  - District heating
  - Boiler (gas, oil, coal, biogas, condensing)
  - Electrode boiler (with storage)
- Indirect
  - Active solar technology
  - Combined heat and power plant (CHP)
  - Heat pump (absorber, solar, bore holes)
  - Flue heat exchanger

Cooling energy
- Direct
  - Electrically driven chiller
  - Absorption chiller
  - Gas-motor driven chiller
  - Cooling towers (free cooling)
  - Tandem systems
- Indirect
  - Cold storage in building
  - Cold storage in terrain
  - Bore holes

Electrical energy
- Mains supply
  - Commercial power supply utilities
- Self supply
  - Combined heat and power plant
  - Emergency generator
  - Photovoltaics
  - Tandem system
  - Wind generator

Water
- Fresh water
  - Public supply (drinking, cooking)
- Greywater
  - Waste water (condenser water, flushing, cleaning)
- Rainwater
  - Flushing, cleaning, cooling

Building fabric

Facade and roof
- Transparent insulating material
- Photovoltaics
- Absorber surface
- Storage masses
- Planted surfaces
- Rainwater
- Daylight elements
- Collectors

Construction
- Storage masses
- Passive solar absorber
- Heat exchanger elements
- Night cooling by outside air

Atria
- Green zones
- Evaporative cooling
- Passive solar energy
- Heat buffer

Fresh air

Fresh air, wind flow and the inherent energy potential they contain are essential parameters for future integrated planning concepts and should be used accordingly. In principle, all buildings should be designed for natural ventilation unless extraordinary hygiene requirements indicate otherwise. Such concepts, which have become more frequent in recent years, are not only accepted by users, but are very popular. Studies of relevant projects have shown that given the choice between natural and mechanical ventilation, users will only resort to mechanical ventilation during one third of the total operation period (above all in winter), preferring to work with open windows during the rest of the time.

Even when a building is located in an environment with less than ideal conditions and is subjected to high emission rates (sound, air pollution, dust, strong winds etc.) natural ventilation is still the most desirable solution. For isolated cases, options must be developed to protect fresh air from emissions and transport it to the usable areas inside the building. There are a number of options for natural ventilation, depending upon the shape and height of the building. The primary sources are negative and positive pressures on the building skin, often supported by thermal lift or buoyancy, which help to promote natural ventilation **(Figure 1.2)**.

The soil

In addition to free cooling, the use of thermal energy (for cooling or heating) is a suitable alternative to active cooling or heating systems. Deep and shallow geothermal installations, aquifer reservoirs, seasonal soil reservoirs etc. are used to extract heat from the building in summer and store it in the soil, and transport heat from the soil to the building in winter. In future, the use of thermal energy will become an essential component in integrated planning, with building engineers and building systems engineers working hand in hand, because large foundations and pile foundations are particularly suited to these types of systems **(Figure 1.3)**.

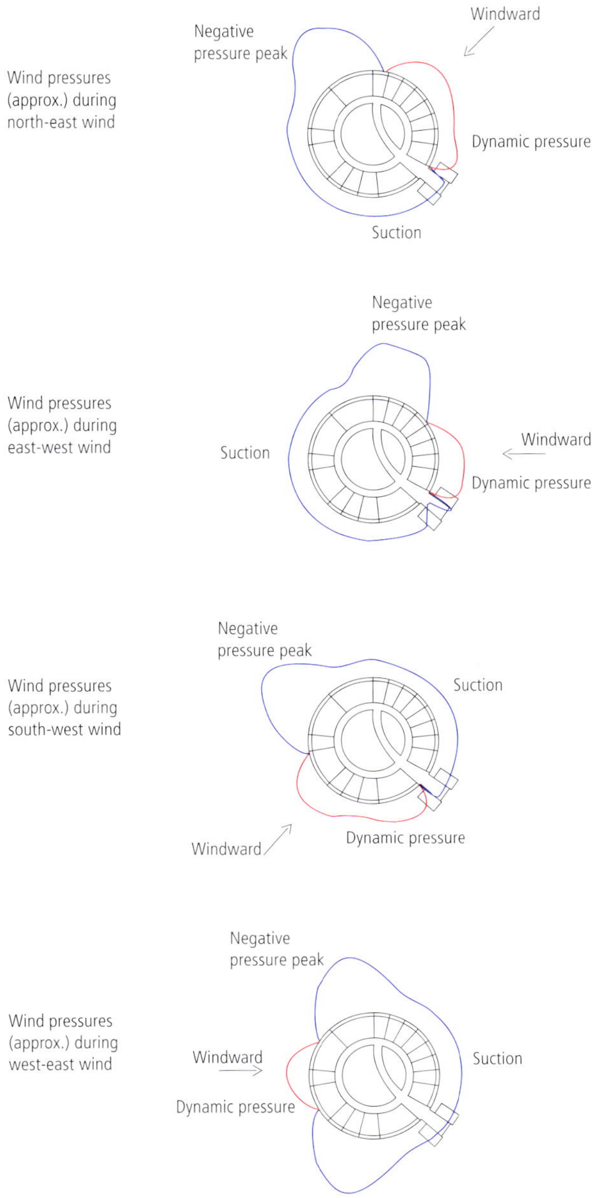

Figure 1.2
Pressure distribution in the event of the wind blowing from several directions

Figure 1.3
Schematic representation of geothermal energy use

Rain and surface water

Fresh water is one of our most precious resources; there is no adequate substitute. Water suppliers process it at considerable expense and market it to consumers under stringent hygiene quality controls. Few people realize how much of this valuable resource is literally being flushed down the toilet at home and at work (approx. 33 %). Cleaning, car washing and plant watering waste large amounts of potable water.

Relevant studies show that nearly 55 % of valuable potable water could be replaced with rainwater – an enormous savings potential. Utility water, i.e. processed rainwater, can be used at the very least for sanitation (flushing), watering gardens and lawns, cleaning and laundry. In future, potable water should be used primarily for food preparation, dishwashing and personal hygiene.

Rainwater use should always be restricted to water collected on rooftops to ensure that heavy pollutants are prevented from entering into the water cycle. Prior to use, rainwater is processed and disinfected, exposed to UV radiation, refiltered and dosed.

But rainwater is not only useful for cleaning and flushing; it can also cool buildings and the areas near buildings. Some of its uses are:

- Evaporative cooling near buildings using the water surface of an artificial lake
- Evaporative cooling near buildings using fountains set into lakes, cascading waterfalls, etc.
- Direct cooling of the ambient atmosphere through evaporation from finely sprayed mist
- Direct building component cooling (on glass surfaces)
- Indirect building component cooling as a result of watering, partial evaporative cooling etc. **(Figure 1.4)**.

Planted surfaces

Suitably designed planted surfaces provide natural shading for external walls exposed to the summer sun; this prevents the surfaces from overheating and results in a massive reduction of the cooling loads. Planted areas should be designed to shed their foliage in autumn to enable passive solar energy use in winter. Evergreen areas are therefore only partly desirable, since they offer limited benefits (collecting dust, oxygen production).

Most users of a building experience trees, shrubs and groundcover as attractive and positive features. These plants also improve the microclimate near the building, which clearly has a positive effect on building performance **(Figure 1.5)**.

Figure 1.4
Water-cooled glass roof surfaces of a trade fair hall

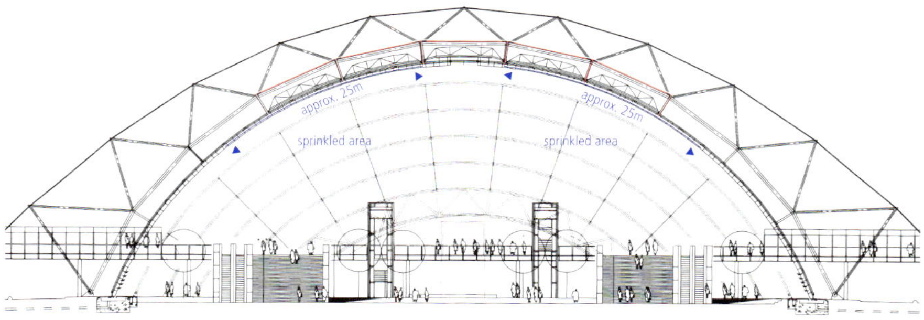

Figure 1.5
Building in green belt

When developing urban concepts, essential ecological aspects should be taken into consideration, chiefly with the aim of promoting natural ventilation and minimizing overheating in the urban area. As density in urban areas becomes increasingly desirable to protect the surrounding countryside, it is essential to chart the natural, diurnal and nocturnal air currents and to design building concepts that do not impede natural air circulation. In addition to fresh air currents flowing into an urban area, nocturnal cooling of building masses also plays an important role. Individual scenarios may call for planted barriers as a protective screen against unwanted currents, for example in winter. All these parameters should be considered and integrated into urban planning concepts **(Figure 1.6)**.

Step 3
Building form and orientation

The heat loss and heat gain in buildings changes automatically according to orientation (north-south versus east-west) and according to their surface-to-volume ratio.

These topics are not only interesting, they are essential if the goal is to develop an integrated planning concept that will achieve consistent savings in terms of the initial investment and operating costs of building systems. This includes the aspect of facade and window design with regard to orientation. While the specific annual heat requirement changes only insignificantly with larger window sizes in north-, east- and west-oriented facades, it is reduced by as much as 30 % in south-facing facades when window surfaces are increased. It is important to determine, for each individual case, whether existing heating energy consumption codes provide truly effective guidelines to reduce heating energy consumption, or whether they are too rigid with regard to building use. This may result in concepts that deliver more heating energy than is necessary through excessive insulation, which in turn must be compensated for through cooling energy. Current heating energy consumption codes are a typical example of linear instead of integrated thinking, since the heating consumption factor for residential structures is applied to all other structures – even those with high internal heat gains – "painting them all with the same brush."

Natural ventilation

In addition to all other positive aspects of minimizing heating energy consumption through the form and orientation of buildings, special attention should be given to their suitability for natural ventilation.

In principle, all buildings should be conceived to produce sufficient differences in the pressure coefficients and, in the case of operable windows, sufficient cross ventilation as a result of the pressure differences.

The collaboration between architect and aerophysicist as members of an integrated planning team is especially thrilling when it comes to designing building volumes and surfaces for optimum natural ventilation. Of particular interest in this context are very tall and very low buildings – building types that were previously considered unsuitable for natural ventilation. For tall buildings the field has widened from single to double-leaf facades with a variety of facade designs (box window, ribbon window, curtain wall facades etc. with circular pressure distribution channels or shaft- and well solutions). For extremely low buildings such as large sports and fair halls and central vestibules or lobbies, curved or pitched roof structures play an essential role by facilitating natural evacuation and fresh air influx across large areas. Several such examples have been realized in recent years – here architecture truly lives up to the maxim "form follows function." There are many applications for corresponding buildings, especially those located near heavy traffic routes with high noise levels; in this case the architecture can respond to high emission rates with double-leaf facades, if necessary supported by thermal towers **(Figure 1.7)**.

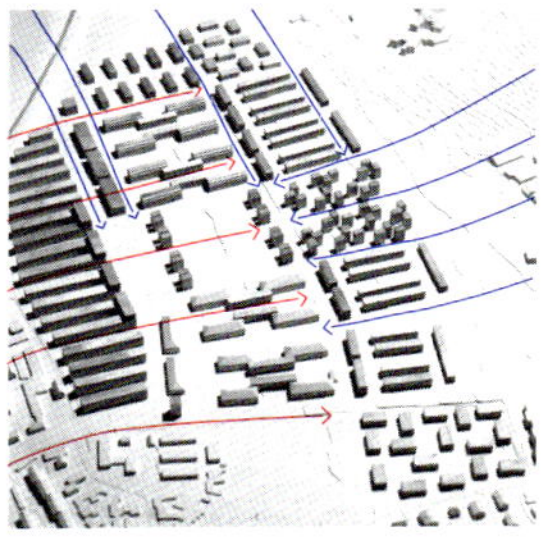

Figure 1.6
Wind flows in urban area (model)

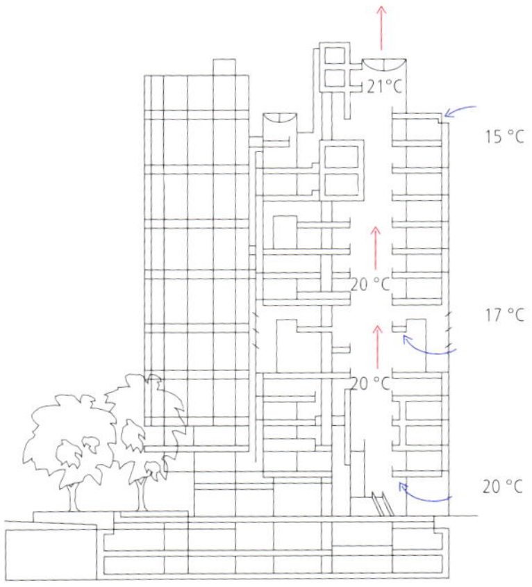

Figure 1.7
Thermic principle of an office building with double-leaf facade and multi-storey hall

Step 4
Buildings and building structure
Building structure, facade design and the use of solar buffer zones (halls, circulation areas, etc.) are another important step on the path towards correct building planning.

Thermal storage in building mass
Thermal storage in the concrete building mass is the logical answer for buildings where room temperatures need to be reduced in summer as naturally as possible. This approach reduces cooling loads by over 30 %; that is, thermal storage reduces room temperatures by more than 5 K in comparison to low thermal mass buildings.

When architects, building climatologists, structural engineers and building technicians work together, the correct use and design of necessary building masses can be an important factor in developing a total thermal storage concept that will effectively reduce investment and operation costs for mechanical or technical systems. At the same time, this approach subjectively and objectively enhances the quality of the space by improving both the actual room temperatures and the perceived temperatures. "Building with mass" can be extraordinarily advantageous and lead to building concepts that require a minimum of technical installations. The theory has been confirmed as correct over the past 25 years in numerous built examples, which have indeed lived up to the promised performance expectations **(Figure 1.8)**.

"Building without weight" or lightweight building may be just as important, especially when local building materials can be used. Lightweight construction and high levels of insulation can achieve equal or similar results, although protection against external heat gain becomes an essential factor for the success of solutions of this kind.

From facade to polyvalent wall
In the past and to some degree even today, facade requirements were defined exclusively by the architect.

Facades should be intelligently designed to satisfy all requirements of future users while ensuring optimum use of the environmental resources available to current occupants, as long as this is compatible with the overall building program. Facades are more than a simple physical barrier between interior and exterior; they must fulfil a multitude of differing functions.

The principal factors are:

- Visual contact between interior and exterior
- Daylight penetration into interior
- Natural ventilation
- Optimal insulation
- Optimal sun protection
- Comfortable surface temperatures
- Adequate glare protection

Facades are an essential component with regard to a building's thermal, hygienic and visual comfort. They should simulate the function of human skin, that is, they should respond to changing interior and exterior conditions.

Facade planning requires a high degree of creativity and innovation from all those involved, since facade solutions that satisfy all economic and ecological demands are rarely achievable by an architect alone. Aerophysicists, building climatologists, building physicists and designers are needed to develop truly optimum solutions in close collaboration.

Highly innovative facades at a cost of approximately 2,050 €/m² are not economical and exceed the allowable limit by far. Such high investment costs more than double the price for each square metre of usable office space and the entire building services that lie behind this type of facade. Projects where the architecture is brilliantly conceived but uneconomic are ultimately unsatisfactory. Such solutions not only risk rejection from the client, they may also compromise the continued success of the relevant planning offices or even entire sectors in the industry **(Figures 1.9 and 1.10)**.

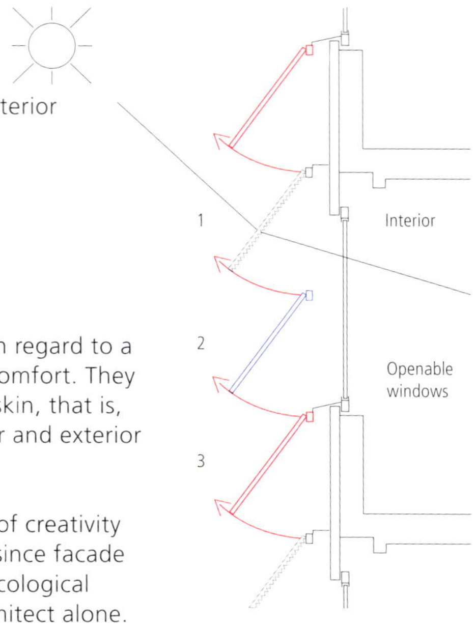

1 Top band:
Light deflection system with prismatic panels, computer-controlled, at right angle to sun position

2 Middle band:
Ventilation panel, manually openable

3 Bottom band:
Balustrade solar collector, temperature-dependent, computer-controlled

Figure 1.10
Principle diagram of a highly flexible facade structure

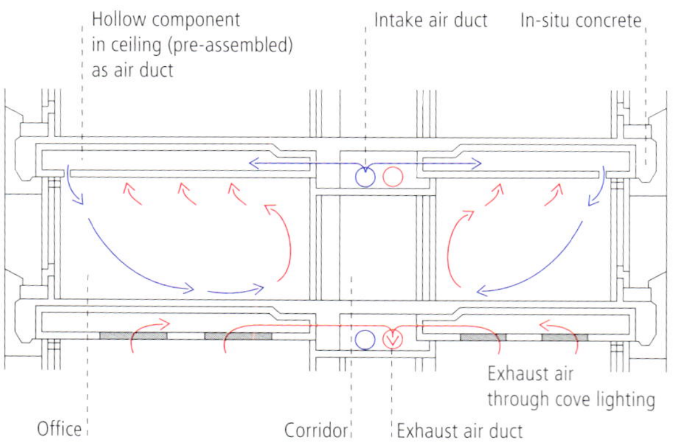

Figure 1.8
Diagram showing building with difficult storage conditions, with night-time cooling by external air currents

Figure 1.9
Overview and flow chart of possible active/passive measures for load compensation

Q Heat quantity
T_R Room temperature
T_{Ou} Air temperature outside

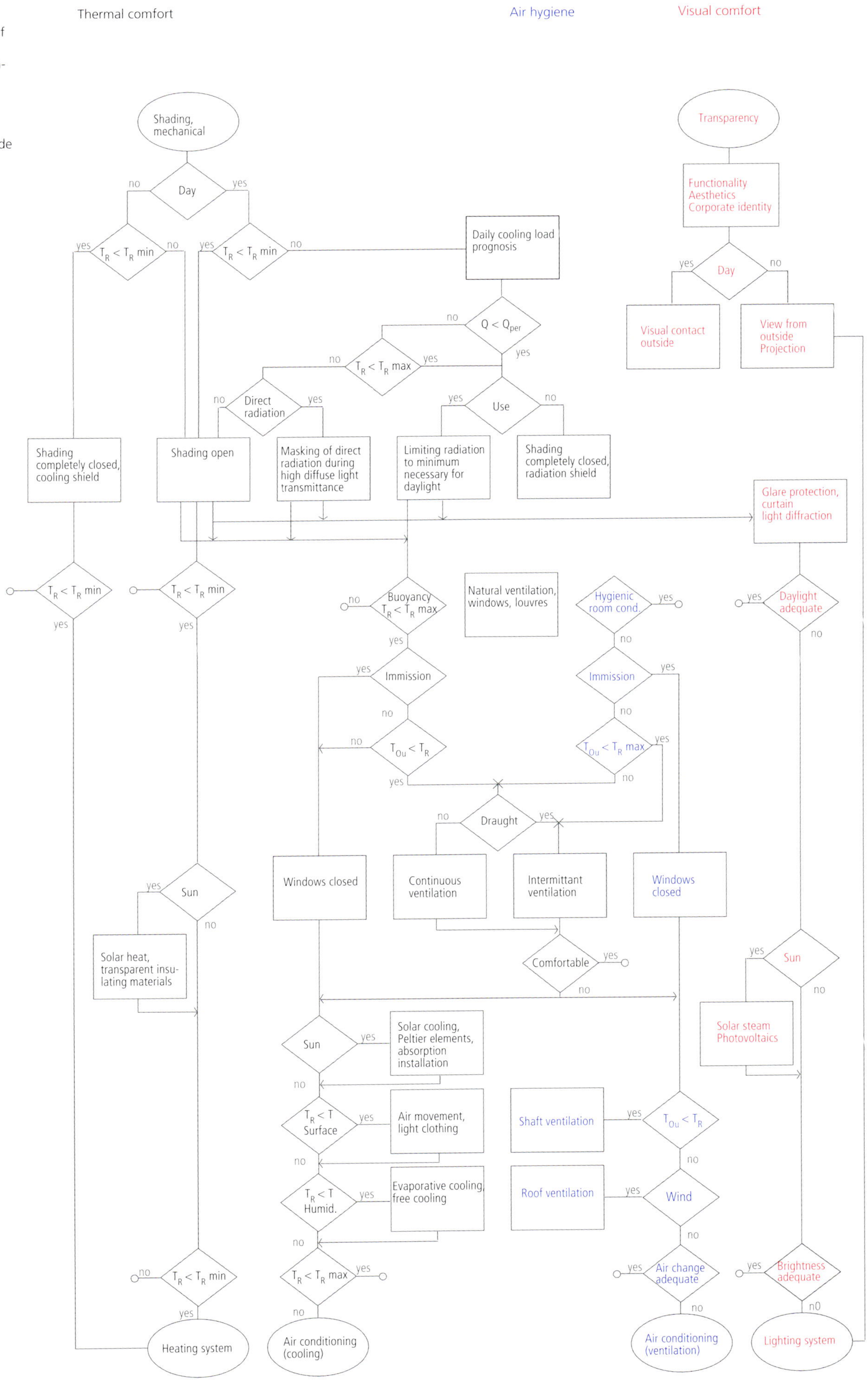

Energy saving, active measures on facade

Passive, zero-energy measures, independent of facade

Active technical measures

Figure 1.11
Example of a hall structure and design of a lightweight resource-saving roof

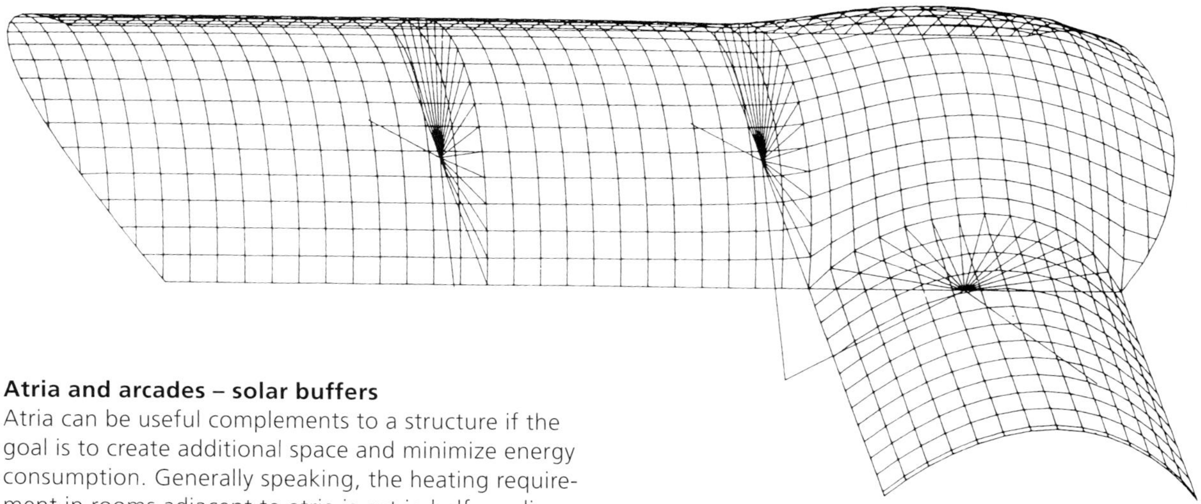

Atria and arcades – solar buffers

Atria can be useful complements to a structure if the goal is to create additional space and minimize energy consumption. Generally speaking, the heating requirement in rooms adjacent to atria is cut in half, cooling loads are reduced and light incidence is diminished. The overall balance sheet of (energy) consumption tends to yield positive results for such glass-covered spaces. Building performance is improved even more when atria are planted or partially planted, and potentially further enhanced with water surfaces to introduce a minimum of evaporative cooling.

Atria should always be designed with the needs of the user(s) in mind, avoiding mechanical ventilation and cooling whenever possible, while working with as light an inner facade as possible to avoid exacerbating the costs for customized roof structures. Net-like roof structures (building without weight) are generally able to meet these demands, although they require a close collaboration between architects and building engineers in order to achieve overall economical solutions (Figure 1.11).

Step 5
Once all resources of the exterior space and the building fabric itself have been integrated into the planning concept, the process of applying building systems as complementary components in the creation of comfortable living spaces can begin. Active building systems should be understood to play a supplementary role in areas where passive building concepts cannot achieve the desired goals, for example, in areas where room temperatures are uncomfortably high, or in case of insufficient extraction of contaminants and odours, unfavourable surface temperatures etc.

Building systems should be integral components of any concept that is open to a variety of innovative approaches for maximum reduction of energy consumption and emissions. Familiar state-of-the-art systems include the so-called combined systems, total energy installations and others, if necessary supplemented by active measures to tap into environmental resources such as the sun, water, rainwater, and the thermal energy potential of the soil.

Combined heat and power systems, heat pump systems, storage techniques, condensing boiler installations and similar systems, as well as building component cooling are standard practice for today's architects.

Integrated planning concepts are absolutely imperative for the future, even if we can make only a small contribution at local level towards solving the predicted global problems. Despite lower energy and operating costs, a more acute sense of the environment and an awareness of resource conservation in some European countries have made Europe a place where the right approaches are being implemented, providing an important stimulus for development elsewhere.

A comparison of per-capita energy consumption in North America and Western Europe reveals that consumption is nearly twice as high in the United States. Several countries have some way to go. This not only true for the United States but also for several European countries as well as nearly all threshold countries, where massive environmental pollution and poor infrastructure are already approaching levels that were unimaginable only a few years ago.

Buildings that represent the right form as a result of integrated planning strategies can serve as models for developing and threshold countries; given the appropriate marketing strategy, they may even become an export item.

Integrated planning methods presuppose that a more abundant catalogue of ideas and innovations translates into the realization of planning concepts that ultimately result in buildings, which can be competitive in a tight market.

Architects and engineers, in particular, are challenged by integrated planning concepts to move away from linear thinking and turn towards integrated thinking. Professionals will need additional training, literature and other tools to upgrade their knowledge and learn how to apply the new methods to their own practice.

There is no doubt that integrated planning processes are more labour-intensive than linear thought processes, especially in the areas of building climate and aerophysics. This greater planning effort is rarely remunerated in today's tight market.

Through the slow combustion of protein, fat and carbonhydrates using inhaled oxygen, humans produce heat ("basal metabolic rate"), which must be dissipated in order to keep the interior body temperature relatively constant.

Heat dissipation occurs in several ways:
- Convection and conductivity
- Heat radiation
- Evaporation
- Exhalation
- Excretion

When the body is comfortable and at rest, minimum heat dissipation is approximately 80 W.
Figure 2.1 illustrates the changes in heat dissipation with increasing activity levels or increasing ambient temperatures. The corresponding skin and surface temperatures are dependent upon the ambient temperature, shown in **Figure 2.2**.

Table 2.1 indicates human heat dissipation during different activities (metabolic rate), while **Table 1.2** provides a record of perceived heat (dry heat) and transpiration (latent heat) in W, i.e. g/h.

As **Table 2.2** demonstrates, perceivable heat diminishes slightly with rising air temperatures, while latent heat (transpiration) remains constant at first. When the room temperature rises above 20 °C the perceived heat begins to fall more noticeably; this is linked to the fact that the human body perspires as room temperatures rise in order to dissipate the heat energy from the body via evaporation from the surface of the skin. The heat dissipation of perceivable heat (dry heat) is effected to approximately 50 % by convection and approximately 50 % by radiation.

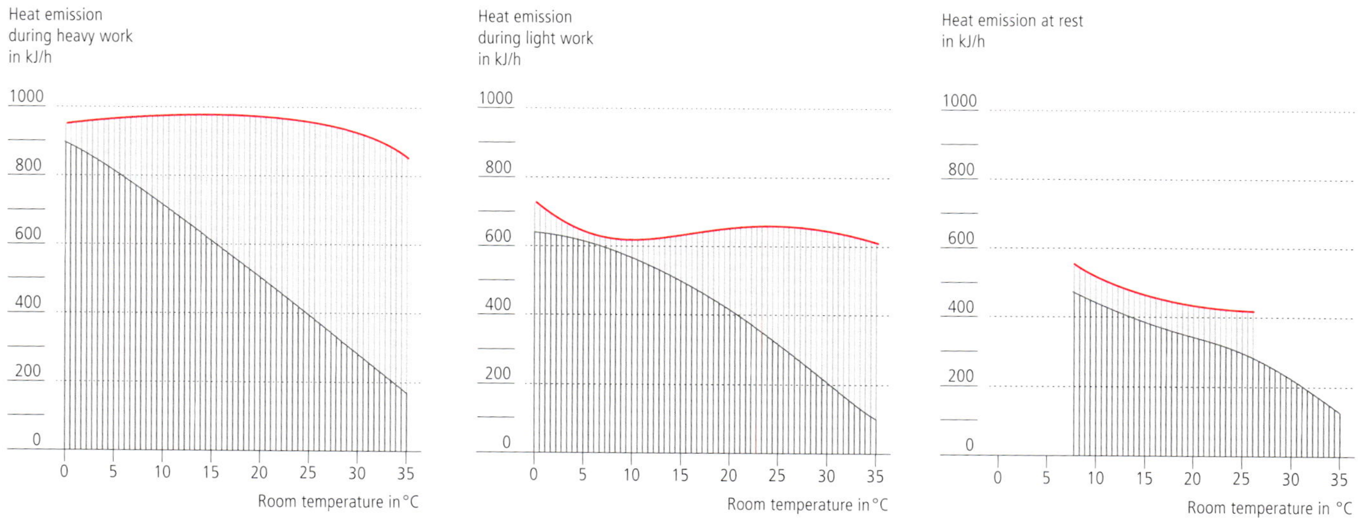

Figure 2.1
Heat emission (wet and dry) of a human being as a function of the room air temperature when performing different activities;
1 kJ/h = 0.278 W.

Wet heat emission $\dot{Q}_f$
Dry heat emission $\dot{Q}_{tr}$
Total heat emission $\dot{Q}_{ges.}$

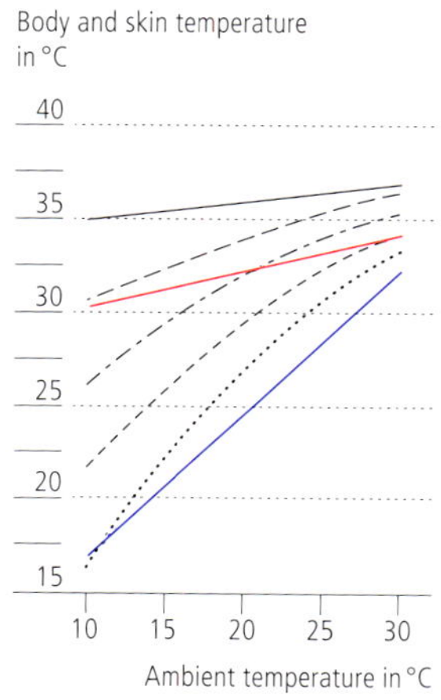

Figure 2.2
Skin and surface temperature of a human being at rest, as a function of the ambient temperature

Mean skin temperature clothed
Mean skin temperature unclothed

Rectum
Head
Trunk
Hand
Foot

Comfortable rooms generally lead to optimum performance conditions for the occupants. Ergonomists have undertaken many studies on this topic, producing performance curves according to effective temperatures shown in **Figure 2.3**. This simplified representation is based on a mean air velocity of 0.15 m/s and a mean relative humidity of 50 %. As **Figure 2.3** shows, optimum performance occurs within a narrow margin of temperatures. This can be achieved with relevant constructional means and types of room ventilation – and **Figure 2.4** indicates different options with maximum room temperatures in summer. As additional information, **Figure 2.5** documents experimental results with regard to room temperatures and performance (based on D. P. Wyon).

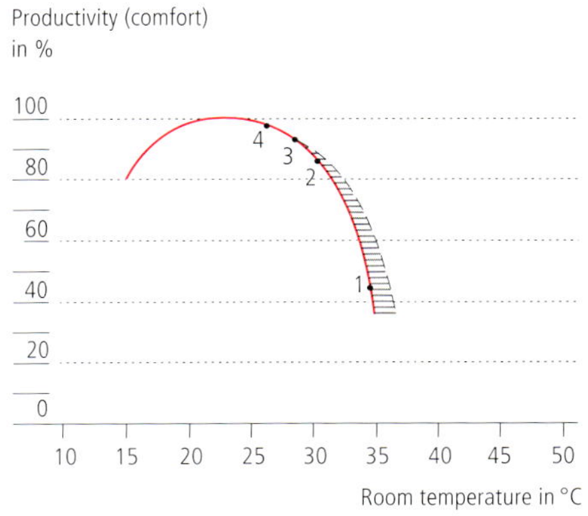

Figure 2.3
Human performance as a measure of "climate comfort" for different ventilation systems and temperature conditions

Activity	**Degree of activity** DIN 1946 T.2	**Metabolic rate = heat emission** W/m²	met	≈ W
At rest		46	0.8	80
Sitting, relaxed	I	58	1.0	100
Standing, relaxed		70	1.2	125
Sitting, light activity (office, home, school, laboratory)		70	1.2	125
Standing, light activity (drawing-board activity) (shopping, laboratory, light industry)	II	81 93	1.4 1.6	145 170
Moderate physical activity (housework, operating machinery)	III	116	2.0	200
Heavy physical activity (heavy machinery operation)	IV	165	2.8	300

Table 2.1
Total heat emission of human being when performing different activities (as per ISO 7730, 1 met = 58 W/m²)

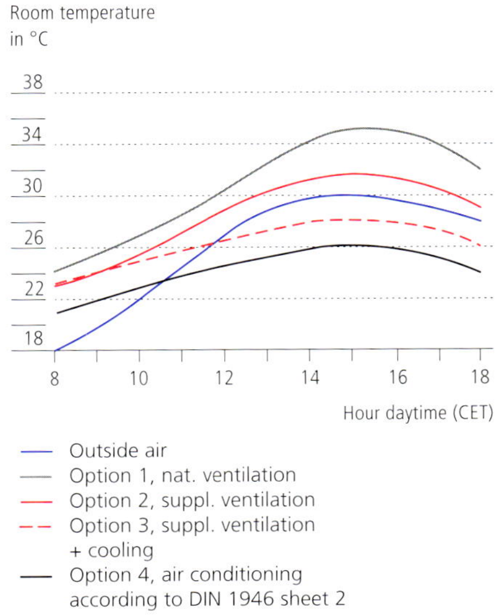

Figure 2.4
Maximum room temperatures with different ventilation and cooling options in general office areas

Air temperature °C	**Perceptible heat** W	**Latent heat** W	**Total heat emission** W	**Water vapour emission** g/h
10	136	21	157	30
12	126	21	147	30
14	115	21	136	30
16	106	21	127	30
18	98	23	121	33
20	92	27	119	38
22	85	33	118	47
24	77	41	118	58
26	69	49	118	70
28	58	59	117	85
30	47	69	116	98
32	33	81	114	116

Table 2.2
Heat emission and water vapour emission of human being (normal clothing, seated man with light activity and still air; air humidity 30 to 70 %).

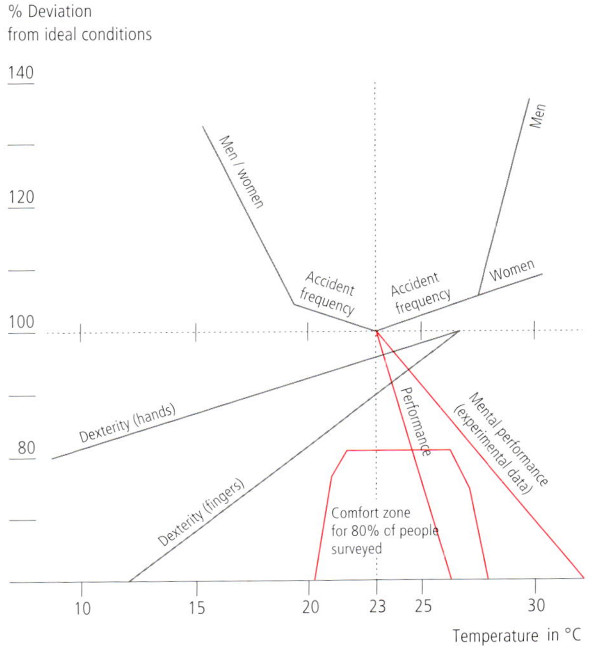

Figure 2.5
Experimental results of studies relating to room temperatures and comfort (D. P. Wyon)

2.1 Thermal comfort

Thermal comfort is defined by the comfort zones within which people are most at ease. In these zones, thermal balance in the body is maintained despite differing physiological influences from the environment.

Many factors influence comfort and it is therefore impossible to establish definitive comfort boundaries. Rather, comfort is the product of the interaction between various influencing factors.
These include:
- Gender
- Health
- Nutrition
- Age
- Season
- Type of work

There are also several factors determined in part by the construction itself, as well as by ventilation and room temperature.
These are:
- Air temperature
- Ambient temperatures
- Humidity
- Air velocity
- Clothing

Air temperatures in a room should ideally range between 20 °C and 24 °C, where the factor of clothing appropriate to outside temperatures has already been taken into account. However, room temperatures of 24 °C in summer require a high investment for cooling and the temperature limit is accordingly set somewhat higher, at 27 – 28 °C, accepting that the performance level will diminish slightly. As **Figure 2.6** illustrates, room temperatures rise when outside temperatures reach approximately 22 – 26 °C and higher.

In addition to room temperature, the average temperature of surrounding surfaces (including heating surfaces) plays a fairly important role through the average radiation temperature, since the body warms up by radiation.

The average radiation temperature t_R is the product of:

$$t_R = \frac{\sum_n (A \cdot t)_n}{\sum_n A}$$

A = surface
t = surface temperature
t_R = radiation temperature
n = number of different surfaces

A stands for the individual surfaces (walls, heating elements, windows etc.), while t stands for the corresponding temperatures.

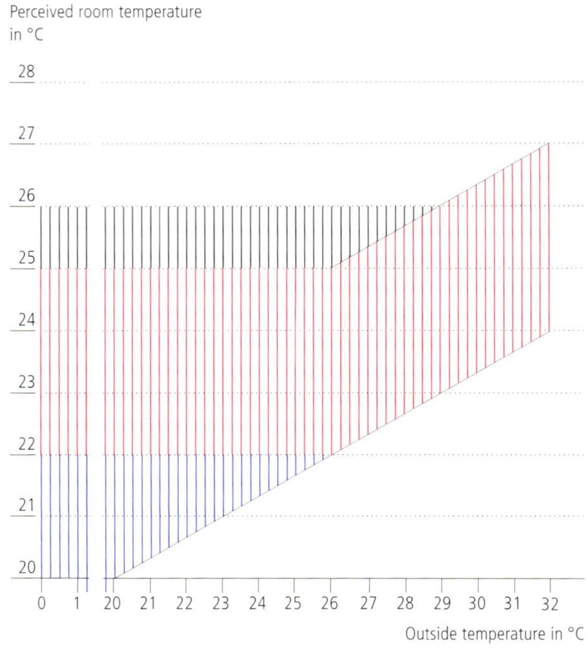

III Acceptable during short-term additional cooling loads
III Recommended range
III Acceptable e.g. for displacement

Figure 2.6
Range of preceived temperatures (DIN 1946, German code for room ventilation, Part 2)
Prerequisites: Activity levels I and II, light to medium clothing

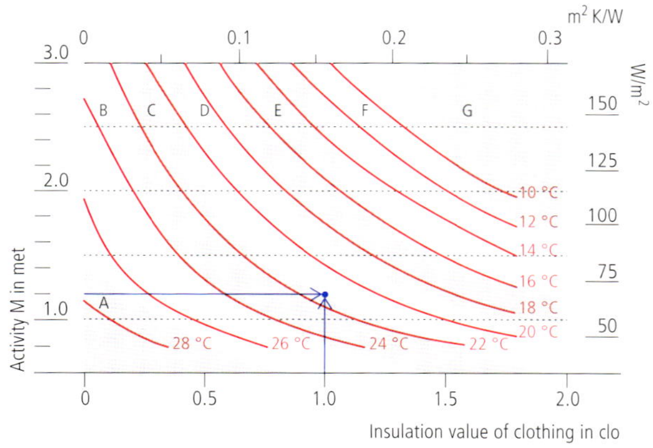

PMV = 0.0, i.e. 5% dissatisfield
PMV = ± 0.5, i.e. 10% dissatisfield
(PMV – predicted mean vote)

A ± 1.0K
B ± 1.5K
C ± 2.0K
D ± 2.5K
E ± 3.0K
F ± 4.0K
G ± 5.0K

Figure 2.7
Optimum operative temperature for PMV = 0 depending on activity and clothing (as per ISO 7730), air humidity 50 %

For example, if the average wall temperature falls by 1 K, this is equivalent to a 1 K air temperature reduction for the person at rest. Thus, air and wall temperatures equally influence the loss of warmth from the human body. For thermal comfort, one must therefore always consider the air temperature in a room as well as the average radiation temperature, which combine to produce an optimal "operative" temperature. The expression "operative temperature" has often been replaced by the term "perceived temperature," although the latter does not take draught-induced temperature decreases into consideration. **Figure 2.7** indicates optimal operative (perceived) temperatures depending on activity and the insulating values of clothing (PMV = 7 point measure valuation of perception, on a scale of -3 to +3).

As **Figure 2.7** shows, clothing is now always taken into consideration in addition to activity; the insulation values of different types and items of clothing are listed in **Table 2.3**.

Figure 2.8 illustrates yet another factor in the evaluation of thermal comfort: the comfort zone according to Grandjean.

The diagram illustrates that only a very small field of specific surrounding and room temperatures results in comfortable conditions. Areas near windows, in particular, require additional and often extensive efforts to achieve favourable surface temperatures, as illustrated in **Figure 2.9**.

Clothing	**Insulate value I_{cl}** $\frac{m^2 \cdot K}{W}$	clo
Unclothed	0	0
Shorts	0.016	0.1
Tropical clothing Short open shirt, short trousers, lightweight socks, sandals	0.047 – 0.062	0.3 – 0.4
Light summer clothing Short open shirt, long lightweight trousers, lightweight socks, shoes	0.078	0.5
Light work clothing Short underpants, open work shirt or light jacket, work trousers, woollen socks, shoes	0.093	0.6
Overalls (cotton) Shirt, short underwear, socks, shoes	0.124	0.8
Lightweight outdoor sports clothing Short underwear, training shirt, trousers, socks, gym shoes	0.140	0.9
Rainwear, two-piece suit (polyurethane) Shirt, short underwear, socks, shoes	0.140	0.9
Strong work clothing Long underwear, single-piece work garment, socks, strong shoes	0.155	1.0
Light street clothes Short underwear, closed shirt, light jacket, long trousers, socks, shoes	0.155	1.0
Foundryworker clothing (flame-resistant equipment) Shirt, short underwear, socks, shoes	0.155	1.0
Leisurewear Short underwear, shirt, pullover, robust jacket and trousers, socks, shoes	0.186	1.2
Foundryworker clothing and heat protection coat Shirt, short underwear, socks, shoes	0.217	1.4
Light street suit with light coat	0.233	1.5
Robust street suit Long underwear, long closed shirt, robust jacket and trousers, waistcoat of cloth or wool, woollen socks, shoes	0.233	1.5
Clothing for wet, cold weather Long underwear, long closed shirt, robust jacket and trousers, pullover, woollen coat, woollen socks, strong shoes	0.233 – 0.310	1.5 – 2.0
Arctic clothing	from 0.465	from 3.0

Table 2.3
Insulation values of selected items of clothing in dry state

*) Parameter for clothing insulation value in "clothing units" [1 clo = 0.155 (m² · K)/W].

Figure 2.8
Comfort zone in the t_r/t_A-diagram after Grandjean
t_r = mean radiation temperature
t_A = room air temperature
||| comfortable

Based on air velocity 0 to 0.2 m/s; relative humidity 30 to 70%

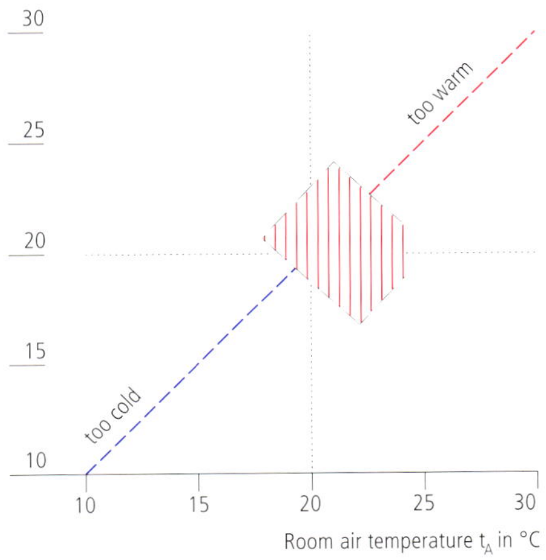

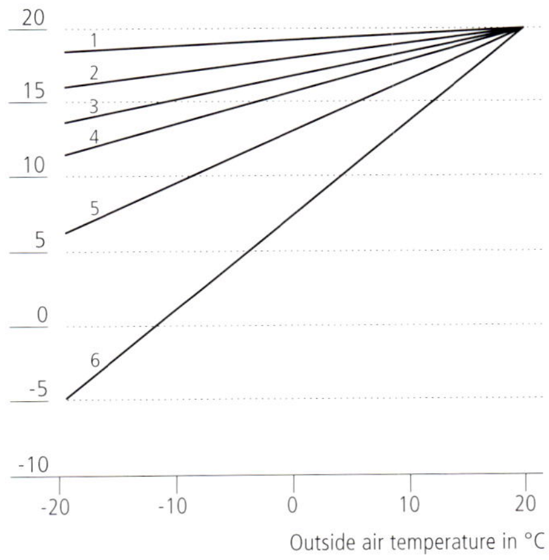

Figure 2.9
Inner surface temperature for exhaust air, standard and at windows with thermal insulating glazing – room temperature 20 °C

1	Exhaust air window	k ≤ 0.4 W/m²K
2	Thermal insulating glazing	k = 0.7 W/m²K
3	Thermal insulating glazing	k = 1.3 W/m²K
4	Thermal insulating glazing	k = 1.6 W/m²K
5	Double glazing	k = 3.0 W/m²K
6	Single glazing	k = 6.0 W/m²K

In terms of temperature influences, thermal comfort is principally determined by the average calculated from the air temperature and the mean temperature of all surrounding surfaces. The less these two temperature values diverge and the closer they are to the average value of 20 – 24 °C, the more evenly distributed is the heat loss in the human body. The maximum difference between air temperature in a room and surrounding temperatures should not exceed 3 K. Moreover, differences in individual surface temperatures should be kept to a minimum to ensure that the body is warmed evenly. This applies especially when chilled or heating ceilings are used or radiant heating walls.

Humidity – in particular, relative humidity – is another important factor for comfort, because the strength of evaporation depends largely on the vapour pressure difference of water on the skin surface and the vapour in the air. For normal room temperatures around 20 °C, heat dissipation by evaporation plays only a minor role and therefore has only negligible influence. Even when temperatures rise, humans have no direct perception of humidity; hence the comfort zone for room temperature and relative humidity is quite generous, see **Figure 2.10**.

When humidity falls below 35 %, experience has shown that more dust is trapped, which may irritate the respiratory organs. Relative humidity values below 45 % may cause slight electric charges, especially in plastic components, which may result in problems from (electric) discharges. High relative humidity can easily lead to mildew and fungus formation, when dew point temperatures fall short as a result of high humidity (building and material damage).

High room temperatures with high humidity levels generally cause skin evaporation to increase drastically in order to cool the body. Experiments have shown that an absolute humidity of more than 12 g water content per kg of air produces a humidity level that is experienced as uncomfortable. In general, it can be stated state that the relative humidity in a room should diminish as room temperatures rise in order to achieve comfortable room conditions. At the same time, relative humidity should be slightly lower for strenuous activities than for sedentary activities.

Figure 2.10
Comfort zone for the pair of variables:
t_A = room air temperature
j = relative room humidity
(Leusden and Freymark)

Air movement in a room strongly influences comfort; in general, increased air movement is still experienced as comfortable as room temperatures rise, and vice versa. As **Figure 2.11** illustrates, few people express either comfort or discomfort when air movement lies below 0.1 m/s, a value that should be considered both for correct natural ventilation and for air conditioning or mechanical ventilation. It is important to avoid constant air movement either in natural or mechanical ventilation, because the turbulence quotient of velocity greatly influences the feeling of comfort in people. Since changes in pressure occur in both mechanical systems and natural ventilation, air velocities can reach more than four times the desired level for very short periods. The air velocities defined in the DIN 1946 standard according to air temperature or operative temperature are generally set too high and produce dissatisfying results. Especially in the case of sedentary activity and light clothing, high air velocities are experienced as uncomfortable and lead to rapid cooling. In the case of pronounced physical activity and warm clothing, both low room temperatures and higher air velocities are acceptable. In spaces where most work is performed while sitting down, air velocities should rarely exceed 0.1 m/s, based on an insulation value of 0.5 for the clothing and a total heat dissipation value of 1.0 met.

Correct clothing has an enormous impact on comfort, since appropriate clothing can help people to quickly react to temperatures that are either too high or too low. The insulation value of clothing is listed in **Table 2.3**, where the clothing factor (m^2 K/W) increases in relation to the density and warmth of a particular item of clothing.

In conclusion, **Figure 2.12** illustrates the recommended comfort zone in an h, x diagram: this is the comfort zone that should exist or be created in a room to provide the best conditions for optimum performance. In addition to the temperature and humidity values given in this diagram, the corresponding recommendations with regard to air velocity and surface temperatures must also be taken into consideration.

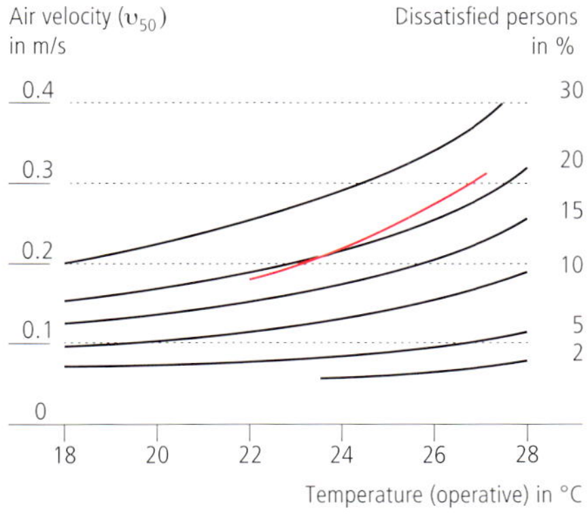

Figure 2.11

Percentage of dissatisfied persons (PD), depending on air velocity and air temperature, after Fanger.

0.8 clo, 1 met,
Degree of turbulence Tu = 40%

$$Tu = \frac{v_{84} - v_{50}}{v_{50}} = 40\%$$

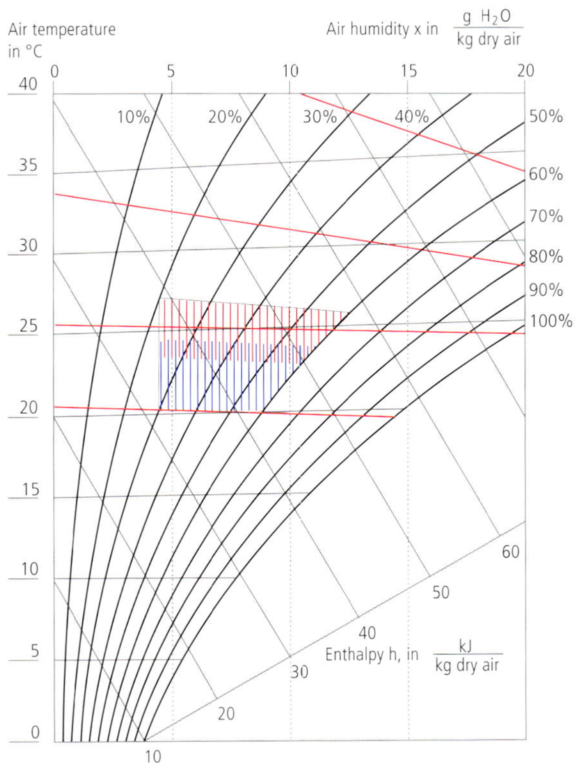

Figure 2.12
New comfort table with "effective temperature" curves, after ASHRAE.

||||| Summer
||||| Winter
— Effective temperature

2.2 Hygienic comfort

Hygienic comfort is not defined in DIN or other standards. It describes air quality and the influence of several factors on the level of comfort in a particular space.

Generally speaking, dust has little adverse effect on comfort in well-cleaned living and working spaces, perhaps with the exception of extremely dry air in winter, which tends to make dust levels more noticeable through the smell created by dust carbonization on overheated heating elements. This is experienced as unpleasant, becoming more noticeable the dustier a room is.

Significant negative impact can occur at production and manufacturing sites, where fine dust is produced in high concentrations (for example, grinding shops, dressing shops, cement works etc.), in which case special extraction fans should be installed to keep dust levels as low as possible (shop hygiene/protective measures).

In living and working areas human exhalation and perspiration are only one factor in addition to evaporation from furniture, carpets, wallpaper, wall paint and building materials, as well as airborne particles from combustion and heating processes, cleaning, the influx of unfiltered outside air especially in industrial areas and near high-traffic roads, and finally odours from food preparation etc.

Table 2.4, maximum workplace concentration of air pollutants, lists a number of air pollutants and their upper threshold limits, which must be maintained to avoid health hazards.

Table 2.4
TLV 1997
(Threshold limit values, substances constituting a health hazard, selection)
(as per TRg S900 1997)

Chemical	Formula	TLV ppm	mg/m³
Acetic acid	$CH_3 \cdot COOH$	10	25
Acetone	$CH_3 \cdot CO \cdot CH_3$	500	1,200
Ammonia	NH_3	50	35
Arsenic hydride	AsH_3	0.05	0.2
Asbestos dust *) **)			15,000 fibres/m³
Benzene *)	C_6H_6	1 – 2.5	3.2 – 8
Bromide	Br_2	0.1	0.7
Butane	C_4H_{10}	1,000	2,350
Cadmium oxide (suspected carcinogen)	CdO		0.015 – 0.03
Calcium oxide	CaO		5
Carbon dioxide	CO_2	5,000	9,000
Carbon disulphide	CS_2	10	30
Carbon monoxide	CO	30	33
Chlorine	Cl_2	0.5	1.5
Chlorine dioxide	ClO_2	0.1	0.3
Chlorobenzene	$C_6H_5 \cdot Cl$	10	46
Chloroform *) (trichloromethane)	$CHCl_3$	10	50
Chloromethane	$CH_3 \cdot Cl$	50	105
Copper (dust)	Cu		1
Copper (vapour)	Cu		0.1
DDT	$C_6H_4Cl_2CH \cdot CCl_3$		1
1,2-dichlor-1,1,2,2-tetrafluoroethane (R-114)	$CF_2Cl \cdot CF_2Cl$	1,000	7,000
Dichlorodifluoromethane (R-12)	CF_2Cl_2	1,000	5,000
Dichlorofluoromethane (R-21)	$CHFCl_2$	10	43
Dichloromethane (suspected carcinogen)	CH_2Cl_2	100	360
Diethyl ether	$C_2H_5 \cdot O \cdot C_2H_5$	400	1,200
Ethanol	$C_2H_5 \cdot OH$	1,000	1,900
Fluorine	F_2	0.1	0.2
Formaldehyde (suspected carcinogen)	$HCHO$	0.5	0.6
Formic acid	$HCOOH$	5	9
Hexane	C_6H_{14}	50	180
Hydrazine *)	$NH_2 \cdot NH_2$	0.1	0.13
Hydrogen bromide	HBr	5	17
Hydrogen fluoride	HF	3	2
Hydrogen chloride	HCl	5	7
Hydrogen cyanide	HCN	10	11
Hydrogen peroxide	H_2O_2	1	1.4
Hydrogen phosphide	PH_3	0.1	0.15
Hydrogen selenide	H_2Se	0.05	0.2
Hydrogen sulphide	H_2S	10	15
Hydrogen tetrachloride	CCl_4	10	65
Iodine	J_2	0.1	1
Iron oxide (fine dust)	Fe_2O_3 ; FeO		6
Lead	Pb		0.1
Magnesium oxide (vapour)	MgO		6
Mercury	Hg	0.01	0.1
Methanol	$CH_3 \cdot OH$	200	260
Naphthalene	$C_{10}H_8$	10	50
Nicotine		0.07	0.5
Nitrobenzene	$C_6H_5(NO_2)$	1	5
Nitroglycerine	$C_3H_5(ONO_2)_3$	0.05	0.5
Nitric acid	HNO_3	2	5
Nitrogen dioxide	NO_2	5	9
Ozone	O_3	0.1	0.2
Phenol	$C_6H_5 \cdot OH$	5	19
Phosgene	$COCl_2$	0.1	0.4
Phosphorus (yellow)			0.1
Pentachloride	PCl_5		1
Propane	C_3H_8	1,000	1,800
Quartz			0.15
Sulphuric acid	H_2SO_4		1
Sulphur dioxide	SO_2	2	5
Styrene	$C_6H_5 \cdot CH = CH_2$	20	85
Toluene	$C_6H_5 \cdot CH_3$	50	190
Trichlorofluoromethane (R-11)	$CFCl_3$	1,000	5,600
Turpentine oil		100	560
Vanadium (V_2O_5-dust)			0.05
Zinc oxide (vapour)	ZnO		5

*) Carcinogenic, special measures to minimize exposure
**) As per TRg 5519

Conversion for 1 ppm (parts per million):

$$1\ cm^3/m^3 \triangleq \frac{\text{molar mass}}{\text{molar volume}}\ mg/m^3$$

In countries where smoking is permitted at the workplace, tobacco smoke is another factor that should not be underrated, although the actual impact on the non-smoker depends greatly on the concentration of smoke. In general, smoke is unevenly distributed in a room, especially when the space is ventilated by thermal currents (ventilation from bottom to top). Statistics show that in offices where smoking is permitted, the average consumption is 1.5 cigarettes per hour, with an average of 40 % smokers among the workforce. Based on these statistics, 100 m³ fresh air is required per cigarette smoked to avoid negative impact on the hygienic comfort (the equivalent of a minimum fresh air rate of 60 m³/h per person).

For this reason it is recommended to introduce at least 55 – 60 m³/h of fresh air into mixed-use rooms (smokers/non-smokers) and in large, open-plan areas at least 75 m³/h of fresh air per person. This helps to extract not only tobacco smoke, but also formaldehyde from interior furnishings and insulating materials, nitrous oxide (heating or cooking) and solvent vapours.

Carbon dioxide scale

The carbon dioxide scale (according to Pettenkofer) indicates the increase of exhaled CO_2 in closed room. This would be of little consequence if carbon dioxide content were not the indicator for impoverished air quality in rooms, notably odours and evaporation, which are especially noticeable in small, densely occupied rooms.

When the CO_2 content rises to above 0.1 %, the air quality is judged to be poor (0 % = 1.000 ppm). Toxic effects of CO_2 occur when levels exceed 0.25 %. As **Figure 2.13** shows, this CO_2 level may rapidly be reached in densely occupied and poorly ventilated rooms, and here ventilation with the highest possible air change rate is the only remedy (on an hourly basis).

Constant air change of 4 ac/h maintains the CO_2 content at approximately 0.14 %, a level that is subjectively experienced as poor air quality, although it isn't harzardous. **Figure 2.14** indicates the fresh air rate per person for different acceptable CO_2 concentrations and different exothermic heat levels according to activity. This diagram also indicates the threshold value 3, mentioned above as the maximum workplace concentration of toxic substances, which must not be exceeded under any circumstances.

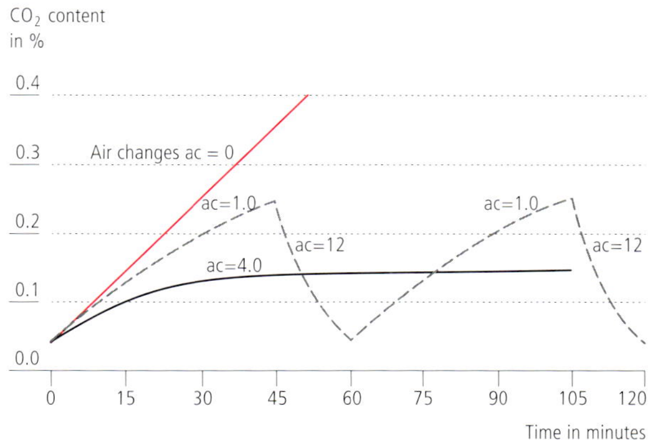

Figure 2.13
CO_2 content in densely occupied rooms, e.g. classrooms (approx. 1 person/m²) for different air change rates

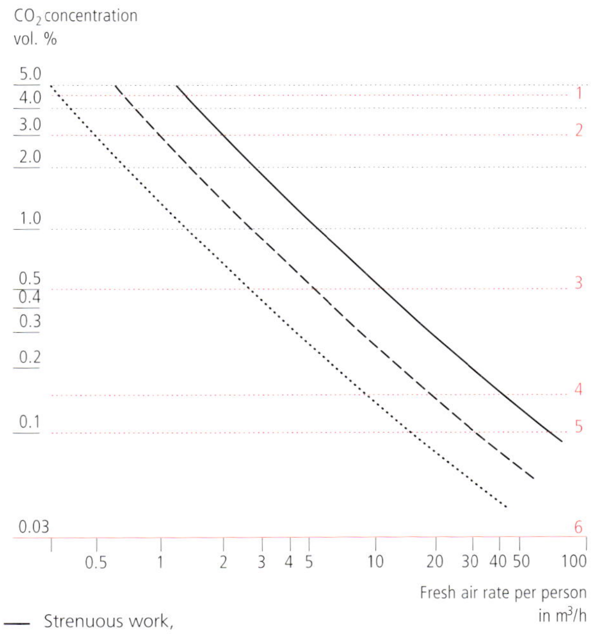

— Strenuous work, 400 W heat transmission
– – Light work, 200 W heat transmission
..... Sitting, 100 W heat transmission

1 Exhaled air
2 Shelters
3 Industrial MAK value
4 Maximum for offices
5 Pettenkofer value
6 Fresh/outside air

Figure 2.14
Fresh air rate per person for different permissible CO_2 concentration levels (Reinders)

2.3 Acoustic comfort

The expression "acoustic comfort" is an uncommon one that appears in neither technical publications or workplace guidelines, nor is it necessarily included in the curricula of polytechnics or universities. This is due mainly to the fact that the term encompasses more than the physical parameters to which we usually refer in the natural sciences. Although acoustic comfort is difficult to define, there are at least some cases where acoustic discomfort can be clearly defined. Acoustic discomfort exists, for example, near low-flying airplanes or at extremely loud rock concerts. These conditions are experienced as jarring by most people; on the other hand, soundless and sound-absorbing rooms are equally uncomfortable acoustically. We usually inhabit rooms at neither one nor the other extreme, that is to say, rooms that may well be acoustically comfortable. Here are some notes on definable physical parameters:

Unpitched sound

Unpitched sound consists of the mechanical oscillations and waves of a medium (in this case, air). When unpitched sound occurs within a frequency range of 16 Hz to 16 kHz, we call it audible sound. Like footfall sound (or impact sound), unpitched sound is partially radiated as airborne sound.

Tone

Tone is a pure sinusoidal oscillation of specific strength and frequency; in contrast to tone, noise (or sound perceived as noise), is the result of random and aperiodic oscillations.

Loudness

The loudness in a room is the perception of tones at frequencies recognized by the human ear. The perceived sound is usually expressed in sound pressure levels, that is, the dual pressure created by sound oscillation in addition to frequency.

Sound pressure

Sound pressure is used to measure loudness.

Sound power level

The sound power level is a parameter of sound energy and is the product of sound pressure, sound velocity and the cross-sectional area vertical to the oscillation direction.

The sound pressure level used to evaluate rooms is the SPL (logarithmic ratio between sound pressure and sound power level), values in the equation are sound pressures. The reference sound pressure for all evaluations has been established to be the lowest sound power level perceived by the human ear for a frequency of 1,000 Hertz. This SPL is 2×10^{-4} microbar. For individual room categories, SPLs generally lie between approximately 25 dB(A) and 65 dB(A). SPLs of 25 dB(A), required for recording studios, concert halls, theatres and opera houses, create absolutely quiet rooms.

Rooms for sedentary activities, such as offices, conference rooms and the like, should have an SPL of approximately 35 dB(A), regardless of room size and number of occupants; these values can be far exceeded for industrial halls. In the latter, physical protection must be provided (headsets, ear plugs etc.). Measurements taken by acoustics experts are alarming since they indicate that the increase in traffic translates into a doubling of noise levels every ten years. Our most sensitive organ is exposed to continuous stress, and this may have serious implications not only for the ear itself, but for the whole organism as well. The consequences have been documented as stress-related illnesses (disorders of the autonomic nervous system, lack of concentration, sleep disorders etc.). **Table 2.5** lists standard values for SPL and reverberation periods.

Type of area	Sound pressure level	Reverberation time in s
Living rooms/bedrooms	35/30	0.5
Hospital: Wards, day/night	35/30	1
Consulting rooms, halls, corridors	40	2
Operating theatres	40	3
Halls: Radio/TV studio	15/25	1/1.5
Theatre/opera house	30/25	1/1.5
Concert hall	25	2
Cinema, auditorium, reading room	35	1
Church	35	3
Offices: Meeting room	35	1
Small office area	40	0.5
Large office area	45	0.5
Restaurants	40 – 55	1
Museums	40	1.5
Reading room/classroom	35/40	1
Gymns/swimming pools	45/50	1.5/2

Table 2.5
Guideline values for sound pressure levels and reverberation time in different areas

2.4 Visual comfort

Visual comfort exists when the perceptive faculties in the human brain can operate without interference. Incorrect distribution of light density in a room, glare, poor colour matching and inappropriate interior design all inhibit perception. On the other hand, when perception is in no way inhibited, the basic senses of the eye, such as vision, speed and contrast sensitivity, are optimized. Optimal working conditions can be achieved by harmonizing the light density conditions in the working environment, starting with the light density conditions at the workstation. A stable perception field for PC workstations, where vision is especially important, should include contrast sensitivity as a criterion.

Visual comfort also depends on sufficient light in the area of visual focus and the avoidance of glare, be it direct, indirect or daylight glare. The light and temperature of colours in lighting systems for rooms are additional criteria. Lit rooms should also exhibit sufficient shadow effects to enhance the plasticity and three-dimensionality of objects and surfaces. Last, but not least, visual comfort means visual contact from the inside out and from the outside in.

2.5 Electromagnetic compatibility

In Europe electromagnetic compatibility is defined by a code for the functional capability of electrical appliances and electrical systems; a conformity approval symbol (CE) identifies appliances that have been approved. The electromagnetic compatibility defines, first, the allowable degree of interference radiating from the appliance itself. Second, it defines the levels of interference from other appliances or environmental influences that the appliance must be able to handle. The aim here is to avoid interference between several appliances or any impediment to their functions, which is especially important when large data-processing systems are used.

While electromagnetic compatibility refers to appliances, electromagnetic environmental compatibility is used in the context of humans and animals. At issue are possible health hazards as a result of electric, magnetic or electromagnetic fields ("electro-smog"). Assessment is hampered by the absence of proven scientific data indicating to which degree humans are influenced by electromagnetic fields. It has been established that high levels are definitely harmful. The current discussions focus on the nature and severity of health hazards that such fields pose. In relation to electricity, data-processing installations, cable systems etc., three different types of fields are described.

Electrical fields
Electrical fields arise from the supply voltage (Volt/m). They have a source and a sink and can be shielded by simple means (protection through plants, metal strips etc.).

Magnetic fields
Magnetic fields form when electricity flows through a conduit (Ampere/m). In contrast to electrical fields, magnetic fields do not have a source or a sink; rather, they are closed circuits and can therefore not be shielded. Shields simply lead to field distortions or to the dissolution of the field. These measures require complex technical efforts (e.g. thick metal covers of soft iron).

Electromagnetic fields
Electromagnetic fields are created by the radiation of electromagnetic waves. Examples are radio reception, television reception via antennae, radio engineering and cell phones. The health risk posed by electromagnetic fields is not yet clearly established; however, recommendations have been formulated, for example, to restrict cell phone use to half an hour per day. Cell phones operate in the microwave range; their transmission performance of approximately 2 W is radiated directly at the head of the user. Wireless telephones have an effect similar to cell phones, but the transmission performance is noticeably smaller and therefore presumably less dangerous. Long-term experience exists only in the area of radiated electromagnetic waves – for example, in the operation of

transmitters. Statistics that there are health hazards when people are exposed to electromagnetic waves over a long period of time or when electromagnetic waves are very strong.

In the area of electrical and magnetic fields, which occur whenever electric energy is transferred, electrical fields can be viewed as relatively unproblematic based on current knowledge, with the exception of the immediate vicinity of high-voltage lines. Magnetic and electromagnetic fields in buildings occur in areas where high electrical performance is transmitted (transformer stations, low-voltage main distributions, busbars, mains supply). These fields are indeed a serious concern; the primary concern is, however, how long a person is exposed to these fields. Today, we think twice before locating a workstation or a desk near a magnetic field. Government guidelines on the subject define boundary values for this scenario in each country.

Boundary values

Biological effects of low-frequency, electric and magnetic fields are primarily assessed by the density of the structure-borne currents flowing through a specific area. Such impact currents occur naturally. Many organs in the human body exhibit impact current densities of approximately 1 mA/m². In the heart and the brain they may reach up to 10 mA/m².
Acute health risks caused by electric and magnetic fields can be avoided when long-term impact current density is no greater than 2 mA/m². This value lies within the range of natural impact current densities. This value is guaranteed in electric field forces below 5 kV/m and magnetic flow densities below 100 mT for 50 Hertz.

The limits listed are general recommendations, primarily for households, and are based on recommendations resulting from international studies. The following overview **(Table 2.6)** shows values for different appliances measured by the Federal Agency for Radiation Protection:

The unit used to measure magnetic fields is the tesla, with most household appliances lying in the microtesla range, as shown in the table.

Representative values of magnetic flux densities emitted by household appliances at varying distances measured in microtesla (μT) standard distances are highlighted

	3 cm	30 cm	1 m
Hairdryer	**6 - 2000**	0.01 - 7	0.01 - 0
Electr. razor	**15 - 1500**	0.08 - 9	0.01 - 0.3
Can opener	1000 - 2000	**3.50 - 30**	0.07 - 1
Drill	400 - 800	**2 - 3.5**	0.08 - 0.2
Vacuum cleaner	200 - 800	**2 - 20**	0.13 - 2
Mixer	60 - 700	**0.60 - 10**	0.02 - 0.2
Electr. discharge lamp	40 - 400	**0.50 - 2**	0.02 - 0.25
Microwave	73 - 200	**4 - 8**	0.25 - 0.6
Soldering iron	105	**0.3**	< 0.01
Radio (portable)	16 - 56	**1**	< 0.01
Stove	1 - 50	**0.15 - 0.5**	0.01 - 0.04
Washing machine	0.8 - 50	**0.15 - 3**	0.01 - 0.15
Iron	8 - 30	**0.12 - 0.3**	0.01 - 0.03
Dishwasher	3.5 - 20	**0.60 - 3**	0.07 - 0.3
Immersion heater 1 kW	12	**0.1**	< 0.01
Toaster	7 - 18	**0.06 - 0.7**	< 0.01
Monitor (colour)	5.6 - 10	**0.45 - 1**	< 0.01 - 0.03
Dryer	0.3 - 8	**0.08 - 0.3**	0.02 - 0.06
Kettle 1 kW	5.4	**0.08**	< 0.01
Computer	0.5 - 3	**< 0.01**	
Fridge	0.5 - 1.7	**0.01 - 0.25**	< 0.01
Clock (AC)	300	2.25	**< 0.01**
Slide projector	240	4.5	**0.15**
Heater	10 - 180	0.15 - 5	**0.01 - 0.25**
Transformer	135 - 150	0.60 - 1.05	**0.24**
Television	2.5 - 50	0.04 - 2	**0.01 - 0.15**
VCR	1.5	< 0.01	

Electrical fields strengths in buildings in V/m; readings taken at 30 cm distance from device

Boundary value est. by radiation protection commission for population at large	5000
Boiler	260
Stereo receiver	180
Iron	120
Fridge	120
Hand mixer	100
Toaster	80
Hairdryer	80
Evaporator	80
Colour TV	60
Coffee-maker	60
Vacuum cleaner	50
Clock (electr.)	30
Fields acting on house from the outside	20
Electr. stove	8
Lamp	5

Table 2.6
Max. permissible field strengths/ magnetic flux densities

2.6 The influence of colours

(Dr Leonhard Oberascher, Ecopsychologist, Salzburg)

Colour, as a basic design element, has a strong influence on the impression and behavioural effect of the architectural environment and thus contributes to the psychological and physical well-being of users in a room.

We experience colour as a fundamental quality in our visual perception: surface colours – mainly perceived as an inherent quality of the object – are the visible "skin" of the environment. From the overall impression (that is, the composite image of texture and colour), the eye derives the condition, material characteristics and utility of an object (thus a yellow-red fruit signifies ripeness and, usually, readiness for consumption).

Colour and form together determine the overall shape, or "Gestalt", of an object (in the sense of Gestalt psychology). Colour is therefore not, as is often assumed, a superficial aspect, but rather a fundamental element in "giving shape". To give shape means to give meaning. Colour therefore influences not only the optical appearance of an object but also its meaning, effect and various uses. The principle design task in this context is to create a harmony between the possible colour functions and the intended object functions. Colour can fulfil the following functions:
- Indicative
- Symbolic
- Aesthetic

By its indicative function, colour (together with the surface texture) clarifies the invitational character, utility and function of an object by indicating its visual validity, condition (e.g. hot or cold), material characteristic (e.g. wood, stone, metal), purpose and functional structure. Colour thus suggests chiefly the real qualities and functions of an object.

By its symbolic function, colour communicates imaginary object qualities. It projects immaterial values (thus in a sports car, the colour red can communicate speed and power); it can also take on a random symbolic meaning (e.g. the same red in another context can represent socialism, postal services or an adult entertainment centre).

By its aesthetic function, colour mainly serves as a decorative element or for formal composition. Since aesthetic behaviour is largely independent of use, the evaluation of colours in this context is less according to function than according to the affect and expression of colours. The decisive aspect herein is the formal aesthetic relationship between different colours, as judged by contrast, harmony, field size and distribution.

All three colour functions listed above are important for the relationship between colour effect and a sense of comfort and well-being. The value is not only determined by the function of the room, but also to a great degree by the relevant needs and preferences of the occupants. Psychological studies have shown that these rarely match the ideas of the architects and designers.
A conservatively inclined user will prefer a beige ambience, not because these colours are intrinsically beautiful or because the composition is balanced in an aesthetic sense, but because these colours express the user's deep-seated associations with "comfort," "reliability" and "earthiness."

The complex influence of colour and its use on psychological and physiological well-being cannot be reduced to simple formulae; rather, it can only be usefully evaluated in the context of a system that encompasses people, social relationships and the spatial and material environment.
A thorough evaluation includes:
- Cultural factors (different colour cultures)
- Milieu factors (group-specific colour preferences and significance)
- Individual disposition and preferences (personal associations with colour)
- Function and meaning of the room (relation to colour function and meaning)
- Colour-form relationships (influence on form effect)
- Colour-material relationships (influence on material impression)
- Factors given by and in the surrounding environment (relationship between colour and landscape, urban surroundings, historic and stylistic references)
- Colour and physical well-being

Empirical studies have shown that colour has immediate physical impact on the human organism. Some of these hypotheses and observations are discussed in the paragraphs that follow.

The colour-excitation hypothesis

Does (room) colour influence the overall physiological state of person? Are warm colours (red, orange, yellow) stimulating, and cold colours (blue, turquoise, green) calming? Are we more stimulated by intense (strongly hued) colours than by subtle colours?

Küller and Mikelides note that the results of studies carried out over the past fifty years give few answers to these questions because they looked mostly at the effect of coloured light or of isolated colour areas, but not at actual rooms.

The physiological effect of the entire visual environment was first studied by Küller (1976, 1986) in two rooms with different degrees of visual complexity. One room was designed in different colours and patterns, while the other was uniformly grey. EEG and heart rate readings were taken for test candidates, who spent three hours in each room. The results showed that the colours and patterns of the first room clearly stimulated brain activity, that is, they produced a measurable physiological state of excitation, while at the same time lowering the heart rate, which was assumed to be a compensatory autonomous-nervous reaction to a high degree of stimulation. These findings were further supported by verbal statements made by the test candidates about a sensory overload in the first room. It took up to three hours before a feeling of balance was achieved.

Mikelides (1989) compared the effect of two rooms uniformly painted in red and blue, respectively. The assumption that red is cortically more stimulating than blue was confirmed by EEG readings. Küller and Mikelides conclude from the impact on heart rate and brain activity that the colour and complexity of the visual environment have a far greater influence on humans than had been previously assumed. This supports the approach of using colour in design as a means of achieving some control over the general excitability of humans (Küller, 1991). However, it would hardly be possible or desirable to use colour as a form of compensation for boring and monotonous tasks. Instead, such use of colour would create additional stress, evidenced by the paradoxical decrease in heart rate.

In conclusion, we can state that intense colours, independent of hue, as well as high visual complexity (e.g. strong colour contrasts) are, generally speaking, physiologically stimulating, while subdued colours, such as grey, have the opposite effect. A comparison of different hues with the same intensity saturation shows that red is far more activating than blue.

The hue-warmth hypothesis

In general, hues between yellow and red are associated with "warmth" and those in the blue to green range with "cold." Can we derive from this that the colours in a room influence our climatic well-being? Does a red room, for example, require less heating for our comfort than a blue room?

Küller and Mikelides note that a general relationship between room colours and climatic well-being could not be confirmed in experiments. According to Mikelides (1989), perceived temperatures in a red room and in a blue room are equally accurate in terms of the actual temperatures.

While ideas about warm and cold colours influence our assessment of interior spaces, they seem to have no impact on thermal comfort. Nevertheless, the characteristics associated with room colours do influence our psychological well-being. It is quite plausible, for example, that we perceive optically "cold" rooms as repellent and unfriendly, while "warm" rooms appear cosy and friendly.

Colour and psychological well-being

In addition to collective and individual colour associations and the resulting psychological impact of a room, there are a number of other factors which are important in the context of colour and psychological well-being:

Colour, used to create order and orientation

Colour generally facilitates the "readability" of the visual environment by imprinting it with a visible structure: the eye automatically links surfaces and areas of the same colour and differentiates areas of contrasting colours. Pronounced contrasts in brightness, hue or saturation create enhancement or isolation; related colours (simillarities in hue, saturation, nuance, black and white content) create homogeneity and integration. Colour as a room-structuring element is principally a means of creating lucidity, safety, orientation and order, as well as helping to represent specific room functions and actions in the space. Visual lucidity, order and comprehensibility promote a feeling of spatial control and are hence an important contribution to well-being.

Colour as territorial marker and status symbol

Colour identification can visually indicate "territorial" claims, hierarchical structures and status. Waiving the use of such symbols appears to have a negative effect (insecurity).

Colour as identifier
As a conspicuous identifying element in all corporate design, colour can be an essential contribution to the creation of a "visual" identity, or "branding." The high recognition value of colour instills confidence and boosts well-being.

Colour as a lifestyle symbol
Next to the colours of clothing and cars, home furnishing and interior decorating colours are important indicators and expressive means of personal lifestyle. Together with material and surface-texture characteristics, colour communicates specific societal value. Positive effects on the psychological well-being of room occupants are more likely when the colour scale reflects personal value perceptions and aesthetic preferences and lies within the acceptable range of group-specific values and styles.

Colour as an expression of trend orientation
Variety and change are a basic human need. Fashionable colours are the result of a collective desire for colour change. They are a symbolic expression of the predominant orientation in taste, and their main purpose is to demonstrate that one belongs to an idealized group (usually promoted in the media), which reflects the *Zeitgeist*. The change in collective colour preferences emphasizes that the relationship between colour and psychological well-being is flexible and can be evaluated in the context of the era and the culture.

It is difficult to give universal recommendations on how colour should be used.

Basically, one should take care that the room or object functions communicated through colour choices support the intended room or object functions or at least avoid being in conflict with them (e.g. if orientation and safety are an important concern, the aesthetic function of colour may have to be secondary to this primary function). With regard to the physiological impact of colour, the range of colours chosen should reflect the occupants and their activities. (The use of strong colour contrasts and highly stimulating hues is probably counterproductive to the well-being of the users in an operating theatre, or for anyone under stress). Furthermore, cultural, group-specific and individual preferences should be considered whenever possible.

Another basis for the formulation of general colour design recommendations is found in the field of environmental psychology where theories have been developed to explain aesthetic environment preferences. Kaplan & Kaplan postulate an accepted evolutionary need for meaning and order. The need for familiarity and the desire for innovation complement one another. The preference for a specific environment depends on the following variables:

- Coherence (degree of cognitive organization of a scene)
- Readability (degree of categorical distinctness of the elements present in the environment)
- Complexity (number and variety of these elements)
- Mysteriousness (degree to which the scene promises the discovery of additional information to the viewer)

Based on this, the following general recommendation can be formulated: the use of colour in design oriented to human needs should principally aim to achieve a balanced relationship between stimulation and reassurance, order and variety, relatedness and contrast. Colour should both integrate, structure and explain, and also offer sufficient variety to stimulate the active interaction with the architectural environment.

2.7 Sick Building Syndrome

Buildings should not cause sickness, but should meet thermal and hygienic requirements that offer an agreeable atmosphere to their users. In the past a number of buildings have been found lacking in this regard. Therefore we must work towards using ecologically sound materials and improving upon basic deficiencies in the areas of ventilation and lighting.

According to Dr P. König's definitions, Sick Building Syndrome reveals itself in various SBS-related symptoms and their possible causes. The individual manifestations and related causes are listed in **Table 2.7**.

Various sources have been identified as factors in causing the Sick Building Syndrome. These are:
– Excessive air velocity or turbulent air flow in a room
– Symptoms arising from microbial allergens or cell toxins
– Malfunction of thermal regulation due to excessively high or low temperatures (poor stimulus climate)
– Problems arising from low-frequency sound (< 100 Hz)
– Odours from poorly maintained humidifiers and filters

These problems can be avoided by ensuring draught-free air supply into rooms, where the velocity of air flow should be slowed sufficiently (< 0.12 m/s) to ensure comfort.

Microbial allergens and cell toxins can be avoided through regular maintenance of humidifiers and filter installations, for instance by using high-performance filters, which can remove a number of pollutants from the air.

It should in principle also be possible to use the buildings without technical air systems for processing ambient air, so that openable windows are an absolute necessity. By the same token, windowless rooms in constant use to accommodate personnel should be avoided as far as possible. The proportion of window area should amount to at least 50 % of the façade, which means that adequate sun protection and large storage volumes are a precondition of avoiding the rooms becoming overheated. It should be borne in mind in this context that the temperatures should lie within an agreeable range (individual temperature regulation for the individual rooms and the possibility of switching off window ventilation).

Rooms should only be humidified if the relative humidity drops below 35 %. Indoor plants are a viable option to support humidification.

SBS complaints	Possible causes
Draught Susceptibility to flues and colds ventilation Rheumatic complaints	Excessive flow velocity Excessive turbulence Intake air temperature too low
Mucous membrane irritation of the upper respiratory tract and eyes, sensation of dryness	Microbial allergy (from air-conditioning system), dust; mites (floor coverings etc.)
Fever Breathing difficulties Rheumatic pains Fatigue	Microbial cell toxins (endotoxins, cytotoxins) from humidifying water, filters and air intake elements
Fatigue Impaired concentration Dizziness Headaches	Poor temperature control: - Temperatures > 23 °C - Physiologically unsatisfactory daily temperature sequence - Increase in relative humidity - No of window ventilation Low-frequency noise (< 100 Hz) Allergens, endotoxins, cytotoxins
	Insufficiency of: - Sun protection (lack/internal) - Window surfaces (too large) - Storage volume (too little) - HVC capacity/maintenance
Poor air quality	Odour from air-conditioning system: - Technical (material, filters) - Microbiological Insufficient air change

Table 2.7
SBS complaints
(after Dr P. Kröling)

In addition to the principal symptoms **Table 2.8** (after Dr P. Kröling) also lists secondary symptoms, causes, origins, and means of avoiding SBS.

SBS symptoms	Secondary symptoms	Causes	Origin	Avoidance
Infectious: Legionella pneumonia	Pontiac fever (similar to influenza)	Legionella bacteria	Aerosol with rich germ burden in air flow (e.g. cooling system)	Hygienic conditions End-stage filters
Allergic: Eye irritation/tears Blocked nose Bronchial disorders Sensation of dryness	Fatigue Headache Impaired concentration Reduced performance	Microbial allergy – Various fungal spores – Various germs – Mites/products – Organic dusts	Air humidification – Humidifying water – Moist system components Filters – Deficient filtration – Washing filters through	Low-germ humidification (ozone/UV/silver ions) – Regular cleaning Renewable filters Electro-filters End-stage filters
Toxic-allergic: Monday fever Humidifier fever Severe fatigue (fatigue syndrome)	Symptoms similar to influenza Breathing impairment Coughing Allergic alveolitis Long term risk: Pulmonary fibrosis (wet lung) Reduced performance	Constituents: Various germs – Cell walls: Endotoxins – Cell content: Cytotoxins - Mycotoxins	Germ-infected humidifier water	Low-germ humidification (ozone/UV/silver ions) – Regular cleaning Renewable filters Electro-filters End-stage filters
Toxic: Fatigue Headache Impaired concentration	Symptoms similar to allergy Reduced performance	External sources of contamination: – CO, NO_x, SO_2, VOC Internal sources of contamination: – Furniture, carpets	External air Building	Pollutant-minimizing filters (external/end-stage) Avoidance of sources Improved ventilation Odour-eliminating filters (external/end-stage)
Olfactory: Unpleasant odours "Lack of oxygen" Need for fresh air Shortness of breath	Dissatisfaction with environment Reduced performance Desire for windows to be opened	Intake air burdened with odours: – Traffic, garages Odour from HVC systems: – Air humidity – Filters, system components	External air HVC system	Odour avoidance Odour-eliminating filters (end-stage)

Tab. 2.8
SBS-Symptoms

3. Integrated planning models

Integrated planning is nothing new. A study of ancient building traditions in all areas of the globe and all climate zones reveals that early building forms were notable for a high degree of integration, e.g. of thermal and ventilation requirements, considering the available technical options. It was only in the 20th century, especially the second half, that client specifications, architectural expression and building technology became emancipated and strove for optimum solutions. In nearly every case, technology delivered the tools to overcome constructional challenges or excessive user requirements. The result was buildings that were functional but only at the cost of high energy consumption and maintenance. The energy and resource crises and above all a greater ecological awareness, especially in Germany and in Europe, have inspired architects and engineers to collaborate in the development of integrated, reciprocal technological concepts. Many clients also have begun to set the bar lower in terms of demands for comfort, giving planners greater scope to develop new concepts. **Figure 3.1** provides an overview of the interdependencies between client, architect and building systems planner in terms of potential savings in building systems.

Figure 3.1
Architecture – Technology – Environment

From the very outset, the architect must consider the conditions in the immediate building environment and the urban planning requirements and seek solutions that minimize the need for elaborate technology.
In the past, the technical solution for many buildings in dense downtown locations exposed to high noise levels was to design hermetically closed facades and fully artificial internal air-conditioning and climate control systems. The contemporary response, by contrast, is a preference for double-skin facades and naturally ventilated rooms, as illustrated in **Figure 3.2.** The fully closed, deep-cavity double facade construction employed for this concept protects the offices against traffic noise, while at the same time allowing for natural ventilation. **Figure 3.3** illustrates the concept in detail. This is an example where the need for elaborate building systems was reduced by means of facade technology and design.

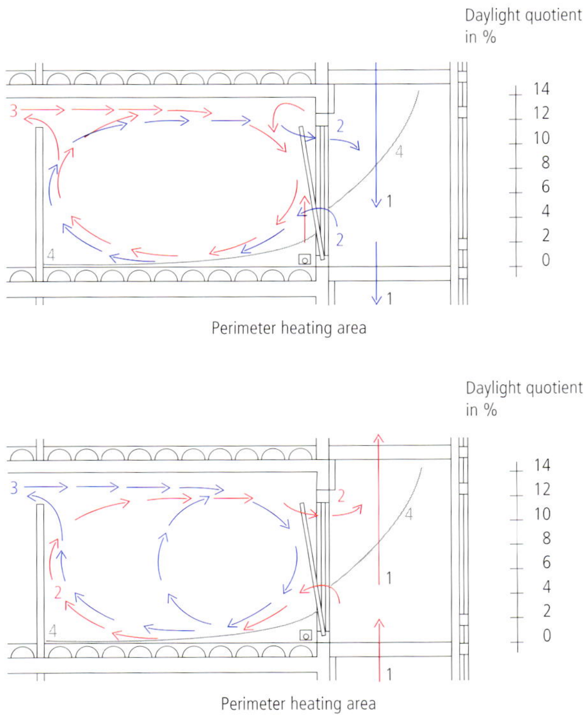

Ventilation of office in winter/ cool transitional season (external temperature −10 °C to approx. +10 °C)

1 Primary fresh air flow (approx. +12 °C to +15 °C)
2 Secondary fresh airflow
3 Mechanical ventilation (warm airflow when windows are closed)
4 Daylight quotients

Ventilation of office in summer/ warm transitional season (external temperature +22 °C to approx. +30 °C)

1 Primary fresh airflow (approx. +26 °C to +30 °C)
2 Secondary fresh air flow
3 Mechanical ventilation (cold airflow when windows are closed)
4 Daylight quotients

Figure 3.3
Detail section of a double-leaf facade

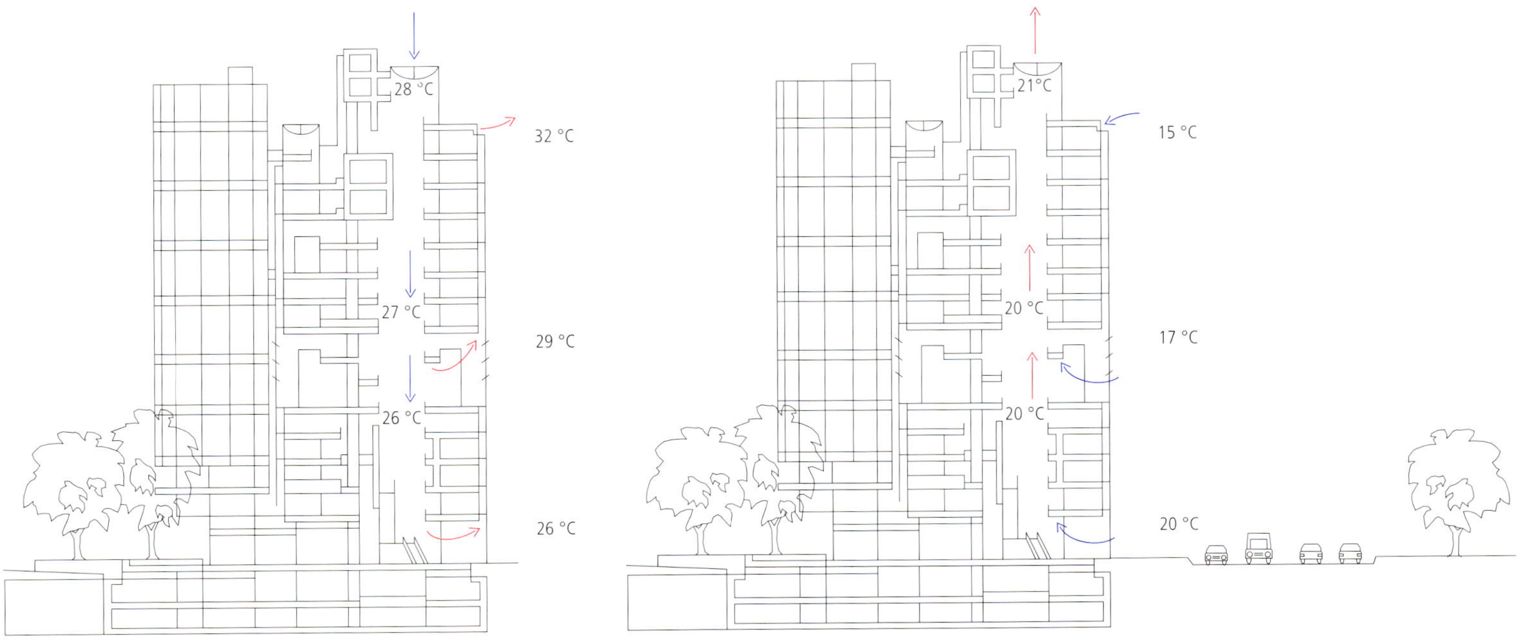

Summer
External temperature 27 °C

Winter and transitional season
External temperature 15 °C

Figure 3.2
Function of a double-leaf facade

Integrated planning methods can, however, achieve far more. The facade cavity, for example, may accommodate the entire communications cable system, which can easily be installed after completion of the construction, or modified or upgraded at a later point without inconveniencing the user. Since the cavity is shielded from exposure to weather conditions, wall penetrations for cable conduits do not pose a problem.

Other opportunities for integrated planning can be explored, for example, in wind canal studies. For one thing, such experiments carefully study and optimize the possibilities for natural ventilation in complex building forms. While the traditional pattern of arranging rooms along the facade and designing the latter with random openings does allow for a direct exchange of vitiated air and fresh air when windows are opened, this method by no means guarantees draught-free airflow throughout interiors and thereby a comfortable internal climate. For another, wind canals can be used for detailed facade studies on appropriate auxiliary construction systems, for example, in order to reduce heat exchange at the surface by decreasing the maximum velocities of the outside airflow. A building that has been optimized in wind canal settings can therefore be improved with regard to heat losses while ensuring practical air exchange by means of intelligently placed openings. Either of these features reduces both energy consumption and the technical installations necessary.

Figure 3.4 illustrates the pressure distribution on different building structures.

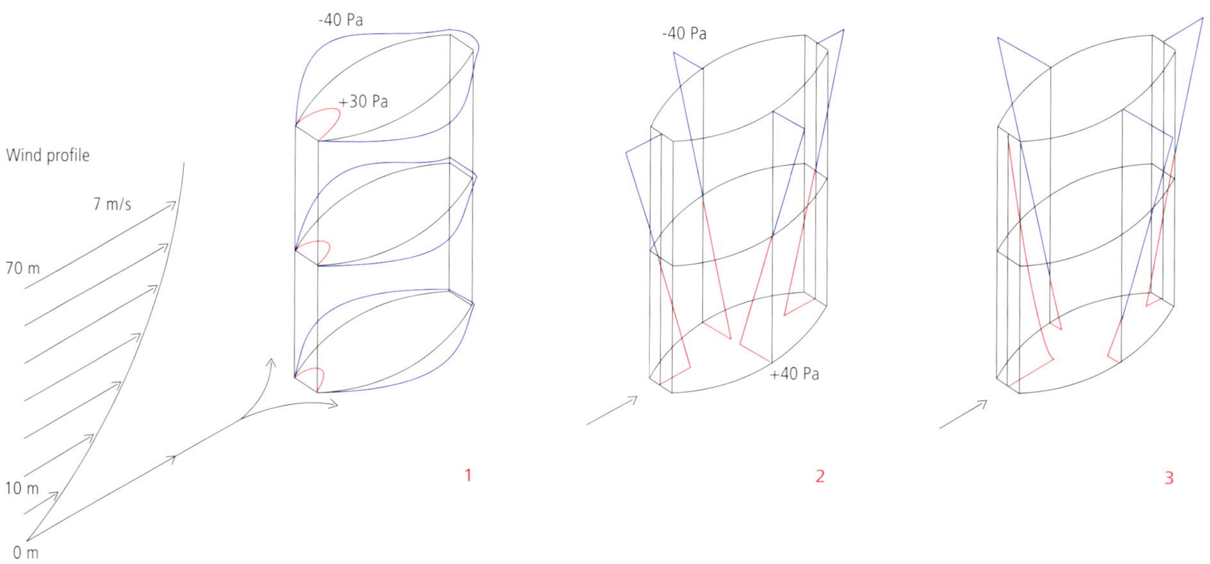

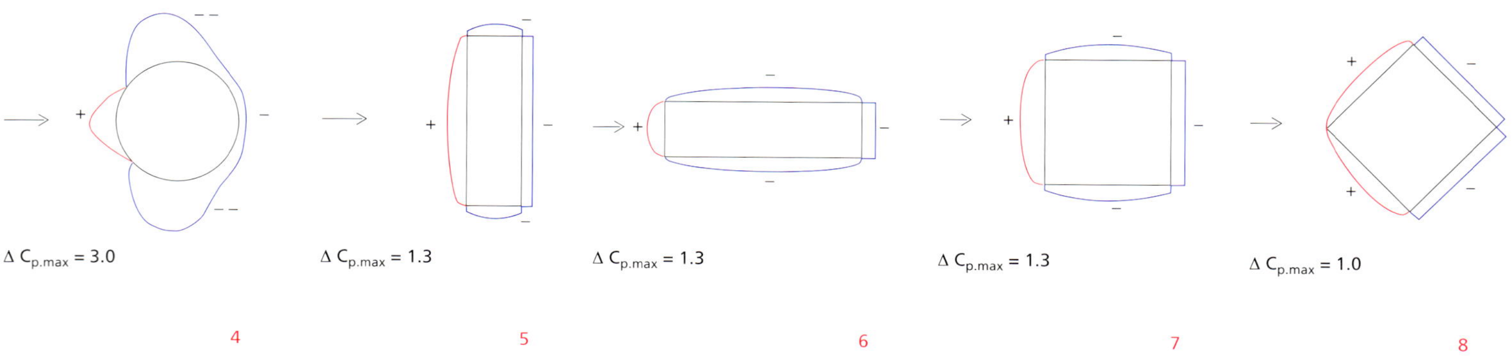

Figure 3.4
Pressure distribution on building envelopes for different parameters

Pressure
Suction

1) Wind-induced pressure distribution on envelope (schematic example)

2) Pressure differences created in winter through internal and external temperature differences (schematic example)
$T_{OU} = 0$ °C $T_i = 20$ °C

3) Distribution of pressure differences on envelope (winter) for average wind and buoyancy

Another common approach involving the entire planning team to improve building performance is to undertake a detailed, total energy simulation of the planned building. This includes temperature performance and the ingress of daylight, as well as air exchange simulation. **Figure 3.5** depicts the model for an unrealized office building. Deliberations on this project focused on either creating a conventional corner block building with a diagonal wing or encasing the entire structure in a glass skin, finishing the street frontage with a double-skin facade and the courtyard side with a 5-storey glass atrium. The atrium was to be naturally ventilated in summer and the transitional seasons and sealed in winter to harness the resulting solar heat gains for heating the building. **Figure 3.6.1/2** presents a sectional view of both options under discussion. A total building simulation is essential in order to arrive at a statement that is conclusive and convincing for the client. The simulation is vital to compare the many individual aspects of the different construction methods, and as a decision-making tool on a cost-benefit basis.

The diagrams in **Figure 3.7** provide a comparison of the simulation results. The calculated operating cost savings can be compared against the additional effort required for facade and roof construction in order to arrive at a final analysis. Results of this nature are possible only in an integrated planning process. They offer all participants the opportunity to develop new methods of improving buildings.

Figure 3.5
Model for administrative building Badenwerke AG, Karlsruhe (Architects: Auer + Weber, Munich)

Figure 3.6.1
Section of building with conventional facade solution, open courtyard / option 1

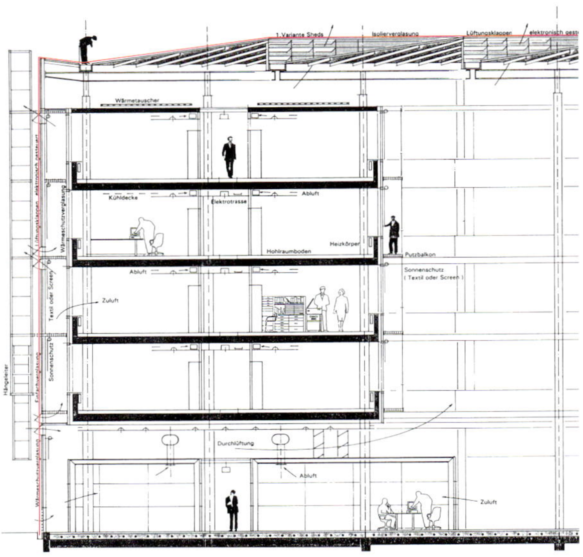

Figure 3.6.2
Section of building with double-leaf facade and glass roof / option 2

Figure 3.7
Cost comparison of building systems for conventional building (1) and fully glazed fabric (2)

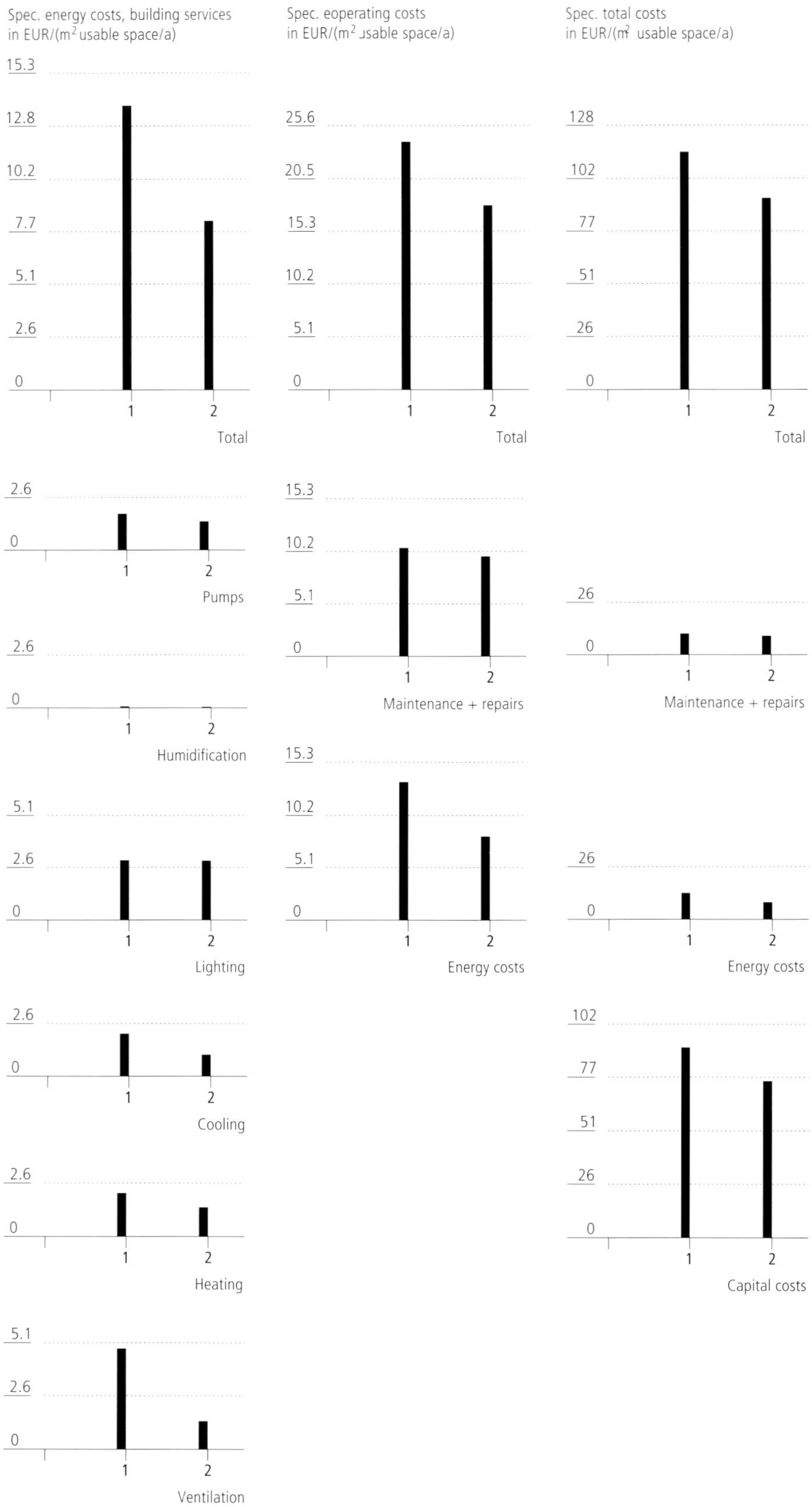

4.
Heating systems

4.1
Heat output requirements

(with Herbert Mudrack, Hamburg)

There are standard guidelines and methods (e.g. Germany DIN 4701/ Switzerland SIA 384/2) for calculating heat output requirements in rooms.

The total heat output requirement Q_S (standard heating load) is composed of:
– Thermal transmittance heat load Q_T
  = Heat loss via surrounding surfaces
– Ventilation heating load Q_V
  = Heating of incoming outside air

The heating load calculation is based on:
– Standard internal temperatures
– Standard external temperatures
– Design room temperatures for different types of rooms (attic rooms, basement rooms, rooms abutting the soil etc.)

Guidelines recommend standard internal temperatures; the perceived temperature (operative temperature), the air temperature and the median temperature of the surrounding areas must be taken into consideration.

Table 4.1 presents some of the recommended room temperatures commonly used for the calculation.

The selection of the standard external temperature is based on defaults and guidelines, although the standard external temperatures fluctuate greatly depending on location and altitude (roughly -8 to -26 °C). Standard external temperatures are generally based on the lowest two-day average values reached a total of ten times over a 20-year period.

In addition to the temperatures mentioned above, room temperatures may also be chosen for different types of rooms (e.g. attic rooms, rooms abutting the soil, room temperatures in atria etc.).

Since extremely cold temperatures generally occur over short periods of time only, the design external temperatures are adjusted accordingly, taking the storage capacity of the building into consideration. The storage capacity delays the effect of extremely low external temperatures, that is, the influence on the heating load is attenuated. The standard external temperature values are therefore diminished (or increased, respectively) depending on the storage capacity (0 to 4 K).

Even well-insulated external walls cannot deliver the desired internal surface temperature in cold seasons, leading to lower perceived temperatures and hence diminished comfort. The standard practice is to compensate for this heat loss in the thermal transmittance coefficient or U-value (also called U-factor or heat flow).

Heat loss through windows is similarly compensated as a result of the diffuse heat radiation penetrating through the window surface. Again, the thermal transmittance coefficient is adjusted accordingly. Direct solar incidence is not taken into consideration in calculating heat output requirements, although it does diminish the annual energy consumption, that is the ultimate performance value is more advantageous than the theoretically computed value.

Table 4.1
Standard indoor temperatures
for heated rooms

Type of room	**Temperature** °C
Residential buildings:	
Living and bedrooms, kitchens, WC	20
Bathrooms	24
Hallways, heated utility rooms	15
Stairwells	10
Administration buildings:	
All rooms except WC and utility rooms	20
Utility rooms	15
Commercial buildings:	
Sales areas, retail shops	20
Grocery retail, general storage	18
Meat and sausage storage	15
Hotels:	
Hotel rooms, halls, ballrooms, conference rooms	20
Schools and teaching facilities:	
Classrooms (general), gyms	20
Teaching kitchens	18
Theatres, concert halls	20
Churches (general)	15
Hospitals (see also DIN 1946, Part 4):	
Operating theatres, prenatal units	25
Other	22
Workshops:	
Minimum temperature	15
For sedentary activities	20
Barracks	20
Pools:	
Halls	28
Showers	24
Changing rooms	22
Museums, galleries, airports	20

4.1.1 Transmission heat output requirement

The transmission heat output requirement (standard transmission heat load) is calculated as follows:

$$\dot{Q}_T = U \cdot A\,(\vartheta_i - \vartheta_e) = \frac{A\,(\vartheta_i - \vartheta_e)}{R_k} \quad [W]$$

where
A = area of building component in m^2
U = thermal transmittance in W/m^2K

$\frac{1}{U} = R_U$ = resistance to thermal transmittance in m^2K/W

ϑ_i = internal temperature in °C
ϑ_e = external temperature in °C.

4.1.1.1 Thermal transmittance (U-value)

The thermal transmittance (U-value) or resistance to thermal transmittance 1/U is calculated as follows:

$$\frac{1}{U} = \frac{1}{\alpha_i} + \frac{1}{\alpha_e} \sum \frac{th}{\lambda} \qquad \text{or} \qquad R_U = R_i + R_e + \sum R_\lambda$$

α_i = internal heat transfer coefficient
α_e = external heat transfer coefficient
th = thickness of layer
λ = thermal conductivity
U = thermal transmittance coefficient
R_U = resistance to thermal transmittance = 1/U
R_λ = thermal conductivity resistance = th/λ.

Accepted standards no longer indicate the thermal transmittance (U-value) of different building components; instead, the conductivity resistance R_λ must be determined for each layer of the component based on the thermal conductivity of the component. The sum of the thermal conductivity resistance values of the individual layers in combination with the resistance values to thermal heat transfer

$$R_i = \frac{1}{\alpha_i} \quad \text{and} \quad R_e = \frac{1}{\alpha_e}$$

yields the total resistance to thermal transmittance

$R_U = \frac{1}{U}$ of the building component.

Therefore the standard thermal transmittance (U-value) is
$U_n = U + \Delta U_A + \Delta U_S$
(ΔU_A represents the diminished perceived temperature, ΔU_S partially compensates for heat gain through diffuse incident solar radiation).

The values for thermal conductivity are found in standard building material specifications and guidelines.

4.1.2 Ventilation heat output requirements

The ventilation heat output requirement $\dot{Q}_V$ (standard ventilation heat load) is calculated as follows:

$$\dot{Q}_V = \dot{V} \cdot \rho\, c\,(\vartheta_i - \vartheta_e) \quad [W]$$

$\dot{V}$ = air volume flow m^3/s
c = specific heat of air in J/kg K
ρ = air density kg/m^3.

In winter, external air penetrates into interiors through joints in windows and doors, lowering room temperatures or requiring compensation through the heating system. The amount of air penetrating into the building depends on leakage, room location, region and wind exposure, i.e. velocity. The following equation can be used to determine the hourly airflow through windows:

$$\dot{V} = \sum (a \cdot l) \sqrt[3]{\Delta p^2} \qquad [m^3/h]$$

$\dot{V}$ = air volume flow in m^3/h
Δp = pressure differential in Pa
a = coefficient of permeability in $m^3/mh\ Pa^{2/3}$
l = length of joint in m.

The heat output requirement is

$$\dot{Q}_{FL} = \dot{V} \cdot c \cdot \rho \cdot (\vartheta_i - \vartheta_e) \qquad [W]$$

$$= \sum (a \cdot l) \sqrt[3]{\Delta p^2} \cdot c \cdot \rho\,(\vartheta_i - \vartheta_e).$$

Figure 4.1 indicates the airflow through window and door joints according to the length of the joint in meters for various pressure differentials and joint permeability coefficients (a).

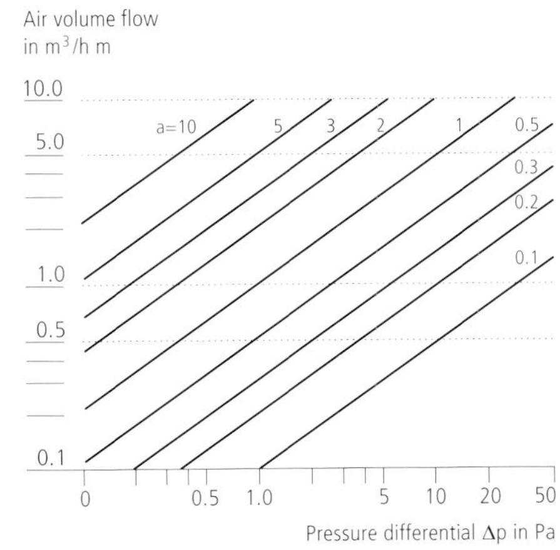

Figure 4.1
Airflow through window and door joints per m of joint length

4.1.3 Earth sheltered spaces

Earth sheltered spaces, that is rooms that are partially or fully below grade, are subject to heat loss partly via the soil to the external air, partly via the soil to the groundwater below or directly to the groundwater. The heat loss in such spaces can be calculated as follows:

$$\dot{Q}_T = A_{total}\ \frac{(\vartheta_i - \vartheta_{OT})}{R_{AL}} + \frac{(\vartheta_i - \vartheta_{GW})}{R_{GW}} \quad [W]$$

where

ϑ_{OT} = mean outside temperature $\approx 0...-5\,°C \approx (\vartheta_e + 15\,K)$

$R_{OT} = R_i + R_{\lambda A} + R_{\lambda B} + R_e$ = equivalent resistance to thermal transmission room-outside air

$R_{\lambda A}$ = equivalent resistance of the soil to thermal conductivity (**Figure 4.2**)

$R_{\lambda B}$ = resistance of building component to thermal conductivity

$R_{GW} = R_i + R_{\lambda B} + R_{\lambda S}$ = equivalent resistance to thermal transmission room-groundwater

$R_{\lambda S} = T/\lambda_S$ = thermal conductivity resistance of soil to groundwater

D = depth of groundwater

λ_S = thermal conductivity coefficient of soil ≈ 1.2 W/mK

ϑ_{GW} = groundwater temperature = 10 °C.

Figure 4.3 contains a standard plan for calculating the heat output requirement. The design room temperatures are indicated next to numbers assigned to each room. **Table 4.2** corresponds to the example illustrated in **Figure 4.3**: the table offers an example of calculating the heat output requirement for a living room (room 401 in the lower right hand corner of the plan illustrated in **Figure 4.3**).

4.1.4 Total heat output requirement

A mean thermal transmittance coefficient U_m and the ratio of building surface to spatial volume are used to arrive at a first estimate of the maximum heat output requirement (**Figure 4.4**). The mean thermal transmittance coefficient represents the average value of all thermal transmittance coefficients of the building. The heat output requirement is therefore calculated as follows:

$$\dot{Q}_n = \{U_m \cdot \frac{A}{V} + 0.34n\} \cdot V \cdot \Delta\vartheta \quad [W]$$

U_m = mean thermal transmittance in W/m²K

A = total enclosed area in m²

V = total building volume in m³

n = air change per hour = 0.5 bis 1.0 in h^{-1}

$\Delta\vartheta = (\vartheta_i - \vartheta_e)$ = temperature difference in K.

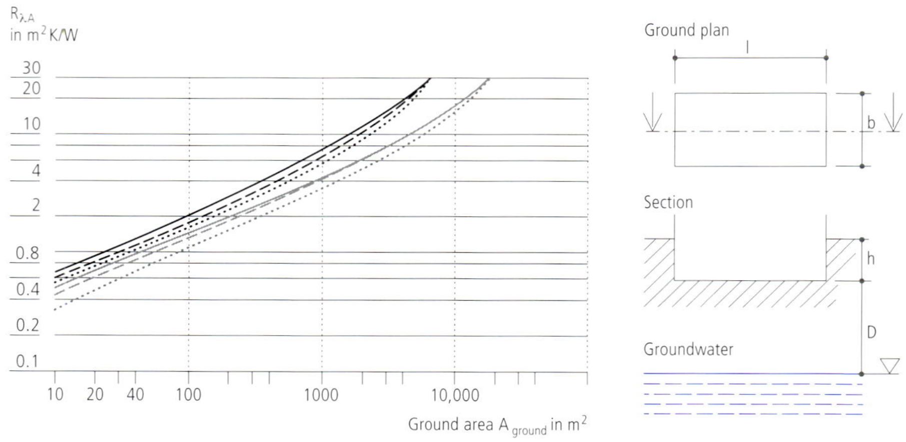

Figure 4.2
Equivalent resistance to thermal conductivity R_A in earth-sheltered building components (after DIN 4701 – Part 2)
$A_{ground} = l \cdot b$

l/b	D = 2m	D ≥ 10m
1	——	——
2	– –	– –
5	·····	·····

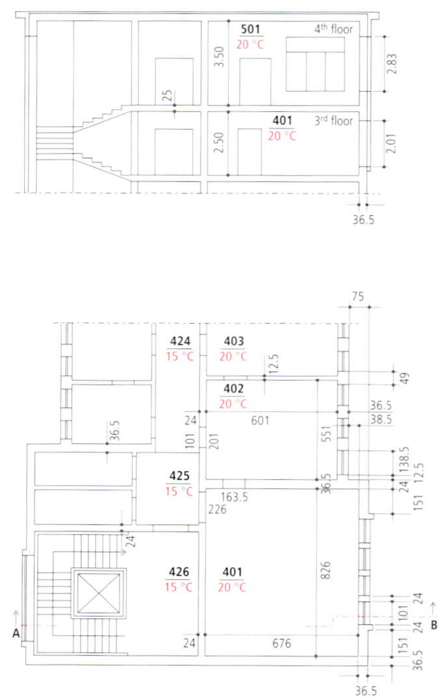

Figure 4.3
Room locations and dimensions for design example after DIN 4701

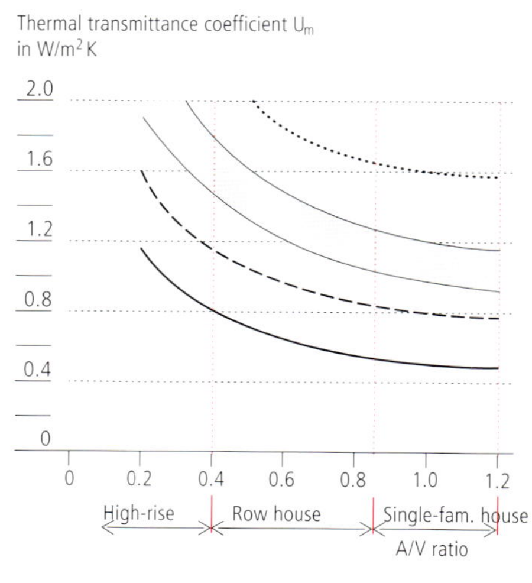

Figure 4.4
Thermal transmittance coefficient according to A/V-value

—— High insulation value
– – Raised insulation value
▭ Buildings prior to 1976
····· Old buildings

The heat requirement can also be determined on the basis of the specific heat output requirement dependant on type of construction and building scale, e.g. in accordance with the SIA 180/1 recommendations (**Figure 4.5**). The goal is to achieve lower values in the diagram.

Figure 4.6 indicates the maximal annual heating requirement permitted for buildings in Germany, from which the maximum heat output requirement can be deduced, assuming roughly 1,700 full-load heating hours.

The specific heat output requirements (according to SIA 180/1) and the permitted maximum annual heating energy requirement is likely to decrease by roughly 30 % in the following two to three years; in other words, the authorities will impose corresponding limitations. Mean thermal transmittance coefficients (U-value) of buildings will therefore settle in a range of 0.5 to 0.8 W/m²K. The insulation codes thus contribute to a massive reduction in heat loss, which may lead to disadvantages in buildings with high internal heat loads, where overheating cannot be ruled out.

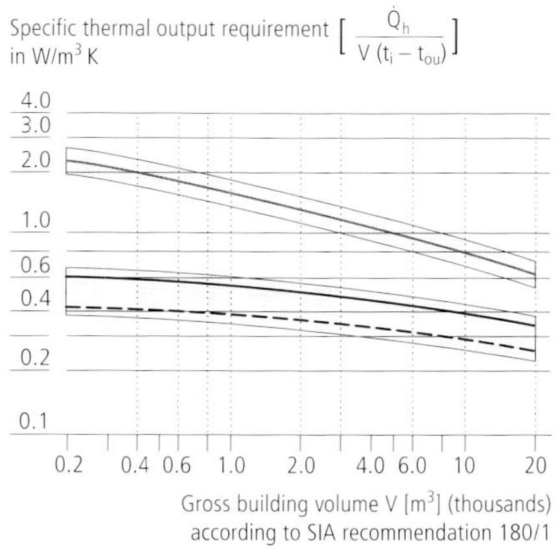

Figure 4.5
Specific thermal output requirement according to type of construction and building scale

Conventional construction
A/V = 0.25 – 1m²/m³
(AC = 0.5 – 1h⁻¹
U = 2W/m² K)

Recommended in SIA 180/1 and 384/2:
A/V = 0.25 – 1
A/V = 0.15 – 0.7
(AC = 0.3h⁻¹,
U = 0.5 – 2W/m² K)

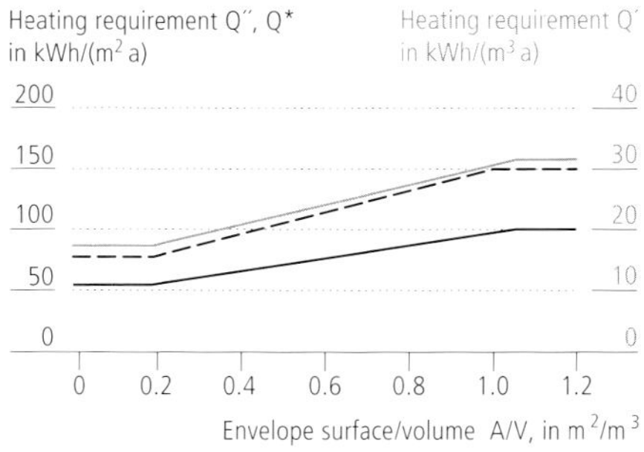

The new thermal insulation guideline (TIG) is based on
Q´ in kWh/(m³ a)
Q´´ in kWh/(m² a) for room heights up to 2.60m

– – Old TIG Q* in kWh/(m² a)
— New TIG Q´´ in kWh/(m² a)
— New TIG Q´ in kWh/(m³ a)

Figure 4.6
Max. permitted annual heating requirement in Germany

Table 4.2
Calculation of standard heat requirement according to DIN 4701

Building project: example | **Room number: 401** | **Room description: living room**

Standard internal temperature:	ϑ_i = 20 °C	House parameter:	H = 0.72 $\frac{W \cdot h \cdot Pa^{2/3}}{m^3 \cdot K}$
Standard external temperature:	ϑ_e = – 14 °C		
Room volume:	V_R = 140 m³	Number of internal doors:	n_1 = 2
Total enclosed area:	A_{total} = 187 m³	Height above ground:	z = 10m
Temperature of ambient air flow:	ϑ_a = – 14 °C	Height correction factor (airflow):	ε_A = 1.0
Exhaust air excess:	$\Delta\dot{V}$ = m³/s	Height correction factor (no airflow):	ε_N = 0

1	2	3	4	5	6	7	8	9	10	11	12	13	14	15	16	17
			Area calculation					Transmission heat requirement					Permeability to air			
Acronym	Direction	Number	Width	Height/ length	Area	Deduct area? (–)	Area invoiced	Standard thermal transmittance	Temperature difference	Transmission heat requirement of building component	Number of horizontal joints	Number of vertical joints	Joint length	Joint permeability factor	Building component permeability	Exposed/ not exposed to airflow (F/N)
–	–	n	b	h	A	–	A'	U_S	$\Delta\vartheta$	$\dot{Q}_T$	n_h	n_v	l	a	$\sum (a \cdot l)$	
–	–	–	m	m	m²	–	m²	$\frac{W}{m^2 \cdot K}$	K	W	–	–	m	$\frac{m^3}{m \cdot h \cdot Pa^{2/3}}$	$\frac{m^3}{h \cdot Pa^{2/3}}$	–
OW	SE	4	1.01	2.01	2.0		8.1	2.2	34	606	8	8	24.1	0.6	14.5	A
EW	SE	1	8.26	2.75	22.7	OW	14.6	1.33	34	660						
EW	SE	1	0.38	2.75	1.0		1.0	1.33	34	45						
EW	SW	1	6.76	2.75	18.6		18.6	1.33	34	841						
ID		1	1.63	2.0	3.3		3.3	2.00	5	33						
IW		1	8.26	2.75	22.7	ID	19.4	1.59	5	154						
										2339						

Permeability exposed to airflow:	$\sum (a \cdot l)_F = 14.5 \frac{m^3}{h \cdot Pa^{2/3}}$	Minimum ventilation heat requirement:	$\dot{Q}_{Vmin}$	= 320 W
Permeability not exposed to airflow:	$\sum (a \cdot l)_N = – \frac{m^3}{h \cdot Pa^{2/3}}$	Standard ventilation heat requirement:	$\dot{Q}_V$	= 809 W
Room parameter:	r = 0.9	Standard transmission heat requirement:	$\dot{Q}_T$	= 2339 W
Ventilation heat requirement through joint ventilation:	$\dot{Q}_{VJ} = \varepsilon_A \cdot 14.5 \cdot H \cdot r \cdot 34 = 320$ W	Critical value:	D	= 0.37 $\frac{W}{m^2 \cdot K}$
Ventilation heat requirement through HVAC systems:	$\dot{Q}_{HVAC}$ = W	Proportional ventilation heat requirement:	$\dot{Q}_V/\dot{Q}_T$	= 0.35
		Standard heat requirement:	$\dot{Q}_S$	= 3148

OW = outside window
EW = external wall
IW = internal wall
ID = internal door

4.1.5 Actual heat output requirement

As a result of more stringent insulation codes, the actual heat output requirement will influence heat requirement calculations in the future. The external and internal heat gains through direct or indirect solar incidence, occupants and lighting, as well as machines and devices will have to be taken into consideration. The following examples illustrate how the annual heating requirement can be reduced, e.g. to remain below a maximum of 60 kW/a. **Figures 4.7** to **4.9** depict different building types and their actual heat balance as a result of architectural measures designed to gain thermal energy in winter and avoid overheating in the transitional seasons or in summer.

Figure 4.7 shows a typical school building, the unique requirements of such a building and the various principles employed to diminish the heat balance throughout the year.

In this school building solar gain is achieved in winter by means of a double-leaf facade, leading to a drastic reduction of the total heat energy balance for the year (reduction of annual heating energy by roughly 50 %).

Figure 4.8 illustrates a facade detail for an office building, again in consideration of the requirements typical for such buildings, the projected periods of use and proposed principles. In this office building, the heating requirement is considerably reduced by means of a double-leaf facade. Rear ventilation is necessary for the facade when no heating energy is required in the building, or, when internal heat radiation from occupants, lighting and machines combine with the external heat gains to fully cover the heat loss of the building.

In the example of a residential building (multi-family dwelling) solar heat gain is achieved by means of winter gardens placed in front of the living areas. The detail sketch in **Figure 4.9** illustrates the use of the relevant room units.

In all these examples the common design element of a double-leaf facade for south- or west-facing rooms leads to a reduction of heat loss and achieves heat gain in winter. It is important, however, to consider that the initial investment costs in the facade component of the building are amortized over a period of eight to ten years through energy cost savings as a result of improved thermal insulation and the passive use of solar energy.

Winter

Transitional season

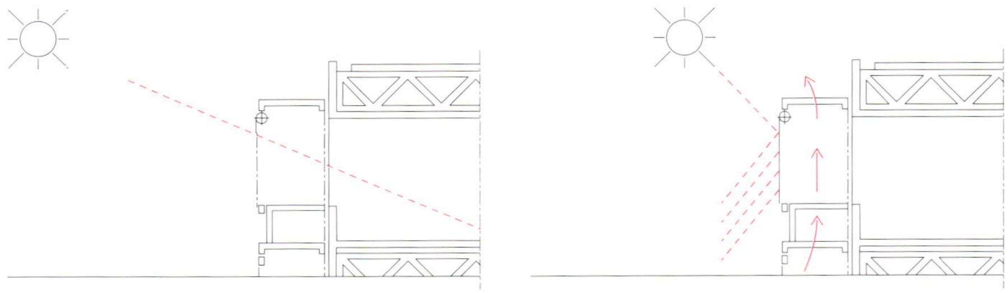

- Optimal use of daylight and solar energy
- Use of waste heat via heat pump

- External shading element with controlled light incidence
- Avoidance of overheating by means of "ventilating" accumulated heat

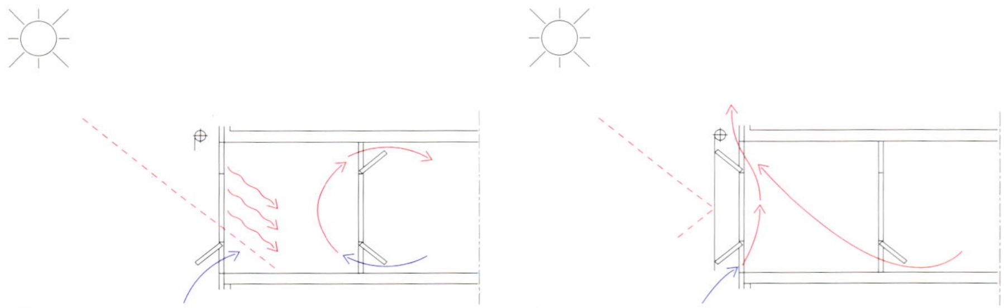

Winter
- Incident sunlight in winter garden
- Storage and distribution of heat in apartments

Summer
- External fabric shade
- Natural ventilation by means of ventilation louvres

Figure 4.7 (top)
Example of projection for a school building

Figure 4.8 (middle)
Example for an administration building with double-leaf facade

Figure 4.9 (bottom)
Example for residential building with winter garden

Table to Figure 4.7

Daylight requirement
Waste heat from humans, light and electrical appliances
Different periods of use (day/night, holidays)

	Classroom	Access	Gym
Luminous intensity:	300 – 500 lux	>100 lux	300 lux
Room temperature:	21 – 26 °C	≈ 18 °C	16 – 22 °C
Fresh air/air change:	>25m^3/h/pers.	1h^{-1}	2h^{-1}

Design principles:
- Good visual contact with exterior and daylight use
- Plan for sun and glare protection
- Use of incident sun especially in the morning and in transitional seasons
- Keep storage mass in interior to a minimum to accelerate warm-up period in classrooms
- Good insulation
- Correct sizing of heating system (take waste heat into consideration)

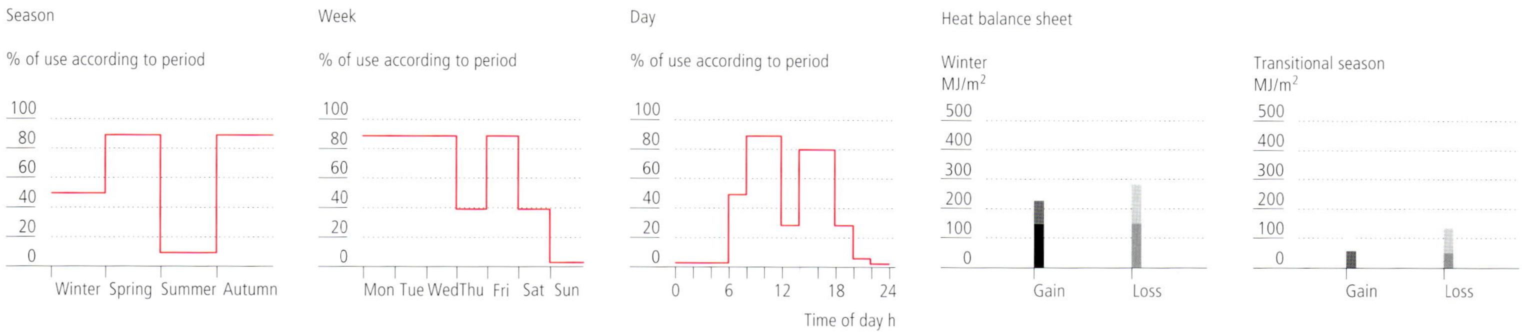

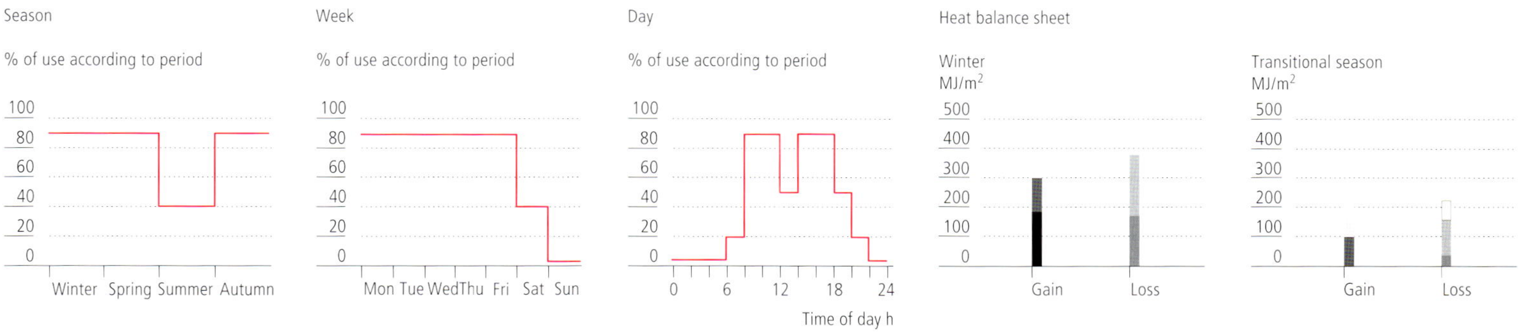

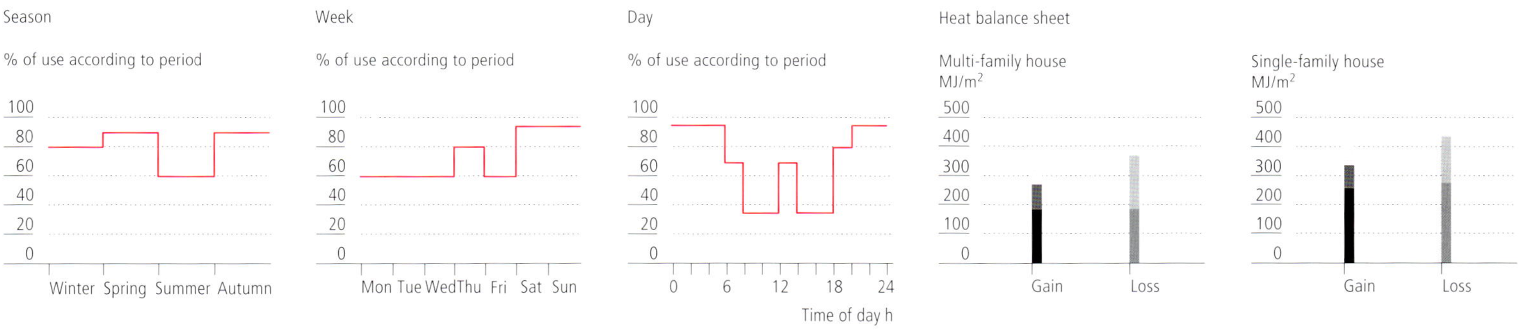

Table to Figure 4.8

Large internal loads (occupants, lighting, computers)
Daylight requirements (excellent light quality)
Excess heat in summer (air conditioning)

	Office	Access	Archives
Luminous intensity:	400 – 1.000 lux	100 lux	100 lux
Room temperature:	21 – 26 °C	18 – 26 °C	≈ 19 °C
Fresh air/air change:	>25m³ /h/pers.	≈ 1h⁻¹	0.3h⁻¹

- Avoid risk of overheating in summer
- Plan for good sun and glare protection
- Achieve cooling predominantly by natural ventilation and partially by means of air-conditioning system
- Aim for good natural light
- Close attention should be paid to interaction between internal loads, solar gains, heating and air-conditioning system in relation to heat balance sheet
- Avoid large areas of tinted glass
- Activate building mass
- Plan with great degree of flexibility in mind to allow for future changes in use

Table to Figure 4.9

Large amounts of heat loss through transmission and ventilation
Solar energy can replace (mechanical) heating to a large degree

	Living room	Bedroom	Bathroom/WC
Luminous intensity:	100 – 500 lux	100 – 500 lux	≈ 250 lux
Room temperature:	19 – 26 °C	14 – 24 °C	≈ 22 °C
Fresh air/air change:	≈ 0.4h⁻¹	≈ 0.4h⁻¹	0.3 – 2h⁻¹

- Aim for compact building form and enclosed types of construction
- Ensure good insulation
- Take course of the sun into consideration
- Sensible disposition of rooms
- Plan for large window areas on south side and fewer openings on north side
- Generally construct mostly buildings with sluggish thermal response (with heavy materials) to store solar energy and achieve natural temperature control

■ Heating energy
■ Waste heat from humans, electrical energy
■ Transmittance
■ Ventilation
Solar energy
□ AC system

In future, the heating energy requirements in buildings for standard uses should not exceed the following values:

Single and duplex houses
approx. 78 kWh/m^{2}a
approx. (280 MJ/m^{2}a)

Multi-family dwellings, retirement homes, hotels and similar buildings
approx. 70 kWh/m^{2}a
approx. (259 MJ/m^{2}a)

Office buildings, schools, retail shops, museums, libraries etc.
approx. 60 kWh/m^{2}a
approx. (220 MJ/m^{2}a)

Industrial buildings, warehouses, distribution centres, sports halls, workshops, railway stations etc.
approx. 66 kWh/m^{2}a
approx. (240 MJ/m^{2}a)

Buildings with special requirements, such as research facilities and laboratories, lecture halls, department stores, restaurants, hospitals, indoor pools, theatres, radio and television studios as well as buildings with other special criteria
approx. 90 kWh/m^{2}a
approx. (330 MJ/m^{2}a).

4.2 Primary energy sources

Currently, most energy is generated through fossil fuel combustion, with coal, crude oil and natural gas being the principal energy sources. They still cover nearly 90 % of the global energy consumption, although nuclear energy is gaining in importance.

Alternative energy sources (renewable energy sources) such as sun, wind, water, rainwater and biomass constitute only a small portion of the total energy supply, although there are some positive indicators that this sector is clearly growing. **Figure 4.10** indicates how the global energy requirements have been met between 1900 and the present day, as well as a projection up to the year 2060. The focus in primary energy sources will shift from coal, crude oil and natural gas to renewable energy sources. Wind and solar energy, especially, as well as geothermal energy and biomass will take on greater importance.

Primary energy sources exist in a variety of forms. These are, among others
- Solid fuels
- Liquid fuels
- Gaseous fuels
- Solar energy
- Wind energy.

Figure 4.10
Sources for global energy requirements, 1900 to 2060
1 exajoule = 31.12 million tCE
(source Energie & Management, annual journal, 12/99)

1 Coal
2 Crude oil
3 Natural gas
4 Nuclear power
5 Biomass (traditional)
6 Hydro
7 Wind energy
8 Biomass (new)
9 Solar energy
10 Geothermal/oceanic energy
11 Undetermined

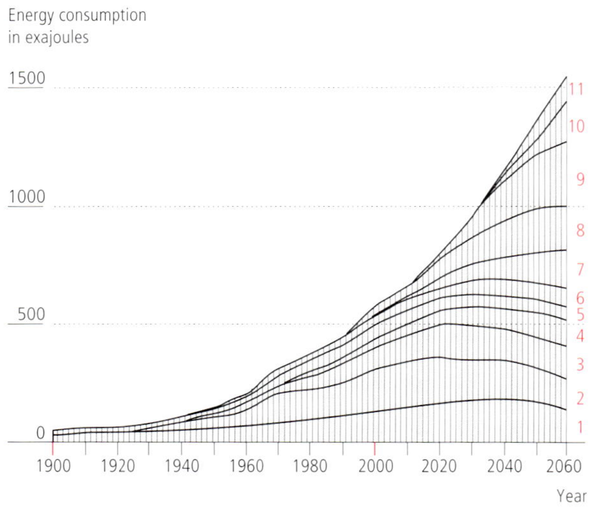

4.2.1 Solid fuels

Solid fuels are defined as natural or refined fuels, according to the method of exploitation.

Natural solid fuels are bituminous coal, brown coal, peat, wood and straw. Coal and peat are residuums produced by the decomposition and disintegration of plants from early periods of the earth, under high pressure and oxygen deficiency.

Bituminous coal is the oldest, natural geological fuel, found worldwide at various depths.

Brown coal dates from a much more recent period and contains fragments of wood-like inclusions. The water content of brown coal is approximately 45 to 60 %; it is generally exploited through open pit mining. Brown coal has a very low calorific value and is hardly worth transporting across great distances.

Peat is formed from decomposed plants and water; it is therefore high in water content and is exploited by cutting large peat blocks on the surface and drying the blocks prior to combustion.

In Europe, wood is used as a combustion fuel mostly in forestry and sawmill operations, generally in chip form (wood chips, sawdust).

Bituminous coal, brown coal bricks, coke and charcoal are **refined solid fuels**. The bricks are manufactured from crushed and dried bituminous and brown coal in briquetting presses. They are available in a variety of shapes.

Coke is produced in special ovens at approximately 1,200 °C during a process of degasifying bituminous coal, brown coal or coal blends; (degasification is the process whereby gaseous components are driven out of solid material in an airtight heating chamber).

Charcoal is a product of wood that has been charred in charcoal kilns or other specialized kilns. **Table 4.3** lists the average compositions and properties of solid fuels.

The percentage of various substances such as carbon, hydrogen etc. in the total weight, as well as the lower calorific value, are important parameters in how solid fuels are used. Combustion plants can only be fully analysed by establishing the required theoretic air volumes, exhaust or waste gas volumes, ash accumulation and amount of fuel. In determining the environmental compatibility of solid fuel installations, it is vital to observe the air pollution regulations in each state. These regulations limit the emissions of combustion plants. In the main, these limitations apply to:
- Dust
- Carbon monoxide (CO)
- Sulphur oxide (SO_X)
- Halogenated hydrocarbon (HCl, HF etc.).

Table 4.3
Average composition and properties of solid fuels

Fuels	Carbon	Hydrogen	Oxygen	Nitrogen	Sulphur	Water	Ash	Net calorific value	Theor. air volume	Theor. dry waste gas volume	Vapour	Theor. humid waste gas volume	Max. carbon dioxide content in waste gases
	**) for raw/crude fuels												
	c	h	o	n	s	w	a	n.c.v.	A_{min}	V_{atr}	V_{H_2O}	V_{af}	$CO_{2\ max}$
	Weight/%	Weight/%	Weight/%	Weight/%	Weight/%	Weight/%	Weight/%	kJ/kg	m_n^3/kg	m_n^3/kg	m_n^3/kg	m_n^3/kg	%
Carbon (pure)	100	–	–	–	–	–	–	33,820	8.9	8.9	–	8.9	21.0
Coal (Ruhr)													
Long-flame gas coal	77	5	8	1	1	3	5	30,100	7.9	7.7	0.6	8.3	18.5
Gas coal	80	5	5	1	1	3	5	31,400	8.3	8.0	0.6	8.6	18.5
Bituminous coal	81	5	4	1	1	3	5	31,800	8.4	8.1	0.6	8.7	18.5
Forge coal	82	4	4	1	1	3	5	31,800	8.3	8.0	0.5	8.5	18.8
Lean coal	84	4	2	1	1	3	5	31,400	8.5	8.2	0.5	8.7	18.8
Hard coal (US), anthracite	85	3	2	1	1	3	5	31,400	8.3	8.1	0.4	8.5	19.3
Coke (blast furnace)	83	0.5	0.5	1	1	5	9	28,900	7.7	7.5	0.1	7.6	20.5
Pitch coal (Upper Bavaria)	58	4.3	10	1.2	5.5	10	11	22,930	3.0	2.95	0.6	3.55	18.2
Brown coal (Rhineland), *)													
raw	30	3	10	1	1	50	5	9,630	3.1	3.0	0.9	3.9	17.2
Briquettes	55	5	18	1	1	15	5	19,250	5.6	5.4	0.7	6.1	17.2
Peat, air-dry *)	38	4	26	1	1	25	5	13,800	3.6	3.5	0.7	4.2	19.8
Wood, air-seasoned *)	42	5	37	–	–	15	1	14,600	3.8	3.8	0.7	4.5	20.4

*) The composition, especially in terms of water content, of brown coal, peat and wood fluctuates across a wide range.

**) Conversion of composition in relation to pure fuel subsequent to multiplication by 100/(100-w-a).

4.2.2 Liquid fuels

Liquid fuels are:
- Mineral oils as distillation products from crude oil
- Creosote as low temperature carbonization and distillation product from coals
- Synthetic oils from coal liquefaction
- Other liquid fuels, for example from plants (potatoes, rape etc.)

Mineral oils were created millions of years ago in sedimentary rocks from residual animal and plant material exposed to high temperatures and partial bacterial influence. Mineral oil is drilled to gain crude oil in the familiar composition. Chemically, natural oil is a blend of different hydrocarbons (paraffin/olefin/aromatics etc.). It is processed by means of fractioned distillation (separation into distillation tails) and refined into light, middle and heavy oils or treated in cracking processes (in which larger hydrogen molecules are split into smaller units by heating the oil under pressure).

Light crude, especially petroleum with a boiling point of 50 to 200 °C, consists primarily of paraffin or limit hydrocarbons.
Middle oils are differentiated as petroleum (boiling point 200 to 250 °C) and gas oil (boiling point 200 to 350 °C), previously used in oil gas production and now mostly known as diesel oil. Extra-light heating oil also belongs into this category.

Heavy oil (boiling point above 350 °C) is generally used a lubricant oil and as heating oil for power stations, and as a fuel for heavy oil machines.

Creosotes, distilled tar products, do not play a role as primary energy sources for heating. Tars are produced in the course of distilling (coking) and low temperature carbonization of the fuels. Tars are processed into light. middle and heavy oils (gasoline, diesel, heating oil) through distillation, cracking and hydrogenation.

Synthetic oils from bituminous and brown coal, and crude oil tailings are not yet used for heating with conventional boilers, although they may become more important in future. Synthetic oils are produced, for example, by liquefying coal in high-temperature reactors.

Heating oils are generally distillation products of natural oil. It is important to differentiate between shale oil, mineral oil and lignite tar oil in consideration of the differing types of oil burners. Extra-light heating oil (EL heating oil) is especially important as a primary energy source; the same is true for heavy heating oil (H heating oil) in industrial applications. Light and medium heating oil have largely fallen into disuse.

The emissions that result from burning heating oil are largely composed of carbon dioxide and water.
Table 4.4 lists the minimum requirements for heating oil, again with particular emphasis on sulphur content, water content, lower calorific value and ash production.

		EL heating oil	L heating oil *)	M heating oil *)	H heating oil	Tested according to
Density for 15°C maximum	max. g/ml	0.860	1.10	1.20	must be indicated	DIN 51757
Flash point in closed crucible	above °C	55	55	65	80	DIN 51755 DIN 51758 DIN EN 57
Kinematic viscosity maximum	mm^2/s (cSt)	for 20 °C 6 (≈ 1.5 E)	for 20 °C 17 (≈ 2.5 E)	for 50 °C 75 (≈ 10.E)	for 100 °C 50 (≈ 7 E) for 130 °C 20 (≈ 3 E)	DIN 51366 DIN 51550 DIN 51562
Pour-point	max. °C	– 6	–	(≈ 45)	40 **)	ISO 3016
Coked residue after Conradson	Weight/%	max. 0.1	2	12	15	DIN 51551
Sulphur content max. weight/% — mineral oils lignite-tar oils coal-tar oils	 max. % max. %	0.2 – –	– 2.0 0.8	1 2.0 0.9	2.8 – –	DIN 51400 T. 1, 2, 3 and 6 DIN EN 41
Water content, non-removable,	max. weight/%	0.05	0.3	0.5	0.5	ISO 3733
Sediment content	max. weight/%	0.05	0.1	0.25	0.5	ISO 3735
Salt-freedom in coal-tar heating oils	°C	–	must be indicated	must be indicated	–	DIN 51603
Net calorific value n.c.v. min.	kJ/kg	42,000	37,700	37,700	39,500	DIN 51900
Ash (oxide ash) max.	weight/%	0.01	0.04	0.15	0.15	DIN EN 7

Table 4.4
Minimal requirements for heating oil after DIN 51603 Part 1 (12.81) and Part 2 (10.76)
(1 kWh = 3600 kJ)

*) Manufactured from brown coal and (stone) coal, but also from mineral oils.

**) When > 40°C; must be indicated by supplier.

4.2.3 Gaseous fuels

The properties of the available heating or combustion gases are very different, and most heating gases are a blend of combustible and non-combustible gases, with the predominant combustible components being hydrocarbons, hydrogen and to a lesser degree carbon monoxide. Heating gas is either natural gas (natural gas or petroleum gas) or residual gases from technical processes, such as coke oven gas, blast furnace gas, suction gas etc.

Heating gases are differentiated by their upper calorific value (H_U) into

Lean gases	$H_U < 2.5$	kWh/m^3_n
Medium gases	$H_U = 2.4 - 4.0$	kWh/m^3_n
Strong gases	$H_U = 4.0 - 6.0$	kWh/m^3_n
Rich gases	$H_U > 6.0$	kWh/m^3_n.

Today, natural gases are the most important primary energy source. All gaseous, generally impure hydrocarbon compounds, which are extracted from the ground and combustible, are categorized as natural gases. They are odourless and contain a multitude of admixtures.

Like crude oil and coal deposits, natural gas fields evolved from simple organisms, whose deposits were transformed through exposure to high pressures and temperatures. Natural gas deposits have accumulated in porous rock formations, which were formed by tectonic shifts and covered in top layers that are impermeable to gas. The composition of natural gas varies greatly from one exploration area to another. It must therefore be processed prior to use in order to remove undesired components such as hydrogen sulphide, water and others. Processing generally occurs at the extraction source, with sulphurous compounds being especially problematic since combustion of these compounds produces the noxious sulphur dioxide (SO_2).

Tables 4.5 / 4.6 list the principal technical data of relevant combustion gases.

4.2.4 Solar energy

The sun radiates a total of 180×10^{12} tCE (tons of coal equivalent) onto the Earth on an annual basis, or approximately 15,000 times more than the total global energy consumption. In Germany, for example, the annual sum of radiation lies in the region of 1,000 kWh/m^2a; this value is two to three times higher in sunny regions. In reference to the total surface of the Earth, consumption, on the other hand, is no more than 0.15 kWh/m^2a.
Solar energy has been harnessed for heating in the sunny regions of the Earth since time immemorial and, more recently, also for cooking and cooling.
The use of solar energy is therefore hardly a new idea. What is new, however, is that even countries where periods of sunshine are relatively brief are employing fairly elaborate means to harness the sun's energy in order to replace conventional energy sources at least to some degree. Solar energy will remain a supplementary energy source in Central and Northern Europe for many years to come, supplying only a fraction of the necessary heating energy. There are two principal reasons. First, the annual availability of solar energy is too low in the northern latitudes. And second, solar energy is most abundantly available when heating requirements are insignificant. Furthermore, the brief periods of sunshine in Central and Northern Europe mean that high investment costs are disproportionate to subsequent energy savings, since nearly 70 % of the radiation occurs between April and September. The amortization calculations of solar energy installations are much more favourable for Southern Europe, because the annual sum of radiation is noticeably higher (e.g. Marseille 1,860 kWh/m^2a, Zurich 1,160 kWh/m^2a, Hamburg 930 kWh/m^2a).

All the same, the time has surely come to harness solar energy to the best of our ability, for solar energy can help in part to solve our energy problems in the long term.

4.2.5 Environmental compatibility

The principal focus of environmental protection targets the following areas:
- Keeping the air clean
- Water protection and waste water/sewage treatment
- Waste disposal
- Noise reduction

With regard to generating heat for heating purposes, air purity is of primary importance. The other environmental protection issues contained in the list are relevant in the area of heat-and-power generation. The emissions resulting from generating heating energy are in the main emissions of contaminants.

The principal emissions are:
- Sulphur dioxide (SO_2)
- Nitrogen oxide, (NO, NO_X)
- Carbon monoxide (CO)
- Organic compounds (hydrocarbons)
- Heavy metals, CFCs (chlorofluorocarbons)
- Carbon dioxide (CO_2) is increasingly regarded as a contaminant linked to the greenhouse effect
- Dust

Table 4.7 lists the specific emissions for different fuels.

No.	Fuel gas	Volumetric composition %/vol.							Gross calorific value	Net calorific value	Density ratio	Wobble index
		H_2	CO	CH_4	(C_3H_6) C_nH_m	other hydro-carbons	CO_2	N_2	H_o kJ/m_n^3	H_u kJ/m_n^3	d_v (Air = 1)	W_o (H_o:$\sqrt{d_v}$)
1	Blast-furnace gas	2	30	–	–	–	8	60	4,080	3,975	0.99	4,100
2	By-product coke-oven gas	12	28	(<) 0.5	–	–	5	54.5	5,340	5,025	0.88	5,700
3	Bituminous coal generator gas	12	29	2	–	–	3	54	5,965	5,650	0.86	6,400
4	Brown coal generator gas	15	27	2	–	–	7	49	6,070	5,760	0.86	6,500
5	Mixed gas (12 + 1)	19.3	22.2	8.4	0.6	–	6	43.7	9,125	8,370	0.80	10,200
6	Coke water gas	50	40	(<) 0.5	–	–	5	4.5	11,510	10,460	0.55	15,500
7	Coal water gas	50	35	5	–	–	5	5	12,770	11,615	0.53	17,550
8	Sewer gas (12 + 6)	51	18	19	2	–	4	6	18,000	16,120	0.46	26,540
9	Sewer gas (12 + 2)	44	12	22	2	–	4	16	18,000	16,120	0.51	25,200
10	Propane – air (17 O_2)	–	–	–	–	18	–	65	18,000	16,740	1.10	17,160
11	Oil carburized coal water gas	37	28	15	5	–	8	7	18,840	17,370	0.64	23,550
12	Coke-oven gas (grid gas)	55	6	25	2	–	2	10	19,670	17,370	0.39	31,500
13	Oil carburized coal water gas	45	35	1	10	–	4	5	20,090	18,420	0.63	25,300
14	Coal gas	52	8	28	2.5	–	2	10	20,930	18,840	0.41	32,700
15	Low-temperature carbonization gas (from bituminous coal)	25	5	45	5	10	5	5	33,500	30,350	0.62	42,550
16	Natural gas type L	–	–	82	–	3.0	1.0	14	35,200	31,800	0.64	44,000
17	Natural gas type H	–	–	93	–	5.0	1.0	1.0	41,300	37,300	0.61	52,880
18	Methane	–	–	100	–	–	–	–	39,850	35,790	0.55	53,750
19	Oil gas	20	5	40	20	10	1	4	45,210	41,230	0.74	52,550
20	Propane C_3H_8	–	–	–	–	100	–	–	100,890	92,890	1.562	80,730
21	*n*-butane C_4H_{10}	–	–	–	–	100	–	–	133,870	123,650	2.091	92,600

Table 4.5
Composition, density and calorific value of industrial gases, after F. Schuster et al.

Substance	Symbol	Molecular mass	Density	Content		Combustion/calorific value				Theoretical combustion air volume		Waste gas volume, moist		Dry waste gas	Moist waste gas
			ρ	c	h	H_o	H_u	H_o	H_u	L_{min}	L_{min}	V_{af}	V_{af}	$CO_{2\,max}$	H_2O
			kg/m_n^3	Weight/%	Weight/%	kJ/kg	kJ/kg	kJ/m_n^3	kJ/m_n^3	m_n^3/kg	m_n^3/kg	m_n^3/kg	m_n^3/kg	Vol.-%	Vol.-%
Acetylene	C_2H_2	26.04	1.17	92.5	7.5	49,910	48,220	58,470	56,490	10.2	11.9	10.6	12.4	17.5	8.1
Benzene	C_6H_6	78.1	3.73	92.2	7.8	42,270	40,580	157,970	151,650	10.2	35.7	10.6	37.2	17.5	8.1
Butane (n)	C_4H_{10}	58.1	2.71	83	17	49,500	45,715	134,060	123,810	11.4	30.9	12.4	33.4	14.1	15.0
Butylene	C_4H_8	56.1	2.60	85	15	48,430	45,290	125,860	117,710	11.6	28.9	12.4	30.9	14.9	12.9
Ethane	C_2H_6	30.1	1.35	80	20	51,880	47,490	70,290	64,345	12.3	16.7	13.4	18.1	13.2	16.5
Ethanol	C_2H_5OH	46.11	2.19	52	13	30,570	27,710	67,070	60,790	7.0	14.3	8.0	16.4	15.0	18.4
Ethylene	C_2H_4	28.05	1.26	85.7	14.3	50,280	47,150	63,410	59,460	11.3	14.3	12.1	15.3	15.1	13.1
Carbon monoxide	CO	28.01	1.25	42.9	0	10,100	10,100	12,630	12,630	1.91	2.38	2.88	34.7	0	
Methane	CH_4	16.04	0.72	75	25	55,500	50,010	39,820	35,880	13.3	9.52	14.6	10.5	11.7	19.0
Methanol	CH_3OH	32.04	1.52	37.5	12.5	23,840	21,090	36,200	32,030	5.0	7.15	6.0	8.6	15.1	23
Propane	C_3H_8	44.09	2.01	81.8	18.2	50,340	46,350	101,240	93,210	11.8	23.8	12.8	25.8	13.8	15.5
Propylene	C_3H_6	42.08	1.91	85.7	14.3	48,920	45,780	93,580	87,575	11.2	21.4	11.9	22.9	15.1	13.1
Toluene	C_7H_8	92.11	4.87	91.2	8.8	42,850	40,940	208,890	199,570	10.4	42.8	10.9	44.8	17.1	8.9
Hydrogen	H_2	2.016	0.090	0	100	141,800	119,970	12,745	10,780	26.4	2.38	32.0	2.88	0	34.7

Table 4.6
Combustion of gaseous and vaporous fuels
(1 kWh = 3600 kJ)

Combustion and calorific values based on 25°C and 1.013 bar, volumes based on 0°C and 1.013 bar (DIN 51850)

Fuel	Emissions in kg/MWh					Calorific value
	SO_2	NO_x	CO	Dust	Org. comp.	kWh/kg
Coal	1.80	0.36	23.00	0.90	0.90	8.35
Brown coal	0.83	0.05	25.00	1.26	0.54	5.3
EL heating oil	0.47	0.18	0.18	0.00	0.04	11.9
H heating oil	1.76	0.65	0.04	0.11	0.03	11.0
Gas	0.01	0.18	0.22	0.00	0.01	10.0

Table 4.7
Specific fuel emissions (kg/MWh)

4.3 Heat generating systems

Heat is usually generated through primary energy in solid, liquid or gaseous form. However, heat energy can also be made available through district heat or electrical energy. **Figure 4.11** gives an overview of the diverse forms of heat generation, i.e. means of making heat energy available.

4.3.1 District heat

District heating installations can be divided into the following principal subgroups, differentiated by the temperature levels of the heat transfer medium:
- Warm-water district heating (≤ 120 °C)
- Hot-water district heating (> 120 °C)
- Steam district heating (> 120 °C)
- Cold district heat (= 25 – 35 °C).

The principal elements of a district heating installation are:
- Heat generation in gas-, liquid- or solid-fuel fired boilers
- Heat distribution via long-distance pipelines
- Heat transfer stations in the buildings to be heated

Other types of heat generating installations are combined heat and power stations, where heat is extracted after (electric) power generation has occurred, and garbage incinerating plants. In the so-called cold district heat, relatively cold water (25 to 35 °C) is separated from the cooling tower cycle of a power plant, conducted via a single pipeline to the maximum consuming point, where it serves as a heat source for a heat pump. The chilled water from the cooling tower is then diverted into a river.

In the past, heating energy was generally distributed through steam. In recent years, these systems have been converted to hot-water installations or otherwise refitted, since condensate return proved difficult. Today, most district heating installations are operated on a twin- or triple-pipeline basis. In the twin-pipeline system, heat is distributed at a sliding flow temperature adjusted to the outside temperature (winter 130 °C/summer 70 °C). The triple-pipeline system offers two different temperature ranges, one remaining constant to supply high temperature subscribers throughout the year.

Figure 4.12 depicts a district-heating pipeline being laid underground to DN 900 for an urban area with a variety of energy consumers. The pipelines can also be installed above ground.

Heat generation (Supplying thermal energy)

District heat – Steam, hot water, warm water	Heat exchanger	Condensate (steam systems)
Solid fuels – Wood, coal, peat, waste, biomass	Boiler installation	Storage (bunker) Environmental impact
Liquid fuels – Heating oil (EL/L/M/H), benzene, benzole, bio-oils	Boiler installation, CHP-systems, conversion boilers	Storage (tanks)
Gaseous fuels – Natural gas (town gas), liquid gas (propane, butane), methane-containing gases, biogas	Boiler installation, CHP-systems, fuel cells, heat pumps	Part. storage (tanks)
Electrical energy	Storage tanks, storage heaters, heat pumps	Storage
Geothermal energy – Geothermal heat from: earth coils, collectors, bore holes, groundwater		Bore holes, regeneration of soil
Environmental energy – Air, surface water, waste water		
Solar energy – Solar radiation	Flat, evacuated tube collectors, absorber mats, parabolic-tube collectors	Storage, seasonal dependency

Figure 4.11
Heat generation
(Supplying thermal energy)

Figure 4.12
District heat installation in Wedel, Germany
In-ground installation, DN 900

Figure 4.13 shows the schematic layout, while **Figures 4.14** and **4.15** depict typical installations of district heating transfer stations in buildings. The district heating transfer consists of components, which are primarily installed by the energy supplier, as well as heat exchangers in an additional heat exchanger unit. While tubular heat exchangers were the standard in the past, plate heat exchangers are the standard form used today, preferred for their smaller dimensions and high efficiency. **Figure 4.16** shows a plate heat exchanger in a central heating station. Here, primary and secondary mediums (district heat/internal water cycle) are bypassed between rib plates to promote heat exchange (not water exchange) between the plates, resulting in the transfer of the heating energy from the district heating system to the internal water cycle. **Figure 4.17** once again demonstrates the installation and function of a plate heat exchanger.

District heating has the following advantages:
- No fuel demand
- No fuel storage
- No waste disposal
- Use of inexpensive fuels (e.g. garbage/low grade coal), although garbage incineration would lead to higher heating costs as a result of elaborate measures for flue gas collection
- Excellent operational dependability through alternating use of several boilers
- Space-saving, since the consumer does not need to provide a boiler room, a fuel storage room or chimney
- Low operating and maintenance cost
- Central heating systems produce less contamination than individual boiler
- Electricity and heat can be coupled in the plant district; simultaneous improvement of the thermodynamic process and use of waste heat
- Improved fire protection for heated buildings

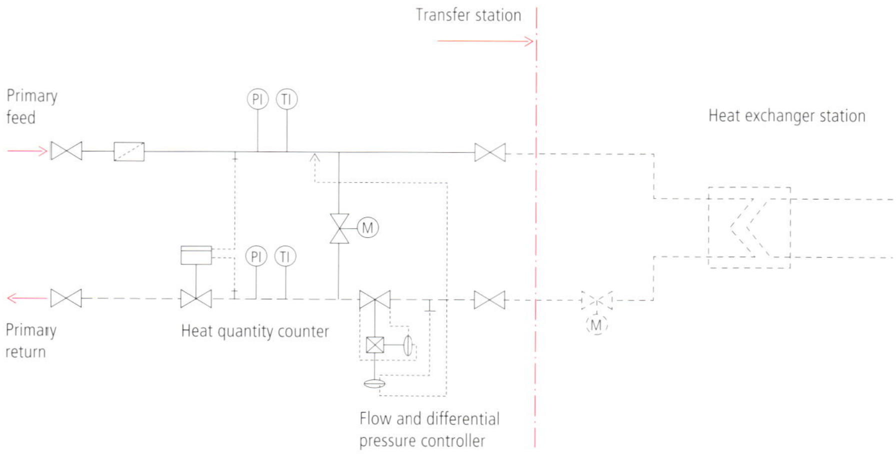

Figure 4.13
District heat transfer station, diagram

Figure 4.14
Small compact district heat station, Minidomo type
(Image: Cetetherm)

Figure 4.15
Compact Cetecom district heat station for heating and warm water with Cetesystem D storage load system
(Image: Cetetherm)

Figure 4.16
Panel heat exchanger in central heating installation

4.3.2 Sensible heat generation

When the primary energy is converted into heating energy in a building, we speak of sensible heat generation. Sensible heat can be generated in individual rooms or centrally within a building, resulting in a wide variety of technical solutions.

4.3.2.1 Individual heating units

Individual heating units in a room are:
- Chimneys
- Open fireplaces or airtight stoves
- Tiled stoves
- Individual stoves fired with coal, wood, oil or gas
- Gas space heaters
- Electrical heating units (e.g. fan heaters)
- Electrical thermal storage heaters

The individual heating units listed above are widely known and usually require a chimney connection (oil, gas and solid fuel heating) or an electrical connection.

Figure 4.17
Bolted panel heat exchanger (image: Cetetherm) and connection options

1 Fixed panel on stand
2 Connecting socket
3 Bottom guide bar
4 Panel set
5 Support
6 Tension rod
7 Moveable pressure panel
8 Top supporting rod

Simple connection

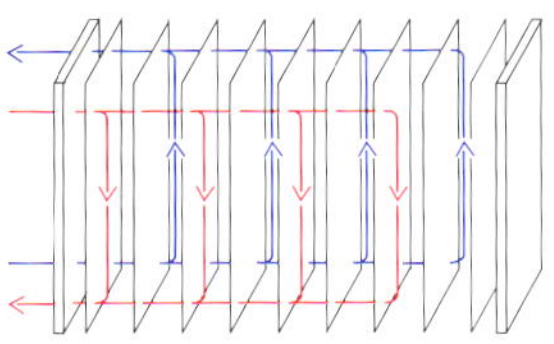

Multiple connection

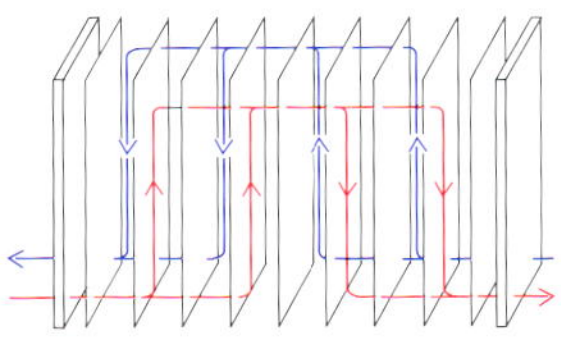

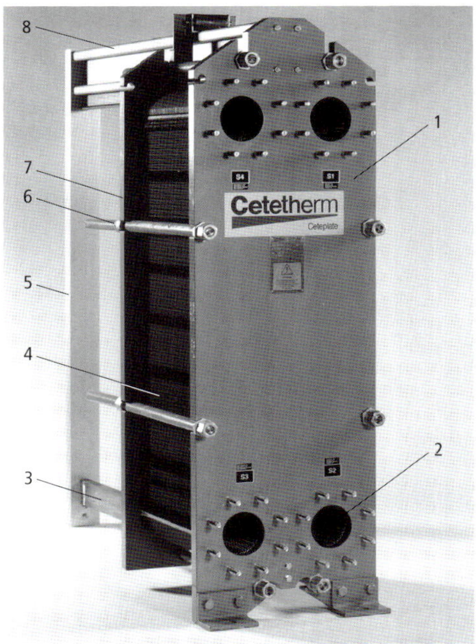

4.3.2.2 Boiler plants (solid fuels)

Boiler size
The hourly heat output required for generating heat in a system is calculated from the sum of the simultaneous user demand volumes. These demand volumes are:

$$\dot{Q}_k = \dot{Q}_{tr} + \dot{Q}_L + \dot{Q}_b + \dot{Q}_{kL} + \dot{Q}_s\ (+\ \dot{Q}_{WWB}).$$

Where
$\dot{Q}_k$ = max. heat output for heat generator
$\dot{Q}_{tr}$ = transmission heat requirement
$\dot{Q}_L$ = ventilation heat requirement
$\dot{Q}_{WWB}$ = heat requirement for service water processing (added only in special cases)
$\dot{Q}_{kL}$ = heat requirement for ventilation and air-conditioning systems
$\dot{Q}_b$ = heat requirement due to operational loss
$\dot{Q}_s$ = heat requirement of other heat consumers

According to heat consumer and use, maximum demands tend to occur at different times of the day. In the interest of efficient operation, it is therefore necessary to determine the peak periods, which define the dimensions of the boiler plant. Factors to be considered include the diurnal curve of temperature-energy of all individual systems for heating, ventilation, air conditioning, warm-water processing etc. It is vital not to install additional reserves, since boiler systems with unnecessarily elevated performance levels result in additional investment costs for boilers, firing and flue gas components, as well as higher readiness-to-operate losses, which have a negative impact on the annual efficiency of the boiler plant.

Readiness-to-operate losses occur when air is continually evacuated through the idle boiler as a result of the draught from the warm chimney, cooling the heating surface in the boiler. (Diminution through positive pressure combustion in larger boilers or motor-driven flue-gas butterfly valves.)

The boiler plant is designed to meet peak demands, which are limited to a few days throughout the year. Consequently, larger plants (> 300 – 500 kW) should be distributed across several boilers with differing heat output levels to ensure optimal response to demands at any given time and hence efficient operation. Precise knowledge of the annual efficiency coefficient is vital for dimensioning a boiler plant, since the goal should be to adapt it to the optimum boiler efficiency.

Boiler plants are necessary when buildings are fitted with central heating systems or many heat energy consumers are being supplied from a central location. **Figure 4.18** illustrates a firing installation with coal as the primary energy source as an example for solid fuels. The coal is transported on a worm drive from the coal bunker to the flame tube boiler, where it burns on a grate and releases thermal energy to heat water. Heat is transferred via hollow bodies containing the heating water and surrounded by the hot flue gases, which flow along them on the outside. The design of such boilers is primarily aimed at achieving a high degree of efficiency by means of a large heat exchange surface. The flue gases are scrubbed after exiting from the boiler and evacuated via a chimney. Both cast iron sectional boilers and steel boilers are used for solid fuel combustion.

Figures 4.19 and **4.20** show smaller boilers suited to conversion from solid fuels to oil or gas. The boiler in **Figure 4.19** is described as a dual-chamber boiler.

The furnace in **Figure 4.20** features two fully separated combustion chambers, thus possessing a higher degree of firing efficiency.

When solid fuels are used to generate heating energy, the entire process (fuel intake as well as ash removal) must be fully automated to ensure that the system can be operated with as little personnel input as oil- or gas-fired systems.

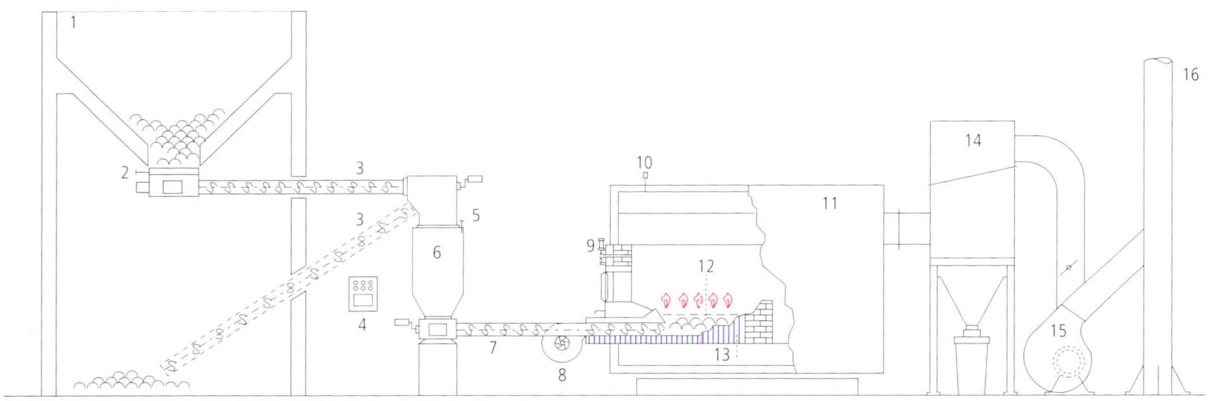

Figure 4.18
Coal-fired installation

1 Coal bunker
2 Bunker slide gate
3 Feed worm drive with drive and alternative
4 Control panel
5 Fill level indicator
6 Feeding funnel
7 Underfeed worm drive with drive
8 Down-draught fan
9 Negative pressure control
10 Thermostat or pressure gauge
11 Flame tube boiler
12 Fire pit and grate
13 Air chamber
14 Flue gas dust collector
15 Induced draught ventilator
16 Chimney

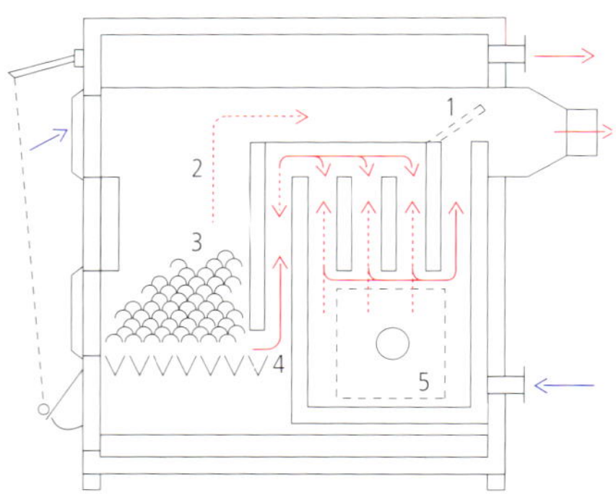

Figure 4.19
Alternating combustion steel boiler with two combustion chambers

1 Reversing valve
2 Flue gas oil
3 Coke
4 Flue gas coke
5 Oil combustion chamber

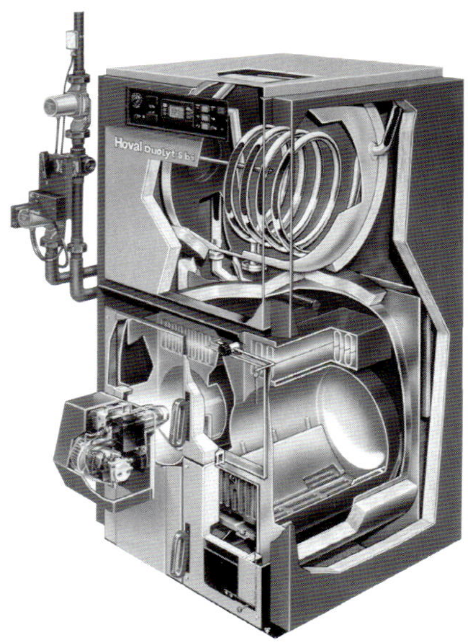

Figure 4.20
Steel double boiler with two separated combustion chambers, attached heat exchanger (Hoval DuoLyt LN)

4.3.2.3 Boilers (oil-, gas-fired)

In reference to the calorific value of the fuel (H_u), modern heat generators, especially low temperature heat generators, achieve annual utilization ratios of 91 – 93 % at relatively low performance levels. Heat generators with greater performance levels achieve 94 – 95 % with two-step or modulating burners. This means that there is little room left to improve the energy efficiency of conventional heat generators, especially since the flue gas temperatures must not fall below minimum values (e.g. in accordance with DIN 4702).

Specialized burners were developed in the past years for low-emission combustion of gas or oil. These are illustrated in **Figures 4.21** and **4.22**.

The Bluetwin oil burner (**Figure 4.21**) combines the advantages of tried and tested components with technical innovation. The Bluetwin achieves a NO_x-decrease through innovative, directed air conduction. By reducing the oxygen content in the hot flame core and simultaneously recirculating the waste gas, the oil-air mixture is combusted in a blue flame in the Bluetwin burner.

The MatriX radiant burner shown in **Figure 4.22** – a gas boiler – reduces nitrogen oxides by considerably lowering the combustion temperature. A homogeneous gas-air mix is prepared in two wire-netting hemispheres and evenly distributed before it reaches a fine stainless steel mesh (hemisphere) in which the gas-air mix is ignited. The gas flows out gradually and burns flamelessly. The glowing wire mesh hemisphere exudes a large part of its heat in the form of radiation, thereby cooling the flame to a clear margin below 1,300 °C.

In addition to better flue gas values, the only approach to considerably improve utilization ratios is to use the evaporation enthalpy of the vapour contained in the heating gas/flue gas and to lower the flue gas temperature even more drastically in order to further reduce the sensible flue gas loss. In conventional heat generators, the evaporation enthalpy is simply evacuated through the chimney, without deriving any use from it.

This goal is achieved with condensing value boilers, as well as with conventional furnaces and retrofitted flue-gas heat exchangers for condensation.

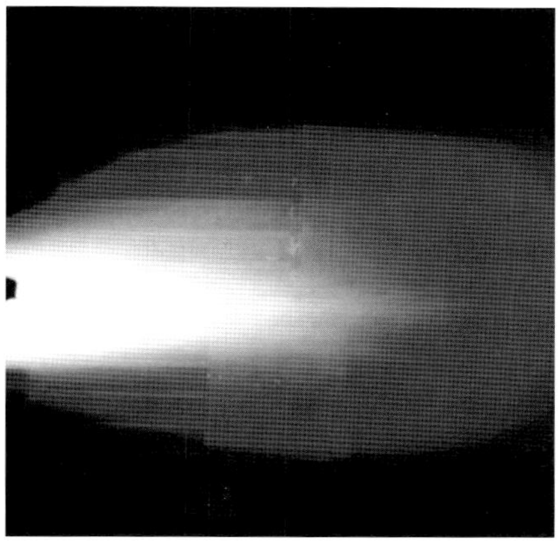

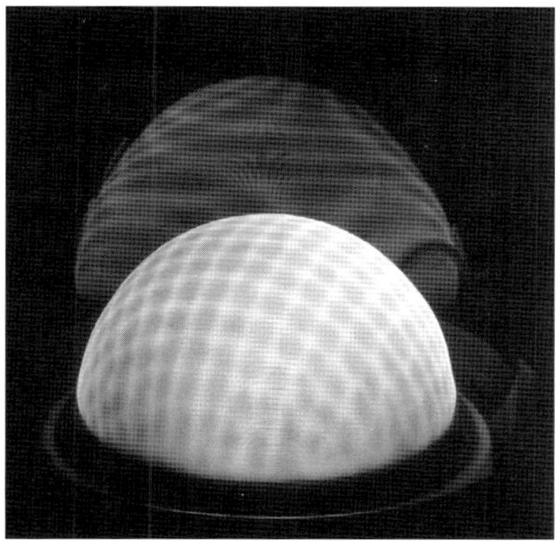

Figure 4.21 (left)
Viessmann Vitoflame oil burner

The Duozon-combustion principles is an innovation. The oil-air mixture combustion is poor in oxygen at the hot core of the flame with low NO_x-formation near the dust shield. Output range of 18 – 27 kW (Product image: Viessmann)

Figure 4.22 (right)
MatriX radiant burner

Viessmann has taken a step towards achieving the goal of "pollution gas free combustion" with its MatriX radiant burner. It is widely recognized as a milestone in gas heating technology, avoiding pollutant emissions at the very point of origin, i.e. at the point of combustion (Product image: Viessmann)

Today, boilers/furnaces are commonly differentiated according to:
- Size
- Small boilers up to approx. 50 kW
- Medium boilers up to approx. 350 kW
- Large boilers from approx. 350 kW upwards

- System design
- Type of firing
- Material

In terms of system design, a differentiation is made between:
- Gas or oil special purpose boilers
- Changeover boilers
- Two-component boilers and
- Condensing boilers.

Gas or oil special purpose boilers are only suitable for use with gas or oil and cannot be converted for use with other fuel types.

Changeover boilers are suitable for oil, gas and solid fuels. The changeover is effected manually by converting the firing installation.

Two-component boilers are used for oil/gas combustion and can be switched to solid fuel combustion without component conversion.

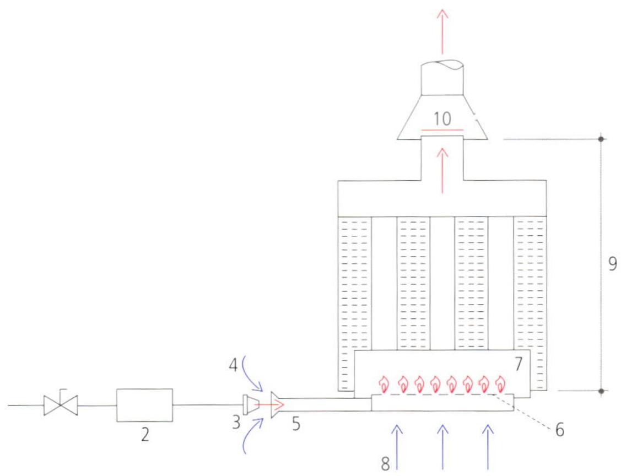

Figure 4.23
Firing process without fan

1 Gas
2 Gas valve
3 Main gas jet
4 Primary air
5 Injector pipe
6 Grate
7 Combustion chamber
8 Secondary air
9 Connection height
10 Draught diverter

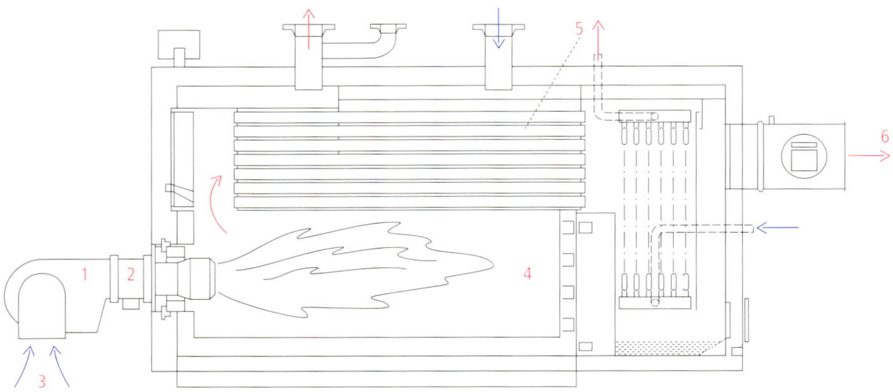

Figure 4.24
Firing process with fan

1 Fan/burner
2 Gas feed
3 Air for combustion
4 Combustion chamber
5 Ancillary heating surface
6 Flue gas

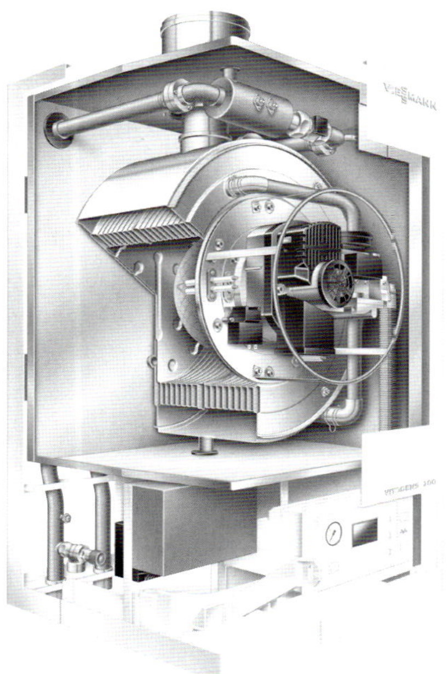

Figure 4.25
Vitodens 200, wall-mounted gas condensing boiler with modulating MatriX radiant burner 7 to 66 kW.

In conjunction with the cascade control unit, rated outputs are available up to as high as 265 kW. (Product image: Viessmann)

The firing processes are differentiated as follows:
- Gas firing with fan burner
- Gas firing without fan burner (atmospheric burner)
- Oil firing with fan burner
- Oil firing without fan burner

In the gas firing process in burners without fan (**Figure 4.23**), the boiler is operated with natural draught, i.e. the combustion air is drawn into the boiler (up to the flow control) solely by convection. As a result of the integrated flow control, the chimney draught has little or no impact in the flue gas duct. The flow control interrupts the rigid connection of the flue gas pipe between boiler and chimney, preventing undue influence on the firing process through draught, back-draught or excessive convection.

In firing processes with fans (for oil and gas, **Figure 4.24**), the air volumes required for combustion are mechanically supplied via a fan (or blower) built into the burner. At the same time, positive pressure is created within the combustion chamber in the case of positive pressure firing to compensate for the boiler resistance on the gas supply side. This allows for stack-independent heating operation, which can be regulated to achieve optimal combustion values.
In this design, the zero pressure point for the blower is located in the exhaust gas connection piece at the end of the boiler.

The advantages of gas firing with fan burners are:
- Higher firing efficiencies
- Better heat transfer to areas to be heated and
- Compact boiler construction

Based on these advantages, boilers with fan burners have become the standard today.

The cutaway image in **Figure 4.25** illustrates a wall-mounted gas-furnace with modulating Matrix-radiant burner (output 7 – 66 kW) for extremely low-emission combustion and energy efficient heating.

Figure 4.26 presents a cutaway image of a low-temperature furnace. The bifilar heating surface consists of a steel cylinder and fitted radial-ribbed cast iron components. It is notable for its outstanding thermodynamic properties. The inner, cast-iron casing exposed to the heating gases takes on a higher temperature than the steel cylinder, surrounded by the boiler water. This inhibits condensation of the combustion gases and condensate accumulation and thus prevents corrosion damage.

The heat flow diagram of a heat generating system in **Figure 4.27** shows that the bulk of the generated energy is directly available as usable heat, while some portions are lost via the chimney and the boiler surface.

Figure 4.28 presents an exemplary solution of a central heating installation with four large boilers (output approx. 800 kW each) and one small boiler (output approx. 110 kW), for improved response to constant partial-load operation.

To reduce flue gas losses and contaminant emissions of a boiler system, the trend is to use gas- or oil-fired boilers as condensing boilers (boiler with internal flue gas exchanger) or boilers with retrofitted catalytic converters and heat recovery components. The diagram in **Figure 4.29** indicates the typical construction of this type of boiler installation with an integrated catalytic converter and flue gas exchanger in the flue gas flow. The design of the catalytic converter is illustrated in detail (on the left).

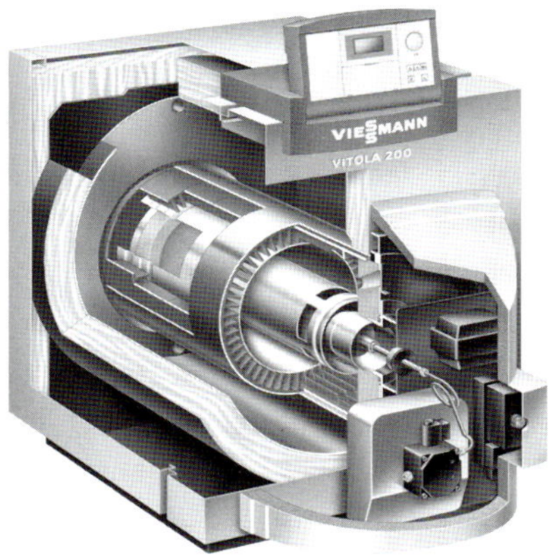

Figure 4.26
Cutaway view of Vitola 200 low temperature oil-/gas furnace (Product image: Viessmann)

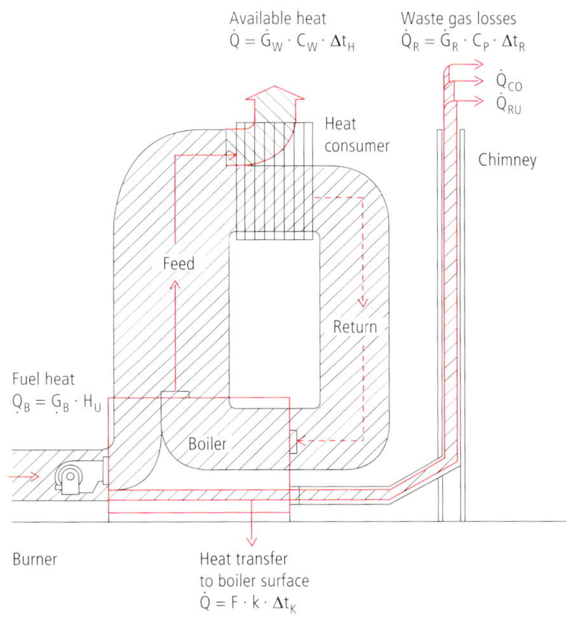

Figure 4.27
Heat flow diagram of a heat generating system

$\Delta t_H = t_{VL} - t_{RL}$
$\Delta t_K \sim t_{OK} - t_L$
$\Delta t_R \sim t_{AG} - t_L$

Figure 4.28
Heating installation with large and small boilers, output per boiler 720 to 895 kW and 105 to 130 kW each, respectively. (Viessmann)

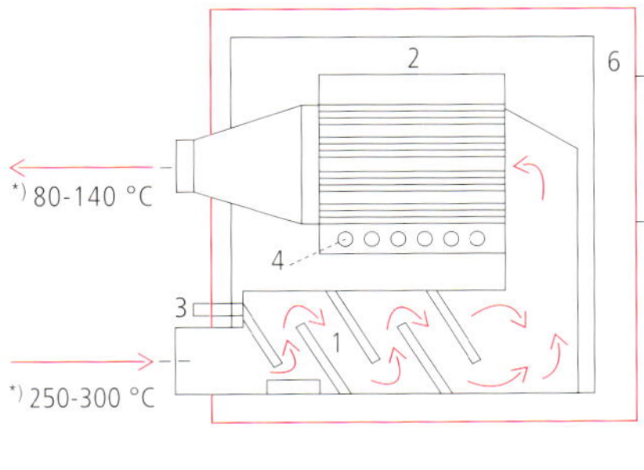

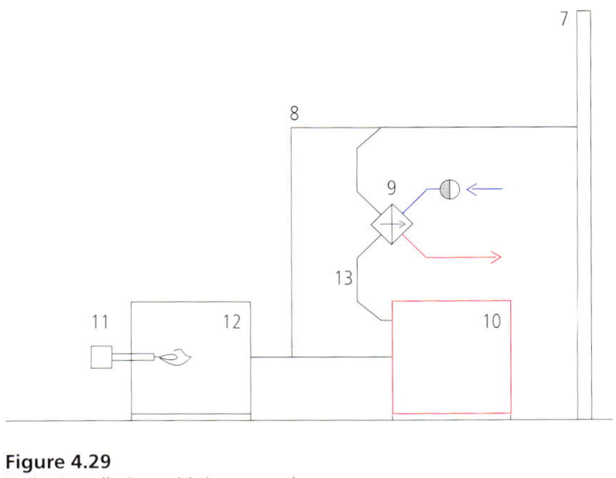

1 SO_2 neutralization
2 Cat-element
3 NH_3 injection
4 Maintainend high temperature
5 Control and monitor
6 Insulating material
*) according to system and layout
7 Chimney
8 Bypass (inspection cycle)
9 Heat recovery
10 Catalytic converter
11 Gas/oil burner
12 Boiler
13 Flue gas

Figure 4.29
Boiler installation with integrated catalytic converter and flue gas exchanger in flue gas flow.

Figures 4.30 to **4.32** depict a "family of combustion devices" for the generation of warm water, i.e. heating.

Figure 4.30 depicts the appearance and design of the output device. **Figure 4.31** shows a wall-mounted gas-condensing furnace that operates on sliding temperature scale without defined lower temperature (ceramic-surface burner/aluminium-silicon heat exchanger).

The final image in the series, **Figure 4.32**, portrays a standard condensing boiler for natural gas.

Since combustion technology is based on the premise that the fuel contains hydrogen, gas lends itself more easily to this application than oil: the water data for condensation in natural gas result in an attainable energy gain of 11 %, while oil only achieves 6 %. Nevertheless, there are combustion boilers and exhaust gas/water heat exchangers for condensation de-signed for use with oil. The dew point temperatures of the exhaust gases for heating oil lie at approx. 48 °C, and at approx. 58 °C for natural gas. The almost ten degrees difference in dew point temperatures, make gas the preferable fuel in this regard as well. The heating water temperatures in combustion systems reach maximum levels of 60/40 °C (supply / return flow temperature), which can be reduced to 40/30 °C without any notable loss in terms of utilization. Combustion technology achieves energy savings of 15 – 20 % to conventional gas-fired boilers and 8 –15 % by compared to low-temperature boilers.

4.3.2.4 Boiler systems (electric)

In countries and regions where electricity is inexpensive, boilers with direct electric heating can be used as attic heating systems. A cost-efficient slip stream tariff is an important prerequisite: electrically operated boilers are supplied with heat overnight and the stored heat is then distributed during the day. Water or solid materials (for example magnesite) are used as a storage medium – and additional storage mediums are currently under development.

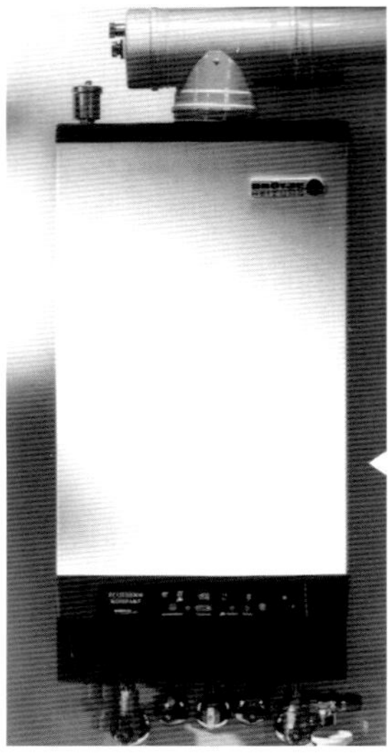

Figure 4.30
Condensing gas heater (wall-mounted) for heating and warm water generation,
output range
6.5 – 18 kW for heating,
6.5 – 22.1 kW for warm water
(Product image: Brötje)

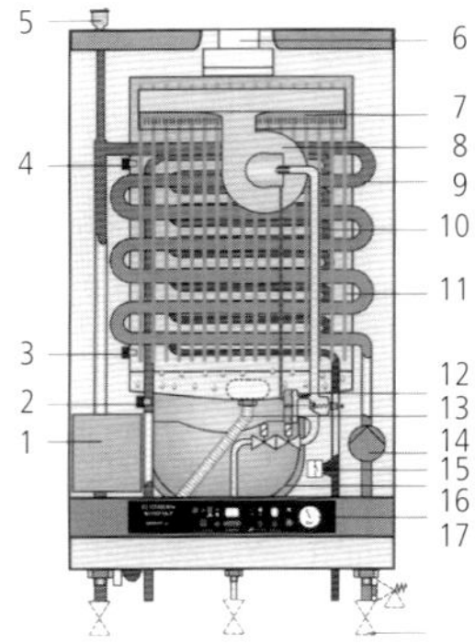

1 Electronic control
2 Warm water NTC
3 Return NTC
4 Feed NTC
5 Quick-action ventilating valve
6 Connection to flue gas system
7 Modulating, premixing metal-mesh burner
8 Speed-controlled fan
9 Heat exchanger coil for heating
10 Heat exchanger coil for warm water
11 Al-Si heat exchanger
12 Modulating gas valve
13 Condensation water pipe
14 Heating cycle pump
15 Flow switch warm water
16 Expansion tank 12 l
17 Control panel
18 Maintenance valves

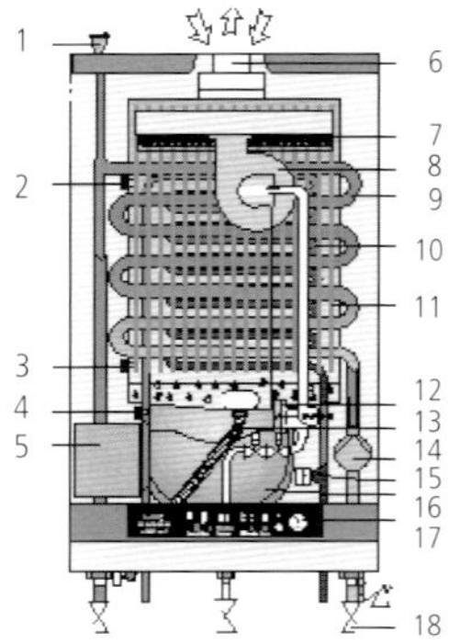

Figure 4.31
Wall-mounted gas boiler Ecotherm Kompakt, with condensing technology for heating a building with warm water storage,
output range 3.5 – 70 kW
(Product image: Brötje)

1 Quick-action ventilating valve
2 Feed, temperature messurement
3 Return, temperature messurement
4 Warm water, temperature messurement
5 Electronic control
6 Connection piece Waste gas system 80/125
7 Modulating, pre-mixing metal-mesh burner
8 Variable-speed blower
9 Heat exchanger coil for heating
10 Heat exchanger coil for warm water
11 Al-Si heat exchanger
12 Modulating gas valve
13 Condensation water run-off
14 Heating cycle pump
15 Flow control switch, warm water
16 Expansion tank, 12 l capacity
17 Control panel
18 Maintenance taps for gas, heating and water (fittings)

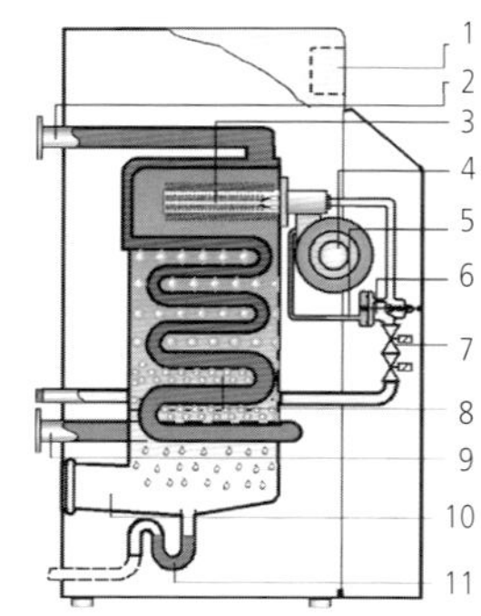

Figure 4.32
Condensing gas boiler SGB 2, for heating a building,
output range 16 – 120 kW
(Product image: Brötje)

1 Regulation and control unit
2 Boiler feed
3 Modulating blower burner
4 Combustion air blower
5 Control unit / blower pressure
6 Gas volume control valve
7 Double-seat gas valve
8 Al-Si heat exchanger
9 Boiler return
10 Condensate collecting basing with waste gas connection piece
11 Condensation water siphon trap

Direct heating through electrical energy is provided in both smaller and large installations, whereby water is stored in hot water or solid material reservoirs scaled to the daily heating requirements. Recharge is generally automated according to outside temperatures and programmed to occur only at low tariff periods. The water circulating in the heating cycle is mixed in a three-way mixing valve with return water depending on the outside temperature to achieve the most favourable heating water temperature.

Figure 4.33 illustrates a central magnesite reservoir for warm-water heating installations where the reservoir core is electrically charged. For the discharge cycle, the air circulated via the reservoir core transfers the thermal energy to the circulating heating water via an air-water heat exchanger.

In indirect heating with electrical energy, shown in Figure 4.34, hot water is generated with the help of a special electrode boiler which transfers its thermal energy to a reservoir. The hot water reservoir in turn is generally heated during the night (low tariff period) and discharged during the day.

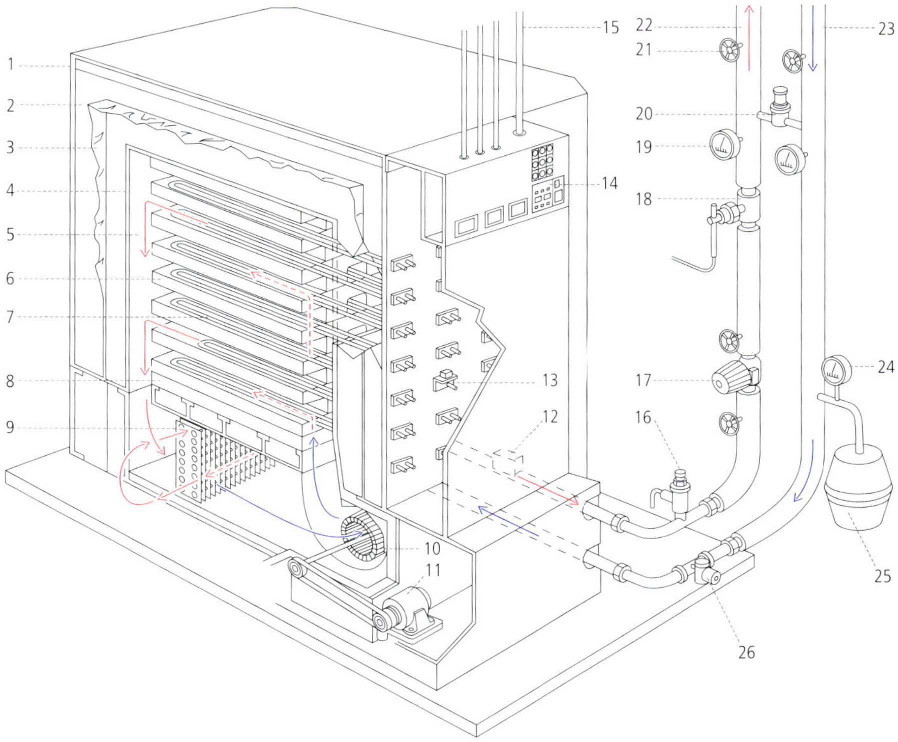

Figure 4.33
Ceramic central reservoir for warm-water heating installations, with magnesite storage material

1 Housing
2 Mineral wool insulation
3 Aluminium foil
4 Core box
5 Air channel
6 Storage core
7 Heating element
8 Hard insulation
9 Air/water heat exchanger
10 Radial ventilator
11 Ventilator motor
12 Feed temperature sensor and delimiter
13 Core temperature sensor and delimiter
14 Control panel with regulating and control switches
15 Electric connection
16 Safety valve
17 Circulation pump
18 Water-flow detector
19 Thermometer
20 Bypass with overflow valve
21 Shutoff valve
22 Feed
23 Return
24 Pressure gauge
25 Expansion tank
26 Fill and drain cock

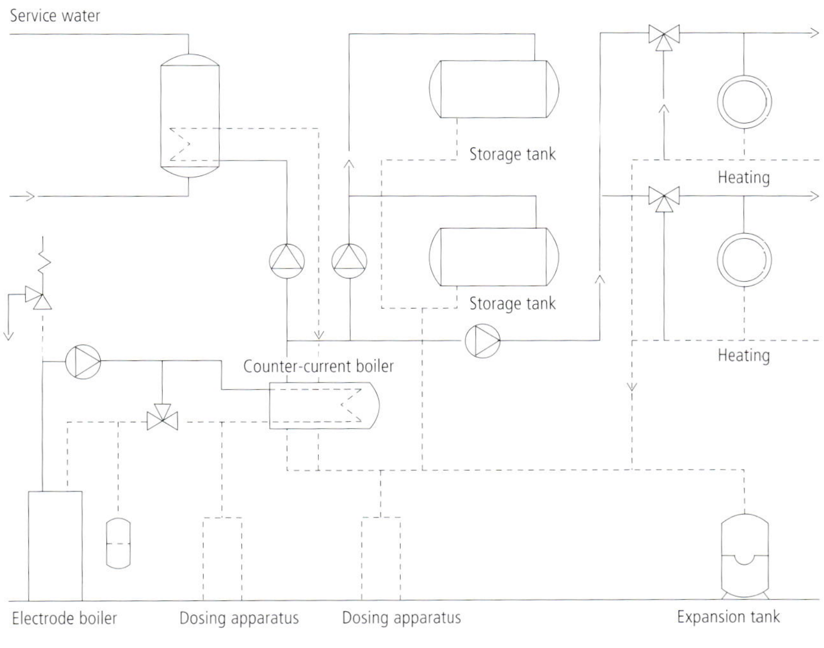

Figure 4.34
Piping diagram of a heating installation with electrode boiler and heat storage

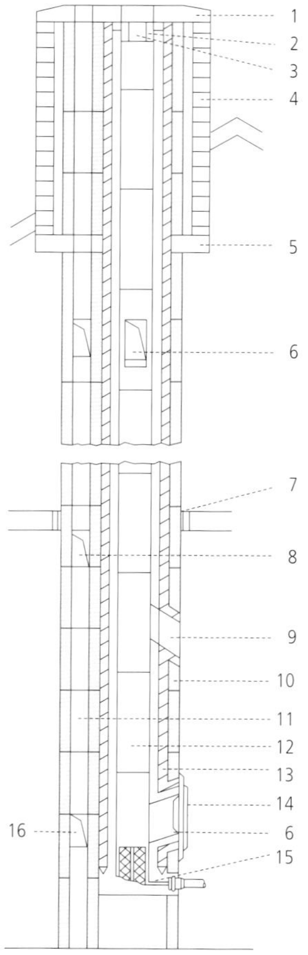

Figure 4.35
Triple-shell Plewa "S+S" chimney
The ceramic inner glazing is mirror smooth and acid-proof.
This chimney construction is also suitable for use with condensing boilers

1 Reinforced concrete cover plate
2 Expansion joint
3 Expansion joint sleeve
4 Sheathing – chimney top (head)
5 Reinforced concrete cantilever plate
6 Casing with rectangular connection piece for installation of cleanout door
7 Mineral fibre slip joint
8 Exhaust air opening for heating chamber ventilation
9 Casing with round connection piece for fireplace link up
10 Light-weight concrete casing bricks
11 Tip-stretched exhaust air shaft in casing brick
12 Ceramic glazed casing
13 Mineral fibre insulating panels
14 Cleanout door
15 Condensate run-off
16 Control door for exhaust air shaft

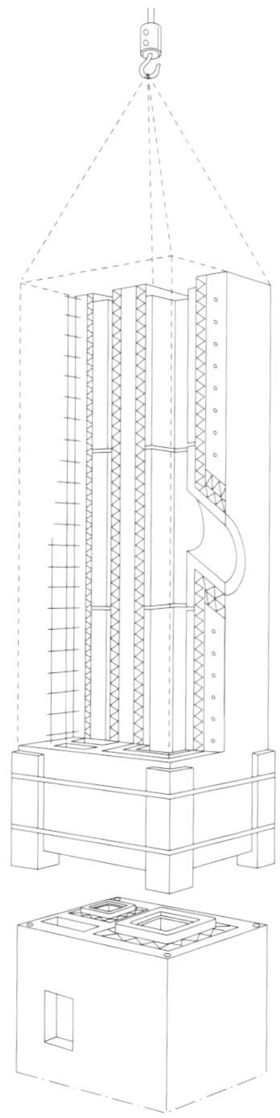

Figure 4.36
Prefabricated chimney

4.3.3 Chimneys and flue stacks

Chimneys have the task of cleanly evacuating flue gases from boiler systems. The draft required to achieve this is produced by the weight differential of the hot flue gases in the chimney and a cold air column of equal height. In firing installations with natural chimney draught, the negative pressure in the combustion chamber of the heating boiler must be strong enough to enable the air volume required for combustion to flow freely into the combustion chamber. In firing installations with positive pressure firing, the chimney draft is only required to overcome the resistance of the flue gas duct and the chimney flow resistance.

Chimneys are differentiated into several types of construction. These are:
- Multi-shell assembly chimneys
- Prefabricated chimneys
- Air-flue gas chimneys
- Moisture-resistant chimneys
- Freestanding chimneys
- Steel chimneys

Figure 4.35 illustrates a multi-shell assembly chimney with a chamotte casing, glazed on the inside, a light-weight concrete casing and mineral fibre insulation. Assembly chimneys are lightweight and easy to erect in very little time.

Figure 4.36 illustrates a prefabricated chimney, whose construction is similar to the chimney shown in **Figure 4.35**. Prefabricated chimneys are delivered in floor-high sections and are very cost-efficient owing to the standardization of the components (prefabrication).

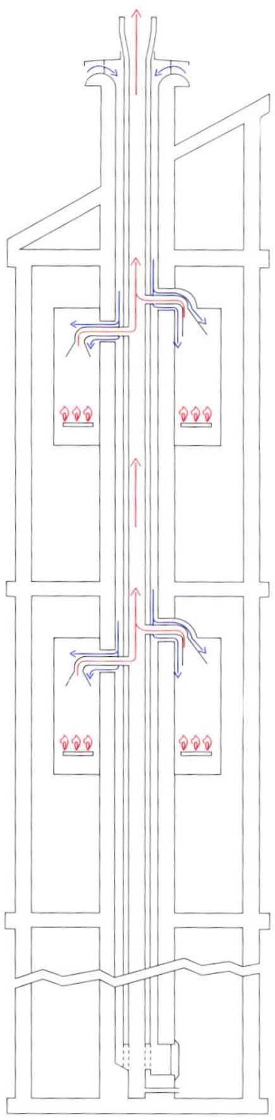

Figure 4.37
Air-flue gas chimney, shown in combination with gas outlets to connected individual furnaces/boilers

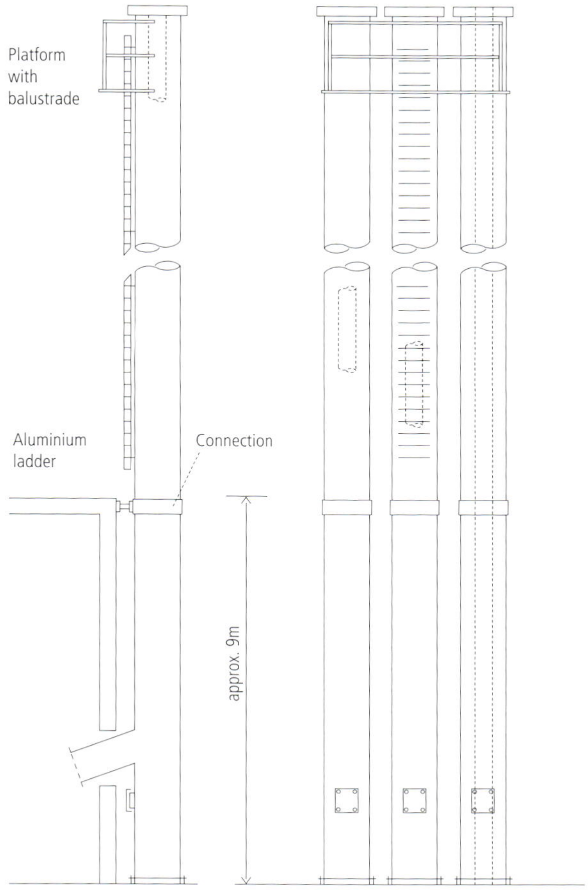

Figure 4.38
Stack installation – threepass, freestanding

Figure 4.37 shows an air-flue gas chimney for individual firing installations. This chimney fulfils a dual role: it evacuates the flue gases of the individual installations connected to it and simultaneously supplies them with the required combustion air. Both functions are arranged either side by side or concentrically integrated. The design of the air-flue gas chimney solves the problem of sufficient air supply in residential objects with individual firing installations and airtight windows and doors.

Moisture-resistant chimneys were developed to link up to low temperature heat generators or condensing boilers, whose flue gas temperatures lie far below 100 °C and therefore produce considerable amounts of condensate, which could damage and even destroy a conventional chimney. In moisture-resistant chimneys air is conducted upward around the gas-conducting inner pipe. Supply air enters in the plinth area of the chimney. It absorbs any occurring moisture and transports it to the outside at the mouth.This prevents the risk of moisture saturation when flue gas temperatures are low, although a water- and acidproof inner glazing is essential to protect the chimney.

Freestanding chimneys consist, as **Figure 4.38** illustrates, of one or more gas evacuating inner pipes, an insulating layer and an external casing in reinforced concrete, steel or masonry. Freestanding chimneys or stacks are only common for large central heating plants. The dimension (height) of these stacks must meet local air pollution control standards.

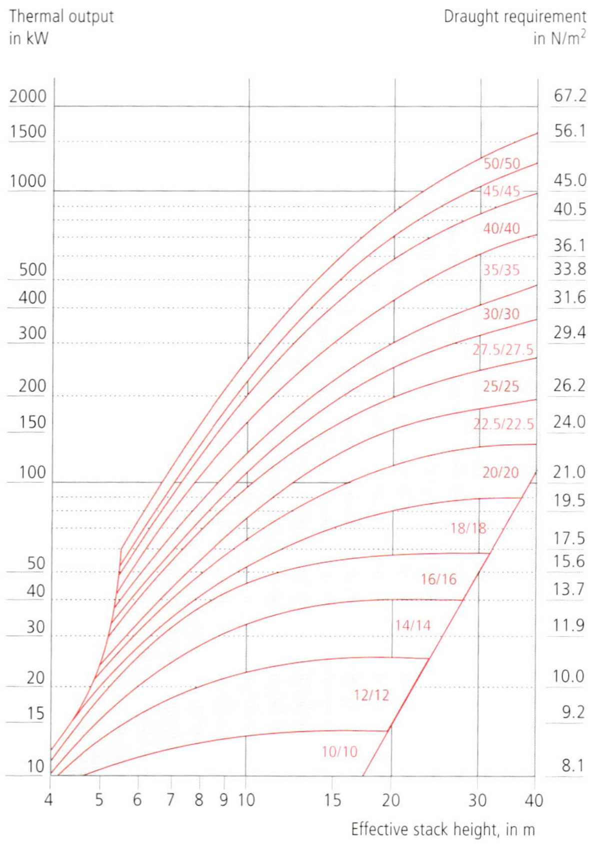

Figure 4.39.1 (left)
Diagram for dimensioning chimney cross-sections, oil burners without fan

Boiler with draught requirement
EL heating oil
TW = 120 °C

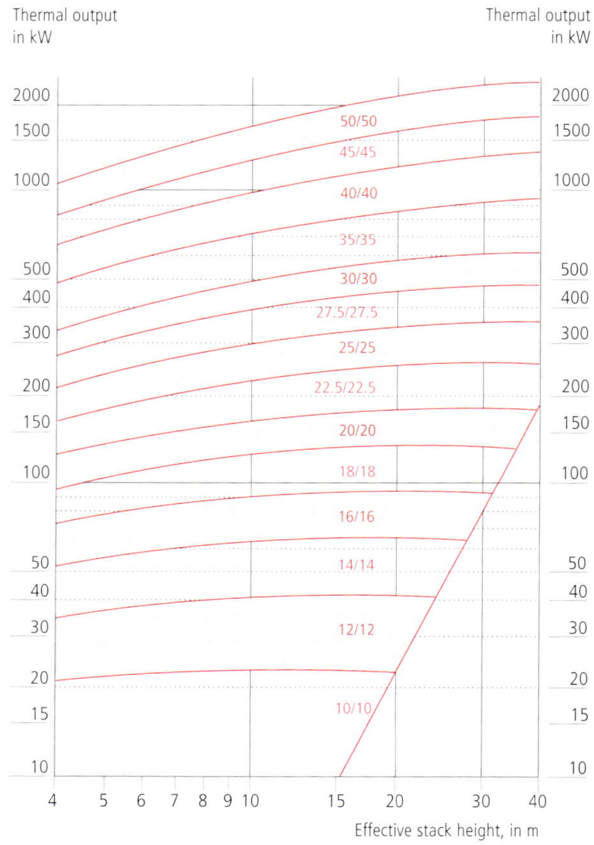

Figure 4.39.2 (right)
Diagram for dimensioning chimney cross-sections, oil burners with fan

Boiler without draught requirement
(Overpressure combustion)
EL heating oil
TW = 120 °C

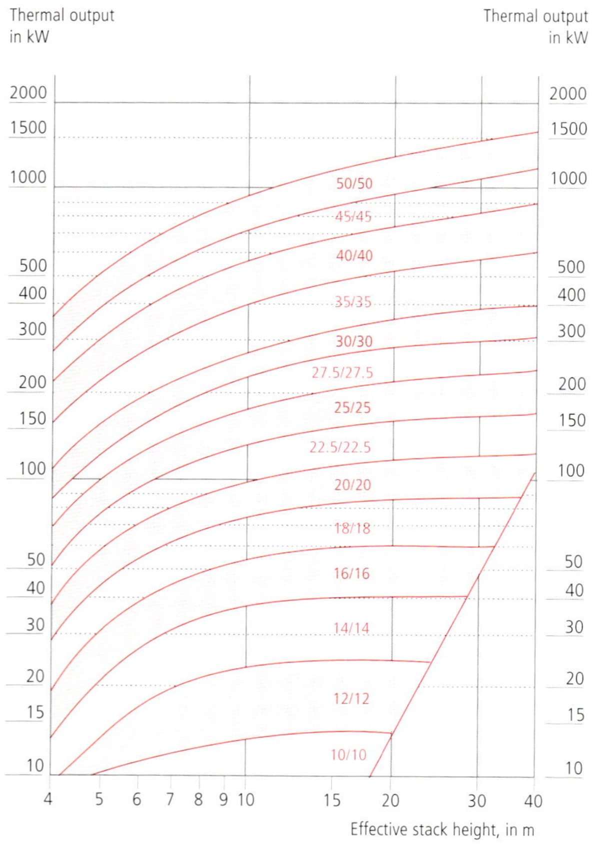

Figure 4.39.3
Diagram for dimensioning chimney cross-sections, gas burners

Boiler with atmospheric gas burner
Natural gas, draught requirement = 5 N/m²
TW = 80 °C

Chimney evaluation studies may be required to determine the chimney height according to contaminant loads. For smaller installations, the dimension guidelines provided by the manufacturers are sufficient. For the purpose of illustration, the diagrams provided by the manufacturing firm Plewa (**Figures 4.39.1** to **4.39.3**) demonstrate the measurements for chimney cross sections according to the effective chimney height, thermal output and boiler construction. The initial values indicated in each diagram must be taken into consideration.

The individual diagrams indicate the measurements for chimney cross-sections for heating boilers (fired with natural gas or oil) and for different draught requirements:

Up to three firing installations for solid or liquid fuels and a thermal output of 20 kW each can be linked to a single chimney for central heating systems. For gaseous fuels, the thermal output lies at 30 kW each.

The effective chimney height must be a minimum 4 m in each instance, if the uppermost firing installation is gas operated, otherwise it must be 4.5 m high. Height above roof: min. 0.4 m. If the aforementioned nominal thermal output is exceeded at each firing installation, each of these must be fitted with a separate chimney.

The same applies to
- Buildings with more than five full storeys
- Flue gas fired systems with open combustion chambers (e.g. open fireplaces)
- Gas firing installations, whose flue gas temperatures exceed 400 °C.

The standard guidelines from local authorities must be adhered to in each individual scenario.

4.3.4 Oil storage

Storage in oil tanks depends on the availability of the heating oil, the costs of tank and oil as well as other factors. Generally, a tank should be scaled to accommodate the heating oil for an entire heating season; this means, however, that the tank is large and expensive. Without taking the energy required to heat utility water into consideration, the approximate heating oil consumption for residential buildings for one entire heating period is:

Fuel consumption/a (Fa), approx.:

Fa = 15 to 20 l/m² use area a
(= 540 to 720 MJ/m²a) for a multi-family dwelling, i.e.

Fa = 20 to 30 l/m² use area a
(= 720 to 1080 MJ/m²a) for a detached single-family home.

The dimensions of standard oil tanks are listed in **Table 4.8**.

Oil tanks can be installed horizontally/vertically above or below ground, or as battery or basement tanks in buildings.

Underground, cylindrical steel tanks must be double-walled and are designed accordingly. Aside from underground steel tanks, there are synthetic tanks manufactured from fibre-reinforced polyester resin. They are single-walled and do not require leakage safety devices. Spherical glass-fibre laminate tanks can also be installed above or below ground for volumes up to 12 m³.

Above-ground heating oil tanks are available in the following models:
- Steel tanks for smaller installations, combined into a battery of tanks (content 1,000 to 2,000 l)
- Synthetic tanks up to 10,000 l, again suitable for set-up as a battery of tanks
- Steel cylinder tanks
- Basement welded tanks in a variety of geometric constructions, also suitable for retrofitting

All above-ground tanks must be installed in an oil drip basin or in a room serving as an oil drip basin.

Figures 4.40/4.41 illustrate the schematic layout of oil supply and storage for boiler installations.

In addition to heating oil, there are instances where liquid gas must be stored as well. Planning for such cases must be carried out with great diligence due to the high risk of explosion and include consultation with all relevant safety authorities. Special standards apply to liquid gas storage and these must be met without fail.

Content	External diameter	Total length
m³	mm	mm
1	1,000	1,510
3	1,250	2,740
5	1,600	2,820
7	1,600	3,740
10	1,600	5,350
(13)	1,600	6,960
16	1,600	8,570
20	2,000	6,960
25	2,000	8,540
30	2,000	10,120
40	2,500	8,800
50	2,500	10,800
60	2,500	12,800
80	2,900	12,750
100	2,900	15,950

Table 4.8
Measurements of oil tanks according to DIN 6608 – Part 1 (10.81)

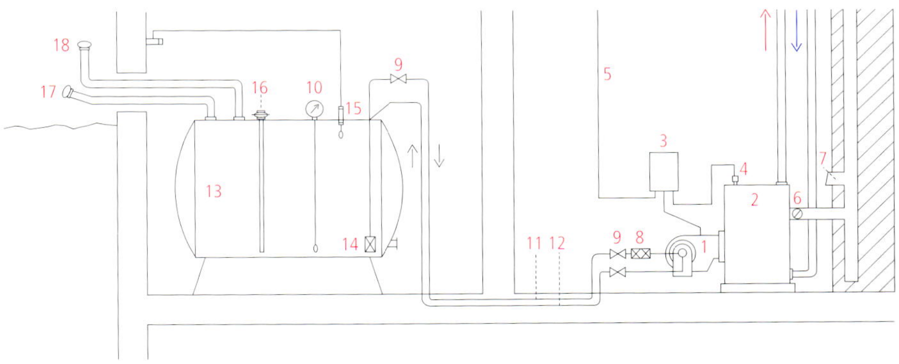

Figure 4.40
Above-ground oil storage

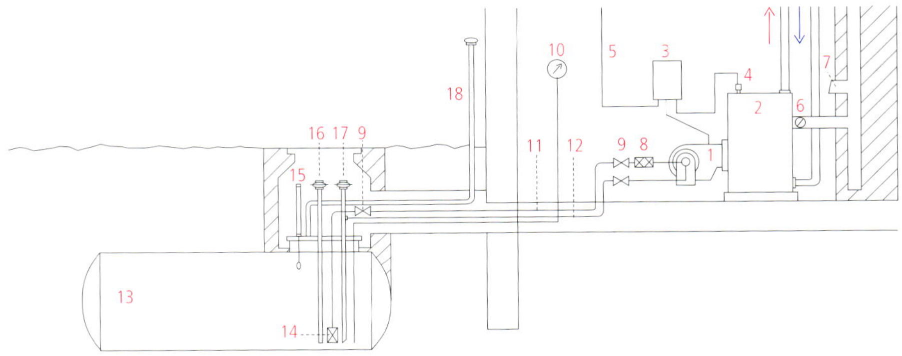

Figure 4.41
In-ground oil storage

1 Burner
2 Boiler
3 Control panel
4 Safety- and regulating thermostat
5 Connection to room temperature control
6 Butterfly valve (throttle valve)
7 Damper
8 Filter
9 Quick-action stop valve
10 Oil level indicator
11 Oil feed pipe
12 Oil return pipe
13 Heating oil storage tank
14 Upstroke valve (approx. 100mm distance to tank floor)
15 Boundary value indicator
16 Fuel dip stick
17 Filling pipe line
18 Aeration according to DIN 4755 (terminating 2.5m above ground level)

4.3.5 Collectors

Solar collectors are available in a wide variety of models for generating thermal energy (warm water/hot water). The following types are differentiated:
- Solar absorbers
- Flat plate collectors
- Evacuated tube collectors
- Focal line systems

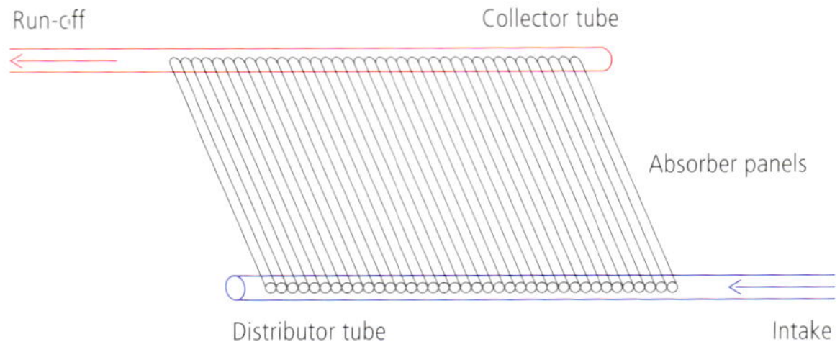

Figure 4.42
Structure of solar absorber

Solar absorbers

Solar absorbers are simple absorber mats consisting of high-grade rubber with moulded-on distributing and collecting tubes, resistant to cleansing agents and a temperature range of -50 to +120 °C. They are also permanently elastic and remain flexible in the cold. Solar absorbers (**Figure 4.42**) can be installed on roofs and on the ground, covering large areas. They generate warm water temperatures up to a maximum of 50 °C. Generally used for the heating of swimming pools (**Figure 4.43**), the surface of the absorber mats should be roughly 50 – 80 % of the surface water to be heated. Solar absorbers can also be used to heat household water. In this case, the circulation system between absorber mat and swimming pool in commercial applications must be replaced by a system linked to a storage tank.

Figure 4.43
Solar absorber for swimming pool
300 m² absorber surface

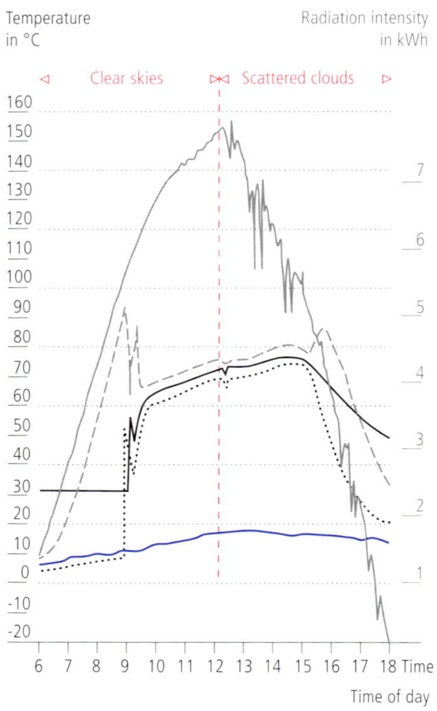

Figure 4.44
Energy curve of flat plate collector over the course of one day (measurements Bell + Grossett)

— Radiation intensity
– – Surface temperature of collectors
— Water temperature (exiting collector)
····· Water temperature (entering collector)
— Outside temperature

Flat plate collectors

In thermal solar collectors, incident energy is transformed into heat and, thus, increases the temperature of the radiation absorber. A liquid thermal medium is used to regulate the temperature of the released heat output.

The following losses occur in solar collector installations:

- Optical losses (reflection)
- Thermal losses (system-dependent)

To keep such losses at a minimum, solar collectors are covered in glass with high light-transmissive properties (light transmission factor > 92 %). The coating on the collector, which absorbs the solar radiation, is highly absorbent with low emission rates. The covering of a collector usually consists of 4 mm thick, pre-tensioned low-iron safety solar glass, with very low reflection properties.

The thermal energy gained by the collector is transported to a storage tank by water circulation. The entire system must be frost- and corrosion-proof. The system is controlled and regulated by measuring temperature differences. **Figure 4.44** depicts temperatures and radiation intensities measured on a single day in a flat plate collector installation.

The relationship between radiation intensity and collector temperature is easily visible in the graph. **Figure 4.45.1** shows a vacuum flat plate collector, which has a higher heat-insulating efficiency than flat plate collectors because of the vacuum.

Improved coatings for the radiation absorbers (TiNOx) in the collectors have increased the efficiency coefficients of the collectors (**Figure 4.45.2**).

To derive the best possible efficiency coefficient from the solar system, the collectors must face south. In some instances, increasing the size of the total collector surface can compensate for lower efficiency coefficients when this ideal orientation is not feasible. **Figure 4.46** shows the reduction of the maximum efficiency coefficient for deviations from a 180° azimuth angle. Based on 50° N latitude, the optimal angle for the collector installation is roughly 45°. The efficiency coefficient of the collector installation diminishes when other angles are used.

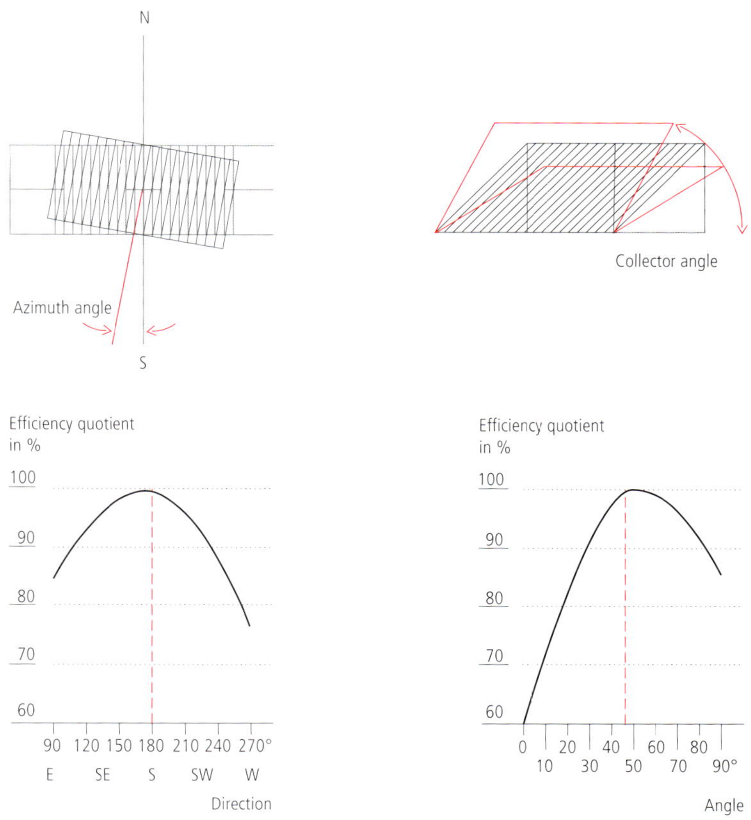

Figure 4.46
Adjustment of collector installation in case of less than optimum placement

Period of active use	Optimum angle
Jan. – Dec. (year round)	30 – 50°
April – Sept. (seasonal)	25– 45°
May – August (seasonal)	20 – 40°
Sept. – April (heating period)	50 – 70°

Figure 4.45.1
Vacuum flat plate collector (Thermosolar)

Figure 4.45.2
Installation-ready solar system with high-performance flat plate collector (TiNOx-coating), (Reflex Solar-Unit)

Evacuated tube collectors

Evacuated tube collectors, like flat plate collectors, transform incident solar energy into heat. The solar radiation permeates the evacuated glass tube and falls onto the absorber surfaces inside the tube, where it is transformed into heat. There is virtually no heat loss to the surroundings as a result of the high-grade selective coating on the evacuator surface and the vacuum. The heat energy collected on the absorber surface can be transferred to a heat tube, located on the underside of the absorber. As a result of the heat transfer process, the liquid in the heat tube evaporates, arriving at the condenser in the form of steam. In the condenser, the thermal energy is transferred to a water stream, where it condenses and returns in the heat tube to evaporate once again due to incident heat radiation. **Figures 4.47** and **4.48** show a typical evacuated tube collector and the working principle. The same placement (azimuth angle/installation angle) as for flat plate collectors applies.

The efficiency coefficients of relevant collectors are indicated in **Figure 4.49**. The graph shows that vacuum tube collectors have noticeably better efficiency coefficients than flat plate collectors when temperature differences are high (mean collector temperature/outside space).

Since most collector installations are components of a two-way heat-supply system, the financial viability should be established for each individual case. In other words, it is important to verify whether the cost-benefit-ratio meets the requirements and, above all, whether the heat gained in summer can be raised to temperature levels that are sufficiently high to supply energy to absorption chillers.

Figure 4.47
Evacuated tube collectors
(Image: Viessmann)

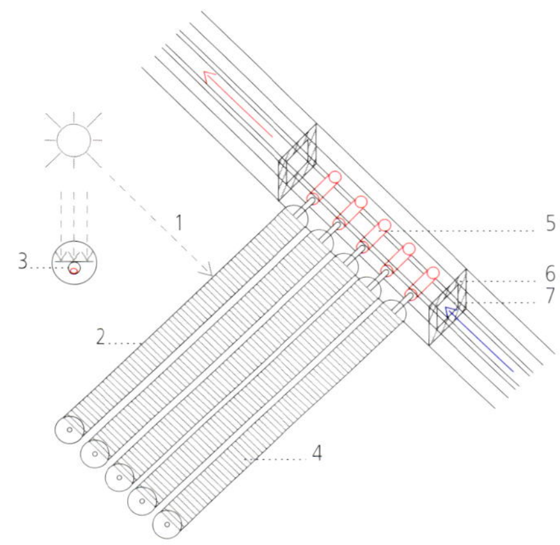

Figure 4.48
Principle of evacuated tube collector

1 Heat radiation
2 Glass tube
3 Conducting tube with filling
4 Absorber, selectively coated
5 Condenser
6 Heat insulation
7 Heat exchanger tube

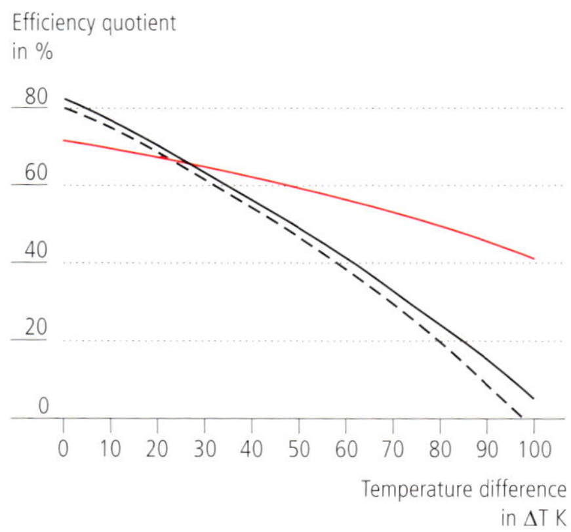

Figure 4.49
Efficiency quotient of flat plate collector/vacuum flat plate collector and evacuated tube collector (Stiebel-Eltron)

-- Flat plate collector
— Vacuum flat plate collector
— Evacuated tube collector

ΔT = Temperature difference between surroundings and average collector temperature (Flow/return)

Indoor measurement;
wind velocity 4 m/s and radiation strength 730 and 750 W/m² respectively

Figure 4.50 shows a system where the solar energy delivers energy for heating and simultaneously thermal energy for the operation of a refrigerating system. To achieve supply temperatures well above 100 °C in such systems, a line-focused installation may be required instead of the aforementioned installations, which operate on the flat plate principle. The system illustrated in this schematic diagram has the definite advantage that periods of high radiation intensity, and thus high cooling loads, coincide with the availability of the required thermal energy necessary for an absorption refrigerating system, keeping energy costs low.

Figure 4.50
Solar energy heating and cooling installation

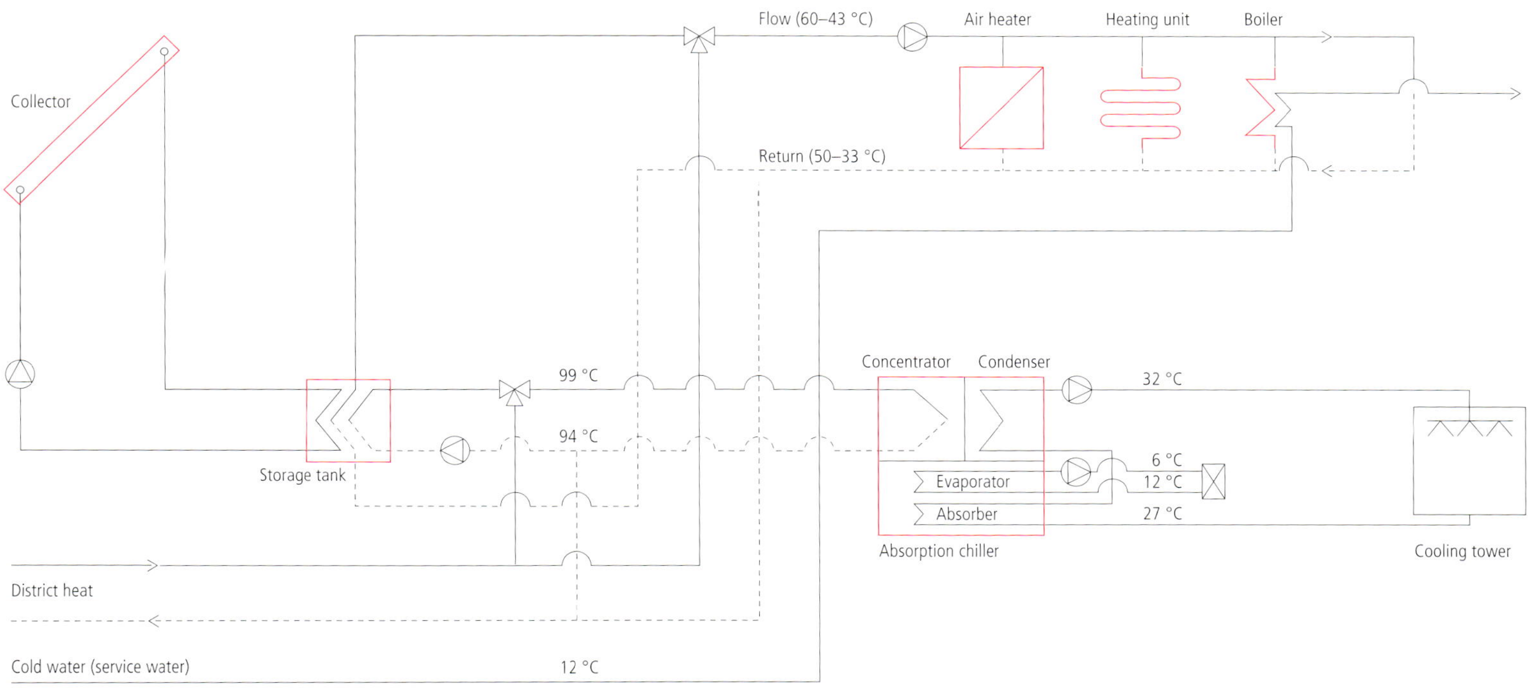

4.3.6 Fuel cells

Fuel cells are a promising form of energy production for the future. The high-temperature fuel cell is of particular interest for building systems. It transforms natural gas with an efficiency quotient greater than 50 % into electricity, producing high-temperature waste gases, which can be used for heating purposes. Several systems are already available on the market, while others are still being tested. They represent an alternative solution to combined heating and power stations. Unlike conventional motor-driven systems, these systems do not emit nitrogen oxide, operation is nearly noiseless and total efficiency quotients are comparable to other systems with power-heat coupling.

The operating principle of fuel cell technology is flameless combustion of natural gas or hydrogen in an electrolyte. Fuel cells are generally differentiated by the electrolyte used. Occasionally they are differentiated according to operating temperature (high- and low-temperature fuel cells). The following types of fuel cells are suitable for use in power-heat coupling: the phosphoric acid fuel cell (PAFC), the molten carbonate fuel cell (MCFC) and the solid oxide fuel cell (SOFC). MCFCs and SOFCs are high-temperature fuel cells (600 °C and 800 °C, respectively).

Although PAFCs are the most technically advanced fuel cells, they have the lowest electric efficiency quotient (system efficiency quotient 40 %). Fuel cells of this type are available on the market.

MCFCs are currently operated in some pilot installations, although some problems with leakage and corrosion remain to be solved. SOFCs are currently undergoing field tests. In contrast to PAFCs and MCFCs, where the liquid electrolyte is contained in a solid body matrix, the solid ceramic component serves as the electrolyte in SOFCs. A cell consists of a ceramic electrolyte (0.2 mm ion-conducting zirconium oxide) and porous electrodes (0.03 mm) consisting of electroconductive metallic ceramics joined both ways (**Figure 4.51.1**). The oxygen of the supply air is ionized in the cathode at roughly 800 °C: in other words, the oxygen absorbs electrons. Next, the oxygen diffuses, due to the resulting concentration quotient, through the electrolyte to the anode. There, the oxygen reacts to the blend of natural gas (CH_4), hydrogen (H_2) and carbon monoxide (CO), forming CO_2 and water and releasing heat. The redundant electrons of the oxygen ion are also set free and return to the cathode via an external electric circuit. The dc voltage created between the two electrodes is less than 1 V, so that many cells must be connected in series to achieve sufficient output voltages. To this end, the fuel cells are stacked. Optimizing the current in a module allows the released thermal energy to be used for preheating supply air, processing natural gas (re-forming natural gas and water into CH_4, H_2 and CO), for heating and for post-combustion of unused natural gas (**Figure 4.51.2**). The latter is employed in the Hexis concept (developed by Sulzer Innotec, Winterthur, Switzerland). The schematic of such an installation is shown in **Figure 4.52**. At present, these systems are still very expensive and only mass series will deliver a favourable cost-benefit ratio. This is nevertheless a very promising technology for the future.

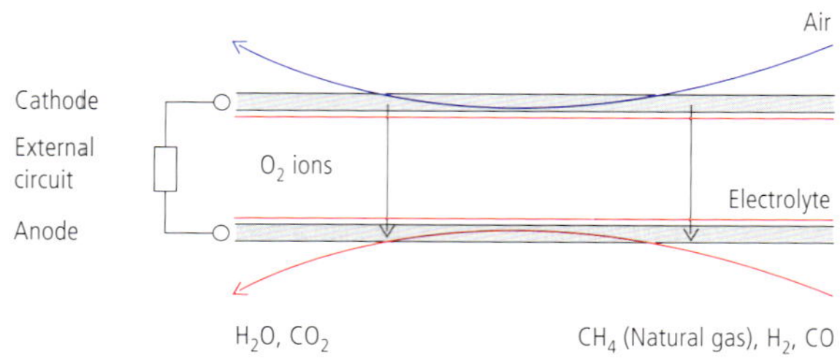

Figure 4.51.1
Principle of fuel cells

Oxygen ions from the air travel from the cathode via the electrolyte to the anode and react with the treated natural gas. This creates electricity and heat.

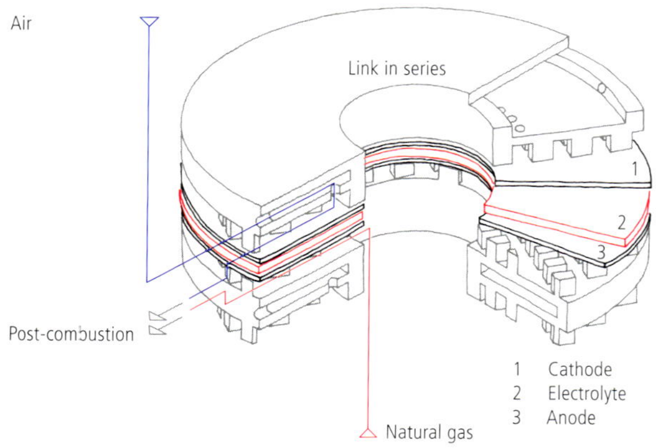

Figure 4.51.2
Section of cell stack

Only when several cells are stacked are units created which meet the required performance in the kW range.

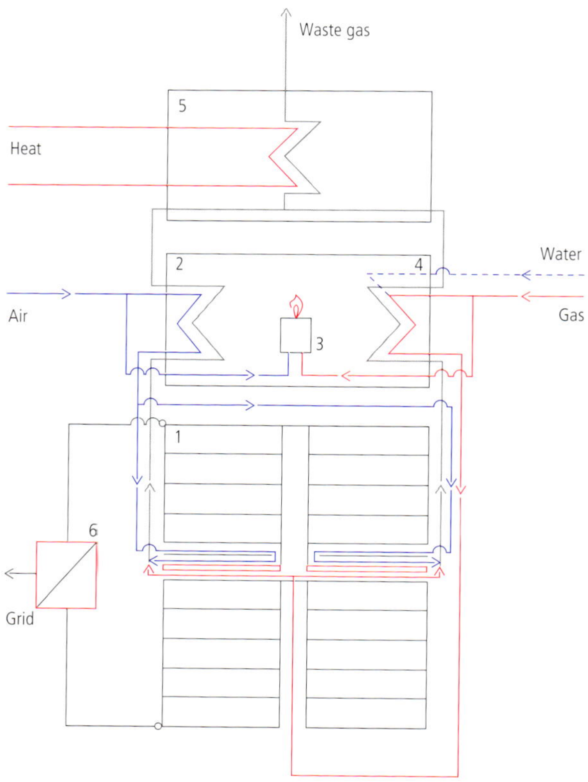

Figure 4.52
Diagram of fuel cell installation

1 Fuel cell stack
2 Air preheating chamber
3 Startup and auxiliary burner
4 Natural gas treatment
5 Flue gas heat exchanger
6 Controls, d.c./a.c. inverter

4.3.7 Heat pumps

In terms of construction, heat pumps are cooling machines that draw energy from the environment in a cyclical process, raising it to a higher temperature level, which is then usable for heating purposes. In this process, the usable thermal volume is a multiple of the thermal equivalent of the work expended.

In electricity-powered heat pumps, for example, the thermal output is three to four times each kW of motor performance (3 to 4 kW heating performance). The available thermal energy is therefore composed of two parts: the energy that is "pumped uphill" from the lower to the higher temperature and the thermal equivalent of the work required.

Heat pumps are either primarily designed for cooling and deliver the thermal energy generated in the process, or primarily designed for generating heat. **Figure 4.53** illustrates the principle of a gas-motor heat pump in a schematic diagram.

The diagram shows the gas motor and attached coolers (motor cooler/exhaust gas heat exchanger) as well as the link up to the cooling cycle with compressor, condenser (for heat generation) and evaporator for natural energy use (harnessing the cooling water of a cooling tower in the example shown).

Heat pump installations can be operated with the following heat sources:
– Groundwater
– Soil
– Outside air
– Exhaust air from ventilation systems
– Surface water
– Wastewater

Figure 4.53 indicates the path of the water flowing back from the heating network through the condenser (the heat pump), the motor cooler and finally the waste gas heat exchanger – a process that raises the supply water for the heating network to a very high temperature level (approx. 95 °C).

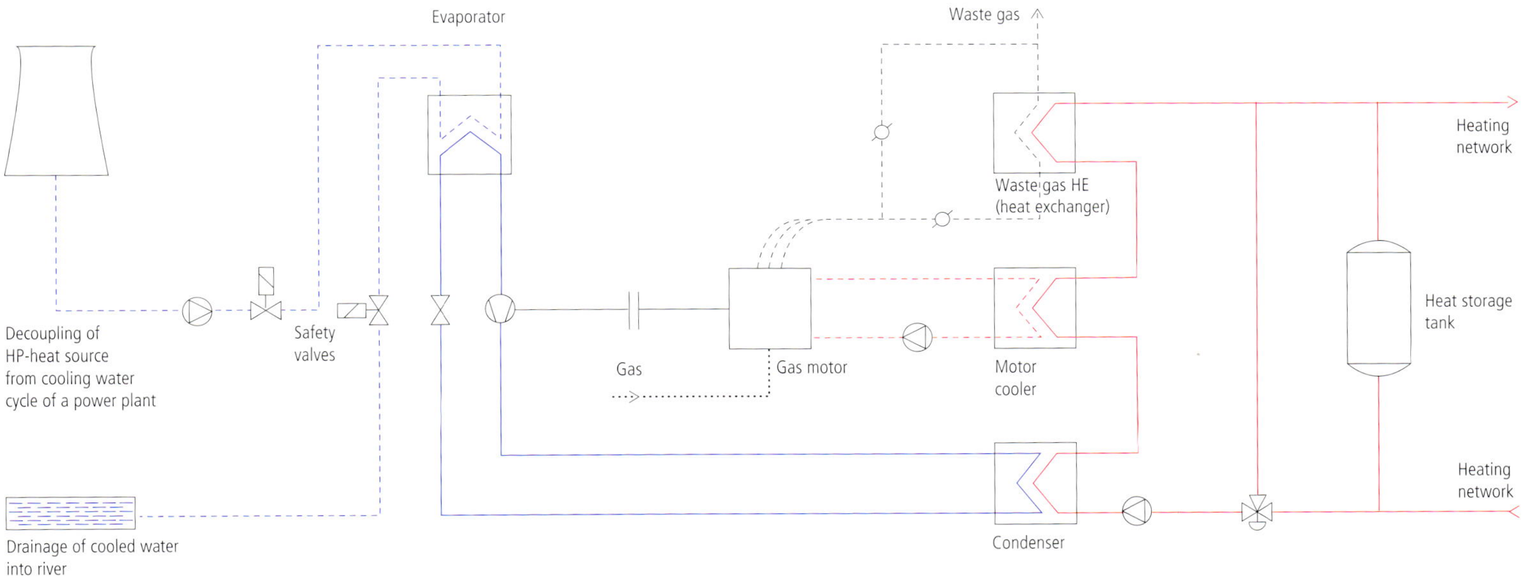

Figure 4.53
System diagram of gas motor heat pump

—— Hot water cycle
—— Chilling medium cycle
- - - Heat source cycle (cooling tower)
····· Gas
- - - Waste gas
- - - Cooling water cycle

The thermal energy output of the condenser is composed of the motor performance of the condenser as well as the thermal output extracted from the cooling water cycle via the evaporator. This is therefore a water-to-water system, since water serves as the heat source, unlike the system in **Figure 4.54**, which depicts the schematic of a gas heat pump designed to supply heating energy to houses. In this system, the outside air serves as a heat source, although, in contrast to the previous system, it can only be operated when outside temperatures are -5 °C or higher. Moreover, it is supplemented by thermal energy from heating boilers.

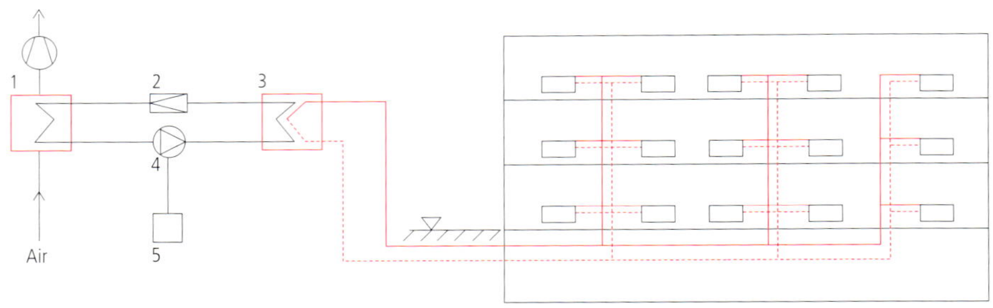

Figure 4.54
Diagram of gas heat pump

1 Evaporator
2 Expansion valve
3 Liquefier
4 Condenser
5 Gas motor
Heating network

The warm wastewater in sewers is another heat source that has rarely been harnessed until now. Heat is extracted from the wastewater via a heat exchanger and raised to a usable level with the help of a heat pump. The heat exchanger is integrated into the dry-run channel (**Figure 4.55.1**).

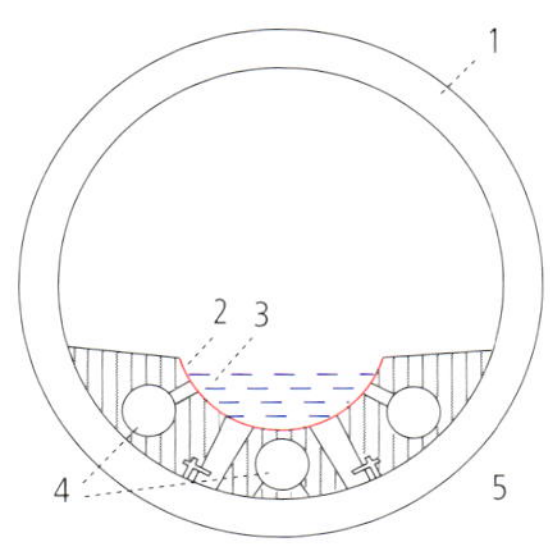

Figure 4.55.1
Cross-section of sewer with Rabtherm heat exchanger
(Drawing: Studer + Partner)

1 Sewer
2 Heat exchanger half shell
3 Sewage water
4 Intermediate medium, feed
5 Intermediate medium, return

Figure 4.55.2 contains the schematic diagram of a groundwater heat pump installation with heat source and heat sink, in which groundwater serves as an ideal heat source since it is available at a constant temperature (approx. +12 °C). Environmental concerns, however, have greatly restricted the use of groundwater today and applications for special permission are required for such systems, which are consequently rarely used. Since this system, too, relies upon water as a heat source it is also described as a water-to-water heat pump system.

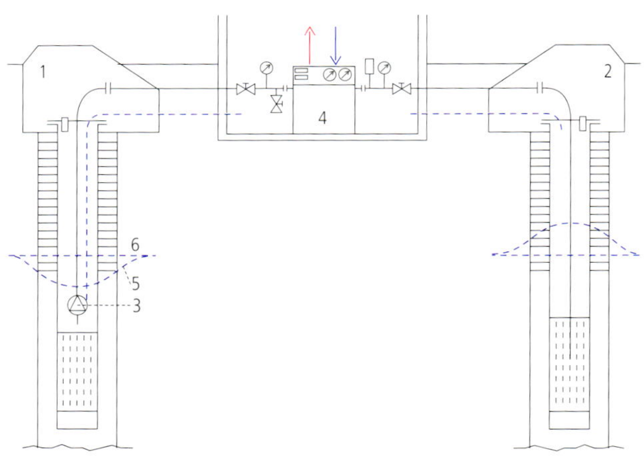

Figure 4.55.2
Diagram of groundwater heat pump installation with heat source, heat sink, submerged or deep-well pump, water/water heat pump

1 Heat source
2 Heat sink
3 Submerged pump
4 Water/water heat pump
5 Feed water level
6 Rest water level

Notation for heat pumps

X – Y – HP
(Source) (Sink)

The diagrams in **Figure 4.56** illustrate examples of using the soil as a heat source for a heat pump installation. There are several methods of harnessing geothermal heat, for example with the help of earth collectors (earth coils etc.).

Figure 4.57 depicts another application of collectors (heat absorbers). This image shows synthetic pipes installed on the roof (roof heat absorbers), which not only absorb incident solar energy but also extract energy from the surrounding air and from rain. In addition to roof-installed absorber surfaces, energy fences (vertical installation on the grounds) and energy stacks are other available options.

Figure 4.58 shows a gas heat pump in a central heating plant for a building. This gas heating plant is integrated into a heating network with additional heat pump systems (gas or electricity powered) and delivers 800 kW heating output per machine. At the same time, the gas heat pump system is used as a cooling machine supplying a bank of coolers – an optimum use of the machine.

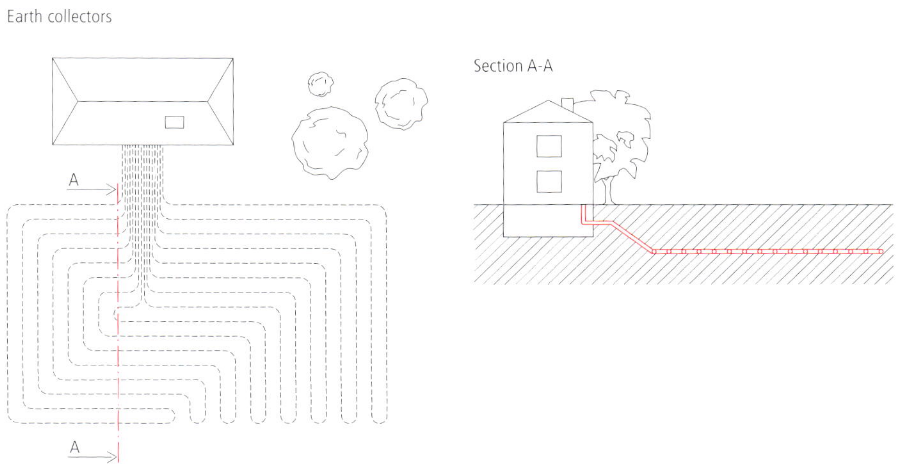

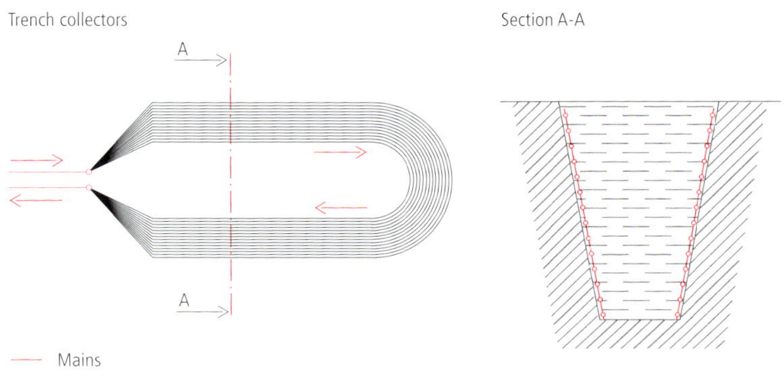

Figure 4.56
Examples: earth collector, trench collector and vertical bore hole

Figure 4.57
Roof heat absorbers

Figure 4.58
Central heating installation with gas heat pumps

4.3.8 Power-Heat Coupling (PHC)

Table 4.9 presents an overview of the various systems for power-heat coupling.

Until recently, power-heat coupling was used exclusively in power plants (heat power plants – HPP) operated with a variety of turbines. As their name indicates, heat power plants were primarily designed to produce heating energy in addition to electric power and to feed the energy into a district-heating network.

In response to the energy crises of the 1970 s and 1980 s, the monopoly of energy producers was softened to allow building users to construct their own heat-power coupling systems. These systems are so-called combined heat and power plants (CHPs), whose performance is clearly inferior, both in terms of power generation and distribution, to the performance of heat power plants (see Table 4.9). The principal aim was, and is, to achieve a balanced, interlinked approach to energy generation and supply.

Combined heat and power plants (CHP)
In principle, combined heat and power plants are small power plants with combustion motors, which power a generator to produce electricity and utilize the waste heat of the motors from cooling water and exhaust gas for heating. In terms of fuel consumption, they achieve electricity efficiencies of roughly 30 to 40 % and a thermal efficiency of roughly 55 %, resulting in a total efficiency quotient of 85 to 95 %.

Figure 4.59 shows the three CHP modules in the heating centre of a pharmaceutical plant with a shaft output of 315 kW each and a waste heat output of 520 kW each. The remaining thermal energy required for supply to the manufacturing and laboratory areas of the plant, as well as its absorption cooling machines and heating in general is produced by means of gas and steam boilers. Heat is withdrawn from the exhaust gases, whose temperature is reduced from roughly 430 to 120 °C in the process. In other words, nearly three quarters of the thermal energy contained in the exhaust gas is used.

CHP modules are expensive and more elaborate to maintain than heating boilers. The financial viability of CHP installations must be established for each instance. No doubt, such installations promote a more rational use of energy than conventional energy supplies and also reduce the environmental burden caused by harmful emissions.

System	Heating power stations (HPS)			Combined heat and power station (CHP)		
	Heating power station with steam turbines	Heating power station with gas turbines	Combined heating power station	Combined heat and power station, "to scale"	Standard combined heat and power station	Small combined heat and power station (TOTEM)
Drive system	Steam turbines	Gas turbines	Gas and steam turbines combined	Gas motor with three-way catalytic converter	Gas motor with three-way catalytic converter	Gas motor
Fuel	Coal, heavy oil (fluidized combustion), natural gas, heating oil (conventional steam boiler)	Natural gas, EL heating oil, gasified coal (in future)	Natural gas, EL heating oil, gasified coal (in future)	Natural gas, biogas (e.g. in sewage treatment plants)	Natural gas, biogas (e.g. in sewage treatment plants)	Natural gas, biogas (e.g. in sewage treatment plants)
Principal application (examples)	Integrated district heating	Process heat for industry, hospitals (steam, hot water)	Integrated district heating	Integrated local heating for larger single buildings	Integrated local heating for larger single buildings	SFH development, single buildings
Performance range	5 MW and up	0.5 MW and up	20 MW and up	50... 1'000 kW	150... 200 kW [2] [3]	15... 50 kW [3]
Current parameter[1]	0.30... 0.60	0.40... 0.70	0.80... 1.20	0.45... 0.65	0.45... 0.65	0.35... 0.45
Efficiency	0.85	0.75... 0.85	0.75... 0.85	0.85... 1.00 [4]	0.85... 1.00 [4]	0.85... 1.00 [4]

Table 4.9
Power-Heat Coupling (PHC)

[1]) Current parameter = electricity production/heat production
[2]) Good performance range in terms of economic efficiency and potential for application
[3]) Several units can be coupled to increase performance
[4]) Efficiencies of 1.00 and above (in relation to the net calorific value) are possible in theory when relevant measures are taken (use of latent heat by means of waste gas condensation)

CHP systems are a good choice for buildings or building complexes with constant, high electricity consumption (e.g. data processing centres, technical installations operated round the clock etc.) where the goal is to simultaneously utilize the thermal energy in a sensible manner.

Figure 4.60 shows a very large CHP system consisting of five modules. Clearly visible in the image, the gas-powered machines operate generators for power generation and simultaneously utilize the thermal energy. Depending on the total machine configuration, the thermal energy can be transformed into cooling energy in summer (via absorption cooling machines) or harnessed to serve other heat energy consumers.

CHP systems are ideal in cases where the installation replaces an emergency power diesel aggregate, selling heat and, possibly, feeding power into the grid. It is important to note, however, that the reimbursements vary widely from region to region and are dependent on the conditions imposed by the utility suppliers.

Capital and operating costs, as well as incidental operating costs, are fixed costs and therefore dependent on the annual operation period of the installation. The key is to ensure maximum operation periods, for as operating time diminishes, the specific costs of the energy produced rise proportionately. Figure 4.61 contains the annual graph for a housing scheme indicating the total heating energy requirement throughout the year in dependence of outside temperatures. The CHP modules marked in the graph (M1/2/3) have annual operating periods of 8,000 to 4,000 hours per year, which translate into high returns. Running times below 4,000 hours for such CHP modules should be avoided, since the energy costs rise too high as the graph demonstrates.

Figure 4.59
Central heating installation: combined power and heat plant CHP, Salutas Pharma GmbH (thermal output approx. 3 x 520 kW) (Image: m+w Zander)

Figure 4.60
Power room with CHP (thermal output approx. 7 MW)

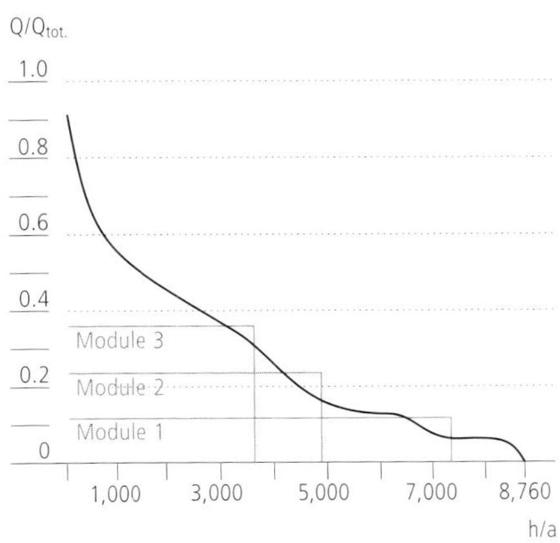

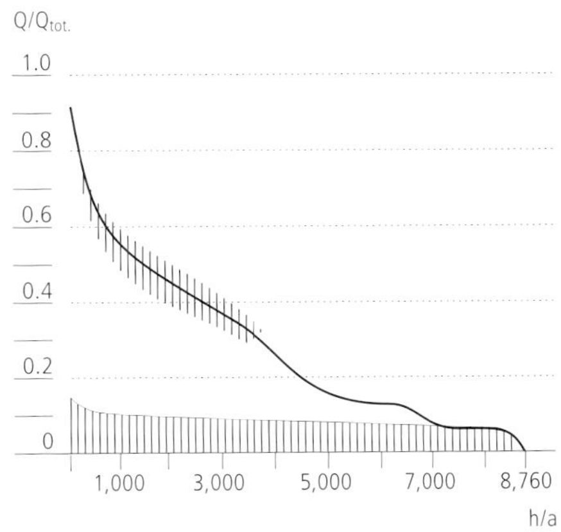

Figure 4.61
Graph charting the annual heating requirements of a residential development and interpretation of graph for CHP sizing

Heating requirement of a residential development

||||| Tolerance range determined by external temperature
||||| Warm water

4.3.9 Wood chip combustion

In contrast to wood combustion in an open chimney or old wood-burning stoves, wood today is burned in chips for environmental reasons, attaining the level of automation required for large systems. Low polluting levels depend largely on complete combustion through very high temperatures and prolonged retention of the gases in the combustion zone. This requires considerable blending of gases and combustion air. Modern wood-chip combustion systems, therefore, separate gasification, combustion and heat utilization, as shown in **Figure 4.62**.

When the wood is almost dry, very little carbon monoxide and carbon hydrogen are emitted. The temperature increase should be quite rapid; otherwise more contaminants are created. The problem in wood combustion is not the burning of new wood but rather of old pieces of pressure treated or impregnated timber. In these cases, contaminants must be removed from the flue gas with the help of catalytic converters.

Great efforts are being made (especially at the ETH in Zurich) to develop clean-burning wood-chip combustion plants, wood being an indigenous, plentiful and renewable resource in Central Europe.

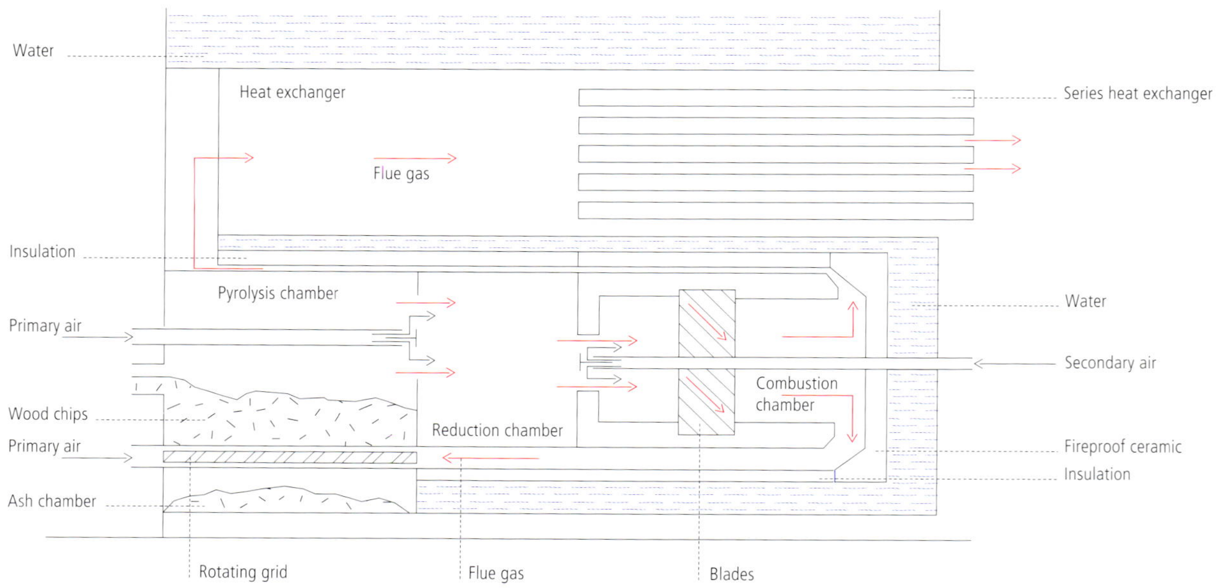

Figure 4.62
Diagram of wood chip combustion installation

The wood chips are transported to the pyrolysis chamber on a spiral conveyor and then slowly transported farther on a rotating grid. In the reduction chamber, the nitric oxides are reduced through preheating. The (combustion) gases created by this reduction enter a combustion chamber, where they are intensively mixed with fresh air to create complete combustion. The hot combustion gases then flow through a heat exchanger and an economiser and transfer thermal energy to the warm water heat cycle.

4.4 Thermal energy distribution

The following systems are differentiated in thermal energy distribution (heating):
- Split system with bottom distribution and vertical branches
- Split system with top distribution and vertical branches
- Split system with horizontal distribution on each floor
- One-pipe systems
- Single-storey heating

4.4.1 Split systems

Split system with bottom distribution

In the split system (**Figure 4.63**) each heating unit in the system is supplied with heating water of equal temperature in a flow pipe; the cooled water is returned to the heat generator via a return pipe.

In larger objects, this kind of a distribution system also contains several subgroups, divided according to orientation or use, which can be regulated individually.

Split system with top distribution

In this system (**Figure 4.64**) heat is distributed as in the split system with bottom distribution. The only difference is that the main distribution pipes are installed on the top floor and in the attic. In addition to the flow and return pipes on the top floor or in the attic, a return pipe is also installed in the basement or in a floor duct in the lowest storey of the floors to be heated.

The system is used when the heat generating installation is located on the roof or if bottom distribution is precluded for constructional reasons.

Split system with horizontal distribution on each floor

In this system, the distribution pipes are installed either beneath the ceiling – in the cavity above the suspended ceiling – or in a double floor and the radiators/convectors above or behind are directly linked to the distribution pipes (**Figure 4.65**).

The system is used in buildings where vertical branches cannot be executed in the facade area and where vertical distribution must therefore be restricted to a few, centrally located branches.

In buildings without double floors or suspended ceilings the system can only be used if the pipes are installed visibly on the unfinished floor, thus accepting the disadvantages of distribution pipes in the screed or if the pipes are installed on the external wall behind the radiator.

In residential buildings, especially, radiators are also connected to a central distributor via floor pipes in the subfloor.

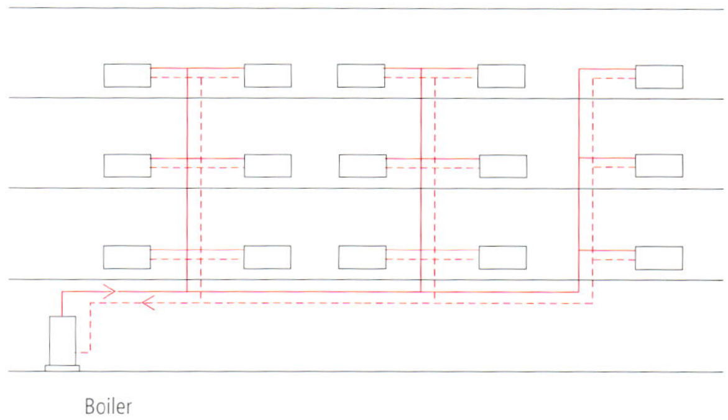

Figure 4.63
Split system with bottom distribution and vertical branches

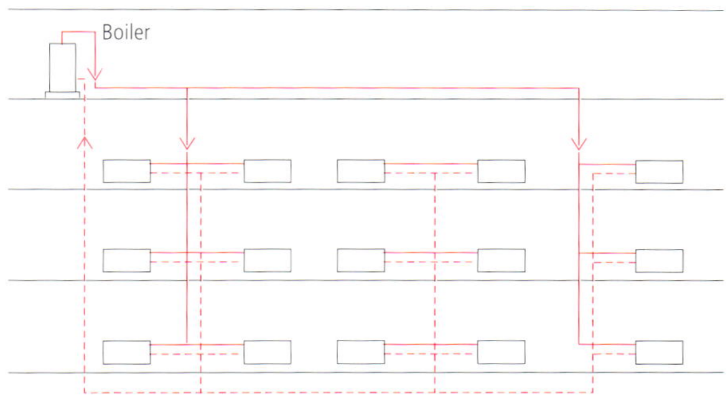

Figure 4.64
Split system with top distribution and vertical branches

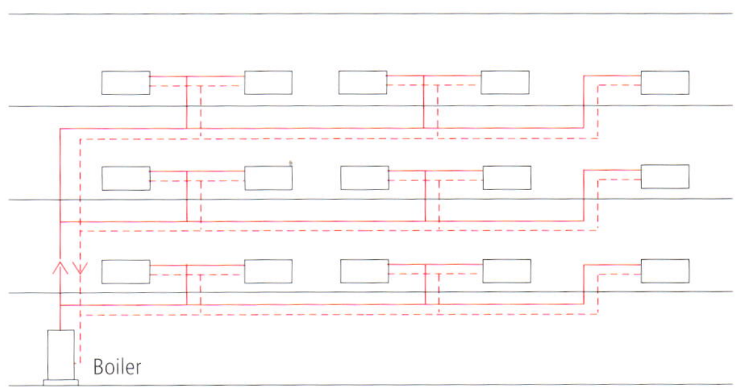

Figure 4.65
Split system with horizontal distribution on each floor

4.4.2 One-pipe system

In one-pipe systems, radiators are supplied with heating water via a single pipe; in other words, flow and return water are transported in a joint pipe. The outcome is that the temperature of the heating water diminishes from radiator to radiator due to the gradual admixture of the cooler return water. To adjust the radiator output, the corresponding heating surface or the volume of water flowing to the relevant radiator must be increased in order to maintain a constant, average temperature in each radiator.

In large one-pipe systems, distribution originates from the plenum. The pipes to the radiators in the one-pipe system are connected to this distribution system.

A differentiation is made between horizontal and vertical one-pipe systems (**Figure 4.66**). In the horizontal one-pipe system, the pipe leading off from the vertical branches is installed horizontally – generally below the ceiling or in the double floor. Heating surfaces located on the same floor are linked one beside the other to this horizontal pipe.

In the vertical one-pipe system, the pipes, branching off from the horizontal distribution pipes in the basement or attic, are installed vertically in shafts or channels. Heating surfaces located above one another on different floors are linked to this pipe. The advantage of a horizontal one-pipe system versus a vertical one-pipe system is that floor-to-floor division of the heating groups allows for individual, separately measurable and controllable zones, which is compatible with building use both in office/commercial and in residential construction.

4.4.3 Single-storey heating

Single-storey heating, **Figure 4.67**, which is almost exclusively employed in refurbishing older buildings, provides a separate heat generator (special gas boiler or gas-circulation boiler) for each floor or usage zone. The individual radiators must be supplied with heating water either in a split-system or in a one-pipe system.

The pipes are predominantly installed on the unfinished floor or behind special baseboards. When floor installation is chosen, the aforementioned disadvantages of pipes in screed must once again be tolerated.

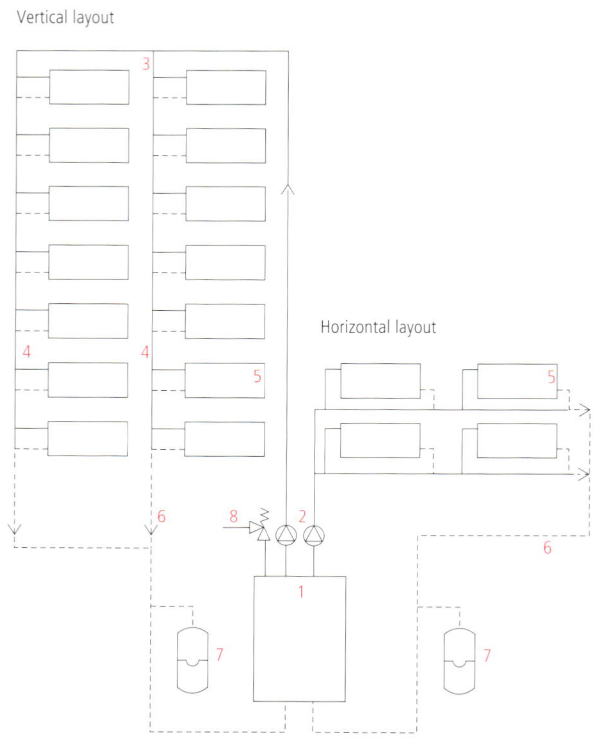

Figure 4.66
One-pipe system

1 Boiler/furnace
2 Circulation pump
3 Distribution pipe on or above top floor
4 Fall pipe strings
5 Radiators with manual or thermostatic valves
6 Return collecting pipe
7 Pressure expansion tank
8 Safety valve

Figure 4.67
Single-storey heating

Feed
Return
Gas pipes

4.4.4 Piping

4.4.4.1 Materials

The following materials are commonly used for piping systems:
- Black steel
- Galvanized steel
- Copper
- Plastic
- Compound materials (plastic/metal)

Black steel pipes
Seamless or welded black steel pipes are used almost exclusively in heating installations. Water quality must be carefully monitored when using black steel pipes. The water may require processing in some cases.

Galvanized steel pipes
Galvanized steel pipes are rarely used in heating installations. They are, however, a frequent choice for sanitary installations, since zinc protects the iron surfaces. Zinc also has the property of forming protective layers in combination with the components of water (calcium). The quality of these protective layers, however, is dependent on the composition, temperature and flow velocity of the water. In principle, galvanized steel pipes can be used for both cold and warm water, although the maximum temperature must not exceed 60 °C. The risk of corrosion increases when this temperature is exceeded and the effectiveness of the protective layer, which is especially good for temperatures below 60 °C, decreases.

Both black and galvanized steel pipes should always conform to DIN standards and be identified as quality (or high-grade) pipes.

Copper pipes
Copper is a suitable material for use in cold and warm water piping. It is therefore employed in both heating and sanitary systems. A protective layer also forms inside copper pipes and once again it is dependent on the quality of the water composition. The formation of the protective layer is especially good at high temperatures.

Although copper pipes can be somewhat problematic with some types of water, copper is by far the best choice for sanitary installations. Preferably one should only use copper pipes from manufacturers who can guarantee a degree of copper percentage purity that corresponds to standard codes.

Plastic pipes
Due to their corrosion-resistant properties, plastic pipes are employed whenever metal cannot be used, although it is important to limit their exposure to high temperatures and pressure.

In heating installations the oxygen diffusion through the pipe wall constantly enriches the heating water with oxygen, which can lead to corrosion in furnaces, boilers and radiators etc. It is important to take the permeability of plastic pipes to diffusion into consideration.

4.4.4.2 Corrosion

Corrosion is the destruction of materials caused by chemical or electrical processes. The principal cause of corrosion is the dissolution of gases, oxygen or carbon dioxide in water. Chemical composition as well as water temperature, flow velocity and pressure are also important contributing factors.

Electrochemical processes (element formation) also play an important role, especially when several metals are used (mixed installation). Additional damage may result from matte formation, which is caused by the separation of the carbonate hardness from the pipe walls when water temperatures rise above approx. 50 °C. The degree of matte formation depends on the salt and mineral contents of the water. Corrosion always sets in at the metal surface. In order to select the correct material it is therefore important to understand the reactions between water components and material and to explore which factors promote, hinder or prevent the formation of protective layers.

In residential installations, potable water and the materials that come into contact with it should meet DIN 2000 standards to avoid damage by corrosion.

In heating systems, where water flows in a constant cycle, corrosion and matte formation are rare in steel piping, since the initial oxygen content is expelled through the rise in temperature and removed from the system by ventilation. If any oxygen is prevented from entering into the system through appropriate measures (e.g. sealed expansion valve) and expert installation, corrosion is more or less eliminated altogether. Heating water should also be slightly alkaline to neutralize the carbon dioxide contained in water, which promotes corrosion.

If corrosion cannot be prevented even through the use of appropriate materials as a result of the water quality, various protective devices or water processing systems can be installed to remove gas, acid and hardness from the water and to promote the formation of a protective layer.

A water analysis is essential in order to determine the appropriate measures.

4.4.4.3 Sizing

Piping is sized with the help of a friction diagram. An example is shown in **Figure 4.68.1**. This pipe friction diagram contains the following parameters:
- Thermal output (in kW) for ΔP (net pressure) = 20 K
- Loss of pressure (pressure drop, head loss)
- Velocity
- Pipe size

The example given is for a thermal output of roughly 400 kW for a maximum pressure drop of 100 Pa/m. This results in a sizing of DN 80 for roughly 0.92 m/s flow velocity.

All piping is sized with the help of such diagrams, with the water volume calculated on the basis of the thermal output and the temperature spread between feed and return flow temperature.

In large buildings with elaborate and widely spread piping networks, the pressure losses in the individual subsections must be calibrated with the help of control valves to ensure that all users are supplied with adequate volumes of water (**Figure 4.68.2**).

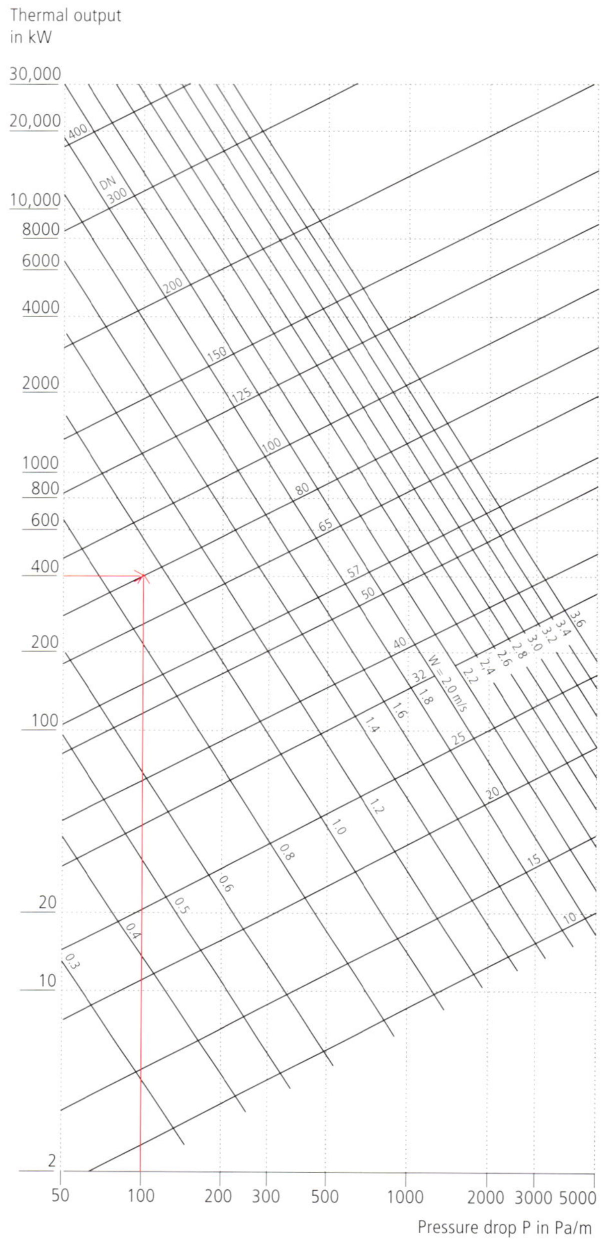

Figure 4.68.1
Pipe friction diagram for steel pipes

Pump warm water heating, water temperature 80 °C, roughness g = 0.045 mm, ΔP = 20 K

When the water temperature is 150 °C, P is roughly 4 % less; it is roughly 4 % greater for a water temperature of 50 °C.

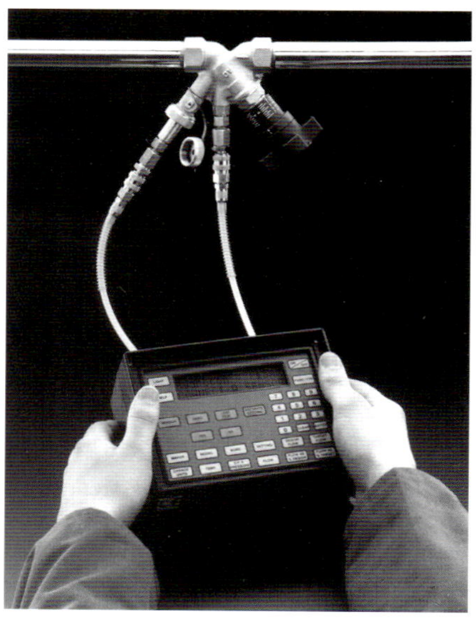

Figure 4.68.2
Control valves for heating and cooling systems and flow measuring computer for precision control of mass flows (System Oventrop)

Figure 4.68.3
Pump-driven pressure maintaining device for larger, closed heating systems
(Image: Reflex)

To ensure that the control valves can operate properly in the heating system, it is important to maintain the operating pressure necessary to do so. This is achieved with the help of expansion tanks or, in the case of large piping systems, with the help of pump-driven pressure maintaining devices (**Figure 4.68.3**).

4.4.5 Insulation

Insulation plays an important role in contemporary installations because it is key to a number of important functions. It prevents energy losses, maintains maximum or minimum temperatures, ensures uninterrupted operation, provides protection against burns from hot pipes and prevents corrosion, especially in cold piping systems whose surface temperatures lie below the dew point temperatures of the surrounding areas.

Insulation on building systems must be installed in an expert manner, making sure that the relevant prerequisites are fulfilled.

Table 4.10 lists the efficient insulation layer thickness for different standard piping sizes depending on the conductivity 8 of the thermal insulation.

To install insulation properly, it is important to maintain minimum distances between the pipes, objects and other building components (**Figure 4.69**). When low-pressure insulation is used (mineral wool with gross densities less than 75 kg/m^3) and operating temperatures are 250 °C or higher, supports must be added for all straight insulated sections, we well as for elbows and fittings.

Conductivity λ of thermal insulation in W/(m·K)	**Nominal diameter of pipes** 10	15	20	25	32	40	50	65	80	≧ 100
	Thickness of insulation layer in mm									
0.035	20	20	20	30	30	40	50	65	80	100
0.040	25	25	25	40	40	50	65	80	100	125
0.045	35	35	35	50	45	60	80	100	125	155
0.050	45	45	40	60	55	80	100	120	150	190
0.055	60	55	50	75	70	90	120	150	185	230
0.060	70	65	60	90	80	110	140	180	225	280

Table 4.10
Insulation for heat distribution installations

Insulation with G+H ISOVER shells IS-HI, IS-HF or IS-HA/WLG 035
Insulation with G+H ISOVER-aminated matting ML 3/WLG 040

Min. distance to	Nominal diameter of pipes (a) up to 32	40 – 50	65 – 100
Pipes (c)	80 / 100	120 / 160	220 / 280
Wall and ceiling (b)	50 / 60	70 / 90	120 / 150

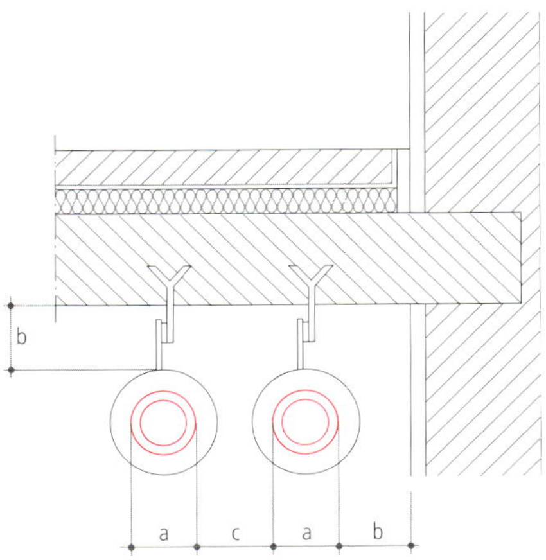

Figure 4.69
Recommended distances between pipes according to heating system guidelines

Insulating materials should have as little impact as possible on the base on which they are installed. It is therefore important to take the material properties of mineral wool into consideration (e.g. sulphide impact on steel, chloride impact on nickel chromium steel). Sulphide-free and low-chloride materials are preferable.

Supplementary heating systems, installed externally, are required for down times and longer periods of suspended operation when there is a risk that the liquids contained in the piping and tanks might freeze. This is especially important in cases where the insulation alone cannot prevent the heat losses that occur during down times.

Diffusion-proof insulation
By reversing the temperature drop in cold water pipes, water vapour from the relatively warm surrounding air can be transported to the cold air at the pipes. Depending on the temperature difference and the relative humidity, this may lead to the formation of condensate. A vapour barrier must be installed to prevent condensate formation and to protect the permeable insulating material against moisture saturation. The type and thickness of the materials used for the vapour barrier is based on the object temperature, ambient temperature, relative humidity and the vapour-resisting effect of the entire insulating system. For the above conditions, aluminium foil with insulating materials on the inside or double layers surrounding insulating material are suitable as insulation.

The vapour barrier must fully enclose the insulating layer and it is important to seal joints, termination points and penetrations with insulating sealant tape.

Additional jackets
Insulation jackets protect insulation against damage, exposure to weather and chemicals. In industrial applications jackets with galvanized or plastic-coated sheets are used (steel, aluminium or austenitic sheeting). The jacket material must not diminish the fire performance of the insulating material. In addition, insulating materials should be non-flammable and not give off smoke or toxic gases in case of fire.

Insulating materials are therefore rated as "noncombustible" (A1/A2), "hardly inflammable" (G1), "inflammable" (B2) and "easily inflammable" (B3). In addition to these basic classifications, fire resistance ratings have been established for structural components (S), ventilation and air conditioning ducts (HVC), pipes (P) and fire doors (D).

Sound insulation
Containers, pipes and ducts through which gases, vapours or liquids flow, can represent a considerable source of noise. The cause may be the high flow velocity of the transported mediums, but also barriers inside the piping and duct systems, which become sound sources. It may therefore be necessary to take measures to suppress sound or sound radiation. In many cases, the required insulation can simply be coupled with soundproofing. Insulating materials with increased pressure resistance tend to decrease noise levels. External or additional jackets (with sealed joints) can also improve the situation, requiring the use of sound absorbing sheeting in sandwich construction in front of and behind existing sound bridges.

For pipes and ducts etc., appropriate impact sound stoppers must be installed at all penetration points on walls and ceilings to prevent the oscillation of the system from being transferred to the building fabric.

In conclusion, **Figures 4.70** to **4.72** indicate reference values for the correct selection of insulation in terms of different operating conditions and requirements.

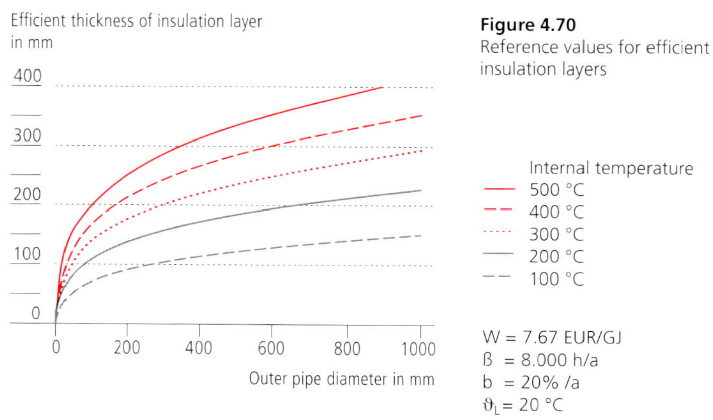

Figure 4.70
Reference values for efficient insulation layers

Figure 4.71
Thickness of insulation in mm required to prevent condensation in in-situ polyurethane foam inside aluminium sheet casing,

– 85 % relative humidity
– 20 °C ambient temperature

Fundamentals:
Operational thermal conductivity for in-situ PUR foam outdoors with
– Pipe densities of 45 kg/m³ to 60 kg/m³ according to Q 138
– Transmittance value 5.0 W/(m² K)

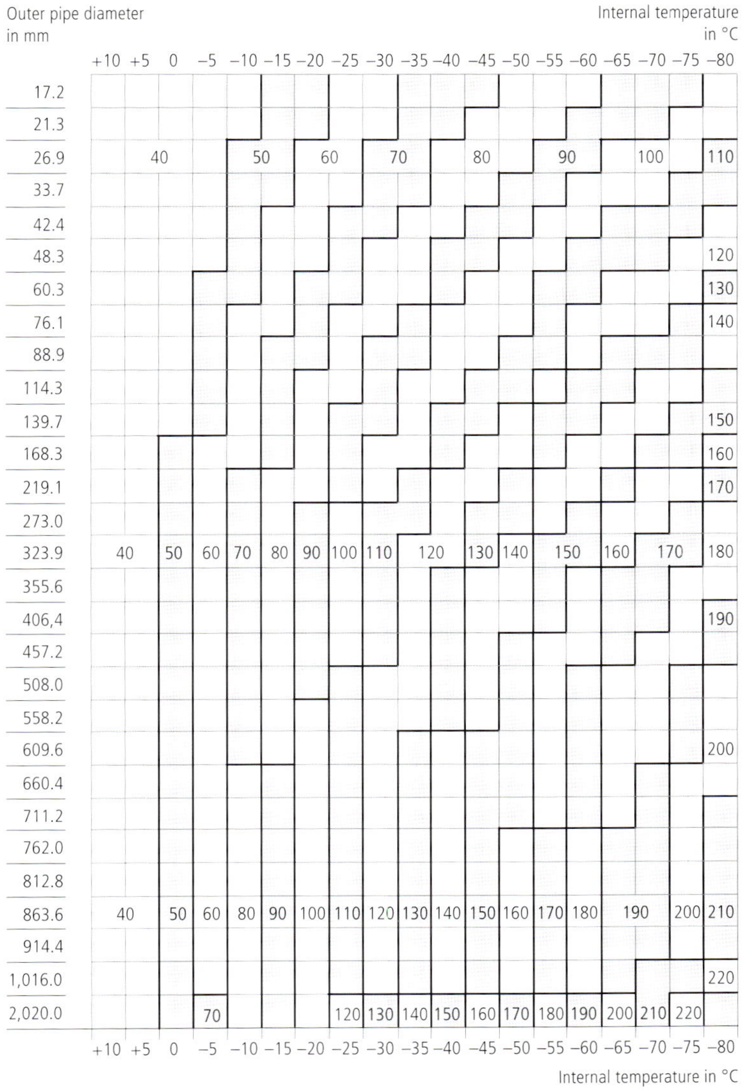

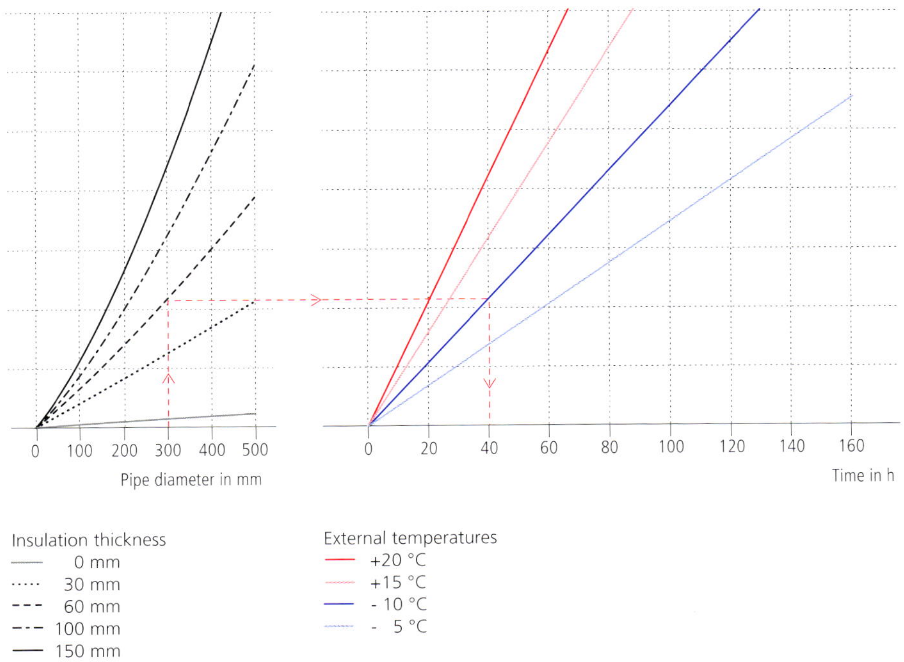

Figure 4.72
Rest period to prevent cold water pipes from freezing (without ice formation)
Water temperature: +5 °C;
Wind velocity: 5 m/s;
Insulation: ISOVER shells or PS (polystyrene) HR foam shells

4.5
Radiators and heating surfaces

Types of space heating

Radiators and heating surfaces are differentiated by construction (e.g. radiators, convectors, area-, ceiling-, floor- and wall-heating units) and material (e.g. cast iron, steel, copper, aluminium and plastic). They are used to maintain thermal equilibrium in an enclosed space (**Figure 4.73**).

International standards have been developed for the measuring methods and test conditions applied to compare the thermal output of heating surfaces produced by different manufacturers. In Germany DIN 4707, Sheet 2 and DIN EN 422/2 set the standard.

This ensures that the heating surfaces correspond to the heat requirements. The standard thermal output is set for a temperature spread of 20 K between feed and return flow and excess temperature of 60 K above a room temperature of 20 °C. The conversion method for different conditions is described in DIN 4703, Part 3.

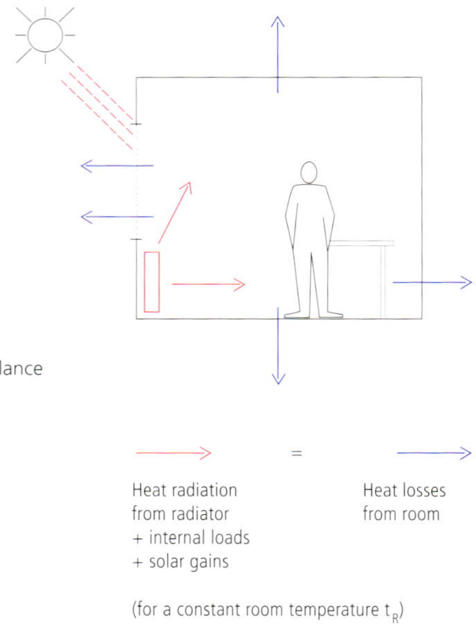

Figure 4.73
Maintaining thermal balance in a room

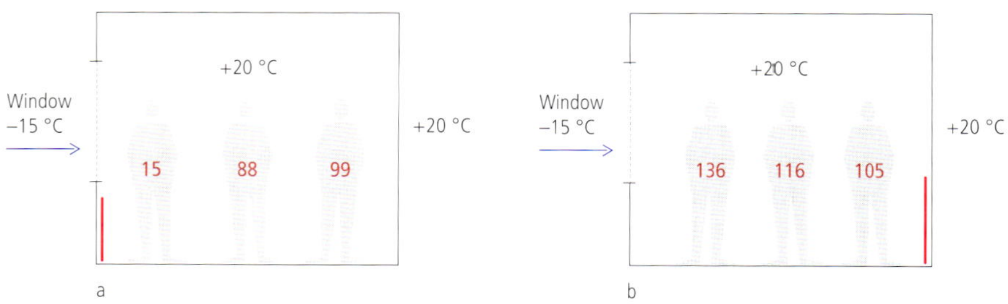

Figure 4.74
Schematic illustration of heat radiating from occupants (in W) in rooms with radiators (after Kollmar)

a Radiator below window
b Radiator on lateral internal wall

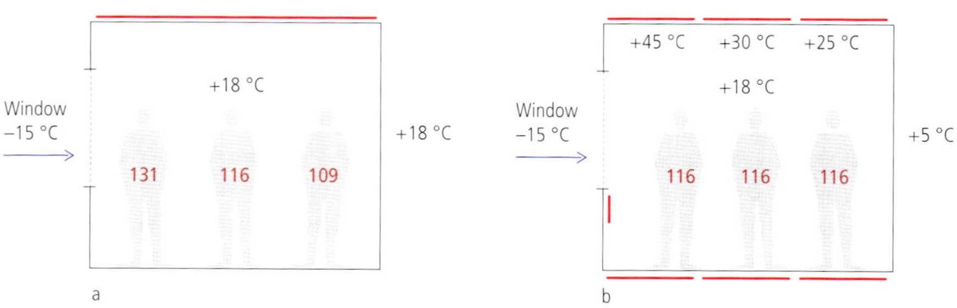

Figure 4.75
Schematic illustration of heat radiating from occupants (in W) in rooms with radiant panel heating systems (after Kollmar)

a Radiant ceiling heating
b Radiant ceiling and floor heating as well as window heating panel (radiant panel heating)

The heat emitted by occupants fluctuates depending on radiator location and type of heating. **Figure 4.74** illustrates that human heat emission changes depending on the type of the heated room (location of radiators) and the position of the occupant in the room. When the radiator is located on the internal wall and the occupant close to the window at the external wall, heat emission to the outside will always result in a feeling of discomfort ("radiation draught"). Moreover, the downflow of cold air on the window surface causes an additional draught effect. Conversely, when heat is provided by means of radiators below the entire window front, the heat radiating from the radiator compensates for the impact of the cold external wall and the window surface. The ratio of heat radiating into and out of the space changes depending on the distance to the window. The radiated thermal energy may even be too great and lead to discomfort, in which case occupants will naturally move away from the radiator. Both scenarios place an asymmetrical thermal burden on the body with negative effects. Differences of 20 to 30 W/m² are especially noticeable and uneven cooling of the body, for example of the head, by more than 40 W/m² causes discomfort.

Additional burdens can be the result of excessive thermal radiation, exemplified in **Figure 4.75**. Where more than half of the human body is exposed to thermal radiation of more than 300 W/m² (e.g. commercial kitchens, ovens etc.), shields must be provided to protect the body from the radiating heat.

The same is true for incident solar radiation through windows, where heat loads of 350 to 450 W/m^2 are quite possible unless shading elements are used to prevent them.

Floor heating can have a similarly undesirable effect, when surface temperatures rise above 27 °C in a constantly occupied space. In areas along the edge, where occupants are unlikely to step, temperatures may be increased to 35 °C to relieve the burden from the occupied zones. On the other hand, floor temperatures should not fall below 17 to 18 °C because an accumulation of heat-sensitive nerves in the ankle area makes people prone to discomfort.

In the case of radiant ceiling systems, the radiation of heat onto the occupant's head in a room with an air temperature of 20 °C should not exceed 12 W/m^2 to avoid discomfort. It is important to remember that the room height and ceiling temperature must be proportionate (i.e. the lower the room height, the lower the ceiling temperature).

Figure 4.76 indicates the layers of temperatures in the occupied area for different types of heat influx. The "ideal heating" is indicated for the purpose of comparison.

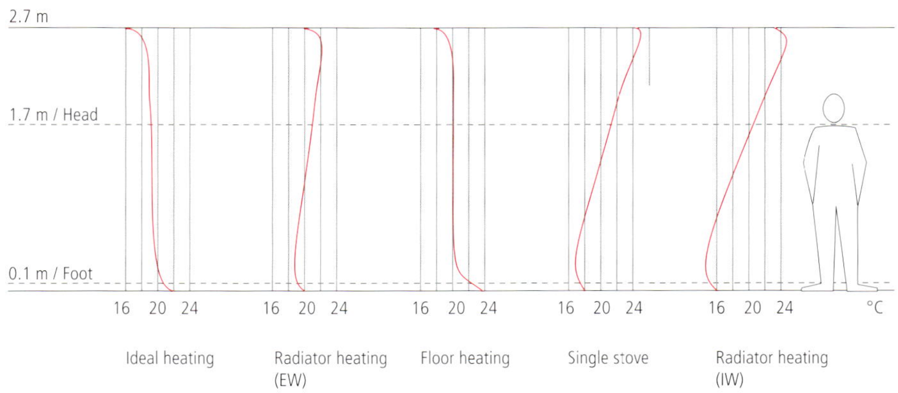

Figure 4.76
Temperature layers in indoor space for human occupation

EW EW External wall location of radiators
IW Internal wall location of radiators

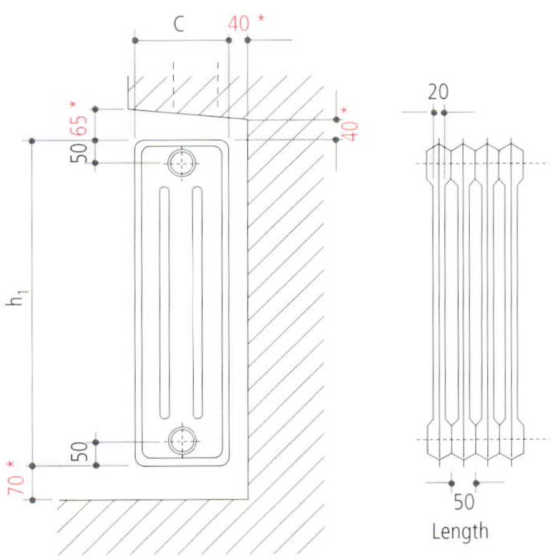

--- Slits are recommended for deep radiator niches
* Minimum distances
h_1 Height
C Depth

Figure 4.77
Cast-iron radiator dimensions according to DIN 4720

4.5.1 Radiator heating

Radiators are the most common heating units. They are manufactured in cast iron, steel, aluminium and plastic. Owing to the relatively small amount of water contained in radiators, they are quick-response heating surfaces, especially desirable in combination with passive solar use. Depending on the total radiator surface and model depth, 60 to 70 % of heat is emitted by convection and 30 to 40 % by radiation. Each radiator consists of individual sections through which water or steam flows. Depending on the material, the sections are assembled with R 1" nipples with right- and left-handed threads or joined by welding. Manufacturers offer a wide range of heights and depths, making it easy to adapt radiators to the layout and thermal requirement of individual rooms.

4.5.1.1 Cast iron radiators

The cast iron radiator (**Figure 4.77**) is composed of individual sections joined in series by nipples until the desired length is achieved.

Radiators can be delivered fully assembled. Weight and the desired output determine the length of each radiator unit. Once the overall length of the unit has been determined, connector, blind or ventilation plugs are attached at both ends depending on the manner of connection and set up. All cast iron radiators are classified as normal or high-pressure models.

Aside from standardized cast iron radiators, some manufacturers also produce flat cast iron radiators with face surfaces. They are not as deep as standard radiators and are manufactured in a variety of heights. Their shallow design makes them particularly suitable for retrofitting in spaces with limited installation depths. Measurements and other technical data are contained in the literature provided by the manufacturers and some examples are given in **Table 4.11**.

4.5.1.2 Steel radiators

By comparison to cast iron radiators, steel radiators are lighter, less expensive, more resistant to fracture, shorter and weldable in case of damage. They are, however, more susceptible to corrosion and for this reason steel radiators are only used for warm water heating, which today is generally employed with closed loop designs, but not for steam heating. Each individual section is welded together from two half shells formed from 1.25 mm thick radiator sheeting. As many as 20 sections in turn are welded into blocks at the headers. As with cast iron radiators, nipple joints are required to connect the blocks, to add individual sections or attach the connector, blind and ventilation plugs. To this end, sockets with right- and left-handed threads (R 1 1/4") are installed in the heads. These sockets are punctured in the centre to permit water circulation and ventilation.

Steel radiators, too, come in models designed for normal and for high-pressure operation. In high-pressure models, stabilizing sheets (tension rods) are welded into the top and bottom head for safety reasons (**Figure 4.78**).

DIN 4703 Part 1 and DIN 18380 specify the applications and technical data (**Tables 4.12.1/2**).

In addition to standardized steel radiators, there are "narrow columns" with depths of 70 to 75 mm and various heights. The same variables as discussed above for flat cast iron radiators apply.

4.5.1.3 Aluminium radiators

In addition to cast iron and steel radiators, there are aluminium lamellae (or fin) radiators. They are manufactured in compression and extrusion presses. Since they are nearly twice as expensive as steel radiators, they have not captured a sizeable market share despite their aesthetic form. Difficult water conditions may also lead to corrosion (risk associated with mixed installations with copper).

Radiator dimensions										
Height in mm	280	430	430	580	580	580	580	980	980	980
Depth in mm	250	160	220	70	110	160	220	70	160	220
Standard thermal output per section in W	92	93	122	68	92	126	162	111	204	260

Table 4.11.1
Standard thermal output of cast iron radiators after DIN 4703 per section in W for warm water with t_m = 80 °C and room air temperature of 20 °C, after DIN 4703 Part 1

Model	Abbr.	Heating medium	Permissible operational overpressure (bar)	Max. operation temperature (°C)	Inspection overpressure (bar)
Standard model	SW	Warm water	4 (static pressure plus pump pressure)	110	7
	SS	Steam	2	133	7
Custom model	CW	Warm or hot water	6 (static pressure plus pump pressure)	140	12
	CS	Steam	4	151	12

Table 4.11.2
Applications for cast iron radiators after DIN 4703

Radiator dimensions								
Height in mm	280	430	430	580	580	580	980	980
Depth in mm	229	139	200	90	139	200	89	139
Standard thermal output per section in W	46	43	58	42	58	73	70	90

Table 4.12.1
Thermal output of Sanaplan (Baufa) steel radiator per section in W for warm water with t_m = 62.5 °C and room air temperature of 20 °C

Model	Abbr.	Heating medium	Permissible operational overpressure (bar)	Max. operation temperature (°C)	Inspection overpressure (bar)
Standard model	S	Warm water	4 (static pressure plus pump pressure)	110	7
Custom model	C	Warm or hot water	6 (static pressure plus pump pressure)	140	10

Table 4.12.2
Applications for steel radiators after DIN 4703

4.5.1.4 Plastic radiators

Plastic radiators can only be operated with maximum operating temperatures of 80 °C and a maximum operating pressure of 2 bar above atmospheric pressure. Other disadvantages include considerable expansion, oxygen diffusion, low mechanical strength and combustibility. Plastic radiators are also more expensive than either cast iron or steel radiators. Advantages include light weight and easy installation.

4.5.1.5 Pipe radiators

Like steel radiators, steel pipe radiators are welded into blocks from individual sections. Each section is manufactured from two head pieces, consisting of two pressed and welded half shells, and the water-conducting precision steel pipes that are welded in between. The pipes have no edges that could lead to injuries. The tubular form allows for operating pressures as high as 10 bar above atmospheric pressure. Since they are manufactured in a variety of depths and heights, they are easily adapted to any spatial situation. Steel pipe radiators can thus also become an element of spatial design for architects (**Figures 4.79/4.80**). Additional measurements and technical data are contained in the literature provided by manufacturers (some examples are given in **Table 4.13**).

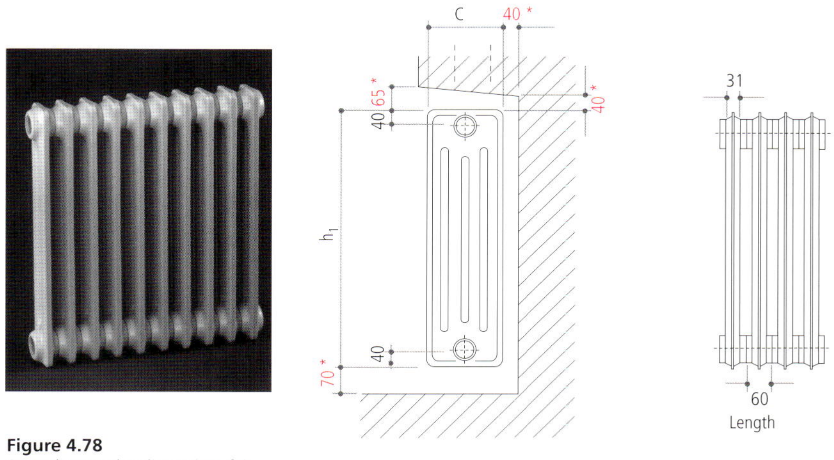

Figure 4.78
Sanaplan steel radiator (Baufa) and steel radiator dimensions according to DIN 4722

--- Slits are recommended for deep radiator niches
* Minimum distances
h_1 Height
C Depth

Figure 4.79
Modern pipe radiator Decor-S (Kermi)

Figure 4.80
Example of using pipe radiator in interior design

Dimension in mm	**Standard thermal output** in W 2 columns 65*	3 columns 100*	4 columns 140*	5 columns 178*	6 columns 215*
300	29	41	53	64	76
350	34	48	63	75	88
400	38	55	70	86	100
450	44	62	78	97	113
500	49	69	87	107	126
550	55	74	95	117	138
600	59	81	103	127	151
750	74	101	128	157	184
900	90	119	150	185	219
1000	98	131	166	203	241
1100	107	143	181	221	263
1200	116	156	197	240	280
1500	142	192	242	288	335
1800	167	228	287	343	399
2000	188	253	318	378	442
2500	233	316	395	465	541

* Depth in mm

Table 4.13
Standard thermal output of pipe radiators per section in W for warm water with tm = 80 °C and room air temperature of 20 °C, 25 mm pipe diameter and 45 mm section length after DIN 4702 Part 1

Figure 4.81
Towel radiator
(Credo-IR, Kermi)

Figure 4.82
Towel radiator
(classic, Brötje)

Figure 4.83
Towel radiator
(Credo-Techno, Kermi)

Figure 4.84
Modern, flat compact radiators obviate the need for niches and deliver high thermal output even for low-temperature systems
(Image: Kermi)

Figure 4.85
Typical smooth-walled flat radiator
(Combiplan Elegance, Baufa)

4.5.1.6 Towel radiators

Towel radiators are more frequently installed in wet cells and kitchens. They come in a wide variety of shapes and colours that offer architects a broad range of design options. Electric heater band inserts can be used to adjust temperatures even in summer, making it possible to dry towels on these units all year round (**Figures 4.81/4.82/4.83**).

4.5.2 Flat radiators

Flat radiators (or plates) are another form of heating surface. Their measurements are not standardized and manufacturers offer different shapes, heights and depths. The variety of shapes and the shallow depth make them suitable for use as interior design elements and installation away from walls. Heat output from the smooth or profiled surfaces is by radiation (**Figures 4.84 and 4.85**).

The feed and return flow fittings as well as the ventilation connections consist of sockets with customized measurements welded to the unit. Brackets of flat steel bent into U-shapes are welded onto the back for attachment; the number of brackets depends on the size of the radiator.

Other designs consist of grooved panels, which are either welded onto a flat steel plate or welded directly to each other. This creates channels through which the heating water can flow. A distribution or collecting channel is located at the top and bottom and connected to the heating water feed (Figures 4.86/4.87).

In terms of heat output, flat radiators are the least expensive heating surfaces. They also have a low water content and thus good regulating properties. These heating surfaces can also be installed one behind another, whereby the heat output diminishes by 40 % per metre since heat emission by radiation diminishes after the second row; heat emission is solely by convection from this point onward. Cover plates are available for flat plate radiators installed in rows to give them the appearance of a single radiator. Examples of thermal output for different models are given in Tables 4.14.1/2.

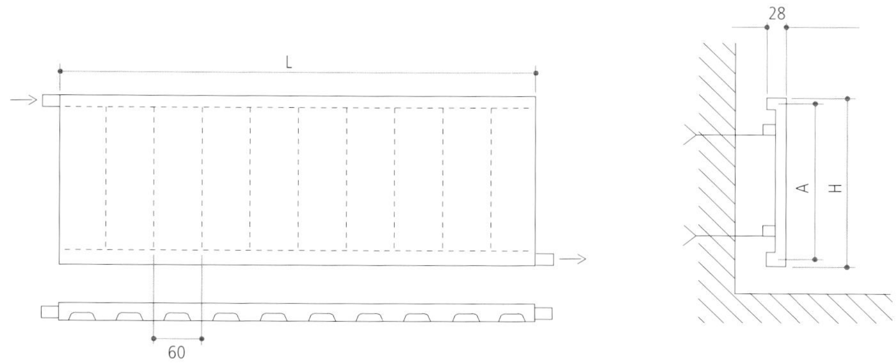

Figure 4.86
Flat radiator with smooth front and grooved back

Length L = number of sections x 60 mm
Height H = A + 28 mm

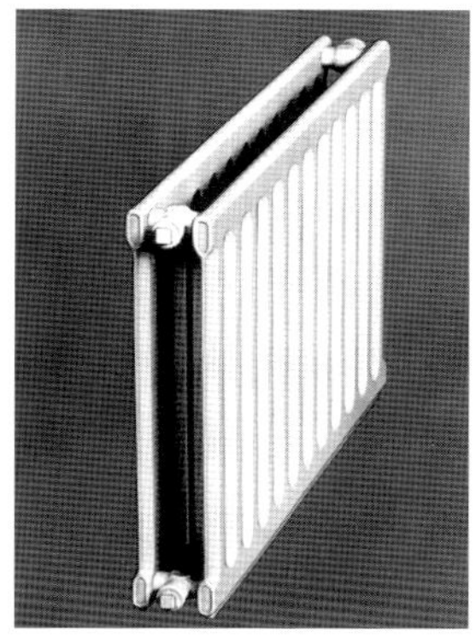

Figure 4.87.1
Grooved flat radiator (Model Classic, Baufa), without housing, accident-proof

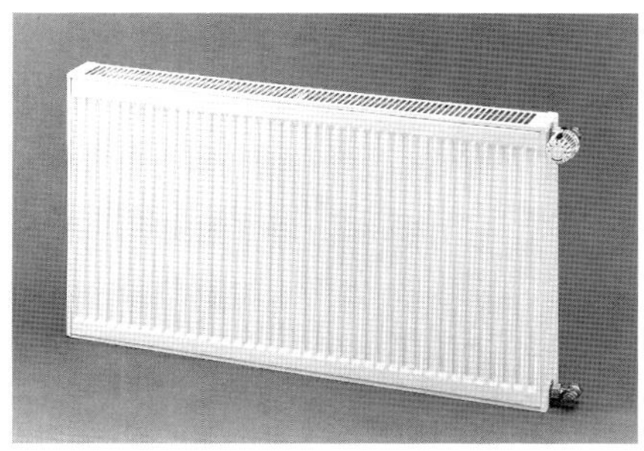

Figure 4.87.2
Compact model of grooved flat radiator (Model DiaPlus, DiaNorm Wärme)

4.5.2.1 Radiant panels

Radiant panels are used for large spaces such as factories and warehouses. These units consist of parallel tubes linked with sheet metal plates. To achieve good heat transfer, the sheet metal plates are welded or clamped onto the tubes. The heating tube may lie above or below the sheet metal plate. Radiant panels are installed on walls or ceilings, even in shed roof constructions. A layer of insulation is installed above the radiant panel, which is generally operated with heating water at a temperature of 180 °C and installed in the working area of the occupants, for example in a long row of so-called ribbon radiant panels.

The literature provided by manufacturers contains information with regard to thermal output, which must be measured by recognized institutes, and design measurements.

Table 4.14.1
Standard thermal output of smooth-walled flat plate radiators according to length, indicated in W/m for mean warm water temperature t_m of 80 °C and a room temperature of 20 °C. Plate thickness 25 ± 3 mm, single row as per DIN 4703, Part 2

Dimensions in mm	200	300	400	500	600	700	800	900
Standard thermal output in W/m	267	400	525	650	773	893	1010	1125

Table 4.14.2
Standard thermal output of flat plat radiators with grooved panels according to length for mean warm water temperature tm of 80 °C and a room temperature of 20 °C. Plate thickness 18 ± 3 mm, profile section at least 10 % greater than plate length, profile height ≥ overall height less 100 mm, single row as per DIN 4703, Part 2

Dimensions in mm	200	300	400	500	600	700	800	900	1000
Standard thermal output in W/m	294	425	556	684	810	935	1058	1180	1300

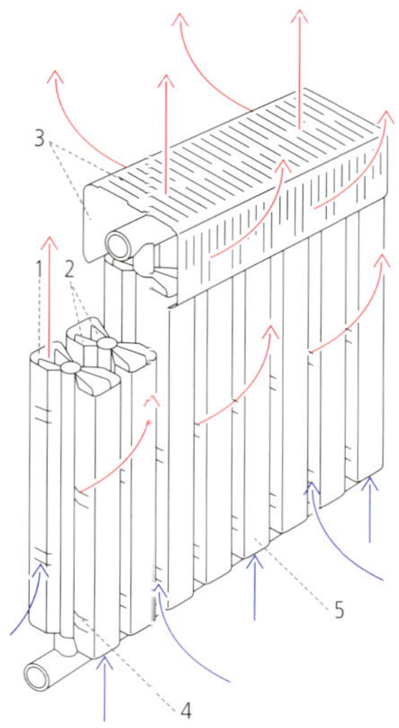

Figure 4.88
Tube radiator
(Gerhard & Rauh)

1 Housing
2 Ribs
3 Warm air openings, top and sides
4 Air intake and venting slits
5 Bottom air intake

4.5.3 Tube radiators

Tube radiators are manufactured from seamless steel tubes and equipped with a feed pipe and a return flow header. Ribs are welded onto the vertical tubes to increase heat emission. Tube radiators are designed to withstand high operating pressures, which makes them suitable for use in high-rises, with hot water systems and for direct feed from the district heating network (see **Figure 4.88**).

Smooth steel tube radiators are available as register and ribbed tube models, although they are increasingly being replaced in the market by the multitude of other heating surfaces.

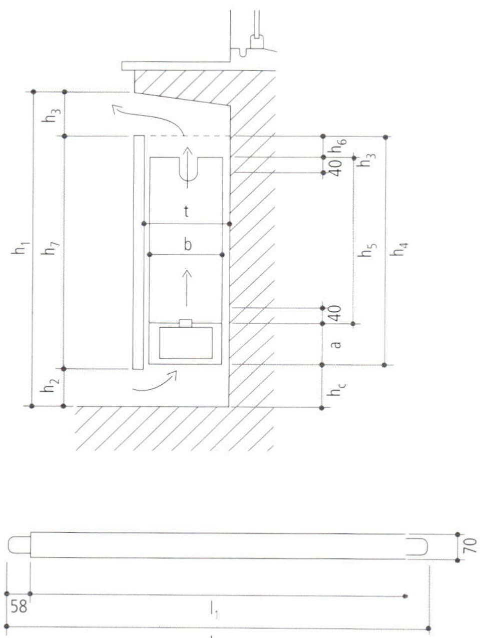

Figure 4.89
Principal measurements for a convector and niche installation according to GEA

h_1 Niche height
h_2 Air intake height
h_3 Air venting height
h_4 Effective stack height
h_5 Height of convector bracket
h_6 Screen projection
h_7 Height of front screen
h_c Convector installation height ($h_c = h_2 + 10$ mm)
a Convector height
b Convector depth
t Depth of convector bracket = b + 4 mm

4.5.4 Convector heating

As the name indicates, heat emission in these units is almost exclusively by convection. There are two principal designs. In the first, the convector consists of a copper tube onto which aluminium lamellae are attached to ensure good heat transfer between pipe and lamella. In the second, the convector consists of an oval steel pipe on which steel lamellae are arranged in rows. To achieve good contact between pipe and lamella, both are galvanized after assembly. The convector must be installed in a shaft or cavity (**Figure 4.89**). Air flows from below through the lamellae of the convector and is heated. The cavity produces a stack effect: the heated air rises and exits through perforations at the top. The thermal output of the convector depends on the shaft height. The higher the shaft, the greater the thermal output. This is true for a shaft height of up to one meter. There are a variety of options for convector installation (**Figure 4.90**). The principal advantage of convectors is that they are small and light; the low water content is another advantage, resulting in rapid response to the heating water volume.

Disadvantages include difficult access for cleaning because the lamellae are housed in a cover and placed very close together. The cover should be easy to remove. Another disadvantage lies in the absence of a component for radiant heat emission. Moreover, convectors cannot be used at low temperature ranges. The cost advantage of the convector partially compensates for the additional expense incurred for the required cover. The heating performance is not defined in any standard and must be established from the lists provided by the manufacturers. The thermal output of the convector can be controlled by damping the water volume and/or the airflow in the shaft.

Figure 4.90
Various installation options for convectors according to GEA

1 Below window
2 In front of smooth wall
3 Detached
4 Built into wall
5 Built into wall
6 Subfloor convector with room air intake
7 Subfloor convector with cold air intake
8 Subfloor convector with intake on both sides
9 Convector behind bench

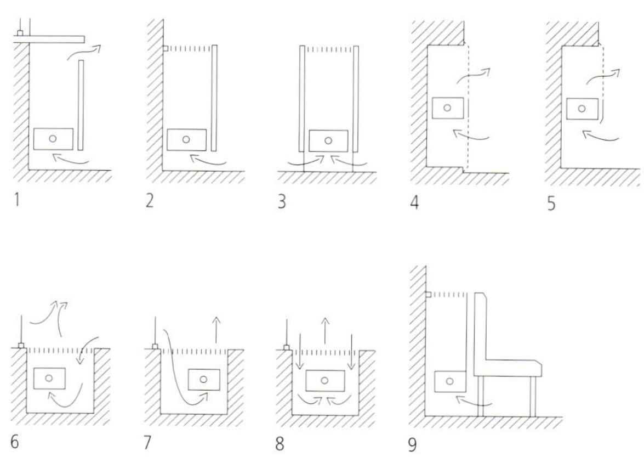

When convectors are used in combination with other heating surfaces it is important to consider that the thermal output of convectors, when the temperatures of the heat medium are low drops more rapidly than the output of other heating surfaces. Whenever possible, convectors should therefore have an autonomous heating cycle.

A variety of materials (hard particleboard, wood, metal sheeting or similar) are suitable for the casing. When installing and encasing a convector, great care should be taken to ensure that there are no gaps at the front and at the back to prevent secondary air from flowing past it.

Aside from convectors with natural air circulation, there are designs with forced airflow. The airflow is produced either by means of a central air distribution network and air discharge below the convector or by means of a fan for each individual convector (**Figure 4.91**). The higher the velocity in the blower stream, the higher the thermal output of the convector.

Fan convectors have an air output of roughly 200 to 2,000 m^3/h and a heating performance of up to 20 kW. They are therefore suited for medium-size spaces such as offices, restaurants, boardrooms, classrooms and also large living rooms. Fan convectors for floor installation are a variation of the standard model, designed to operate exclusively with recirculating air systems (**Figure 4.92**).

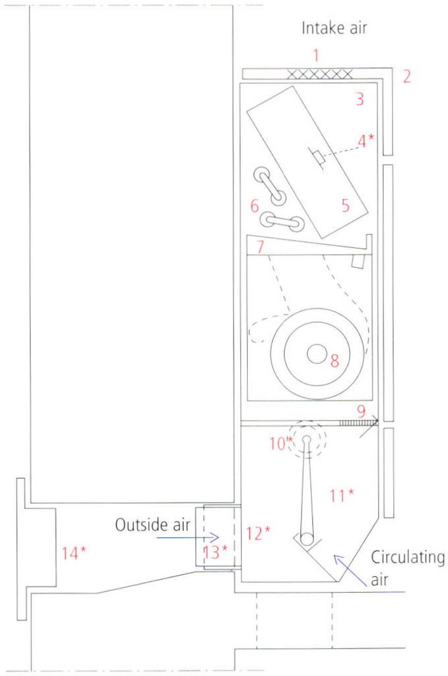

Figure 4.91
Section of fan convector according to GEA

1 Air venting grid
2 Housing
3 Basic structure
4 Antifreeze thermostat *
5 Heat exchanger
6 Supplementary electric heating
7 Condensation water basin
8 Fan
9 Filter
10 Servomotor *
11 Secondary air flap *
12 Outside air intake stack *
13 Wall frame *
14 Weather-proofing mesh *

* only for mixed air units

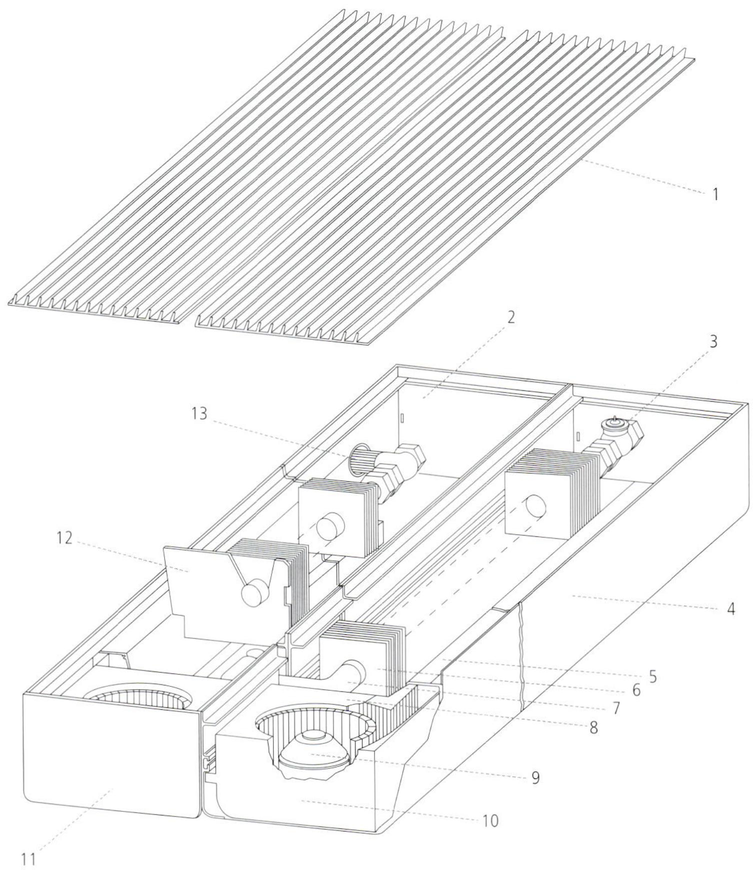

Figure 4.92
Fan convector for floor installation
(Emco)

1 Grating
2 End piece
3 Foot valve
4 Tub
5 Convector
6 Duct
7 Convector elbow joint
8 Cover
9 Motor
10 Fan housing
11 End piece
12 Convector bracket
13 Control valve

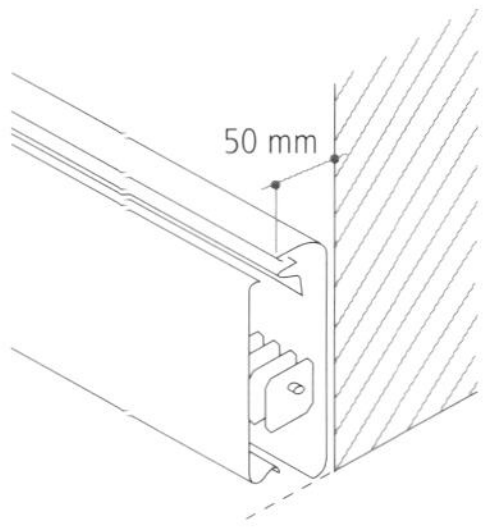

Figure 4.93
Baseboard radiator
(Evitherm)

Baseboard convectors are another variation. Baseboard units consist of a copper tube with aluminium lamellae and a prefabricated tin plate cover. In some models the convection opening at the top can be closed off to regulate thermal output (**Figure 4.93**).

When baseboard convectors are installed along the full length of the exterior walls of a room, they deliver good heat distribution, since the warm air rising from the units flows up along colder exterior walls. Due to the placement of the copper tube, this type of unit is ideally suited for single-pipe heating since it does not require a distribution of its own. Its shape, small size, good regulating properties and easy installation also make it an interesting solution for retrofitting in renovation projects. Disadvantages are that baseboard convectors limit the options for arranging furniture in a room and that the convection almost inevitably leads to dirt accumulating above the baseboard (dust accumulation).

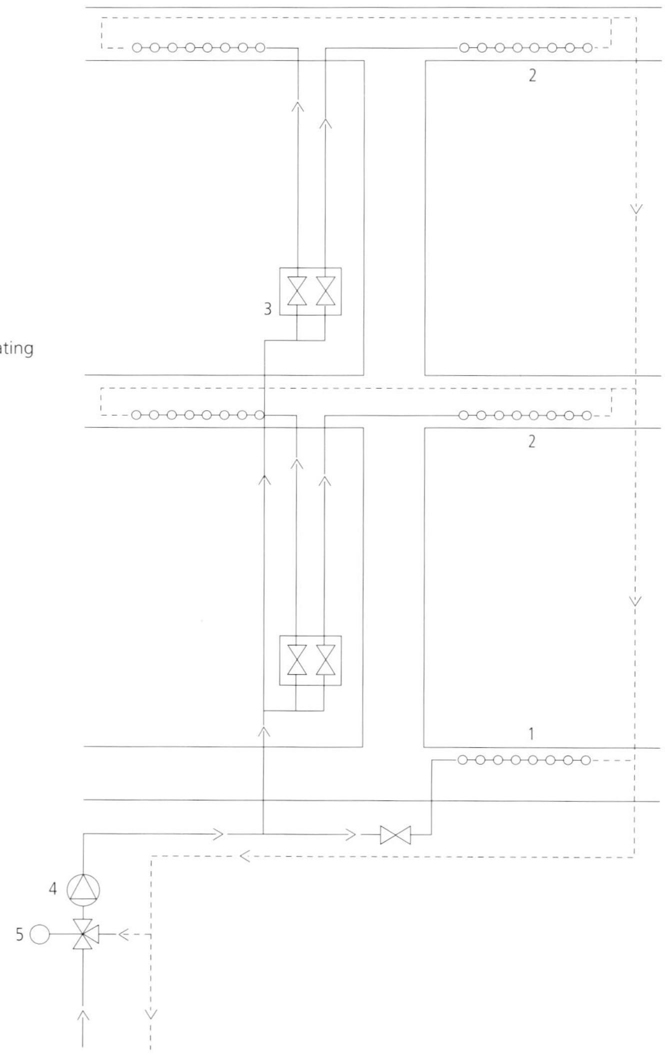

Figure 4.94
Radiant ceiling and floor heating
(radiant panel heating)

1 Floor heating
2 Ceiling heating
3 Control valves
4 Circulation pump
5 Mixing valve

4.5.5 Radiant panel heating

While systems described thus far deliver the required heat through heating units, radiant panel heating operates on the principle of warming the surfaces that enclose the space. This is achieved by integrating or suspending heating tubes. Depending on the placement of the heating tubes, we speak of ceiling, floor or wall heating (**Figure 4.94**). Since these solutions function largely by radiation in comparison to conventional heating units, they are also often referred to as radiant heating.

The radiation emitted from the surface lands on the objects in the space and on the other surfaces enclosing the space. These are warmed and in turn emit heat by convection or radiation.

Since radiant panel heating is usually operated with warm water that reaches a feed water temperature of 50 °C, it is especially well suited for the operation of heat pumps.

There are a number of advantages and disadvantages associated with radiant panel heating.

The **advantages** are that this system does not call for localized heating surfaces, allowing for greater freedom in terms of spatial requirements and reducing the risk of burn injuries. Moreover, the heating surfaces are invisible and generate only negligible amounts of air movement, that is, they do not stir up dust or disperse germs. In addition to the even temperature distribution, this system also offers a cooling option for summer operation when equipped for cold water operation (conversion system).

The **disadvantages** are the higher costs, which result from the necessary co-ordination in the construction phase. Moreover, the system tends to respond sluggishly, especially with large heated masses, and offers only limited options for adjusting the thermal output or room layout. Repairs can be labour-intensive and difficult. Radiant panel heating is very sensitive to cold air influx and to cold exterior and wall surfaces, especially in the ceiling heating variation. The system must be assembled and installed with great care.

4.5.5.1 Floor heating

Floor heating consists of pipes laid in the flooring screed or concrete topping, through which warm water flows. The floor emits heat partially by convection and partially by radiation. To avoid undesired heat emission below the floor, an insulating layer is installed beneath the pipes. In the past, the aim was to allow heat to be emitted downward from floor slabs in apartment buildings, so that the floor heating on one storey would act as a ceiling heating on the storey below. The idea was that this would achieve an even more balanced temperature profile and increase comfort. It had the disadvantage, however, of occupants being dependent upon the heating preferences of their neighbours. Floor heating is an increasingly popular choice, especially in the single-family home market, despite the higher costs associated with it.

Floor heating systems are classified into wet or dry installation systems.

In **dry installation**, the pipes are installed in prefabricated rigid foam panels, manufactured with matching grooves or channels. These rigid foam panels are generally used to serve as thermal and impact sound insulation. Since the pipes are surrounded by air in the rigid foam panels, aluminium lamellae are sometimes attached to the pipes to improve heat transfer to the floor screed. After the pipes have been pressed into place, a film is laid across the panel and the screed or flooring is then dry installed. One advantage of this method is that the pipes can expand freely, preventing thermal stress and thus fissures (**Figure 4.95**). PVC floor heating is a special variation of this type of heating. It is installed in cavities and equipped with special lamellae, which can be adjusted to the construction height of the cavities. This floor heating can be connected to an existing heating network with a feed temperature of 90 °C and a temperature spread of 20 K because the PVC pipes can expand freely (**Figure 4.96**). This also allows for a combination with other heating units, especially convectors, without requiring separate thermostat controls.

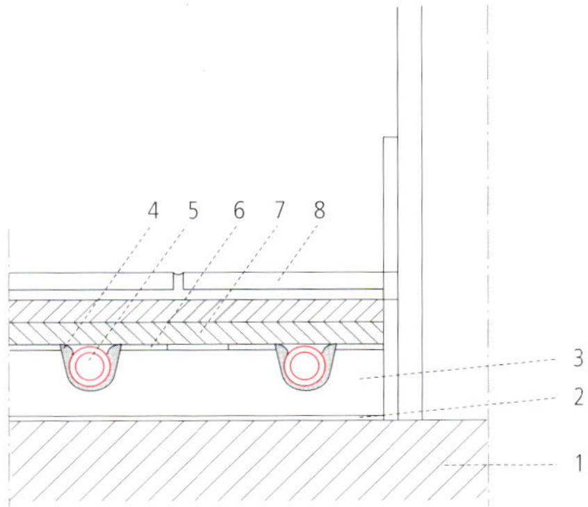

Figure 4.95
Floor construction for dry installed floor heating
(System Purmo)

1 Loadbearing floor structure
2 Impact sound insulation (if required)
3 Polystyrene installation component
4 Heat conducting plate
5 Floor heating pipe
6 Heat distribution plate
7 Dry floor screed (here: 2 layers of Fermazell)
8 Floor covering

Figure 4.95.2
Example of dry-installed floor heating
(velta siccus, Velta)

Figure 4.95.1
Example of dry-installed floor heating
(TS14 Floor heating, Purmo)

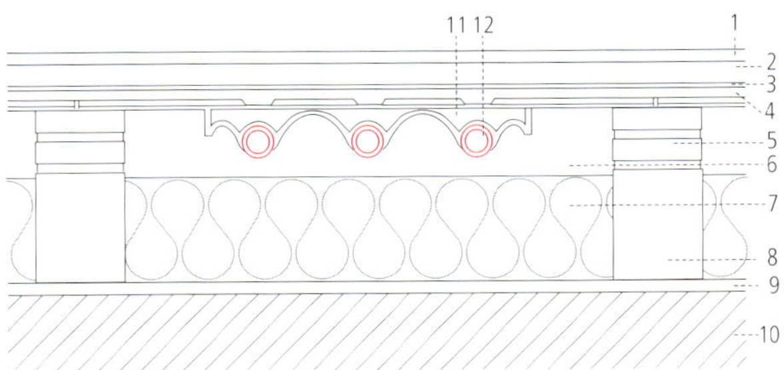

Figure 4.96.1
Floor construction for the Velta System

1 PVC or linoleum flooring
2 Wood-chip board
3 PE sheet
4 False floor
5 Double rocker beam
6 Air cavity
7 Insulation
8 Fill block
9 Waterproofing
10 Concrete
11 velta pipe bracket
12 velta plus pipe PE-Xa 110

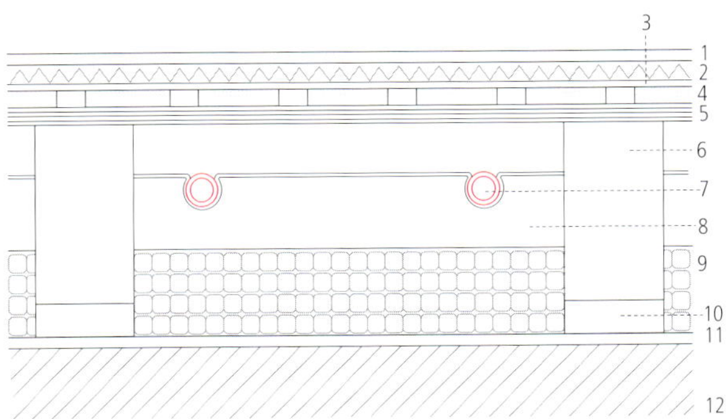

Figure 4.96.2
Floor construction in a sports hall, Polytherm System

1 Parquet flooring, 10mm
2 Load distribution panel, 16mm
3 PE sheet
4 False floor, 18mm
5 Rocker beam
6 Fill block
7 Airtight PE-Xc heating pipe, 20 x 2mm
8 Guide rail, 35mm
9 Insulation, 85mm, next to soil PUR (λ = 0.03 W/mK)
10 Permanently elastic spring pad
11 Waterproofing, approx. 10mm
12 Foundation/concrete slab

In **wet installation**, a film is laid on top of the thermal and impact sound insulation, followed by a reinforcement steel mesh or a grid to which the fasteners are attached.

The pipes are then laid and connected to the fasteners to prevent them from shifting. Once the pressure and leakage test has been completed, the floor screed can be installed; this must be executed with great care to ensure that no air is trapped as the pipes are covered (**Figures 4.97.1/2**).

Since the pipes and the floor screed have different expansion coefficients, various additives are mixed in with the floor screed. The relevant expansion factors must be taken into account because the entire screed is warmed by the system. The screed should be allowed to dry for at least four weeks prior to the first heating cycle, which should be started gradually to prevent detachment between pipes and screed. The feed cycle temperature must not exceed 55 to 60 °C to avoid thermal stress and fissuring.

While steel and copper piping was used for earlier floor heating installations, the majority of such installations are currently executed with plastic piping. In damp rooms the sealing film must be installed above the pipes to prevent water from coming into contact with the pipes. For this reason, expensive copper piping is generally used for floor heating in swimming pools.

The pipes can be laid in a variety of patterns. The simplest form is a serpentine installation, which has the disadvantage, however, of producing uneven floor temperatures. To compensate for this shortcoming, feed and return pipes are laid next to one another, which results in a spiral pattern. The two patterns can also be combined. Along the edges of external walls and beneath windows the standard pipe distance of 15 to 30 cm is usually diminished (**Figures 4.98, 4.99**).

For obvious physiological reasons, surface temperatures in floor heating should not exceed 24 °C in the accessible areas. Along the edges, surfaces temperatures up to 29 °C are acceptable, since it is safe to assume that no one will spend a long time there. In the bathroom, where occupants are barefoot, the surface temperature should be set to 32 °C.

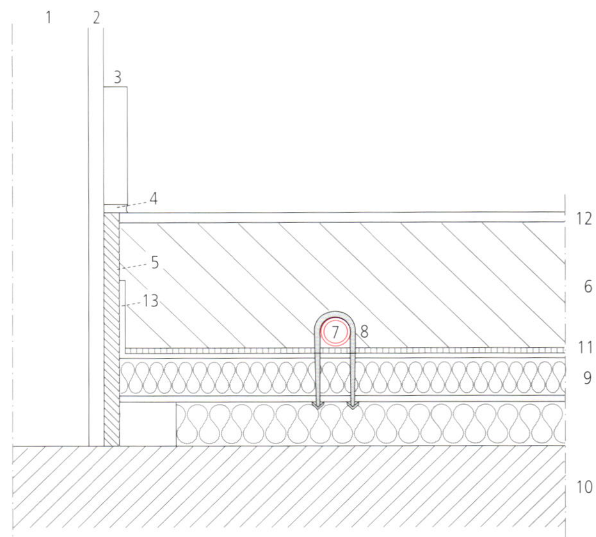

Figure 4.97.1
Floor construction for wet-installed floor heating

1 Wall
2 Plaster
3 Baseboard
4 Elastic joint sealer
5 Edge insulating strip
6 Screed DIN 18560/18353
7 PE heating pipe
8 Pipe-bracket, type A
9 Insulation
10 Loadbearing base
11 Loadbearing structure
12 Floor covering
13 PE sheet

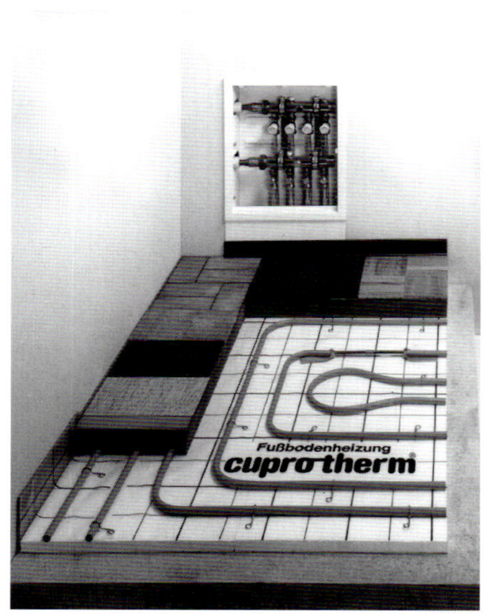

Figure 4.97.2
Floor construction and floor covering options

Figure 4.98.2
Example of installed floor heating with decreasing distance between pipes near walls
(Image: Velta)

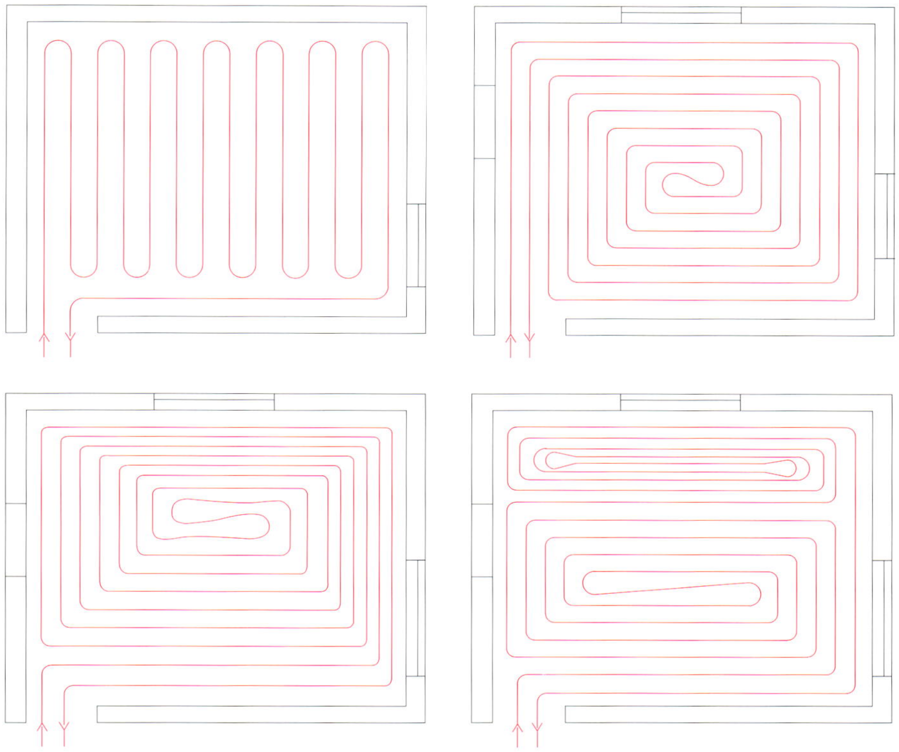

Figure 4.98.1
Various floor heating installation options
(after Velta)

Floor heating delivers a thermal output of roughly 80 W/m^2 for 20 °C room temperature. Since only those buildings constructed after the new insulation guidelines came into effect generally achieve a specific heat requirement of less than 80 W/m^2, additional heating units are sometimes necessary in combination with floor heating. This is especially true in rooms with built-in furniture and appliances (e.g. kitchen and bath).

Since the entire floor screed with a minimum thickness of 45 mm is heated above the top edge of the heating pipe, the response time of floor heating tends to be sluggish. Thermostats should therefore be designed to respond to weather rather than room temperature. The estimated cooling period of floor heating per 1 K is roughly one hour for each cm of screed thickness. Pure floor heating is therefore not suited for passive solar use. The argument that floor heating has a natural self-regulating effect is only true from a thermodynamic perspective when the room temperature is equal to or greater than the surface temperature, which should be 26 °C.

Figure 4.99
Example of installed floor heating with parallel feed and return pipes, large area heating
(Image: Polytherm)

To compensate for the disadvantage of this sluggish response, 40 to 50 % of the thermal requirement can be delivered by floor heating and the remaining demand can be met with conventional heating surfaces. The latter have a rapid response time due to the smaller water volumes that flow through them and the small heated masses. The heat supply to the heating unit can be diminished to compensate for heat gain through incident solar radiation or heat emission from occupants. The same approach can also be used to compensate for the impact of furniture, which diminishes the thermal output of floor heating. This array of dual heating makes it possible to operate floor heating on a continuous basis.

One variation of floor heating is to combine static and hot air systems. The "air conomy" system unifies these methods to keep operating temperatures as low as possible. The underlying warm air movement provides very a comfortable warmth as well as clean, fresh air and a constant evacuation of contaminants and odours. The high degree of comfort which this system solution delivers is the result of even heat distribution and constant renewal of air, which means that air contaminants (pollen, dust, odours etc.) are evacuated. The system also reacts quickly to partial warming of the air, thus overcoming the sluggish response of the floor heating system. **Figures 4.100.1/2** show a section and a functional diagram of the system solution, respectively.

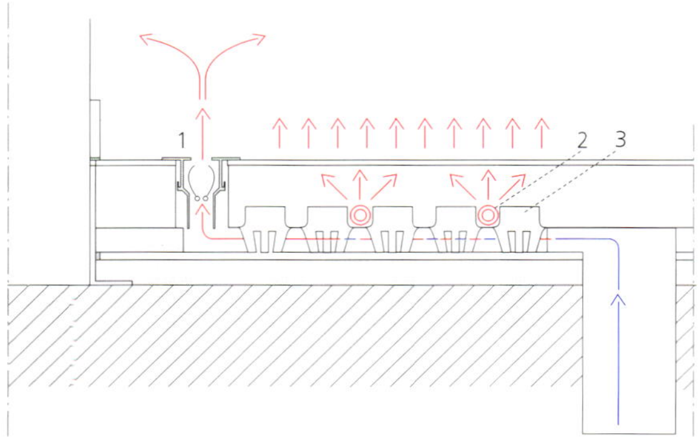

Figure 4.100.1
Section of floor construction,
air conomy system®
(Schütz eht)

1 Air vent on floor or wall
2 Water-conducting heating pipes
3 Spacing cone

4.5.5.2 Ceiling heating

The Crittall ceiling, named after its inventor, is the oldest style of ceiling heating. Steel tube coils, consisting of 3/4" seamless or special-rolled Fretz-Moon pipe, are installed above the bottom reinforcement. The concrete covering of the tubes should be 1 to 3 cm thick on the bottom and 6 to 7 cm thick on the top (**Figure 4.101**). The coils are individually connected to the secondary distribution, which is generally located in the corridor.

The top reinforcement can be installed once the pressure and leakage test has been complete, and the ceiling can be cast in a single piece. Installation on the bottom reinforcement promotes thermal conductivity and thus even temperature distribution within the ceiling.

Since concrete and iron have more or less identical expansion coefficients in the temperature range of up to 60 °C, there is no risk of stress fissures. To protect the pipes against external corrosion, they, like the reinforcement, should be embedded in the concrete without air inclusions. Insulation is added to the unfinished ceiling at a later point to limit heat transfer towards the top. Since this ceiling has an even temperature distribution, it can be finished with standard plaster without additives. For large ceilings, expansion joints must be included and the statical engineer must be familiar with the ceiling temperature to ensure that appropriate measures are taken. Heavily reinforced ceilings are best able to absorb any thermal stresses that might occur.

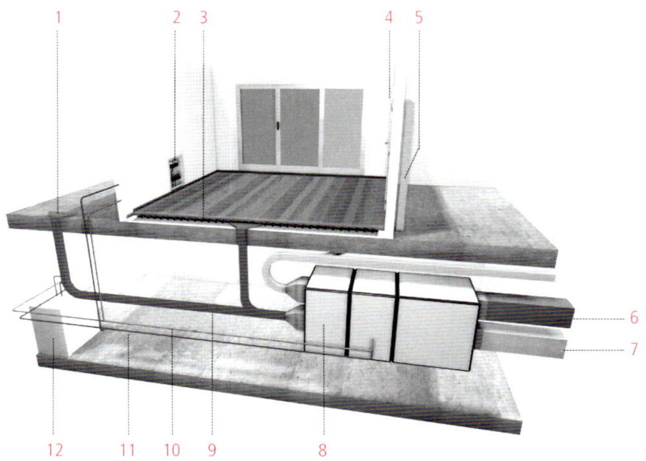

Figure 4.100.2
Functional diagram
Air conomy system®
(Schütz eht)

1 Injection module
2 Heating cycle distributor
3 Air conomy® hollow floor system
4 Exhaust air disk valve
5 Exhaust air
6 Escaping air
7 Outside air
8 Ventilator
9 Intake air
10 Return
11 Feed
12 Heat pump, oil furnace or similar

The pipes are installed in distances of 15, 20 and 25 cm. To compensate for heat lost through walls and window surfaces, the tube coils are spaced more closely along edges. However, the first pipe should lie at a minimum distance of 0.3 m from the external wall to avoid excessive heat loss to the outside. In the load-bearing area, the insulation should therefore be as thick as it is on the ceiling. Finally, the feed pipe is installed near the external wall with the result that the temperature of the heating water diminishes in the direction of the centre of the room.

The pipe registers should be distributed across the space to allow for individual room temperature control in response to user behaviour.

One approach aimed at improving the considerable sluggishness of this system is to reduce its mass. Hollow brick ceilings and suspended heated ceilings are two solutions worth mentioning in this context.

Depending on the specific purpose, however, ceiling heating should react more or less sluggishly. The firm Stramax has developed a ceiling heating system with lamellae for this reason (lamellae ceiling heating).

In this system 0.7 to 1 mm thick aluminium lamellae with lock beading are placed over the heating pipe. For smooth lamellae an additional latchwork is required. Other styles of lamellae are such that they are simultaneously plaster support elements. The plaster must come into full contact with the aluminium lamella to ensure good thermal conductivity (**Figure 4.102**).

In coffered plaster ceilings (**Figure 4.103**) the aluminium lamellae are integrated in the plaster panel. The panels are available with smooth or perforated undersides. Depending on the model, they are either clamped onto the heating pipe from below or slid onto the pipe with the help of a special section. Gypsum plasterboards are suspended from wires or special fastening systems, familiar from conventional suspended ceilings. The boards leave sufficient space along the edges to accommodate the heat expansion.

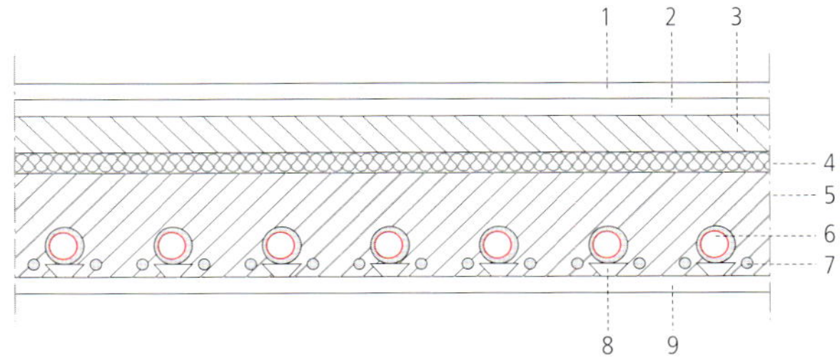

Figure 4.101
Ceiling heating with heating pipes in concrete (Crittall System)

1 Floor covering
2 Screed
3 Pumice stone concrete
4 Impact sound and thermal insulation
5 Loadbearing ceiling structure
6 Ceiling heating pipes
7 Armouring
8 Concrete blocks as spacers
9 Plaster

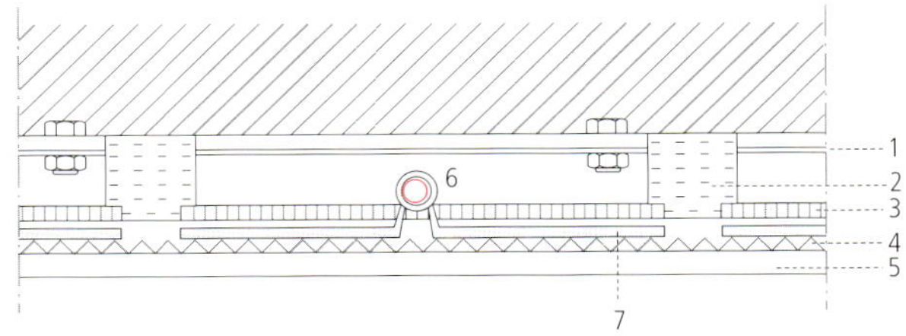

Figure 4.102
Ceiling heating and lamellae ceiling heating (Stramax System)

1 Insulation with aluminium foil
2 Floor grid
3 Loadbearing structure for ceiling heating pipes
4 Expanded metal
5 Plaster
6 Ceiling heating pipes
7 Lamella

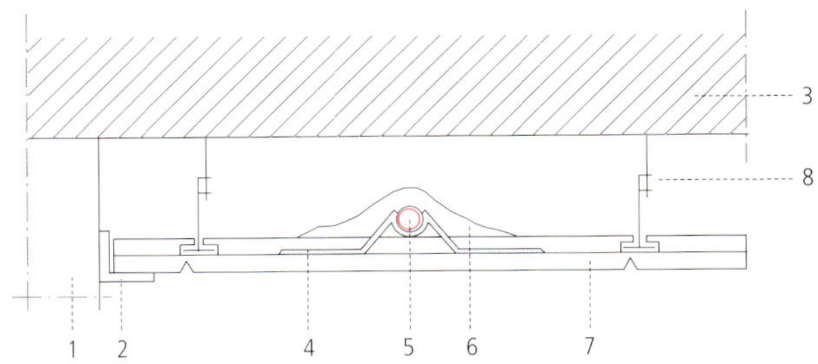

Figure 4.103
Ceiling heating with coffered ceiling (Destra System)

1 Wall
2 Trim
3 Loadbearing ceiling structure
4 Lamella in coffered ceiling
5 Ceiling heating pipe
6 Insulation
7 Plasterboard coffer
8 Suspension

Prefabricated lacquer-finished steel or aluminium panels with a width of 85 mm and a thickness of 0.75 mm are used for the Zent-Frenger ceiling (Figure 4.104). The panels are fastened with steel brackets to the pipes, which are spaced 300 or 600 mm apart according to the required heating capacity. A layer of insulation above the panel acts as a sound barrier. Since no plaster is used, the ceiling can be operated with standard warm heating water of 90/70 °C.

While cooling at foot level plays an important role in floor heating, the temperature range in the head area determines the surface temperatures in the case of ceiling heating. The desired approximate value for standard room dimensions and clear heights of 2.7 to 3.0 m is an average ceiling temperature of 35 °C. At this temperature level, the thermal output is roughly 140 W/m^2.

Ceiling heating has not become a viable choice for reasons of cost and, in part, thermal-physiological reasons. The limited air movement associated with it has, however, made it fairly popular in hospitals, clinics and sanatoriums, where hygiene is the uppermost priority. Evenly distributed temperature, supported in part by floor heat, is the desired effect in these environments.

4.5.5.3 Wall heating

Radiant wall heating systems are designed in the same manner as floor heating systems, that is, they are integrated into the masonry or plastered. Insulating layers should be sufficiently thick on external walls to avoid excessive heat loss to the outside (Figure 4.105).

Since lateral radiation is most comfortable for people, radiant wall heating systems should be the ideal solution. Unfortunately they are often impractical because they leave little area for furniture. The surface temperature can be set higher for radiant wall heating than for radiant ceiling heating because people are less sensitive to heat radiation from the sides.

All radiant heating systems should be controlled by thermostats that respond to outdoor rather than indoor temperatures. In larger building ensembles, temperature controls must be installed for all directions.

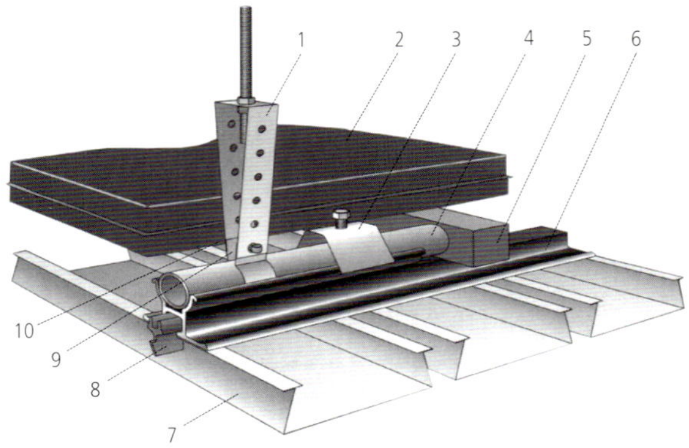

Figure 4.104.1
Ceiling heating as panel ceiling (Zent-Frenger System)

1 Punched strip
2 Insulating pad
3 Tension spring
4 Register pipe
5 Distribution pipe
6 Contact bar
7 Panel
8 Serrated slat
9 Shackle
10 Splint with U-shaped disks

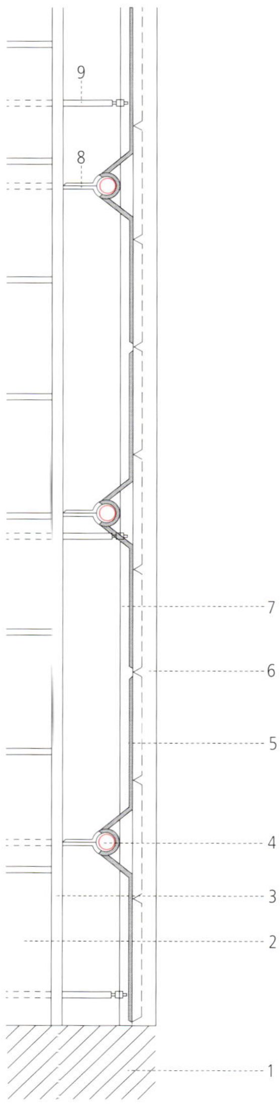

Figure 4.104.2
Ceiling heating as panel ceiling (Zent-Frenger System)

Figure 4.105
Wall heating with plaster wall (Deria System)

1 Loadbearing ceiling
2 Wall
3 Insulation
4 Heating pipe
5 Lamellae
6 Expanded metal plaster wall
7 Angle
8 Pipe bracket
9 Angle bracket

4.5.5.4 Facade heating

Facade heating is a variation of radiant heating systems (**Figure 4.106**). The floor-height posts, required for structural soundness, and the lateral girders of each inward-facing facade component are filled with warm water at temperatures up to a maximum of 45 °C. The window panes with double glazing are installed on the outside, and thermally separated from the facade heating components. This thermal separation also serves to diminish heat loss to the outside. Heat radiates from the posts and girders onto the pane, which consequently has a surface temperature between 16 and 20 °C.

This surface temperature creates a tremendous sense of comfort and transforms window areas into agreeable indoor areas, a characteristic that is especially appealing in office buildings. Corrosion problems are rare, since the heating system is operated in a closed loop (no mixed installation).

This facade also offers further advantages in terms of fire protection. Since the metal components contain water, there is no need for additional fire protection. The middle girder can be fitted with a convector pipe that is heated separately.

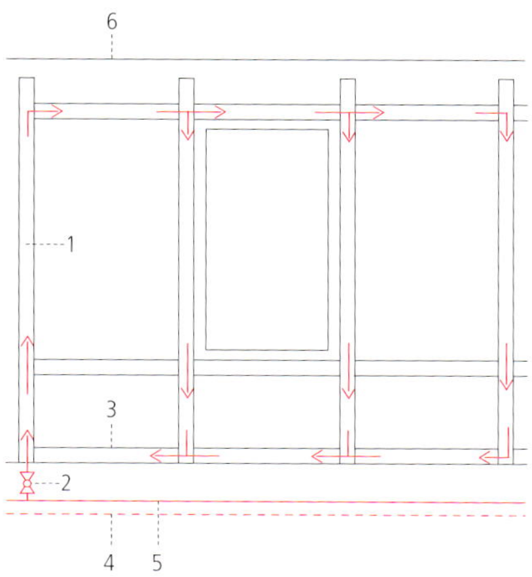

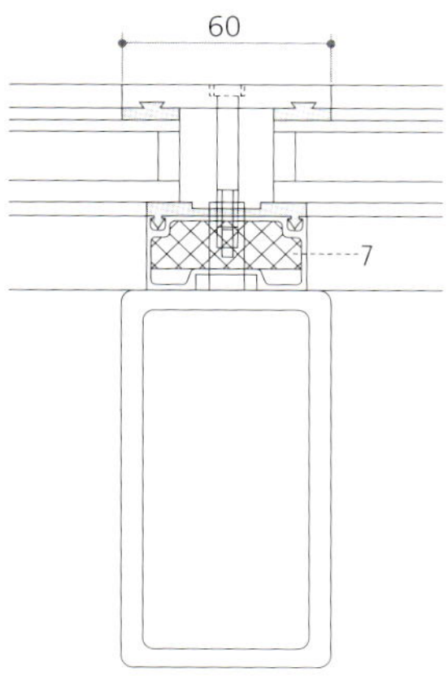

Figure 4.106.1
Facade component and cross-section of facade post, after Gartner

1 Heated hollow section
2 Thermostatic valve
3 Electrical channel
4 Return
5 Feed
6 Ventilation
7 Plastic section

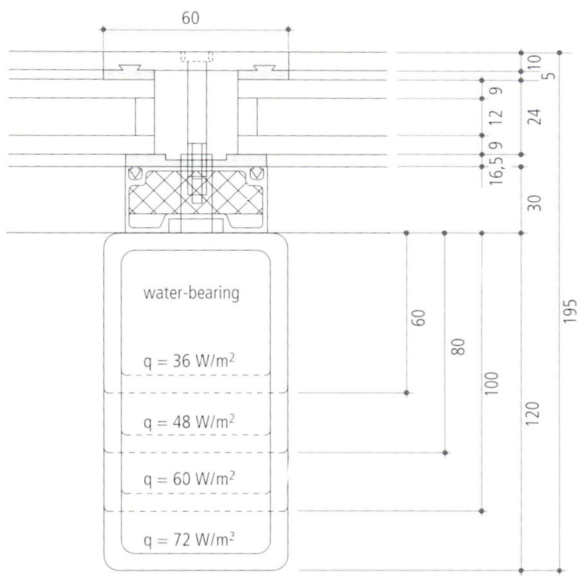

Figure 4.106.2
Detail of a facade heating system

Thermal output of section for mean heating water temperature of 36°C
q = 200 W/m² active heating surface

4.5.6 Room temperature control

All heating systems must be equipped with thermostats to control the temperature in individual rooms.

Thermostatic valves are solid, gas or liquid filled temperature sensors, whose spring pressure mechanism throttles the mass flow rate of heating water when temperatures rise or releases it when temperatures fall. Calibration is not required for temperature sensors; regulators on temperature sensors are marked with numbers one through five. The number three setting corresponds to a room temperature of roughly 20 °C. **Figure 107.1** illustrates different thermostatic valves. The placement of the sensor plays a decisive role in the effectiveness and precision of the temperature control (**Figure 4.107.2**).

Electric single-room controls with electrothermal (**Figure 4.107.3**) or motor-driven valves are preferable. They allow for programmed temperature reduction at night, with the option of excluding individual rooms if desired. With the help of a central thermostat or a link to a bus system (**Figure 107.4**), temperatures can also be lowered in daytime, either in each individual space or pre-programmed with a timer. This system design can help to save energy.

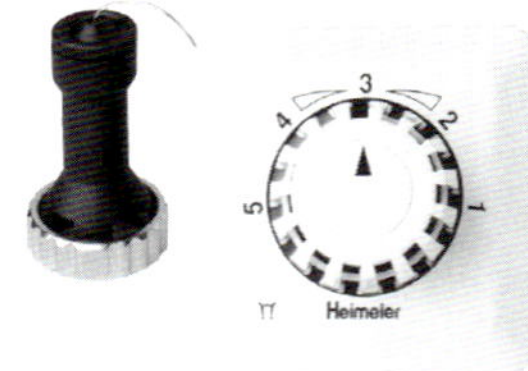

2

3

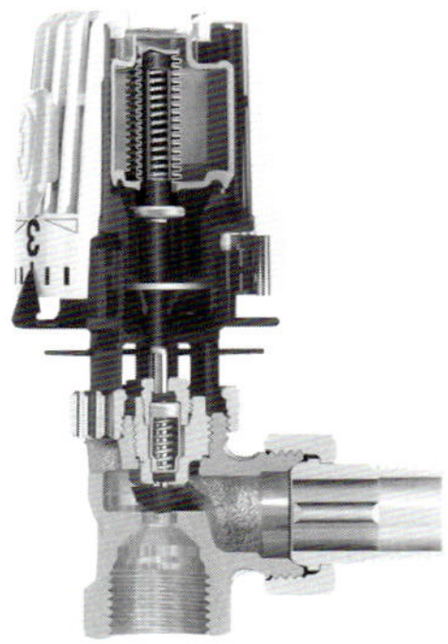
1

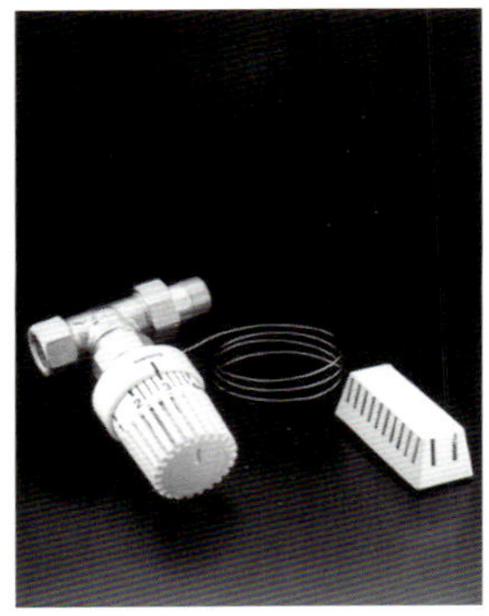
4

Figure 4.107.1
1 Cutaway view of regulated thermostatic valve (Image: Heimeier)
2 Thermostat remote adjusting device with built-in sensor (Thermostat (Type) F, Heimeier)
3 Thermostat, modern design (Thermostat head Uni C Colani, Oventrop)
4 Thermostat with remote sensor (Thermostat head Uni L Oventrop)

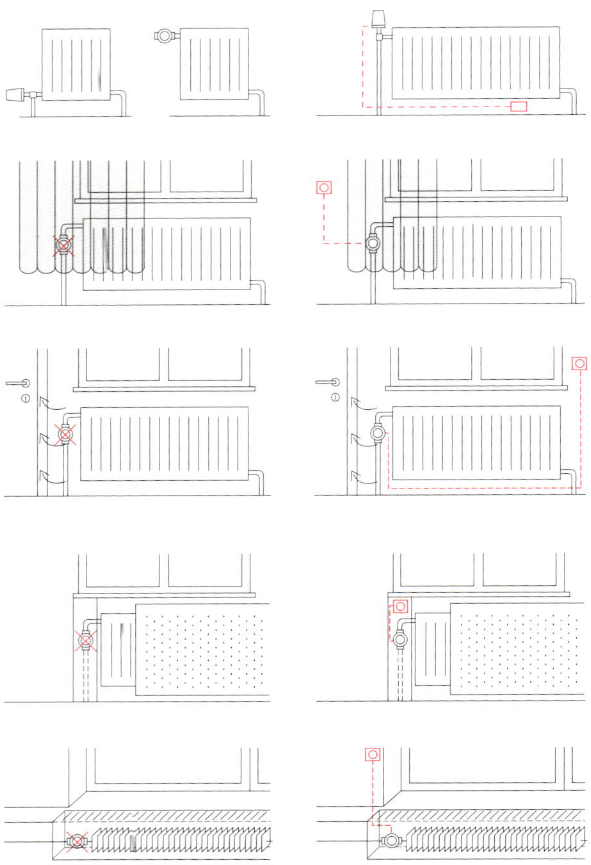

Figure 4.107.2
Correct and incorrect placement of thermostatic valves in various radiator/convector installations (after Danfoss)

Figure 4.107.3
Electrothermic actuator for heating, ventilating and air-conditioning systems (Type EMO T, Heimeier)

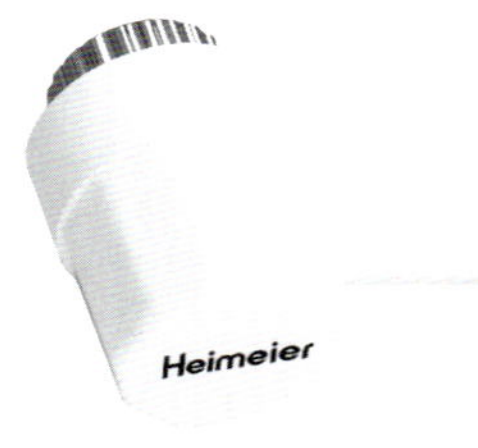

Figure 4.107.4
Electric motor-driven radiator valve for direct connection to a bus system (EIB) (EIB System, Oventrop)

Today, high-tech has also become a factor in temperature control technology for heating systems. Infrared control systems are used to pre-program savings measures for warmth/comfort/energy. Relevant valve heating units (IR) are programmed to set the correct room temperature constantly. Infrared control systems have a programmable transmitter, a built-in receptor device with an electronic actuator, which influences the heating system. The series of images in **Figure 4.108** displays the individual components of the integrated solution.

Operating panel with text display: the display explains each step

Clear and simple: commands are entered with a few actions

Figure 4.108
The full array of connection and control devices is elegantly hidden behind a screen. Since the valve is controlled via an actuating drive, there is no need for a wall penetration to accommodate the thermostat knob (Kermi, Credo-Ventil IR)

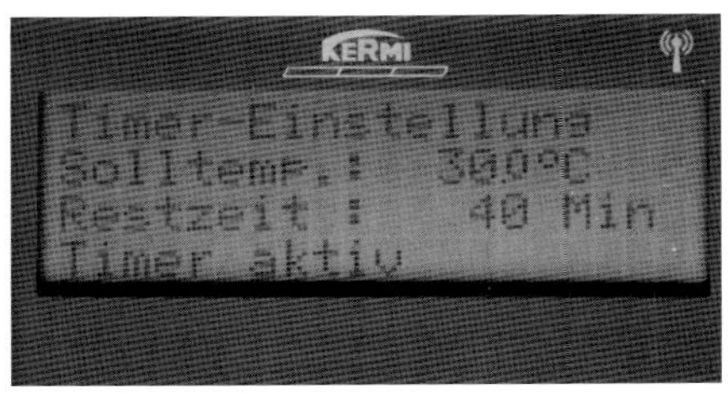

Everything presented at a glance: day, time, target temperature and programming periods are indicated on the screen

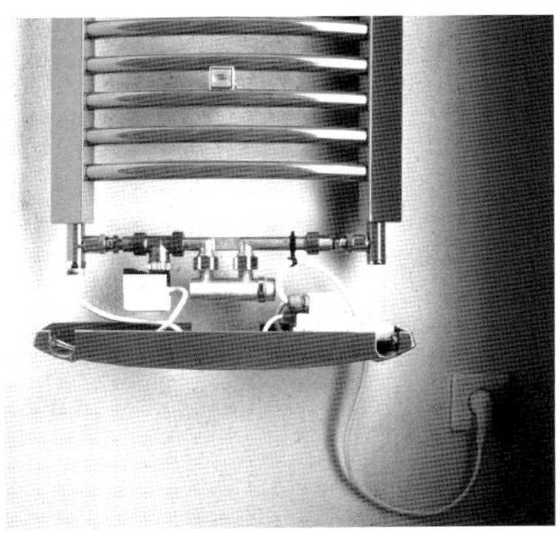

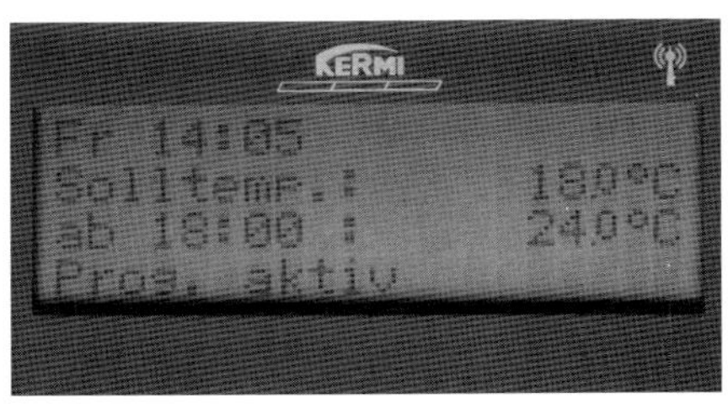

Truly user-friendly: once the batteries are loaded, time and date are set automatically

4.6 Central heating systems/shafts/horizontal installations

Installation site
The installation site should be as close as possible to the largest heat consumers and easily accessible for maintenance and repair work. Firing systems must be installed, located and operated at all times according to code. We recommend that the relevant authorities be consulted in good time for all systems for which notification, i.e. permission, is required.

There are two principal sites for installing central heating systems in buildings:
- Basement installation
- Roof installation

Attic installation may also be an option for technical and financial reasons.

The technical advantages are:
- Short chimneys
- Standard model boiler (no high-pressure boiler required)
- No fresh air and ventilation ducts required

The financial advantages are:
- Freeing up useable space in the basement
- Low building costs for boiler room
- Slightly higher boiler efficiency due to lower flue gas temperature

Disadvantages:
- Additional roof load
- Increased safety requirements to protect against gas and oil leaks
- Increased sound insulation required

When installing boiler rooms in the attic, it is important to provide a waterproof, tub-shaped floor and a sealed door threshold. Oil and gas supply pipes to roof (or attic) installations must be laid in accordance with standard specifications. Fuel pipes in shafts or ducts must be double-walled, and the shafts or ducts must be ventilated.

Appropriate sound insulation features and measures must be implemented to prevent acoustic irritation in the rooms adjacent to the boiler (including pumps, piping etc.).

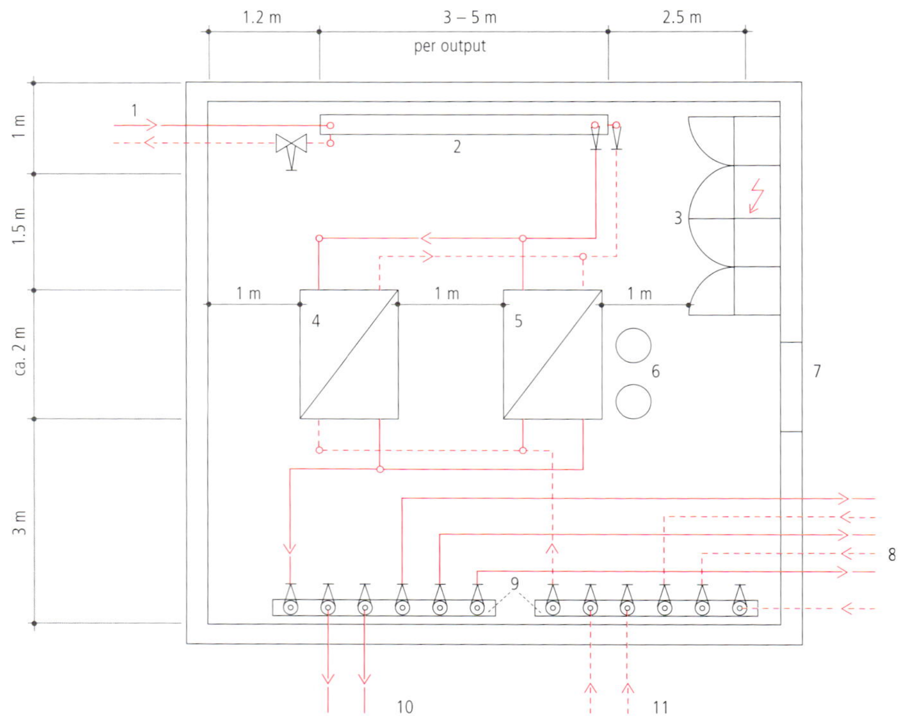

Figure 4.109
Large district heating centre

1 District heating service lines
2 Measuring and control station (transfer station)
3 Control panel
4 Heat exchanger 1
5 Heat exchanger 2
6 Pressure expansion tanks
7 Access
8 to individual consumer groups
9 Heat distributor
10 Feed
11 Return

The following examples provide an overview of the amount of space required for central heating systems by presenting several typical central systems and their principal measurements. Clearance for installation (assembly/disassembly) has been taken into consideration, since most boilers and furnaces must be replaced after an operating period of roughly 20 to 25 years. The same applies to large control systems, pumps and similar systems (within a shorter period of time).

Figure 4.109 illustrates a central district heating installation with transfer station, reverse flow devices, principal distributors and principal collectors, as well as switch and control systems.

Figure 4.110 illustrates a central heating system with electric boilers and heat storage tanks for heating a large building project with electrical energy.

Figure 4.111 illustrates a smaller central heating system, typically used for oil and gas furnaces in single- or multi-family houses.

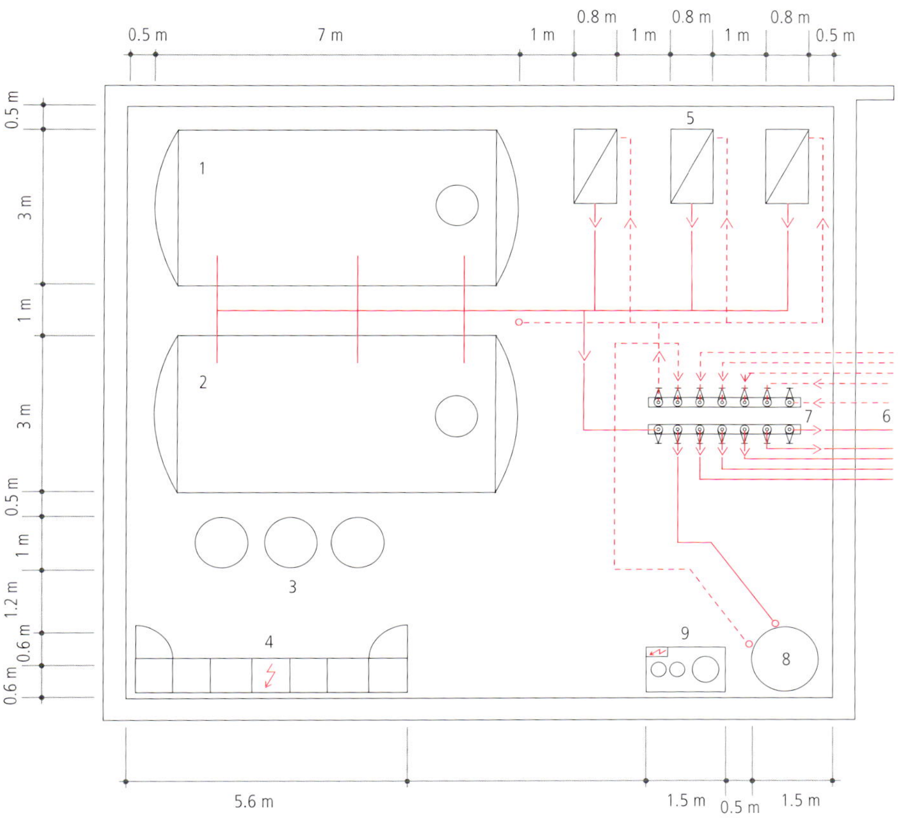

Figure 4.110
Central heating installation with heat storage tanks

1 Storage tank 1
2 Storage tank 2
3 Pressure expansion tanks
4 Control panel
5 Electric boiler
6 to consumers
7 Distributors
8 Warm water generation
9 Water processing

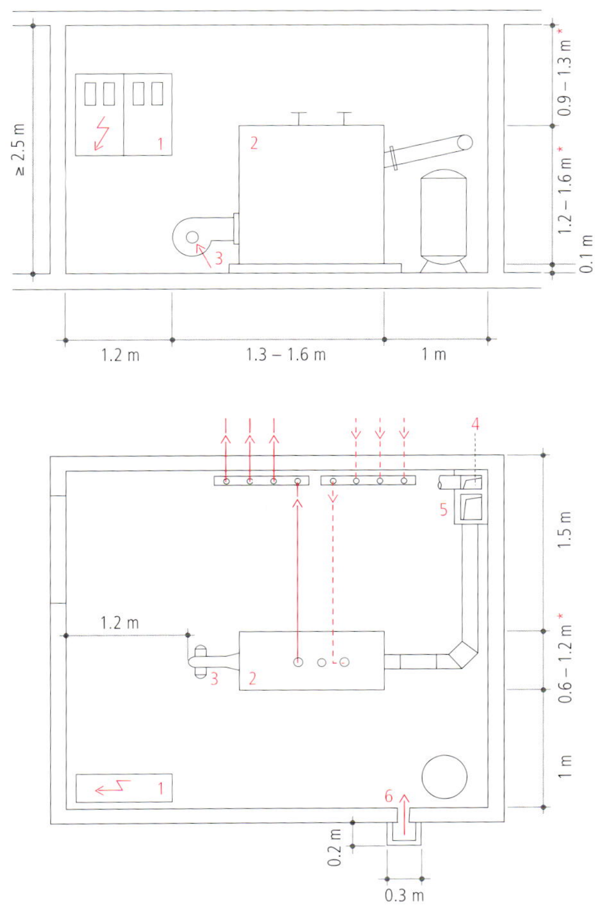

Figure 4.111
Small oil/gas central heating installation

1 Control panel
2 Boiler with warm water generation
3 Fuel feed
4 Combustion chamber exhaust air
5 Exhaust gas chimney
6 Combustion chamber air intake

* according to heat generation type

Figure 4.112.1/2 illustrates a central oil- or gas-fired system for a larger complex supplied from this energy centre.

Figure 4.113 illustrates a central roof heating system for a large shopping centre with a total heat output of roughly 12 MW, distributed across six, oil-fired and equally large cast iron boilers.

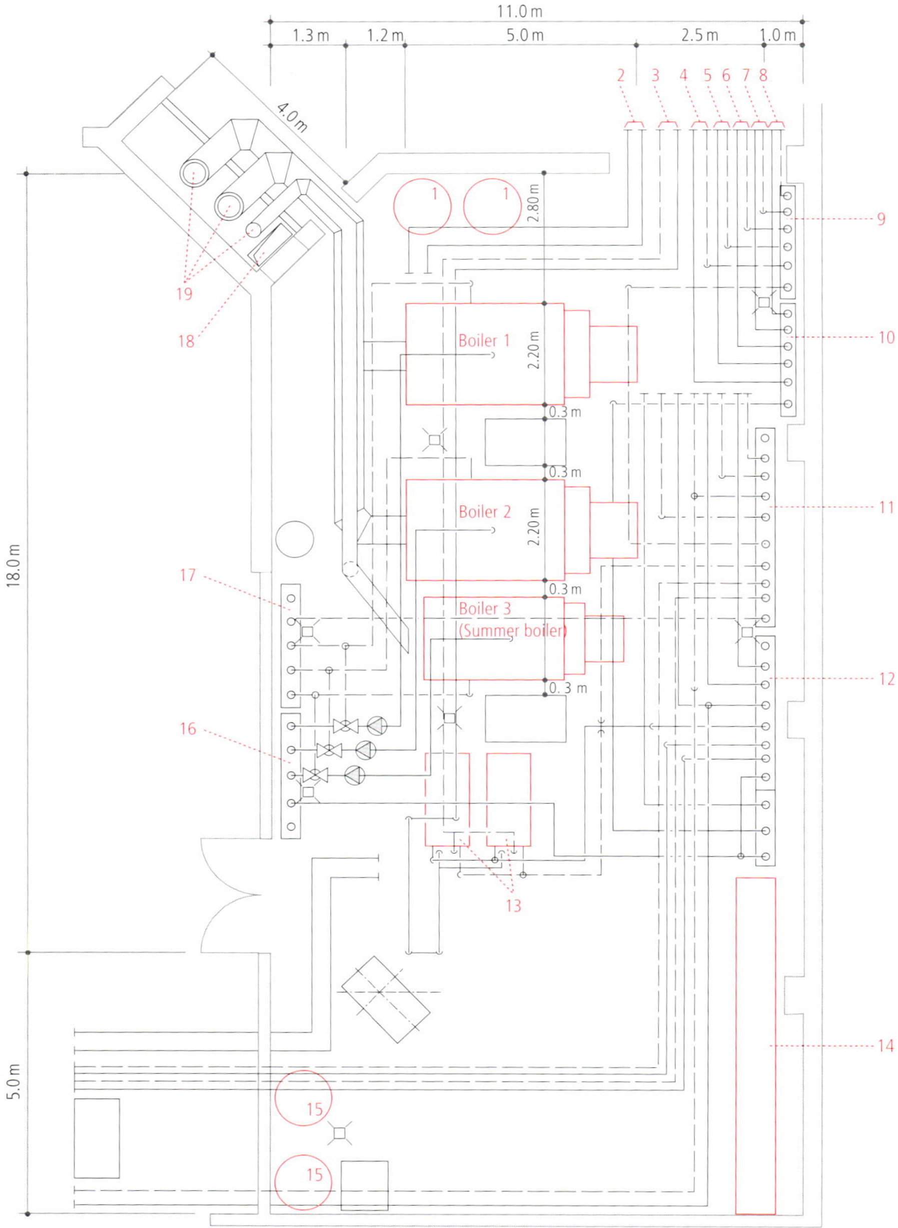

Required room height = 4.0 m

1 Expansion tanks, q 1.2 m, h = 2.0 m
2 Cold water (DN 150)
3 Tower (DN 125)
4 WWG (DN 80)
5 HVAC Café 1 (DN 50)
6 HVAC Café 1 (DN 50)
7 Static heating Café 1 (DN 32)
8 Static heating Café 2 (DN 32)
9 Return/bottom collectors
10 Sub-distributors/feed
11 Return collectors
- Kitchen DN 125
- Tower (DN 125)
- Bank DN 25
- Gym DN 25
- Main return DN 200
- Reserve DN 100

12 Feed distributors
- Tower (DN 125)
- Bank DN 25
- Gym DN 25
- Main feed DN 150
- Kitchen DN 125
- Main feed DN 200

13 Heat exchanger for proportioning pressure separation
14 Control panel (l = 7.2 m)
15 Expansion tanks with air compressor, q 1.2 m, h = 2.0 m
16 Main collectors boiler installation
- Feed boiler 1, DN 125
- Feed boiler 2, DN 150
- Feed boiler 3, DN 150
- Reserve DN 150

17 Main collectors boiler installation
- Return boiler 1, DN 125
- Return boiler 2, DN 150
- Return boiler 3, DN 150
- Reserve DN 150

18 Combustion chamber air vent above roof, 1.0 x 0.4 [m]
19 Chimneys (q 65 cm/30 cm)

Figure 4.112.1
Large oil or gas central heating installation with distributors and collectors

Section

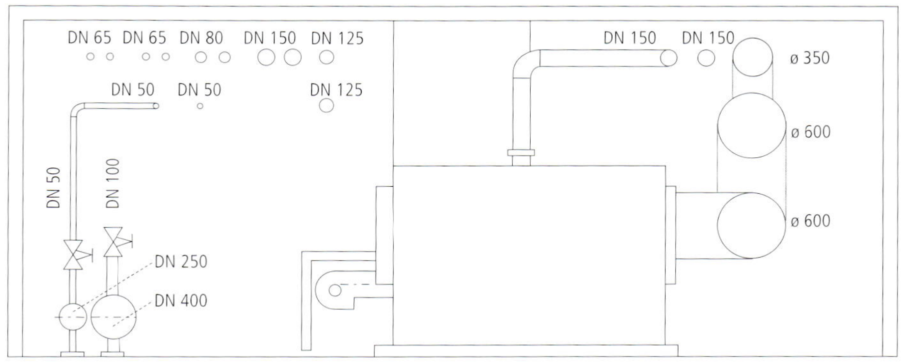

Figure 4.112.2
Large oil or gas central heating installation with distributors and collectors

Section

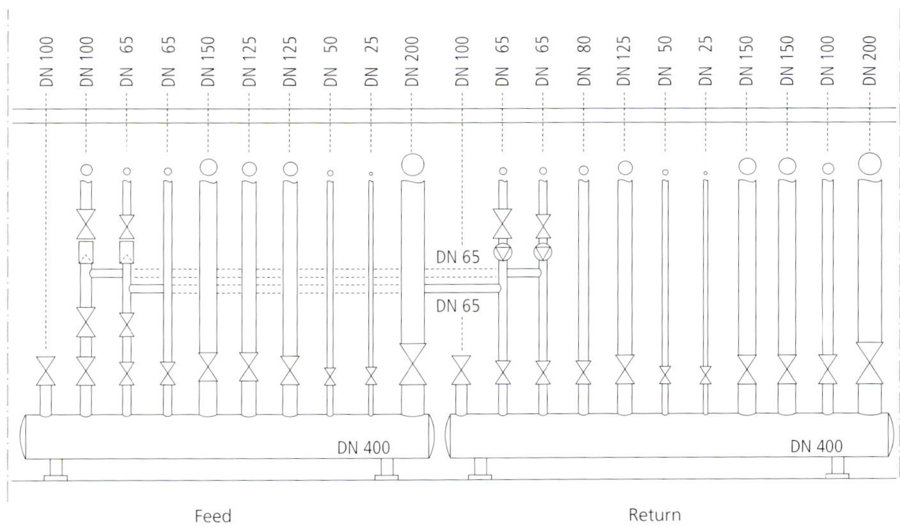

Figure 4.113
Example of a roof-installed central heating system Holstein-Center complex, Itzehoe, Germany
Manufacturer: Detlev Andres, Kiel

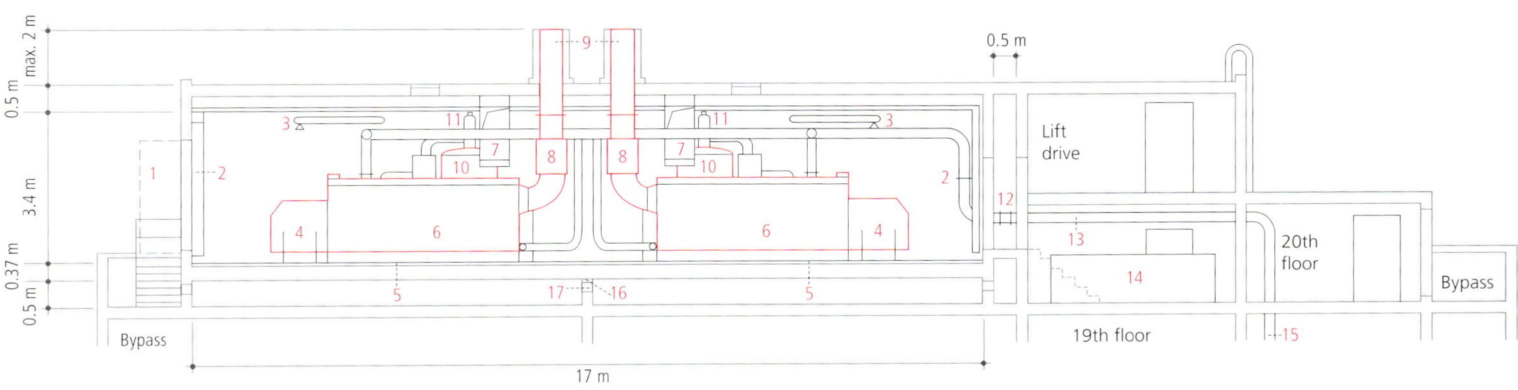

1 Emergency exit
2 Intake air ducts
3 CO_2 fire extinguishers
4 Two-stage oil burners with combustion sound insulation cones and oil sump pans
5 Sound-absorbing boiler pads
6 Boilers
7 Air evacuation ducts with built-in sound absorbers
8 Exhaust gas sound absorbers
9 Chimneys (6 in total)
10 Expansion tanks (4 x 2,500 l)
11 Sound absorber compensators
12 Connection piece
13 Expansion equalizing section
14 Tank for daily heating oil supply (5,000 l)
15 Heating ducts (approx. 70m high)
16 Sound insulating layer
17 Pressure equalizing bore holes for air cavity beneath heating installation (air cavity as sound insulation)

The photographs in **Figures 4.114, 4.115** and **4.116** show views into a central heating installation with large boiler systems, distributors and circulating pumps, electronically equipped for remote infrared operation and supervision.

Figure 4.114
Central heating installation with boilers

Figure 4.116
Static frequency changer driven circulating pumps with infrared remote control
(Image: Grundfos)

Figure 4.115
Static frequency changer driven (pressure) intensifier system for water distribution as required
(Image: Grundfos)

Table 4.15 indicates the area required for each central heating system in relation to heat output.

Shafts are required to furnish the necessary vertical links between the central heating system and the floors to be heated. The shafts (water conducting shafts) may house chimneys or feed and return flow pipes. Large shafts are generally designed to provide access for a variety of necessary tasks (repair/retro-fitting/conversion) and equipped with steel grid platforms. Roughly 5 % of total area for these central system areas should be estimated as shaft areas for all piping and chimneys.

The breeching (horizontal distribution) of heating pipes is largely dependant on the heat quantities, with installation heights for the principal breechings (top/bottom distribution) ranging from 20 cm (single-family house) to one meter (large central heating systems). Within each floor, the standard installation height ranges from 3 – 5 cm, either in the ceiling or the floor area. Radiators are located according to the options, that the model and the available space allow.

Boiler housing **Boiler output**	**Heating chamber**		**Pump and distribution room with WW generation**	**Space for expansion tanks**	
	Area m^2	Height m	Area m^2	Area m^2	Height m
up to 25 kW	6				
25 – 45 kW	6 – 11				
45 – 70 kW	11 – 14	2.50			
75 – 95 kW	14 – 16			2.00	1.60
95 – 116 kW	16 – 18			2.00	1.60
for multiple boiler installations					
0.12 – 0.35 MW	24 – 30	3.00	10 – 12	4.00	2.00
0.35 – 0.95 MW	30 – 60	3.00	12 – 20	7.00	2.20
0.95 – 1.75 MW	60 – 75	3.50	20 – 37	9.00	2.50
1.75 – 5.80 MW	75 – 160	3.50	37 – 80	18.00	2.80
5.80 – 8.70 MW	160 – 200	4.00	80 – 110	20.00	3.80
above 8.70 MW	Plan for custom boiler sheds or building components				

District heat transfer **Thermal output**	**Transfer station and distributor for district heat link with**				**Substation**		
	Steam Area m^2	Hot water Area m^2	Steam Height m	Hot water Height m	Area m^2	Steam Height m	Hot water Height m
up to 0.12 MW	6	10	2.40	2.40	5	2.40	2.40
0.12 – 0.35 MW	6 – 12	10 – 15	2.40	2.40	5 – 8	2.40	2.40
0.35 – 0.95 MW	12 – 24	15 – 40	2.50	3.00	8 – 18	2.50	3.00
0.95 – 1.75 MW	24 – 36	40 – 58	2.80	3.50	18 – 32	2.80	3.50
1.75 – 5.80 MW	36 – 60	58 – 120	3.00	3.50			
5.80 – 11.60 MW	60 – 120	120 – 200	4.00	4.00			
11.60 – 17.40 MW	120 – 200	200 – 300	4.00	4.00			

Table 4.15
Central heating system dimensions

Radiator placement

Radiators are usually placed beneath windows to compensate for the influx of cold air and to render these areas comfortable for use. Experts are currently investigating options for placing radiators along internal walls, in combination with triple glazing or insulated glazing, because this allows for shorter and simpler piping routes. A central heat consumption sensor can generally be installed.

Radiators and other heating units are installed on wall brackets and fasteners, footings or upright brackets.

A variety of adjustable brackets are available to facilitate installation in all building heights, including dowel fasteners for conventional or light-weight constructions (**Figures 4.117.1/2**). When radiators are installed in niches or in front of floor-to-ceiling windows, the relevant insulation specifications must be followed. In niches, this means installing a layer of insulation and in the case of placement in front of room height windows, this means custom radiator design.

Radiator casing

Whenever possible one should avoid covering heating units. If a casing is desired at all costs, it must be easily removable for cleaning the radiator and perforated to allow air to circulate freely. In order to meet these requirements, the cover perforations should be as large as possible along the top and the bottom of the entire length of the unit. Covers diminish the standardized thermal output by 10 to 30 %, depending on the design of the cover.

Flat bracket for fixing in a wall

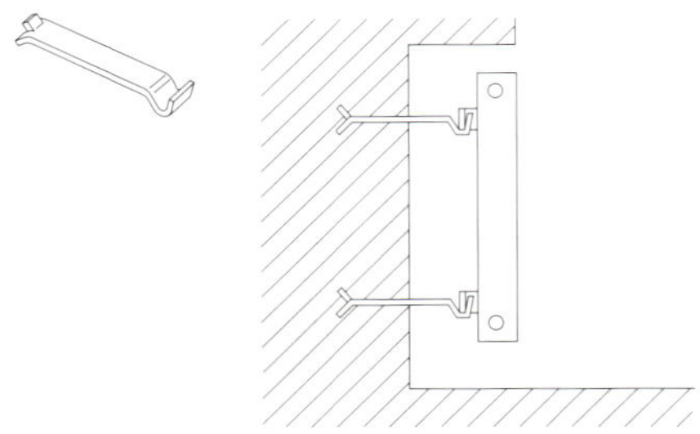

T-shaped bracket for fixing in a wall

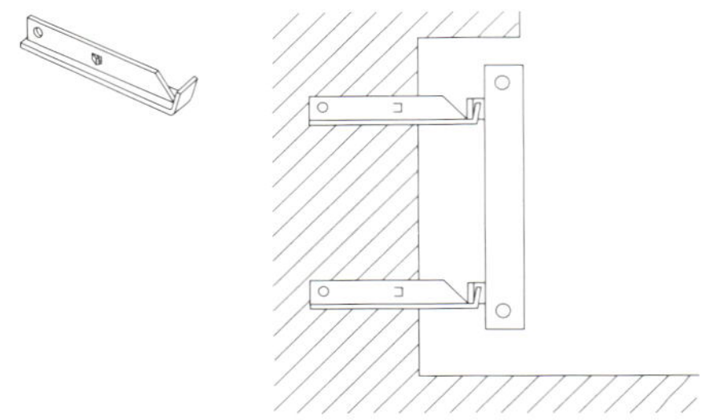

Flat bracket for plugging

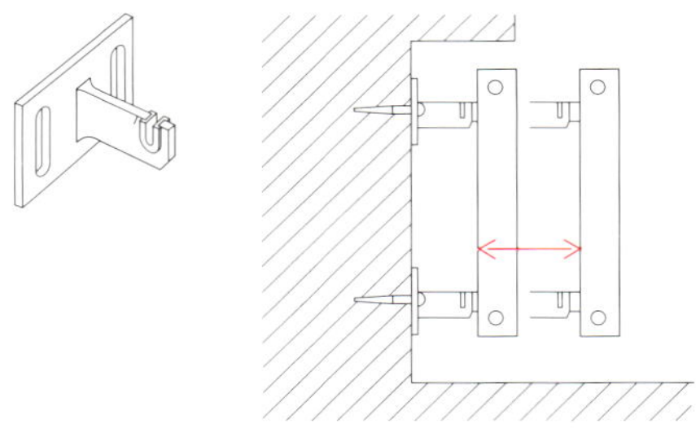

T-shaped bracket for plugging

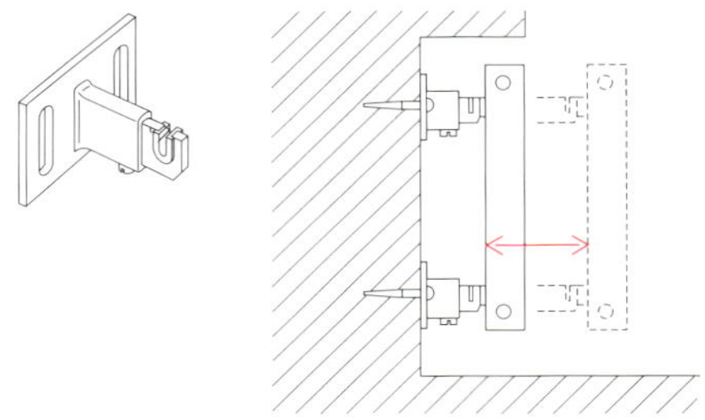

Figure 4.117.1
Brackets for flat radiators

1 and 2:
Flat bracket and T-shaped bracket for fixing in a wall
3 and 4:
Flat bracket and T-shaped bracket for plugging

Height-adjustable upright bracket suitable for any plate height

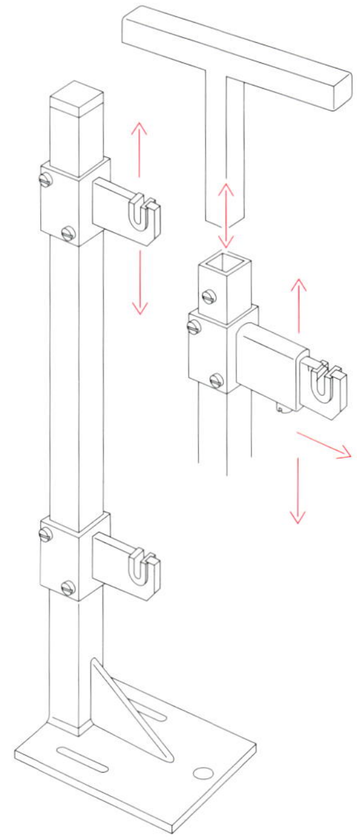

Upright bracket with additional adjustment option for depth and use as window sill bracket

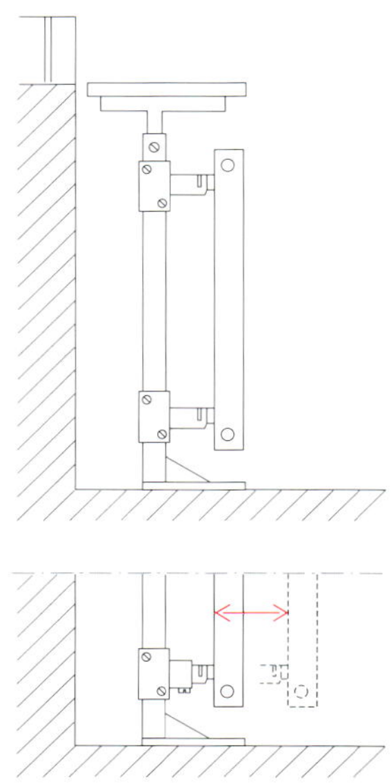

Figure 4.117.2
Universally applicable upright bracket

Left:
Adjustable to any plate height
Right:
Additional adjustment option for depth and window sill bracket

5. Sanitation and fire protection

(with Manfred Woiwode, Düsseldorf)

Sanitation primarily concerns water supply and drainage. Human beings have appreciated water as a life-sustaining element since the beginning of time, as a refreshing beverage and for health care and hygiene. In recent years, however, this gift of nature has become a hot political issue. Water shortages and pollution are gradually turning into nightmare scenarios for the future.

We must tap into the knowledge and experience gained from the natural sciences, different modes of living and building systems, to counteract this ominous development and aim for optimum efficiency when planning sanitary installations.

The relationship between sanitation technology and building systems
In pondering this relationship one should think of the sanitary installations as a cohesive component that is fully integrated from a functional and a structural perspective.

Fixtures that must fulfil a variety of purposes form the core of this component. These fixtures must be supplied with appropriate operating "fuels", which must then be evacuated once the operation is complete. The principal operating "fuels" is water, and a number of physical laws must be considered when planning the piping system.

Unlike furniture, sanitary fixtures and appliances cannot be moved within a room as the need arises. It is important to remember that each fixture is always connected to two or more pipes, which would need to be moved along with the fixture.

Hygiene
Clean fixture surfaces, ease of cleaning, avoidance of dirt-collecting corners, odourless evacuation of human excrement, prevention of pest infestation and the formation of sewage gases as a result of installation inside a building, prevention of drinking water contamination and disinfection of water for public pools – these are standard demands made on modern sanitary installations.

Design
Sanitary fixtures and appliances come in a wide variety of designs. The simple, clean and deliberately utilitarian models of recent years reflect the preferences and fashions of our time.

Function and purpose
The functional form, design and layout of sanitary installations must always answer to human requirements. When manual functions are replaced by automated mechanical options, material considerations should have priority. Ultimately, all functions must be subordinate to the purpose of the sanitary system.

Operational safety
Negligent endangerment of human health and life is a criminal offence and the prevention of such negligence is a matter of public interest. To this end, guidelines and codes have been formulated with regard to the manufacture and operation of sanitary installations. Public water and sewage utilities are trustees charged with representing these interests. The obligation to obtain a licence is therefore a necessary requirement for all gas fitters and plumbers.

The principal risks are:
- Water damage caused by pipes
- Explosion of connected warm water fixtures
- Gas explosions
- Breakdown due to system component failure
- Clogged pipes
- Infections due to contaminated drinking water
- Aggravated illnesses due to unhygienic installations

Economic efficiency
The minimum investment for capital costs and operating costs determines the ratio of lifecycle to acquisition costs. Selecting suitable materials for the fixtures, piping and insulation is the principal criterion for economic efficiency.

Sound insulation
Sanitary installations should be usable at any time, day and night, without creating disruptive noise for users or occupants of other rooms.

Assembly and installation
This point addresses all measures for simplified, clean assembly, uniformity, standardization, rationalization and prefabrication.

5.1
Cold water supply

5.1.1 Cold water demand

A steady rise in water consumption has been clearly linked to an increase in the number of sanitary installations and greater demands with regard to comfort. Water consumption and its distribution in private households is shown in **Figure 5.1**.

Table 5.1 provides a list of water demands, an important prerequisite in the computation of water supply installations.

The average, current water consumption rates are:
- Bachelor apartment, approx. 80 m^3/a
- 1-bedroom apartment, approx. 100 m^3/a
- 2-bedroom apartment, approx. 120 m^3/a
- 3-bedroom apartment, approx. 160 m^3/a
- 4-bedroom apartment, approx. 200 m^3/a.

5.1.2 Procurement of water

Environmental pollution and increased demands make the procurement of water a challenging task. In the past, well water played an important role and today most groundwater sources are already widely exploited. Consequently, lake and river water is becoming an increasingly important source.

Each country has special requirements for the quality of potable water: in Switzerland, for example, these requirements are specified in the Federal Food Act. The act stipulates that potable water supply must meet "the standards for hygiene and microbial purity established for food, articles for daily use and consumer products" and also correspond to "the regulations on foreign matter and ingredients in foodstuffs". This applies to both cold and warm water.

Water utilities are obliged to deliver water that meets these requirements. Their sphere of accountability reaches as far as the water meter, or in the absence thereof, as far as the main shut-off valve at the connecting pipes.

Potable water must correspond to the general requirements for hygiene and in particular to the food guide requirements in terms of appearance, odour and taste. These requirements form the basis of the criteria that potable water must meet:

- Absence of morbific bacteria or other harmful substances
- Clear, colour- and odourless
- Flawless taste
- Average hardness of approx. 15 for 20 °C (based on French hardness rating)
- Refreshing, with a temperature range of approx. 7 – 12 °C

Figure 5.1
Domestic water requirements – today.
In the Federal Republic of Germany, the average personal water consumption is roughly 140 litres per day. Only some 3 litres are used as drinking water. The bulk is consumed for washing and cleaning and flushing human waste

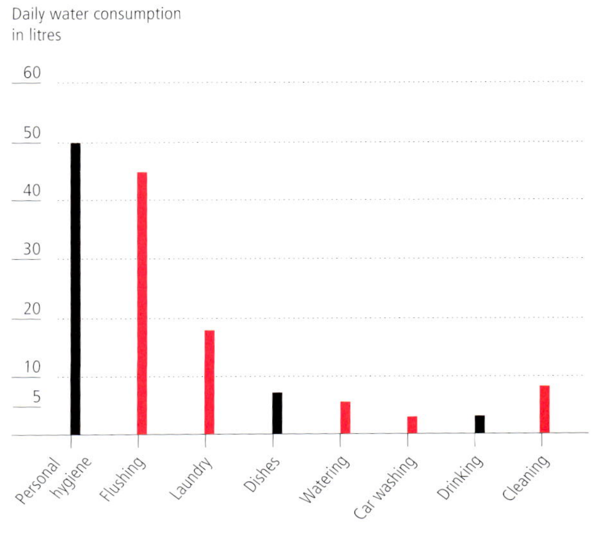

Table 5.1
Parameters for water demands (cold and warm)

Use	**Unit** Consumer	Time	**Demand in litres** Min.	Average	Max.
Residential use:					
Drinking, cooking, cleaning	Person	Day	60	90	130
Laundry	Person	Day	15	30	50
Bathing, showers	Person	Day	60	110	190
Toilet flushing	Person	Day	40	50	60
Car washing	Family car		150	250	400
Housing:					
Simple housing	Resident	Day	110	170	230
Medium comfort	Resident	Day	170	225	330
Elevated comfort	Resident	Day	275	440	660
Public use:					
Schools	Student	Day	10	12	15
Hospitals	Patient	Day	350	500	650
Public baths	500 – 600 baths		500	550	600
Restaurants	Guest	Day	60	100	150
Barracks	Person	Day	70	120	180
Market halls	m^2	Day	4	5	6
Abattoirs	Large livestock	animal	400	450	500
Abattoirs	Small livestock	animal	300	350	400
Laundromats	kg dry laundry		50	60	70
Parks	m^2 area		1	1.5	2
Street cleaning	m^2 area		1	1.25	1.5
Car repair shop	Employee	Day	30	40	50
Baker	Employee	Day	120	130	140
Hairdresser	Employee	Day	160	170	180
Photo studio	Employee	Day	280	300	320
Administration building	Employee	Day	40	55	70
Department stores	Employee	Day	60	70	80
Industrial:					
Brewery (without refrigeration)	100 l beer		600	700	800
Brewery (with refrigeration)	100 l beer		1,400	1,700	2,000
Dairy	100 l milk		400	500	600
Paper mill	kg/fine paper		1,500	2,200	3,000
Tannery	Skin		1,000	1,500	2,000
Mining	t/coal, ore		1,000	1,500	2,000
Raw iron production	t/raw iron		7,000	10,000	13,000
Agricultural:					
Horses	1 animal	Day	60	70	80
Cows	1 animal	Day	60	65	70
Young cattle	1 animal	Day	40	45	50
Pigs	1 animal	Day	15	20	25
Sheep, goats	1 animal	Day	5	6	7

Stringent safety measures are therefore enacted for the procurement, storage and distribution of potable water; these are set forth in detail in the laws and standards of each country.

The procurement of water is a vast field in its own right. For the purpose of this volume, the topic is therefore only discussed in the most general terms.

5.1.2.1 Spring or well water

Natural springs are sites where groundwater rises to the surface; they are tapped into as shown in **Figure 5.2**. The exploitation of springs is determined by the yield, temperature, chemical and bacteriological properties. The principles for designing spring water gathering facilities are as follows:

- Sufficient protection of the water from external pollutants/contaminants
- Protecting the temperature of procured water from external air (external temperature)
- Preventing backpressure at the water gathering site

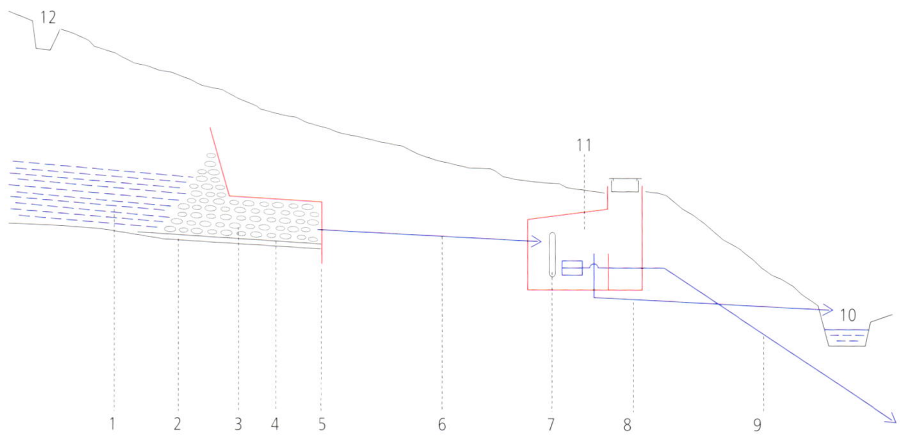

1 Spring water stream
2 Compacted stratum
3 Stone and coarse gravel
4 Waterproof seal
5 Concrete barrier
6 Spring water feed
7 Baffle
8 Outflow
9 Water to reservoir
10 Bottom hill drainage
11 Spring water well chamber
12 Top hill drainage

Figure 5.2
Principle of procuring well water

Figure 5.3.1
Submersible motor-driven pump Wilo Drain model for house and estate drainage
(Model image: Wilo)

5.1.2.2 Groundwater

Groundwater is surface water that is stored beneath the earth's surface above impermeable layers, after having seeped through sandy-gravely stratum and naturally or in some cases artificially enriched.

Where flow velocities of groundwater are low (a few metres per day), appropriate distances to the seepage site and suitable ground materials exist, one can expect to extract high-quality potable water.
Spring and groundwater do not require processing when gathered with the correct means. Lake and river water, on the other hand, must always be processed; i.e. filtered and sterilized. **Figure 5.3** illustrates the principle of gathering groundwater; **Figure 5.3.1** shows a submersible motor-driven pump.

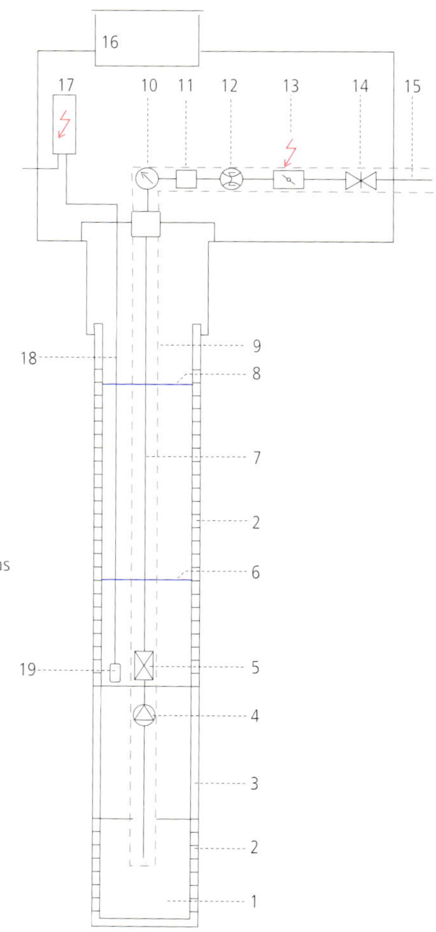

Figure 5.3
Schematic diagram of a groundwater gathering installation

1 Mud catch pit
2 Perforated filtering tube
3 Filtering tube without perforations
4 Submerged motor-driven pump
5 Backpressure valve
6 Minimum water level
7 Riser
8 Operating water level
9 Tube mounting
10 Manometer
11 Steel regulator
12 Optional water meter
13 Automatic throttle flap
14 Gate valve
15 Pressure line (delivery pipe)
16 Inspection shaft (manhole)
17 Electric panel
18 Electric cable
19 Contact electrode

5.1.2.3 Lake water

In regions with water shortages, e.g. in densely populated areas and industrial zones, lake water becomes an essential source that can provide nearly limitless volumes at constant temperatures and low hardness when the gathering site is carefully chosen (at a depth of approx. 30 m) and the procured water is appropriately processed (**Figure 5.4**).

5.1.2.4 Rainwater for industrial use

The mean precipitation in Central Europe is roughly 0.03 l/s m^2/a; annual precipitation is approximately 600 – 800 mm/m^2.

People are generally unaware that a large amount of precious potable water is being literally flushed down the toilet in housing (approx. 33 %). Cleaning, car washing and watering the garden consume needlessly large amounts of drinking water. **Figure 5.5** shows water demands for private households and areas where water of potable quality and water of rainwater quality can be used respectively. In other words, nearly half of the drinking water could be replaced with rainwater – an enormous savings potential. Only 3 % of drinking water are normally used for food preparation or human consumption.

In Central Europe, where precipitation levels are sufficiently abundant, rainwater can be used as a suitable and efficient resource.

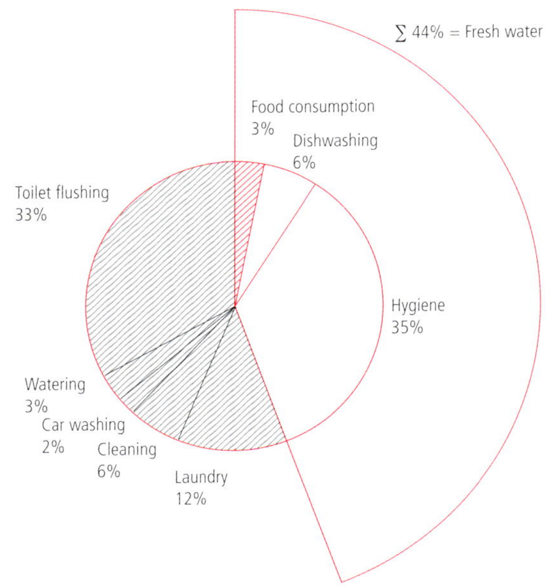

Figure 5.5
Water requirements in private household:
56 % of potable water consumed in private households could be replaced with greywater

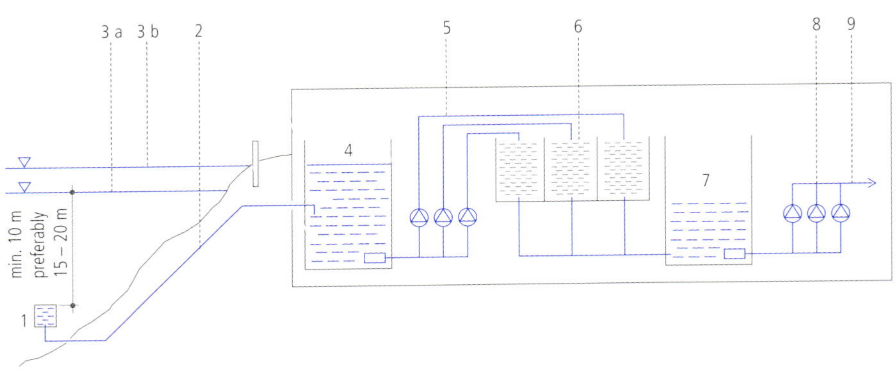

Figure 5.4
Schematic diagram of lakewater gathering installation

1 Catchment point with sieving filter
2 Suction tube on lake-bottom, covered
3 a Minimum water level
3 b Maximum water level
4 Equalizing basin
5 Pump station
6 Open (or closed) filters
7 Pure water reservoir
8 Pump station
9 Distribution pipe

When rainwater is used as service water, such uses should always be restricted to rainwater collected on roofs to limit the amount of pollutants in the rainwater cycle. Planted roofs offer an even better solution as they filter heavy metals, soot and dust. Rainwater is generally collected in a cistern, purified and exposed to UV rays before it is piped into the storage tank proper (**Figure 5.6**).

This tank is the actual water reservoir from which the rainwater consumers are supplied. Prior to use, the rainwater is run through a filter system and then dosed to the appropriate degree of hardness. After the filtering and dosing process, the water is pumped from the cistern to a distributor, from where it is passed on to the individual consumers. Naturally all service water processed from rainwater must be kept separate from the potable water system at all times to prevent any potential contamination.

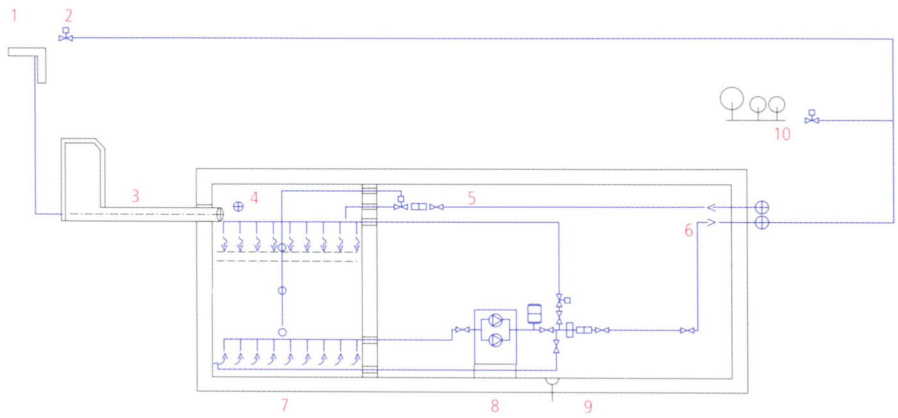

Figure 5.6
Diagram of rainwater processing installation

1 Roof
2 Roof irrigation
3 Stormwater
4 Overflow
5 Secondary feed potable water or well water
6 Service water for grounds and roof gardens
7 Collecting basin
8 Pressure booster installation
9 Filter installation
10 Garden irrigation

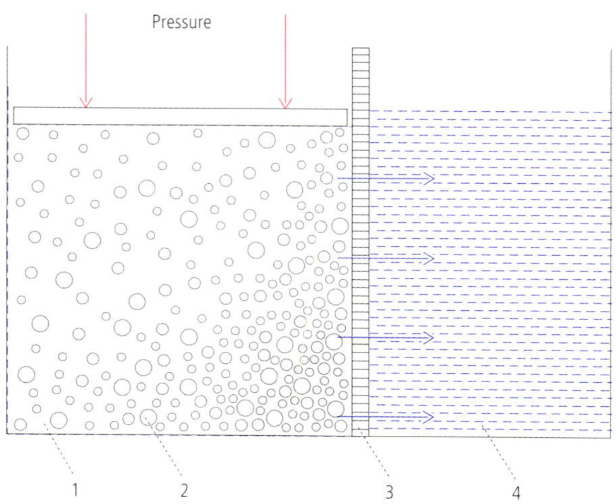

Figure 5.7
Principle of reverse osmosis

1 Saline-containing water
2 Dissolved salines, organic components, bacteria and suspended agent
3 Semi-permeable (fine pored) membrane
4 Permeate/pure water

5.1.2.5 Seawater

While desalinization is not pertinent to the issue at hand, it is very much an essential theme in terms of water procurement. Seawater is processed either by means of evaporation or by reverse osmosis.

In reverse osmosis pressure is applied to the saline-containing water, thus transporting water molecules through a semi-permeable membrane. The molecules of the dissolved salines, organic substances, bacteria and suspended agents are too large to pass through the fine pores of the membrane. They accumulate in the residual water, which is siphoned off as concentrate. The quality of the filtered water is suitable for immediate consumption, albeit of a slightly lesser quality than the potable water we are generally used to consuming. **Figure 5.7** illustrates the principle of reverse osmosis for extracting potable water from seawater.

5.1.3 Cold water distribution

Potable water supply must be guaranteed 24 hours a day at all tapping locations. Consumption is subject to enormous variations. Water reservoirs compensate the difference between these two parameters, also taking into account the additional storage of potable water for fire extinguishing. To supply buildings with natural (water) pressure, the water level in the reservoir should never fall below 30 m above the highest tapping station (**Figure 5.8**).

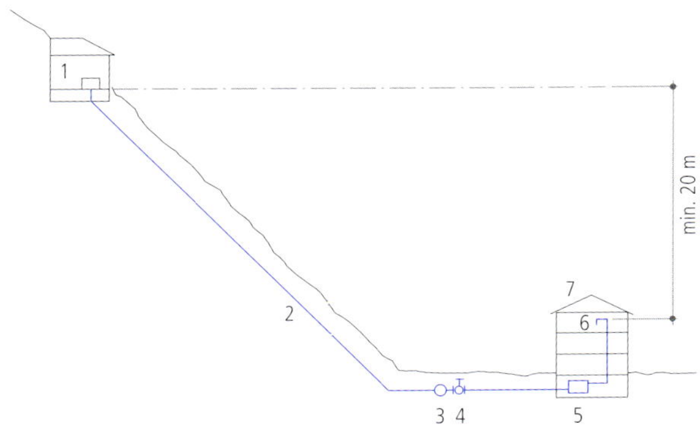

Figure 5.8
Schematic diagram of cold water distribution system

1 Reservoir
2 Main feed
3 Ring or main distribution line
4 Main feed shut-off
5 Distribution battery in building
6 Uppermost tap position
7 Building

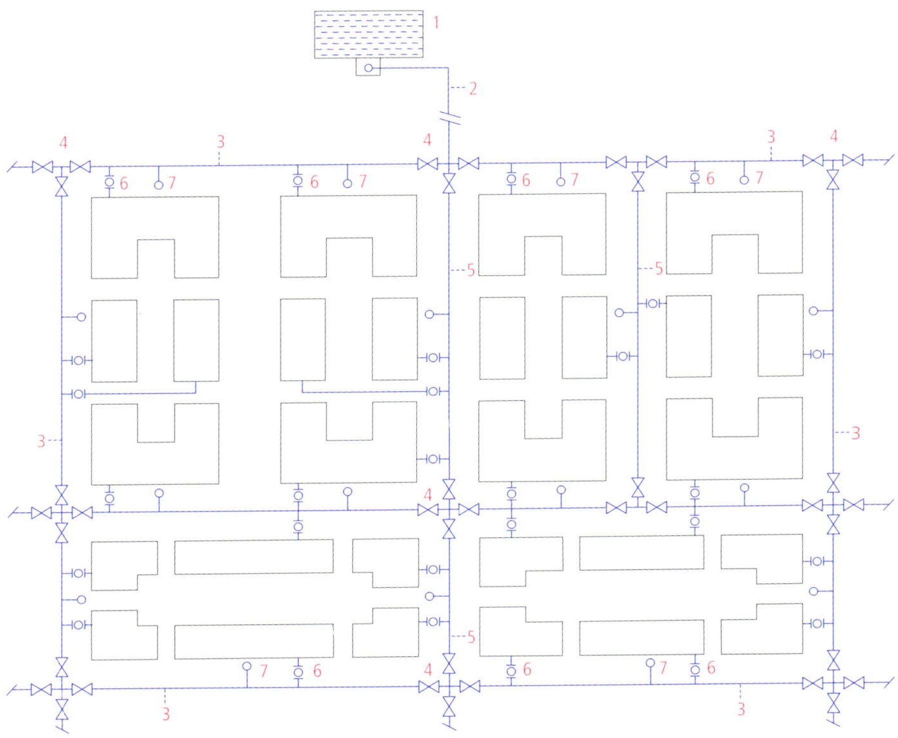

Figure 5.9
Schematic diagram of cold water distribution system
(Plan of ring lines)

1 Water reservoir
2 Main feed
3 Ring or main distribution line
4 Gate valve
5 Connecting lines
6 House hook-up lines
7 Hydrants

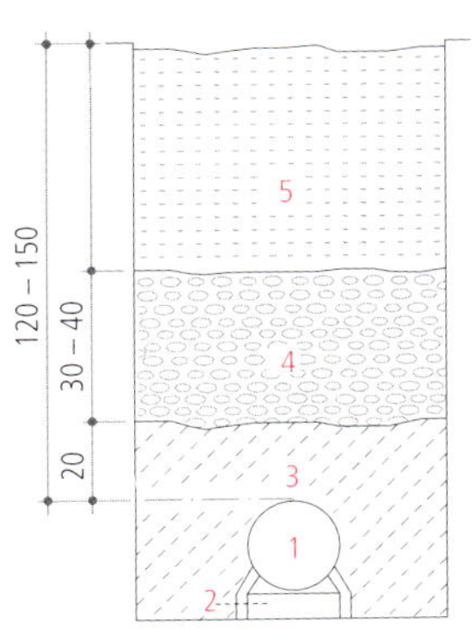

Figure 5.10
Schematic diagram of piping trench

1 Pipe
2 Concrete berm
3 Sand filling
4 Loose gravel and fine soil
5 Standard soil

5.1.3.1 External distribution pipelines

Distribution pipes from the reservoir to buildings are always laid in public ground (preferably streets) to facilitate supervision, maintenance and future expansion. For reasons relating to even pressure distribution and operational safety, the distribution pipes are gathered into so-called rings (**Figure 5.9**).

In case of defective connections or the installation of new hook ups, gate or stop valves can be activated at branch-off points to limit the suspension of supply to small intermediate sections of piping at any given time.

Installation of distribution pipes
Depending on the region, the pipes should be buried at least 1 to 1.5 m below ground to prevent freezing. This is especially important in cases where the pipes run behind or parallel to a retaining wall, a deep trench or a light well. Pipes should generally be installed in washed sand or gravel (**Figure 5.10**). Passages through unstable ground must be reinforced with appropriate support structures to prevent sagging.

Water and sewage pipes should not be installed in the same ditch. If this is unavoidable, water piping must always lie above sewage piping. Berms must be constructed to prevent caving in or sagging.

Potable water pipes must be covered in order to:
- Protect water in the pipes against excessive heat in summer
- Protect water in the pipes against freezing in winter
- Protect the pipes from mechanical loads exerted by heavy rolling stock

Stepped trenches are a suitable solution for piping for different media that must be installed in parallel.

5.1.3.2 Hydrants

Hydrants are taps or filling stations for drawing off water in case of fire, for road cleaning and similar tasks. The following criteria govern hydrant configuration and use:
- Above-ground hydrants for rural conditions (rapid readiness for use, easy to locate)
- Underground hydrants for urban conditions (no traffic obstruction)
- Distances between hydrants range roughly from 60 to 100 m
- Type and location of hydrants must be defined in accordance with official regulations

5.1.3.3 Residential hook-up

This is generally the responsibility of the water supplier or in special cases the plumber commissioned or licensed by the relevant city/community.

Type and size of the building, number of taps i.e. sanitary fixtures, separate guidelines for special water consumers such as swimming pools, cooling plants and other installations for commercial uses determine the ultimate sizing. **Figure 5.11** illustrates the principal layout of water supply piping.

The minimal pipe diameters are:
- Steel piping: 1 1/4"
- Copper piping: 35 mm
- Plastic piping: 40 mm

Water utilities are free to specify smaller pipe diameters for lesser connections.

The wall penetration of the hook-up pipe must be designed in a manner to prevent damage to the pipes when the structure subsides. **Figure 5.12** illustrates one example.

The entire length of the water supply pipe up to the water meter must be exposed on the interior.
With permission from the water utility company, it may be installed inside an accessible duct or shaft.

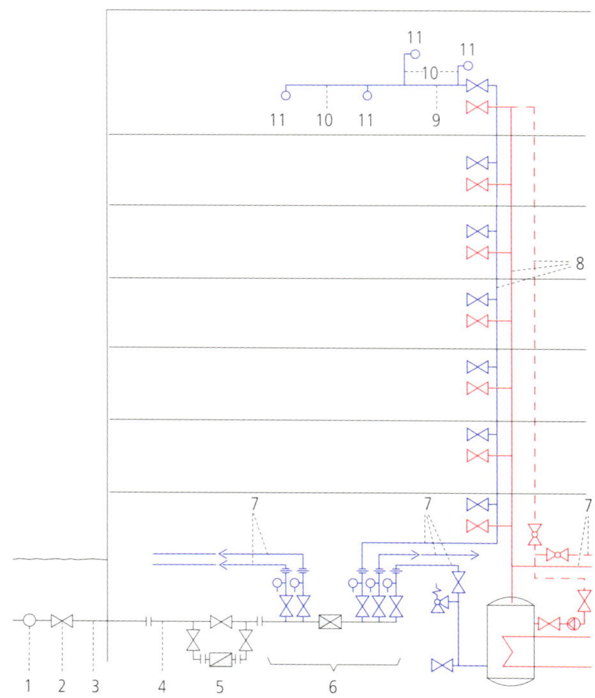

Figure 5.11
System components of a water installation in a building

1 Supply pipe
2 Building shut-off device
3 Building connection pipe (external)
4 Building connection pipe (internal)
5 Water meter installation
6 Cold-water distribution battery
7 Horizontal distributions
8 Vertical piping and warm-water circulation pipe
9 Branch pipes
10 Appliance/fixture connection pipes
11 Tapping positions

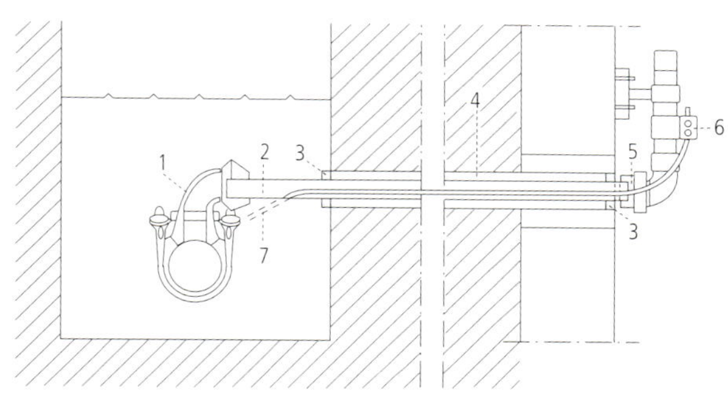

Figure 5.12
Example of house connection (plastic connection)

1 Tapping sleeve with shut-off device
2 PE seal/protective conduit
3 Protective conduit depending on subsoil
4 Transition steel/PE
6 Contact clips
7 50 mm² grounding T-cable, installed between polyethylene and protective conduit

5.1.3.4 Water metering

The water utility company is responsible for installing water metering devices (water meters, calibrating faucets). The utility also determines the size and location of all meters. Water meters must conform to the test and requirement standards of the SVGW (Switzerland), DVGW (Germany) or ÖVGW (Austria) or the equivalent standards in other countries.

The water meter must be installed in a location that is accessible to the consumer at all times, where temperatures are constant and the meter is protected against frost, heat and other influences. Moreover the meter should be readily accessible for taking readings and for periodic renewal.

Depending on application (e.g. commercial operations or many apartments linked to the same meter) it is a good idea to supply a bypass with a lead-sealed valve to avoid interrupting the water supply when the meter is being exchanged.

5.1.3.5 Distribution

The distribution battery is the heart piece of the cold water plumbing system (**Figure 5.13**).

The division into individual branches is based on the following criteria:
– Pressure ratios
– Fixture groupings
– Operational safety
– Efficiency

5.1.3.6 Materials / Sizing

The materials must comply with the recommendations formulated by the gas and water experts in the corresponding country (SVGW, DVGW or ÖVGW, as above, or others).

They must be hygienically safe and exert absolutely no negative influence on the water quality.

To ensure the quality safety of the water that will subsequently flow through them, pipes must be protected against any kind of contamination during storage, assembly, insulation, coating etc. Plastic pipes, connection pieces and fixtures must be protected against harmful exposure (e.g. to heat, UV rays etc.).

The following criteria determine the selection of materials in terms of type and grade:
– Type of installation: in-ground or in-house
– Surrounding conditions: ground conditions, room use, mechanical loads, operating pressure (generally a nominal pressure of 10)
– Water temperature and composition etc.

Depending on material selection, type of installation, surrounding conditions as well as water composition, the pipes must be protected internally and externally (against heat, mechanical influences and corrosion).

Buried pipes
The following materials are suitable for buried pipes:
– Ductile cast iron
– Steel
– High- and low-alloyed steel
– Fibrated cement
– Plastic and other certified materials

Pipes for in-house installations
The following materials are suitable for in-house installation:
– Ductile cast iron
– Steel
– Copper
– Copper alloys
– High- and low-alloyed steel
– Plastic and other certified materials

Waterlines, that is pipes, form pieces, accessories and fittings must be protected against external and internal influences, for example corrosion, in keeping with the relevant material properties.

For new installations, especially when galvanized piping is used, it is a good idea to integrate replaceable regulating components at a suitable location: this makes it possible to easily test the condition of the piping.

Where condensation is a risk factor, piping must be insulated accordingly. Hidden piping generally requires special corrosion protection whenever there is a possibility that it may be exposed to moisture. Any integration into or contact with materials that exert corrosive action, such as gypsum, is not allowed. Steel pipes must be fire-galvanized to protect against internal corrosion. The pipe quality and galvanization must conform to relevant standards, e.g. DIN 2440, 2441 and 2444. Mixed use of metals should be avoided for pipes, connections and other fittings.

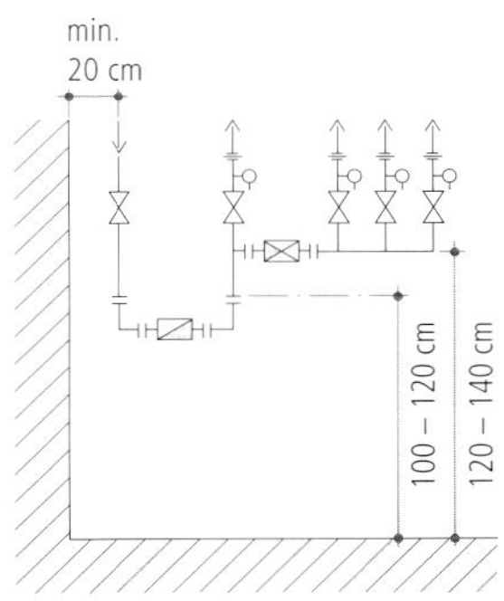

Figure 5.13
Sanitation distribution battery with water meter (direct installation)

Copper must never precede steel in the flow direction of the water. Contamination with and accumulation of sand, rust etc. in the pipes must be prevented. Copper alloys such as brass, gunmetal and similar materials, as well as stainless steel, generally do not require additional corrosion protection for cold or warm water installations. The risk of corrosion tends to increase as hardness decreases due to subsequent water processing and appropriate measures must be taken to counter this risk.

Only certified systems may be used for pipe connections.

Removable connections such as bolts, flanges and clamps must be visible and accessible in building interiors. The same applies to plastic piping. The fundamental rule is that connections must never compromise the mechanical durability of the pipes. When metallic piping is employed for grounding purposes it is vital to ensure that the electrical conductibility is guaranteed at all times (especially in connection pieces, integrated into the bridging or in the connecting element).

Fittings in sanitary installations are divided into the following categories:
- Flow fittings
- Tap fittings
- Safety fittings

Open fixtures are those whose water surface is in direct contact with the atmosphere, such as
- Cold water reservoirs
- Intermediary containers (e.g. rinse container)
- Warm water containers with open outlet
- Pools
- Fountain basins, fish basins

Closed fixtures are those whose water volume is subject to overpressure, as is the case in
- Water storage
- Water transport and pressure increase
- Water heating
- Water processing

Each fixture must be equipped with a shut-off and bleed-off device and a safety device against possible backflow.

Sizing

Basics

Pipe diameters are determined using simplified methods that take the pressure conditions and the type of installation into account.

Pressure conditions

When static pressure is 2 bar or less, computational proof must be established that a minimum flow pressure of 1 bar is guaranteed at each tap position. Flow pressure may need to be adjusted to ensure faultless operation of special fittings or fixtures.

The static pressure should not exceed 5 bar at all tap positions. In cases where the grid pressure is higher, the pressure must be reduced. The water utility company may issue permission for exceptions in the case of specialized installations. The static pressure at garden and garage draw-off taps, as well as fire extinguishing hook-ups, should not exceed 10 bar.

Instead of the pressure reduction valve, which may not be necessary, a bolster or regulating wedge can be integrated to counteract possible pressure fluctuations in the grid. The maximum permissible pressure loss for the entire installation from the water meter onward, i.e. from the central pressure reduction valve onward is 1.5 bar according to SVGW (Swiss standard), while the German standard, DVGW, leaves it variable. An internal pressure increase device is generally required in cases where the minimum flow pressure cannot be guaranteed at all taps.

Computing the pipe network

The sizing of drinking water pipe networks is generally based on mass flow rates according to DIN. Load computation is another simple method of sizing pipelines.

A load value (LV) corresponds to a volume flow of 0.1 l/sec. **Table 5.2** lists fittings and fixtures according to function and performance.

The pipe diameters of the individual subsections for both cold and warm water pipes are derived from the load values and the pipe materials listed in **Tables 5.3.1** to **5.3.3**.

The pipe diameters of standard installations are determined using the load value methods given in **Tables 5.2/5.3**. The load values of the individual subsections are first determined for the most remote tap; the pipe diameters are then determined according to these values. If a minimal flow pressure of 1 bar at each tap or the maximum permissible pressure loss of 1.5 bar for the entire installation from the pressure reduction valve to the most remote tap (pipe length including all individual resistance points such as form pieces and fittings) cannot be maintained, computational analysis is required to determine the appropriate sizing. The load values and drawing-off volume flows (**Table 5.2**) or the effective drawing-off volume flows in l/s are added.

A simultaneity factor can be established on the basis of empirical values or from data provided by the system operator for each subsection. The sum of all load values per subsection times the volume flow value per load value and the simultaneity factor per subsection gives the peak flow that must be taken into consideration.

If the situation does not allow for computation with load values, the partial volume flows per subsection are added and the sum of the drawing-off flows are multiplied by the simultaneity factor per subsection.

Max. number LV	**6**	**16**	**40**	**160**	**300**	**600**	**1600**
DN (mm)	15	20	25	32	40	50	65
Threaded tube (inch)	1/2	3/4	1	1 1/4	1 1/2	2	2 1/2
d_i (mm)	16	21.6	27.2	35.9	41.8	53	68.8

Table 5.3.1
Load values and pipe diameters for galvanized steel pipes DIN 2440/44

Max. number LV	**2**	**5**	**10**	**16**	**40**	**160**	**300**	**600**	**1600**
d_o (mm)	12*	15	18	22	28	35	42	54	76.1
d_i (mm)	10	13	16	20	25	32	39	50	72.1

* 1 LV: max. permissible pipe length 15m
2 LV: max. permissible pipe length 6m

Table 5.3.2
Load values and pipe diameters for copper pipes VSM 211 641

Max. number LV	**4**	**8**	**16**	**40**	**160**
d_o (mm)	16*	20	25	32	40
d_i (mm)	11.6	14.4	18	23.2	29

* 2 LV: max. permissible pipe length 15m
4 LV: max. permissible pipe length 6m

Table 5.3.3
Load values and pipe diameters for PE-X (crosslinked polyethylene) pipes DIN 16893

Connection parameters for fittings and fixtures

Use	**Drawing-off volume flow per connection**		**Number of load values per connection**
	l/s	l/min	LV
Sinks, washstands, bidets, sluice sinks, water tanks	0.1	6	1
Sinks, sink basins, school wall-mounted sinks, hair salon showerheads, domestic dishwashers, instantaneous gas heaters, washtubs	0.2	12	2
Medium performance shower batteries, instantaneous gas heaters	0.3	18	3
Large sinks, floor-mounted sink basins, wall-mounted sink basins, tub batteries, launderettes up to 6 kg, instantaneous gas heaters	0.4	24	4
Draw-off taps for yard and garage	0.5	30	5
3/4"connections: – Sinks – Large tubs – Showers	0.8	48	8

Heating valves are not factored into calculating the pipe sizing.

Table 5.2
Connection parameters for fittings and fixtures

The pipe diameter computation of special installations occurs in two stages, the first setting forth a default selection of pipe diameters and the second carrying out a detailed calculation of the pressure losses. The default pipe diameters can be established using pressure loss diagrams (**Figures 5.14.1 – 3**) based on a flow velocity within the indicated boundaries. Here, too, the calculations are generally based on the values measured at the most remote taps.

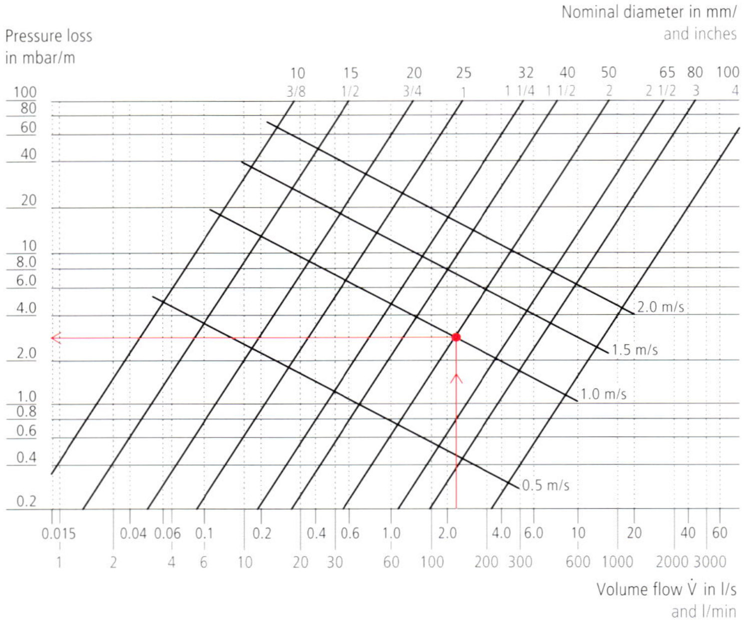

Pipe roughness k = 0.15 mm
Reference temperature 10 °C

Figure 5.14.1
Pressure losses in galvanized steel pipes

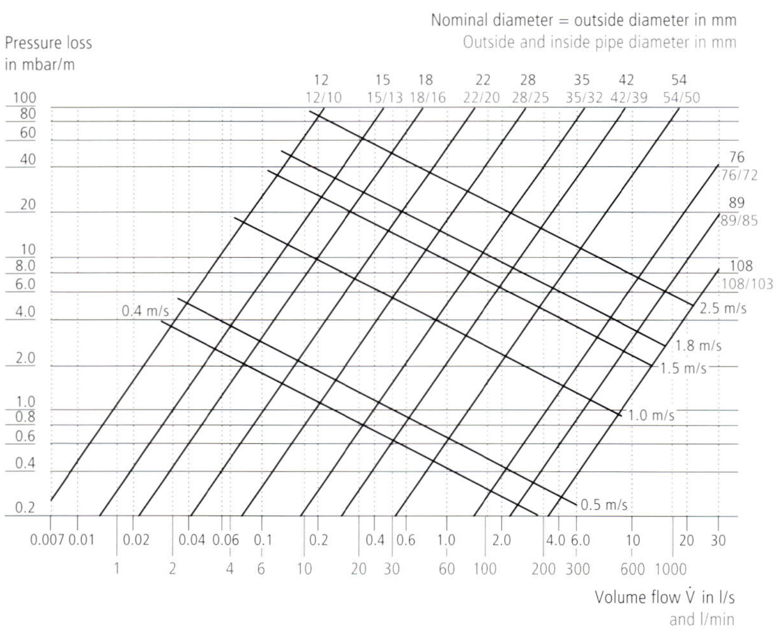

Pipe roughness k = 0.0015 mm
Reference temperature 10 °C

Figure 5.14.2
Pressure losses in copper pipes

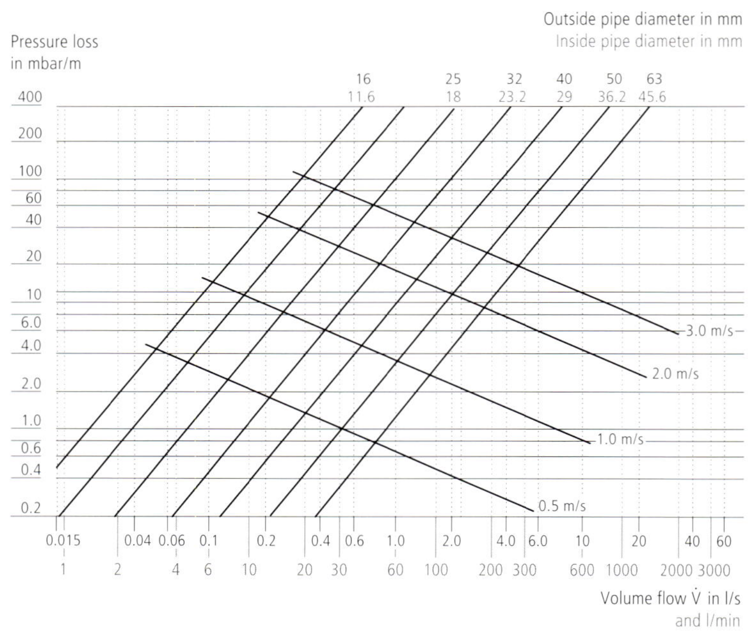

Pipe roughness k = 0.007 mm
Reference temperature 10 °C

Figure 5.14.3
Pressure losses in PE-X pipes

The detailed calculation must deliver analytical proof that the "pressure conditions" are being fulfilled. The standard approach is to determine the pressure losses for the longest pipe (measured from the pressure reduction valve or the water meter onward) up to the most remote tap by adding the pressure losses of each individual subsection through which water flows in a direct line. **Figure 5.15** shows an example for a standard installation in a single-family house, complete with sizing details and load values (LV).

Acoustic insulation requirements
These requirements are founded on universal recommendations and standards. In principle no pipe may come into direct, sound-conducting contact with any building component; this also applies to pipe brackets and fasteners. Pipes containing flexible or moveable parts (compensators, elastic connection pieces) must be properly supported and attached to prevent the transfer of vibration and "hammering" effects. Pipes should never be embedded in concrete in massive wall and ceiling constructions. If pipes must be integrated into ceilings for other compelling reasons, they must be placed inside ceiling channels above the armouring, always taking all safety measures with regard to expansion, corrosion, heat and acoustic insulation into account.

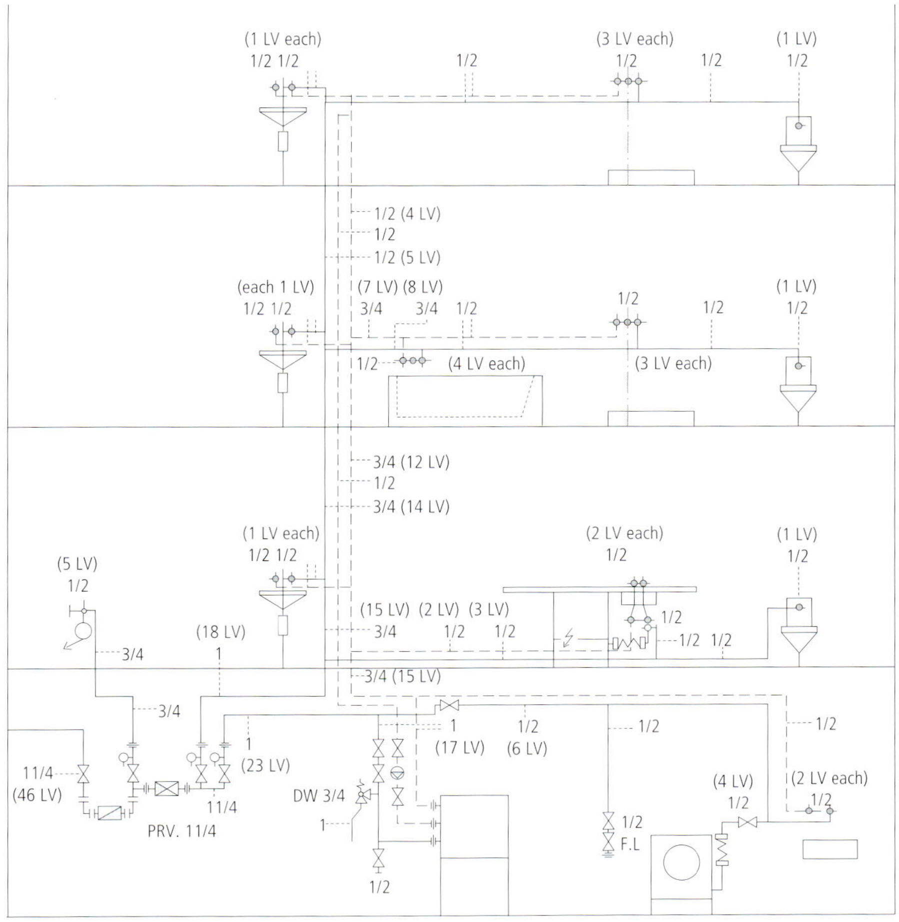

Figure 5.15
Sizing model for standard installation in a single-family house

Figure 5.16
Pressure booster system

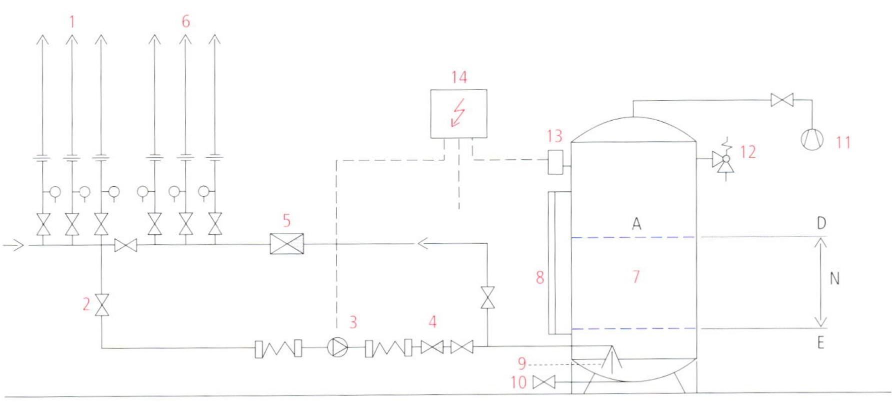

1 Grid pressure distributor
2 Pump feed
3 Pressure booster pump
4 Back pressure valve
5 Pressure reduction valve for variable admission pressure and constant secondary pressure
6 High pressure distributor
7 Hydropneumatic cushioning tank
8 Water level pipe
9 Feed and tapping throat
10 Outflow
11 Air compressor
12 Safety valve
13 Manometric switch
14 Electrical panel

D Disconnect level
E Engage level
A Air cushion
N Net volume

5.1.3.7 Pressure booster systems

Booster systems are employed to supply water to buildings (**Figure 5.16**) where the grid pressure is insufficient or subject to fluctuations or in tall buildings (high-rise). Booster systems are also required in cases where direct link-up to the potable water grid is not allowed (for example, fire protection systems with large water demands). Pressure booster systems (**Figure 5.16.1**) consist of a group of pumps with fittings, a closed membrane pressure tank, a control panel for automatic pump regulation or a closed container (Hydropneumatic cushioning tank) for smaller systems and an air compressor to replenish or renew the air cushion in the tank. A default pressure is set for the air cushion and regulated by the automatic control or maintained by the group of pumps. A pressure reduction valve with constant secondary pressure must be installed after the pressure booster system.

In situations where even volumes are required over a longer period (irrigation systems) the pressure-boosting pump can be integrated into the piping.

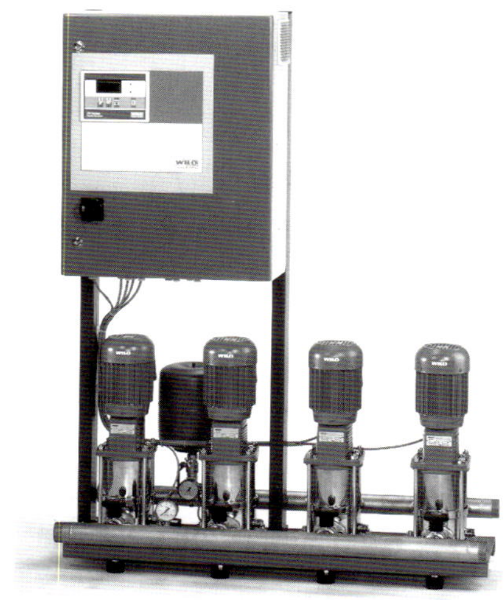

Figure 5.16.1
Comfort-N-Wilo pressure booster system with wet running technology (Image: Wilo)

5.1.3.8 Distribution systems

Instead of using many distribution branches with only one shut-off at the distribution battery, it is preferable to supply entire building blocks or several groups of residential units via a single distribution branch fitted with several shut-off valves for smaller units, e.g. an apartment shut-off valve. The diagram in **Figure 5.17** shows a system solution. Operational safety is primarily based on anticipated problems as a result of a long interruption to the water supply. The most common reason for shutting the supply off is to repair fittings. One economical and practical solution is to replace the group valves (apartment shut-off valves) with individual shut-off valves at each outlet.

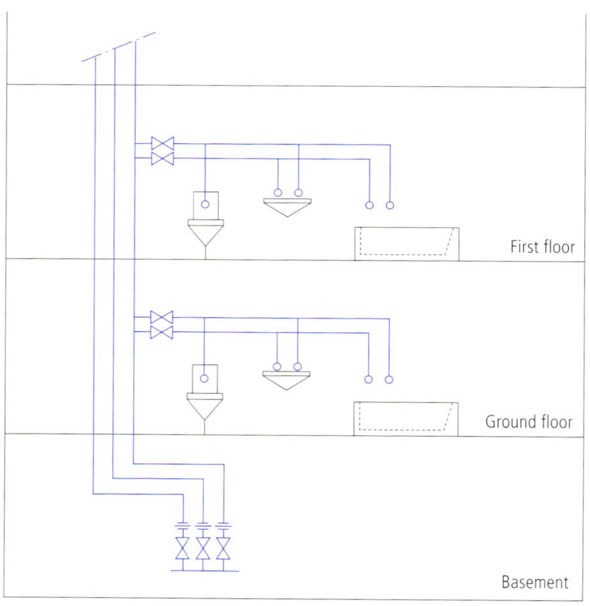

Distribution branches for apartments on separate floors, shut-off option at distribution battery and for each apartment

Figure 5.17
Example of a distribution system

5.1.4 Water processing

Water processed by waterworks contains salts, gases and traces of organic substances. Water hardness is formed by dissolved alkaline soils.

Total hardness is the term used to describe the sum of all calcium and magnesium bonds. Calcium and magnesium are primarily bonded with carbon dioxide but also with chloride, sulphate and nitrate. Expressed in milligram per litre of water, 10 mg CaO/l = 1 °dH (dH = *deutsche Härte* or German hardness) and 10 mg $CaCo_2$/l = 1 °fH (fH = French hardness). The carbonate hardness (temporary hardness) is the part of calcium and magnesium bonded to the carbon dioxide. When water is boiled, the carbonate hardness can be eliminated as white residue (calcium).

The noncarbonate hardness (permanent hardness) is the part of magnesium and calcium bonded to chloride, sulphate, nitrate and silicic acid. Parts remain in the water during boiling.

The salt content of water is the sum of all alkaline soils and salts divided into total cations and total anions. In addition to the alkaline salts, which are factors in cation and anion exchange (desalinization), water contains dissolved alkaline soils, which cause hardness.

The acid capacity is the amount of acid with a concentration of hydrogen ions of 0.1 mol/m^3, which must be added to 100 ml of a water sample to achieve a pH value of 4.3. The acid capacity is thus an indicator of the water's alkalinity caused by the Ca and Mg hydrogen carbonates.

The total hardness of water is only important in situations where water comes into contact with saponifiable material, or where it is steamed or evaporated near such materials. Carbonate hardness is a factor when water is warmed.

Chlorides, sulphates and nitrates have a corrosive effect; nitrates present a health hazard in concentrated form.

Water hardness can be roughly classified as follows:

Total hardness mol/m^3	°fH	°dH	Description
0 – 0.7	0 – 7	0 – 4.0	very soft
0.7 – 1.5	7 – 15	4.0 – 8.5	soft
1.5 – 2.2	15 – 22	8.5 – 12.5	medium soft
2.2 – 3.2	22 – 32	12.5 – 18.0	fairly hard
3.2 – 4.2	32 – 42	18.0 – 23.5	hard
above 4.2	above 42	23.5	very hard

The following processes are employed to treat water for a variety of purposes:

Dosage (inoculation, coagulation)

Dosage systems (inoculation systems) are principally employed to provide protection against corrosion in order to prevent scale formation due to cold and for sterilization. **Figure 5.18** illustrates such a system. Softening equipment should be integrated into the supply pipes to the water heater if the total hardness of the water exceeds 3 – 3.5 mol/m^3. Untreated water should be added to achieve a value of approximately 1.5 – 2 mol/m^3 (roughly 8.5 – 10 °dH), that is, the water is diluted until it reaches that value.

Softening through ion exchange

The Ca and Mg ions in the water are replaced with Na ions, a process that preserves the salinity but ensures that the salts are in solution. Softening by means of ion exchange is mainly used for commercial laundries, cooling towers supplied with water that is otherwise too hard, and for warm water supply. Regeneration is achieved through common salt (NaCl), whereby the wastewater is drained directly to the wastewater treatment plant without undergoing any preliminary treatment.

Partial desalinization through ion exchange

Desalinization through ion exchange is a process of de-alkalization with subsequent softening. The salt content of the water is reduced by the amount of carbonate hardness. This process is mainly used for industrial water, cooling towers, low-pressure steam boilers and air washers. Regeneration is achieved through H-exchangers by means of HCl (hydrochloric acid), whereby the wastewater streams must be neutralized (Na-exchanger through NaCl).

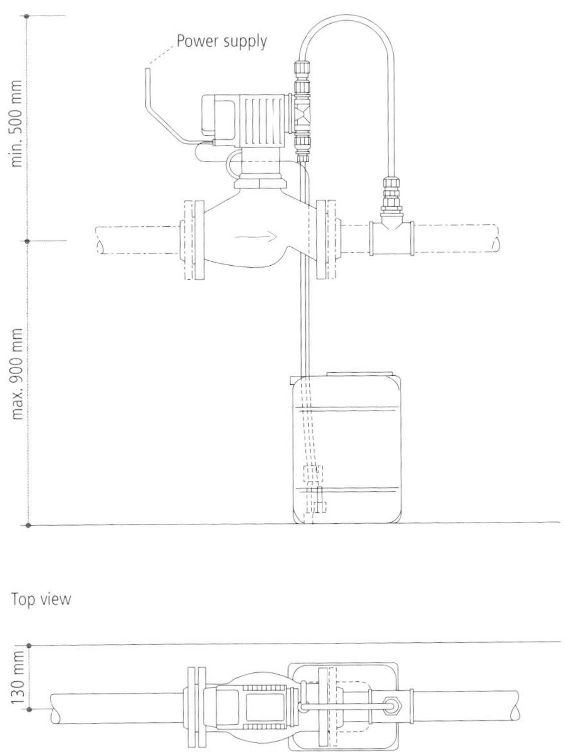

Figure 5.18
Inoculation system

Softening (cation exchange)

This is a softening process with filter-like masses of artificially manufactured Permutit, which attracts the hardness-producing agents. The mass requires regeneration from time to time, a process achieved by adding common salt. Following the regeneration, the hardness-producing agents are flushed out with water. The degree of softening can be selected. This softening process is used for dishwashers in commercial kitchens, humidifiers for air-conditioning systems, commercial laundries and similar operations. **Figure 5.19** illustrates the schematic layout of such a system.

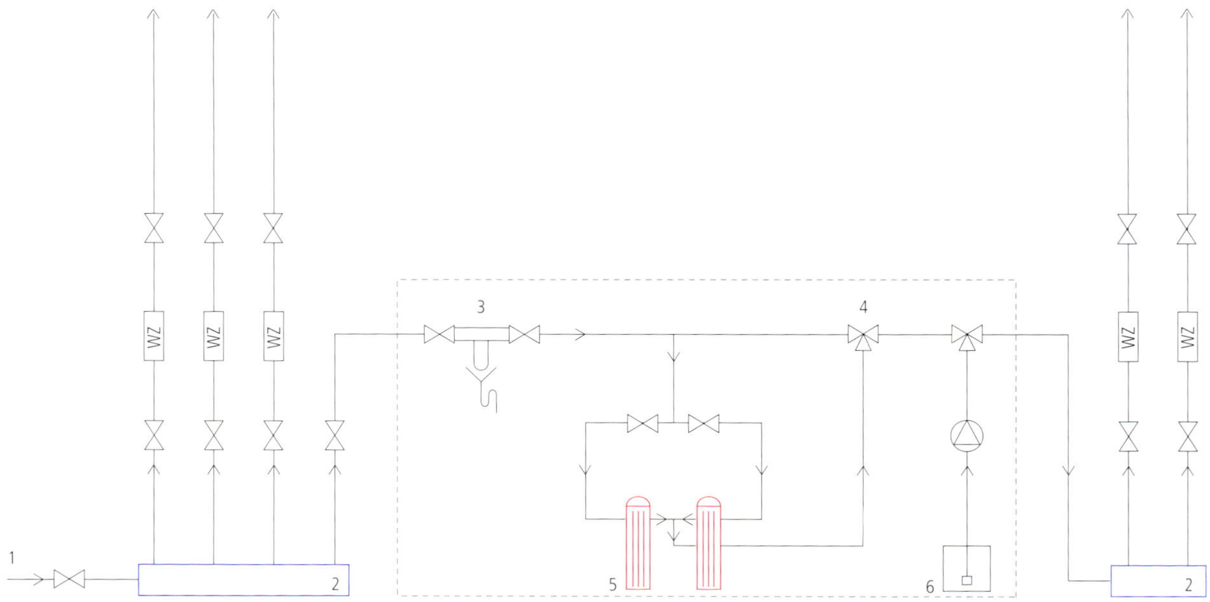

Figure 5.19
Softening system

1 House hook-up
2 Cold-water distributor
3 Protective filter
4 Mixing fitting
5 Softening station
6 Dosing station

Two-column total desalinization with subsequent mixed bed

Total desalinization produces virtually pure, residue-free water required for laboratories, high-pressure boilers and similar installations where residue-free water is an essential requirement. Two-column total desalinization is regenerated by means of a cation exchanger with common salt, anion exchanger with soda lye or buffer filter (mixed bed) with hydrochloric acid or soda lye. All wastewater must be neutralized prior to drainage.

Reverse osmosis

The quality of water treated with reverse osmosis lies between that produced by partial desalinization and total desalinization. In combination with a mixed bed, the result is pure water. The pretreatment (softening) must be carried out flawlessly because the very expensive permeators are otherwise damaged within a short period and rendered non-operational. The same is true for an active-carbon filter within the system when chlorinated water is used. Water treated with reverse osmosis is used in laboratories and for processing feed water for steam boilers. The concentrate of the dissolved salts can be drained directly into the sewage.

Application	Water treatment process	Types of water 1	2	3	4
Domestic and similar applications	Total hardness °fH	< 15	15 – 25	25 – 35	> 35
	°dH	< 8.4	8.4 – 14	14 – 19.5	> 19.5
	mol/m³	< 1.5	1.5 – 2.5	2.5 – 3.5	> 3.5
Cold water: Drinking/cooking	Fine filter	◑	◑	◑	◑
	Inhibitory processes:				
	Corrosion inhibitor process	◑	⌽	○	○
	Furring inhibitor process	○	○	⌽	◑
	Partial softening	○	○	○	⌽
Cleaning Toilet flushing		○	○	○	○
Warm water: Cooking/dishwashing/ personal hygiene	Inhibitory processes:				
	Corrosion inhibitor process	●	◑	⌽	⌽
	Furring inhibitor process	○	⌽	◑	●
	Partial softening to 12 –18 °fH 1.2 – 1.8 mmol/l	○	○	⌽	◑
Water heater Warm water tank	Cathode protection Electrolyte protection	●	◑	⌽	○
Commercial applications					
Laundry (textiles)	Partial softening to 3 ° max. 5 °fH 0.3 – 0.5 mmol/l	○	⌽	●	●
Commercial kitchens	Partial softening to 10 ° – 15 °fH 1 – 1.5 mmol/l	○	○	◑	●
Dishwashing	Softening 0 °fH 0 mmol/l dealcalization	○	◑	●	●
Coffee makers	Partial softening to 8 – 10 °fH 0.8 – 1.0 mmol/l	○	○	◑	●
Pools see SIA 385	Partial softening	○	○	⌽	◑
	Filtering/disinfection	●	●	●	●

Table 5.4
Water treatment for sanitary installations, non-binding approximate values

○ No water treatment
⌽ Optional water treatment
◑ Water treatment recommended
● Water treatment required Application to be determined

Degasification

Water is degasified to remove oxygen and carbon dioxide. In non-thermal processes (hydrazine/diamide) only oxygen is extracted, in thermal gasification processes both oxygen and carbon dioxide are extracted. Degasification is employed chiefly in the treatment of water for boilers and recooling plants.

Table 5.4 lists the types of water treatment that may be required for sanitation fixtures. The degree of hardness is indicated in °fH and in mol/m³ (fH = French hardness/dH = German hardness).

Central disinfection

Disinfection techniques are widely used today in hospitals, pools, dry cleaners and commercial laundries, in the food and beverage industries and in related fields. Their principal functions are to maintain human health, to provide hygienic conditions in food preparation and to extend the shelf life of foodstuffs. The principle is based on the automatic addition of disinfectants into a separate pipe system for which the industry has developed a variety of admix devices. According to DIN 1988 and DVGW, Sheet W 503 (German standards) or Swiss guidelines, for example, only smaller systems of this type may be connected directly to the potable water piping. (For example: a differential-pressure controlled system with a capacity of 5 to 15 l supplying a maximum of 5 to 20 outlets). Disinfection systems are connected directly to the drinking water system via a safety loop with pipe aerator and backflow stopper or a pipe separator.

Cabinet dosage stations are available for applications where several rooms must be supplied; they can be delivered with the mandated system separation, that is, water flows first into an open container, whence it is pumped into the disinfection loop, thereby ensuring that there is no direct contact with the drinking water system.

Large systems of this kind are generally installed in building complexes with their own dedicated system separation. In these cases, a dosage station without system separation is used, as shown in **Figure 5.20**.

Galvanized steel or PVC is used for the piping. Connections at outlet sites are provided in the form of hose nozzles (descriptive labels required) or fixed spraying nozzles with pressure valves.

Spray lances with hoses can be hooked up to the hose nozzles for the purpose of disinfecting rooms or objects or to draw solutions off in buckets. The fixed nozzles are used, for example, in pools, public showers etc. to guard against athlete's foot, and also for shoe disinfection in food preparation plants.

The following should always be connected to disinfecting systems: operating theatres, including scrub rooms, dissection rooms, infectious disease wards, human waste disposal rooms and the scouring and flushing fixtures, bathing facilities, WCs, bedding closets, clean-up and laundries in hospitals. Kitchens and ward kitchens are equipped with separate disinfecting systems operating with odourless and non-toxic disinfectants.

Spray installations are required in pools and public baths to prevent athlete's foot and to disinfect the rooms and objects.

In other areas such as industry, food industry and abattoirs etc., the use and design of disinfecting systems must be decided for each case. However, all personnel and common areas in such plants require disinfecting stations as well as nozzles for room and object disinfection.

Ion content / pH-value
Chemically pure water is by no means neutral, because a fraction of water molecules (H_2O) is always split into electrically charged H^+ and OH^- ions.

When H^+ ions are more prevalent, water has the effect of a mild acid. If, on the other hand, OH^- ions are more prevalent, water produces an alkaline reaction; that is, it has the effect of a mild lye solution. The hydrogen ion concentration is expressed as the pH-value, where:

– chemically pure water	pH = 7
– acidic water	pH < 7
– alkaline water	pH > 7

Corrosive action on iron metals only ceases to exist at a pH-value of roughly 9.6, measured at 23 °C.

Figure 5.20
Central disinfection plant

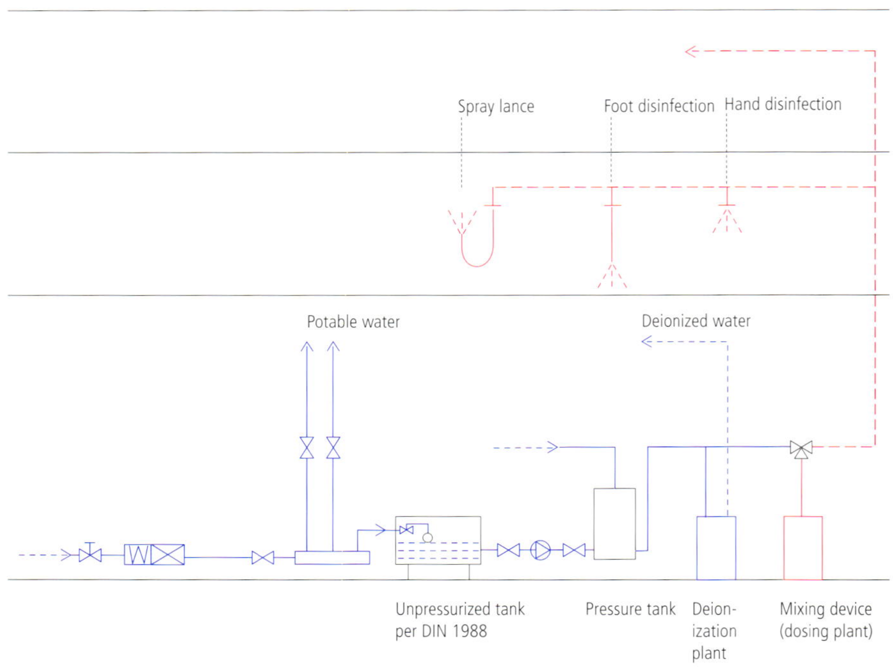

Table 5.5
Water treatment processes

Application	Purpose	Available treatment options		
Cold water General purposes	Prevention of mechanical impurities	1.1	1.3	
	Corrosion protection Scale protection	2.1	2.4	
	Potable water quality	5.1	5.2	
Manufacturing	Specific water quality to match manufacturing	3.1	3.9	
Dilution of product rinse	process	4.1	4.2	
Warm water	Scale and corrosion protection	2.1	2.4	3.7
Heating systems 4 system types	Scale and corrosion protection	2.2 3.8 4.2	2.4 3.9	3.4 4.1
Steam boiler 4 system types	Scale and corrosion protection	3.7 4.1	3.8 4.2	3.9
Air-conditioning systems	Scale protection Bonding of salt deposits in air-conditioned rooms	2.1 4.1	3.8 4.2	3.9
Cooling towers	ditto plus prevention of micro-organism growth	2.5 4.2	3.8	3.9
Commercial laundries	Soft water	3.7		

1. Filtering
Removal of separable and filterable matter
1.1 Sedimentation process
1.2 Upflow sludge blanket process
1.3 Silica sand and gravel filter
1.4 Diatomite filter etc.

2. Inoculation / Dosing
Admixing of conditioning chemicals as protection against crystalline scale and corrosion
2.1 Phosphate to stabilize the carbonate hardness
2.2 Hydrazine (diamide) for oxygen absorption
2.3 Sodium hydroxide (caustic soda) for alkalization
2.4 Sodium silicate as corrosion protection
2.5 Anti-algae additives etc.

3. Removal of dissolved matter
3.1 Neutralization (deacidification)
3.2 Iron extraction
3.3 Demanganization
3.4 Partial softening, dealkalization
3.5 Precipitate softening
3.6 Ion exchange
3.7 Softening cation exchanger
3.8 Dealkalization
3.9 Deionization

4. Degasification
4.1 Removal of oxygen
4.2 Removal of carbon dioxide

5. Disinfection
5.1 Chlorination processes (various)
5.2 Oxygenation
5.3 UV radiation
5.4 Oligodynamic processes (catadyne process) etc.

6. Wastewater treatment
6.1 Detoxification
6.2 Neutralization
6.3 Mechanical clarification
6.4 Biological clarification
6.5 Treatment of radioactive wastewater

The pH-value plays an important role in all situations where the acidic, alkaline or neutral properties of liquids must be supervised and controlled. This is the case for concentrates and solutions in the chemical industry for bleaching, dyeing and washing baths, in electroplating baths, film baths, sugar solutions, brew irrigation, mashing, seasoning, in condensates, boiler feed water, wastewater, potable water, in medical and pharmaceutical applications etc.

The pH-value (hydrogen ion concentration) is measured by:
- Measuring electrodes for continuous measuring
- Metal electrodes (antimony, bismuth, tungsten)
- Reference electrodes
- Calomel electrodes

A variety of water treatment processes are available for different applications (**Table 5.5**).

Consumers	Demand	Temperature
Hospitals	100 – 300 l/day, bed	60 °C
Barracks	30 – 50 l/day, person	45 °C
Office buildings	10 – 40 l/day, person	45 °C
Therapeutic baths	200 – 400 l/day, patient	45 °C
Department stores	10 – 40 l/day, employee	45 °C
Schools (250 days/yr)		
without shower facilities	5 – 15 l/day, student	45 °C
with shower facilities	30 – 50 l/day, student	45 °C
Sports facilities with showers	50 – 70 l/day, athlete	45 °C
Bakeries	105 – 150 l/day, employee	45 °C
	10 – 15 l/day, for cleaning	45 °C
for baking production	40 – 50 l/100 kg flour	70 °C
Hair salons (incl. customers)	150 – 200 l/day, employee	45 °C
Breweries incl. production	250 – 300 l/100 l beer	60 °C
Laundries	250 – 300 l/100 kg laundry	75 °C
Dairies	1 – 1.5 l/l milk	75 °C
	av. 4,000 – 5,000 l/day	
Butchers		
without production	150 – 200 l/day, employee	45 °C
with production	400 – 500 l/day	

Table 5.6.1
Warm water demand and temperatures for commercial applications

Tap point	One-time drawing off Litres	Temperature t_w °C	Duration Minutes
Outlet valves:			
DN 10, half opened	5	40	1
fully opened	10	40	1
DN 15, half opened	10	40	1
fully opened	18	40	1
DN 20, half opened	25	40	1
fully opened	45	40	1
Scullery sinks:			
single	30	55	5
double	50	55	5
Sinks:			
Hand sink	5	35	1.5
Sink	10	35	2
Washbasin, single	15	40	3
Washbasin, double	25	40	3
Bathtubs:			
small (size 100)	100	40	15
medium (size 160)	150	40	15
large (size 180)	250	40	20
Shower	50	40	6
Sitzbath	50	40	4
Bidet	25	40	8
Total consumption (60 °C)			
modest requirements	10 – 20 l/day and person		
higher requirements	20 – 40 l/day and person		
top requirements	40 – 80 l/day and person		

Table 5.6.2
Warm water demand in apartments

5.2 Warm water supply

5.2.1 Warm water demand

Warm water demands for a wide variety of uses are subject to considerable fluctuations. In housing, demands are linked not only to the size of the house or apartment and the number of residents but also to lifestyle, age, installation of water meters, profession, season and other circumstances. In hotels, warm water consumption depends on the number of tubs and showers and on the hotel rating. Luxury hotels consume considerably more warm water than economy rated hotels, with morning and evening peak times. Warm water consumption follows similar patterns in factories, sportshalls etc., where considerable volumes of water are required during brief peak periods. **Tables 5.6.1/5.6.2** contain details on warm water demands and temperatures for buildings and apartments.

The data provided in these tables are complemented by comparative empirical values for restaurants and hotels, pools and washing and shower facilities in industrial operations etc.

5.2.2 Supply categories

Supply is generally divided into:
- Individual supply
- Group supply
- Central supply

Figure 5.21 contains a schematic illustration of a central warm water supply system.
The principal components of the warm water supply installation are:
- Primary energy supply
- Service water heating
- Warm water distribution
- Warm water delivery/output

Figure 5.22 illustrates the layout of a warm water supply system with indirect heating.

Primary energy supply
There are various standard and alternative types of energy for warm water generation.

Standard primary energy sources
The following sources are available for warm water generation:
- Heating oil
- Coal
- Electric power
- Gas (city gas, natural gas and liquid gas)

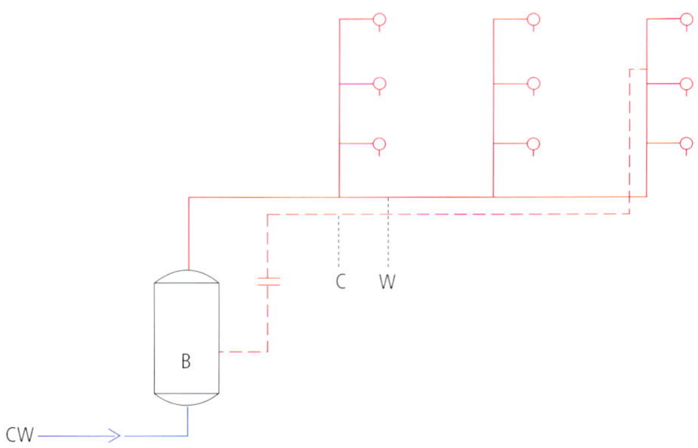

Figure 5.21
Central warm water supply with direct-fired boiler

B Boiler
CW Cold water
W Warm water
C Circulation

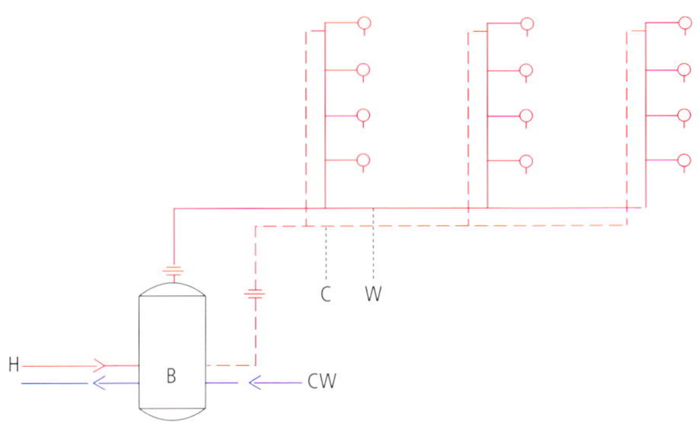

Figure 5.22
Central warm water supply with indirect heating

H Heat supply
B Boiler
CW Cold water
W Warm water
C Circulation

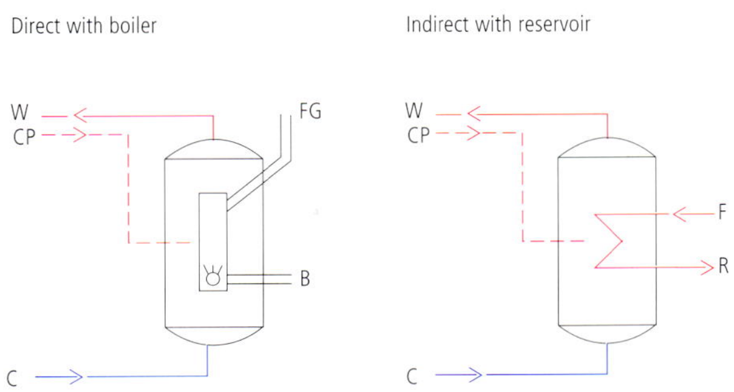

Figure 5.23
Warm water production

C Cold water pipe
R Return
F Feed
W Warm water pipe
CP Circulation pipe
FG Flue gases
B Burner

Alternative energy sources

The following energy sources are used in conjunction with new technologies:

- Water
  River, lake, ground and wastewater in combination with heat pumps
- Sun
  Solar collectors, energy roof, facade and fence, in part combined with heat pumps
- Earth/soil
  Earth collectors with heat pumps
- Air
  Ambient air, exhaust air in combination with heat pumps

Separating warm water generation from the building heating system is often a good solution for economic reasons; this means that a dedicated warm water generation system is created for summer operation, preparing the required warm water. It is important to prevent the occurrence of stagnant water areas when switching operations from one water heater to another since this would severely compromise the water quality in terms of hygiene and bacteria. Stagnation would also alter the water composition due to oxygen depletion and lead to corrosion in the piping.

5.2.3 Warm service water generation

With regard to operation, a differentiation is made between boiler and storage tank.

Water is heated directly in the boiler, in contrast to the storage tank, where water heated elsewhere is merely stored (**Figure series 5.23**).

The principal criteria for designing installations to generate service water are:

- Hygiene and mechanical stability of materials and coatings
- Temperature stability
- Corrosion resistance
- Pressure ratios
- Test and operating pressure (1.5 times grid pressure = at least 12 bar)
- Thermal output
- Temperature control
- Hydraulic and thermal safety
- Configuration

Storage tank design

Storage tanks for warm water storage and tanks for heating cold water should be designed as upright components with the smallest possible diameter and a diameter to height ratio of 1:3. This ratio has proven advantageous since the small diameter minimizes tensile stresses produced by the internal pressure, which means that the plate thickness of the tank walls can be reduced (keeping manufacturing costs low). This design also minimizes heat losses (storage tank surface in relation to volume).

1 Handhole cover with magnesium anode
2 Warm water
3 Insulation
4 Circulation
5 Heating feed
6 Heating spiral
7 Heating return
8 Adapter
9 Cold water outflow

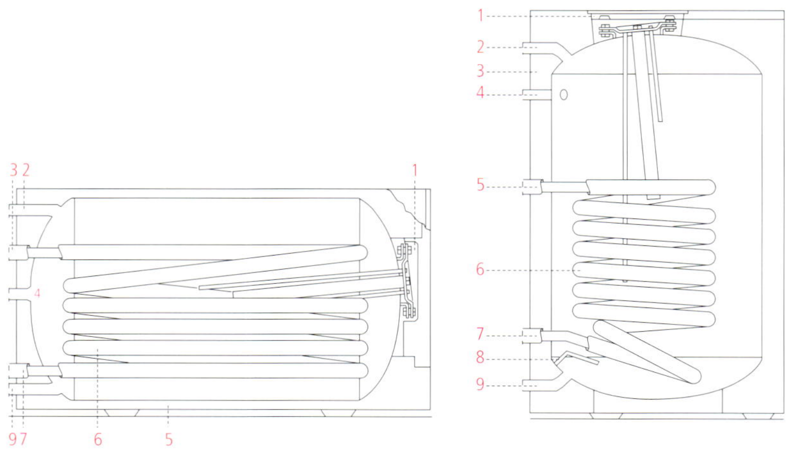

Figure 5.23.1
Construction of horizontal and vertical warm water storage systems
EAS T deep storage system
EAS 120 – 250 vertical storage tank
(Drawings: Brötje)

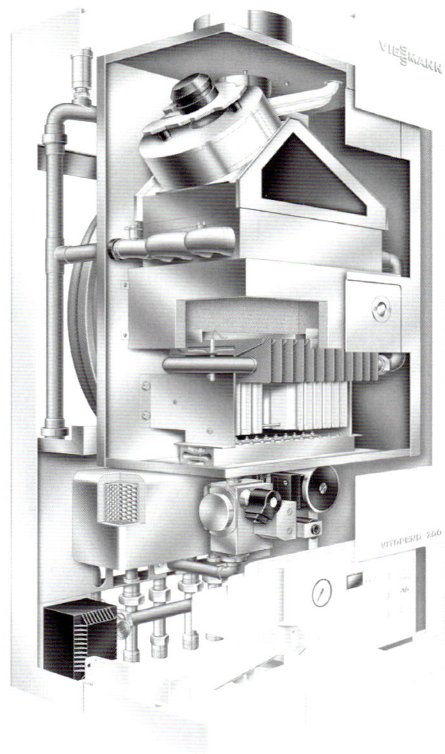

Figure 5.23.2
Vitopend 200
Gas-circulation water heater and gas-combination water heater, 10.5 – 24 kW
(Image: Viessmann)

Figures 5.24 and 5.24.1 illustrate the principal system components of a service water heating system with storage tank based on the assumption that heat is supplied from a boiler installation.

Storage tanks for service water heating can be installed in series, which means that temperatures increase from tank to tank (e.g. 10/40 °C and 40/60 °C). They can also be set up in parallel configuration for increased operational safety.

When no storage tanks are used, the service water is heated directly via the furnace of a heat generating plant (regardless of the type of firing) with an added boiler inside the furnace. However, this form of service water heating is linked to a considerable disadvantage: operation must continue even in summer (when there is zero demand for heating) or, for example, for industrial processing; operation during this time is still exclusively dedicated to generating warm water and compromised by high operational losses.

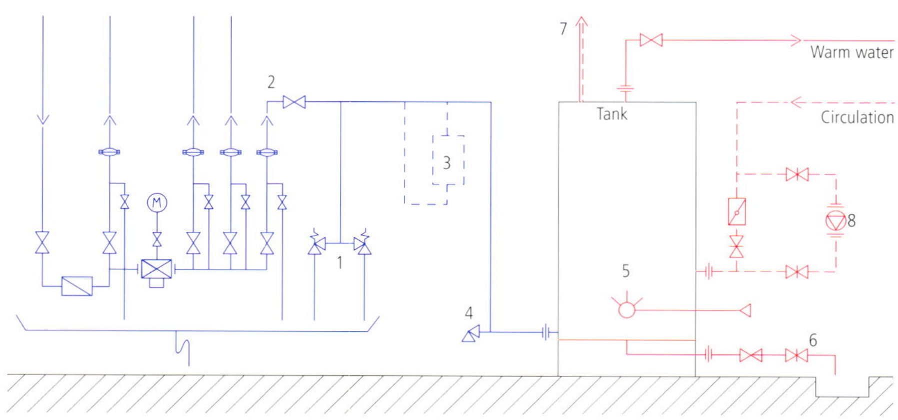

Figure 5.24
CW and WW installations, central with WW tank

1 Safety valve
2 Back pressure valve
3 Pressure chamber (feed content must be 4% of tank)
4 Outflow
5 Heating source
6 Emptying with valve and quick opening valve
7 Aeration and dearation
8 Circulation pump with timer

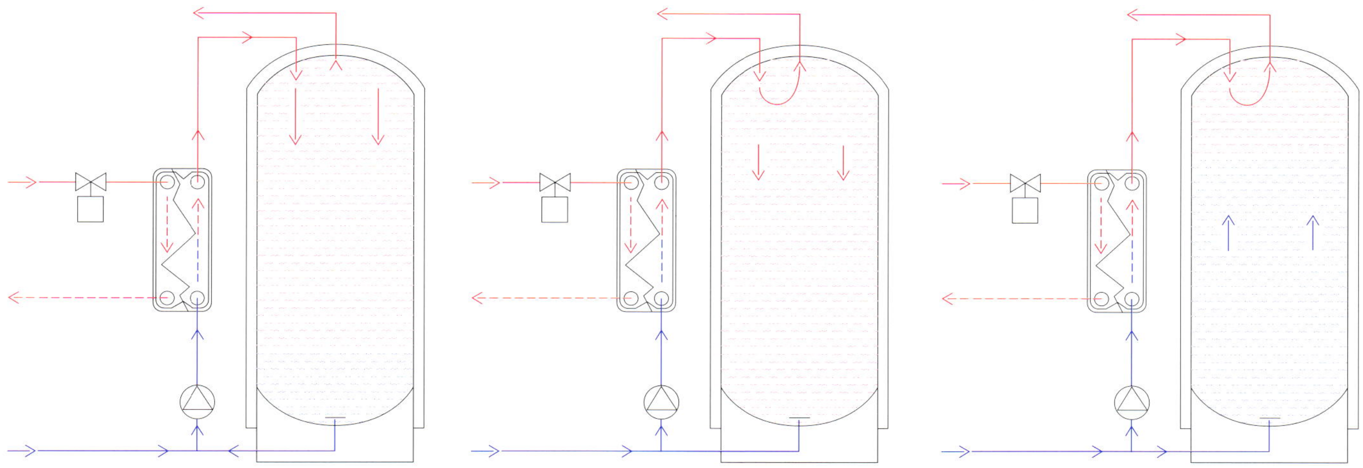

Figure 5.24.1
Feed and drawing-off process in Cetesystem D potable water storage loading system (Drawing: cetetherm)

Figure 5.25
Warm water distribution systems

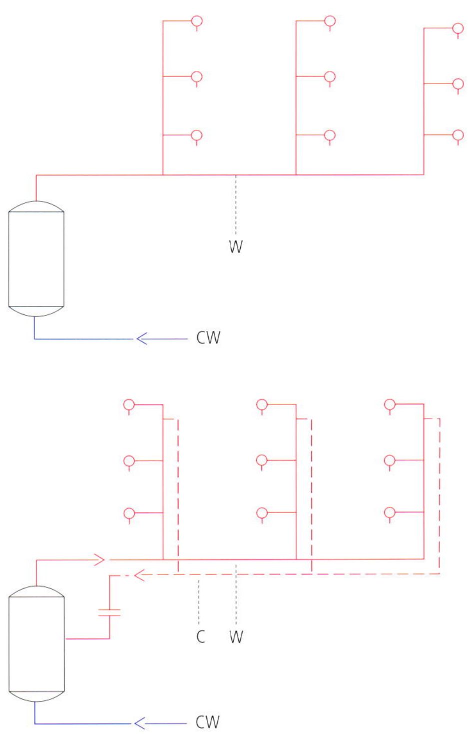

Distribution system without circulation

Distribution system with circulation (consumption is calculated on the basis of area or number of taps)

5.2.4 Warm service water distribution

Since warm service water is not tapped on a continuous basis, considerable stagnation losses are inevitable, and may represent 10 to 30 % of the energy used in the distribution systems. The following criteria must be taken into consideration to avoid these immense losses:

- Shortest possible distribution system
- Concentration of warm water tapping points
- Adjusted warm water temperatures (maximum 60 °C)
- Efficient warm water pipe insulation
- Interrupt circulation during long downtime periods (e.g. more than four hours)

Figure 5.25 illustrates different types of warm water distribution systems for various uses. Warm service water distribution systems without circulation are ideal in situations with short piping distances and where immediate flow at the tap is not required (e.g., bathroom sink 8 – 12 sec., kitchen sink 5 – 10 sec., shower 15 – 20 sec. and bathtub 15 – 20 sec.). Warm service water distribution systems with circulation are suitable when warm water piping lengths to the tapping points exceed 30 m and where immediate flow at the tap is required at all times. Thus the circulation piping system has the task of transporting warm service water to the tapping point and, conversely, providing the return link to the heat generating plant whenever warm service water is not drawn off.

In circulation systems (large piping system extension) maximum temperature drops of up to 5 K (between feed and return) are permissible. This can be achieved by augmenting the rate of circulation and providing appropriate insulation.

Pump-operated circulation piping should be no less than 15 mm (1/2 inch) clear width and laid so as to avoid "air pockets". Where air pockets are unavoidable, the pipes must either be vented at the tapping points or fitted with ventilation valves at the most elevated circulation points.

5.2.5 Warm service water delivery

Warm service water is delivered via standard fittings in the form of individual taps or combination sets. The following options are available for combination sets:
- Simple two-lever combination set with valves (cold/warm) and joint tap
- Mechanical mixer with single-lever or turning handle control
- Thermal mixer with thermostat-controlled, constant mix temperature and open or closed tap

5.2.6 Sizing

Sizing for warm water piping is determined in the same manner as for cold water piping since the pipe materials are nearly identical, with similar flow velocities and water output parameters (l/s).

5.2.7 Insulation for warm service water systems

Insulation guidelines for warm water piping for warm service water and circulation piping address the same conditions explained in the previous chapter with regard to heating systems. Efficient insulation parameters can therefore be based on the corresponding diagrams and data. In addition the thickness of insulation for warm water storage tanks should be approximately as follows:

Small tanks up to 100 l capacity = approx. 60/80 mm insulation

Tanks for 250 l, 1000 l = approx. 60/100 mm insulation

Tanks for 1000 l, 3000 l = approx. 100/150 mm insulation

Tanks with more than 3000 l capacity = at least 150 mm insulation

Local energy regulations may specify thicker layers of insulation as minimum requirements for both piping and storage tanks. These regulations must be adhered to in all cases.

For savings in both energy consumption and initial investment costs, circulation piping should be installed simultaneously with warm service water pipes (pipe-to-pipe system) as shown in **Figure 5.26**. In well executed pipe-to-pipe systems a thermostat-controlled circulation pump is only required during brief periods and the circulation pipe can be smaller than for individual piping systems.

As an alternative to the circulation system, a temperature-dependent, self-regulating, electrical heating strip can be installed on top of the warm water distribution piping. This strip constantly compensates for the heat losses incurred in the distribution piping. There is no definitive preference for either of these systems and the selection should be made on a case-specific basis.

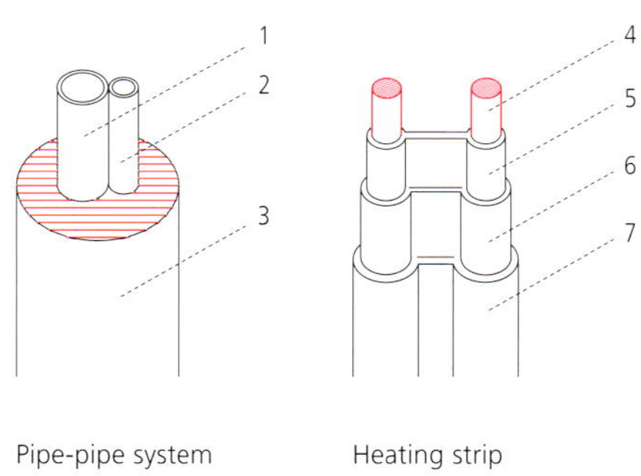

1 Warm water distribution
2 Circulation return
3 Insulation
4 Live strand
5 Self-regulating and semi-conducting plastic strip
6 Electrical insulation
7 Protective sleeve

Figure 5.26
Steam trace for pipelines in warm service water systems

5.3
Gas supply

With more readily accessible natural gas supplies, gas heating has assumed a more important role. Gas heating units are used to heat homes, offices, retail stores and schools etc. The advantages of gas space heating are especially appreciated for renovation projects.

Gas heating has the following specific advantages:
- Simple controls
- Rapid warm-up period
- Clean operation
- Constant readiness
- No consultation required
- Heating costs are easily established
- Low air pollution

Gas can be used in combination with a wide variety of standard appliances such as refrigerators, boilers, dish-washers/washing machines, stoves and instantaneous water heaters.

The example shown in **Figure 5.27** indicates various gas consumption points in a building with several itemized definitions. Gas must always be handled with the greatest care since unprofessional use or inexpert tampering with gas installations has caused severe accidents. All gas installations must be absolutely gas-tight, resistant and durable. Gas-operated fixtures and appliances must be placed to avoid any risk of fire and explosion and the total combustion of gas must be ensured at all times. The required volume of combustion air must flow unimpeded and combustion gas must be ventilated safely.

The entire installation must be executed with materials that have been tested and certified. Binding standards apply. New installations should be designed to accommodate the future use of other types of gas. The local gas utility must be notified of each new installation, expansion or modification of existing gas-supplied installations.

Gas piping is laid in ground using black or galvanized, seamless or welded steel or ductile cast iron pipes. Plastic pipes should be used only upon approval from the gas utility and the relevant fire department, in conjunction with documented proof of the material's resistance to the gases, their conditioning substances and other additives. In buildings, gas lines are installed in black, seamless or welded steel (galvanized) pipes, or in seamless or welded copper, brass or aluminium pipes.

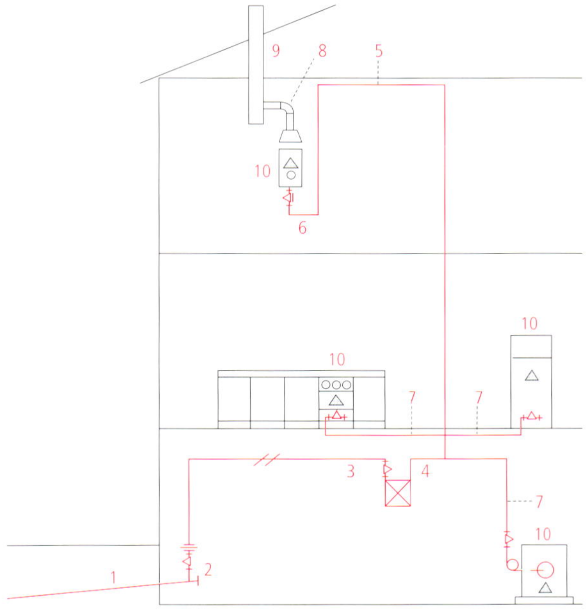

1–2 House feed
2–3 Internal pipe up to gas meter (riser)
4–5 Internal pipe after gas meter (outgoing line)
6 Appliance connection pipe
7 Branch pipe
8 Flue gas pipe
9 Flue gas chimney
10 Gas appliances

Figure 5.27
Diagram of gas supply in residential house

The use of lead or plastic pipes is prohibited in buildings. All connection hoses to gas appliances must conform to certified standards. Pipe connections must not weaken the gas pipes in any undue fashion. For this reason screw, clamped, welded, soldered and bolted or flanged joints, as well as screwed-sleeve joints, are used for all connections (bell-and-spigot or lead joints and glued joints only for in-ground connections). All piping must be protected against corrosion; in-ground piping must be fully embedded in sand and protected against damage. If there is a risk of corrosion for pipes installed inside buildings, additional protective measures must be employed (protective sleeves, hot-galvanizing, coatings and insulation) or the pipes themselves must consist of corrosion-proof materials (copper, brass, aluminium or rust-proof steel). When pipes are integrated into concrete or brick masonry, it is vital to ensure that cement additives will not cause corrosion.

The sizing of the pipes is determined using either tables or computations. The following pressure losses (decrease of pressure in gas flow due to friction with internal pipe wall, changes in direction and cross section) can be assumed for standard scenarios:

- Feed A-B 5 mbar
- Internal piping up to meter B-C 5 mbar
- Gas meter C-D 5 mbar
- Internal piping after gas meter D-E 5 mbar
- Total pressure loss: 20 mbar

The flow pressure in piping leading to gas appliances (pressure of flowing gas) should be as follows:

- City and grid gas 60 mbar
- Natural gas 180 mbar
- Propane-air mixture 60 mbar

Depending on the specific gas type, the connection load (load unit) and pipe material, gas pipes can be sized in a simplified manner based on the values given in **Tables 5.7/5.8**.

Appliance	**City gas**	**Natural gas** Propane-air Ho,n = 56.5 MJ/m³	Propane-air Ho,n = 27.2 MJ/m³
	Supply value m³/h i.e. load units	Supply value m³/h i.e. load units	Supply value m³/h i.e. load units
Refrigerators *	0.04	0.02	0.03
Boilers and other appliances	0.5	0.25	0.3
Small washing machine	1	0.5	0.6
Domestic dishwasher	2	1	1.2
Domestic dryer	2	1	1.2
Domestic stove	2.5	1.25	1.5
Instantaneous heater:			
0.5 MJ/min; 8.7 kW	2.5	1.25	1.5
1.4 MJ/min; 22.6 kW	7	3.5	4.3
1.5 MJ/min; 24.4 kW	7.5	3.75	4.6
1.7 MJ/min; 27.9 kW	8.5	4.25	5.2
Other appliances corresponding to supply value			

* If another gas appliance is present in the same room, refrigerators do not need to be considered in sizing the gas supply

Table 5.7
Supply values for the most common appliances

City and district gas

Loading case					**1**	**2**	**3**	**4**	**5**	**6**	**7**
Max. unfurled pipe length in m					5	10	15	20	30	40	50
Steel pipes			Copper pipes		Max. permissible load units						
Medium threaded pipes VSM 11520		Thin-walled steel pipes	VSM designation mm	Clear width mm							
Nominal width inch	Clear width mm	Clear width mm									
			7 x 1	5	0.2						
			8 x 1	6	0.25						
			10 x 1	8	0.4	0.1	0.1				
3/8 "	12	12	*12 x 1	10	1	0.5	0.5	0.3	0.3	0.2	0.2
1/2 "	16	16	*14 x 1	12	2.5	2	1.5	1	0.8	0.6	0.5
3/4 "	21	21	*17 x 1	15	6	5	4	3.5	2.5	2	1.5
1 "	27	27			15	12	10	9	7	6	5
1 1/4 "	36	36			30	25	20	18	15	12	11
1 1/2 "	42	42			60	50	40	36	30	26	24
2 "	53	53			120	100	80	75	60	52	48
2 1/2 "	69	69			260	220	180	170	140	120	110
3 "	81	81			440	360	300	280	240	220	200
4 "	105	105			1000	850	680	650	500	460	420

* For appliance connections

Table 5.8.1
Design table for pipes for house installations, city and district gas

(For pipes leading to the gas meter or pipes after the gas meter) Fittings and other individual resistance factors and the probable maximum load have been taken into consideration

Table 5.8.2
Design table for pipes for house installations, natural gas, propane-air

(For pipes leading to the gas meter or pipes after the gas meter) Fittings and other individual resistance factors and the probable maximum load have been taken into consideration

Natural gas Propane-air Ho, n = 56.5 MJ/m³

Loading case					1	2	3	4	5	6	7
Max. unfurled pipe length in m					5	10	15	20	30	40	50
Steel pipes			Copper pipes		Max. permissible load units						
Medium threaded pipes VSM 11520		Thin-walled steel pipes	VSM designation mm	Clear width mm							
Nominal width inch	Clear width mm	Clear width mm									
			7 x 1	5	0.15						
			8 x 1	6	0.2						
			10 x 1	8	0.35	0.08	0.08				
3/8 "	12	12	*12 x 1	10	1.0	0.5	0.4	0.3	0.2		
1/2 "	16	16	*14 x 1	12	2.3	1.5	1.0	0.8	0.5	0.4	0.4
3/4 "	21	21	*17 x 1	15	6.0	4.5	3.5	3.0	2.2	1.8	1.5
1 "	27	27			14	10	8.0	7.0	6.0	4.5	3.5
1 1/4 "	36	36			38	28	22	20	18	15	14
1 1/2 "	42	42			58	45	35	30	28	24	22
2 "	53	53			110	90	70	65	60	50	45
2 1/2 "	69	69			250	200	160	150	130	120	100
3 "	81	81			400	340	260	250	200	180	160
4 "	105	105			850	750	600	550	450	380	360

* For appliance connections

Table 5.8.3
Design table for pipes for house installations, propane-air

(For pipes leading to the gas meter or pipes after the gas meter) Fittings and other individual resistance factors and the probable maximum load have been taken into consideration

Propane-air Ho, n = 27.2 MJ/m³

Loading case					1	2	3	4	5	6	7
Max. unfurled pipe length in m					5	10	15	20	30	40	50
Steel pipes			Copper pipes		Max. permissible load units						
Medium threaded pipes VSM 11520		Thin-walled steel pipes	VSM designation mm	Clear width mm							
Nominal width inch	Clear width mm	Clear width mm									
			7 x 1	5	0.12						
			8 x 1	6	0.16						
			10 x 1	8	0.26	0.06	0.06				
3/8 "	12	12	*12 x 1	10	0.65	0.32	0.32	0.2	0.2	0.13	0.13
1/2 "	16	16	*14 x 1	12	1.6	1.3	1.0	0.65	0.5	0.38	0.32
3/4 "	21	21	*17 x 1	15	3.8	3.2	2.6	2.3	1.6	1.3	1.0
1 "	27	27			10	8.0	6.5	5.8	4.5	3.8	3.2
1 1/4 "	36	36			20	16	13	11.6	10	8.0	7.0
1 1/2 "	42	42			40	32	26	23	20	17	15.5
2 "	53	53			80	65	52	49	40	34	31
2 1/2 "	69	69			170	140	116	110	90	80	70
3 "	81	81			280	230	200	180	155	140	130
4 "	105	105			650	550	440	420	320	300	270

* For appliance connections

Figures 5.28 – 5.30 give values for more precise sizing. Figure 5.31 provides a simplified example for executing a gas installation based on natural gas and the corresponding, simplified sizing for the pipes.

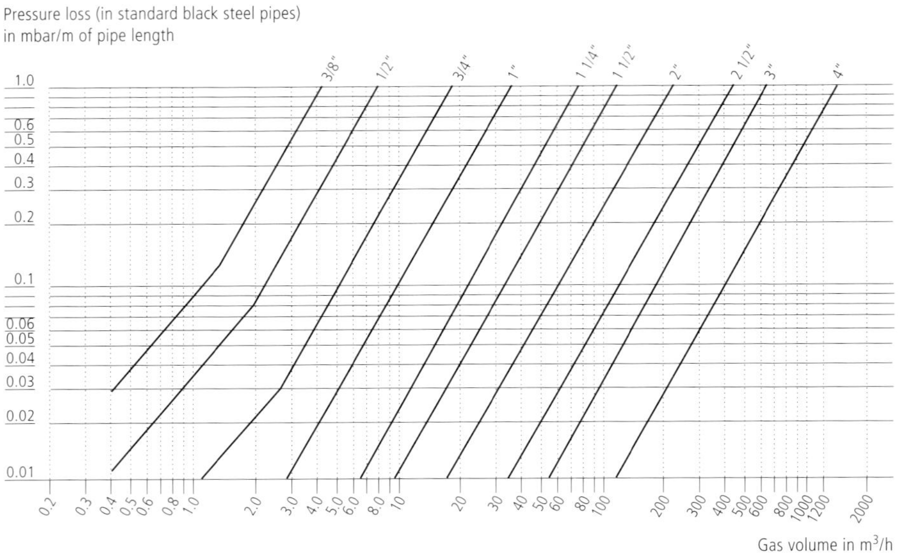

Figure 5.28
Pressure loss, city gas-district gas

Density: 0.5 (air = 1)

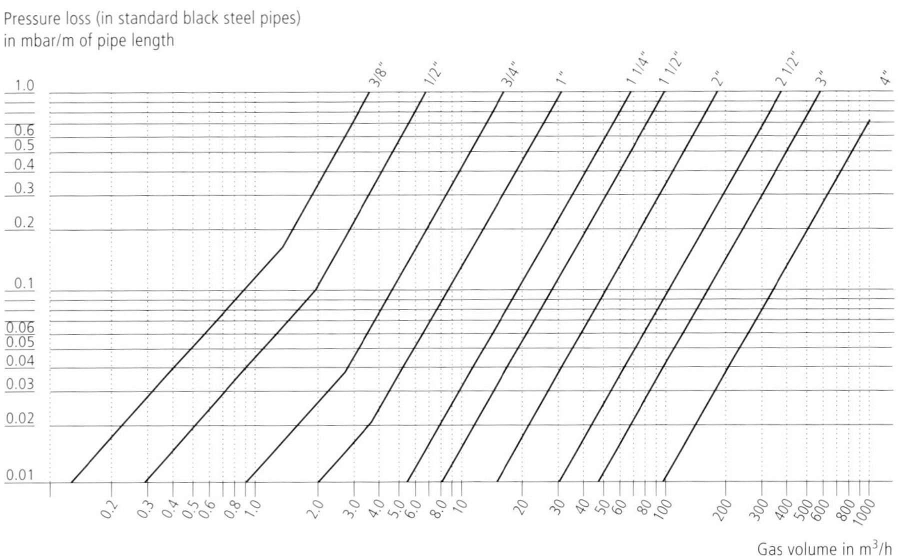

Figure 5.29
Pressure loss, natural gas, propane-air

$H_{o,n}$ = 56.53 MJ/m³ = 15.7 kWh/m³

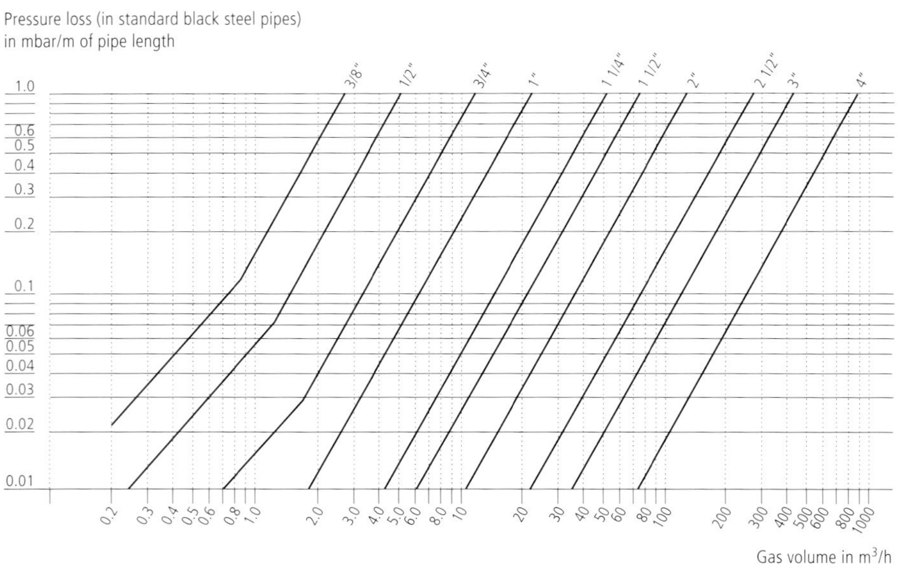

Figure 5.30
Pressure loss, propane-air

$H_{o,n}$ = 27.21 MJ/m³ = 7.56 kWh/m³
Density: 1.14 (air = 1)

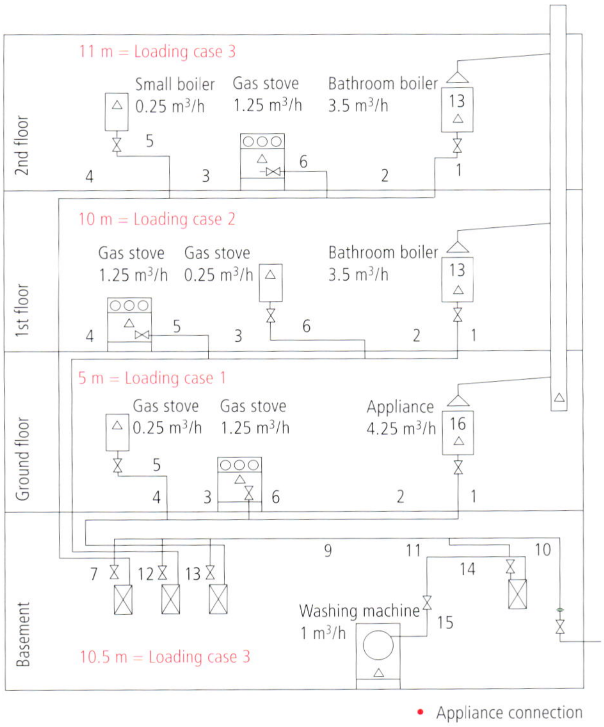

1	0.8 m	3.5 m³/h = 1/2 "	•
2	1.5 m	3.5 m³/h = 3/4 "	
3	1.2 m	4.75 m³/h = 1 "	
4	7.5 m	5.0 m³/h = 1 "	
5	1.2 m	0.25 m³/h = 3/8 "	•
6	2.0 m	1.25 m³/h = 1/2 "	•
1	0.7 m	3.5 m³/h = 1/2 "	•
2	2.3 m	3.5 m³/h = 3/4 "	
3	1.0 m	3.75 m³/h = 1 "	
4	6.0 m	5.0 m³/h = 1 "	
5	1.4 m	1.25 m³/h = 1/2 "	•
6	2.0 m	0..25 m³/h = 3/8 "	•
1	0.6 m	4.25 m³/h = 3/4 "	•
2	1.5 m	4.25 m³/h = 1 "	
3	0.5 m	5.5 m³/h = 1 "	
4	2.4 m	5.75 m³/h = 1 "	
5	1.8 m	0.25 m³/h = 3/8 "	•
6	0.4 m	1.25 m³/h = 1/2 "	•
7	1.0 m	5.0 m³/h = 1 "	
8	0.5 m	10.0 m³/h = 5/4 "	
9	6.0 m	15.75 m³/h = 5/4 "	
10	3.0 m	16.75 m³/h = 5/4 "	
11	1.0 m	1.0 m³/h = 1/2 "	
12	0.5 m	5.0 m³/h = 1 "	
13	0.5 m	5.75 m³/h = 1 "	
14	2.5 m	1.0 m³/h = 1/2 "	
15	1.5 m	1.0 m³/h = 12x1	•

Figure 5.31
Design example
Gas installations with natural gas, simplified sizing

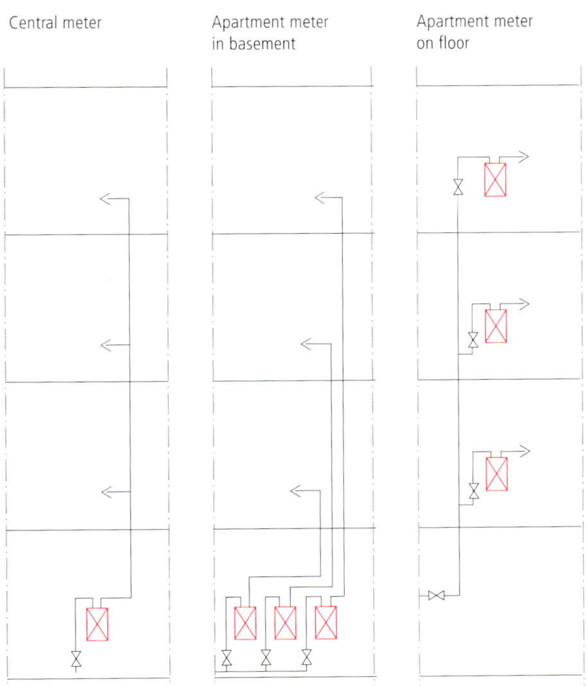

Installation:
a shut-off device must be installed in front of each gas meter

Figure 5.32.2
Meter locations, placement examples

5.3.1 Domestic delivery pipes

The domestic delivery pipe, the section of piping that leads from the public main to the first shut-off device in the building should be laid on a slope towards the public main so that it can be drained in that direction. If this is not feasible, a frost-proof condensate collector must be installed after the main shut-off device (**Figure 5.32.1**).

The main shut-off device of the gas pipe must be installed in a fully accessible location immediately after the entry point into the building with the delivery pipe being no less than 1 1/4 inch in diameter. Domestic delivery pipes must always be installed by the gas utility or authorized installation contractors. The gas meter is located right behind the main shut-off device (pipe no less than 3/4 inch).

The piping inside the building (internal piping) connects the gas meter to the consumption points. These should be steel pipes no less than 1/2 inch in diameter.

Gas utilities will install pressure control stations within the gas supply in the building only if large pressure differences are anticipated in the grid. Pressure control stations are fitted with additional safety features, which protect the connected gas supply appliances against excessive pressure in case of pressure controller failure.

The gas meter(s) must be installed in a dry, frost-proof room and protected from direct heat radiation, corroding agents and mechanical damage. The room must be well ventilated. These prerequisites preclude positioning the gas meter in a furnace room. **Figure 5.32.2** depicts examples of meter installation with the required shut-off devices.

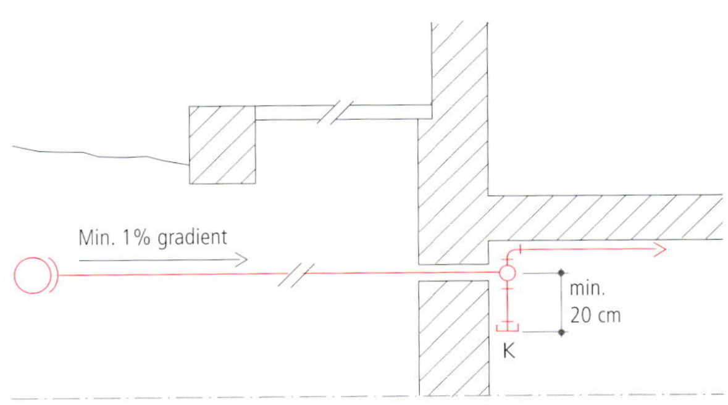

Figure 5.32.1
Gas supply pipes with drip pot

5.3.2 Gas burning installations / placement

Gas burning installations must be sited so as to ensure the unimpeded flow of the necessary combustion air into the room and the installation fixture. Unimpeded evacuation of waste or flue gases must also be ensured. Combustible materials must never be used in immediate proximity to gas-fired installations. Gas burning installations for propane-air mixture should not be placed in underfloor spaces unless they are fitted with additional safety features to prevent leakage of unfired gas and the relevant spaces are sufficiently ventilated.

Since the following volumes of air are required for the combustion of 1 m^3 gas
- 5 m^3 for city gas
- 11 m^3 for natural gas
- 8 m^3 for propane-air mix

the space may well be sufficiently ventilated if it is vented to the outside or to a sufficiently ventilated adjacent space.

Figure 5.33 indicates the fresh air openings in a bathroom with a gas-fired warm water boiler.

In addition to the gas-burning installations mentioned above, gas heat radiators are used in outdoor spaces and in large factory halls. As their name indicates, these devices emit most of their heat by radiation. Convection heaters and gas-fired air heaters (e.g. in fair halls), which transfer thermal energy to the space via air currents (mechanical ventilation), can serve as supplementary heating units.

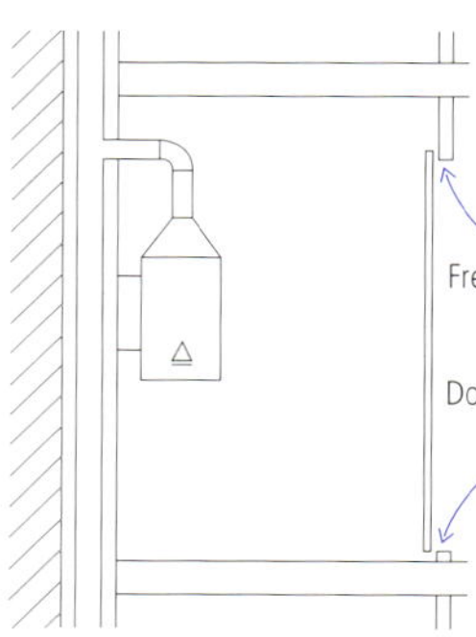

Figure 5.33
Placement of fresh air openings

5.4 Building drainage

Building drainage is always subject to binding regulations. The regulations apply to drainage installations in the building and to estate drains connected to the sewer. Sewage water from sanitary installations and rain runoff from roofs etc., is transported through the drainage system to the public sewage drain. The system is self-cleaning provided it is subjected to standard use and has been properly installed. The drainage system should protect against:
- Sewage gas leakage
- Water and gas pipe leaks
- Emptying of odour traps
- Piercing of odour traps through over pressure in the pipes
- Pipe blockage, especially against backup from the public sewer
- Use of sewage system under normal conditions in case of mechanical or chemical loads

Any direct linkage between potable water and sewage water pipes is prohibited on principle. Separate installations must be provided for the drainage of acid-laden, alkaline or toxic materials or liquids, as well as materials that spread harmful or bothersome evaporations or others that can damage drainage installations. These separate installations must be fitted with treatment systems to neutralize such wastewater (biological, mechanical systems in industrial operations etc.).

Rainwater and sewage must always be drained separately from buildings and may only be drained jointly outside of buildings if a mixing system exists in the planning area (mixing of rain and sewage water).

Sewage water requires pre-treatment only where the materials given in **Table 5.9** are drained.

5.4.1 Terms and definitions

Wastewater

Grey water
Rainwater/cooling wastewater/hill water or groundwater/purified wastewater/seepage

Stormwater
Rainwater runoff that flows into the seepage

Sewage water
Domestic black water/heavily contaminated rain runoff/pre-treated industrial wastewater

Industrial wastewater
Wastewater that has been specially treated prior to being diverted to the sewage system

Sewage water coefficient (SW)
Sewage water coefficient corresponds to a sewage water volume flow of 1 litre per second

Hydraulic load capacity
Allowable volume flow for a drainage system

Sanitation fixtures/appliances
Tub/shower/bathroom sink/bidet/WC/urinal/drain/sluice/kitchen sink/dishwasher/washing machine/laundry tub etc.

Table 5.9
Pretreatment of rain and sewage water

Substances	Countermeasures
Floating matter and/or heavy suspended matter in larger volumes	Collectors, sand catchers, filters or sieves, rakes
Oil and grease (possibly in emulsified form) in larger volumes	Oil and grease interceptors (if necessary with enhanced cleansing effect)
Toxins	Detoxification
Acids and alkaline matter Condensate from heating plants	Neutralization
Dangerous amounts of pathogens	Disinfection
Radioactivity	Deactivation

Drainage components

Special drainage appliances
Commercial kitchen and laundry appliances/laboratory hydrotherapy machines etc.

Subsurface water inlet
Installation with/without drain trap for gathering and draining subsurface water

Rainwater inlet
Installation with/without drain trap for gathering and draining rainwater

Overflow
Device to prevent overflow

Fixture trap
Device to prevent gases produced in sewage pipes/drains/individual purification systems from escaping

Sludge trap
Container with drain trap in which wastewater is collected and sand, gravel, sludge etc. are trapped and separated

Sand catcher (also: grit chamber)
Container without drain trap in which wastewater is collected and heavy materials such as sand, gravel etc. are separated

Grease interceptor
Installation with drain trap to separate mineral oil, petrol, grease and starch from wastewater

Wastewater conveying installation
Components such as pumps, vacuum-tank installation etc.

House trap
Trap to prevent wastewater from flowing back from the sewage system into the building

Drainage system

Building drainage
All drainage systems within a building

Estate drainage
Full drainage system within the boundaries of a property

Branch waste
Pipe leading from the last clean-out on the property to the public sewer

House drain
Wastewater pipe beneath basement floor or buried in ground

Branch soil
Horizontal, projected pipe for the collection of wastewater from down, branch and connecting pipes

Soil stack
Vertical pipe across one or several storeys, roof vented and conducting wastewater to subsurface or collecting piping

Branch soil
Joint pipe for several connecting pipes leading to the down, collecting or subsurface piping

Connecting pipe
Pipe connecting drainage object to subsurface or down piping (DIN 4045)

Pump pressure pipe
Pipe conducting wastewater to a collecting or subsurface pipe via a pump

Fall route
Vertical route section in a connecting or branch soil, max. length 0.2 m

Down route
Route section in a connecting or branch soil with more than 10 % or more than 5.8 ° gradient and more than 0.2 m height

Moving the centre line
Moving of the centre line with the help of a pipe fitting or two elbows up to 45° in the vertical section of a connecting, branch or soil stack

Loop
Horizontal link of two soil stack sections up to a max. length of 8 m

Ventilation pipe
Pipe dedicated exclusively to air circulation

Stack vent
Continuation of all soil stacks without change in cross-section from the uppermost intake across the roof

Secondary ventilation
Ventilation pipe parallel to soil stack

Collecting ventilation
Collecting pipe of two or several soil stacks, secondary pipe, branch soil or connecting stack vents

Branch vent
Ventilation of a branch soil

Connection ventilation
Ventilation of a connecting pipe

Pressure pipe ventilation
Ventilation at the highest point of the pump pressure pipe

Ventilation
Ventilation of a ground or collecting pipe above the roof or into a soil stack

Shaft ventilation
Ventilation of a shaft above the roof or into a soil stack

These terms are illustrated in part in **Figure 5.34**. **Figures 5.35/5.36** show examples of a sewage system (building drainage) for a fictitious residential building.

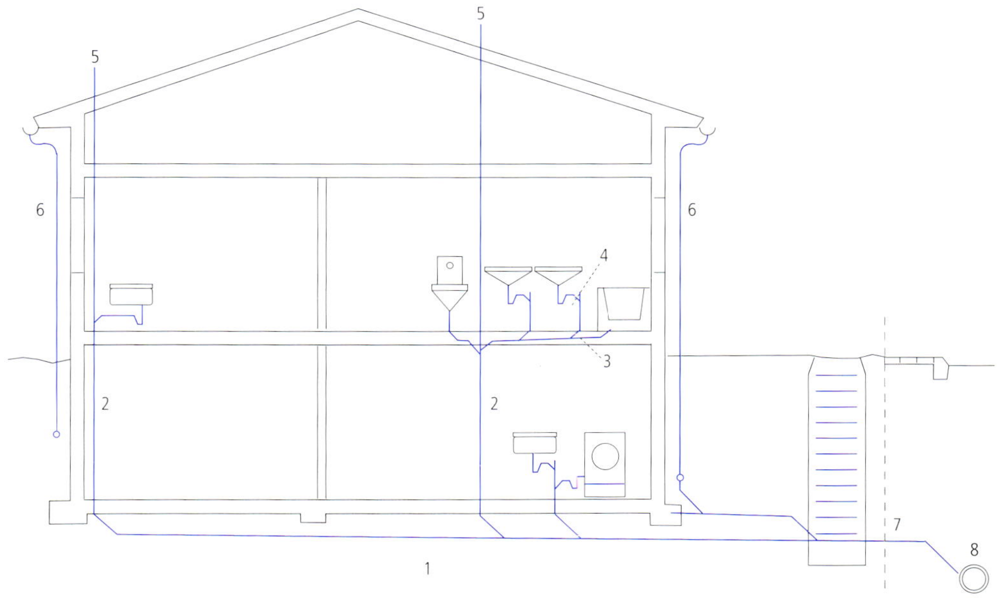

Figure 5.34
Individual components of a drainage system

1 House drain
2 Soil stack
3 Branch soil
4 Branch waste
5 Stack vent
6 Storm leader
7 Estate connecting pipe
8 Public sewer

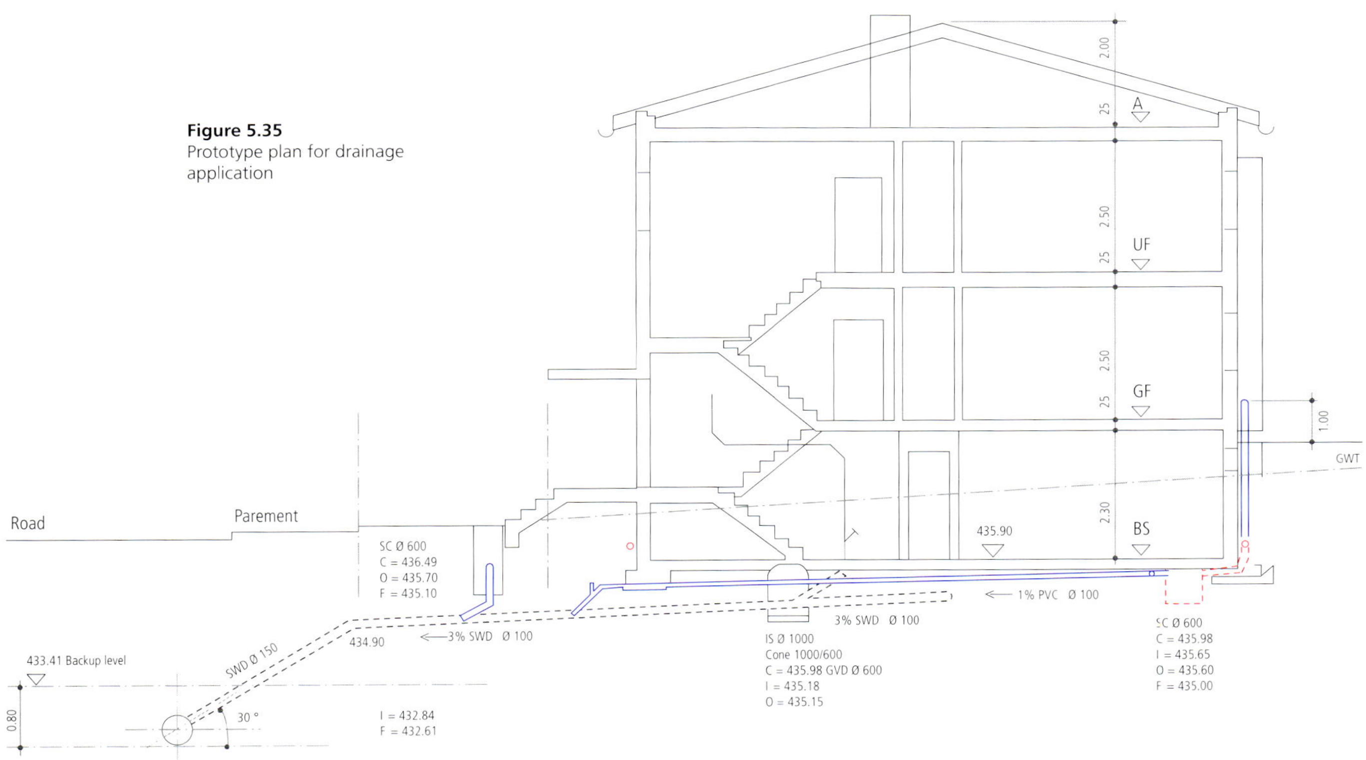

Figure 5.35
Prototype plan for drainage application

When plastic pipes are used, shaft lining must be installed

--- Sewage
—— Stormwater

WAR	Stormwater runoff	F	Floor elevation
WAS	Black water	WC	Water closet
SWD	Stoneware drain	Bt	Bathtub
PVC	Polyvinylchloride drain	Ws	Washstand
PVCS	Polyvinyl chloride drain/ seepage drain	Ss	Service sink
		S	Sink
IS	Inspection shaft	WM	Washing machine
SC	Sludge collector	FD	Floor drain
Dq	Dip quadrant	FDI	Floor drain inlet
C	Cover elevation	FTC	Fixture trap cover
I	Inlet elevation	DG	Drip gutter
O	Outflow elevation	CO	Clean-out

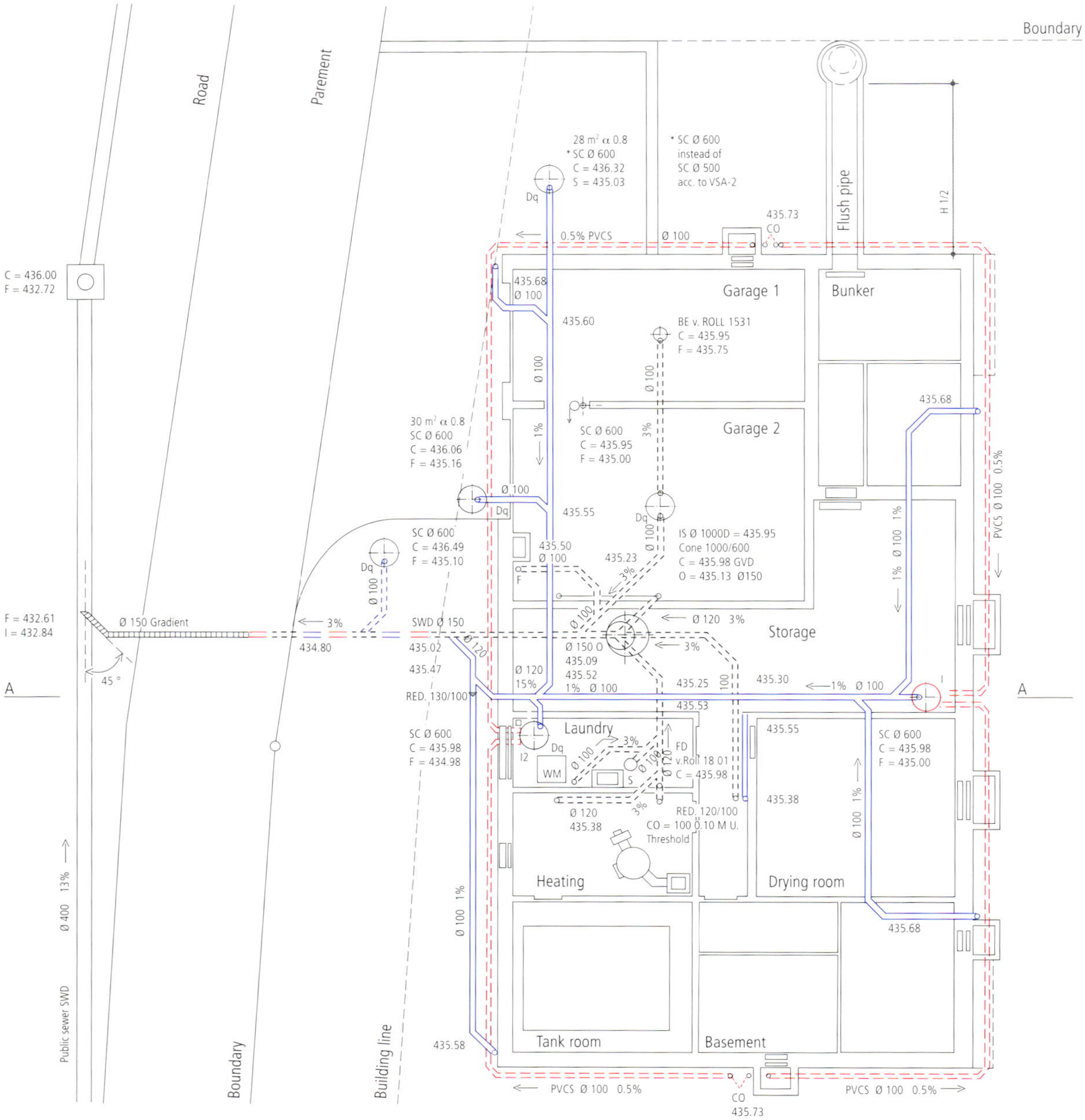

Figure 5.36
Prototype plan for drainage application
(Ground lines)

--- Sewage water
— Stormwater
— Drainage

WAR Stormwater runoff
WAS Black water
SWD Stoneware drain
PVC Polyvinylchloride drain
PVCS Polyvinyl chloride drain/ seepage drain
IS Inspection shaft
SC Sludge collector
Dq Dip quadrant
C Cover elevation
I Inlet elevation
O Outflow elevation

F Floor elevation
WC Water closet
Bt Bathtub
Ws Washstand
Ss Service sink
S Sink
WM Washing machine
FD Floor drain
FDI Floor drain inlet
FTC Fixture trap cover
DG Drip gutter
CO Clean-out

5.4.2 System characteristics

The purpose of the drainage system is to carry off waste and stormwater. It must be resistant to domestic waste and black water and sized to handle the volumes of wastewater to be carried off. The system must be designed to prevent leakage of wastewater and gases from the piping as well as backup from the main sewer into the building. Sewage pipes and drains must be protected against mechanical stresses of all kinds, external temperatures and corrosion. In principle, drainage systems should be channelled and installed in rooms that do not require elevated sound protection.

It is best to avoid long, horizontal connecting pipes (more than 5 m), since they are likely to cause functional and constructional problems. Horizontal collecting pipes and house drains must be laid on a gradient of at least 2 cm per 1 m of horizontal pipe length (1:50) to promote the unaided drainage of the wastewater.

Chemically contaminated wastewater must be drained to a separate pretreatment plant. Wastewater containing floating matter and heavy suspended matter, oils and fatty matter, strong toxins, acids or matter that has a strong alkaline reaction also requires pretreatment. The aforementioned sand catchers (or grit chambers), filters, sifting plants, oil, grease and starch interceptors are all pretreatment methods. Detoxification is required for wastewater containing toxins and neutralization for acid-carrying wastewater. Wastewater exposed to radiation must be collected in cooling containers.

Drainage piping is sized to ensure air circulation inside the pipes even as water is drained. The pipe diameter must not diminish in the flow direction.

Access to drainage systems for cleaning must be provided by installing relevant openings; connection to house drains is equally essential. All drainage objects must be equipped with fixture traps.

Where there is a risk of backup from the public sewer, house traps, pumps etc. must be installed to prevent flooding of affected rooms. All drainage objects located above the backup level must be designed for direct drainage via a natural gradient.

Stormwater must be carried off from roofs, balconies and other extensions via gutters and storm leaders. It is important to ensure that calcium and other precipitate matter flaking off from coated surfaces exposed to rain does not clog the storm pipes.

Permits must be obtained for all drainage systems; moreover, any drainage installation must be constructed solely with certified materials and parts.

5.4.3 Materials

All materials and products used for building drainage must comply with relevant national or international standards. The criteria established for approved materials are listed in **Table 5.10**.

Pipe connections for sewer pipes must be executed according to approved standards. The same is true for the type of fixture traps, which should be self-cleaning through the draining water and are generally installed immediately behind the drainage object outlet. The seal water level for fixture traps should be roughly 50 mm for toilets and at least 70 mm for all other fixture traps. Typical fixture traps are shown in **Figure 5.37**.

Trap types

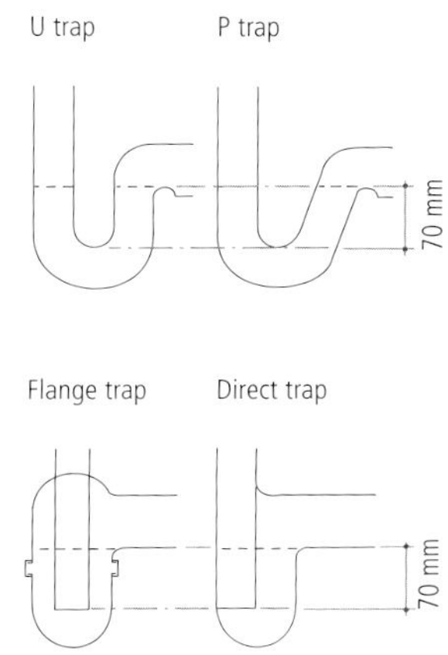

Figure 5.37
Traps

Criteria	Cast iron	Steel		Asbestos cement	Plastic		Stoneware (vitrified clay)
	sleeveless	stand. corrosion protection	plasticized	(Eternit)	PE	PVC	
Corrosion/ageing stability:							
1 Acid resistance	◑	○	○	○	●	●	●
2 Alkali-proof	●	●	●	●	●	●	●
3 Solvent	◑	○	○	○	●	○	●
Mechanical properties:							
4 Impact strength	◑	●	●	◑	●	○	◑
5 Bending strength	○	●	●	○	●	◑	○
6 Temperature stability	●	●	●	◑	●	○	●
7 Expansion	●	●	●	●	○	○	●
8 Sound-absorbing	◑	◑	◑	◑	○	○	●
9 Weight	○	◑	◑	◑	●	●	○
Assembly characteristics:							
10 Flexible application of form pieces, i.e. construction components	◑	●	●	◑	●	◑	○
11 Prefabrication	◑	●	●	◑	●	◑	○
12 Rational assembly	◑	●	●	◑	●	●	◑

● advantageous
◑ sufficient
○ detrimental

Table 5.10
Criteria for approved materials for sewer pipes

5.4.4 House drains

House drains are sewer pipes that are laid more or less horizontally beneath a building or underground, to which all storm leaders are connected (see **Figure 5.34**).

House drains should be laid out in a straight line and parallel to the building walls whenever possible. Directional changes in house drains must be realized with fittings of max. 45° and pipe connections must never be used for changing direction. At least one inspection chamber must be created per estate; in the case of long house drains, there should be an inspection chamber every 40 metres (**Figure 5.38**).

Separate inspection chambers are required for separating systems and the drainage of different types of wastewater (storm/black/industrial wastewater). Several soil stacks can be connected by means of certified fittings (**Figure 5.39**) without having to consider special connection distances. Subsidence (of soil and buildings) is a factor that must be considered when installing house drains; penetrations of house drains through building components must be flexible in design.

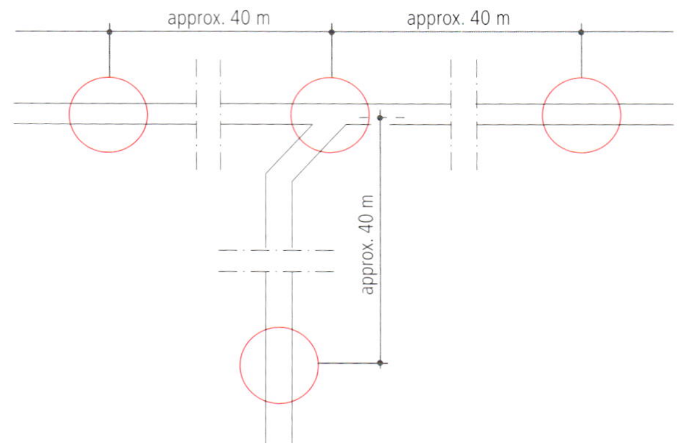

Figure 5.38.1
Placement of inspection chambers

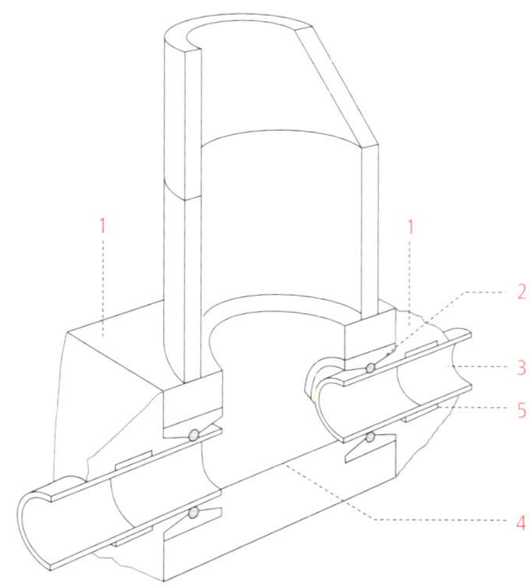

Figure 5.38.2
Inspection chamber

1 Concrete
2 Lining
3 Connecting piece
4 Berm
5 Anchor if required

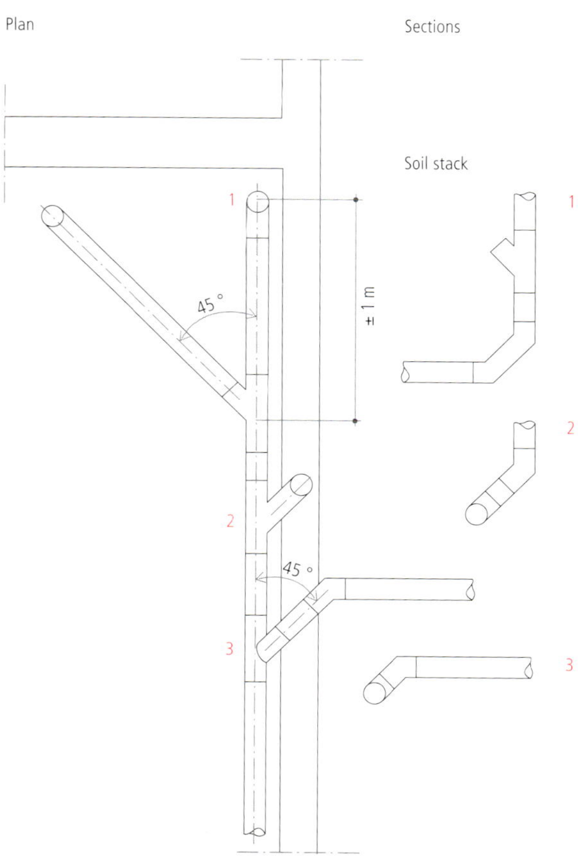

Figure 5.39
Connecting soil stacks to a house drain

5.4.5 Branch soils

Branch soils should be laid out in a straight line and parallel to the building walls whenever possible. Directional changes must be realized with fittings of up to 60°. Soil, branch and connecting pipes are joined at 60° angles; the same applies to the union of two branch soils.

Clean out openings are required for the inspection and cleaning of branch soils; they should be installed at intervals of 40 m and in cases where several directional changes occur (total of 180°). Otherwise, branch soils are subject to the same rules as house drains.

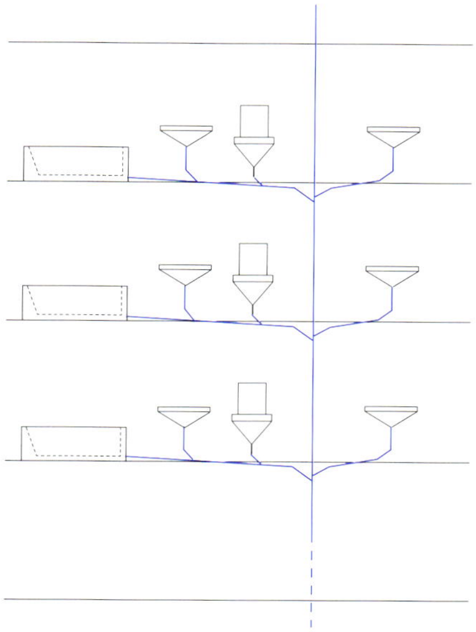
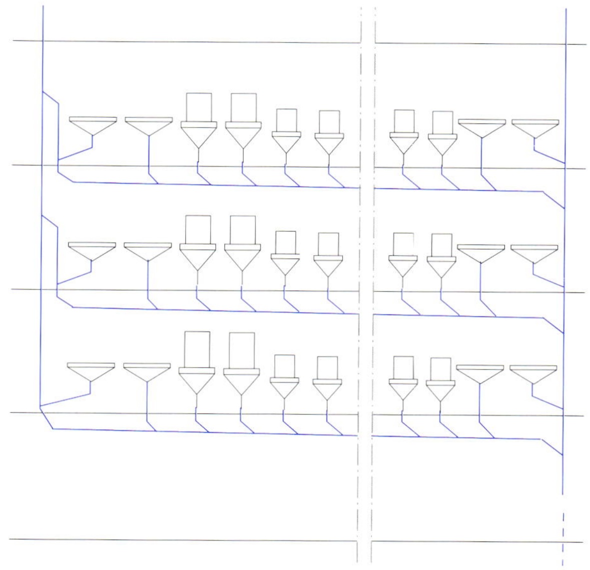

Figure 5.40
Options for soil stacks with stack ventilation (and sizing)

Primary ventilation system

Max. permissible number SW Total	Largest individual SW	WC Total	WC Per floor	Permissible load $\dot{V}_S$ (l/s)	Soil stack Ø in mm
3 1)	1.0	–	–	1.3	57/ 63
7	1.0	–	–	2.0	69/ 75
20	1.5	–	–	3.0	83/ 90
70	2.5	14	6	4.2	101/110
100	2.5	20	7	5.0	115/125
400	2.5	80	22	10.0	147/160

1) max. 2 fixtures per 1 SW

Soil stack with direct and indirect stack ventilation

Max. permissible number SW Total	WC Total	WC per floor	$\dot{V}_S$ 1) allowable (l/s)	Fall pipe Ø in mm	Secondary ventilation 2) Ø in mm
64	–	–	4.2	83/ **90**	57/ **63**
150	30	6	5.9	101/**110**	83/ **90**
200	40	7	7.0	115/**125**	83/ **90**
800	160	22	14.0	147/**160**	101/**110**

1) $\dot{V}_S$ permissible values roughly 40% higher than for primary ventilation system

2) The secondary ventilation can be linked to the soil stack with a 50/**56** dia. connection

5.4.6 Soil stacks

Soil stacks are designed to receive wastewater gathered on different floors of a building and drain it downwards. They should run in as straight a line as possible from floor to floor to avoid the need for soundproofing as a result of directional changes. Soil stacks are vented above the roof and require clean out openings above the connecting points to house drains and branch soils. **Figure 5.40** shows the most common system solutions for venting soil stacks.

5.4.7 Storm drains

Rainwater runoff (from roofs, balconies, loggias etc.) is drained via a separate drainage system, which must be fitted with fixture traps placed in frost-proof locations. Naturally, spouts and safety overflow drains should not be located near footpaths or street spaces.

For roof areas with more than average dirt accumulation, mud/sludge collectors must be employed.
To avoid annoyance caused by bad smells, all storm drains are fitted with fixture traps; this does not apply to drains leading directly into draining ditches or seepage trenches and in the case of planted roofs.

The amount of roof water (stormwater volume) must be established in order to size storm leaders. Stormwater volume is calculated with the following equation:

$$\dot{V}_r = A \cdot r \cdot \alpha \quad \left[\frac{m^2 \cdot l}{s \cdot m^2} = \frac{l}{s}\right]$$

Where:
A = area exposed to rain (only the projected area in case of slanted roof surfaces)
r = rain intensity (depending on location – average = 0.03 l/s m²)
α = delay coefficient

Storm leaders can be sized roughly according to the dimensions listed in **Table 5.11**.

Figures 5.41.1 to 3 give schematics of different roof constructions with roof inlets for flat roofs.

Stormwater leaders must be constructed from durable pipes above ground, especially where they connect to stormwater gutters on the outside of the building. In each of these cases it is important to decide whether pipeline heating is necessary to prevent downpipes from freezing.

Figure 5.42 complements the previous images by illustrating a type of floor drain suited for use on patios or large balconies.

Area exposed to rain A		**Pipe dimension** 83/**90**	101/**110**	115/**125**	148/**160**
	α	Permissible area exposed to rain m^2			
Slanted and flat roofs, independent of roof skin material	1.0	153	296	430	873
Humus-covered areas	0.3	–	998	1.433	2.911
$\dot{V}$ in l/s for a 3% gradient		4.6	8.9	12.9	26.2

Table 5.11
Dimensions for storm leaders

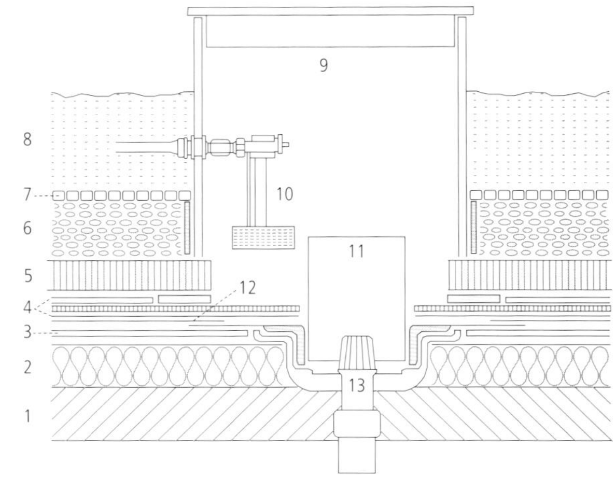

Figure 5.41.1
Roof outlet for planted roof

1 Concrete ceiling
2 Insulation
3 Roof membrane
4 Buffer and protective layers
5 Drainage pipes
6 Drainage layer
7 Filter mat
8 Soil
9 Inspection chamber
10 Float valve for irrigation
11 Standpipe
12 Foil ring
13 Pluvia rainwater inlet

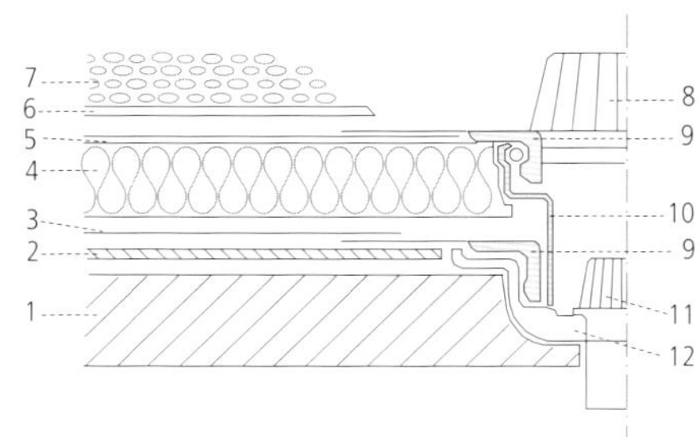

Figure 5.41.2
Construction of insulating roof (massive construction)

1 Concrete construction
2 Levelling mortar
3 Vapour barrier
4 Insulation
5 Sliding sheeting
6 Waterproofing (roof membrane)
7 Gravel fill
8 Leaf trap
9 Sheeting joint
10 Spacer
11 Inlet strainer
12 Roof water inlet

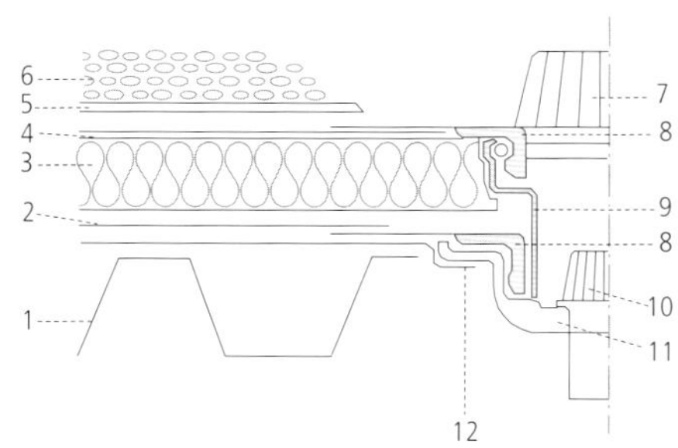

Figure 5.41.3
Insulating roof construction (light construction)

1 Trapezoidal sheet metal
2 Vapour barrier
3 Insulation
4 Sliding sheeting
5 Waterproofing (roof membrane)
6 Gravel fill
7 Leaf trap
8 Sheeting joint
9 Spacer
10 Inlet strainer
11 Roof water inlet
12 Reinforcing plate

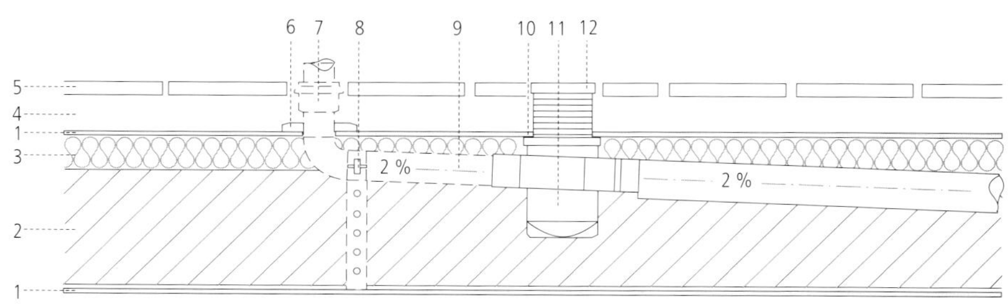

Figure 5.42
Floor drains

With gasket and dampproofing

1 Insulating sheeting
2 Massive floor construction
3 Insulating course
4 Overlay
5 Floor slabs
6 Weatherproofing
7 Mounting sleeve
8 Pipe clamp
9 Lateral inlet connection piece
10 Cup rim
11 Floor drain
12 Grate and frame

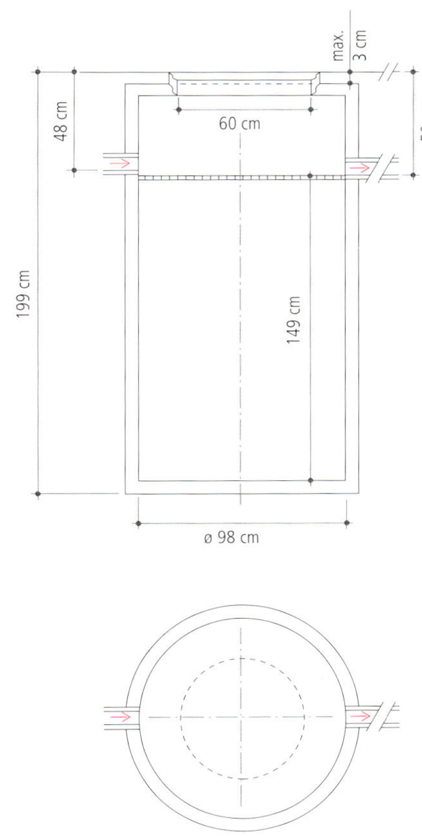

Figure 5.43
Diagram of sand trap construction

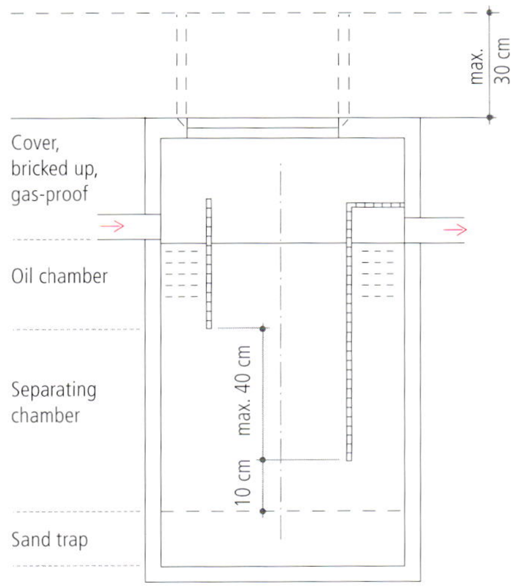

Figure 5.44
Petrol or oil interceptor

5.4.8 Interceptors

Interceptors (also called separators) are required to separate sludge, sand, mineral and organic fats and oils, and chemical additives. **Figure 5.43** illustrates a sand trap for separating solid matter (sand or sludge). When required, this type of trap is combined with petrol and oil interceptors preceding the sand trap in the direction of flow. Petrol or oil interceptors (**Figure 5.44**) are containers with closed lids and deep dip walls at the inlet and outlet. Petrol or oil collects at the surface in the space between the dip walls and is siphoned off from there. Solid matter settles at the bottom, where it, in turn, is siphoned off. Petrol and oil interceptors are used in auto body shops, car-wash facilities and gas stations, and in operations where mineral oil products are processed.

Grease and starch interceptors must be installed in operations that generate large amounts of organic fats, oils and starches (food product manufacture, commercial kitchens etc.). The bacterial transformation (decomposition) in grease interceptors can produce a foul odour; such interceptors must therefore be emptied to prevent any waste gases from entering into the building. Domestic sewage must never be drained to the grease interceptor and must therefore be measured only after the point where fat-containing wastewater joins in the stream (**Figure 5.45**).

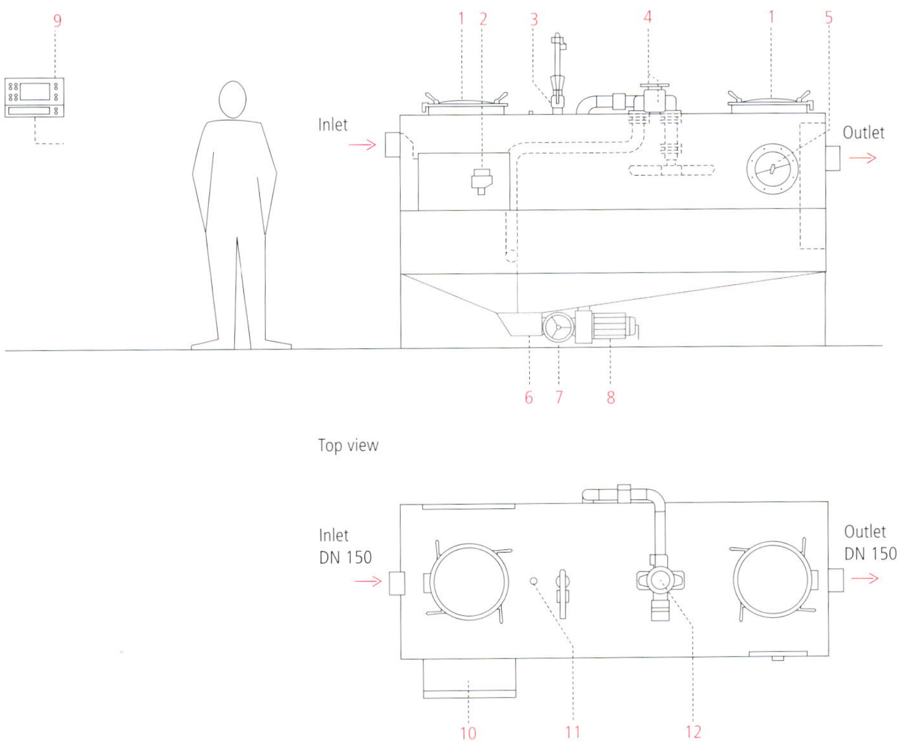

Figure 5.45
Grease interceptor installation (basement installation) for institutional applications, free-standing NG 7

1 Dome lid
2 Power supply
3 Fill unit
4 Distribution stopcock with actuator
5 Inspection glass
6 Floor outlet
7 Pump maintenance gate valve
8 Pump
9 Remote control
10 Program device
11 Low level measuring device
12 Flange connection for emptying pipe DN 65, PN 10

Catch pits (**Figure 5.46**) are used in situations where sewage cannot be drained directly and intermediate storage is required. These pits have different settling basins in which solid matters of varying consistency can accumulate.

Neutralization and detoxification plants are required when industrial or commercial wastewater contains toxic materials. Contracting firms with authorized personnel design system solutions that correspond to the specific wastewater volume and toxic gas concentration in each case.

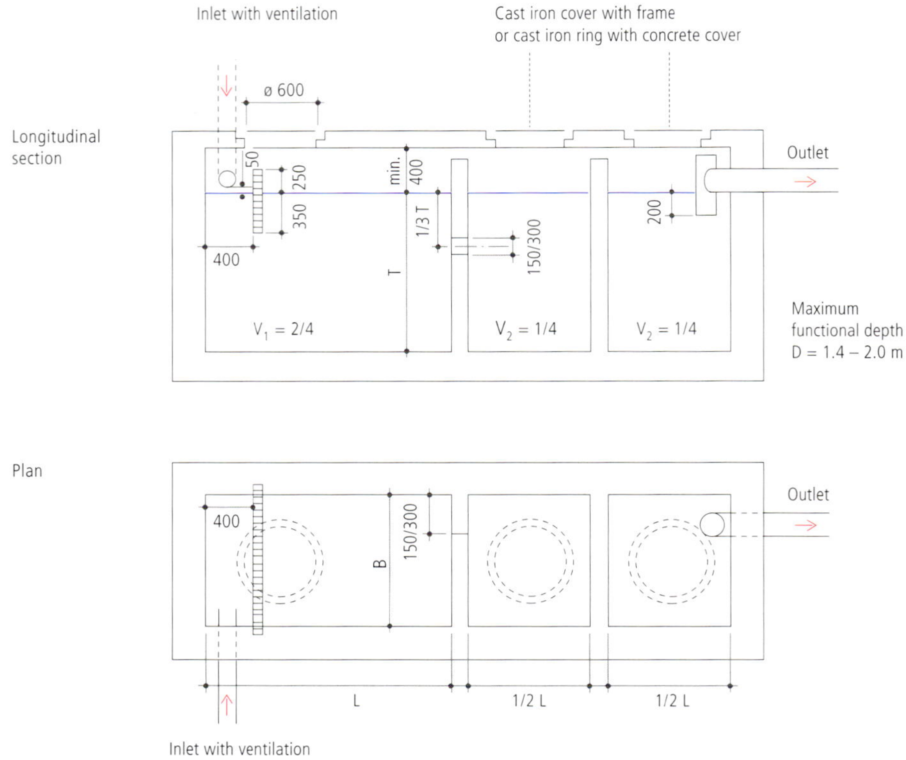

Figure 5.46
Catch pit with digestion tanks

5.4.9 Sizing sewage drainage systems

Sewage drainage systems are sized on the basis of a sewage water coefficient (SW/sewage water volume in l/s). If the sum of all sewage water coefficients lies below 60 SW, the piping is sized according to **Tables 5.12 – 5.13.3**.
If the sum of sewage water coefficients is greater than 60, the probable peak load (maximum simultaneous wastewater load) is calculated with the following equation:

$$\dot{V}_{s\,max.} = 0.5\sqrt{\sum SW}\ [l/s]$$

The sewage water coefficients of sewage drains must always be added. Based on the diagrams, the anticipated maximum load can be determined for each section based on each partial sum.

The decisive factor for measuring soil stacks, branch soils and house drains is the probable peak load, derived from the sum of all objects (SW) connected prior to this point and taking simultaneity into consideration. The allowable load of the soil stack differs depending on the vent stack and must be sized according to **Tables 5.13.1/2**.
Pipe dimensions for house drains and branch soils are calculated on the basis of the maximum anticipated load capacity $\dot{V}_{s\,max.}$.
For a house drain in a mixed system the maximum anticipated load $\dot{V}_m$ is determined from the sum of $\dot{V}_{s\,max.} + \dot{V}_r$.

$$\dot{V}_m = \dot{V}_{s\,max.} + \dot{V}_r$$

$\dot{V}_{s\,max.}$ = max. simultaneous wastewater volume

$\dot{V}_r$ = amount of precipitation

Diagram 5.47 can assist in sizing soil stacks, branch soils and house drains.

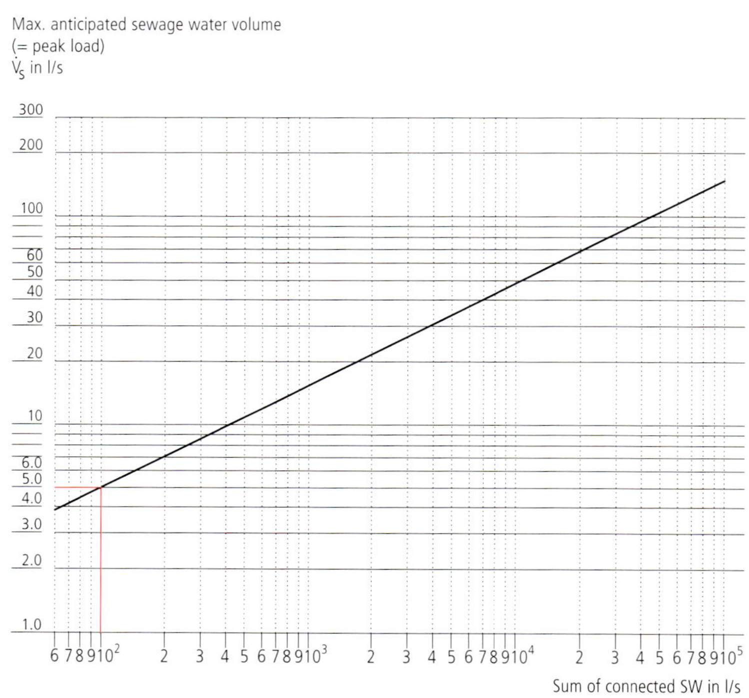

Example
Given:
20 apartments at 5 SW each = 100 SW

Problem:
probable peak load
Solution:
$\dot{V}_S = 0.5\sqrt{100\ [l/s]} = 5$ l/s

Figure 5.47
Probable peak load, housing

Drainage object or type of line	Connection value AW_s/ SW	Nominal width of individual connecting line DN
Handbasin, washstand, bidet, institutional sink	0.5	40
Kitchen drain (kitchen sink, single and double) including dishwasher up to 12 standard place settings, drain, domestic washing machine up to 6 kg dry laundry with individual drain trap	1	50
Washing machine for more than 6 – 12 kg dry laundry	1.5 *)	70 *)
Commercial dishwasher, refrigerating machine	2 *)	100 *)
Urinal (individual units) **)	0.5	50
Floor drain DN 50 DN 70 DN 100	1 1.5 2	50 70 100
Water closet, plug-in rinse bowl	2.5	100
Shower tub, foot tub, shower stall	1	50
Bathtub with direct connection	1	50
Bathtub with direct connection via a connection line of up to 1 m laid above floor level and connected to a ≥ DN 70 pipe	1	40
Bathtub or shower tub with indirect connection (bath drain), connecting line behind bath drain up to 2 m	1	50
Bathtub or shower tub with indirect connection (bath drain), connecting line behind bath drain longer than 2 m	1	70
Connecting line between tub drain valve and bath drain	–	≥ 40
Laboratory drain	1	50
Drain in dentistry fixture according to DIN 13937 or amalgam interceptor	0.5 *)	40 *)

*) When manufacturer values are provided, the sizing must be based on real values

**) Bank of urinals, see Table 5.12.1

Table 5.12
Connection values for drainage objects and basic values for the nominal width of individual connecting lines

Number of urinals	Connection value AW_s	Nominal width of connecting line DN
up to 2	0.5	70
up to 4	1	70
up to 6	1.5	70
more than 6	2	100

Table 5.12.1
Connection values for a bank of urinals and nominal widths of corresponding connecting lines

SW/AW		max. number of closets		Nominal clear width CW	Load
Total	Largest individual SW	Total	Per floor	Soil stack	max. permissible $\dot{V}s$ (l/s)
3 *)	1.0	–	–	57	1.3
7	1.0	–	–	69	2.0
20	1.5	–	–	80	3.0
70	2.5	14	6	100	4.2
100	2.5	20	7	118	5.0
150	–	30	10	125	6.1
400	–	80	22	150	10.0

*) max. 2 fixtures per 1 SW

Table 5.13.1
Soil stack, main venting system

SW/AW	max. number of closets		Nominal clear width CW		Load
Total	Total	Per floor	Soil stack	Secondary venting system [2]	max. permissible $\dot{V}s$ (l/s) [1]
64	–	–	80	57	4.2
150	30	6	100	80	5.9
200	40	7	118	80	7.0
300	60	10	125	80	8.5
800	160	20	150	100	14.0

[1] $\dot{V}s$ allowable values raised by 40% by comparison to main venting system

[2] The connection to the fall pipe can be sized with the CW

Table 5.13.2
Soil stack, secondary venting system direct or indirect

$\dot{V}$ permissible in l/s [1]

Gradient in % [3]

1	1.5	2	3	4	5	CW in mm
2.8	3.4	4.0	4.9	5.6	6.3	80
5.0	6.2	7.2	8.9	10.2	11.5	100 [2]
8.0	9.8	11.3	13.9	16.0	17.9	118 [2]
9.2	11.3	13.1	16.0	18.6	20.8	125
15.0	18.4	21.3	26.1	30.2	33.8	150
27.0	33.1	38.1	47.0	54.3	60.8	187
32.3	39.7	45.8	56.2	64.9	72.6	200
58.4	71.7	82.8	101.6	117.3	131.2	250
94.7	116.2	134.2	164.6	190.1	212.6	300

[1] For all raw material after Prandtl-Colebrook:
Operating roughness k = 1.0 mm
Kinematic viscosity ν = 1.31 10

[2] Minimal pipe width for house drain CW = 100
at WC connection CW = 118

[3] Standard gradient and minimal gradient must be considered
Minimal pipe width for house drain CW = 100
at soil stack connection to WC CW = 118

Table 5.13.3
Branch waste and house drain

5.4.10 Wastewater conveyance systems

The type of drainage system shown in **Figure 5.48**, which lies below the backup level of a main drain, must be linked to the estate drainage system via a pump installation. Stormwater falling above the backup level flows into the estate drainage system as a result of the natural gradient. If stormwater accumulates below the backup level, it must be drained as follows (priorities):

1. Receiving water (main drain)
2. Seepage
3. Pump installation

The required pump installation to transport waste- or sewage water, in particular, (also called pump station) consists of:
- Pump shaft with accessories
- Submerged pump with controls
- Pump pressure pipe
- Ventilation pipe

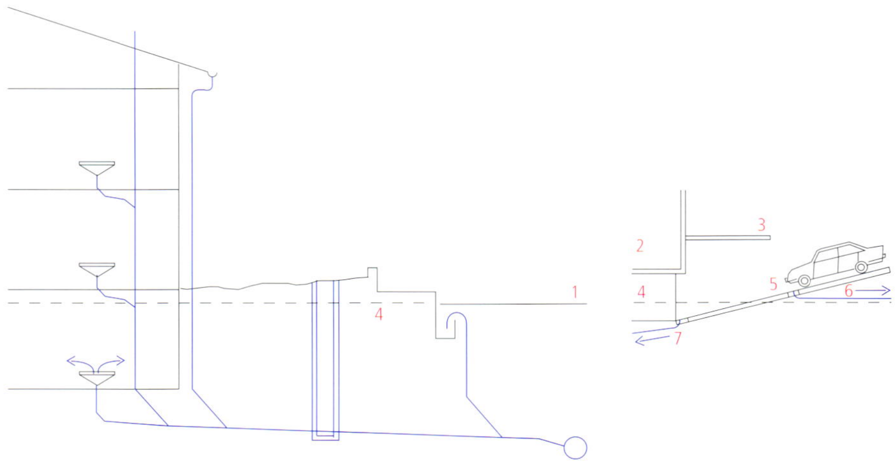

Figure 5.48
Drainage below backup level

1 Road
2 Building
3 Canopy roof (recommendation)
4 Backup level
5 Gutter
6 to estate drainage
7 to pump

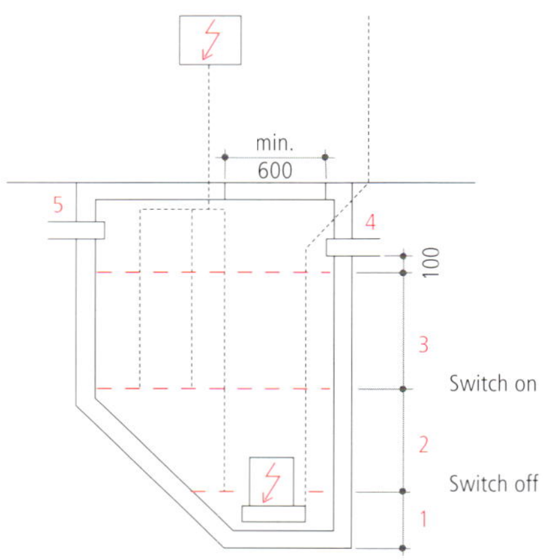

Figure 5.49
Pump shaft for hoisting equipment

1 Pump well
2 Net volume
3 Reserve volume
4 Bottom, lowest inlet
5 Ventilation (min. Ø 100)

The pump station must be accessible for maintenance and modification at all times. **Figure 5.49** illustrates this type of pump shaft with pump well, net volume and reserve volume as well as the submerged pump and inlet. Pump shafts must be designed to be impermeable to sewage, sealed against gas leaks and ventilated (clear width 100 mm). The pump shaft access should be at least 600 mm in diameter. Small pump shafts, designed exclusively for wastewater without faeces and without kitchen wastewater, can be installed without ventilation pipe and fixture trap.
A rustproof ladder must be installed in large shafts with manholes and shaft depths of 1.5 m or more.
All inlets must be merge into the shaft at least 0.05 m above the maximum water level (SEV codes must be fulfilled). **Figures 5.49/50** illustrate different approaches to employing a submerged pump.

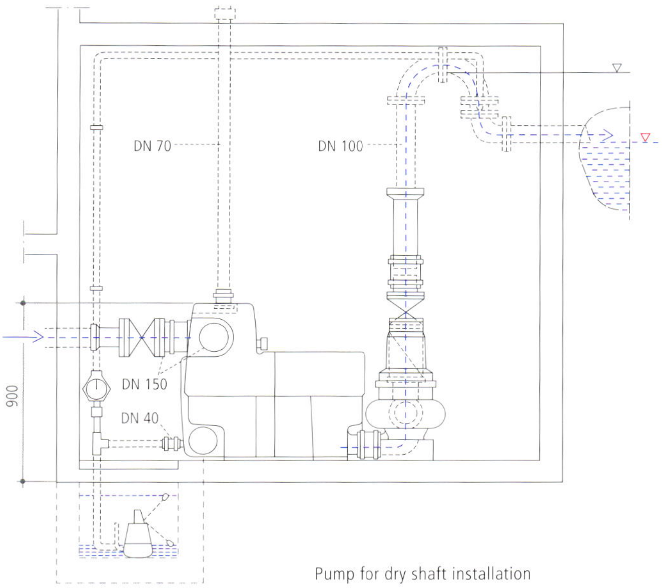

Pump for dry shaft installation

Figure 5.50
Wastewater/sewage pumps
Sectional drawings

Figure 5.50.1
Sewage pumps, type Ama-Drainer
(Product image: KSB)

1 Sewage pump, complete
2 DN 100 pressure pipe
3 Plungers for floater control
4 Back pressure valve
5 Stop valve
6 Elbow
7 Pressure pipe connection with flange
8 Underground conduit
9 Control panel
10 Outlet chain
11 Shaft cover

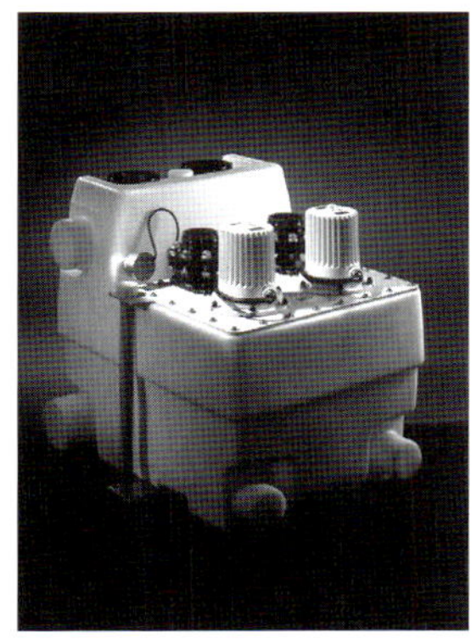

Figure 5.50.2
Twin-hoisting system Compacta
(Product image: KSB)

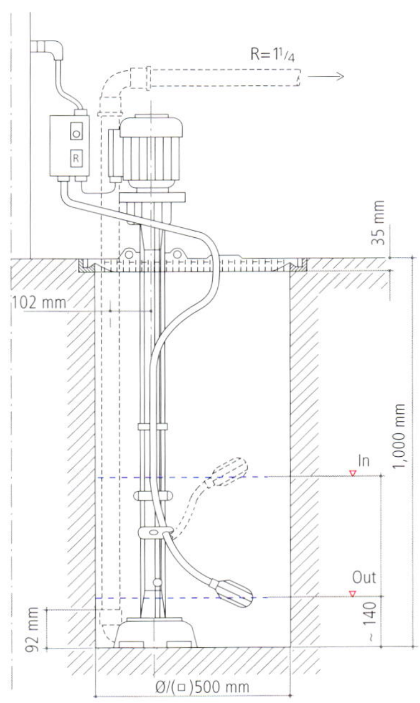

Vertical submerged pump

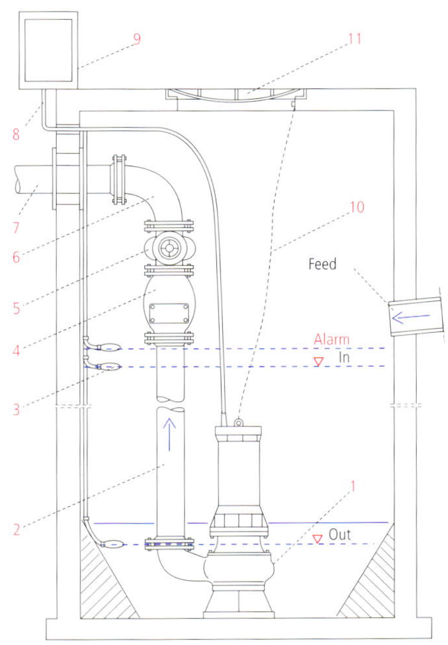

Submersible motor-driven pump

The pressure pipe of the wastewater pump is generally identical in size to the pump connection. The pipe must be laid above the lowest drainage point with natural gradient. No drainage object may be connected to the pressure pipe itself, and it is important to avoid corrosion and vibration transfer. The pump size is determined on the basis of flow rate coefficients or the data contained in the diagram in **Figure 5.51**. The minimum flow rates are as follows:

- Minimum flow rate without faeces = 2 l/s
- Minimum flow rate with faeces = 5 l/s
- Max. permissible flow rate = 10 l/s

The flow rate of the pump station is not identical to the peak performance indicated in **Figure 5.47** because the operation cycle of the pump station is greater than the period of maximum wastewater load. Stormwater and cooling wastewater must be considered as additional factors in the effective flow rate coefficient.

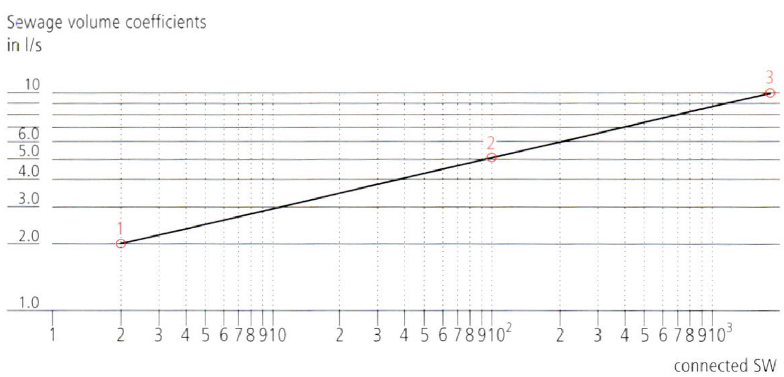

Figure 5.51
Dimensioning diagram for hoisting installations

5.5 Estate drainage

The estate drainage system gathers all wastewater from the building and transports it together with all other wastewater occurring on the estate to the public sewer. Estate drainage therefore includes all drains either buried on a private property or installed in the basement ceiling as well as all components required for operation and maintenance, such as elevated collectors, roof water collectors, sludge collectors, inspection shafts, oil and fat separators, wastewater pretreatment systems and, where necessary, wastewater pump stations. Private estate drainage ends at the connection point to the public sewer and the property owner is responsible for maintaining the system. Each property should be drained independently without encroachment onto the neighbouring property. If this is not feasible, relevant easements must be established and entered into the deed. **Figure 5.52** provides an overview of all types of local drainage with a mix or separating system. The different types of wastewater can be drained into a stormwater or sewage mains depending on origin and composition.

Table 5.14 provides insight into the drainage of different types of wastewater for various sewage or drainage systems.

Figure 5.52
Types of local drainage

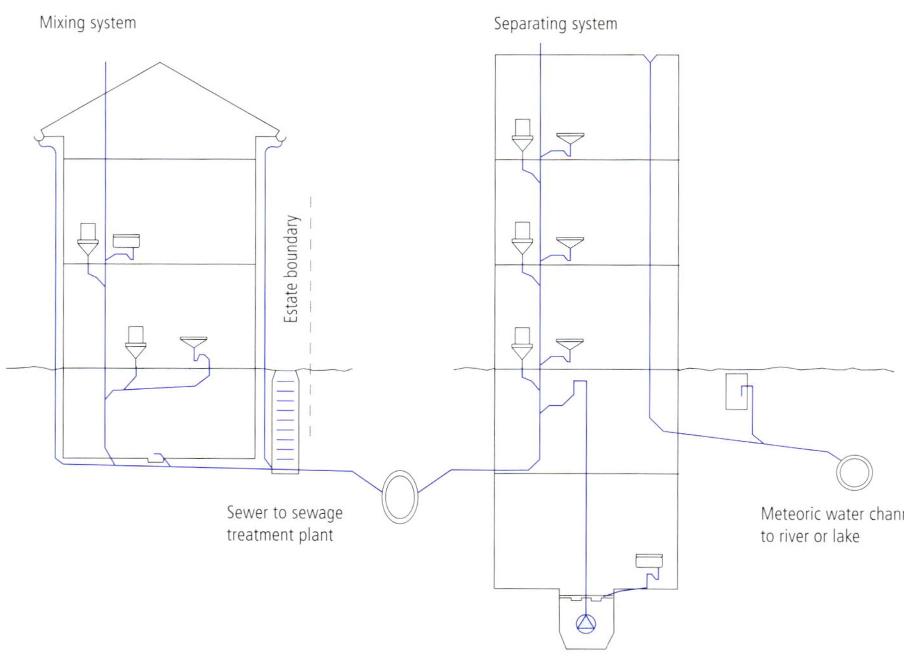

Type of wastewater	Separating system		Mixing system	Pure water channel
	Rainwater channel	Sewage channel		
Roof runoff [2]	X	O	X	O
Private road runoff	X	O	X	O
Stagnant water standard special cases	 X O	 O X	 X X	 O O
Well water [2]	X	O	X[1]	X
Seepage water [2]	X	O	X[1]	X
Ground and spring water [3]	X[1]	O	O	X
Cooling water [2]	X	O	X[1]	X
Sewage (without stormwater)	O	X	X	O
Treated wastewater	X	O	O	O
Streams (brooks)	X	O	O	X

X Permitted
O Not permitted
1 Only allowed with special permission
2 Seepage allowed with special permission, maybe be subject to regulation
3 Drainage requires permission granted by authorities

Table 5.14
Different systems of wastewater drainage

Application	Pipe material						
	Asbestos cement	Standard concrete	Special concrete	Cast iron	PE-h	Hard PVC	Stoneware
Rainwater [2]	X	X	X	X	X	X	X
Domestic and industrial wastewater [3]	X	O	X	X	X	X	X
Sewage/wastewater aggressive to cement	O	O	O	O	X	X	X
Acid ground	X[1]	O	X[1]	X[1]	X	X	X
Pressure pipes	X	O	O	X[2]	X	O	O
Standard seepage water	X	X	X	X	X	X	X
Strongly alkaline seepage water	O	O	O	X	X	O	O

X May be used
O Should not be used
1 Special protective measures required
2 Ductile cast iron
3 Provided the conditions of the Sewage Drainage Act are met

Table 5.15
Suitability of pipe materials for different types of wastewater

5.5.1 Materials

The materials listed in **Table 5.15** are deemed suitable for the various purposes indicated. Spigot and socket joints must be executed to manufacturer specifications and with the appropriate materials; they must also be installed without visible cracks or roughness in the pipe interior, which would impede the wastewater flow rate. Exposed pipes should be avoided and all sewage drains must be fully sealed and impermeable since small roots quickly grow into even the smallest and most minute cracks, blocking the pipes.

5.5.2 House drains

House drains should be laid in a straight line whenever possible and follow the shortest route to the public mains. The sewage pipes must be laid on a sufficient gradient, always in adherence to approved plans. Subsequent modifications must be based on the established plans for the wastewater pipes. The best scenario is to lay house drains on a concrete base to prevent subsidence and fissuring. Flexible pipe connections must therefore be used for wall penetrations (basement areas) into the open ground.

As for house drains in the building, it is important to make ample provision for clean-out with particular attention to inspection shafts in the area of sledge collectors, separators etc.

Figure 5.53 illustrates the wastewater hydraulic system of a horizontal house drain. The diagram on the right demonstrates that the wastewater volume follows a different course when the system is only partially filled (Q_t) than the flow velocity in the partially filled pipe (V_t), a fact linked to the filling of a circular cross section.

As previously explained for house drains in building drainage, directional changes of 90° in such drains may only be realized with two 45° elbows and two horizontal drains may only be joined at angles less than 45°.

Similarly, where long down sections or great differences in elevation exist, 90° directional changes should be employed instead of steep gradients, always considering the aforementioned restrictions (breaking the 90° angle down into a min. of 2 x 45°).

5.5.3 Connections to the public sewer system

The design of the connections to the public sewer system is an important prerequisite for drainage and sewage as a whole. The connecting pipe to the public sewer must be laid in specific angles shown in **Figure 5.54**, with large connecting lines joined to the public sewer in an inspection shaft with open flow gutter.

Figure 5.53
Sewage hydraulics, horizontal pipes

Flow rates for partial fill volume in circular profiles, for horizontal pipes (after Prandl-Colebrook) in %

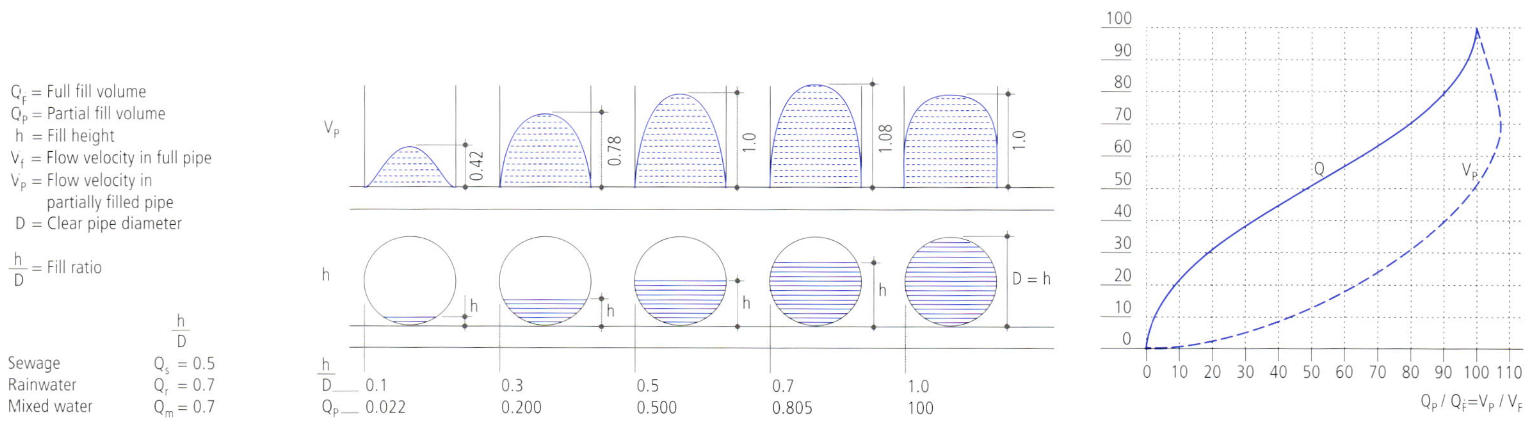

5.5.4 Seepage pipes

Hill and groundwater flows below ground on top of an impermeable layer of soil or within a water-bearing substratum, emerging to the surface in the form of rivulets, natural springs or small brooks. This clear, fresh water is diverted to the public sewer only in rare exceptions.

Seepage water, which occurs only during or after precipitation, can be collected in seepage pipes and diverted to the sewer in order to protect foundations, basement walls and floors, forecourts etc. Seepage pipes are laid as shown in **Figure 5.55**, that is, in a manner similar to ground lines, although the required gradient is no more than 0.5 %. Since seepage pipes are designed to accommodate only seeped rainwater, a pipe diameter of 10 cm is fully sufficient. Seepage water is frequently calcareous and should flow through the pipes as slowly and with as little turbulence as possible to prevent sediment formation. **Figure 5.56** demonstrates the principle of laying seepage pipes around a building.

Figure 5.54
Connection to public sewage system

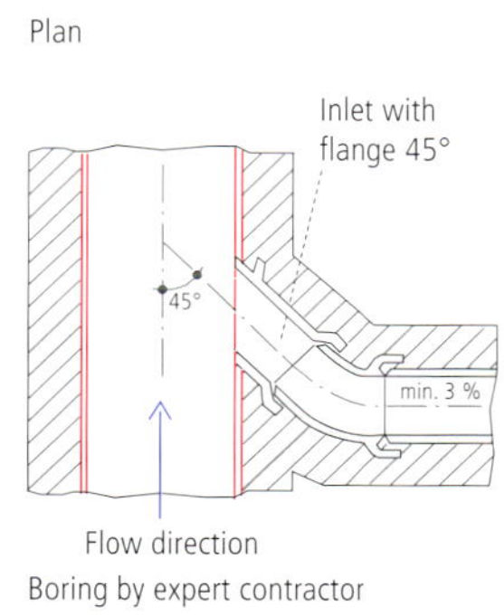

Figure 5.55
Detail of seepage pipe installation

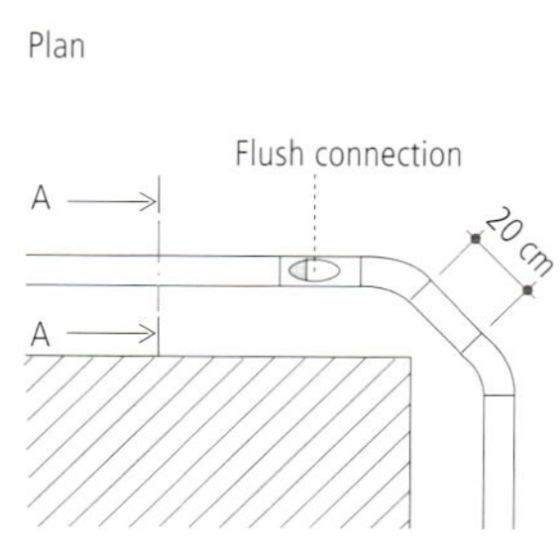

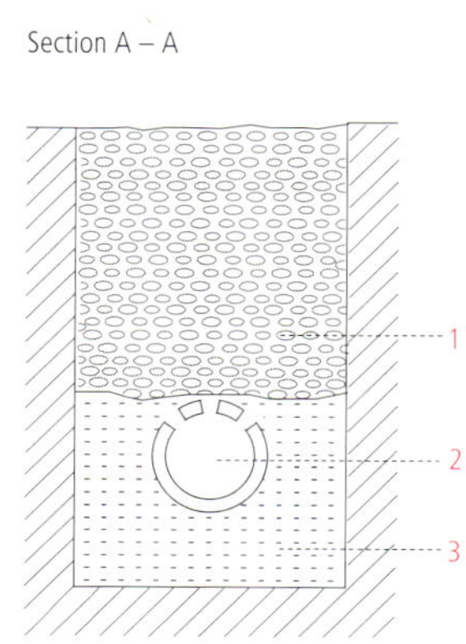

1 Filter material
2 Seepage pipe
3 Pipe bedding

Seepage pipes must never be connected to the domestic drains; instead they must be drained separately via a seepage collector with sludge trap. Seepage pipes must therefore drain neither the stormwater run off of a building nor groundwater.

The plans shown in **Figure 5.57** illustrate the drainage systems for a private home (top) and a commercial garage/auto body shop (bottom); the seepage pipe is clearly visible in the residential plan.

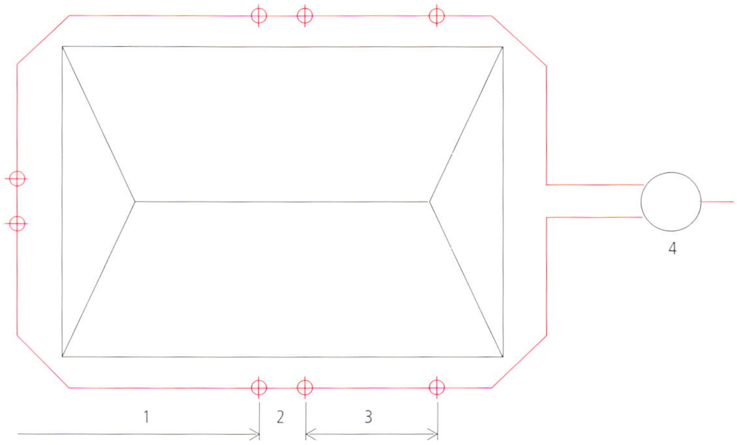

Figure 5.56
Diagram of seepage pipe layout

1 up to 20m
2 Short intermediate piece
3 after 20m/or prior to directional change
4 Collecting shaft

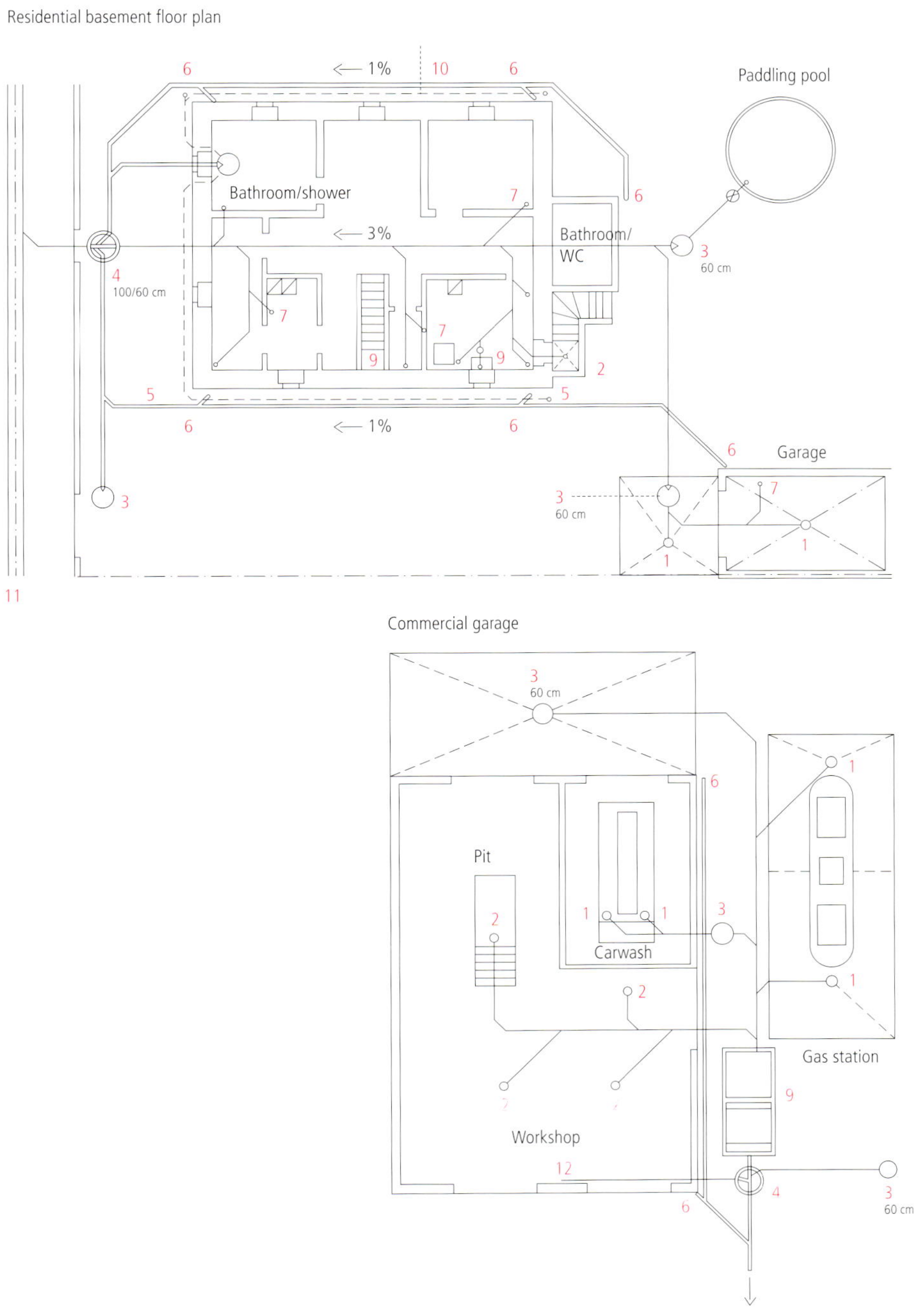

Figure 5.57
Drainage layout

1 Floor inlet
2 Floor outlet
3 Mud/sludge collector
4 Inspection shaft
5 Clean-out
6 Roof water
7 Wall-mounted sink
8 WC
9 Mineral oil interceptor
10 Seepage pipe
11 Sewer
12 Domestic sewage

5.6 Rainwater use

As noted above, freshwater is far too costly and precious a commodity to be used for flushing toilets, watering, car washing and similar purposes. The principal use of freshwater should therefore be restricted to food preparation, cleaning objects and utensils that come into direct contact with food, food consumption and personal hygiene. Service or process water prepared from rainwater is perfectly adequate for all other types of water consumption, especially since the softness of rainwater offers many other advantages (e.g. reducing the amount of detergents required, reducing the need for detergent additives etc.). The two principal categories of rainwater use are process water for rinsing and process water for cooling.

5.6.1 Process water for rinsing

Rainwater use should in principle be restricted to rainwater from roofs, since this ensures that serious contaminants are not introduced into the rainwater cycle, which would otherwise necessitate elaborate sterilization and treatment. The diagram in **Figure 5.58.1** illustrates the principle of a rainwater treatment plant including rainwater collection, a cistern and rainwater treatment via gravel filters and dosage, as well as distribution. **Figure 5.58.2** illustrates a built system with filter shaft and prefabricated storage tank. The degree of treatment required for rainwater depends on the planned uses. Experience has shown that no treatment is required in most cases.

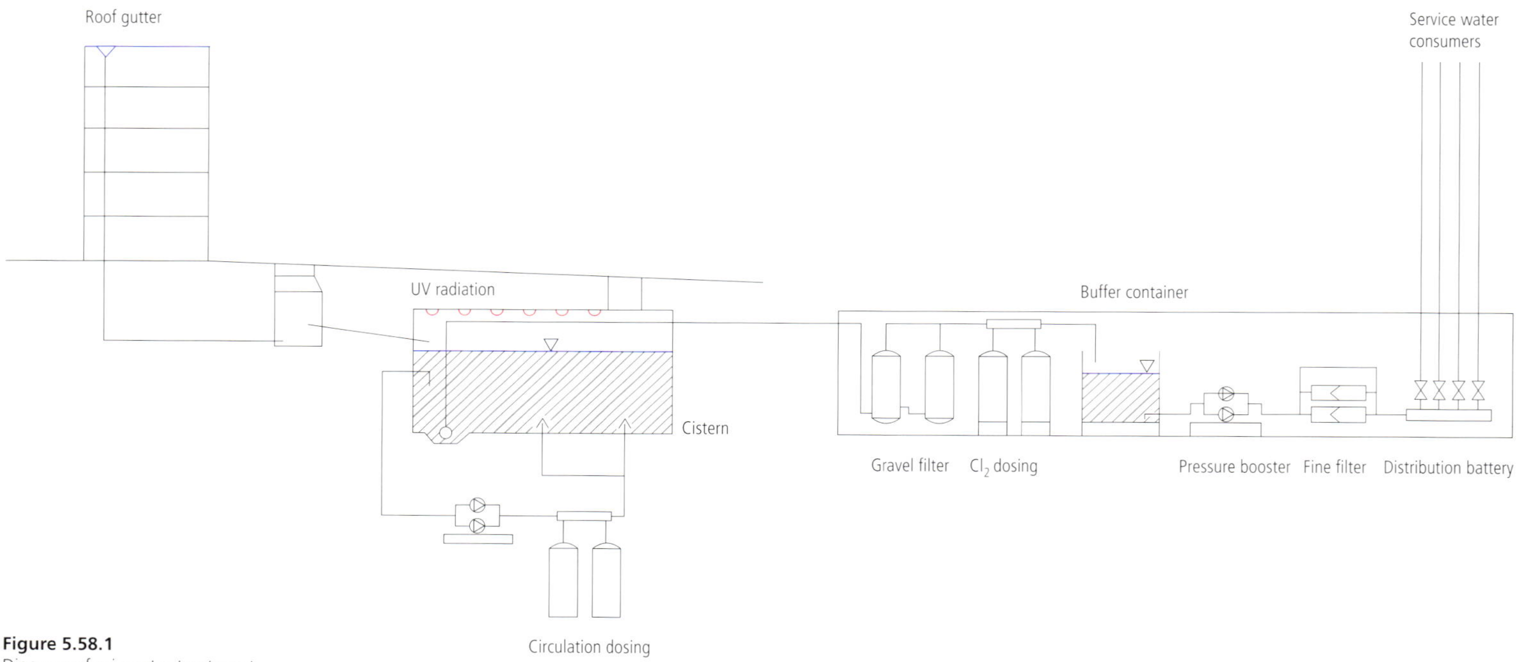

Figure 5.58.1
Diagram of rainwater treatment plant

Figure 5.58.2
Filter shaft in front of rectangular prefabricated storage tank for STEP construction project in Potsdam (Image: Mallbeton)

The cost-efficiency of increased initial investment in relation to decreased water costs must be established for each individual case. In the long term, the use of rainwater is highly recommended in all cases since it is an obvious method of preserving environmental resources. Although rainwater use was still rejected as being too expensive in Central Europe a few years ago, there is a noticeable shift in attitudes, especially in view of the fact that rainwater not only serves to cover the demand for water consumption for cleaning and rinsing, but also for cooling the building skin and the immediate building environment. All other aspects of the operating water system are identical to the parameters already described for water supply in general, albeit strictly separated from the latter. Potable water may only be added via an open outlet according to DIN 1988, Part 4.

5.6.2 Process water for cooling

There are several options for using treated rainwater to cool buildings or building surroundings. These include:

- Evaporative cooling near a building through the water surface of a manmade pool or similar features
- Amplified evaporative cooling near a building by means of installing fountains in a lake, water cascades or similar water features (direct or indirect use)
- Direct cooling of outside air by means of the evaporation from the fine mist of water

For all these options it is important to ensure that cooling by evaporation does not translate into excessive water consumption and to avoid surpassing the comfortable limit of humidity.

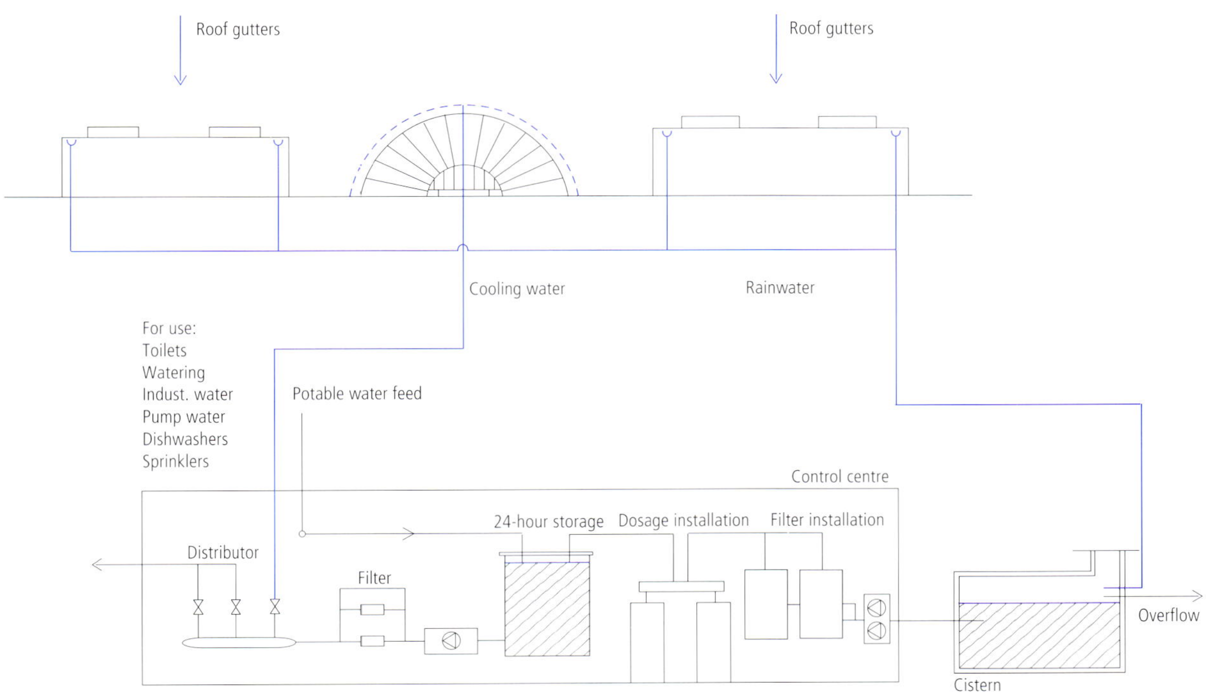

Figure 5.59
Diagram of rainwater usage and reduction of cooling load through envelope cooling

As an alternative, rainwater can also be used for direct building component cooling, an example of which is illustrated in **Figure 5.59**. The diagram shows how rainwater is collected through roof gutters and gathered in a cistern, which is linked to an overflow and to a rainwater treatment plant where the rainwater is filtered, dosed and stored. The process water is distributed across roof areas that require cooling (especially glass roofs) in a manner to create a film of water, which gradually evaporates and thus cools the surfaces.

Table 5.16
Demand statistics for sanitary fixtures/appliances and comfort levels in residential construction

Indications in number of pieces
X option

Persons	Bathtub	Shower	Washstand	WC	Bidet/WC 3000	Bidet/WC 7000	Urinal	Sink	Dishwasher	Washing machine per apartment	Dryer per apartment
	Bt	Sh	Ws	WC	WC	WC	Ur	Ss	Dw	Wm	Dr
Housing											
1 – 2	1		1	1	1			1			
3 – 4	1	1	1	2	1			1			
5 – 7	1	1	2	2	2			2	1		
Luxury housing											
1 – 2	1	1	1	1	X	1		1	1		
3 – 4	1	1	2	2	X	1		2	1		
5 – 7	1	1	3	2	1	1		2	1	1	1
Condominiums, family homes											
1 – 2	1	1	1	1	1	1	1	2	1	1	1
3 – 4	1	1	2	2	1	1	1	2	1	1	1
5 – 7	1	2	3	3		2	1	2	1	1	1

Table 5.17
Demand statistics for sanitary fixtures/appliances and comfort levels in public buildings

Persons	WC ♂	WC ♀	Urinal ♂	Washstand ♂	Washstand ♀	Shower ♂	Shower ♀	Teachers ♂ ♀
Cultural buildings								
10	1 – 2	1 – 2	1 – 2	2	1			
20	2	2	2	2	2			
25	2	3	2	2	3			
35	2	3	2	2	3			
50	2	3	2	2	3 – 4			
75	2	3 – 4	2 – 3	2	3 – 5			
100	2 – 3	5	3	2 – 3	5			
125	3	5 – 6	3 – 4	3	5 – 6			
150	3 – 4	6 – 7	4	3 – 4	6 – 8			
175	4	7 – 8	4 – 5	4	8 – 10			
200	4	8	5	4	10			
250	4 – 6	10	5	4	12			
School buildings								
10	1 – 2	1 – 2	1 – 2	2	2			up to 3
20	2	3	2	2	3			1 WC
25	2	3 – 4	2	2	3 – 4			1Wt
35	2 – 3	4	2 – 3	2 – 3	4			
50	3	4 – 5	3	3	4 – 5	8	8	up to 10
75	3 – 4	5 – 6	3 – 4	3 – 4	5 – 6	8	8	2 WC
100	4	6 – 7	4	4	6 – 7	10	10	1Wt
125	4 – 5	7 – 8	4 – 5	4 – 5	7 – 8	10	10	
150	5	8 – 10	5	5	8 – 10	12	10	
175	5 – 6	9 – 10	5 – 6	5 – 6	9 – 12	12	12	
200	6	10	6	6	10 – 12	12	12	
Administration buildings								
10	1 – 2	1 – 2	1 – 2	1 – 2	1 – 2			
20	2	2 – 3	2 – 3	2	2 – 3			
25	2 – 3	3 – 4	3	2 – 3	3 – 4			
35	3	4	3	3	4			
50	3 – 4	4 – 5	3 – 4	3 – 4	4 – 5			
75	4	5 – 6	4 – 5	4	5 – 6			
100	4 – 5	6	5 – 6	4 – 5	6			
125	5	6 – 7	6 – 7	5	6 – 7			
150	5 – 6	7 – 8	7 – 8	5 – 6	7 – 8			
175	6 – 8	8 – 10	8 – 10	6 – 8	8 – 10			
200	8	10	10	6 – 8	8 – 10			
250	8 – 10	10 – 12	10 – 12	8 – 10	10 – 12			

Persons	WC ♂	WC ♀	Urinal ♂	Washstand ♂	Washstand ♀	Shower ♂	Shower ♀	Foot basin ♂	Foot basin ♀
Sports facilities									
10	1	1 – 2	1 – 2	2	1 – 2	1	1		
20	1 – 2	2 – 3	1 – 2	2	2	2	2		
25	2	3 – 4	2	2 – 3	2 – 3	2	2		
35	2	4	2 – 3	3 – 4	4 – 5	2	4		
50	2 – 3	4 – 5	3	4 – 5	6	4	6		
75	3	5 – 6	3 – 4	4 – 6	8	6	10		
100	3 – 4	6	4 – 5	6	10	10	12		
125	4	6 – 8	5 – 6	8 – 10	12	12	16		
150	4 – 6	8	6 – 8	10 – 12	14	16	18		
175	6 – 8	8 – 10	10 – 12	12 – 14	14 – 16	18	22		
200	8 – 10	10 – 12	12	15 – 16	16 – 18	25	25		
250	10 – 12	12 – 14	12 – 14	15 – 18	18	30	30		
Industrial buildings									
10	1 – 2	1 – 2	1 – 2	3 – 4	3 – 4	1	1		
20	2	2	2	5 – 6	5 – 6	2	2	1	1
25	2	2 – 3	2 – 3	6 – 7	6 – 8	2	2	1	1
35	2 – 3	4	3	7 – 8	9 – 10	3	3	2	2
50	3	4 – 5	3	10 – 12	10 – 12	4	3	3	3
75	3 – 4	5 – 6	3 – 4	15	15	5	4	4	4
100	4 – 5	6 – 7	4 – 5	20 – 24	20 – 24	6	5	5	5
125	5 – 6	7 – 8	5 – 6	25 – 30	25 – 30	7	5	6	6
150	6 – 7	8 – 10	6 – 7	30 – 35	30 – 35	8	6	7	7
175	7 – 8	9 – 12	7 – 8	35 – 40	35 – 40	9	6	8	8
200	8 – 10	9 – 12	8 – 10	40 – 45	40 – 45	10	6	9	9
250	10 – 12	10 – 14	10 – 12	50	50	10	8	10	10
Restaurants									
10	1	1 – 2	1 – 2	1 – 2	1 – 2				
20	1 – 2	1 – 2	1 – 2	1 – 2	1 – 2				
25	2	2	2 – 3	1 – 2	1 – 2				
35	2	2 – 3	2 – 3	1 – 2	2				
50	2	3	2 – 3	1 – 2	2 – 3				
75	2 – 3	3 – 4	3	2	3 – 4				
100	3	4	3 – 4	2	4				
125	3 – 4	4 – 5	4 – 5	2	4 – 5				
150	4	5	5 – 6	2 – 3	5				
175	4 – 5	5 – 6	6 – 7	3	5 – 6				
200	5 – 6	6 – 8	7 – 8	3 – 4	6 – 8				
250	5 – 8	7 – 10	8 – 10	4 – 5	7 – 10				

5.7 Sanitary installations

The sanitary installations are determined by the spatial and functional program. The tables below contain data and planning tools.

5.7.1 Standard values for room installations

Sanitation rooms in different buildings require various installations; these are listed in **Tables 5.16/17** (facing page). Demands as well as comfort levels must be taken into consideration (excerpt from GEBERIT, handbook for sanitation engineers).

5.7.2 Spatial requirements for sanitary installations

The spatial requirements for sanitary installations as well as necessary centre-to-centre and wall distances for sanitary fixtures and appliances are listed in **Tables 5.18/5.19**.

WC installation (two piece)

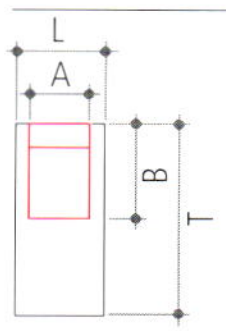

	MA	MI	MC
A	40	38	45
B	65	60	75
L	60	55	75
T	130	110	150

WC installation

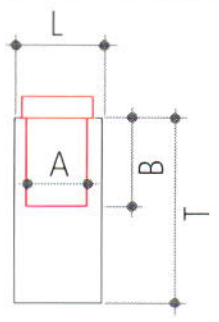

	MA	MI	MC
A	40	38	45
B	60	55	65
L	60	55	75
T	125	105	145

Washstand

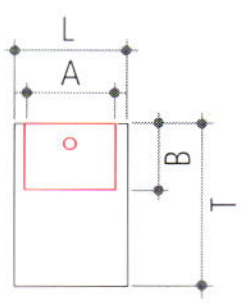

	MA	MI	MC
A	60	50	65
B	45	35	55
L	75	60	90
T	110	90	130

Bidet

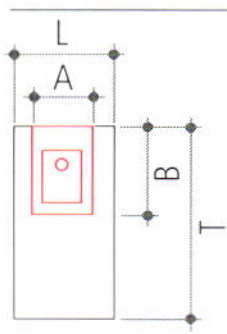

	MA	MI	MC
A	40	35	45
B	60	55	65
L	65	60	75
T	130	115	150

Bathtub

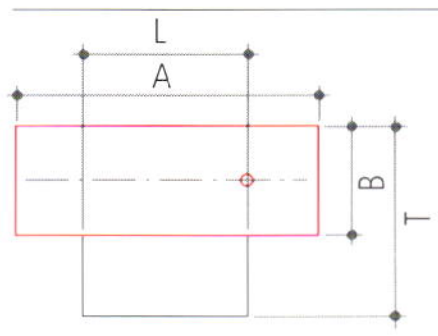

	MA	MI	MC
A	170	160	180
B	75	70	80
L	110	100	120
T	130	120	150

Shower

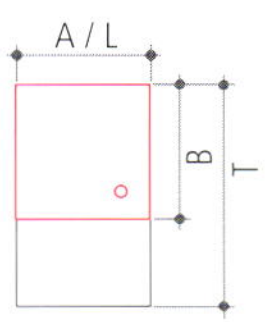

	MA	MI	MC
A	90	80	100
B	90	80	100
L	90	80	100
T	150	130	170

Sink

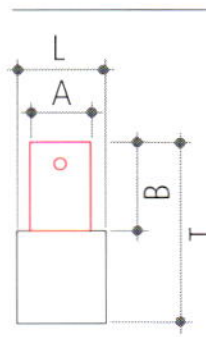

	MA	MI	MC
A	40	35	45
B	60	55	60
L	60	55	60
T	120	105	140

Dishwasher/oven

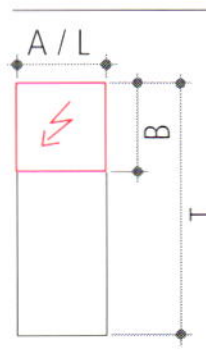

	MA	MI	MC
A	60	55	60
B	60	60	65
L	60	55	60
T	170	160	180

Table 5.18
Space requirements for sanitary fixtures/appliances

A Appliance width
B Appliance depth
L/T Surrounding space
M Distance wall – appliance middle
MA Average measurement
MI Minimum measurement
MC Comfort measurement

WC installation

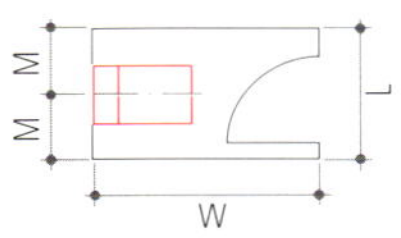

Mass	MA	MI	MC
M	45	40	55
L	90	80	110
W	150	140	160

WC installation and sink

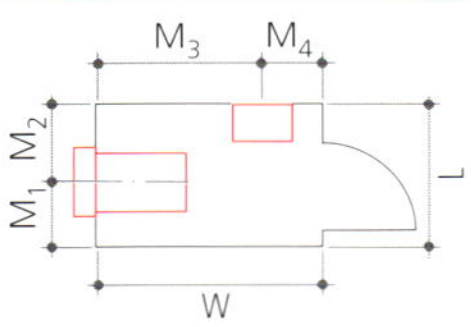

Mass	MA	MI	MC
M_1	45	40	50
M_2	55	50	60
M_3	110	100	120
M_4	40	35	45
L	100	90	110
W	150	135	165

WC installation and washstand

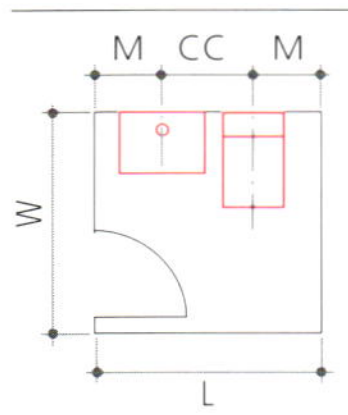

Mass	MA	MI	MC
M	45	40	55
CC	60	55	75
L	150	135	185
W	150	140	160

Shower and washstand

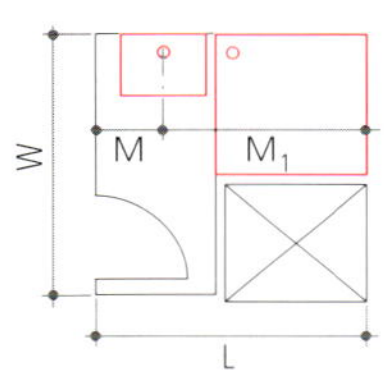

Mass	MA	MI	MC
M	45	40	55
M_1	135	120	150
L	180	160	205
W	160	150	170

Bathtub and washstand

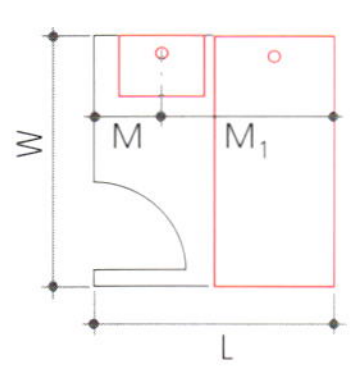

Mass	MA	MI	MC
M	45	40	55
M_1	115	110	125
L	160	150	180
W	170	160	180

W Room width
L Room length
M Distance wall – appliance middle
MA Average measurement
MI Minimum measurement
MC Comfort measurement
CC Center-to-center appliance

WC installation, washstand and bathtub

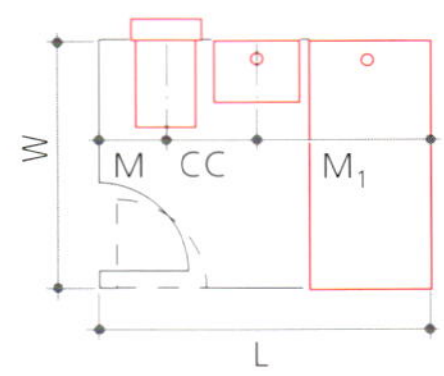

Mass	MA	MI	MC
M	45	40	55
CC	60	55	75
M_1	115	110	125
L	220	205	255
W	170	160	180

WC installation, washstand and shower

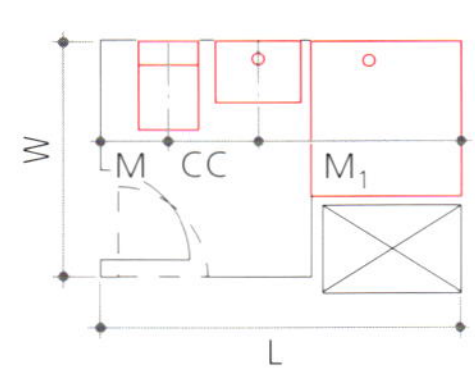

Mass	MA	MI	MC
M	45	40	55
CC	60	55	75
M_1	135	120	150
L	240	215	280
W	160	150	170

WC installation, bidet, washstand and bathtub

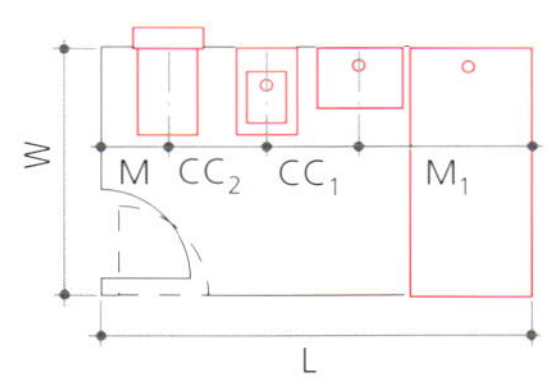

Mass	MA	MI	MC
M	45	40	55
CC_1	65	60	75
CC_2	60	55	75
M_1	115	110	125
L	285	265	330
W	170	160	180

WC installation, double sink and bathtub

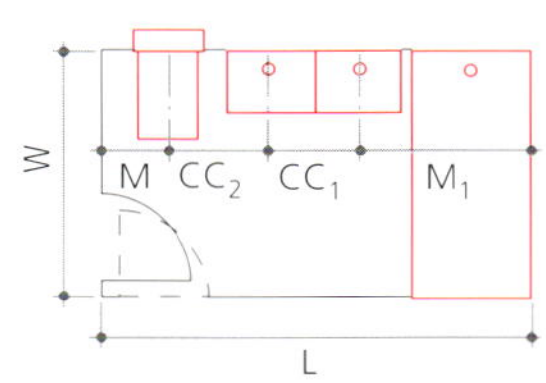

Mass	MA	MI	MC
M	45	40	55
CC_1	65	60	75
CC_2 *	60	55	75
M_1	115	110	125
L	285	265	330
W	170	160	180

* According to model for double sink

Table 5.19
Centre-to-centre and wall distances for sanitary fixtures/appliances

5.7.3 Space requirements for kitchen appliances (residential)

Based on kitchen functions, the following areas are differentiated:
- Rinse and wash area
- Food preparation area
- Cooking area
- Storage area
- Eating area

Most residential kitchens today are designed on the basis of national standards (e.g. the Swiss or European standard). **Figure 5.60** contains typical standard measurements for kitchen installations as well as the necessary face lengths for different apartment sizes. It is important to note that the Swiss standard is based on a module width of 55 cm per component, while the European standard assumes a width of 60 cm.

In addition to the aforementioned criteria, kitchens can also be divided into the following categories:
- Long-front kitchen (single row/double row)
- L-shaped kitchen
- U-shaped kitchen
- Free-form kitchen

Figure 5.61 provides an overview of these different kitchen models. **Figure 5.62** offers a planning aid in terms of calculating the space requirements for the eating area. This figure indicates the space requirement in relation to the number of diners and the number of seats/chairs in the eating area.

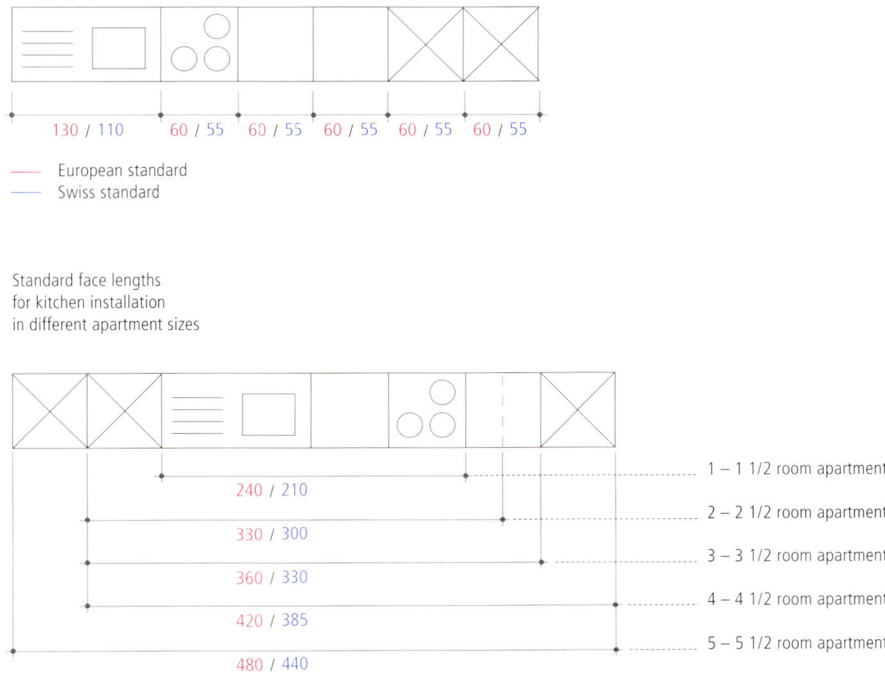

Figure 5.60
Standard kitchen layout (functions: rinsing/food preparation/cooking/storage)

Dining area		Measurements in cm L minimum	L standard	W minimum	W standard
Option a	1 person	150	160	70	80
Option b	2 persons (rectangular)	210	240	70	80
Option c	2 persons (square)	150	160	150	160
Option d	4 persons	150	160	220	240
Option e	6 persons	210	240	240	270

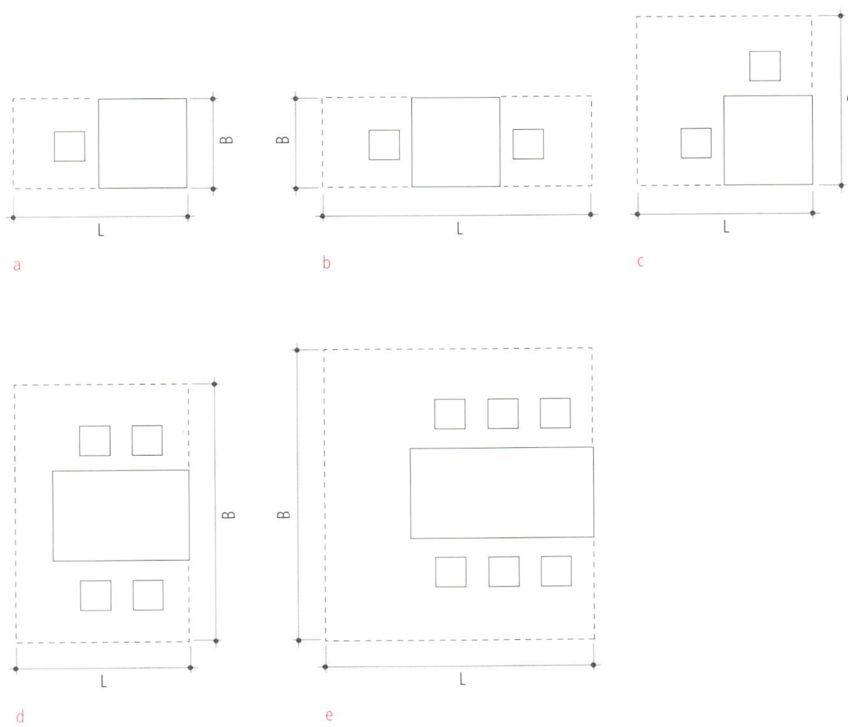

Figure 5.62
Dining area, scale sketch

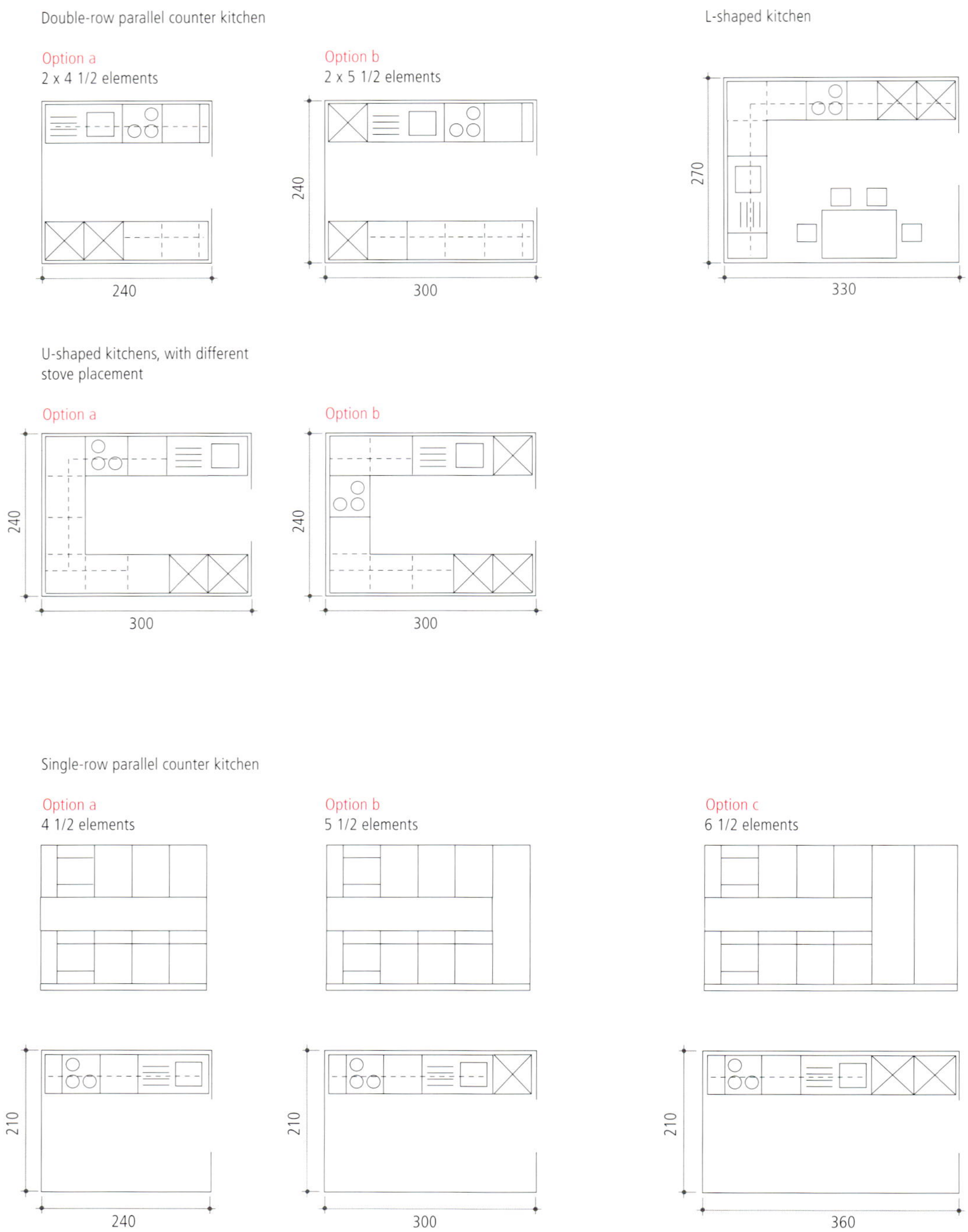

Figure 5.61
Various small kitchen layout designs

5.7.4 Wheelchair access

Systems for wheelchair access should be planned with a view to
- Ease of maintenance
- Splash-proof surfaces
- User-friendly design (wheelchair)
- Usability without outside help

Sanitation areas for people with disabilities must be large enough to be accessible to wheelchair users.

Bathrooms and toilet facilities should be provided in separate rooms with appropriate handlebars to facilitate use by disabled people. Depending on the type of disability, toilet facilities should be planned to offer "left" or "right" WC cubicles, that is, wheelchairs are placed either to the left or to the right side of the toilet. **Figure 5.63** illustrates different heights relevant for wheelchair users and reachable distances for women and men (data taken from: guideline to avoid architectura barriers and impediments published by Procap, the Swiss Association of Disabled People).

The measurements for the WC and shower installations shown in **Figure 5.63** are in part based on the parameters contained in **Figure 5.64**. It is important to remember not to use shower mountings but to install a floor drain with a 2 to 3 % gradient and to consider that most disabled people are seated while showering.

Wheelchair access to the sink must also be provided by using concealed siphons instead of the standard models. In addition to the facilities featured in these drawings, wheelchair-accessible kitchens and sanitary installations are equipped with special fittings whose extended levers facilitate use.

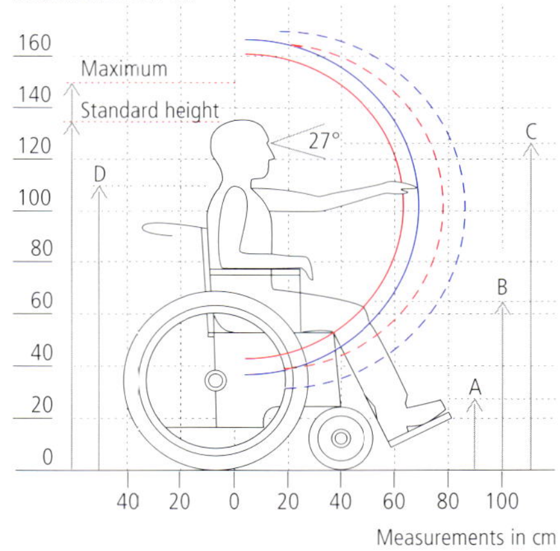

—— Reach for women in upright position
- - - Reach for women, slightly forward-bent position

—— Reach for men in upright position
- - - Reach for men, slightly forward-bent position

A Foot height
B Knee height
C Eye level
D Shoulder height

Figure 5.63
Reach from a standard wheelchair

Figure 5.64
Wheelchair-accessible WC and shower areas

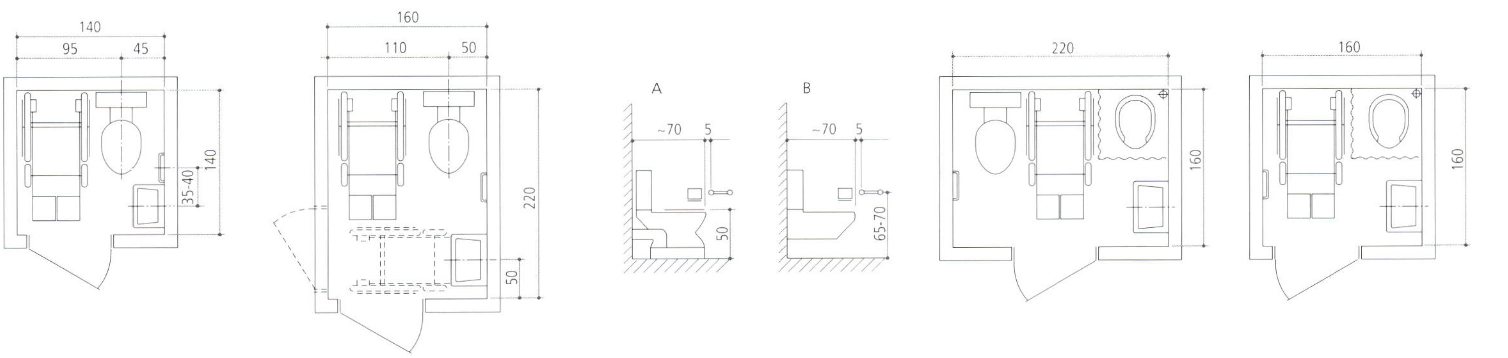

WC, minimum solution where only side access is possible.
Sink next to WC allows for hand-washing at the same time

Doors should always open outwards

Standard WC layout

A Standard toilet, opt. base
B Wall-mounted toilet with visible tanks – excellent solution

Shower, minimum solution with WC

Showers provide excellent amenities for the handicapped and the elderly

Shower, minimum solution without WC

5.7.5 Objects / Fixtures

Despite the multitude of available models, all sanitary fixtures are subject to the same selection criteria described above (efficiency, resistance, durability, maintenance and assembly).

The following materials are used for sanitary fixtures:
- Stoneware (vitrified, impermeable clay)
- Earthenware (non-vitrified, white clay)
- Fireclay (refractory clays)
- Vitreous china
- Steel (enamelled on all sides)
- Stainless steel (rustproof steel/nickel chromium steel)
- Acrylic glass (Plexiglas)
- Plastics (rigid vinyl or PVC or glass fibre reinforced unsaturated polyester resin)

Today, sanitary objects are available in nearly all colours. The following psychological responses have been observed in conjunction with various colours:
- Red to brown scale = warm, comfortable (obtrusive when colours are too intense)
- Blue to green scale = elegant and fresh (cold when colours are too intense)
- Co-ordinated colour scale = restrained, calm

The colours of sanitary objects should naturally harmonize with the tile colours (wall panels), although contrasts may achieve energizing, refreshing effects.

It is important to remember that dark colours make rooms appear smaller than they are, while light colours have the opposite effect. Large mirrored surfaces can create an optical illusion of more space. Sanitary objects are generally accompanied by a range of accessories such as shelves, soap dishes or dispensers, towel racks and towel storage (towel dispenser), disinfectant dispensers, rinse glass holders, mirrored cabinets etc., which are all sensible and necessary features. With the exception of custom installations, sanitary objects include the following:
- Washbasins (bathroom)
- Spit and rinse basins
- Bidets
- Tubs
- Showers
- Footbaths and footwash fixtures
- Sitzbaths
- Baby baths
- WCs
- Urinals
- Drinking fountains
- Sinks (kitchen)
- Drains
- Washing installations (row sinks)

The **Figure series 5.65** shows some examples of these objects, selected from the wide range of available models.

Figure 5.65.1
Washbasin

Figure 5.65.2
Rinse basin

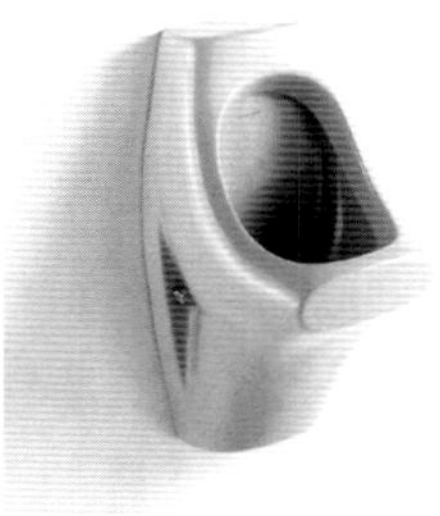

Figure 5.65.3
Urinal

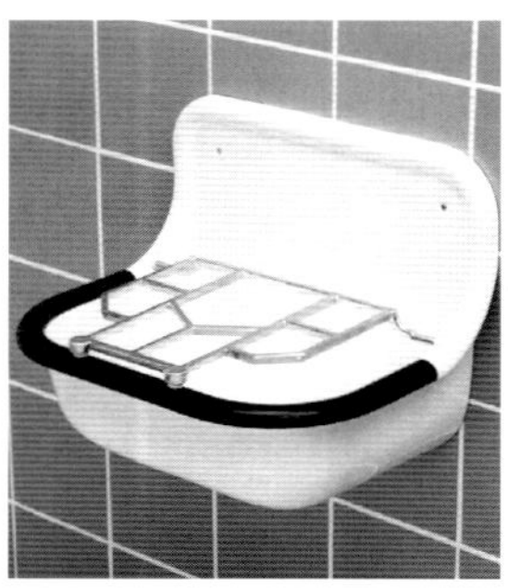

Figure 5.65.4
Utility sink

5.66.9
Concentration on pure essentials: "Atrio" (Grohe)

5.66.10
Floor-mounted bath mixer ("Atrio", Grohe)

Figure 5.65.5
Institutional wall-mounted sink

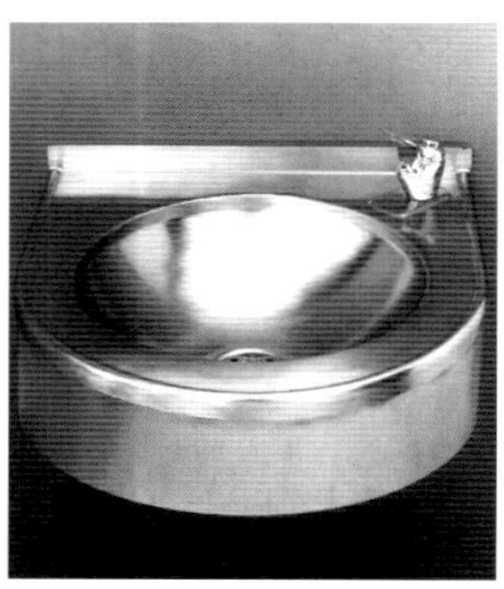

Figure 5.65.6
Drinking fountain

5.66.11
Wall-mounted three-hole wash basin mixer ("Atrio", Grohe)

5.66.12
Wash basin mixer ("Atrio", Grohe)

5.7.6 Fittings

Sanitary fittings are differentiated according to application. They are:
- Water fittings
- Wastewater fittings
- Gas and waste gas fittings

Water fittings are pipe and tap fittings and serve either for use and maintenance or safety and monitoring of water supply installations.

Tap fittings are used at the end points of water piping networks and serve to draw off water for free discharge. These outlet fittings are the most visible fittings and hence of greater interest to architects and clients than those that are used exclusively for technical purposes. The **Figure series 5.66** depicts various tap fittings.

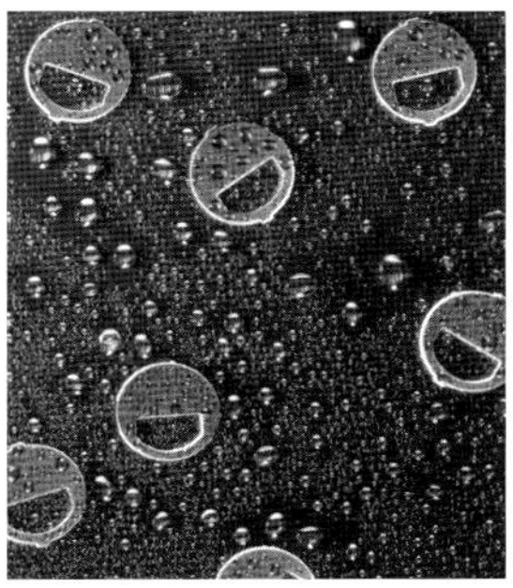

Carbodur ceramic panes with adamantine coating

Figure 5.66.1
Grohe's "Taron" line of shower and tub wall thermostats offer the new "Carbodur" coating technology. The proven ceramic panels are coated with a paper-thin layer of adamantine carbon with the help of a "plasma" process. The result: the fitting operates as smoothly after years of use as it did on the first day; gasket wear is no longer an issue for the lifetime of the fitting. (Image: Grohe)

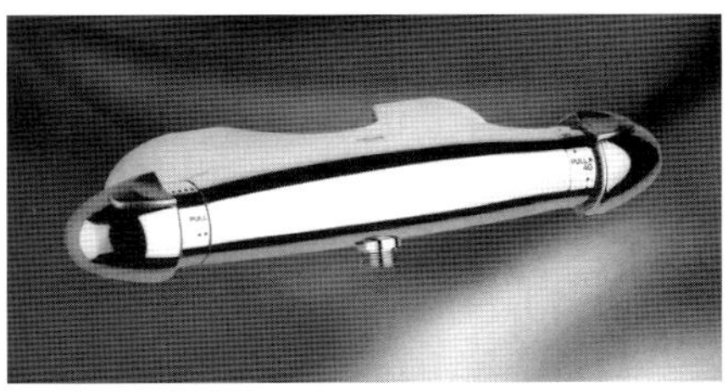

Figure 5.66.4
Thermostat shower mixer ("Taron", Image: Grohe)

Figures 5.66.9 – 5.66.12
"Atrio" of the Grohe "Articulation" product range.

"Articulation" is the name of the company's premium program which comprises selected designer faucets, bathroom accessories, shower systems and sanitary systems. All "Articulation" products meet the very highest standards in terms of product quality and design. In addition, the company continues to market its proven ranges of high-quality bathroom and special faucets, versatile showers and shower systems, convenient kitchen faucets as well as sanitary systems which facilitate creative bathroom design.

Market studies and trend research have shown that today's sophisticated consumers want to articulate their sense of style not only through the design of their living rooms and kitchens but also through individual and harmonious bathroom design. This requires holistic design concepts and products which lend themselves to coordinated interior design schemes.
(Image: Grohe)

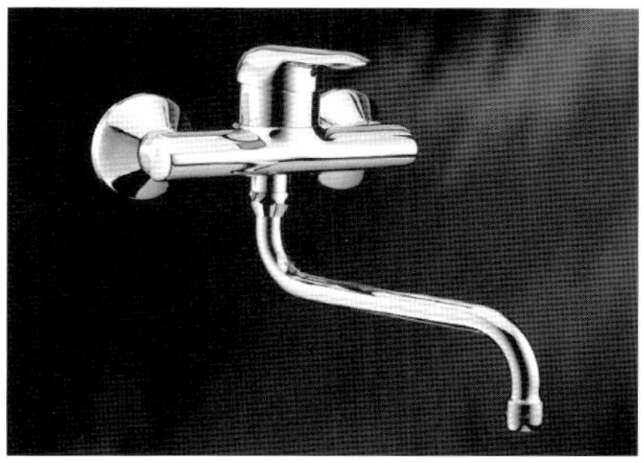

Figure 5.66.2
Sink wall-mounted fitting with swivel spout
(Hansatwist, Hansa)

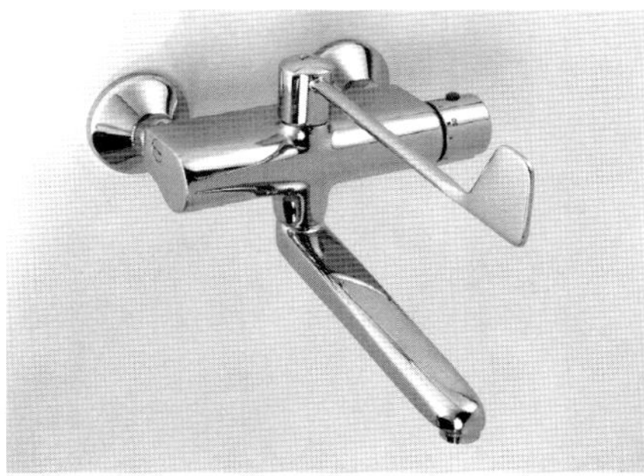

Figure 5.66.3
Thermostat fitting with swivel spout for wall mounting for clinics and hospitals
(Hansatempra, Hansa)

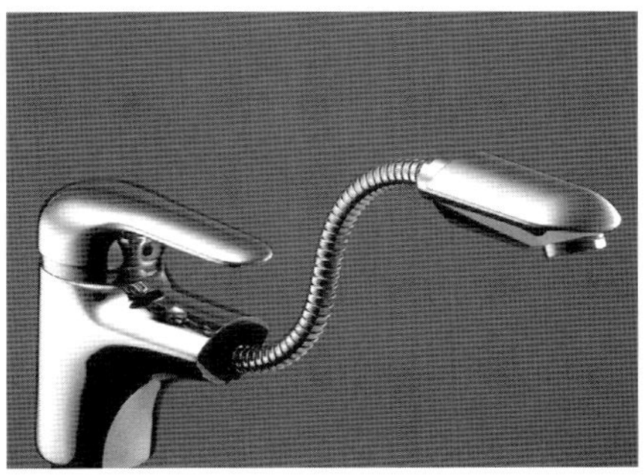

Figure 5.66.4
Single-hand mixer for sink with pull-out spout (Hansaronda, Hansa)

Figure 5.66.5
Single-hand mixer with swivel spout (Hansamix, Hansa)

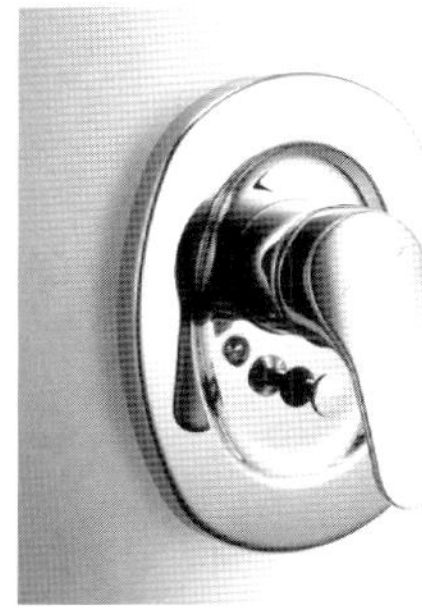

Figure 5.66.6
Single-hand mixer with conversion for wall installation
(Hansamix UP, Hansa)

Figure 5.66.8
Single-hand mixer for sink with fixed spout
(Hansaronda, Hansa)

Figure 5.66.7
Wall-mounted tub fitting with automatic reset and shower head
(Hansalux with Hansaduojet-S, Hansa)

Figure 5.66
Tap fittings

5.8 Fire protection/fire-extinguishing equipment

Fire protection, especially with water as extinguishing agent, is an important theme in the context of sanitation technology. Unlike air conditioning, elevator or WC installations, fire protection systems in buildings are not designed for daily use. They are preventative measures in case of fire, that is, an investment that is not of immediate use to the client. For these reasons, costs for fire protection systems are often questioned. One of the architect's many tasks is therefore to explain to the client that additional costs incurred for an appropriate fire protection system are justified and economical in terms of the greater security they provide. Moreover, appropriate fire protection systems help lower insurance rates, at times even in the case of additional structural measures.

5.8.1 Fire-extinguishing equipment (hydrants)

Building construction and fire-extinguishing equipment must always conform to the fire department codes. Structural measures such as division into fire compartments, use of fire-resistant building materials, placement of doors and stairs etc. affect the architect's planning. Fire-extinguishing equipment must be included in office and industrial buildings, public buildings, department stores, cinemas and theatres etc., in short, wherever large gatherings are anticipated or where an elevated risk of fire exists. The equipment consists of wet, wet/dry and dry piping for fire fighting, rain walls, area sprinklers, automatic sprinklers and other means of direct fire-fighting.

The installation of fire-extinguishing components must be approved by the building authorities or the fire department (this also determines the pipe diameters/water supply etc.). The necessary supply pipes, taps etc. should be integrated into the building's wall systems to ensure sufficient refresh water supply to all wet systems at all times. Piping for fire fighting and fittings must be manufactured from non-combustible materials.

The following fire-extinguishing systems are currently employed:

Stand pipes for firefighting (wet) with wall hydrants
Suitable for use by the fire department and any person available on site to extinguish a fire (emergency firefighting).

The following criteria must be considered when planning for wall hydrants:
- Minimal flow pressure, 3 bar (min. 100 l/minute) at hose connection valve
- Minimal diameter of connecting pipe DN 50
- Diameter of riser i.e. distribution pipe DN 80
- Simultaneity factor 3 (min. 200 l/minute)

The diagram in **Figure 5.67** shows the connection of floor-to-floor piping and wall hydrants via a joint riser as well as the overall system layout.

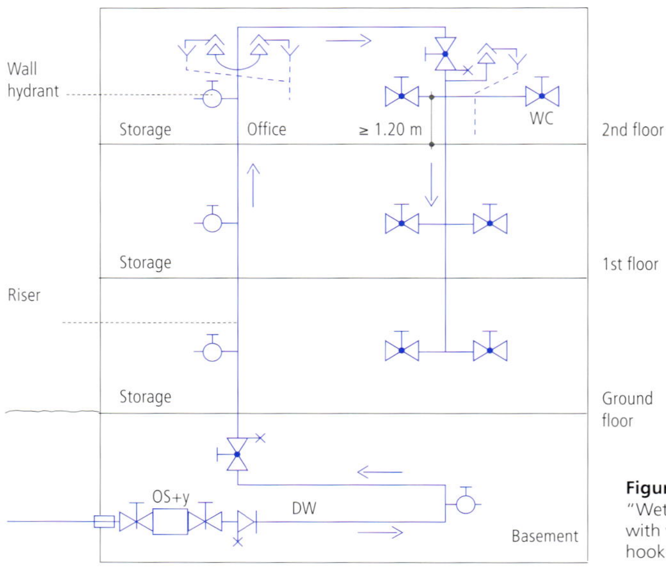

Figure 5.67
"Wet" fire-extinguishing system with wall hydrants and consumer hook-ups

Stand pipes for firefighting (wet/dry) with wall hydrants
Suitable for use by the fire department and any person available on site for emergency firefighting.

Wet/dry water pipes are designed to supply water for fire fighting from the drinking water piping with no or only minimal delay. Moreover, they prevent the formation of stagnant water, which is no longer suitable as potable water. Since the pipes are only filled with water in case of fire and remain dry at all other times, there is no risk of water freezing in the pipes, thereby obviating the need for frost-proof installation of the risers.

Figure 5.68 contains a schematic of a (wet/dry) fire-extinguishing system with valve stations.

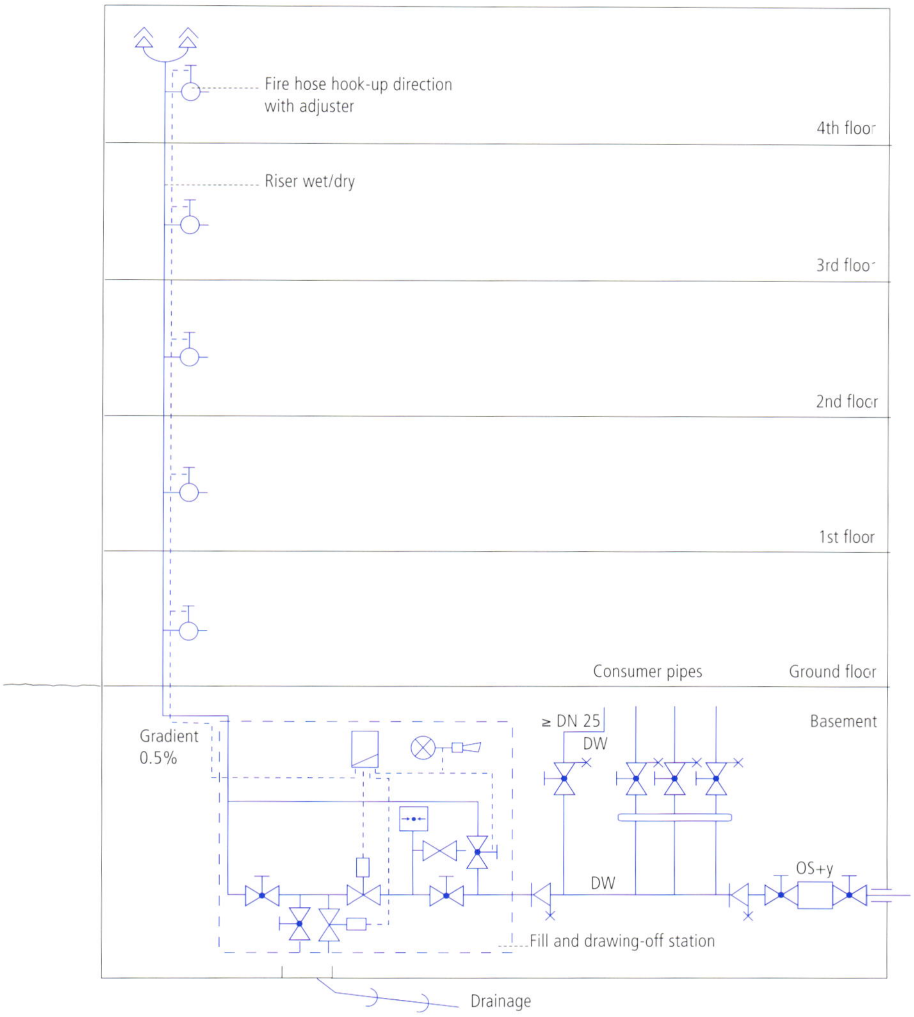

Figure 5.68
"Wet/dry" fire-extinguishing system with valve station

Stand pipes for firefighting (dry) with wall hydrants

Suitable for use by the fire department.

"Dry" pipes for firefighting with wall hydrants are designed for the exclusive use of the fire department. The "dry" riser pipes are water pipes for firefighting, which are supplied with water by the fire department only when needed. In case of fire, water for firefighting can thus be drawn off without the time-consuming process of laying out hose connections. These riser pipes must have no direct link to other water piping. If several riser pipes are installed in a building, each riser must be laid separately and equipped with a dedicated feed.

Figure 5.69 contains a diagram of a "dry" pipe for firefighting and the connection. In case of fire, this system is supplied with water from fire engines or indirectly via hydrants, with access to water for firefighting provided by wall hydrants installed on each floor.

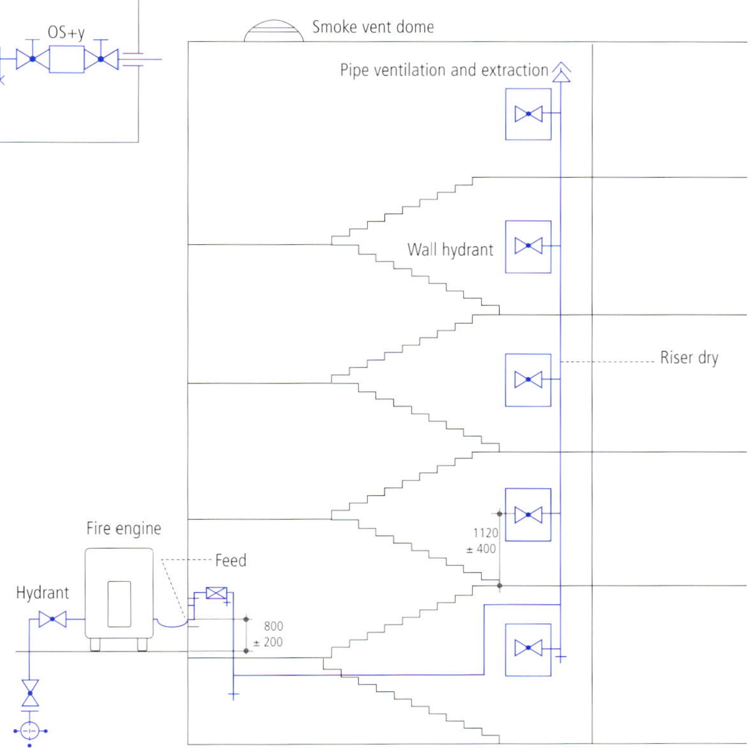

Figure 5.69
"Dry" fire-extinguishing system with feed

5.8.2 Hand-operated fire-extinguishers

Hand-operated fire-extinguishers must be installed in all buildings for a variety of (minor) fire scenarios. Their number, type, size and location must be approved by the local fire department.

Only hand-operated fire-extinguishers suitable and certified for fire rating A, B, C, D (DIN EN2) must be used. These include:
- Dry powder fire-extinguisher with glowing fire powder
- Dry powder fire-extinguisher with metal fire powder
- Dry powder fire-extinguisher with special powder
- Carbon dioxide fire-extinguisher
- Water fire-extinguisher
- Foam fire-extinguisher

Hand-operated fire-extinguishers and wall hydrants are often housed in a joint box in stairwells. Figure 5.70 illustrates this type of box and its measurements.

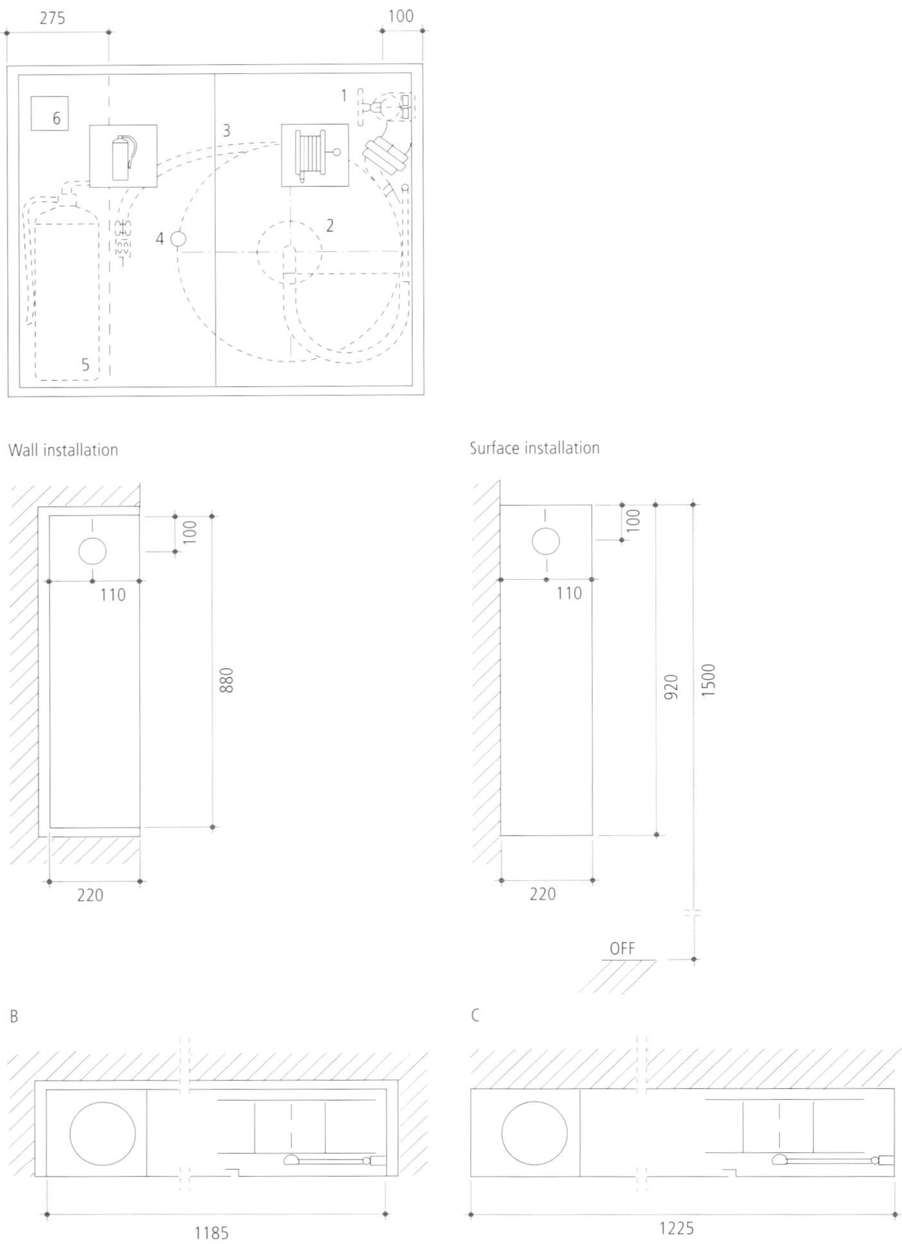

Figure 5.70
Wall hydrant

1 Hose connection valve
2 Water-bearing hose reel
3 Dimensionally stable pressure hose
4 Steel pipe
5 Powder extinguisher
6 Wall cabinet for hand fire-extinguisher

Figure 5.71
Sprinkler pipe and sprinklers
(Image: Krantz – TKT)

5.8.3 Sprinkler systems

Sprinkler systems are the most common, autonomous water-based fire-extinguishing systems. They operate on the principle of early fire detection and targeted firefighting. The sprinklers are screwed onto a pipe distribution network that is linked to the water supply and runs across the targeted building or specific areas within the building. In case of fire, thermal sensors triggered by the incendiary heat open those sprinklers in whose proximity a fire has erupted. Sprinklers disperse fire extinguishing water across an area. Sprinklers located at some distance from the source of fire remain closed. Some 200 l of water are dispersed by an opened sprinkler per minute, which means that in 75 % of all cases a fire is extinguished by four simultaneously opened sprinklers. When sprinklers are set off, the resulting flow of water in the pipe distribution network triggers connected pieces of alarm equipment, which automatically relay alarms to the fire department in the district.

Sprinklers are manufactured and tested according to stringent instructions, e.g. by the Verband der Sachversicherer (or Association of Property Insurers (VdS)/Federal Republic of Germany), with production restricted to licensed manufacturers.

Because of the high effectiveness, reliability and damage control these systems deliver, insurers offer far greater premium discounts than for any other fire protection systems (e.g. fire alarm systems, smoke alarm systems etc.).

Sprinklers are extinguishing nozzles, kept closed by means of a sealant and a heat sensitive breakable bulb. The bulb is filled with a liquid that responds to fire by expanding until it bursts. This opening temperature of the sprinkler should be roughly 30 °C above the maximum anticipated room temperature in order to avoid false alarms. **Figure 5.71** depicts a sprinkler head with baffle plate.

The authorized sprinkler distances (protected area per sprinkler) are adapted to the existing fire risk. In low risk areas, sprinklers are spaced to cover an area of 21 m^2; the distances are decreased to a max. of 9 m^2 each for high-risk situations.

To supply sprinklers with water for firefighting, pipes are installed in the protected building into which the sprinklers are screwed. The nominal diameter of the pipes is determined by means of a hydraulic calculation (similar to the calculation used for cold water pipes).

The pipe network itself is generally filled to full capacity with water and subjected to an operating pressure of approx. 10 bar (wet system).

In regions where there is a risk of frost, dry systems are installed instead of water-bearing sprinkler pipes; in these cases, the sprinkler pipe is filled with compressed air when in a state of readiness. Such system components are only flooded in response to the pressure drop that results from opened sprinklers in case of fire.

The type of water supply is determined by the building's size and fire risk rating. For large objects and in cases of elevated fire risk an "inexhaustible" source of water must be provided in addition to the connection to the drinking water pipes. This additional source supplies water to the sprinklers via pumps linked to a storage tank.

If space is an issue, the storage tank (inexhaustible water source) can be replaced by intermediate containers, which are automatically filled in case of fire. The necessary sprinkler pumps and valve stations **(Figure 5.72)** are housed in a separate room. The sprinkler pumps are either electrically powered or driven by diesel engines; very large sprinkler systems must be equipped with a second, independent energy source (pump with e-motor/diesel-powered pump).

Figure 5.72
Valve station of a sprinkler installation
(Image: Krantz – TK)

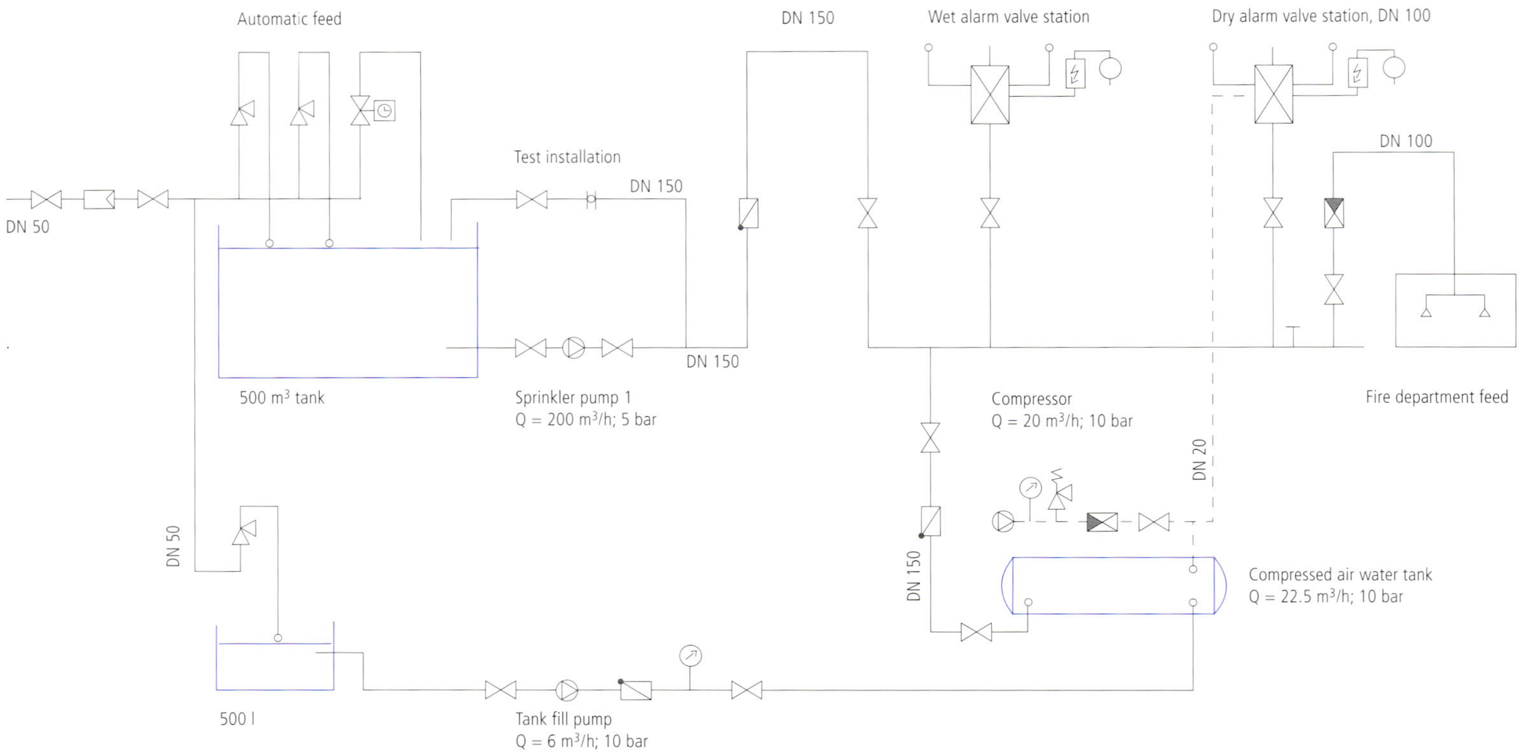

Figure 5.73
Sprinkler system plan

- Main pipe, wet
- Compressed air pipe
- Shut-off valve
- Dirt trap
- Back pressure valve
- Back pressure flap
- Safety valve
- Manometer
- Float valve
- Metering orifice
- Alarm valve station

Water spray fire-extinguishing systems

Water spray or spray-flooding systems are designed similarly to sprinkler systems. The difference is that the standard closed sprinklers (described above) attached to the pipe network above the targeted area are replaced by open extinguishing nozzles. The pipe network is not fed with water in its state of readiness; instead all nozzles are supplied simultaneously with water in the case of fire. Hence spray/flooding systems do not extinguish in a selective manner as sprinkler systems do and are therefore only employed in cases where the risk of a rapid spread of fire exists (flashover).

Figure 5.73 illustrates a typical layout and functional diagram of a sprinkler system.

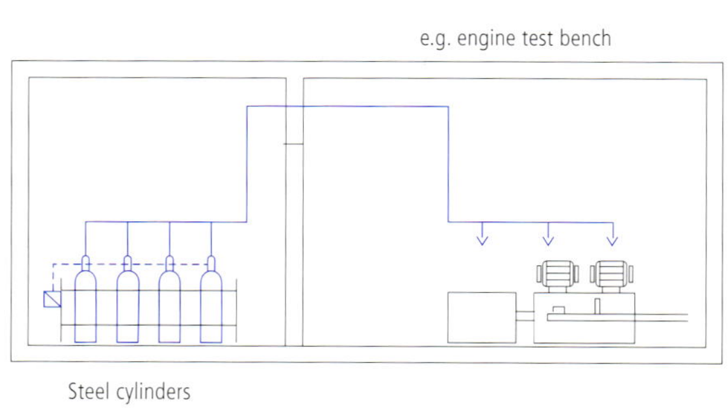

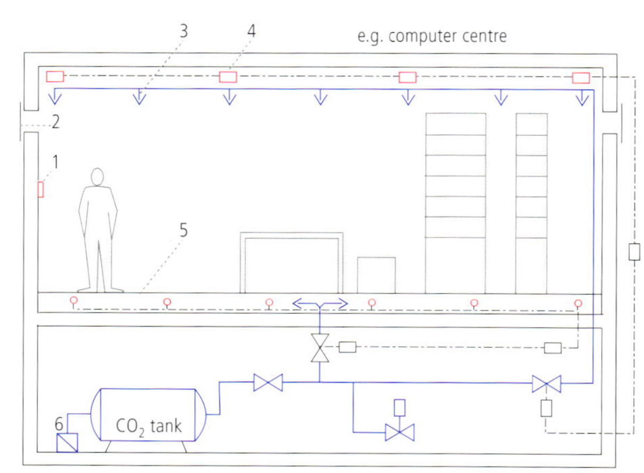

Figure 5.74
Fire-extinguishing systems with CO_2 for object and room protection

1 Alarm device
2 Quick-acting stop valves for pressure equalization
3 Extinguishing nozzles
4 Smoke detectors
5 Ionization detectors
6 Refrigerating set

5.8.4 CO_2 fire-extinguishing systems (Carbon dioxide fire-extinguishing systems)

Fire-extinguishing systems of this type are employed in areas where fire cannot be extinguished with water or where water cannot be used because of the resulting damage to the building and equipment. CO_2 is a colourless gas. It is clean, non-corrosive, non-conductive and does not leave a residue. It extinguishes a fire by means of oxygen depletion. Thus, CO_2 fire-extinguishing systems are employed in commercial paint- and enamelling shops, submersion installations, paint mixing rooms, dry kilns/furnaces, lacquer cabins in lacquer shops, in electrical installations for transformers, generators, IT centres (where protection is primarily aimed at the subfloor), switching stations, radio and television stations, manufacturing rooms, rotation print shops, test benches, paper manufacturing machines, film and foam manufacture, spinning mills, refineries, in warehouses for chemical products and/or combustible liquids, fur storage, fine precision mechanical products and switching devices etc. CO_2 fire-extinguishing systems are divided into systems for space protection and systems for object protection.

Space or area protection is chosen only in cases where entire rooms require fire protection. It is important to note that all openings and doors in such rooms must close automatically in case of fire.

In IT rooms in particular, the fresh air duct in the hollow floor must be automatically locked down in addition to the entire ventilation system. At the same time, overflow pipes must allow the room air to evacuate in sealed rooms (CO_2 displacement).

Object protection is an appropriate choice if only a specific machine or similar equipment requires protection in a given room where it can act as a source of fire or if additional, targeted protection is desired for very fragile equipment or machine parts (generators or transformers, milling machines, film and paper machines, engine test benches, cables in hollow floors in IT rooms, tape storage and similar situations).

The carbon dioxide fire extinguishing system consists of a CO_2 supply system, distribution station, pipe network and custom nozzles. Gas can be supplied from steel bottle batteries or a large special CO_2 deep-freeze containers. For small and medium systems, liquid CO_2 is stored in 30 kg steel bottles at room temperature (not above 45 °C). The economically efficient limit for bottle and tank storage lies somewhere in the region of 3000 kg storage volume.

Heavily insulated tanks in which liquid CO_2 is stored at temperatures ranging from -27° to 20 °C are used for large systems. Small refrigerating machines with direct evaporators provide constant cooling. It is common practice in large-scale industrial operations to draw CO_2 from the same container for purposes other than firefighting, e.g. for inert gas shielded arc welding or in beverage processing plants, provided the required volume of the fire-extinguishing agent is secured at all times.

The volumes of fire-extinguishing agents are calculated according to the fire material to be extinguished and the size of the rooms or objects (according to VdS directive 2093) being protected.

The fire-extinguishing process itself is triggered by smoke sensors located in the relevant room or protected area; the sensors simultaneously emit an acoustic alarm. CO_2 is expelled from the nozzles after a delay to allow occupants to vacate the room, since CO_2 can cause death by asphyxiation .

In some cases CO_2 fire extinguishing systems are triggered by ionization or thermal sensors, instead of smoke sensors, because the former respond more quickly. **Figure 5.74** shows the schematic layout of CO_2 fire-extinguishing systems for object and room protection.

Halon fire-extinguishers were developed in past decades to eliminate the principal disadvantage of CO_2 fire-extinguishing systems (risk of asphyxiation) and to reap the full benefit of the advantages such systems offer. However, these have since been prohibited in the wake of CFC legislation. In the meantime, extinguishing gases such as inert gas and argon have been used as alternatives. Inert gas fire-extinguishing systems are ideal for situations where rapid firefighting is a priority and damage caused by fire-extinguishing agents must be avoided at all cost. They are thus the preferred solution for the protection of computer centres and other electrical and electronic equipment, e.g. control and routing rooms, telecommunications centres, data security centres etc. Another essential advantage of extinguishing electronic equipment by this method rather than with CO_2 fire-extinguishing systems is that the supply of the gaseous extinguishing agent does not produce any evaporation as it flows out of the nozzle, which means that there is virtually no cooling in the affected area. The much-feared "cold shock" to which sensitive equipment may be exposed as a result of dry ice formation is thus avoided.

Besides CO_2 and argon, nitrogen and arginines are also employd as fire-suppressing agents.

5.9 Central installations/shafts/horizontal installations

Central sanitation installations for warm water processing fixtures, perhaps smaller heat generating systems, cold and warm storage, pump groups, control cabinets etc. require approx. 4 m^2 per 1,000 m^2 usable area of the relevant object, without consideration for special fire-extinguishing equipment. The percentile area demand increases dramatically when water demands rise steeply as a result of special equipment and operations, for example in commercial kitchens, manufacturing sites etc.

The scale of the fire-extinguishing equipment with pump stations, valve stations, pressure water tanks and inexhaustible water sources may equal that of the sanitation centre itself, although it may also be considerably larger. In the case of sprinkler systems, especially, it is important to locate the inexhaustible water sources (10 – 150 m^3 water) in areas that have few other uses (e.g. near the departure gate in an underground garage). The necessary shaft areas for sanitation installations are not only a question of dimension but also of correct placement; they generally represent roughly 5 % of the area required for the technical installations.

Detail of apartment plan

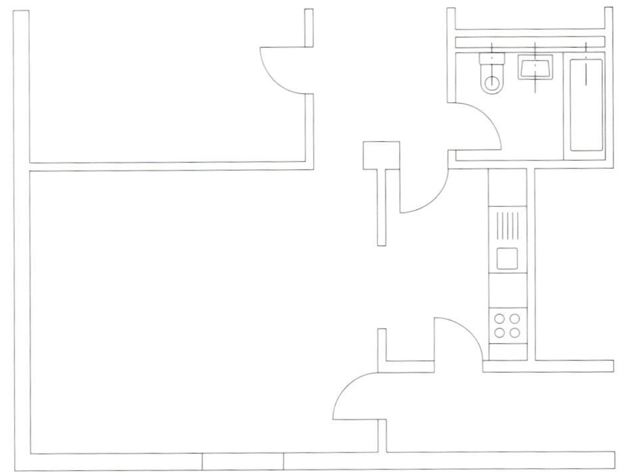

Bathroom detail, step 1

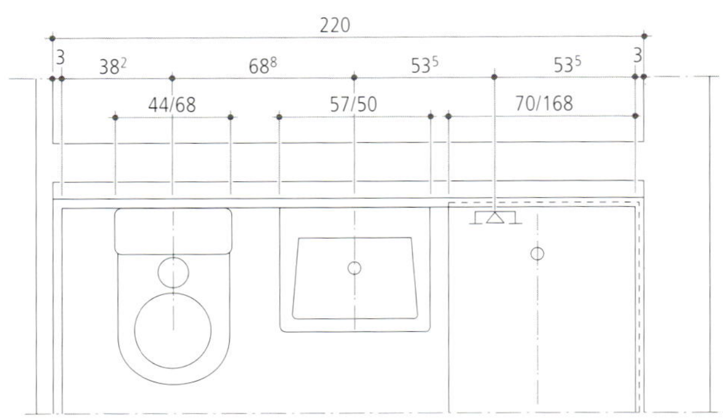

Figure 5.75.1
Planning scheme for a bathroom

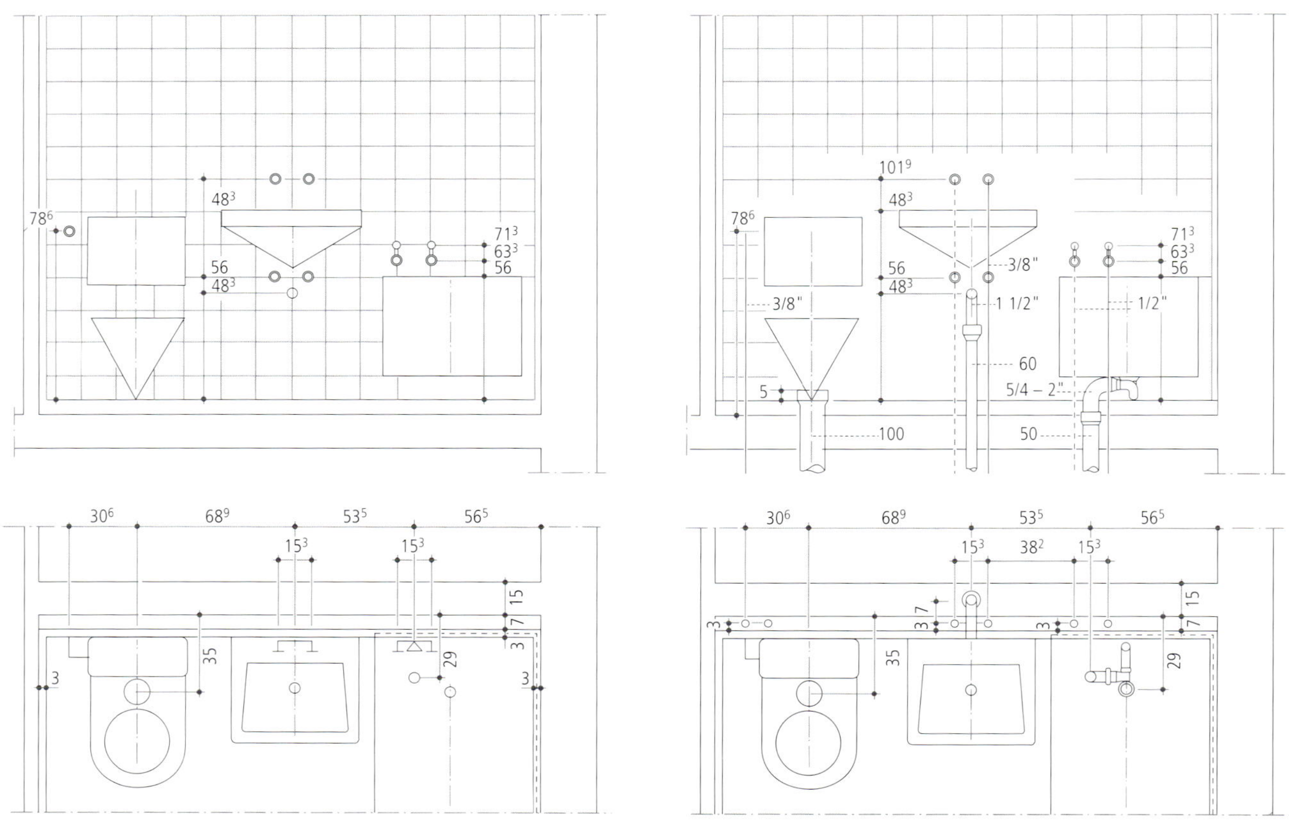

Figure 5.75.2
Bathroom details

Figures 5.75.1/2 illustrate the challenge. The apartment floor plan includes one small sanitary block (bathroom) with only three objects (WC, sink and bathtub).

Lined up in a row, the three objects must be arranged to be user-friendly and also accommodate a flagged mirror. This results in a parallel installation of the waste pipes for warm water, cold water and sewage for the various objects. Even for this small project the resulting shaft width is 2.2 m for a required depth of roughly 0.3 m. Generally, this shaft serves to accommodate the vertical pipes and must also offer sufficient space to allow vertical and horizontal pipes to cross. The pipes, laid one in front of the other and with basic insulation, may nevertheless have an outer diameter of roughly 100 mm despite their minimal sizing. To ensure that insulation can be properly installed, minimal distances must be kept (see chapter on insulation). Although only the WC has a diameter of 100 mm in the configuration shown, the smaller pipe diameters and relevant insulation also contribute to similar measurements.

Installation in front of the wall allows for optimum use of the available space **(Figure 5.75.3)** as well as offering simple and rapid assembly and installation.

For horizontal installations of water and sewage pipes, special attention must be given to soil stacks, which, as mentioned earlier, must fall at least 2 cm per each metre of pipe length. This may quickly lead to problems if longer fall pipes must be laid from appliances to a shaft. The best solution for a flexible plan suited for a variety of requirements is to erect a floor structure of up to 30 cm in which relevant fall pipes can be laid to a maximum of 30 – 35 m.
This demand is still somewhat unusual and is rarely executed, although it will be increasingly necessary to fulfil the requirements of mobile real estate.

Figure 5.75.3
Wall-mounted installation system

The DAL-RAPID wall-mounted installation system allows for highly individualized bathroom planning, as well as rapid and convenient (clean) installation

1. Building performance

2. Comfort

3. Integrated approaches to planning

4. Heating systems

5. Sanitary and fire-extinguishing systems

6. Ventilation and air-conditioning technology

7. Refrigerating and cooling systems

8. Power plants

9. Illumination

10. Daylight technology

11. Low-voltage current systems

12. Conveying systems

6. Ventilation and air-conditioning systems (HVAC systems)

6.1 Natural ventilation (free ventilation)

In the pursuit of fully or at least predominantly natural ventilation for buildings, it is important to remember that there are factors that limit options for natural ventilation. Some of these are:
- Maximum room depth
- Wind forces at the site
- Frequency of calms (no wind)
- Emissions
- Thermal properties in the room

The first step in planning for natural ventilation is therefore to determine the frequency and prevalent wind directions at the site, that is, to evaluate whether the site is suited for natural ventilation at all. Window types and other openings also play an important role in natural ventilation. It is also essential to establish whether emissions present at the site may preclude natural ventilation or whether dust, noise, odours and other factors limit the degree to which natural ventilation is useful.

An interesting topic is the challenge of providing natural ventilation in a building with extreme room depths, or linking or covering building sections with large glass houses or winter gardens. As has been shown in practice, these challenges inspire interesting solutions that address a wide range of individual themes.

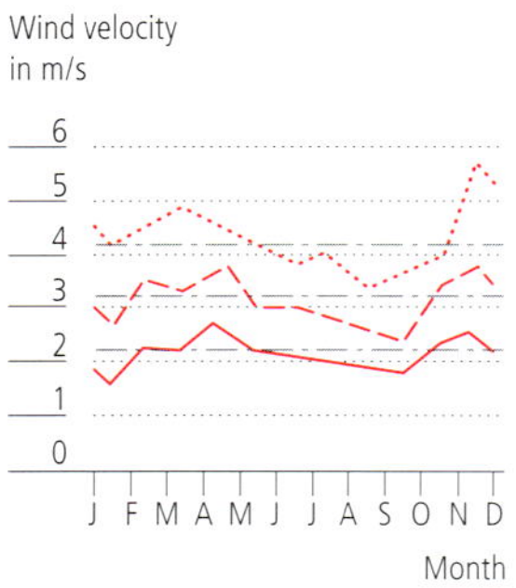

Figure 6.1
Mean monthly wind velocity in different cities, DIN 4710

...... Hamburg
– – Munich
—— Stuttgart

6.1.1 Wind forces

Based on the assumption that it is generally known how wind occurs, the site analysis is therefore focused on establishing the frequency, degree and prevalent wind directions to which the building is exposed.

In Central Europe wind directions, frequencies and velocities are documented in measurements gathered at particular stations (data gathered in adherence to DIN 4710). The differences in the daily changes of air velocities are relatively small and do not need to be considered for the purpose of calculations (HVAC computation). The annual statistics, in contrast, reveal that average wind velocities are greater in winter than in summer in Central Europe, and that velocities are considerably higher in coastal regions than inland.

Figure 6.1 indicates average monthly wind velocities in different cities. The percentages of annual occurrences of calms (no wind) are of interest in this context. The graph in **Figure 6.2** illustrates that calms do not necessarily correspond to the rules for coastal inland regions; instead, the locations of the cities seem to play a considerable role. Thus Stuttgart's 10.4 % annual rate of days without wind is a direct reflection of the city's location in the cul-de-sac of a valley.

It is also interesting to note that the average wind force is generally lower when temperatures are extreme (very high or very low) than during average temperatures. In other words, wind velocities are lower in winter and at the height of summer than during the transitional seasons (**Figure 6.3**).

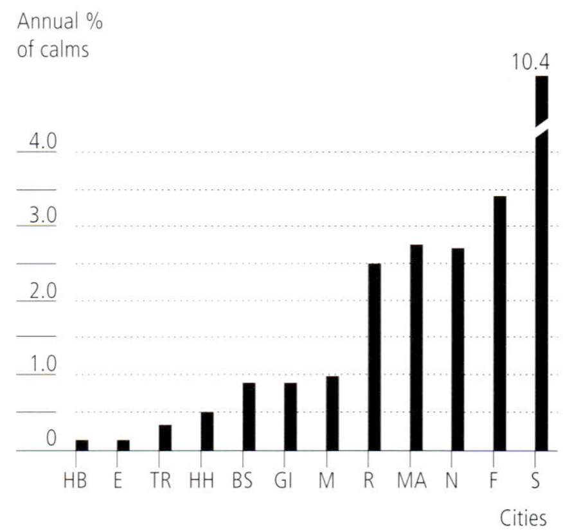

Figure 6.2
Annual % of days without wind in different cities (indicated by licence plate code)

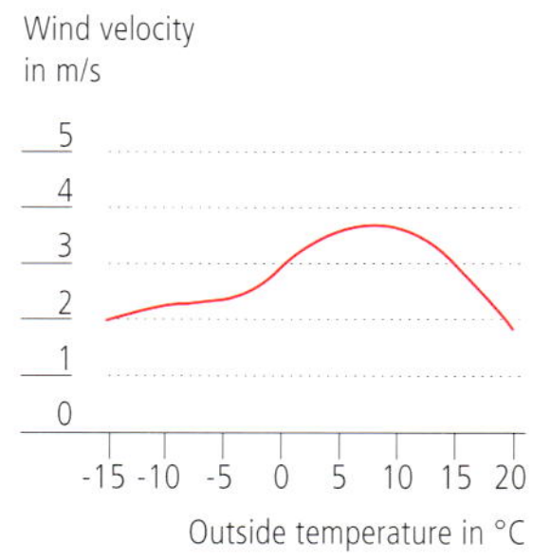

Figure 6.3
Mean wind velocity depending on outside temperature in Hamburg

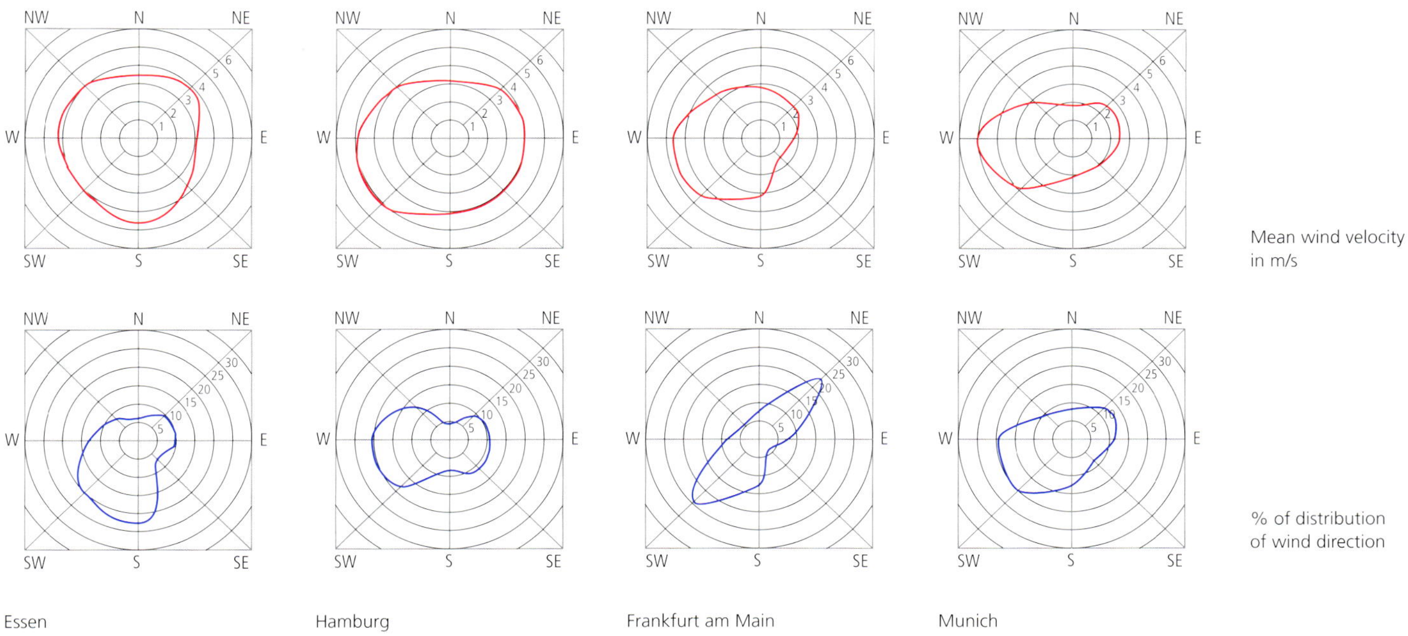

Figure 6.4
Wind statistics for Essen, Hamburg, Frankfurt am Main, Munich

Figures 6.4.1 to 6.4.6 indicate wind velocities, as well as average annual wind directions and their distribution for various locations. The readings at all locations clearly indicate predominant northeasterly and southwesterly winds in Central Europe.

When a building is exposed to wind, positive pressure is created on the side facing the wind (windward or weather side) and negative pressure on the side that is protected from the onslaught of wind (lee). Figure 6.5 also demonstrates that negative pressure fields are created on the sides that are not directly exposed to wind. This realization is vital when planning naturally ventilated buildings and it is especially important when buildings with extreme depths are to be naturally ventilated. Here, the issue is how negative pressure can be increased to improve the effectiveness of natural ventilation.

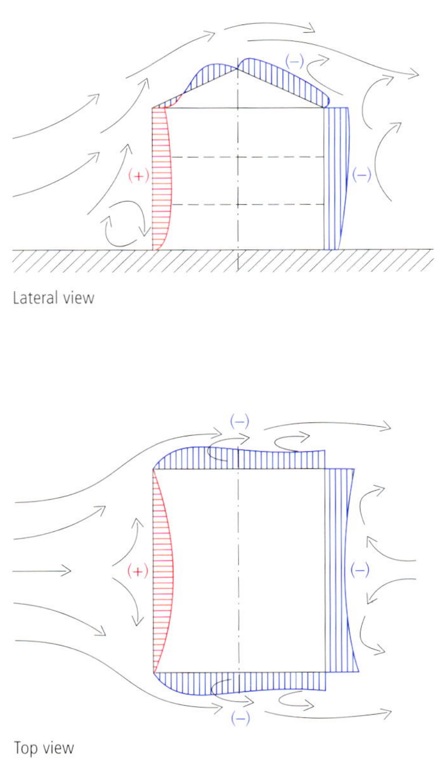

Figure 6.5
Wind forces at building

6.1.2 Window ventilation

In principle, rooms suitable for natural ventilation are those whose maximum depth is no more than 2.5 times the clear height (for ventilation from one side), or 5 times the clear height in rooms where windows on opposite sides allow for cross-ventilation (ventilation from two sides). Permit-granting building authorities generally recognize these maximum room depths as the standard parameters of a room where natural ventilation is feasible. Theoretically, rooms could thus be adapted for natural ventilation by increasing the heights as required.

Wind penetration and air currents in the room change both in response to the ratio between ambient temperatures (external temperatures to room temperatures) and in response to outside air velocity. **Figure 6.6** contains a schematic representation of air currents as a result of natural ventilation in conjunction with a variety of parameters. **Figures 6.7.1** and **6.7.2** offer examples of how room air velocities change for different temperatures and different dynamic pressures on the facade.
As temperature differences increase, air moves more evenly throughout the room, as **Figure 6.7.1** shows. The same is true when the pressure exerted on the facade increases.

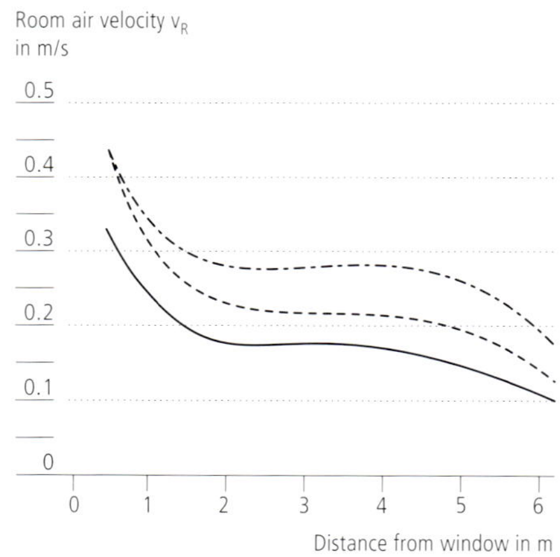

Figure 6.7.1
Velocity V_R plotted to room depth for various temperature differences between inside and outside $t_R > t_O$
(velocities measured roughly at head height)

-·- P_{dyn} = 10 Pa; (≈ 4m/s); Δt = 3K
--- P_{dyn} = 10 Pa; (≈ 4m/s); Δt = 2K
— P_{dyn} = 10 Pa; (≈ 4m/s); Δt = 0K

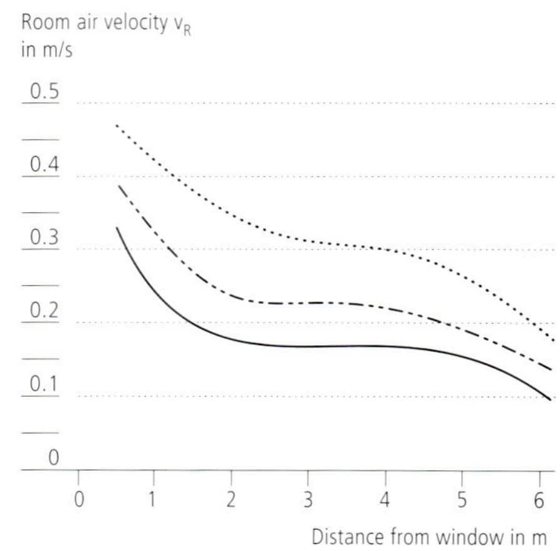

Figure 6.7.2
Velocity V_R plotted to room depth for different dynamic pressures
(velocities measured roughly at head height)

····· P_{dyn} = 30 Pa; (≈ 7.2m/s); Δt = 0K
-··- P_{dyn} = 20 Pa; (≈ 6.0m/s); Δt = 0K
— P_{dyn} = 10 Pa; (≈ 4.0m/s); Δt = 0K

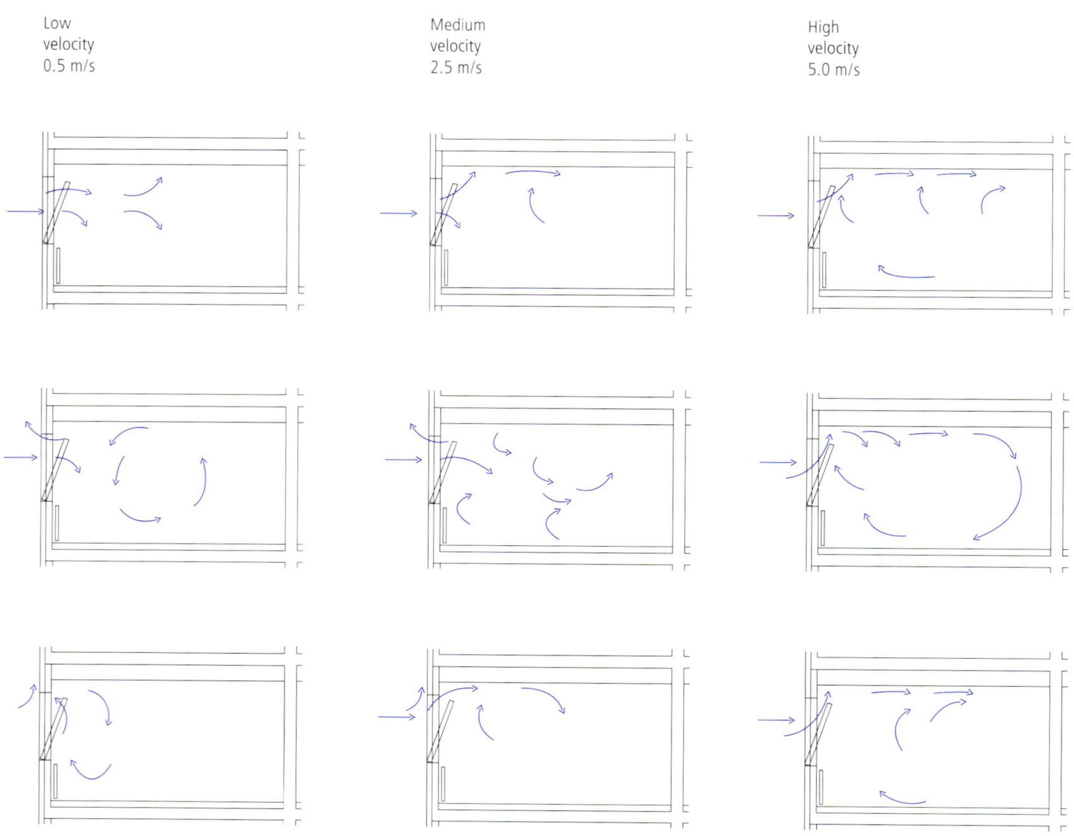

Figure 6.6
Diagram of air movement for natural ventilation and different temperature gradients and influx velocities

The efficacy of window openings for natural ventilation is shown in **Table 6.1** (based on VDI [Association of German Engineers] 2719). Efficacy ratings 1 to 5 are defined as follows:

1. Too little air change – fresh air levels are unhygienic for occupants
2. Marginally hygienic fresh air levels
3. Air change acceptable in all regards
4. High air change rates with strongly diminished CO_2 content (may require adjustment by closing windows)
5. Very high air change rates, suitable for rooms with large crowds and short bursts of ventilation

Finally, the graph in **Figure 6.8** plots the air change in a model room as a function of wind direction, velocity and opened window area. Since a 3 – 4ac/h (air change per hour) is generally sufficient provided there is good natural ventilation, observation has shown that minimal open window areas suffice when wind velocities are high (e.g. 5 m/s) to achieve appropriate air changes. The example documents that 50 % of open window area combined with an oncoming wind velocity of 3 m/s from the east or northwest achieves air changes in the range of 4.5 to 13.3.

Window opening type	Position	Efficiency				
		1 none to low AC ***) 0.2 to 0.8 h^{-1}	2 low to barely sufficient AC ***) 0.8 to 1.5 h^{-1}	3 medium AC ***) 0.8 to 4 h^{-1}	4 high AC ***) 4 to 9 h^{-1}	5 very high AC ***) 9 to 50 h^{-1}
Casement and tilt-and-turn	closed	●				
	turn open *)		●	●		
	turn open				●	
	turn open with opposite openings					●
Tilt-and-turn Hopper Top-hinged	closed	●				
	tilt open		●			
	tilt (top-hinged) open			●		
	tilt (top-hinged) open with opposite openings				●	
Vertical sash	closed	●				
	sash partially opened		●	●		
	sash fully opened				●	
	sash fully opened with opposite openings					●
Sliding (vertical or horizontal)	closed	●				
	slider partially opened		●	●		
	slider fully opened				●	
	slider fully opened with opposite openings					●
Ventilation without fan **)	closed	●				
	open		●	●		
	open with opposite openings				●	
Ventilation with fan **)	closed	●				
	fan operating, depending on setting			●	●	

Table 6.1
Efficiency of window opening types for ventilation

*) So-called partial ventilation positions of window casements are made possible with the help of additional window fittings; these positions often create sufficient air changes, especially in cool seasons.
The following systems are available:
- Scissors that limit the opening width
- Window stopper and window adjuster for casement and tilt-and-turn windows; this makes it possible to fix the casement in any position
- Handles and gears with adjustment devices; this allows for various tilt positions
- Additional window fittings to adjust hopper positions
- Gap limiters for sash windows (with optional locking device for burglar protection)
- Gap limiter (also lockable) for sliding and vertical sliding windows
- Electrical operation of hopper windows (optional control by means of timers, heat sensors, wind measuring devices etc.)

**) Ventilation installations also refer to combination constructions for, e.g. at sound insulation, heat recovery, rain protection, insect protection etc. The requirements should be detailed for each individual case.

***) AC = Air change coefficient in h^{-1}

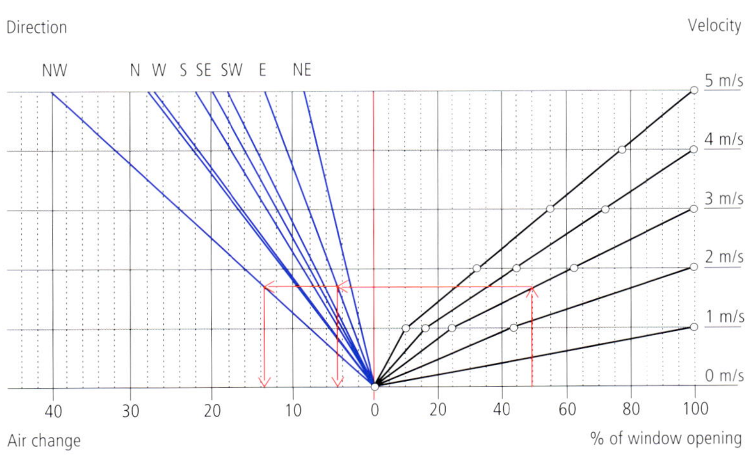

Figure 6.8
Air change as a function of wind direction, velocity and open window area

6.1.3 Stack ventilation

One approach to predominantly natural ventilation for very deep buildings or large open-plan spaces is to design the facade so that it permits a high degree of fresh air influx (window elements). This is complemented by creating shafts (stacks) in the building interior, which generate strong negative pressures in the area beneath the roof, thereby producing a suction effect that ventilates the building. **Figure 6.9.1** depicts a sample floor plan, while **Figure 6.9.2** identifies the peripheral zones, suited to natural ventilation, and the internal zones, where natural ventilation is usually not possible. Wind tunnel studies are an excellent tool to determine the type and manner of air currents acting upon the building from different directions and the resulting pressure coefficients in the facade area.

Figure 6.10.1 depicts the mass model of a building with surrounding built structures, **Figure 6.10.2** a floor model designed to determine and visualize the air currents on a particular building storey (floor model constructed in Plexiglas).

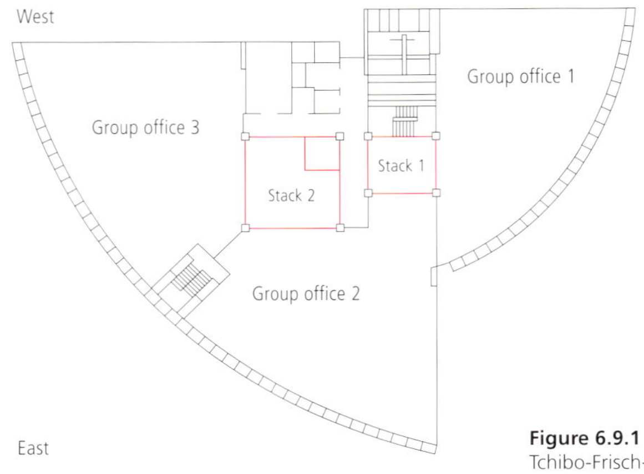

Figure 6.9.1
Tchibo-Frisch-Röst-Kaffee GmbH, standard floor plan
(Architects: Bürgin & Nissen)

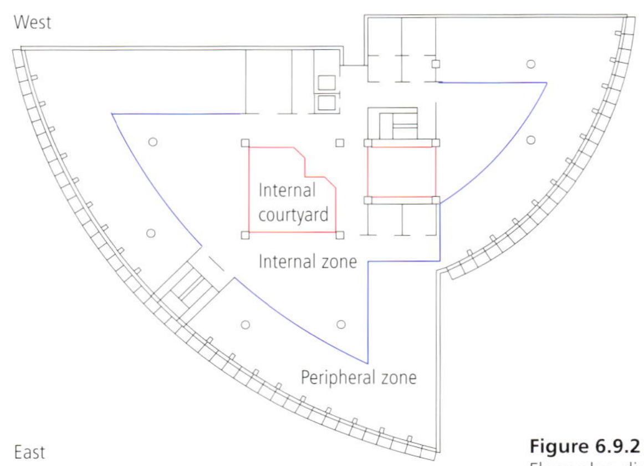

Figure 6.9.2
Floor plan divided into zones for natural ventilation

Figure 6.10.1
Mass model in wind tunnel with built environment

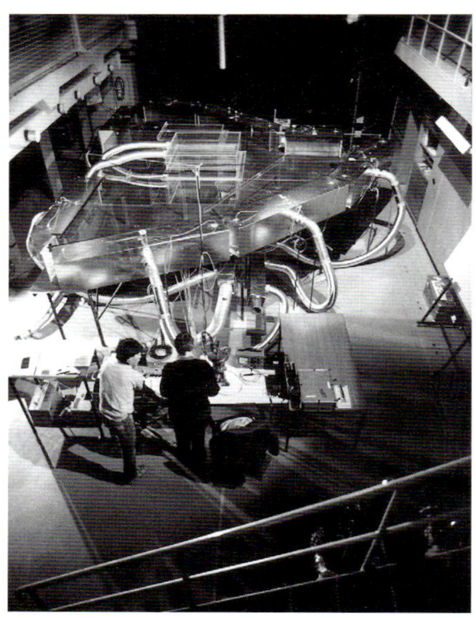

Figure 6.10.2
Large-scale model (1:10) of a floor in flow laboratory

Figure 6.11 illustrates rooftop structures with window sashes that promote negative pressure in the roof area and, depending on oncoming winds, also in the light well, thus creating natural cross-ventilation within the large room depths (flow direction: window – peripheral zone – internal zone – light well – roof).

When differences between inside and outside temperatures are sufficiently high, the interior is naturally ventilated even during periods of low wind velocities in winter. Air change is only achieved, however, when the outside air is colder than the air in the room. When isothermal conditions exist (inside temp. equals outside temp.), air movement cannot be achieved during periods when there is no wind (calms).

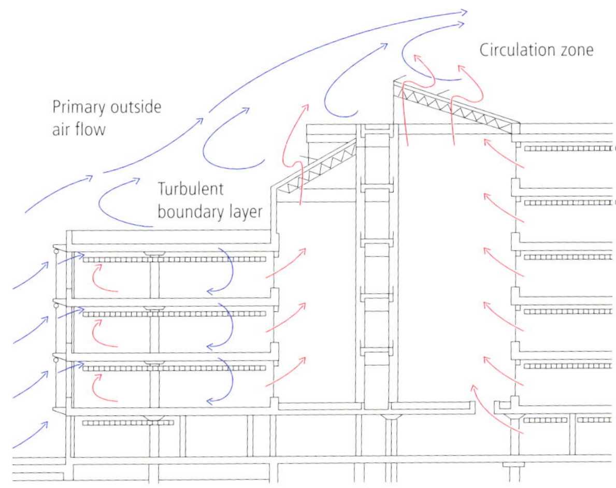

Figure 6.11
Cross-section of building with oncoming wind conditions and cross-ventilation

6.1.4 Ventilation by means of atria and winter gardens

Large winter gardens or glass-covered arcades are part of a growing trend aimed to improve the urban spatial programme in the immediate vicinity of a building. Figure 6.12.1 shows the plan for an administration building in Hamburg with winter gardens and a roughly 150-m-long glass-covered arcade. The isomet-ric drawing in Figure 6.12.2 depicts the architects' vision for the arcade (internal street). As the drawing clearly shows, the full 150 m length of the arcade is covered in glass, linking the adjacent building complexes. Figure 6.13 contains a section through the building with the arcade and adjacent winter gardens, as well as a view into the winter gardens.

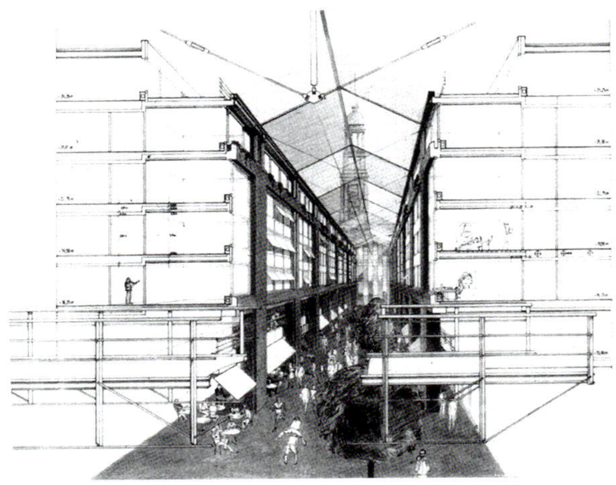

Figure 6.12.2
Isometric drawing of internal street

Figure 6.12.1
Site plan Gruner & Jahr AG (Design by Gerkan, Marg + Partner)

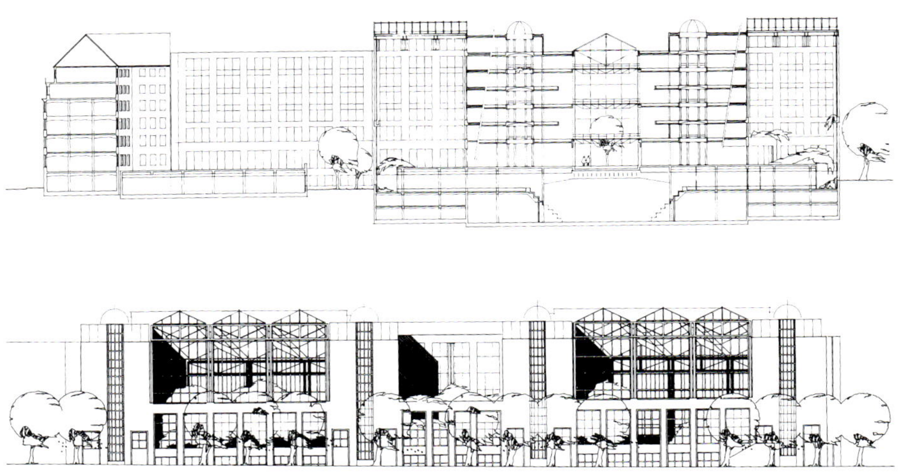

Figure 6.13
Elevation and section of building

Figures 6.14.1 to **6.14.6** contain photographic images and diagrams of the air currents both in the arcade and in the winter gardens for different oncoming winds (all smoke images taken at 2.0 m/s). As the individual images reveal, the documented arcades and winter garden spaces are filled with a fair amount of smoke (an indicator of strong cross-ventilation); the same is true for the adjacent spaces, which are naturally ventilated via the arcade or the winter gardens.

As previously noted, some concepts are not suited to natural ventilation and it may be sensible to use the winter gardens as "thermal buffer zones" (winter operation). Depending on the selection of plants for such winter gardens, the adjacent useable areas (here, offices) may require no or very little additional cooling in summer.

Figures 6.15.1 and **6.15.2** illustrate the desired approach to operating with natural ventilation via the arcades and winter gardens, respectively, for different seasons. It is important to consider that the transitional season represents 60 to 65 % of the entire year; in other words, natural ventilation is feasible for the useable areas during a very long period. A variety of solutions exists for natural ventilation in large glass atria, adaptable to each project and based on its geometry (for example, concave roofs, convex roofs, cupolas etc.).

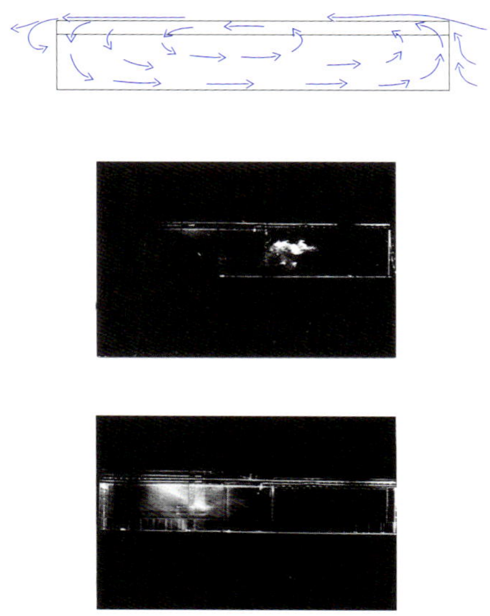

Figure 6.14.1
Wind flow parallel to internal street – longitudinal section

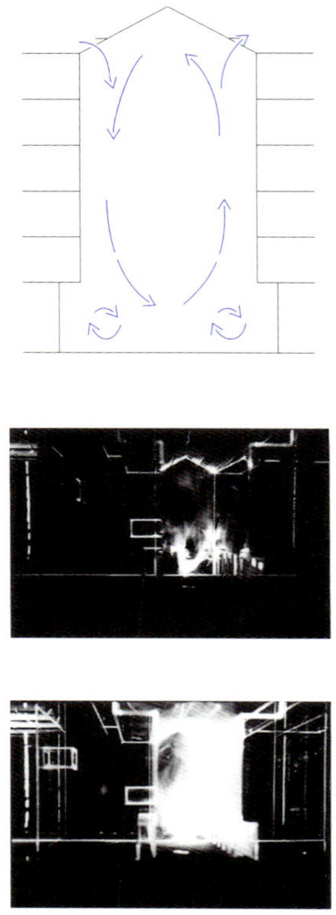

Figure 6.14.2
Wind flow parallel to internal street – cross-section

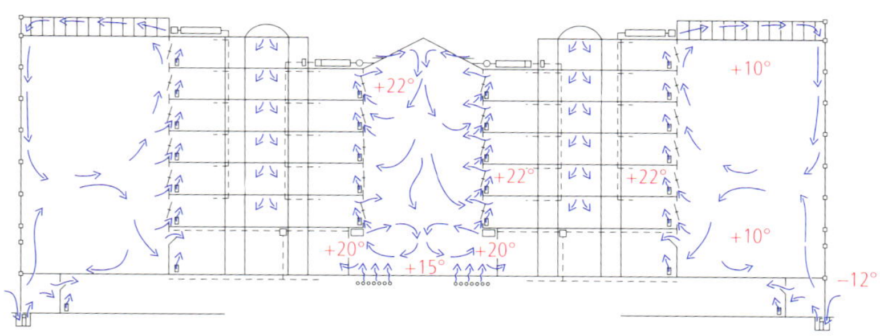

Winter operation
Overcast sky, moderate wind

in °C

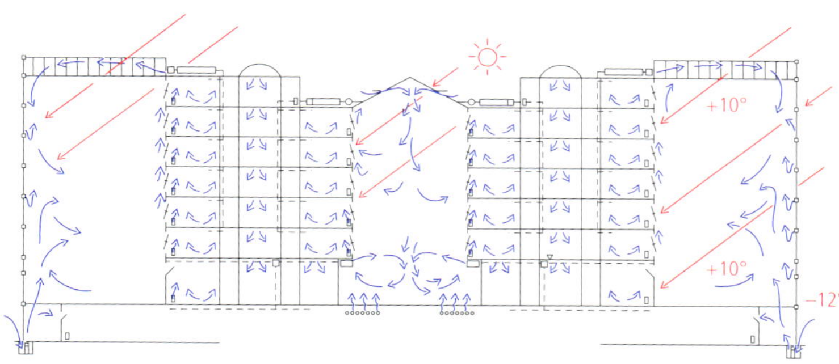

Winter operation
Clear, sunny sky, moderate wind

in °C

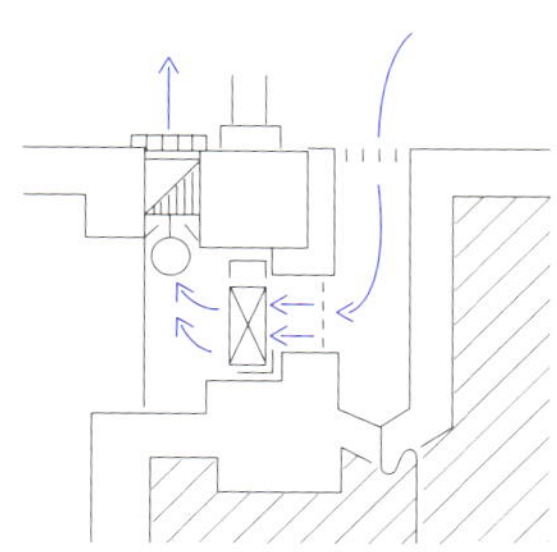

Detail

Figure 6.15.1
Schematic of operating conditions (winter)

Since precise calculation of natural air currents is a very elaborate process, computer simulation based on simplified assumptions, or model studies in the wind tunnel are used. It is always advisable to proceed with great care in designing natural ventilation for such large spaces in order to avoid potential planning errors.

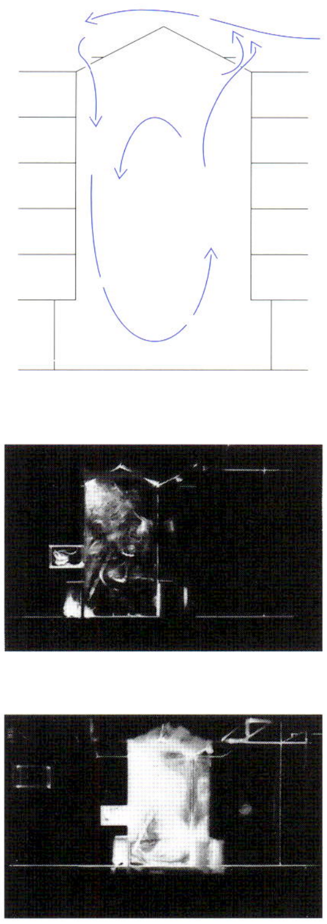

Figure 6.14.3
Wind flow across internal street – cross-section

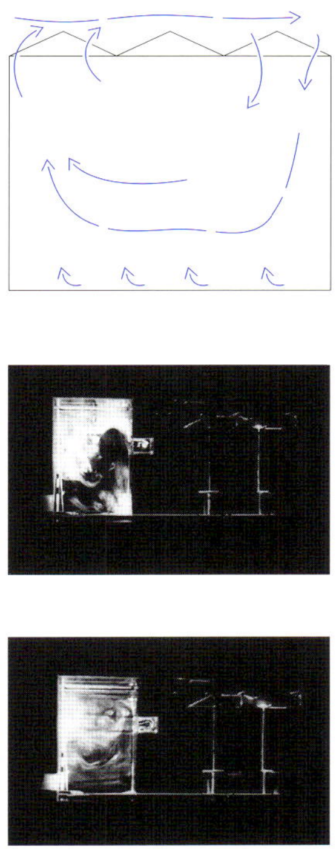

Figure 6.14.4
Wind flow parallel to longitudinal axis of winter garden – cross-section

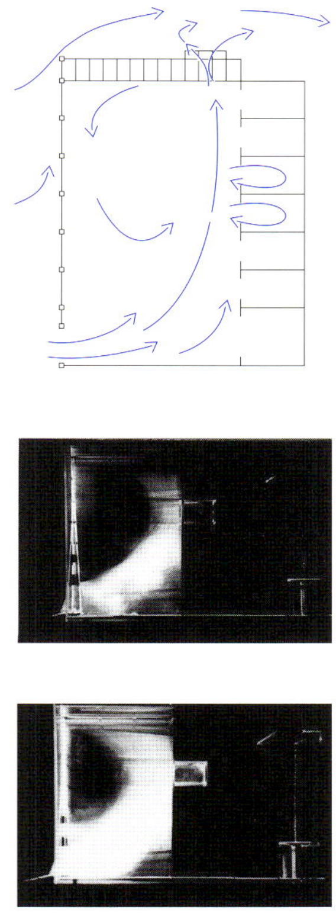

Figure 6.14.5
Wind flow across longitudinal axis of winter garden – cross-section – lee

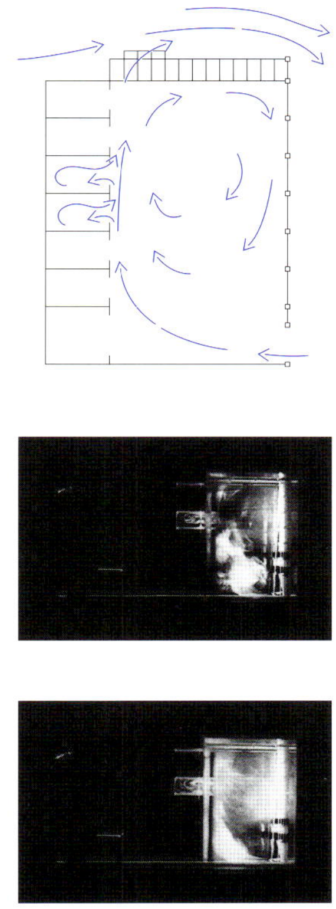

Figure 6.14.6
Wind flow across longitudinal axis of winter garden

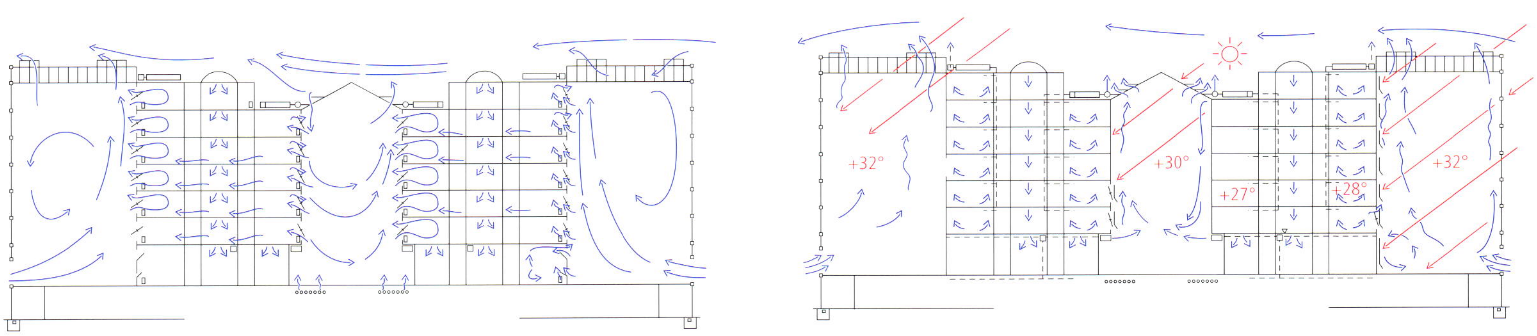

Transitional season
Sunny/overcast sky, moderate to strong wind

Summer operation
Sunny sky, low wind

in °C

Figure 6.15.2
Schematic of operating conditions (summer/transitional season)

6.1.5 Natural ventilation

An analysis of the site conditions is vital in the context of natural building ventilation, especially with regard to sufficient air circulation in protected areas of the structure (e.g. atria).

The wind roses in **Figure 6.16** provide an example of conditions in Frankfurt am Main: the principal wind directions are south west and north east and the highest wind velocities come from a westerly to south westerly direction. The facades exposed to wind from these directions should therefore be designed to ensure that the positive and negative pressures acting on the surface of the building are utilized to promote natural ventilation while avoiding potential problems in the building's interior (for example, as a result of excessive wind pressures on doors, draught etc.).

The architectural concept in this example features perforated facades (southwest) and generous glazing (northeast), an ideal response to the conditions outlined above. The facade and the individual window components should be designed so that initial opening of individual window leaves creates only a narrow gap to minimize wind pressure, while still ensuring air flow throughout the room to evacuate contaminants and odours.

Figure 6.17 indicates the frequency of wind velocities in Frankfurt am Main. Wind velocities above 10 m/s occur during no more than roughly 70 hours per year at this location. Natural ventilation is limited during these periods because high wind velocities can lead to high pressures. In these conditions it is vital to open windows only briefly and then close them again (instantaneous ventilation), unless the relevant area is mechanically ventilated. This limitation – effective only during 1 % of the entire year – imposes only negligible restrictions. Detailed studies should be undertaken of the actual positive and negative pressure quotients, especially with regard to the impact of turbulence on the high-rise. The air currents and pressure quotients for Frankfurt am Main are illustrated in **Figure 6.18**.

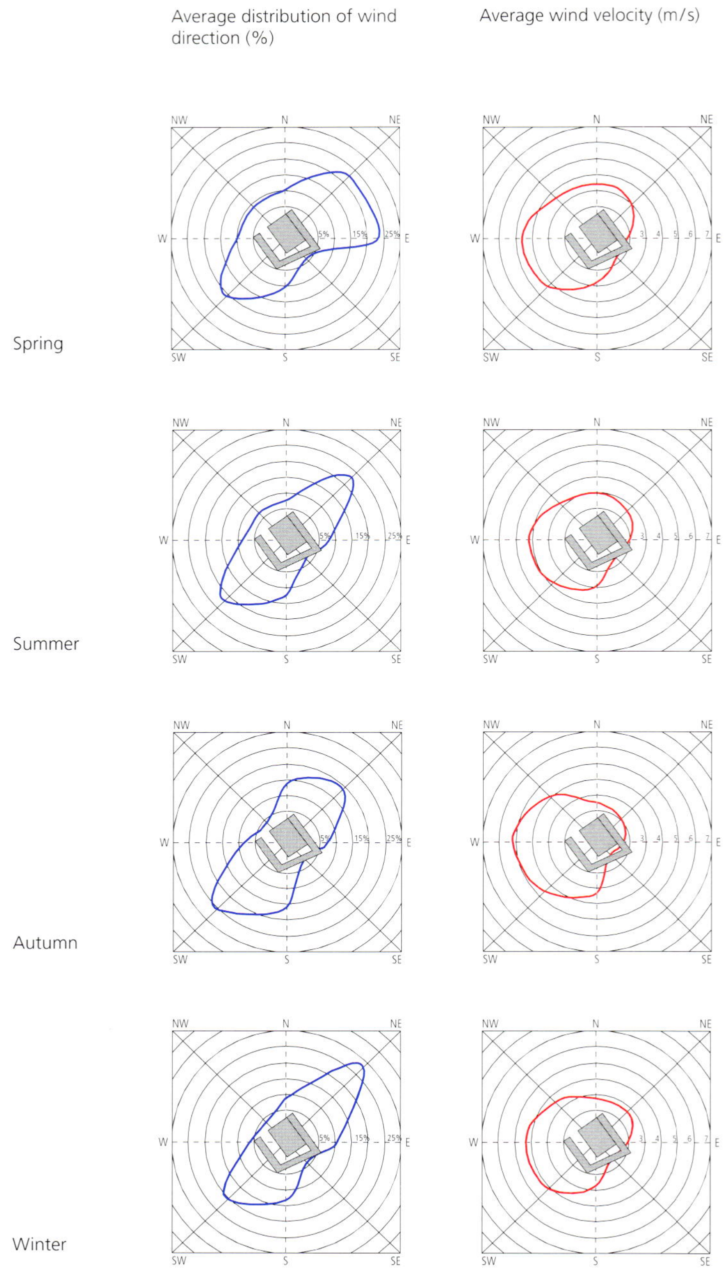

Figure 6.16
Average wind velocity and distribution of wind direction in Frankfurt am Main

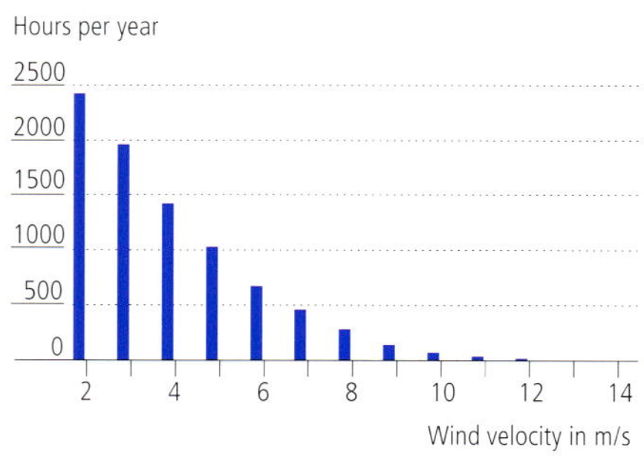

Figure 6.17
Frequency of wind velocities in Frankfurt am Main

Oncoming winds from NE

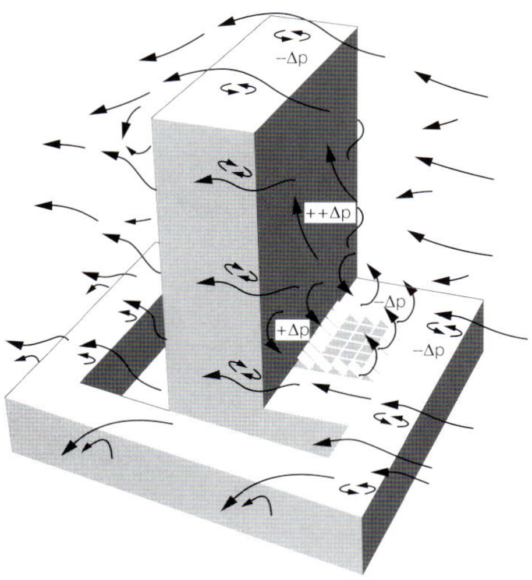

Oncoming winds from SW

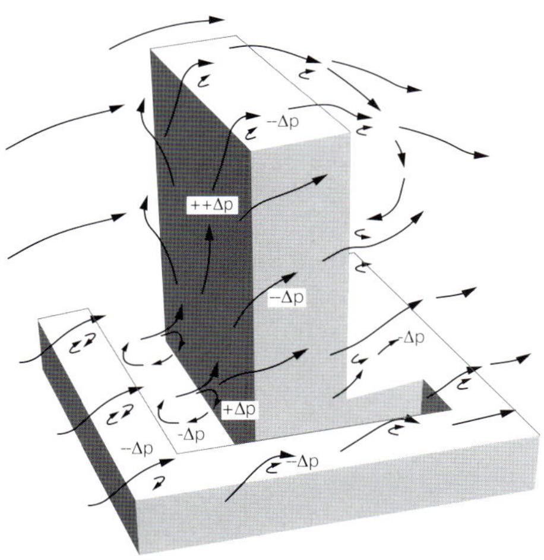

Figure 6.18
Wind directions and pressure coefficients, example Frankfurt am Main

6.1.6 Seasonal ventilation strategies

Natural ventilation of high-rises is especially viable during the transitional season with average outside temperatures, the only limitation being when the outside air velocity exceeds 8 m/s. In Frankfurt am Main, this occurs only during approx. 200 hours per year (see **Figure 6.19**).

When outside temperatures exceed 22 °C, mechanical ventilation a useful or even necessary adjunct to avoid uncomfortably high room temperatures in summer. Mechanical ventilation and heat recovery systems are also recommended when outside temperatures fall below 5 °C.

Figure 6.20 demonstrates that natural ventilation is feasible for more than 80 % of the year, representing circa 1,800 to 3,200 office hours annually. The following measures must be implemented to achieve optimum operation given a variety of wind and thermal conditions:

- Motorized operation of windows and vents
- Automatic adjustment of opening angle according to wind and thermal conditions and to temperatures in each individual room
- Optional individual control, i.e. individual override of automatic settings when feasible
- Use of appropriate bus system with reliable software programs.

Figure 6.19
Frequency of varying outside air velocities per year, Frankfurt am Main

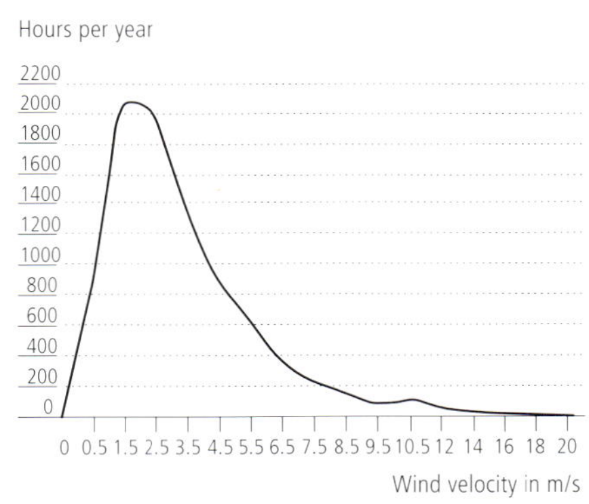

Figure 6.20
Frequencies of outside air temperatures during office hours per year

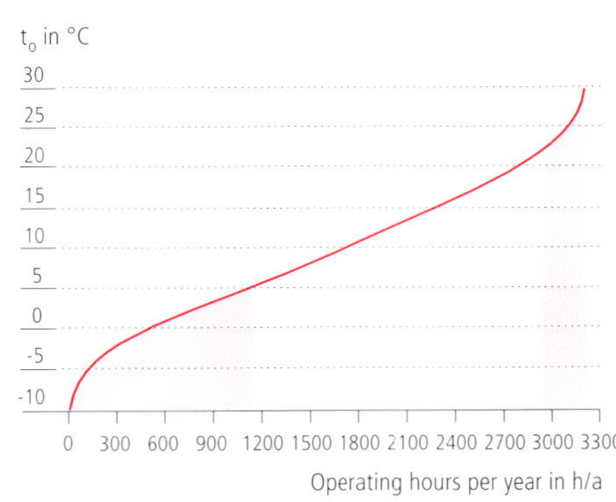

6.1.7 Completion concepts for offices

The goal for the office building featured in our example was not only cost- and energy-efficient operation but also maximum use of natural resources. A variety of completion concepts were developed with a view to building technology and corresponding facade options. The initial task of drafting a completion concept is to determine whether comfortable room temperatures can be achieved without supplementary technical systems, such as ventilation, cooling with water or air etc. **Figures 6.21** and **6.22** illustrate the different approaches with regard to natural ventilation, natural cooling, supplementary (mechanical) cooling etc., with a focus, in each instance, on which completion standards deliver acceptable and comfortable room temperatures during the height of summer.

Option 1

The premise for Option 1 is that all office spaces are exclusively heated with static heating and naturally ventilated through window openings; this option does not include mechanical cooling in summer.

Option 2

The premise for Option 2 is that the static heating and window ventilation featured in Option 1 are complemented by mechanical night ventilation with outside air (operating between circa 7 p.m. to 5 a.m.), activating the storage capacity of the exposed concrete ceilings.

A comparison between Options 1 and 2 reveals that the latter reduces temperatures by approx. 3 K. Additional options are explored because the resulting room temperatures are still too high in summer.

Option 3

Cooling of concrete masses using an active cooling water system, integrated into the ceilings. This cooling water system can also be utilized for heating the rooms in winter. The Batiso system (active building component cooling) is operated as follows:
For outside temperatures ranging from +12 °C to +15 °C, warm water (fluctuating flow temperature) is conducted through the pipe system, warming the concrete masses, which in turn radiate thermal energy into the room and raise the room temperature.
For heating, the maximum flow temperature is +35 °C, resulting in a maximum ceiling temperature of less than +27 °C in extreme winter conditions.
When outside temperatures range from +15 °C to +18 °C, the rooms are neither heated nor cooled, i.e., the pipe system is switched over in readiness for anticipated cooling loads.

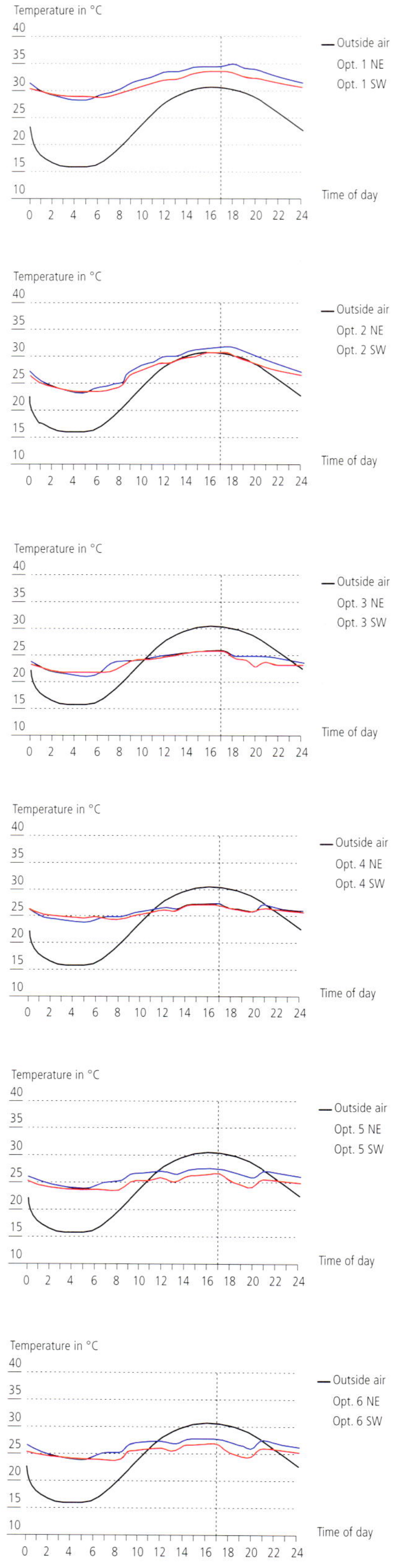

Figure 6.21
Room temperature according to time of day for different completion standards

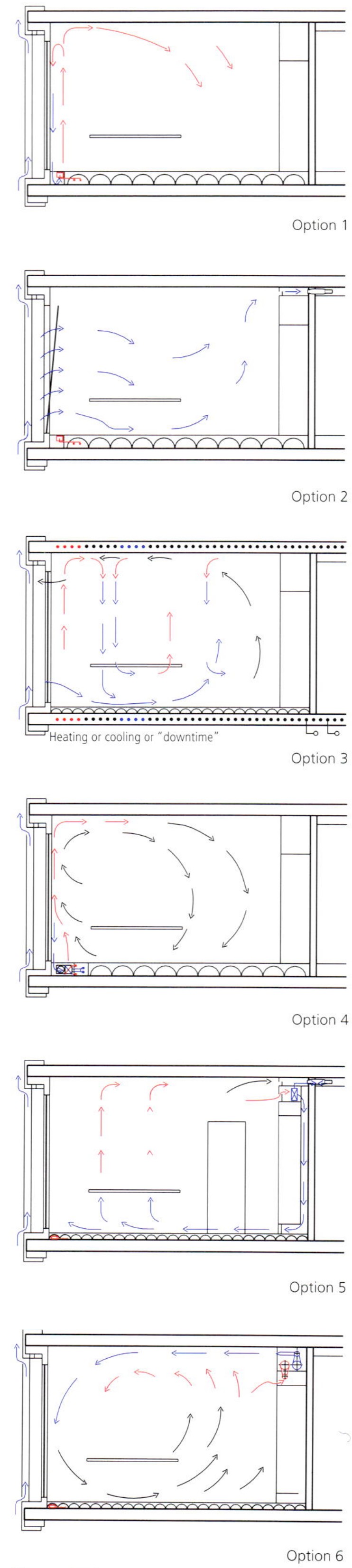

Figure 6.22
Completion and operating options, diagrammatic sections

When outside temperatures rise above + 18 °C, the same pipe system is employed for cooling, drawing the required cooling energy from refrigerating machines (flow temperature approx. + 16 °C), earth coils or a free cooling system.

Although the system structure is simple, it does not offer individual regulation per room and has a sluggish response time (similar to floor heating systems).

Option 4

In Option 4 the rooms are heated and cooled via fan convectors near the windows coupled with natural ventilation. The fan convector need only operate with or without air to heat or cool specific axes in the rooms. The concept is designed to connect the fan convectors to a four-wire installation, allowing for individual heating or cooling operation in rooms depending on cooling loads or heat losses. This system does not have a "down time" as is the case with Option 3. The fans of the subfloor convectors operate only during cooling periods and are regulated continuously.

Option 5

In Option 5 (Gravirent System) also operates with convector heating (subfloor system as in Option 4) and cooling components integrated above the wall units. These components cool the warm air that collects in the ceiling area sufficiently for it to sink to the floor, providing ventilation by means of a quasi stack effect. This solution, also described as "static cooling", offers the considerable advantage that cooling and/or heating can be regulated for individual rooms. It also allows for a high degree of load compensation (70-Watt cooling load per square metre). Another advantage is that the cooling system operates exclusively on a water basis, avoiding undesired noise transmission and increased energy consumption through blowers.

Option 6

This is a well-known ventilation solution (single-duct system, variable volume flow system) in which air is blown into the room at the upper wall unit level on a tangent and then flows through the entire space. The option illustrated for this example is based on the premise of a maximum 2.5 air change rate, achieving a constant room temperature of approx. 27 °C in the height of summer (intake air temperature + 16 °C). Heating in winter is provided by convectors, as in Option 1.

For the purpose of clarification, Options 5 and 6 are subjected to an additional analysis whose principal focus is to determine how they perform during a good weather period (at least seven sunny days in a row).

The diagrams in **Figures 6.23.1 – 6.23.3** and **6.24.1 – 6.24.2** show that exclusively natural ventilation and heating results in room temperatures of roughly 35 °C, while Options 5 and 6 yielded average temperatures of 27 to 27.5 °C.

A temperature graph indicates the frequency of estimated room temperatures for each option in the test office spaces on the northeast and southwest sides, respectively. These graphs show that temperatures above 26 °C are limited to so few hours per year as to be statistically negligible, while Option 1 results in high room temperatures on both the north east and south west sides for longer periods of time. Natural ventilation alone is therefore not recommended. The best solution is a combination of supplementary ventilation and cooling as outlined in the options listed above.

Figure 6.23.
Transient temperature behaviour in rooms with different completion options
(Summer spell of fine weather)

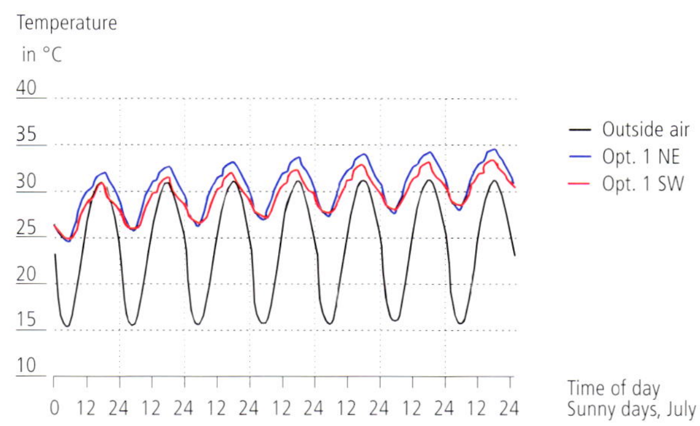

Figure 6.23.1
Option 1

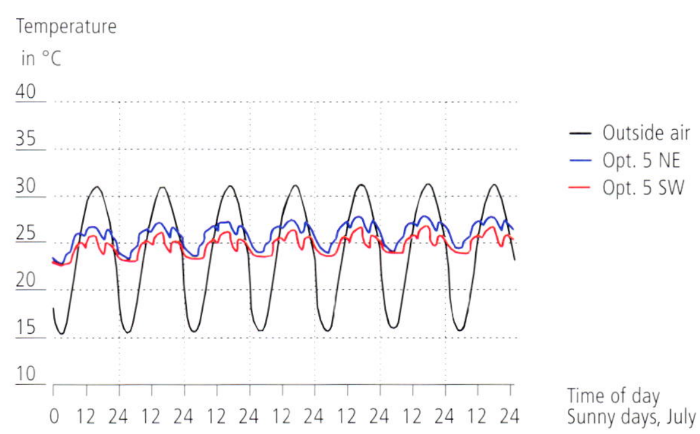

Figure 6.23.2
Option 5

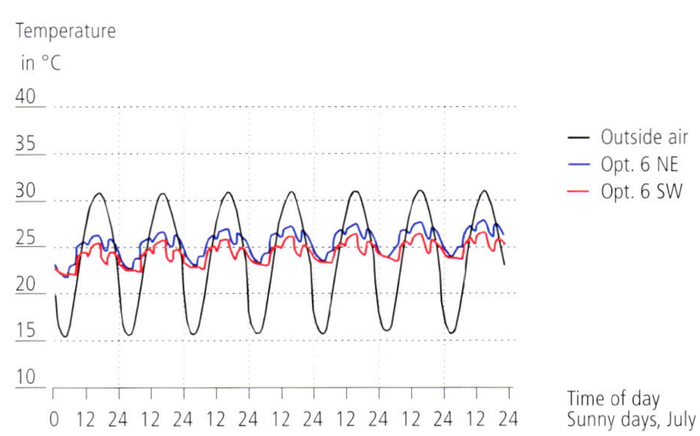

Figure 6.23.3
Option 6

Other technical installations

In addition to the technical systems for general office areas, there are a number of heating and ventilation installations for utility areas, circulation and traffic areas, underground spaces etc.
In terms of usage requirements and operation times, these are:
- Ventilation systems for equipment rooms
  Attic floor/underground levels/mezzanine levels
- Ventilation and cooling systems for conference and meeting areas
- Ventilation systems for archives and storage/warehouse areas
- Air-conditioning systems for computer rooms/central server rooms
- Ventilation and cooling systems for mail rooms and shipping and receiving areas
- Ventilation and smoke exhaust systems for underground car parks
- Ventilation systems for sports facilities
- Ventilation systems with cooling for the print shop
- Pressure cooling systems for stairwells and gates
- Smoke exhaust systems for warehouse areas and equipment rooms, as well as for all rising areas

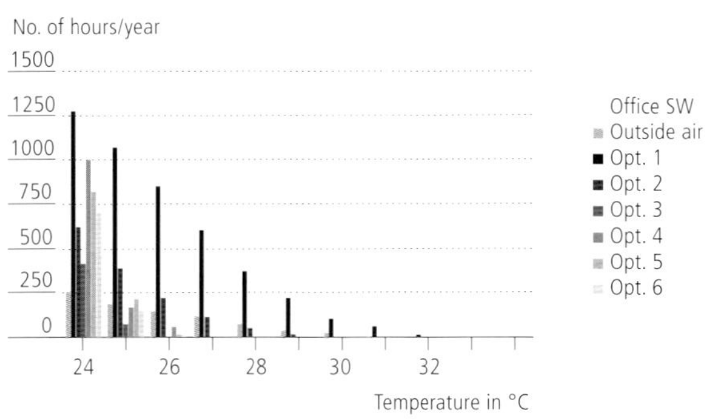

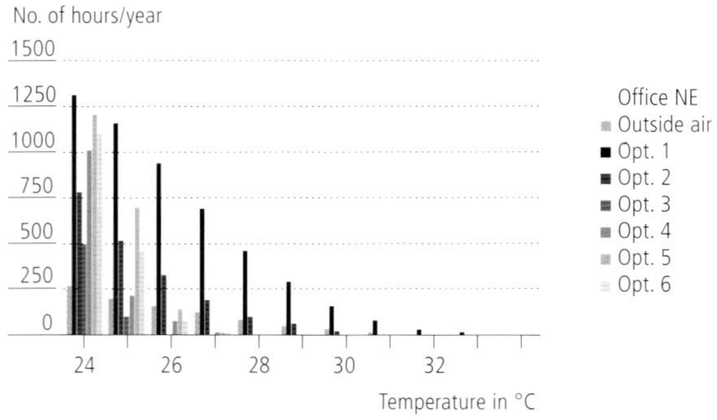

Figure 6.24
Annual distribution of internal and external temperature frequencies. The temperatures recorded between 7 a.m. and 6 p.m. are greater or equal to the temperatures marked on the X-axis.

Figure 6.24.1
Office SW

Figure 6.24.2
Office NE

6.1.8 Completion concepts for high-rise conversions

Buildings should be conceived with future conversion to different uses in mind, for ecological as well as economic reasons. "Flexible real-estate" offers the benefits of a long life cycle, which translates into added value as a long-term investment.

Conversions of apartment buildings into hotels and vice versa are fairly straightforward thanks to the similarities in installations and structural layout. The same applies to converting hotel areas with double floors into office areas. In this case, adjustments are required to the building systems, especially with regard to fresh air supply.

The double floor design provides space for installation of all air, media and energy-related systems in all positions and areas, and supply to rooms at the core is as easy as to rooms at the perimeter. The utility value of the building is thus ensured through its infrastructure, provisions for all kinds of user-completion works and technical options for conversion and different uses. **Figure 6.25** illustrates an example of a high-rise structure with different uses in individual units, which can be divided flexibly.

Offices

Offices on the lower floors of the building are constructed with a basic system comprising convectors, on the floor and in part wall-mounted below the window sills. Fresh air is also injected into the office areas via double or false floors at an air change rate of 1.5 to 2 to achieve guaranteed minimum ventilation for hygienic air conditions. The air volumes supplied in this manner helps to compensate for a cooling load up to a maximum of 10 W/m^2. Based on estimated cooling loads of approximately 50 to 60 W/m^2 in summer, the room temperature is maintained at approximately 32 °C. If users desire a higher degree of thermal comfort or if market demands call for rooms with a more advanced level of technical finishes, retrofit options with capillary tubes (mounted) or suspended cooling ceiling systems (tenant completion) are easily implemented. Both of these retrofit options achieve a total cooling load compensation up to approximately 80 W/m^2. All rooms have flexible access to electrical power via the double floors. All workstations are equipped with connections to information and data communications systems.

Lighting systems may be part of the basic building completion or customized user completions, the latter are generally based on general space lighting with an average illuminance of approximately 500 lux.

The basic interior completion work also includes sprinkler installations in all areas as well as the necessary sanitary blocks. The double floor construction makes it possible to inject air volumes depending on the room division (individual rooms, group rooms or combination solutions etc.) in the areas where they are needed (excluding corridors, for example). Similarly, cooling components (cooling ceiling components) can be installed in those specific areas where cooling loads must be compensated. In addition to fresh air supply, the basic completion also includes ventilation, which results in night-time ventilation in offices constructed with a high-storage capacity (overnight cooling).

Figure 6.26 illustrates air circulation for operation in various seasons.

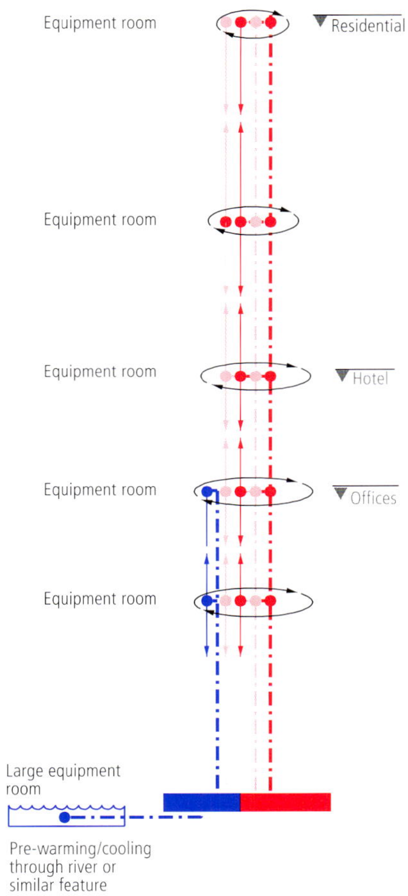

Figure 6.25
Distribution of equipment rooms in a convertible high-rise

Hotels

Hotel rooms and bathrooms are equipped with static heating surfaces, taking into consideration that temperatures in unoccupied hotel rooms are maintained at only +5 to +10 °C. The convectors at floor level beneath windows are therefore used for basic heating. When hotel rooms are occupied, additional heating is provided via fan coil units (systems that function similarly to fan convectors) coupled with regulated fresh air supply and natural ventilation.

In addition to fresh air, the fan coil unit also draws off room air to supply the necessary cooling energy to cool the rooms. Fresh air is injected in the door and ceiling area in the form of downdraught cooling with stack effect ventilation, the air exhaust is located in the bathrooms (see also **Figure 6.27**).

Another basic completion feature in hotel rooms is the lighting in the sanitary blocks – in contrast to the lighting in the hotel room itself. Hotel rooms, like offices, are always equipped with sprinklers (basic completion standard).

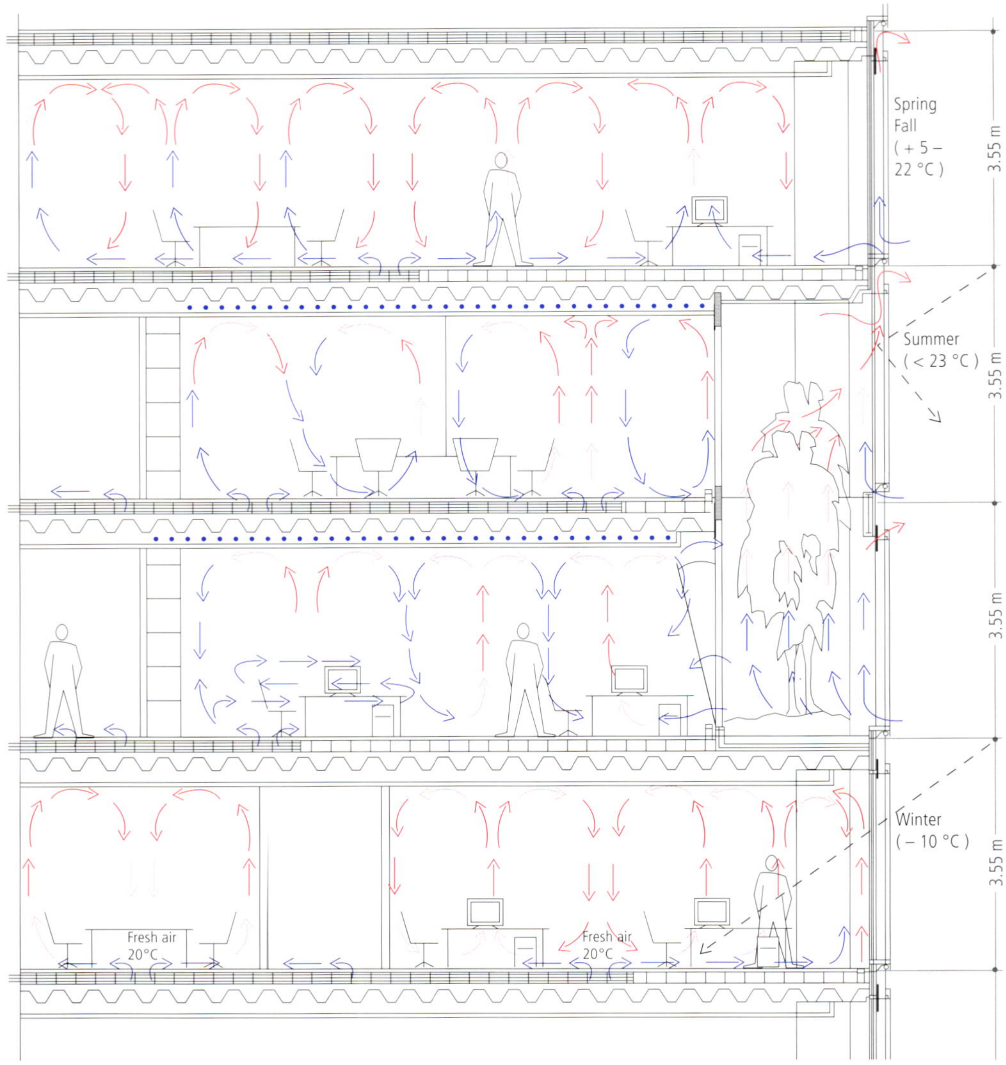

Figure 6.26
Seasonal air circulation (offices)

Apartments

The basic completion standard in apartments (**Figure 6.28**) includes static heating near the windows and fresh air supply at an air change rate of 1.5 to 2, integrated into a suspended ceiling in the entrance area. Depending on tenant requirements, fan coil appliances can also be provided for greater heating and especially cooling performance.

Apartments are constructed with additional floor structures (footfall insulation/total screed approx. 8 cm) in which the cabling ducts are installed to provide floor outlets in all living and lounging areas for all appliances and data/media equipment. Basic completion in the living areas, sanitary blocks and kitchens also includes sprinklers. Stale air from the living areas is vented via WCs, bathrooms and kitchens grouped around the core to avoid odours within the living areas. Lighting systems are only provided in the common areas (bathrooms, toilets, kitchen) – not in the bedrooms and living rooms.

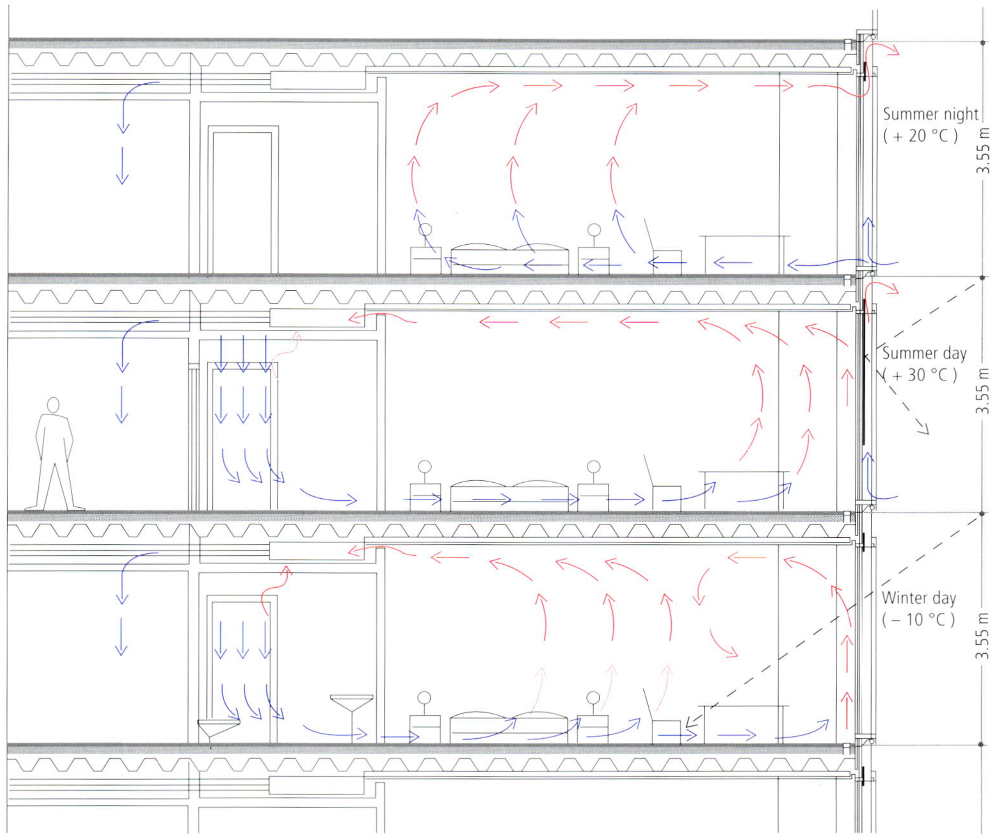

Figure 6.27
Seasonal air circulation (hotel)

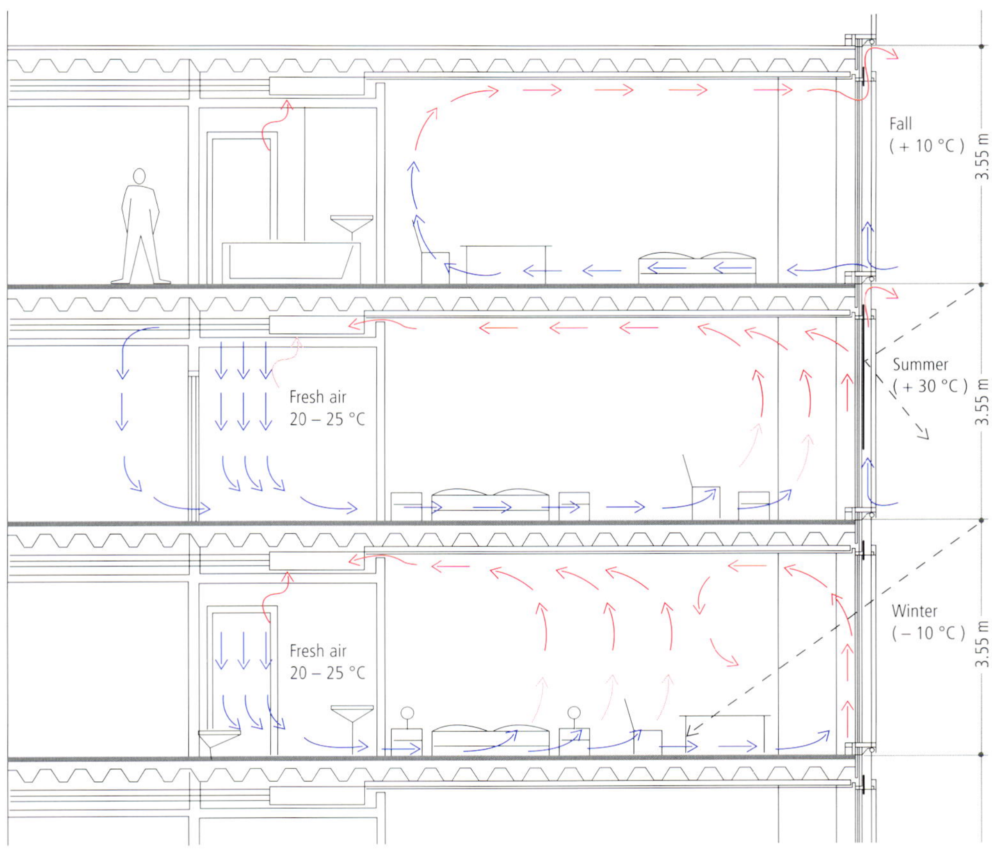

Figure 6.28
Seasonal air circulation (apartments)

6.1.9 Designing natural ventilation systems

Designing a plan for natural ventilation is far more complex than planning for mechanically ventilated or cooled rooms, where air changes as well as the room air variables are predetermined by the system.

In addition to thermal lift caused by temperatures and height, strongly time-dependent, wind-induced pressure differences on the building skin are often important mechanisms of natural ventilation. Each of these parameters is complex in its own right. Thermal and aerodynamic effects often exert equal influence and must be considered together.

As **Figure 6.29** shows, the construction, facade system and impermeability of building components play an important role in addition to the external physical influences.

Window design is a key factor. The frequency with which occupants open windows and how open windows influence internal air currents and thus internal cross-ventilation have considerable impact on the efficacy of natural ventilation.

Odour and contaminant emissions at the building skin, which may originate in the building's HVAC system, and the emissions in the interior space itself are further factors that must be taken into consideration.

Computation methods
In general, room specific air change rates and venting spent air or toxic gases from a building can be calculated by creating a balance sheet of air mass currents that takes all openings and leaks into consideration. All computer programs designed to carry out such calculations operate on the basis of modelling the real building as a network.

Doors, windows, leaks and joints as well as piping and passive HVAC building components can be understood as flow resistances (or "impedance" in analogy to electrical impedance, albeit with non-linear characteristics) as elements of this network. In this model, fans are analogous to terminals in electricity grids; the influence of fans must be taken into consideration for mixed air ventilation systems (characteristic curve of fan performance).

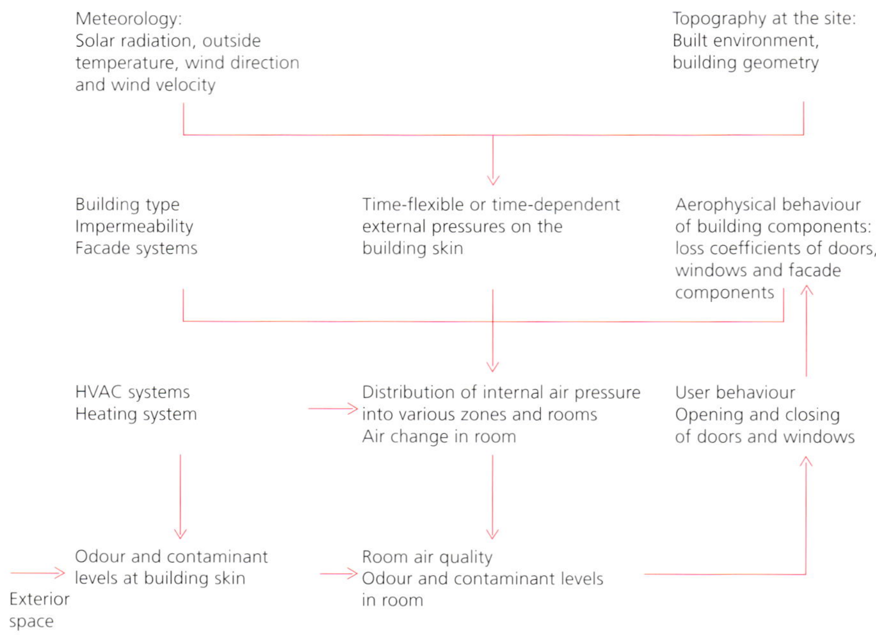

Figure 6.29
Parameters that influence the design of natural ventilation systems

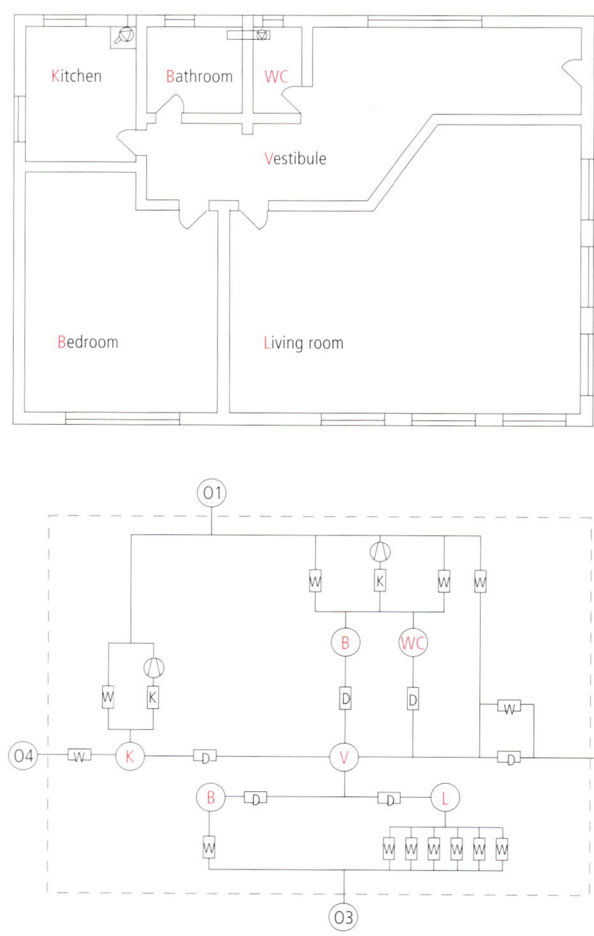

Figure 6.30
Flow grid for the calculation of cross-ventilation in an apartment with natural ventilation, supported by mechanical vents in kitchen, bathroom and WC

The first step in the calculation is to define the location and the precise progression of the coefficient of permeability, i.e. the mass flow rate (m_i) through the relevant building component according to momentary and localized pressure differential (ΔP_i):

$\dot{m}_i = f(\Delta P_i)$.

Figure 6.30 provides a comparison of an apartment floor plan and the abstract flow model. Kitchen, bath and WC are mechanically ventilated; fans are identified as terminals. In this case the external parameters have been summarized for each facade (O1 to O4).

Establishing the external parameters

To deliver practical and accurate predictions, it is vital to correctly establish the external parameters, that is, to make reasonable assumptions with regard to the distribution of temperatures, pressures and contaminant or pollutant emissions, as well as exhaust air attenuation at the building skin.

Temperature distribution can be determined with reasonable accuracy by means of computer simulation referring to location, building site, season and time of day, and construction and facade systems. Collecting data on the external pressures or levels of concentration, however, can be a far more labour-intensive process.

This is due to the complexity of flow conditions near buildings. In addition to the geometry of the building itself, the built environment and statistical frequencies of wind direction distribution, and wind velocity in the region and at the specific site also play an essential role.

At present, precise calculations of these influences on the air currents near the building and hence predictions of the temporal and spatial fluctuations in pressure distribution are still uneconomical.

Physical models of the flow field

It is common in Central Europe to wish to forego mechanical air-conditioning for reasons of economy, ecology and comfort. As regards thermal and hygienic comfort, the predictions concerning the flow conditions in the relevant rooms are essential if fundamental errors are to be avoided. To this end, various computational methods are employed in conjunction with experiments to determine the flow conditions in the surroundings.

The wind-induced, natural air movement in and around a building can be reproduced to scale and with great precision in an atmospheric boundary layer wind tunnel.

The wind-induced air change through large building openings, for example atria or large halls, can be partially determined by means of a direct measuring method. This allows for an iterative approach to optimizing the configuration of the building openings with regard to costs and benefits.

The diagram in Figure 6.31 depicts a typical wind tunnel designed for simulation of wind forces on building models.

Such experiments also yield data that can help in designing intelligent controls that allow the building system to respond optimally to momentary wind conditions at any given time.

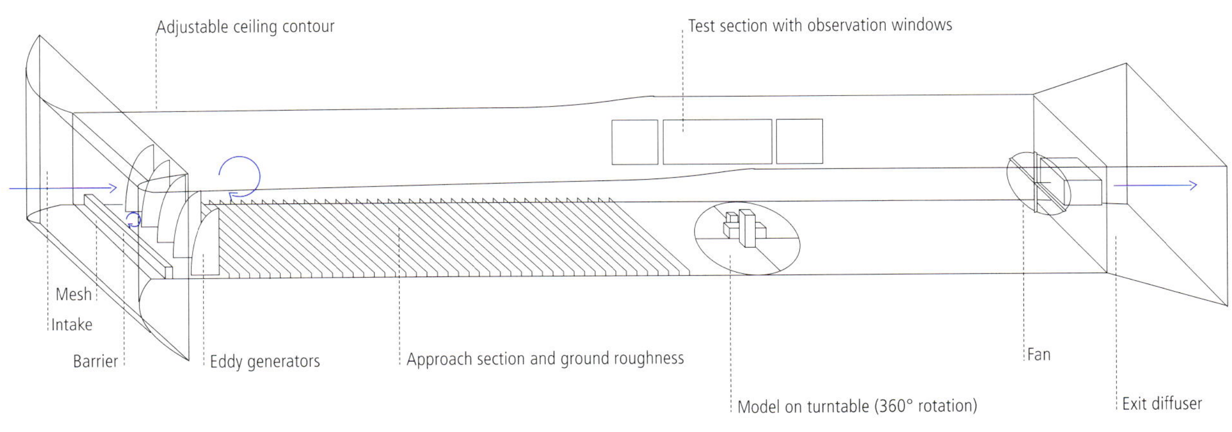

Figure 6.31
Atmospheric boundary layer wind tunnel for aerophysical simulation on building models (schematic, longitudinal section, approx. 3 m depth)

6.2 Calculating the cooling load

The cooling load calculation is analogous to the one for heating load (see **Chapter 4**). It is used to calculate the thermal currents in rooms.

As described for the heating load, the cooling load is calculated on the basis of standard design parameters set forth in relevant regulations, for example, the cooling load regulations in VDI 2078.

Again similar to heating load calculations, cooling loads are now generally calculated by computer to achieve more rapid and more accurate results. The architect should have an excellent understanding of the links between building design, materials and internal thermal transmission or loads.

For cooled or air-conditioned buildings, the cooling loads (thermal currents that must be evacuated) generally set the standard for the definition of necessary air volumes or cooling output in order to achieve cooling load compensation. In all-air systems, the scale of mechanical air-conditioning and cooling is defined on the basis of thermal loads, which must be compensated in order to maintain a set room temperature.

6.2.1 Internal cooling load

The internal heat gain $\dot{Q}_i$, which determines the cooling load, is composed of:

$\dot{Q}_O$ = heat emitted by occupants
$\dot{Q}_E$ = heat emitted by equipment
$\dot{Q}_R$ = heat radiation from adjacent rooms

Heat emissions from equipment may be from the following sources:

$\dot{Q}_L$ = lighting
$\dot{Q}_M$ = machines and devices
$\dot{Q}_G$ = material throughput
$\dot{Q}_C$ = other sources

Thus: $\dot{Q}_i = \dot{Q}_O + \dot{Q}_E + \dot{Q}_R$
with $\dot{Q}_E = \dot{Q}_L + \dot{Q}_M + \dot{Q}_G + \dot{Q}_C$.

6.2.2 External cooling load

The external (or outside) cooling load $\dot{Q}_{OU}$ includes all manner of energy penetrating into the building from the outside (gains from outdoor air), insofar as it has to be evacuated from the room air. A differentiation is made between the following heat flux volumes:

$\dot{Q}_W$ = heat flux through walls (roofs)
$\dot{Q}_F$ = heat flux through windows (fenestration)
$\dot{Q}_{JV}$ = heat flux through joint ventilation

Thus: $\dot{Q}_{OU} = \dot{Q}_W + \dot{Q}_F + \dot{Q}_{JV}$

The energy flux through windows F is differentiated into transmission heat T and radiation heat R.
This results in:

$\dot{Q}_F = \dot{Q}_T + \dot{Q}_R$

Since the cooling load of a room CR is composed of external and internal loads, the equation is:

$\dot{Q}_{CR} = \dot{Q}_i + \dot{Q}_{OU}$

6.2.3 Cooling load of the building

Peak cooling load periods tend to vary and the maximum building cooling load (CB) is therefore defined as:

$\dot{Q}_{CB} = \max \sum \dot{Q}_{CR}$

In other words, as the maximum derived from the sum of all simultaneous cooling loads (not as the sum of peak cooling loads), $\dot{Q}_{CB}$ determines the maximum cooling output required in the building, i.e., in the case of all-air systems, the flow volume of the appliances for the supply area or the entire building.

6.2.4 Thermal loads

Thermal loads in a room are composed of convection loads and radiation loads. Thermal convection loads (TCL) represent heat directly absorbed by the room air and thus immediately transformed into a cooling load. Thermal convection loads also act upon the thermal storage mass of the room enclosure, which have a damping effect depending on storage capacity and heat transmission and are therefore capable of reducing room temperatures (1 – 4 K).

Thermal radiation loads (TRL) are loads that impact the space either through short- or long-wave radiation of light. They do not have a direct effect on the room temperature since the radiation is first absorbed by relevant media and then transformed into a convective (or anabatic) heat flux.

Internal thermal storage capacity is a key factor in planning. It may be sufficiently high to obviate the need for (mechanical) cooling systems or at least greatly reduce their scale.

All standard rules on the calculation of cooling loads (design cooling load) are based on room temperatures of 22 °C (constant condition after initial transient temperature behaviour). The design cooling load is moreover based on the assumption of 24-hour operation of the HVAC system.

6.2.5 Thermal storage in the building mass

As we have already noted, thermal storage in the building mass can play an essential role. Heat gain in buildings is evident when the heat load is a result of solar incidence or when room temperatures change. In these cases the momentary loads are so changed by the storing elements that the heat gain time function changes into a cooling load time function. For calculating cooling loads, thermal storage may be approached in several ways (storage factors in the room, storage factors in external walls). The room, as a unit, absorbs radiant heat at the interior surrounding walls, through windows (diffuse and direct solar incidence), (artificial) lighting, occupants, machines etc., all of which have an effect. How far radiant heat penetrates walls or building storage components depends on the wall construction and the duration of the radiation, thus causing an increase in surface temperatures. The surfaces are in a constant state of reciprocal radiation exchange. Depending on air and surface temperatures, heat is transferred by convection from the surfaces to the air and vice versa.

For variable room temperatures, a rise in temperature triggers a reduction in the cooling load due to thermal storage. A fall in room temperatures triggers an increase in the cooling load due to heat release. Of particular interest here is the heat backflow, i.e. convective heat transfer and heat absorption occurring on the same side of the wall.

The impact of thermal storage processes on fixed load conditions is charted in the diurnal curve of the cooling load factors, which describe fluctuation in the cooling load indexed to damping effects and lag. This correlation is factored into basic calculations of momentary cooling loads by multiplying the maximum values with a storage factor (see also **Chapter 6.2.9**).

6.2.6 Outside temperatures

As a space enclosing surface, external walls are exposed to additional thermal flows and must be treated accordingly, although the cooling loads exerted through the walls from the outside are relatively small. When external surfaces are exposed to direct solar incidence, the resulting surface temperatures can be considerably higher than the outside air temperature.

These high surface temperatures produce additional thermal flows, which, in turn, result in an additional cooling load factor (dependent on the thermal transmittance coefficient, storage capacity and time curve of the radiation load). This thermal transmittance is factored into the calculation using an equivalent temperature difference dependent on damping effects and lag.

To establish the correct peak load temperatures, climate zones are differentiated for the purpose of calculation. These are:

Zone 1 Coastal climate, maximum temperature +29 °C
Zone 1a Slope climate (highlands)
Zone 2 Continental climate I, maximum temperature +31 °C
Zone 3 Continental climate II, maximum temperature +32 °C
Zone 4 River climate in the southwest (Germany), maximum temperature +33 °C
Zone 5 High altitude climate

Figures 6.32.1/2 indicate diurnal outside temperature curves in July and September for the zones listed above.

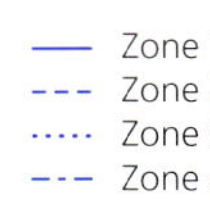

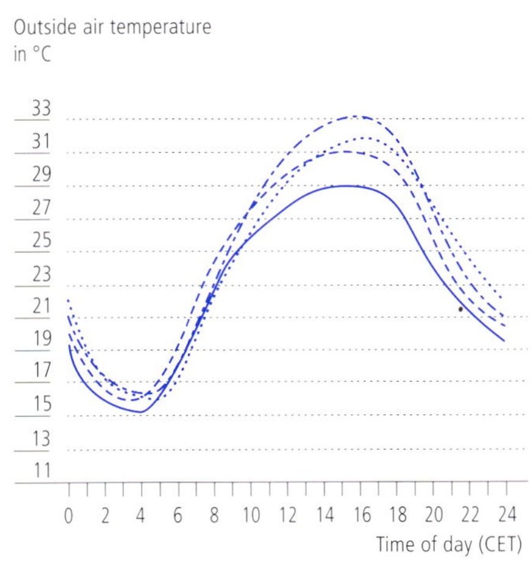

Figure 6.32.1
Diurnal outside temperature variations in July

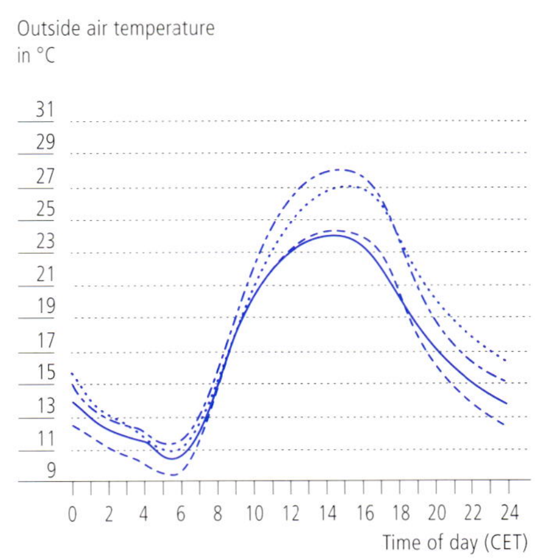

Figure 6.32.2
Diurnal outside temperature variations in September

6.2.7 Solar radiation

An area on the outer atmosphere of the earth with vertical orientation to solar radiation is exposed to a heat-flux density of roughly 1.39 kW/m² at a median distance from the sun (solar constant). This value ranges between 1.35 and 1.44 kW/m², whereby global (or total) radiation is composed of radiation vectors of varying wavelengths and differing intensities. Only a fraction of these heat volumes reaches the earth's surface since the atmosphere surrounding the earth reduces the intensity of radiation. Radiation is diminished through scattering and reflection on air molecules, dust and moisture particles and through absorption as a result of ozone, CO_2 and water vapour, as well as dust and haze. The turbidity coefficient T was introduced for the purpose of indicating by which degree the radiation intensity is diminished. In a clear atmosphere the turbidity coefficient T = 1; the annual average values are:

approx. 5.0 – industrial areas
approx. 3.5 – urban areas
approx. 2.75 – rural areas

As shown in **Figure 6.33.1** the diurnal course of direct solar radiation on walls facing in various directions (lat. 50 °N) with a turbidity coefficient T = 4 (roughly the degree of turbidity present in an urban environment) results in characteristic sine curves. **Figure 60.33.2**, on the other hand, plots the annual and diurnal course of mean global radiation, that is, the sum of direct and diffuse radiation and atmospheric radiation. Cooling load specifications indicate the diurnal course of total radiation and diffuse radiation for the relevant months and times behind a clear double-glazed surface. This approach delivers the required accuracy in terms of calculating incident heat energy in rooms and buildings.

The cooling load calculation takes the glazed area of the window (not the wall penetration) and the type of sunshade system into consideration. The solar energy and diffuse celestial radiation penetrating into the room depends to a large degree on the total solar energy transmission (g-factor), achieved by the window and sunshade combination i.e. the combination's shading coefficient b. It is important to remember that all shading coefficients indicated in the cooling load specifications are based on double-pane thermopane glazing.

For solar radiation, the relationship between the g-value (total solar energy transmission) and the b-factor (shading coefficient) is roughly calculated as:

$$b = \frac{g}{0.8}$$

that is, the b-factors (shading coefficients) are approximately 20 % greater than the g-values (total solar energy transmission). Ideally, calculations should be based on the b-factors given in the relevant product literature prepared by glass manufacturers, since the values listed in VDI 2078 are generally speaking too approximate. The same applies to calculations for shading systems of all kinds (external and internal).

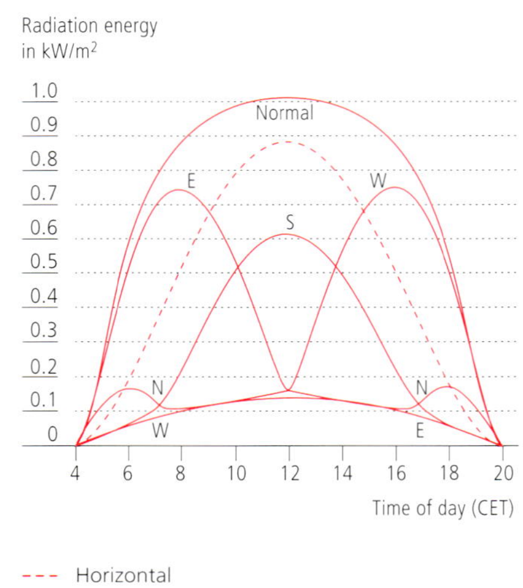

Figure 6.33.1
Total radiation on walls facing in different directions, in July, lat. 50 °N, with turbidity coefficient T = 4 (urban atmosphere)
DIN 4710 (11.82)

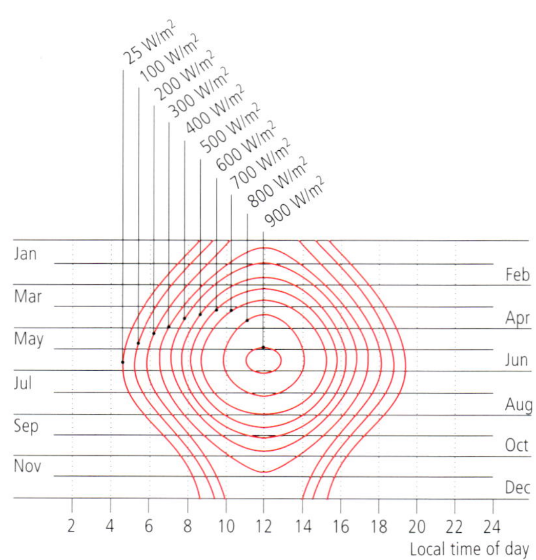

Figure 6.33.2
Mean global radiation in W/m² on radiation days onto a horizontal surface (urban atmosphere)

6.2.8 Facades

(From weather protection to polyvalent walls)

Technical data – the first step

In any study of glass facades on buildings, certain principal factors will be found to influence light and heat incidence, heat emission and colour-reproduction qualities. For climate and light experts working intensively with facades, these factors provide important information for an initial assessment. The principal factors are:

Visible transmittance VT

The visible transmittance VT is the percentage of daylight that penetrates a glass pane from the outside. The visible transmittace VT is indicated in the wavelength of visible light ranging from 380 to 780 nm, related to the light-sensitivity of the human eye.

Total solar energy transmission g-value/b-factor

The total solar energy transmission (g-value) is taken in the wavelength range of 320 – 2500 nm. It is the sum of direct energy transmission and secondary heat transmission to the inside (radiation and convection). The shading coefficient b describes the percentage of energy permeating a double pane (80 %). Therefore, the shading coefficient $b = g/0.80$. Previously, b was related to a single pane (3 mm) and thus $b = g/0.87$. Many data records are still based on this definition (**Table 6.2**).

The transmission value changes depending on the solar altitude and azimuth angle, varying from nearly total transmission (horizontal radiation onto vertical glass surface, $g \simeq 0.9$) to nearly total reflection (total reflection, $g \simeq 0.2$) as shown in **Figure 6.34**.

Diminution factor

The diminution factor of a shading element (from DIN 4108) measures the percentage of incident radiant energy that passes through a shading element and thus contributes to the warming of an interior space.

Product	Appearance (external)	Visible transmittance VT in %	Total solar energy transmission g-value in %	Shading coefficient b-factor	Comments
Standard clear glass	normal transparent				
single		90	83 – 87	1.00	
double 6/12/6		84	78	0.89	
triple per 12 mm air		74	70	0.81	
Calorex	"neutral" mirror effect (enhanced)				Metal oxide coating
single IRO		58	62	0.72	
double IRO		53	48	0.55	
IRA 1 single		42	50	0.57	
IRA 1 double		38	42	0.48	
IRA 1 + IR 2, double		33	29	0.34	
Parsol					Mass-tinted, values dependent on thickness, 2nd pane clear glass
Parsol bronze 6 + 6 mm		44	64	0.59	
Parsol grey 6 + 6 mm		47	47	0.56	
Parsol green 6 + 6 mm		64	49	0.56	
Infrastop					Superimposed gold or silver coating, with or without interference layer, only double pane, internal pane is clear
Auresin 66/44	blue mirror	66	44	0.50	
Auresin 50/36		50	36	0.41	
Auresin 39/28		39	28	0.32	
Bronze 36/26	bronze appears	36	26	0.30	
Gold 40/26	as a light gold-tinted mirror	40	26	0.30	
Gold 30/23	dark gold tone	30	23	0.26	
Silver 36/33		36	33	0.38	
Silver 22/22	silver	22	22	0.25	
Metallic		50	47	0.54	
Parelio	slight bronze tint				Metal oxide coating on clear glass or Parasol, 2nd pane clear glass
Parelio 24 6 + 6 mm		57	56	0.59	
Parelio 50 6 + 6 mm		42	51	0.64	
Parelio 24 green 6 + 6 mm		46	33	0.38	
Parelio 50 green 6 + 6 mm		35	29	0.33	
Parelio 24 grey 6 + 6 mm		27	37	0.43	
Parelio 24 bronze 6 + 6 mm		30	36	0.41	
Thermolux standard		63	52	0.60	
Thermolux insulating glass		53	46	0.52	

Table 6.2
Data for glass with shading factors for near vertical radiation incidence ($b = g/0.87$)

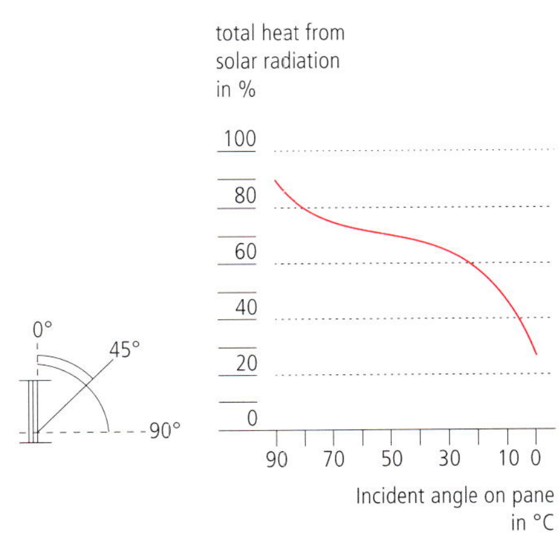

Figure 6.34
Total energy transmission of direct solar radiation through single-glazing (clear) depending on incident angle (angle in relation to vertical pane)

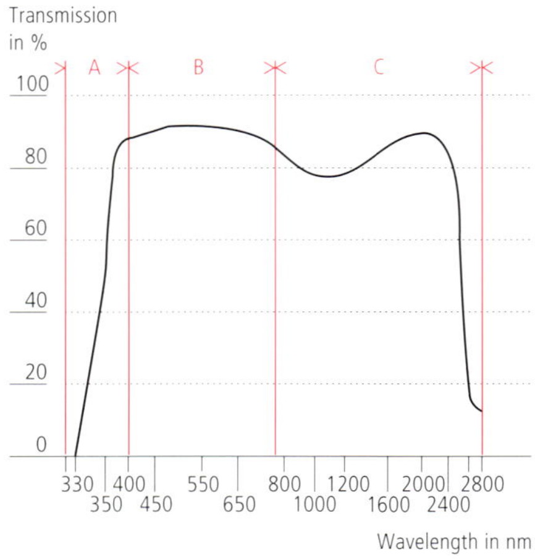

Figure 6.35
Radiation transmission of standard window glass
A Range of ultraviolet radiation
B Visible light
C Infrared range (thermal radiation)

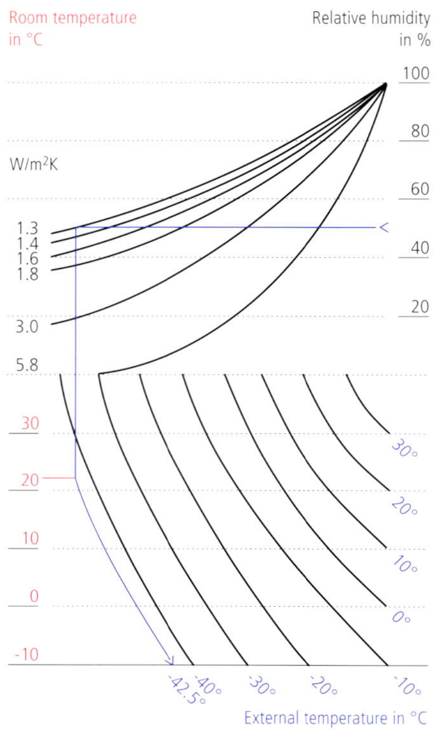

Figure 6.36
Dew point diagram with example

The dew point diagram serves to calculate the temperature necessary for condensation to form on the inside of a window pane (= dew point).
Shown example: pane iplus neutral
U = 1.3 W / m²K,
Room temperature +22 °C, relative humidity 50 %

Result:
Condensation on the iplus neutral pane forms only at a temperature of -42.5 °C

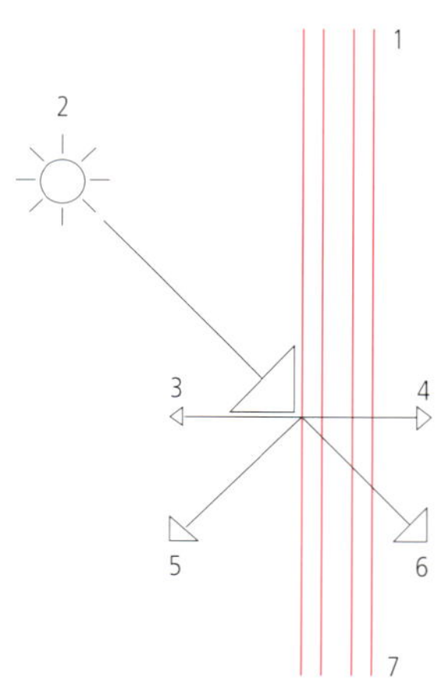

Figure 6.37.1
Total solar energy transmission of iplus neutral R
Distribution of incident energy (g-value from DIN 67 507)

1 Light transmission 75%
2 Solar energy 100%
3 Absorption with secondary heat transfer to outside q_a = 12%
4 Absorption with secondary heat transfer to inside q_i = 16%
5 Reflection ρ = 26%
6 Direct transmission τ_e = 46%
7 Total solar energy transmission g = 62%

Other factors

Other factors in the evaluation of window panes measure:
- Percentage of light reflected towards the outside
- General colour reproduction values (colour reproduction index R_a after DIN 6169)
- Percentage of UV permeability in wavelength range of 280 – 380 nm
- Selectivity factor S, representing the relationship of light transmission to total solar energy transmission etc.

Table 6.2 (previous page) cites data for brands of glass with shading elements (old definition); **Figure 6.35** supplements this with the radiation transmission of standard glazing.

For large glass surfaces and high temperature differences, there is an increased risk of condensation and hence corrosion around the frame. In winter, large glass walls in rooms or atria pose the additional problem that cold air is dumped at the glass surfaces as the air cools, which must be compensated by upward-flowing warm air currents. Another way of avoiding cold air currents is to sufficiently increase surface temperatures. The temperature difference between room air and inner glass surface should not exceed 4 – 5 K. Convectors (warm water, electric) or warm air currents can be used to avoid excessive cooling of the inner surfaces.

Figure 6.36 provides a diagram for determining condensation points on window panes according to the U-value, the relative humidity in the room and the inside and outside temperatures. This diagram is an essential tool for correctly designing relevant (window) pane surfaces.

Figure 6.37.1 shows the physical context for determining the total solar energy transmission or g-value. In **Figure 6.37.2**, the graph illustrates typical ranges of transmission, absorption and reflection on a specific glass pane (Calorex A1), used as an example, and thus indicates the percentages of incident daylight in the room.

Effect of shading on room temperature

The following paragraphs explain the combined effect of the total shading coefficient of a sun protection element and insulating glazing upon a room's cooling loads and temperature. **Figure 6.38.1** illustrates the effect in a room with light to medium storage capacity (in analogy to Option 2 in **Figure 6.43**) fitted with various types of shading. The options are as follows:

Option 2a external louvre
Option 2b closed external louvre (highly reflecting)
Option 2 c open awning
Option 2d indoor shading and glare protection

Figure 6.38.2 plots predictable room temperatures on a warm summer day in a southward-facing room of medium storage capacity based on the varying total solar energy transmission coefficients. The differences between the most effective shading system (lowest total solar energy transmission) and the least effective (internal vertical blinds or curtains) are considerable. The maximum temperature difference resulting from these shading systems may reach almost 10 K, causing a significant change in the thermal comfort experienced in a room. Furthermore, it should be noted that ventilation and air-conditioning systems are sized according to a "peak" temperature during a fair weather period, which may result in higher or lower system investment costs.

The combination of shading, cooling load and temperature behaviour is, however, only one area of influence for defining and designing shading systems. Another, equally important aspect is the incidence of daylight.

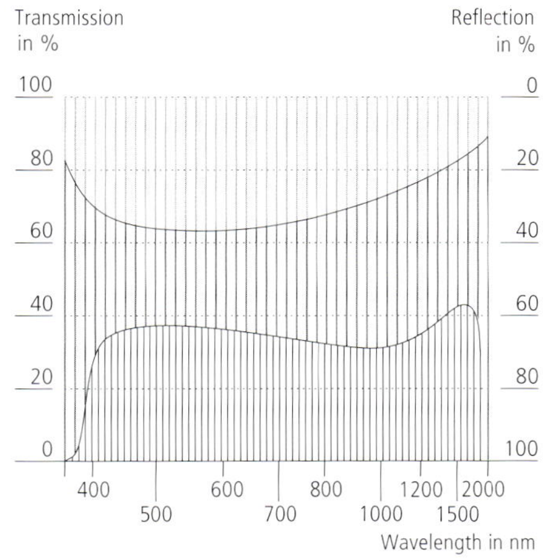

Figure 6.37.2
Schematic of transmission, absorption and reflection on pane (Colorex A1)

Reflection
Absorption
Transmission

Figure 6.25.1
Shading options

Room dimensions:	
Orientation:	South
Floor space:	21 m^2
Occupants:	2
Room height:	2.90 m
Lighting:	158 W
Ceiling suspension:	0.15 m
Machines:	315 W
Balustrade height:	0.70 m

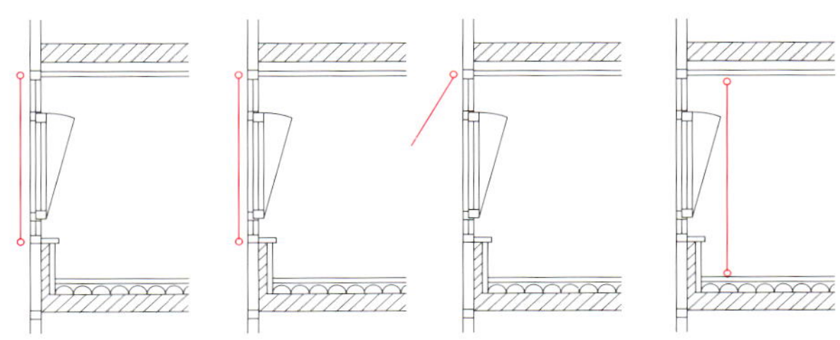

Option	2a	2b	2c	2d
Window	k = 2.0 W/m^2K g = 62%	k = 2.0 W/m^2K g = 62%	k = 2.0 W/m^2K g = 62%	k = 2.0 W/m^2K g = 62%
Shading	external/g = 0.12	external/g = 0.20	external/g = 0.35	internal/g = 0.60

Figure 6.38.2
Room temperatures following a five-day fair weather period in summer;
Weather data from test reference year (TRY region 8)

Fresh air
Option a (g = 0.12)
Option b (g = 0.20)
Option c (g = 0.35)
Option d (g = 0.60)

Room temperatures are indicated as perceived temperatures (from DIN 1946, Part 2), including building component and surface temperatures.

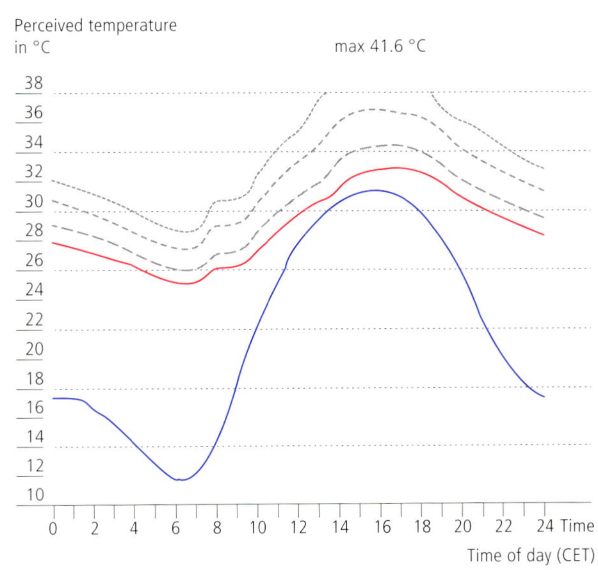

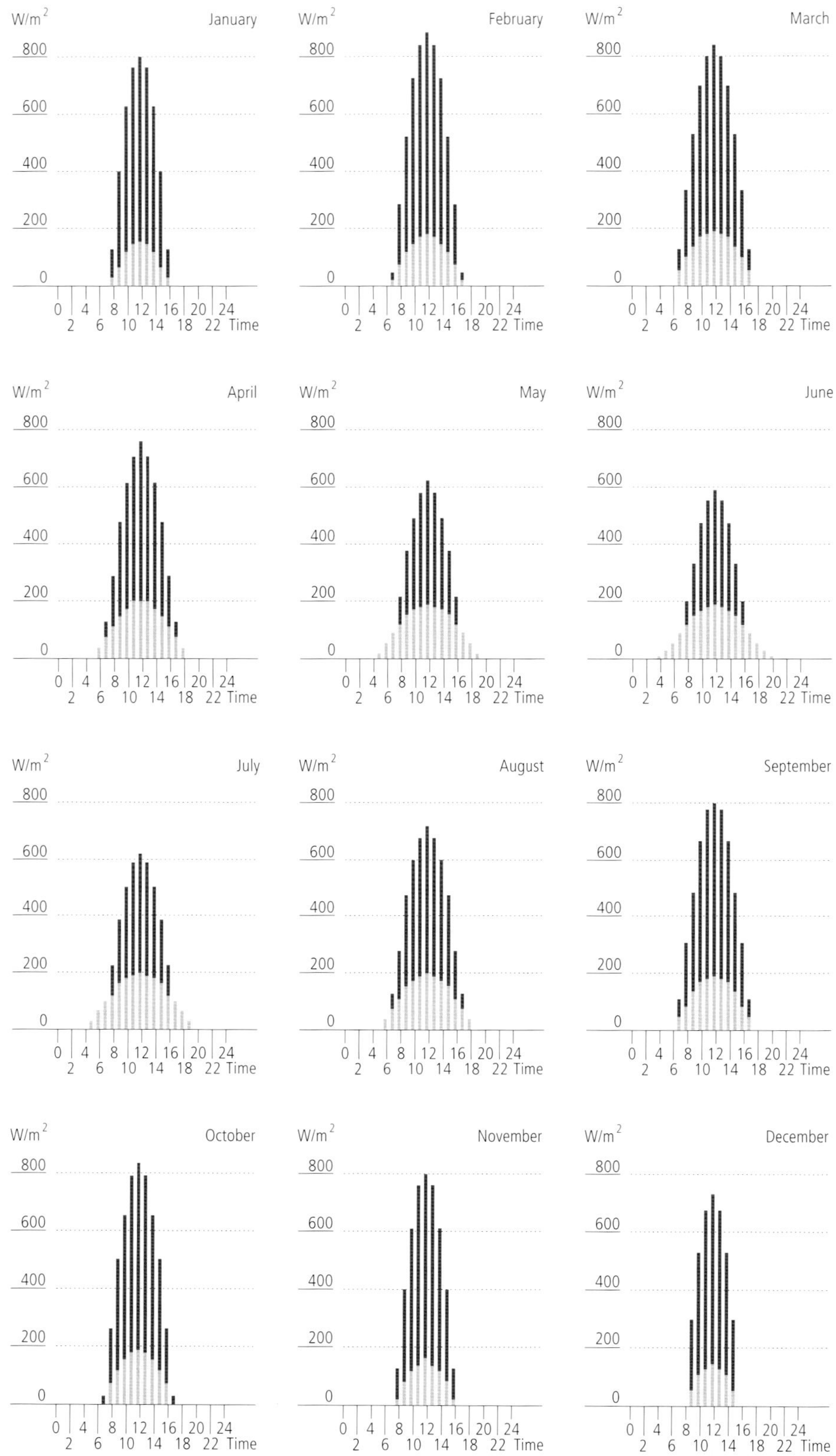

Figure 6.39.1
Incident radiation (vertical) on south facade during a 24-hour period

■ Direct radiation
■ Diffuse radiation

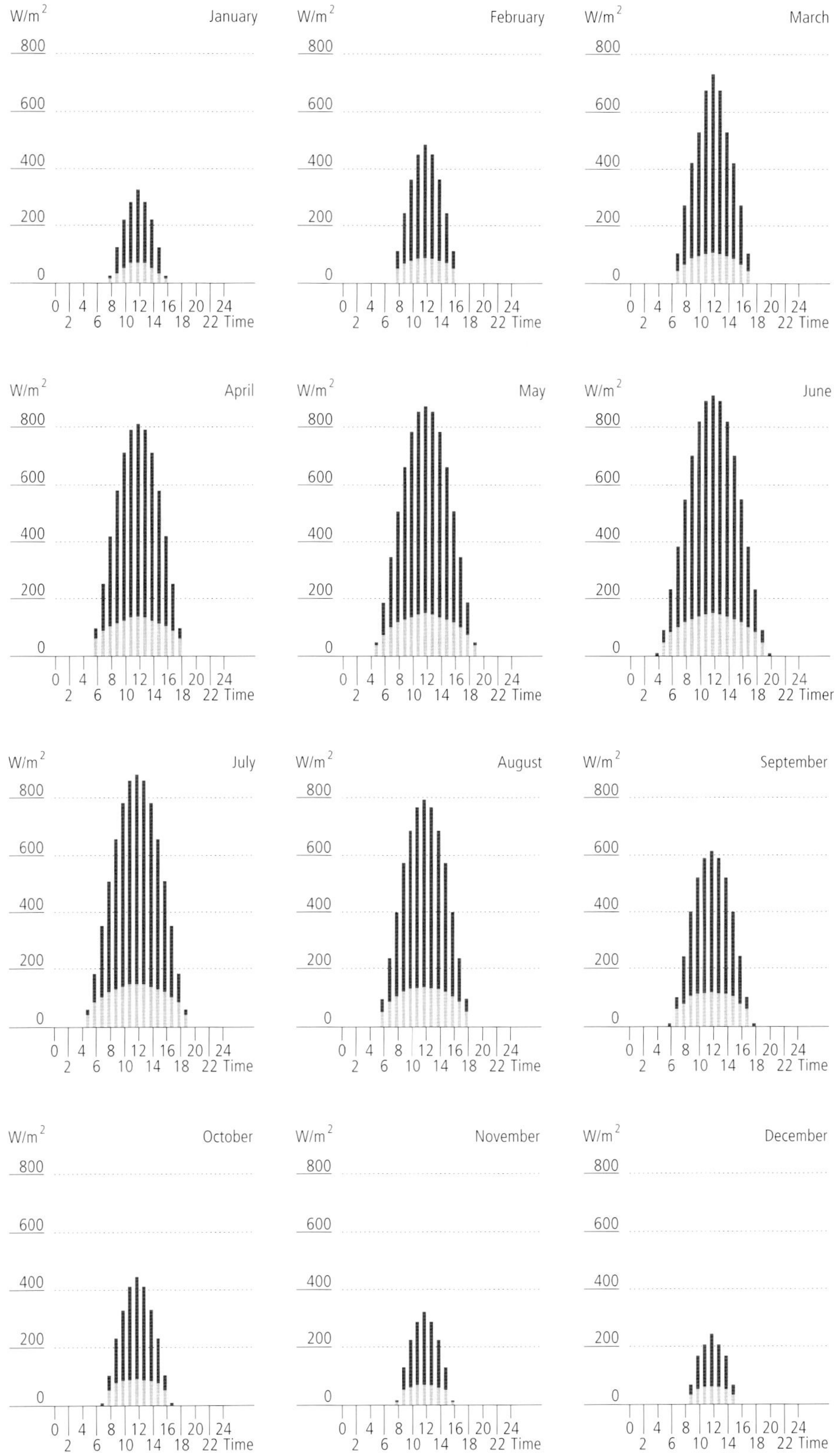

Figure 6.39.2
Incident radiation on roof during a 24-hour period

■ Direct radiation
■ Diffuse radiation

Daylight incidence and total solar energy transmission

External cooling loads can only be minimized when the incident radiation on window and roof surfaces is considerably reduced to diminish additional energy. At the same time, daylight penetration should be optimized to reduce the need for artificial lighting.

Unwanted energy absorption in well-insulated office buildings may occur even when outside temperatures are as low as +5 °C, since the indoor heat sources are likely to provide sufficient heat to the building during the day. The graphs in **Figures 6.39.1/2** serve as a general introduction to the topic by depicting the incidence of radiation (direct and diffuse) on south-facing facades and roof areas. **Figure 6.40** shows the cumulative frequency of radiation on roof surfaces and indicates how much energy is generally available for direct use, for example, in office buildings (passive solar gain), assuming that a building must be heated to +10 °C. The annual timeline of radiation illustrates the dilemma resulting from enormous solar heat gains in buildings with excellent insulation.
Only a small portion of the radiation is actually beneficial for heating the building; most of it is a potential source of discomfort.

The total solar energy transmission is very important for minimizing the incident heat energy produced by direct or diffuse radiation. This coefficient should be as low as possible, combined with maximum light transmission. **Figures 6.41.1/2** show the values for these factors for several examples of neutral and reflective glazing for single-skin facades, and window options for single- and double-skin facades respectively. Optimum window combinations exist for each case.

Different solutions for internal and external sun protection result in different total solar energy transmission coefficients, shown in **Figure 6.42**. These can be improved by removing heat absorbed in a room (secondary heat output/convective heat flow) to prevent undue cooling loads in the room.

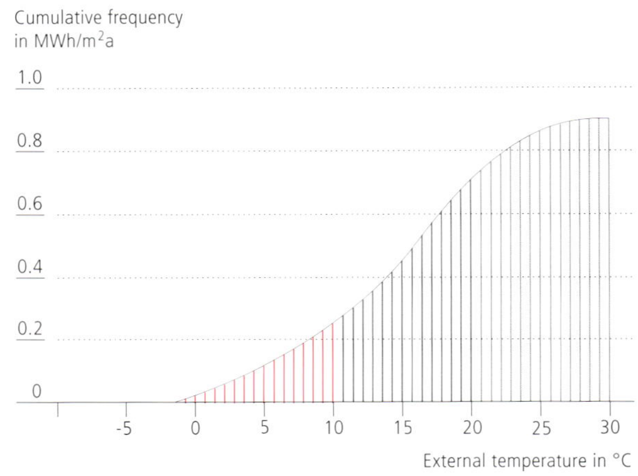

Figure 6.40
Cumulative frequency of radiation on roof surfaces

|||| Direct energy potential
|||| Low energy potential
|||| Barely useable energy potential

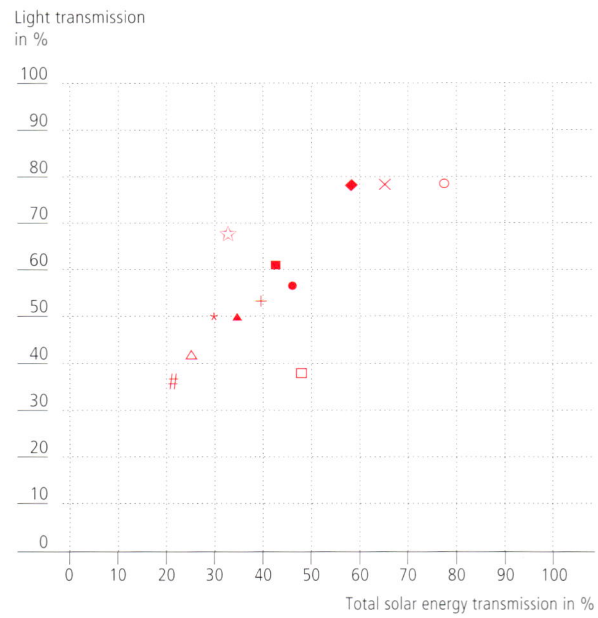

Figure 6.41.1
Selectivity of neutral and reflecting glazing (single-leaf facade)

○ Insulating glazing
× Thermal glazing
◆ Thermal glazing
■ Luxguard Natural
☆ Ipasol Natura
▲ Climasol Neutral
□ Parsol grey
● Antelio Silver
+ Ipasol Silver
* Infrastop Silver (50/30)
△ Luxguard Silver
Infrastop Silver (37/22)

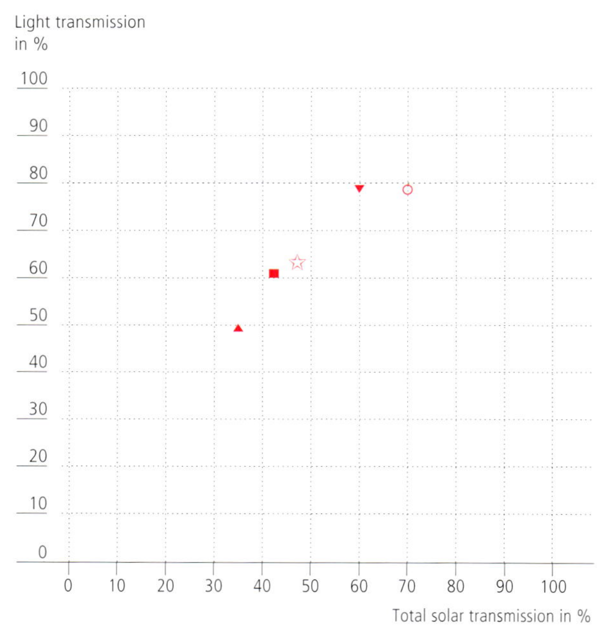

Figure 6.41.2
Selectivity of different window combinations (single-leaf facade)

Double-leaf facade
○ Box window
Insulating glazing/
single glazing

☆ Box window
Single glazing/
thermal glazing

Single-leaf facade
▼ Thermal glazing
■ Luxguard Natural
▲ Climasol Neutral

6.2.9 Computation methods, abbreviated method (manual method)

The standard computation method differentiates, as previously described, between internal and external cooling loads. Hence:

Internal cooling load $\dot{Q}_i$

$$\dot{Q}_i = \dot{Q}_P + \dot{Q}_L + \dot{Q}_M + \dot{Q}_G + \dot{Q}_C + \dot{Q}_R$$

Cooling load generated by persons $\dot{Q}_P$

$$\dot{Q}_P = n \cdot q_P \cdot s_i$$

n Number of persons
q_P Heat emitted by the human body
s_i Cooling load factor for internal loads (50 % convection)

Cooling load generated by lighting

$$\dot{Q}_L = P \cdot l \cdot \mu_L \cdot s_i$$

P = Total connect load of the lamps, in the case of discharge lamps including loss of power supply units (in W)
l = Simultaneity factor of lighting at relevant time
μ_L = Load factor due to lighting
s_i = Cooling load factor for internal loads

Depending on the percentage of convection in the heat emitted by the lamps, the cooling load factor for internal loads can be deduced from the cooling load specifications (e.g. VDI 2078).

Cooling loads generated by machines and devices $\dot{Q}_M$

$$\dot{Q}_M = \sum_{j=1}^{\eta} \left[\frac{P_i}{\eta} \mu_{aj}\right] \cdot l \cdot s_i$$

P_j = Nominal power (spindle shaft power) of the j-machine
η = Mean engine efficiency
μ_{aj} = Load factor of j-machine at relevant time
l = Simultaneity factor
s_i = Cooling load factor for internal loads

Cooling load as a result of rate of flow $\dot{Q}_G$

$$\dot{Q}_G = \dot{m} \cdot c \cdot (\vartheta_I - \vartheta_O)\, s_i$$

$\dot{m}$ = Mass flow of goods brought into or removed from room
c = Mean specific heat capacity
ϑ_I = Inlet temperature
ϑ_O = Outlet temperature
s_i = Cooling load factor for internal loads

The cooling load specifications list types of rooms that refer to their thermal storage capacity. The types are differentiated as:

XL = Very low thermal mass
L = Low thermal mass
M = Medium thermal mass
H = High thermal mass

Rooms can classified into specific types by following these guidelines:

XL = Total mass < approx. 200 kg/m² floor
L/M = Total mass 200 to 600 kg/m² floor
L = Thermal mass is covered (wall-to-wall carpeting/insulation below floor screed/suspended ceiling etc.), total mass 200 to 400 kg/m² floor
M = Thermal mass is exposed (stone floor/plastered concrete ceiling etc.), total mass 400 to 600 kg/m² floor
H = Total mass > 600 kg/m² floor

The classification is based on the assumption that all rooms share the same geometry and insulation but vary in wall construction and plan. In the manual calculation, room types are matched to corresponding cooling load factors (thermal storage factors).

Figure 6.42
Total solar energy transmission coefficients for different shading options in a single-leaf facade

1 DIN 4108
2 Ferrari Soltis 86, silk tinted (external) alu (internal)
3 Agero G-1907
4 MHZ Hachtel No. 400
5 Hüppe No. 10, silver/black
6 Siemens prisms, injected

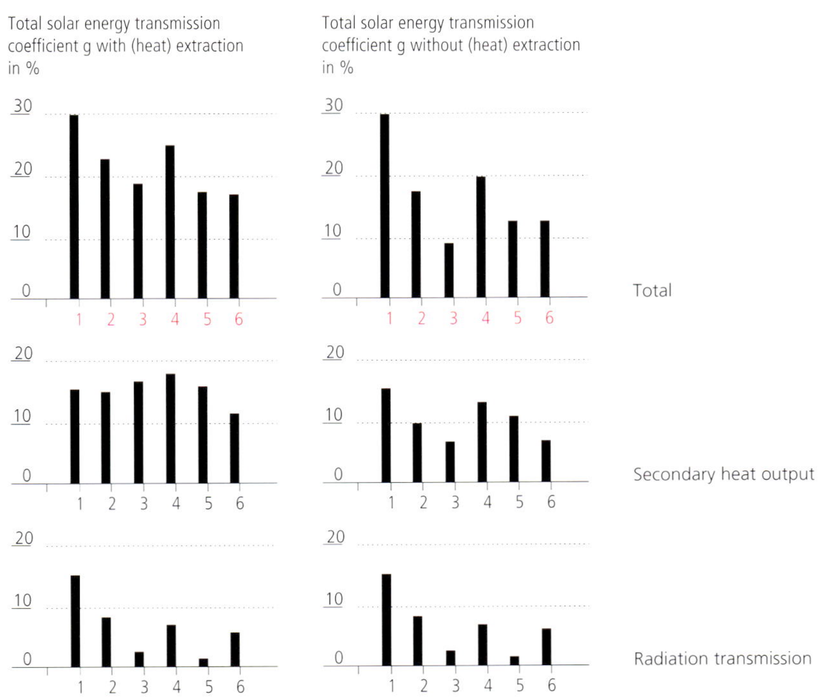

However, care must be taken to base the storage calculation on the different operating times of air-conditioning systems, that is, to ensure that the cooling loads are for the most part applicable to buildings where air-conditioning is used. Hence, the first step is a qualitative comparison of the room to be studied with the room types in the table in order to allocate the space to one of these types.

The following comparative study demonstrates the effects of storage behaviour in a south-facing test room. The room and its construction (Options 1 – 6) are illustrated in **Figure 6.43**.

The essential differences in construction for each option are as follows:

Option 1
Light to medium thermal storage mass with full-height window glazing
Option 2
Light to medium thermal storage mass with window sill (approx. 70 cm high)
Option 3
Medium thermal storage mass with window sill as in Option 2 but without suspended ceiling
Option 4
High thermal storage mass with window sill, without suspended ceiling and with limited thermal mass in floor
Option 5
High thermal storage mass with window sill and brickwork side partitions
Option 6
Medium thermal storage mass with window sill and demountable partitions (metal)

Figure 6.43
Room types with different surrounding structures

Dimensions:	
Orientation :	South
Floor space:	21.0 m^2
Occupants:	2
Room height:	2.90 m
Lighting:	158 W
Ceiling suspension (void):	0.15 m
Machines:	315 W
Window sill height:	0.70 m

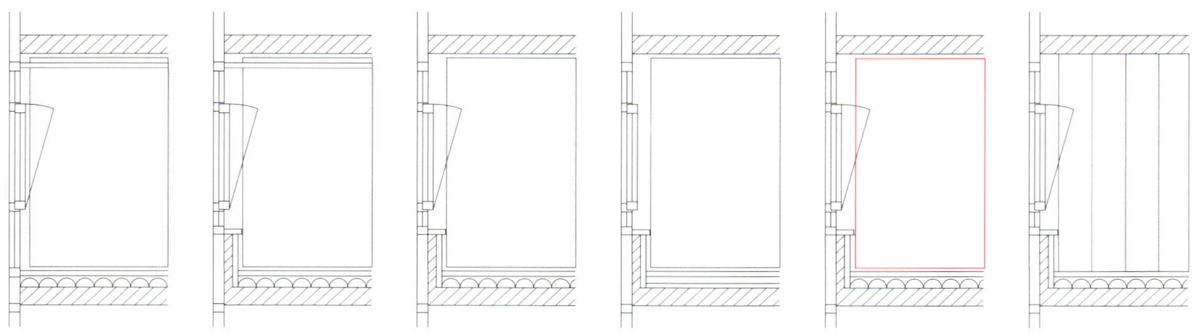

Option	1	2	3	4	5	6
Window	U=2.0 W/m^{2}K g = 62%	k = 2.0 W/m^{2}K g = 62%	k = 2.0 W/m^{2}K g = 62%	k = 2.0 W/m^{2}K g = 62%	k = 2.0 W/m^{2}K g = 62%	k = 2.0 W/m^{2}K g = 62%
Sun protection	external/g = 0.20	external/g = 0.20	external/g = 0.20	external/g = 0.20	external/g = 0.20	external/g = 0.20
Window sill	without	with	with	with	with	with
Suspended ceiling	with	with	without	without	without	without
Floor	Raised floor	Raised floor	Raised floor	Floating screed	Raised floor	Raised floor
Side partitions	Drywall	Drywall	Drywall	Drywall	Brickwork	Insulated metal-stud wall
Rear partition	Drywall	Drywall	Drywall	Drywall	Brickwork	Insulated metal-stud wall
Exterior wall	Louvre 0.6 cm Insulation 8.0 cm	Louvre 0.6 cm Insulation 8.0 cm Concrete 12.0 cm Plaster 1.5 cm	Louvre 0.6 cm Insulation 8.0 cm Concrete 12.0 cm Plaster 1.5 cm	Louvre 0.6 cm Insulation 8.0 cm Concrete 12.0 cm Plaster 1.5 cm	Louvre 0.6 cm Insulation 8.0 cm Concrete 12.0 cm Plaster 1.5 cm	Louvre 0.6 cm Insulation 8.0 cm Concrete 12.0 cm Plaster 1.5 cm
Ceiling/ floor	Suspension 2.0 cm Air 13.0 cm Concrete 25.0 cm Raised floor 15.0 cm Carpet 0.5 cm	Suspension 2.0 cm Air 13.0 cm Concrete 25.0 cm Raised floor 15.0 cm Carpet 0.5 cm	Concrete 25.0 cm Raised floor 15.0 cm Carpet 0.5 cm	Concrete 25.0 cm Insulation 3.0 cm Plaster 8.0 cm Carpet 0.5 cm	Concrete 25.0 cm Raised floor 15.0 cm Carpet 0.5 cm	Concrete 25.0 cm Raised floor 15.0 cm Carpet 0.5 cm
Side partitions	Drywall 1.5 cm Insulation 7.0 cm Drywall 1.5 cm	Drywall 1.5 cm Insulation 7.0 cm Drywall 1.5 cm	Drywall 1.5 cm Insulation 7.0 cm Drywall 1.5 cm	Drywall 1.5 cm Insulation 7.0 cm Drywall 1.5 cm	Plaster 1.5 cm Brickwork 11.5 cm Plaster 1.5 cm	Insulated metal-stud wall 0.1 cm Insulation 9.8 cm Insulated metal-stud wall 0.1 cm
Rear partition	Drywall 1.5 cm Insulation 7.0 cm Drywall 1.5 cm	Drywall 1.5 cm Insulation 7.0 cm Drywall 1.5 cm	Drywall 1.5 cm Insulation 7.0 cm Drywall 1.5 cm	Drywall 1.5 cm Insulation 7.0 cm Drywall 1.5 cm	Plaster 1.5 cm Brickwork 11.5 cm Plaster1.5 cm	Insulated metal-stud wall 0.1 cm Insulation 9.8 cm Insulated metal-stud wall 0.1 cm

The curves in **Figure 6.44** indicate room temperatures after five days of clear weather in summer for a south-facing room. As the temperature chart shows, the storage factor in the external walls (Option 2) causes a temperature drop of roughly 1.2 K (maximum value) compared to full glazing (Option 1). This is caused not only by the external wall's thermal storage capacity but also, especially, by the reduced solar incidence as a result of the decreased glass surface. A comparison of Options 2 and 3 is of greater interest. A suspended ceiling is not used (in Option 3) and the room temperature drops by a further 1 K (approx. maximum value) compared to Option 2; this difference is entirely due to the increased thermal storage capacity.

The difference between Options 3 and 4 is approx. 0.2 K, because the wall-to-wall carpeting almost cancels out the increased thermal storage capacity of the floor compared to the raised floor. A significant improvement could be achieved only if stone flooring were used instead of carpeting. Option 5 reveals the greatest difference due to the extremely high storage capacity of the brickwork partitions. The temperature is lowered by approx. 2 K to 3 K compared to the other room options (with the exception of Option 1).

Figure 6.44
Room temperatures following five days of good weather in summer
Weather data from TRY (TRY region 8)

Room temperatures are indicated as perceived temperatures (from DIN 1946/part 2), under consideration of building component and surface temperatures.

— Fresh air
— Option 1
– – Option 2
--- Option 3
---- Option 4
— Option 5
–·– Option 6

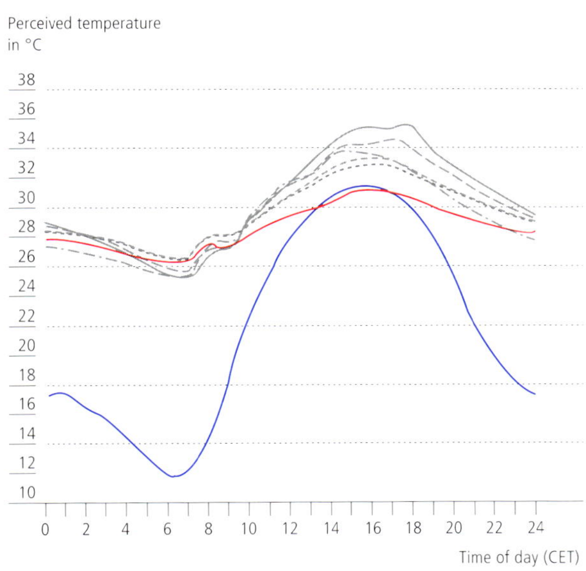

In Option 6, the overall behaviour is similar to that in Option 3, but slightly poorer since the insulated metal-stud walls have an even lower storage performance rating than the drywalls in Option 2.

Judging from how the curves in **Figure 6.44** relate, it is apparent that the full potential of the temperature decrease due to storage in a south-facing room is limited to the period between 4:00 p.m. and 6:00 p.m., that is, four hours after the room has been exposed to maximum solar radiation (external shading). For east-facing rooms, the greatest reduction of the cooling load, and hence in room temperatures, was observed between 12:00 a.m. and 2:00 p.m.; for west-facing rooms, between 10:00 p.m. and 12:00 p.m..

Figure 6.45 contains the temperature statistics for the summer months of May through September for this room and the layout options described above.

Because the room in Option 5 has the greatest thermal storage capacity, its temperatures do not exceed 30 °C, i.e. the temperatures in this room reach a maximum of 29 °C without ventilation and cooling on approx. 20 days of the year. All other options, with the exception of Option 1, exhibit similar behaviour with regard to the predicted number of days when room temperatures range from approx. 22 to 30 °C. Option 1 shows the poorest performance with regard to recorded temperatures, although the increase of incident heat due to the large window surface plays an important role in this model.

The thermal storage capacity of a room can be significantly improved by flushing the space with outside air during the night. **Figure 6.46** represents the temperature behaviour in two test rooms (Options 2 and 5) with and without night cooling. In the case of night cooling, a standard 3 ac/h (air change) is assumed for average wind velocities and with slightly opened windows. A comparison of the curves with and without night cooling for Options 2 and 5 reveals that the temperature is reduced by approx. 2 to 6 K in a room with light to medium storage capacity, whereas in Option 5 it can be reduced by approx. 3 to 4 K.
It is important to note that the temperature level in Option 5 is roughly 4 K less than that in Option 2. The increased thermal storage capacity seems to have an effect, especially during the operating hours of the building in question. Finally, these curves and their comparative values demonstrate that night cooling in a building is as important as the increased storage capacity.

Therefore, the best way to reduce energy consumption and the investment cost in ventilation and air-conditioning systems is to design a room with high thermal storage capacity and efficient night cooling in order to improve the storage behaviour.

The external cooling loads (external heat gains) are composed of cooling loads through external walls and roofs based on the equation described below (see page 214), where the thermal transmittance and an equivalent temperature difference play an essential role.

The cooling loads resulting from incident radiation penetrating through windows in combination with different shading elements also play an important role. Another factor is whether and to what degree windows are partially exposed to the sun or shaded, and whether entire facades lie in shade.

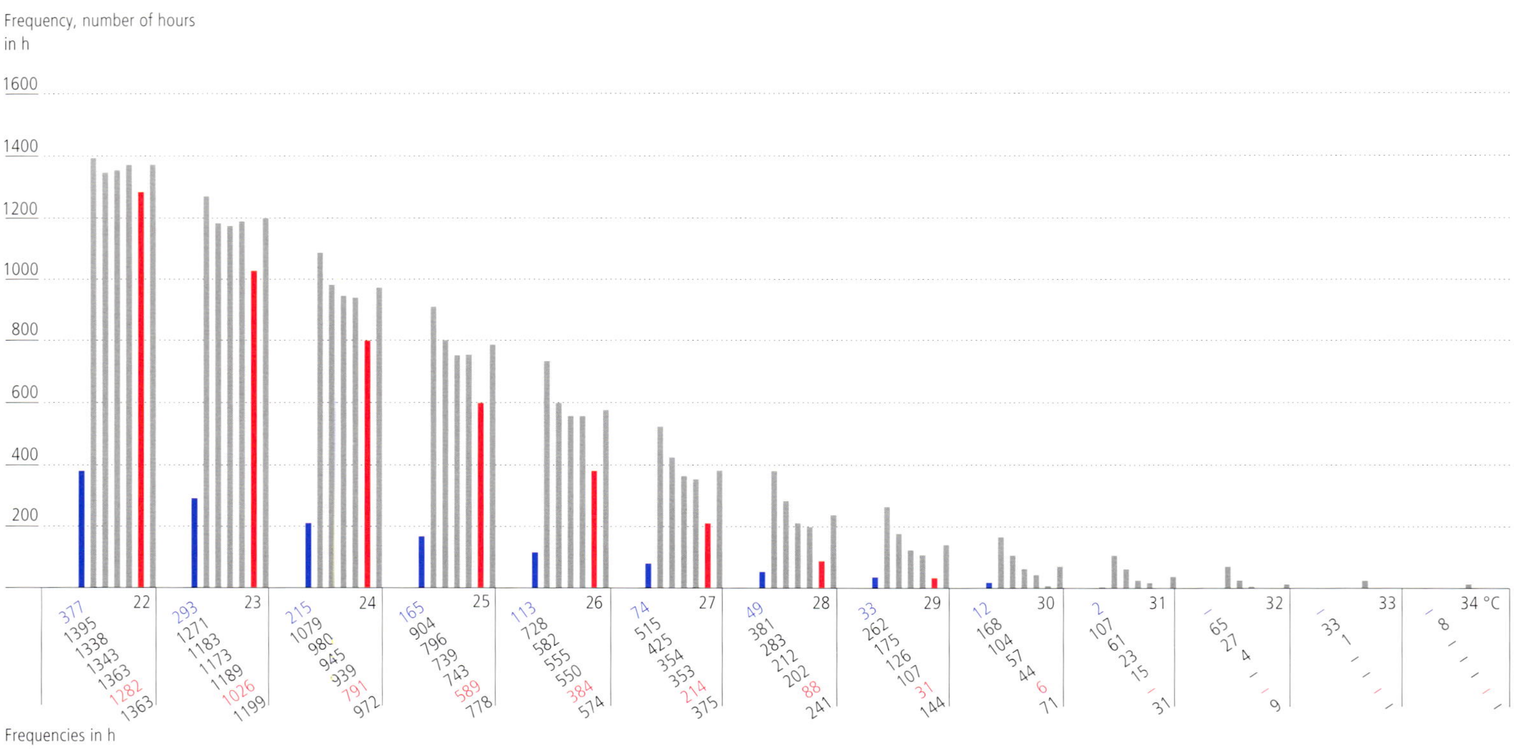

Figure 6.45
Room temperatures from May to September
Weather data from TRY (TRY region 8)

Number of hours when temperature is greater than indicated temperature from 7 a.m. to 6 p.m. (daily)

- Fresh air
- Option 1
- Option 2
- Option 3
- Option 4
- Option 5
- Option 6

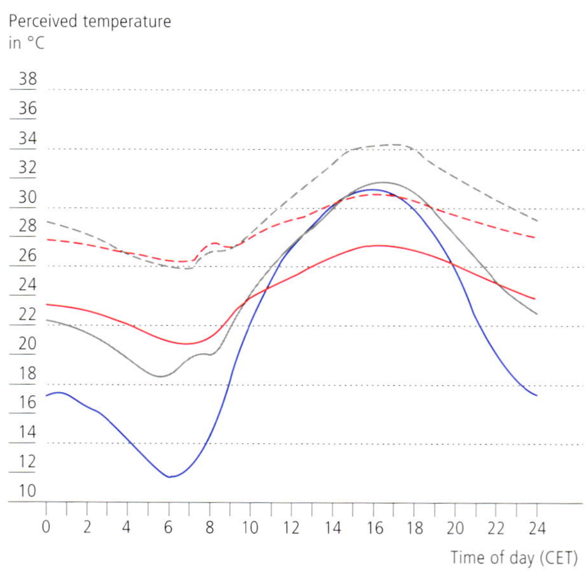

Figure 6.46
Room temperatures following five days of good weather in summer

Comparison: with/without night ventilation
Weather data from TRY (TRY region 8)

- Fresh air
- Option 2 (without NV)
- Option 2 (with NV)
- Option 5 (without NV)
- Option 5 (with NV)

NV = night ventilation (air change 3.0 h from 5 p.m. to 8 a.m.)

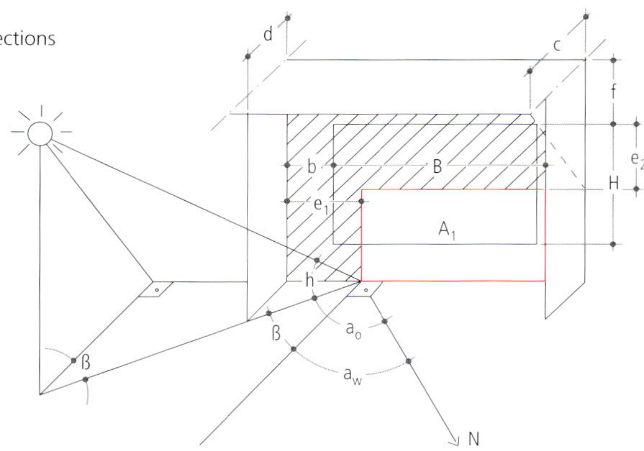

Figure 6.47
Shading from projections

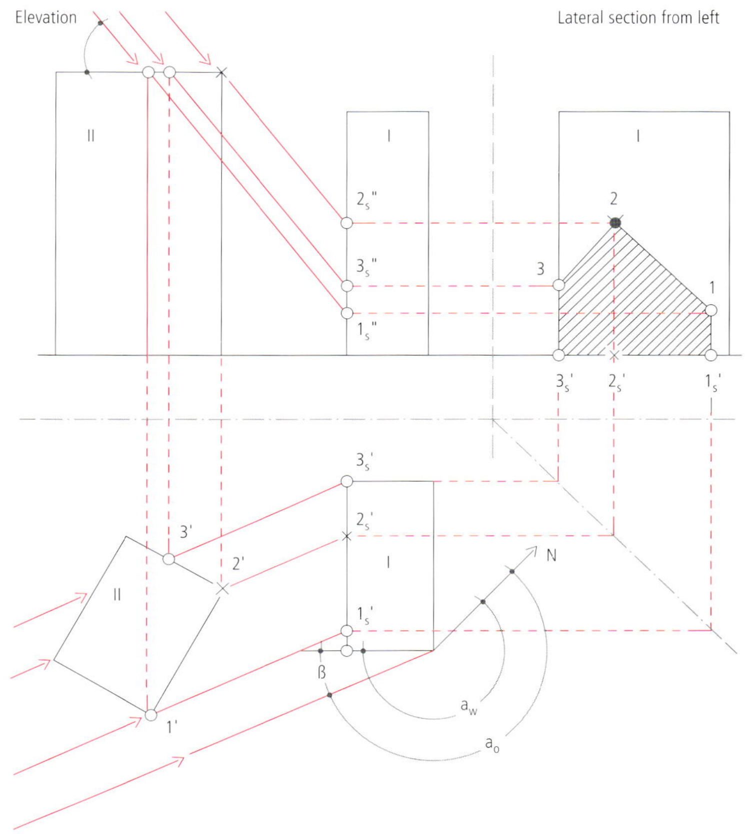

Figure 6.48
Facade shading from neighbouring buildings

a_w	Wall azimuth
a_o	Solar azimuth
h	Solar altitude
A = B · H	Total glass area
A_1	Glass area exposed to sun

Table 6.3
Solar altitude a and solar azimuth a_o

Solar time	20. 12. h	a_o	24. 1. u. 20. 11. h	a_o	20.12. u. 23. 10. h	a_o	22. 3. u. 24. 9. h	a_o	20. 4. u. 24. 8. h	a_o	21. 5. u. 23. 7. h	a_o	21. 6. h	a_o
4													1°	53°
5											6°	66°	9°	64°
6									9°	83°	15°	77°	18°	74°
7					1°	109°	10°	102°	18°	94°	25°	88°	27°	85°
8			3°	125°	9°	121°	19°	114°	28°	106°	34°	100°	37°	97°
9	7°	139°	10°	137°	17°	134°	27°	127°	37°	120°	44°	114°	46°	110°
10	12°	152°	16°	151°	23°	148°	34°	143°	44°	137°	52°	131°	55°	128°
11	15°	166°	19°	165°	27°	163°	38°	161°	50°	157°	58°	153°	61°	151°
12	17°	180°	21°	180°	29°	180°	40°	180°	51°	180°	60°	180°	63°	180°
13	15°	194°	19°	195°	27°	197°	38°	199°	50°	203°	58°	207°	61°	209°
14	12°	208°	16°	209°	23°	212°	34°	217°	44°	223°	52°	229°	55°	232°
15	7°	221°	10°	223°	17°	226°	27°	233°	37°	240°	44°	246°	46°	250°
16			3°	235°	9°	239°	19°	246°	28°	254°	34°	260°	37°	263°
17					1°	251°	10°	258°	18°	266°	25°	272°	27°	275°
18									9°	277°	15°	283°	18°	286°
19											6°	294°	9°	296°
20													1°	307°

Table 6.4
Wall orientation and solar altitude for lat. 50° N

Wall azimuth a_w

N	0°	O	90°	S	180°	W	270°
NNO	23°	OSO	113°	SSW	203°	WNW	293°
NO	45°	SO	135°	SW	225°	NW	315°
ONO	68°	SSO	158°	WSW	248°	NNW	338°
						N	360°

Figure 6.47 illustrates an example of a window shaded by wall projections and the relevant angles; **Figure 6.48** illustrates a facade shaded by a neighbouring building. The relevant solar height and azimuth angles are indicated in **Table 6.3** based on the diagram contained in **Figure 6.49**. **Table 6.4** indicates the wall azimuths in terms of wall orientation and solar angle for lat. 50° N. Hence, external cooling loads are calculated as follows:

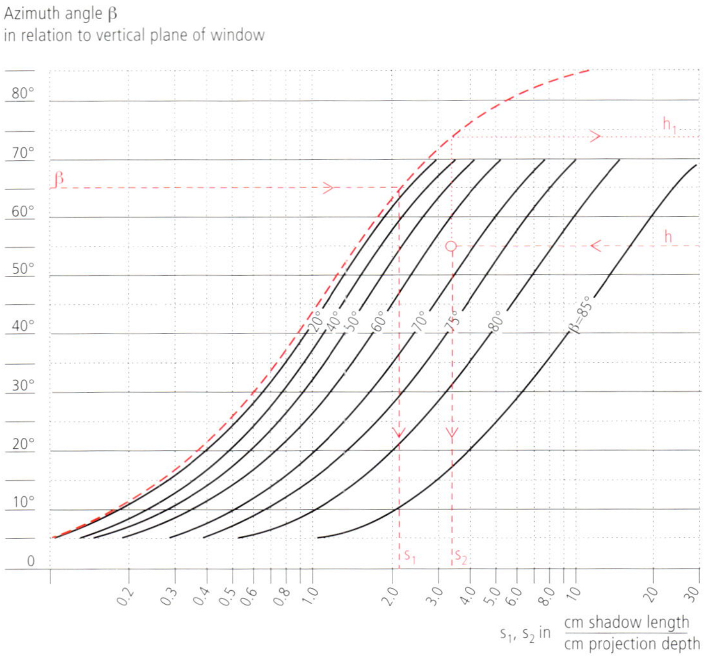

Figure 6.49
Shading diagram

s_1 Specific lateral dimension of projections per unit of length, based on β
s_2 Specific shadow length (from above) per unit of length of projections, based on h
h Solar altitude
h_1 Altitude angle

Example 1
β = 65°
s_1 = 2.15 cm/cm

Example 2
h = 55°
β = 65°
h_1 = 73.5°
s_2 = 3.4 cm/cm

External cooling load $\dot{Q}_E$

$$\dot{Q}_E = \dot{Q}_W + \dot{Q}_T + \dot{Q}_S + \dot{Q}_{LF}$$

Cooling load through external walls and roofs $\dot{Q}_W$

The momentary heat flow through external walls and roofs $\dot{Q}_W$ into a room is the product of

$$\dot{Q}_W = U \cdot A \cdot \Delta\vartheta_{eq}$$

U = Thermal transmittance
A = Area
$\Delta\vartheta_{eq}$ = Equivalent temperature difference

The cooling load specifications indicate the equivalent temperature differences according to orientation and time of day for six classifications of wall and roof constructions.

The same specifications also provide a summary of standard wall and roof constructions. In addition to the allocation to the six options mentioned above, the tables also contain the construction, thermal transmittance (U-value), mass per area (m_a) and the so-called time correction factor (ΔT) for each construction.

The time correction factor takes the lag in reaction on the part of specific building constructions into account in as much as it deviates from the behaviour of the relevant classification. If the time correction factor assumes a value other than zero, the point in time at which the equivalent temperature difference is determined must be modified by the indicated time correction factor and read accordingly. When the time correction factor is zero, the value of the equivalent temperature difference at the definition time must be used:

$$\Delta\vartheta_{eq}\ (T) = \Delta\vartheta_{eq},\ \text{Table}\ (T + \Delta T)$$

where T = Time of day

If the mean value of the outside air temperature deviates from the indicated default in the calculation, $\dot{Q}_W$ is established with an adjusted value.
In summer

$$\Delta\vartheta_{eq1,\ July} = \Delta\vartheta_{eq} + (\vartheta_{OT,\ m} - 24.5\ °C) + (22\ °C - \vartheta_{RT})$$

$\vartheta_{OT,\ m}$ = True mean value of outside temperature in °C
ϑ_{RT} = True room air temperature in °C

Correspondingly, the following applies for a south-facing wall in September:

$$\Delta\vartheta_{eq1,\ Sept.} = \Delta\vartheta_{eq} + (\vartheta_{OT,\ m} - 18.5\ °C) + (22\ °C - \vartheta_{RT})$$

If the walls and roofs feature surfaces whose absorption and emission coefficients deviate from the assumed values (light-coloured wall, dark roof), $\dot{Q}_W$ is also established with an adjusted value $\Delta\vartheta_{eq2}$.

Dark-coloured wall ($\varepsilon = 0.9$, $a_s = 0.9$):
$\Delta\vartheta_{eq2} = \Delta\vartheta_{eq} + \Delta\vartheta_{eq,\,as}$

White wall ($\varepsilon = 0.9$, $a_s = 0.5$):
$\Delta\vartheta_{eq2} = \Delta\vartheta_{eq} - \Delta\vartheta_{eq,\,as}$

Metal, bare wall ($\varepsilon = 0.5$, $a_s = 0.5$):
$\Delta\vartheta_{eq2} = \Delta\vartheta_{eq} - \Delta\vartheta_{eq,\,as} + 2.0$

Light-coloured roof ($\varepsilon = 0.9$, $a_s = 0.7$):
$\Delta\vartheta_{eq2} = \Delta\vartheta_{eq} - \Delta\vartheta_{eq,\,as}$

White roof ($\varepsilon = 0.9$, $a_s = 0.5$):
$\Delta\vartheta_{eq2} = \Delta\vartheta_{eq} - 2\,\Delta\vartheta_{eq,\,as}$

Where $\Delta\vartheta_{eq,\,as}$ is the corrective value in modification of the absorption coefficient by

$\Delta a_s = 0.2$.

The corrective values of the equivalent temperature difference for walls are indicated in the specifications. The corrective values themselves are always read at the hour specified in the table at which the equivalent temperature difference is read.

Cooling load as a result of transmittance through window $\dot{Q}_W$

$$\dot{Q}_T = U_F \cdot A_M \cdot (\vartheta_{AU} - \vartheta_{RA})$$

U_F = Thermal transmittance (U-value) of window
A_M = Total window area (wall opening measurement)
ϑ_{OU} = Momentary outside air temperature
ϑ_{RA} = Room air temperature

Cooling load as a result of incident radiation through window $\dot{Q}_R$

$$\dot{Q}_R = [A_1 \cdot I_{max} + (A - A_1) \cdot I_{dif,max}] \cdot b \cdot R_e$$

A_1 = Glass area exposed to sunshine
A = Total glass area
I_{max} = Maximum total radiation for design month
$I_{dif,max}$ = Maximum diffuse radiation for design month
b = Shading coefficient (b-factor) of windows and sunshade elements
R_e = Cooling load factor for external radiation loads

In cases where the shaded area shifts over the course of a day, the area exposed to sunshine should be determined at the time of maximum total radiation.

The cooling load factor for external radiation loads, R_e, is dependent on the building mass, the shading element (internal, external) and the month that forms the basis for the calculation (design month). If the ratio of A1/A is less or equal to 0.1, R_e must be calculated on the basis of northern exposure.

In scenarios where moveable shading elements are used, the calculation method assumes that the shade is down throughout the day. If this is not the case, that is, if the shade is not pulled down during times where there is no direct incident sun, the following preliminary calculation must be executed in order to determine which radiation values and cooling load factors must be used:

$$\dot{Q}_{R,\,I\,max} = b_1 \cdot b_2 \cdot I_{dif,\,max} \cdot R_{e\,max}$$

$$\dot{Q}_{R,\,II\,max} = b_1 \cdot I_{N\,max} \cdot R_{eN\,max}$$

The radiation values, cooling load factors and shading coefficients must then be applied to establish the greater cooling load value.

Cooling load as a result of infiltration $\dot{Q}_{LF}$

This cooling load factor is only relevant in special cases.

Room cooling load $\dot{Q}_{CR}$

The room cooling load is the sum of the internal and external cooling load factors:

$$\dot{Q}_{CR} = \dot{Q}_I + \dot{Q}_A$$

Building cooling load $\dot{Q}_{CB}$

The building cooling load at the point in time t is the product of the sum of all room cooling loads at t:

$$\dot{Q}_{CB} = \sum_{j=1}^{\eta} \dot{Q}_{CBj}\,(t)$$

Once the timeline is established, the maximum cooling load can be determined.

Clearly, calculating the cooling load is a laborious process. Since this projection not only aims to establish the real cooling loads but also to predict temperature curves under certain conditions – an undertaking subject to many possible variants – most of these calculations are carried out using computer programs, which have been greatly improved and fine-tuned in recent years.

6.3 Air rates and air changes

6.3.1 Air rates

The fresh air required hourly to achieve an hygienic internal climate is defined in relation to the room content, height, location and the degree of deterioration in the internal air quality. The type of fresh air intake also plays a role, a factor whose impact can be diminished when outside air is transported directly to the relevant area (for example, from bottom to top).

To achieve hygienic room air conditions, a minimum degree of outside air should flow into the room per person per hour. The following standards for minimum fresh air rates are based on investigations carried out by medical experts on health at the workplace:

- Theatres, concert halls, cinemas, reading rooms (libraries), fair halls, sales rooms, museums, gym and sports facilities:
  20 m^3/h person
- Offices, resting rooms, canteens, conference rooms, class rooms, auditoriums, lunch rooms:
  30 m^3/h person
- Restaurants: 40 m^3/h person
- Open-plan offices: 50 m^3/h person

When outside temperatures fall below 0 °C or rise above +26 °C, the air rate can be reduced to 50 % to conserve energy. Where unpleasant odours are present (e.g. tobacco smoke) the indicated rates should be increased by 20 m^3/h per person.

6.3.2 Air change

The air change, i.e., the exchange of air in a room per hour (exchange of room volume) results from a wide range of parameters, the most important being:

- Air volume/h required for cooling load compensation
- Necessary intake air volume to achieve hygienic room conditions
- Intake air volume linked to technical reasons (generally coupled with specific types of ventilation)

The air change rates listed below are based on empirical evidence, which is generally a good basis for initial design without having to undertake detailed cooling or heating load calculations etc.

Table 6.5 lists the empirical data for hourly air changes in different types of rooms. **Table 6.6** addresses the detailed requirements for ventilation in hospitals and clinics. **Table 6.7** indicates the fresh air changes in schools (air changes resulting from number of occupants). **Tables 6.8** and **6.9** list air changes in kitchens, where rates are to a great degree determined by the type and number of kitchen appliances. **Table 6.10** offers planning guidelines for sales areas (department stores, retail shops in malls etc.).

Room type	Air change/hour (approx.)
Latrines	6 – 10fold
Battery rooms	4 – 6fold
Bathrooms	4 – 6fold
Pickling plant	5 – 15fold
Libraries	3 – 5fold
Shower rooms	20 – 30fold
Offices	3 – 6fold
Dye shops	5 – 10fold
Paint shops	20 – 50fold
Garages	4 – 5fold
Cloakrooms	3 – 6fold
Guest rooms	5 – 10fold
Auditoriums	8 – 10fold
Canteens	6 – 8fold
Department stores	4 – 6fold
Cinemas and theatres, non-smoking	4 – 6fold
Cinemas and theatres, smoking permitted	5 – 8fold
Hospitals	see Table 6.6
Kitchens	see Table 6.8 / 6.9
Laboratories	15 – 20fold
Shops	6 – 8fold
Operating theatres	15 – 20fold
Plating mills	8 – 10fold
Computer labs	60 – 65fold
Schools	see Table 6.7
Pools	3 – 4fold
Conference rooms	6 – 8fold
Dining rooms	6 – 8fold
Washrooms (toilets)	4 – 6fold
Safes	3 – 6fold
Change rooms in pools	6 – 8fold
Retail areas	4 – 8fold
Meeting rooms	5 – 10fold
Laundries	10 – 15fold
Workshops without unusual air contaminants	3 – 6fold

Table 6.5
Empirical values for hourly air change in different types of rooms

1) Unless otherwise indicated, all values are based on DIN 1946-2.
2) In addition to the reasons relating to climate physiology and infection prevention, HVAC systems may also be necessary for other reasons.
3) + means that the boundary values indicated in DIN 1946-2 must be maintained (30–65% rel. humidity).
4) These values are only applicable to rooms that are constantly occupied.
5) Min. and max. temp. optional throughout year in conjunction with the other rooms in the OR department. An external temperature that is 4K below the temperature indicated in VDI 2078 can be used as a base for sizing the cooling system. The temperatures at the operating table are relevant for the entire operating theatre.
6) For immune-deficient patients (patients w/ weakened immune systems)
7) Room temperature 2 to 4K above water temperature up to a room temperature of 28°C. For water temperatures of 28°C and up, both (water and room) temperatures should be equal.
8) Definitions must be based on tolerability and building requirements.
9) In individual cases, medical and technical equipment may impose the use of HVAC systems and adherence to specific humidity values.
10) The hygienist must determine whether an HVAC system is indispensable for aerogenically transmittable diseases in order to prevent infection.
11) HVAC systems can be omitted, if all premature infants are placed in incubators.
12) Same supply air temperature and supply air humidity as for operating theatres or wards
13) Fresh air volume flow based on contaminant profile
14) Contaminants must be evacuated in case of chemical disinfectants or sterilization; see DIN 58948-7.
15) Construction measures must be taken to prevent air exchange between sterile and non-sterile zones as much as possible.
16) After DIN 1946-7
17) Type A: very high requirement for sterile environment (e.g. heart transplants, heart surgery) Type B: high requirement for sterile environment (even for minimally invasive surgery)
18) Depending on germ concentration in air, see DIN 1946-T4.

Room category	Room type	Room air conditions[1]			Min. fresh air volume relative to			Exhaust air volume flow	Values for max. system sound level [4]	HVAC system indispensable [2]	
		Temperatures		Humidity [3]	1 Person	1 m² Room area	1 m³ Room content (air change)			Climate physiology	Infection prevention
		min.	max.								
		°C	°C		m³/h	m³/h · m²	m³/h · m³	m³/h	dB (A)		
I. High or very high requirements for sterile conditions	Operating theatres Type A or B [17] including trauma and delivery rooms	22 [5]	26 [5]	+		[18]			40	+	+
	Transport corridors or storage areas for sterile goods, wash stations, supply and clear-out rooms, possibly machine rooms in OR departments	22	26	+		15	5		40	+	+
	Recovery rooms in OR departments	22 [5]	26 [5]	+		30	10		35	+	+
	Other rooms, hallways in OR departments	22	26	+		15	5		40	+	+
	Endoscopic procedure rooms					30	10		40		+
	Intensive care units	24	26	+		30	10		30		
	Emergency room for intensive care	24	26	+		30	10		40		
	Patient rooms for special care units [6]	24	26	+		30	10		40		
	Emergency room for special care units [6]	24	26	+		30	10		40		
	Dispensary (sterile rooms)					10	3		45		+
II. Standard requirements for sterile	Delivery rooms	24				15	5		40		
	Endoscopic examination rooms (aseptic, septic)					30	10		40		
	Physiotherapy (baths)	[7]	[7]			[8]			50	+	
	Rooms for minor procedures					15	5		40		
	Recovery rooms outside of OR department		26	+		30	10		35	+	
	X-ray, examination rooms			[9]		15	5		40		
	Wards for intensive observation	24	26	+		15	5		30		
	Wards for infectious disease care, possible incl. anteroom					10	3		35		[10]
	Rooms for premature infant care	24	26	+ (30 – 65% rH)		10	3		35	[11]	
	Rooms for premature infant, infant and general care					10	3		35		
	Other rooms, hallways in intensive care and special care units	[12]	[12]			15	5		35		
	Other rooms, hallways					10	3				
	Other rooms, hallways in patient care areas					10	3		40		
	Sterilization; sterile and non-sterile side, sterile goods storage					[13]			50	[14]	[15]
	Bedding, laundry processing, laundry; sterile and non-sterile side					[13]			50	[14]	[15]
	Pathology/autopsy facilities		22						50		
	Labs					[16]			45		
	Changing rooms							100 per changing room			
	WCs							60 per fixture			
	Ward bathrooms							150 per room			
	Sanitary blocks							100 per block			

Table 6.6
Ventilation requirements in hospitals and clinics
Excerpt from:
DIN 1946, Part 4, Page 10

Room type	Air change
Classrooms	4 to 5fold
Screening rooms	6 to 8fold
Chemistry labs	6 to 8fold
Natural science labs	4 to 5fold
Gyms and school pools	2 to 3fold
Phys.ed. rooms	4 to 6fold
Changing rooms	8 to 10fold
Washrooms incl. anterooms	5fold

Table 6.7
Fresh air changes for HVAC systems in **schools**

Kitchen type	Room height in m	Air change per h
Small kitchens in apartments, homes	2.5 – 3.5	25 – 15fold
Medium sized kitchens in restaurants, hotels, canteens	3.0 – 4.0 4.0 – 6.0	30 – 20fold 20 – 15fold
Large commercial kitchens in hospitals, barracks	3.0 – 4.0 4.0 – 6.0 more than 6.0	30 – 20fold 20 – 15fold 15 – 10fold
Dietary kitchens	3.0 – 4.0	20 – 15fold
Dishwashing kitchens	3.0 – 4.0 4.0 – 6.0	20 – 15fold 15 – 10fold
Cold kitchens	3.0 – 4.0 4.0 – 6.0	8 – 5fold 6 – 4fach
Baking rooms	3.0 – 4.0 4.0 – 6.0	15 – 8fold 8 – 6fold
Sculleries	–	5 – 8fold

Table 6.8
Air changes for **kitchens**

Kitchen appliance	Type or size	Air volume
Stove	Coal-fired, gas-fired, electric	3,000 m³/h per m² hotplate 1,500 m³/h per m² hotplate 1,000 m³/h per m² hotplate
Boiler	100 l 200 l 500 l 1,000 l	300 m³/h 600 m³/h 1,000 m³/h 1,500 m³/h
Tipping frying pan	Gas-fired, electric	1,500 m³/h per m² hotplate 1,000 m³/h per m² hotplate
Warming table	Gas-fired, electric	450 m³/h per m² hotplate 300 m³/h per m² hotplate
Roasting and baking oven	Based on surface	1,000 m³/h per m² hotplate
Coffeemaker	Gas-fired, electric	450 m³/h 300 m³/h
Toaster	2 kW	800 m³/h
Barbecue	Gas-fired, electric	3,000 m³/h per m² hotplate 2,000 m³/h per m² hotplate
Tipping cauldron	10 l 20 l 50 l	200 m³/h 300 m³/h 500 m³/h

Table 6.9
Defining the volume flow based on number and type of kitchen appliances

All room types not listed above require a wide range of air changes as a result of special demands, thermal loads, dust content etc. and can therefore not be included in this summary overview of standard scenarios. This also applies to shelters (e.g. storm shelters), which operate with an air change rate of 9 m^3/h person under normal conditions, and 1.8 m^3/h per person in emergency conditions.

Air change definitions in response to deterioration in air quality are mainly a concern in industrial operations. The sources of deterioration must be known. Once the extent of foreign matter has been determined, we can calculate the volume of air required to achieve a specific level of air purity.

The required volume flow (fresh air volume) is calculated as follows:

$$\dot{V} = \frac{\dot{K}}{k_1 - k_2}$$

Where:

$\dot{K}$ = hourly gas or steam volume in m^3/h

k_i = MAK-value in m^3_{gas}/m^3_{air} **(Table 2.4)**

k_o = gas content in intake air in m^3_{gas}/m^3_{air}

It is important to note that the noxious gas and the gas contained in the outside air must have identical MAK-values; a conversion is otherwise necessary. Here is an example:

Room content 4,000 m^3
Ammonia (NH_3) emitted in room = 1 kg/h

What is the required air change?

– Gas volume arising $\dot{K} = \dot{m} \cdot \frac{V_m}{M_m}$

$\dot{m}$ = mass flow arising (kg/h)

V_m = molar volume (m^3/mol) for 20 °C = 24 m^3/mol
M_m = molar mass (kg/mol) of gas = 17 kg/mol for ammonia

$$\dot{K} = 1\,\text{kg/h}\ \frac{24\,\text{m}^3/\text{mol}}{17\,\text{kg/mol}} = 1.41\,\text{m}^3/\text{h (ammonia flow)}$$

– Fresh air volume flow

$k_i = 50\ ppm = 50 \cdot 10^{-6}\ m^3_{gas}/m^3_{air}$
$k_o = 0$ (clean outside air)

$$\dot{V} = \frac{\dot{K}}{k_i - k_o} = \frac{1.41\ m^3/h}{50 \cdot 10^{-6} - 0} = 28{,}000\ m^3/h$$

– Air change

$$I = \frac{\dot{V}}{V} = \frac{28{,}200\ m^3/h}{4{,}000\ m^3} = 7.05\ 1/h$$

Table 6.10
Planning guidelines for room air conditions, air change/hour (3 m clear room height), occupant density and illumination heat for HVAC systems with air-conditioning in **retail areas**
Excerpt from: VDI 2082

Room categories	**Room temperature** Winter °C	Summer °C	**Room humidity** Winter %	Summer %	**Air change** per hour	**Occupant density** Persons/ 10 m² area	**Illumination heat** W/m²
Category retail and services							
General retail, e.g. textiles, hardware, shoes, jewellery, packaged groceries	19 to 22	22 to 26	–	to 65 … 50	2 to 6	1 to 2	15 to 30
Odour-intensive retail, dry cleaners	19 to 22	22 to 26	–	to 65 … 50	4 to 8	0.5 to 1	15 to 30
Groceries such as meat, fish, cheese, fruit, baked goods	18 to 22	18 to 24	–	to 75 … 65	4 to 8	1 to 2	15 to 30
Retail with low customer frequency, e.g. furniture, household goods	19 to 22	22 to 26	–	to 65 … 50	2 to 8	0.1 to 0.5	15 to 30
Retail with high heat emissions, e.g. lighting/lamps, radio and television, hair salon	20 to 24	22 to 28	–	to 65 … 45	6 to 20	0.1 to 0.5	50 to 200
Integrated restaurants, e.g. refreshment rooms, café, snack bar, casino	20 to 23	22 to 26	–	to 65 … 50	6 to 15	2 to 6	10 to 30
Category ancillary/utility rooms							
Warehouses, general	depending on goods		–	–	1 to 2	0.1 to 0.2	3 to 10
Grocery/food warehouses	depending on goods		depending on goods		1 to 4	0.1 to 0.2	3 to 10
Warehouses with constant staff presence, e.g. goods receiving and shipping	17 to 20	22 to 26	–	to 65 … 50	1 to 2	0.2 to 0.4	10 to 20
Office, also open-plan office	22	22 to 26	–	to 65 … 50	4 to 8	1 to 2	15 to 40
Kitchen based on VDO 2052	to 28	to 28	–	70	15 to 40	–	–
Preparation and processing rooms, e.g. for meat, sausage, cheese, fish, fruit (also internal)	depending on goods		depending on goods		3 to 12	0.5 to 1.5	15 to 30
Workshop, studio	18 to 22	22 to 26	–	to 65 … 50	5 to 12	0.5 to 1.5	20 to 40
Changing rooms *)	21 to 24	–	–	–	4 to 8	–	5 to 10
Training rooms, common rooms and lounges	20 to 22	22 to 26	–	to 65 … 50	3 to 8	4 to 6	20 to 40
Washrooms *)	18 to 21	–	–	–	5	–	5 to 10

*) without air-conditioning

6.4
h/x-diagramm

The h/x-diagram (Mollier diagram) is used to calculate and represent changes in condition with humid air. It is vitally important not only for the climate engineer, who bases essential design components on h/x-diagram calculations, but also in terms of building physics, as this diagram helps to visualize dew points, changes in relative humidity etc. The oblique-angled system of coordinates indicates the x-value (water content in g/kg) as well as the enthalpy value (h) and the dry-bulb temperature t. **Figure 6.50** contains an h/x-diagram for humid air. The saturation curve (relative humidity φ = 100 %) for 1.013 bar is also indicated in the diagram. The range of unsaturated air lies above the humidity curve representing 100 % [humidity], while the range of oversaturated air (fog range) lies below. The graphs representing constant temperatures (isotherms) are moderately rising straight lines in the unsaturated range, which turn right and downwards (fog isotherm) at the saturation curve.

Figure 6.51 illustrates the various changes describing the following condition changes beginning with point 1 (dry-bulb temperature 20 °C / rel. humidity 50 %):

Change in condition 1 – 2
(Dry warming of air)
Dry temperature rises, relative humidity falls, absolute humidity remains constant
Change in condition 1 – 3
(Influx of vapour/steam)
Temperature, relative humidity and absolute humidity of air increase
Change in condition 1 – 4
(Humidification/influx of cold water in unsaturated air)
Temperature falls, relative and absolute humidity rise
Change in condition 1 – 5 – τ
(Dry cooling without dehumidification)
Temperature falls, absolute humidity remains constant, relative humidity rises

If the temperature falls sharply enough to cross past point 5 and beyond the saturation curve, the air emits water (condensation).

Change in condition 1 – 6 – 7
(Change in condition during cooling and dehumidification)
Temperature and absolute humidity fall, relative humidity rises

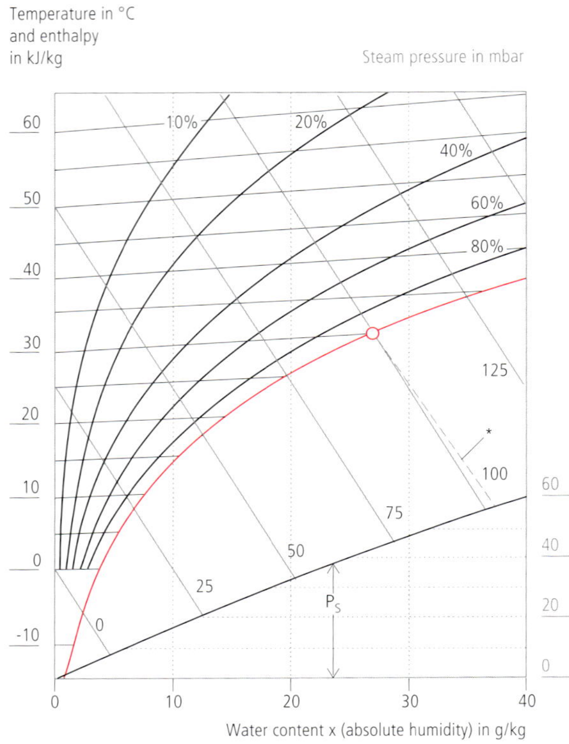

Figure 6.50
h/x-diagram for humid air after Mollier

— Temperature t in °C
— Enthalpy h in kJ/kg
— % relative humidity φ
— Saturation line
– – *) Fog isotherm (t = 30 °C)
P_S Steam pressure

The h/x-diagram in **Figure 6.51** also contains additional parameters, such as the wet-bulb temperature (temperatures indicated next to the saturation curve) as well as the marginal scale relevant for humidification.

The h/x-diagram is also used to determine the temperature and humidity of a glass pane, frame or steel structure, to name but a few examples, which lead to condensation in winter. Assuming a room temperature of 20 °C and 50 % relative humidity (see **Figure 6.51**, point 1) and a pane surface temperature (dry-bulb temperature) of 8 °C, the dew point temperature corresponding to 8 °C dry-bulb temperature is 1.5 °C lower than the dew point temperature in a room that corresponds to point 1. Condensation occurs when the surface temperature of a particular object lies below the dew point temperature.

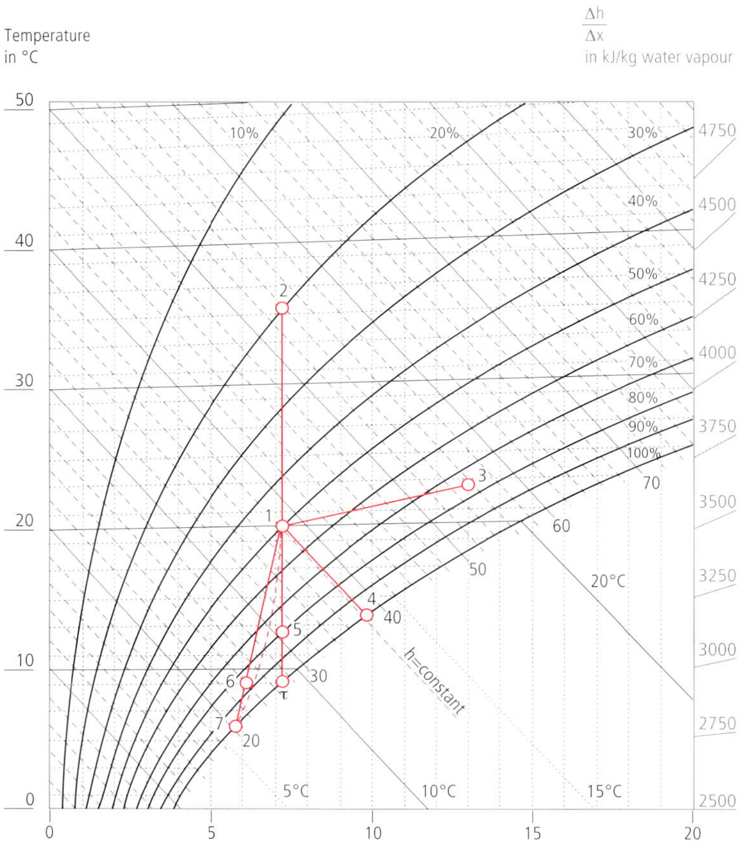

Figure 6.51
Condition changes in h/x-diagram (psychometric diagram)

- —— % relative humidity φ
- Temperature t in °C
- Enthalpy h in kJ/kg
- Dry air
- Absolute humidity x in g/kg dry air

- 1–2 Dry heating
- 1–3 Steam humidification
- 1–4 Water humidification
- 1–5–τ Dry cooling
- 1–6–7 Wet cooling
- 7 Appliance dew point
- τ Dew point of air temperature

6.5 VAC components

VAC systems (ventilation and air-conditioning systems) comprise central units as well as a multitude of components within the entire system. These are briefly explained here as they exert considerable influence of the changes in condition.

6.5.1 Air filters

Air filters in VAC systems are devices or fixtures that filter and separate the particles or gaseous impurities from the air. Atmospheric air contains various matter of differing particulate dimensions, the diameters of the particles ranging from 0.1 to 20 mm (micron) (**Table 6.11**).

Gaseous impurities are separated by means of chemical or physical adsorption processes, that is, pollutants are bonded to adsorptive material.

Test procedures operating either with synthetic laboratory dust or atmospheric dust used to check the separating efficiency and dust storage capacity of the filters have been instituted in order to classify air filters.

Description	**Filter class**	**Mean interception degree** A_m (%)	**Mean efficiency quotient** E_m (%)	**Old descriptions**	**Classification according to StF** [1]
Grit filter	G 1	$A_m < 65$	–	EU 1/A	A Grit or pre-filter
	G 2	$5 \leq A_m < 80$	–	EU 2/B_1	B Particulate filter
	G 3	$80 \leq A_m < 90$	–	EU 3/B_2	
	G 4	$90 \leq A_m$	–	EU 4/B_2	
(High-efficiency) particulate filter	F 5	–	$40 \leq E_m < 60$	EU 5	C High-efficiency particulate filter
	F 6	–	$60 \leq E_m < 80$	EU 6	
	F 7	–	$80 \leq E_m < 90$	EU 7	
	F 8	–	$90 \leq E_m < 95$	EU 8	
	F 9	–	$95 \leq E_m$	EU 9	
			Particle size in μm		
(High-efficiency) submicron particulate airfilter	EU 10	85	0.3 – 0.5	Q	(High-efficiency) submicron particulate airfilter
	EU 11	95	0.3 – 0.5	R	
	EU 12	99.5	0.3 – 0.5	S	
	EU 13	99.95	0.3 – 0.5	S	
	EU 14	99.995	0.05	ST	
	EU 15	99.9995	0.1	T	
	EU 16	99.99995	0.12 – 0.2	U	
	EU 17	99.999995	0.2 – 0.5	V	
	EU 18	99.9999995	0.2 – 0.5	–	

Table 6.11
Air filter classification after DIN EN 779

[1] The classification into "grades based on the guidelines for testing filters for HVAC systems", published by the StF (Staubforschungsinstitut des Hauptverbandes der gewerblichen Berufsgenossenschaften e.V., Bonn, or Particulate Research Institute), was replaced by the air filter classification after DIN EN 799.

Table 6.11 indicates the air filter classification common in Europe and refers to some filtering processes. **Figure 6.52** provides a global overview of air filter classification in relation to kernel size and dust removal efficiency.

Rotating air filters and circulated air belt filters are classified as category A.

Oil-coated rotating air filters (**Figure 6.53**) are used as preliminary filters for strong dust concentrations (99 % dust-collecting efficiency, boundary grain size roughly 5 mm). Built-in wall and duct filters (**Figures 6.54.1/2**) are also used.

These filters are especially suited for preliminary filtering of heavily dust-burdened air, in some cases also as exhaust air filters for industrial operations.

High-performance rotating air filters are frequently equipped with metal mesh filter cells that divide the total air flow into several partial flows; dust is trapped on the filter surface as a result of the directional changes. Metal filters are also ideally suited for intercepting oil and grease, paint mist and similar material.

Figure 6.52
Air filter classification according to characteristic curve determined in test

Classifications:
A Coarse dust or pre-filter
B Fine dust filter
C High-grade fine dust filter

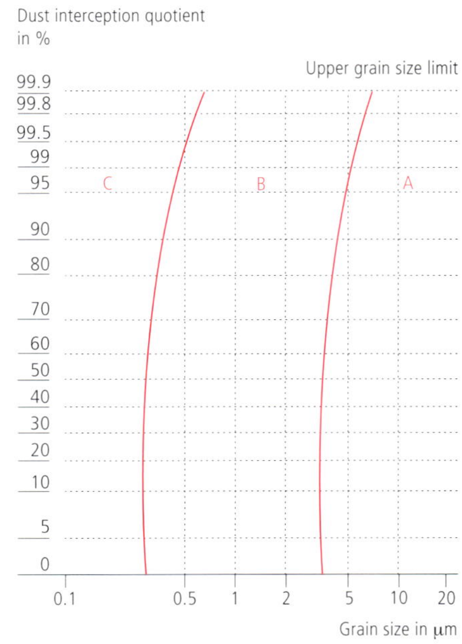

Figure 6.54.1
Dry layer filter in V-form filter design with filter component

Figure 6.53
High-performance rotating air filter with hand-operated mud dredger and catch basin

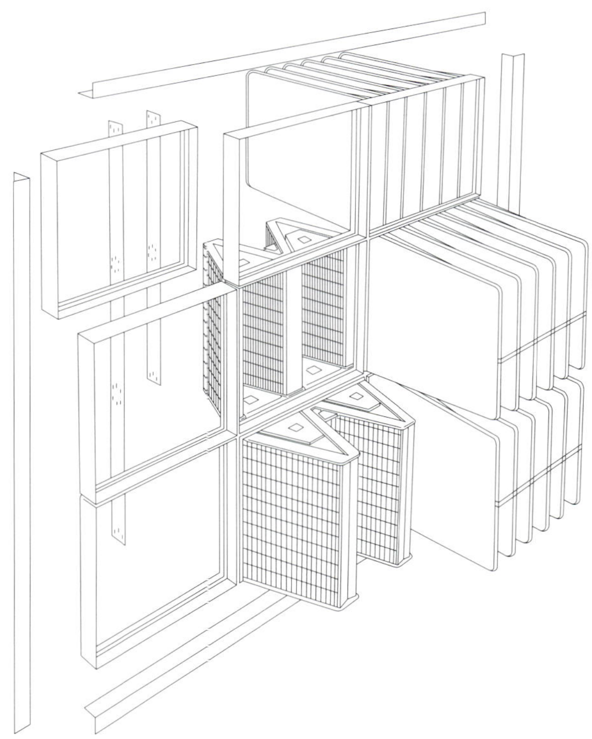

Figure 6.54.2
Dry layer filter in wall cell filter design with metal cells

Figure 6.55 illustrates a dry layer circulated air belt filter, which operates similarly to a roll of film in a camera. The filter is rolled from one spool to another (depending on the set differential pressure of the filter material). The fibre of the filter material may consist of glass, synthetics or natural products. **Figure 6.56** provides a sectional view of the filter medium. The glass fibres are generally wetted with a dust-trapping agent to augment the efficiency of the filter. The fibre density inside the filter increases and the fibre diameter decreases in the direction of the flow (from right to left, as shown in the image).
Other dry layer filter models are built-in wall filters, as shown in **Figures 6.56.1** to **6.56.3**.

Electrostatic filters (**Figure 6.57.1**) are classified as class C filters. In addition to dust and soot they also intercept bacteria, mould fungus, smoke, vapour and oil mist with an efficiency of roughly 85 to 95 %.
In electrostatic filters (with ionization) dust particles in the air flowing through the filter are electrically charged by an ionization component (electrical discharge wires = positively charged tungsten filaments) in order to be extracted in the dust interceptor component with plate condensers (alternatively positively and negatively charged). The operating principle of an electrostatic filter is illustrated in **Figure 6.57.2**.

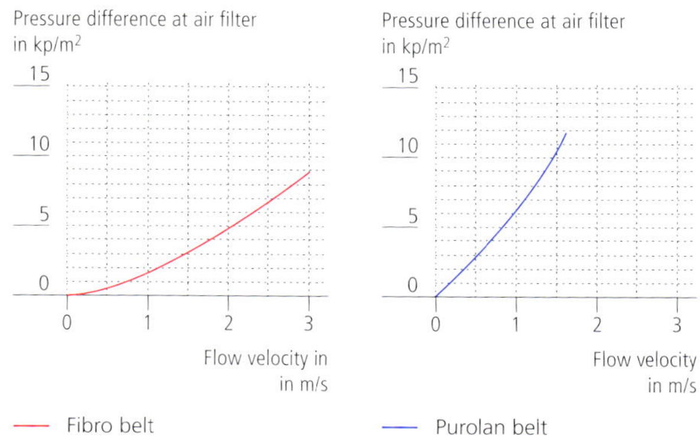

Figure 6.56
Filter medium (glass fibres) for dry layer circulated air belt filter

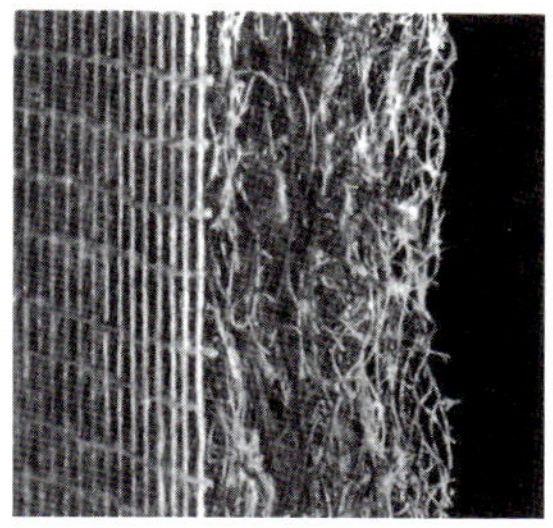

Figure 6.56.1
For intercepting coarse to fine dust

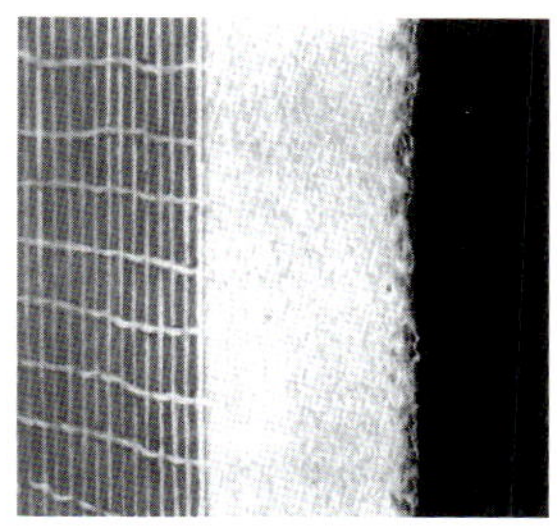

Figure 6.56.2
For intercepting fine to fine dust

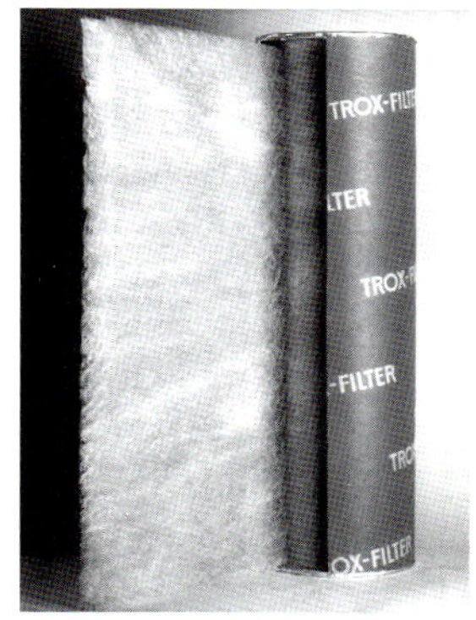

Figure 6.56.3
Rolled belt filter

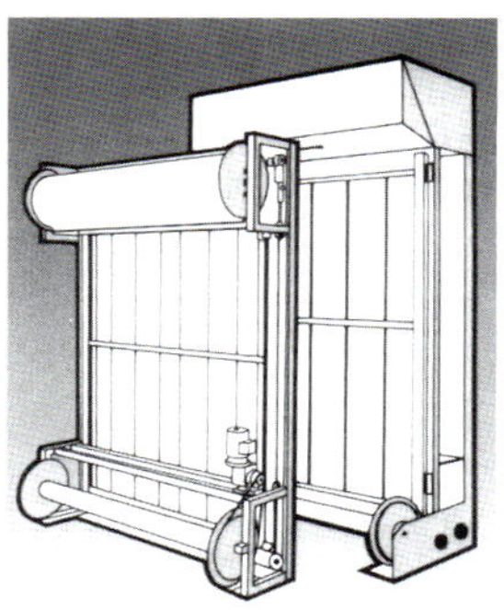

Figure 6.55
Dry layer circulated air belt filter with differential pressure drive

Figure 6.57.1
Electric air filter with collector cell

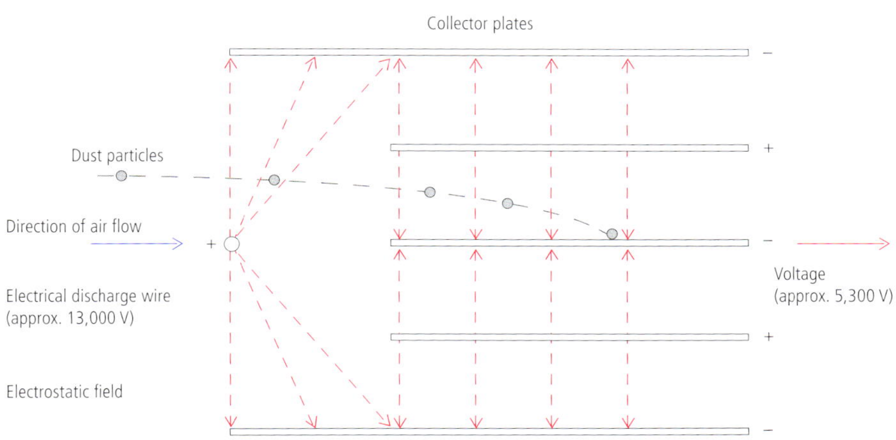

Figure 6.57.2
Schematic and operating principle of electrostatic filter

High-grade fine dust filters and suspended particle filters are used in all situations with stringent requirements for air purity (hospitals/laboratories/chip manufacturing/pharmaceutical industry etc.). Glass fibre, cellulose, paper or blends of these materials are used as filter media in fine dust filters with, for example, filter bags (**Figure 6.58**). The filter medium is either arranged in a zigzag formation inside a frame (**Figure 6.59**) or bundled into filter bags, which are attached airtight onto a front plate. The filter units in the illustrated cell shape are installed as filter walls, filter ceilings or similar components, or installed as individual components inside duct systems. The dust extraction efficiency lies in the region of 90 to 95 % for boundary grain sizes of 0.5 to 0.1 mm. For the highest demands in dust removal (filter class C3, previously aerosol filter, special class S) high-grade fibre air filters are used, which are capable of intercepting dust and aerosols less than 0.5 mm in size. The efficiency of these filters is almost 99 % (dust grain size 0.1 to 0.5 mm). Aerosol filters of this class are employed primarily in chip manufacturing and pharmaceuticals, as well as in areas where radioactive dust particles must be intercepted.

Activated charcoal filters (**Figure 6.60**) absorb primarily gaseous and vapour impurities (odours). Activated charcoal has an extraordinarily large surface in relation to its volume (1 g activated charcoal = 2 cm^3 = 1260 m^2 area). It is thus capable of absorbing gases and vapours when dust-containing air passes through it. The activated charcoal cells are employed in plate or canister like models in combination with preliminary filters to prolong the service life. In large systems, the cells are built up to form walls or ceilings.

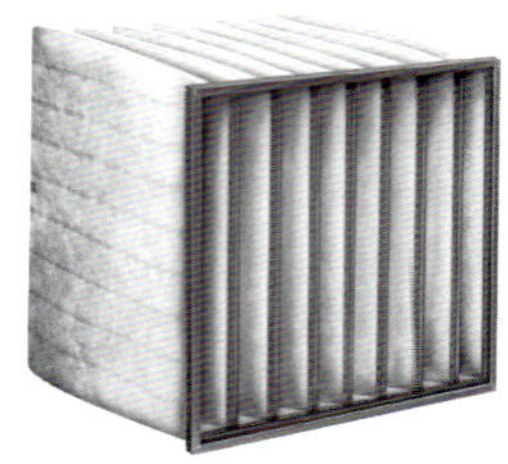

Figure 6.58
Bag filter with high-grade glass fibre mats
(Image: TROX)

Figure 6.59
Ceiling module for RRT suspended matter filter cell
(Image: TROX)

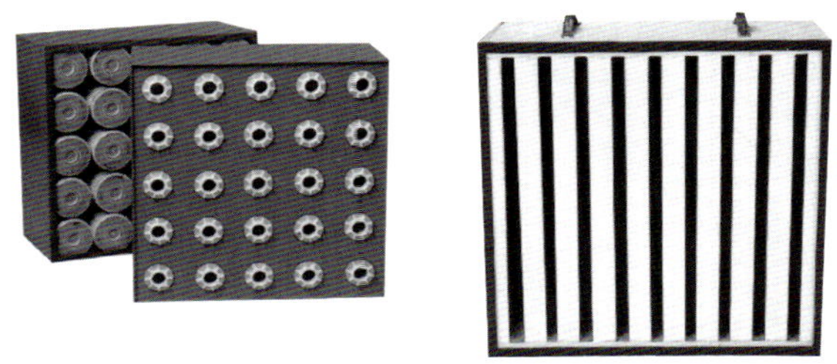

Figure 6.60
Activated carbon filter cells in panel and box construction

Figure 6.61
High-performance filter and filter characteristics

Contaminants, bacteria and particles		Intercepting performance of Human-Air filter
Sulphur dioxide	SO_2	< 90%
Nitrogen dioxide	NO_2	< 80%
Ozone	O_3	< 70%
Benzene compound, toluene, xylene	BTX	< 70%
Polycyclic aromatic hydrocarbons, more than 100 compounds (carcinogenic)	PAHC	< 65%
Particles 3mm		< 95%
Particles 5mm		< 75%

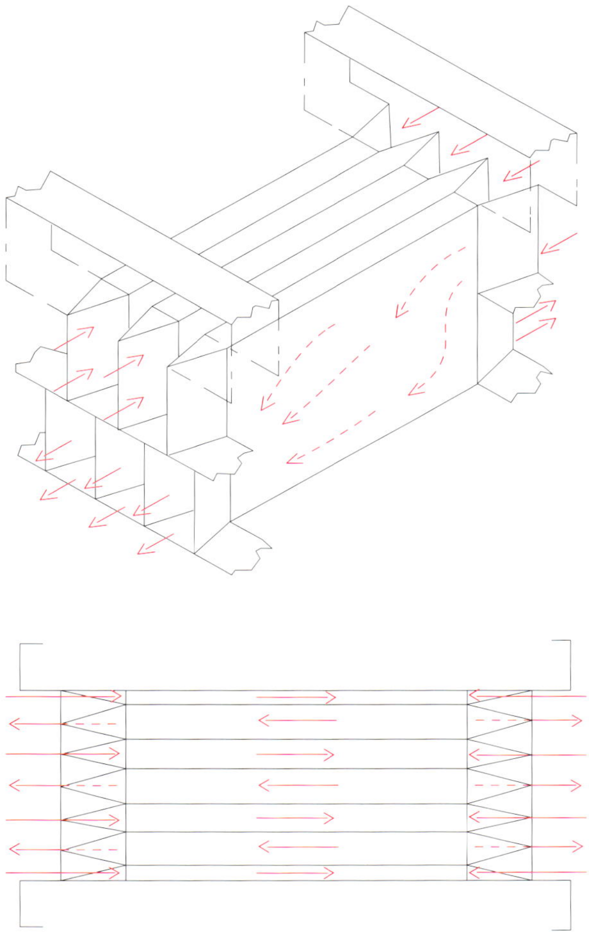

Figure 6.62
Plate heat exchanger

To prevent the so-called Sick Building Syndrome associated with buildings fitted with mechanical air-conditioning and ventilation systems, it makes sense to intercept microbial allergens and cell toxins. In some cases it is necessary to employ terminal filters capable of absorbing a full range of toxins. **Figure 6.61** depicts a high-performance filter as terminal filter installed in front of the air outlet, as well as the filter characteristics. As these characteristics show, the filter intercepts not only pollutants but also bacteria and particles to a large degree, achieving a highly hygienic air quality.

6.5.2 Heat exchanger – heat recovery

For some years, heat exchangers have been installed in air-conditioning and ventilation systems in the interest of energy savings (heat recovery systems). The chief purpose it to recover the thermal energy contained in the exhaust air. Heat recovery systems are differentiated into recuperative and regenerative heat exchange as well as systems that are joined into a cycle.

Plate and bare-tube heat exchangers (**Figure 6.62**) count among the recuperative heat exchangers, while the heating tubes themselves as well as regenerators with rotating thermal storage mass (**Figure 6.63**) are described as regenerative heat exchangers. Heat recovery processes occur in the regenerators, represented in the h/x-diagram (**Figure 6.64**). Rotating heat recovery systems can also lead to material exchange (humidity transfer) in addition to heat transfer, provided the appropriate storage material is used.

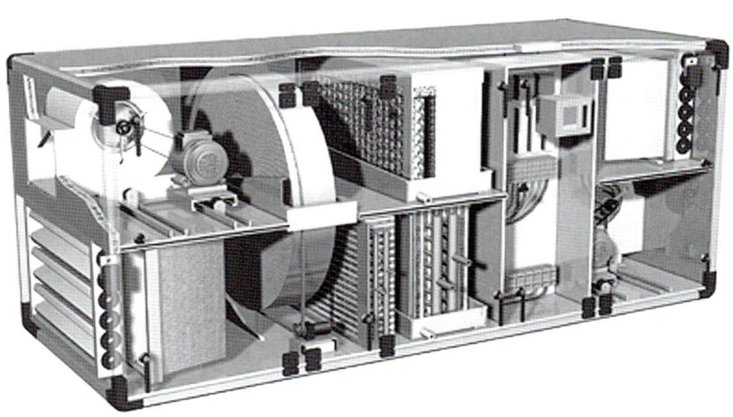

Figure 6.63
Device construction of a rotor system
(Cutaway image: ABB)

The recuperative and regenerative heat recovery systems are somewhat hampered by the disadvantage that fresh air and exhaust air streams must converge. The impact of this disadvantage is quite considerable, since it imposes duct systems, which can lead to large structures in the central installation. Cyclical heat exchanger systems can help bypass this disadvantage, using ribbed pipe exchangers (**Figure 6.65.1**). The cyclical heat exchangers transfer only thermal energy, not matter. Their exchange efficiency is generally lower than that delivered by regenerative heat exchangers.

Recent developments have greatly improved efficiencies in this area. Using highly efficient counter-current thermal layer exchangers and by regulating the mass flow in the cyclical system, the firm SEW has been able to achieve heat recovery efficiency coefficients of 80 – 90 %. By feeding the remaining heating energy directly into the water cycle of the heat recovery, this system can even operate without a separate secondary heater.

In summer, a preliminary measure of using an evaporative cooler can be used for adiabatic cooling of the exhaust air, thereby recovering most of the required cooling energy for the intake air from the exhaust air flow (**Figure 6.65.2**).

Figure 6.64
Air condition changes in heat recovery process in h/x-diagram

— Temperature t in °C
— Enthalpy h in kJ/kg
— % of relative humidity φ
== Sorption regenerator
== Condensation regenerator
A/A Recovered heat (sorption regenerator)
B/B Recovered heat (condensation regenerator)

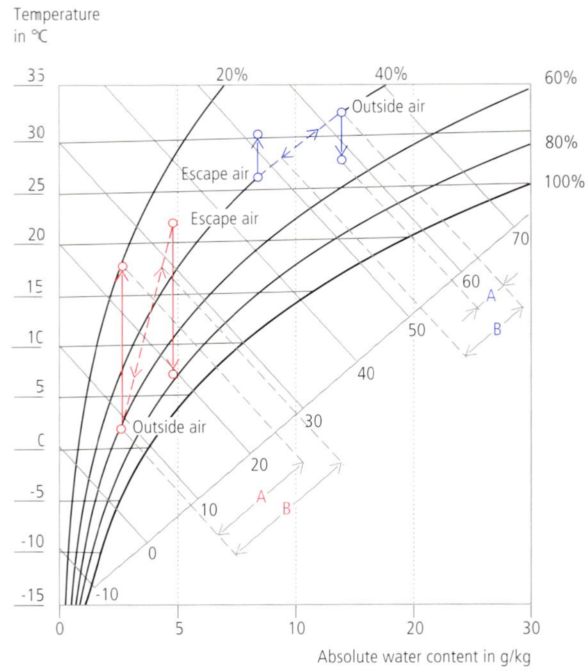

Figure 6.65.1
Cyclical heat exchanger

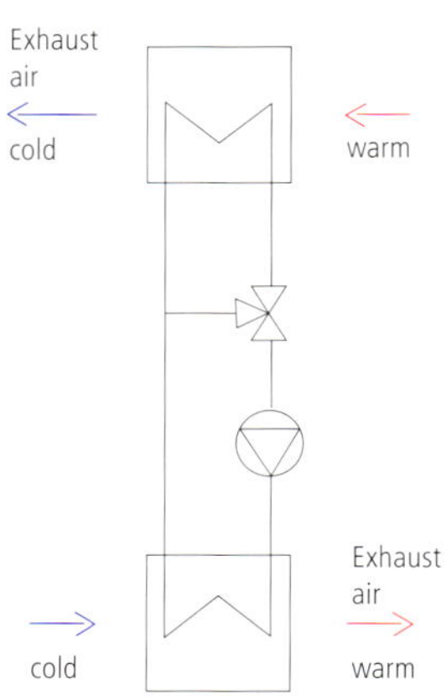

Figure 6.65.2
Cyclical heat recovery with high-efficiency counter-current layer heat exchangers
(SEW system)

System functions:

- Heat/cold recovery
- Integrated night heating
- Integrated night cooling
- Indirect adiabatic evaporative cooling

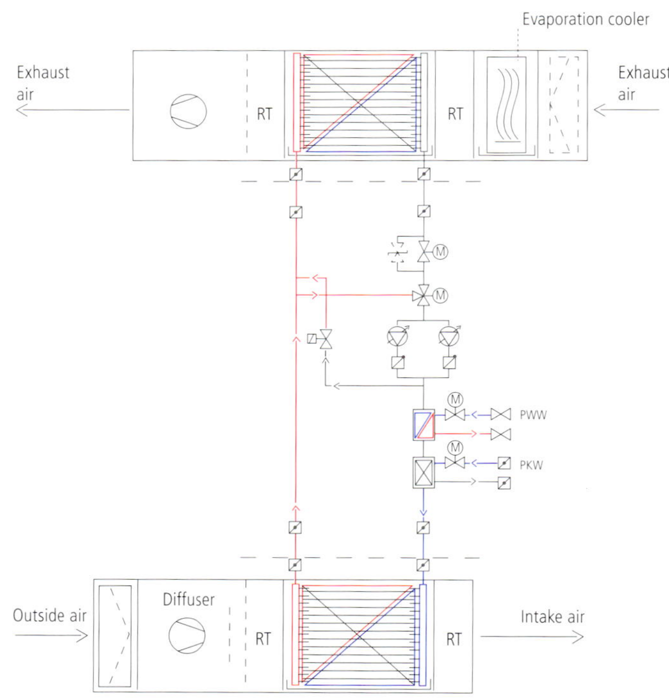

6.5.3 Air heaters and coolers

Air heaters and coolers are used for heat exchange or heat transfer from air to water and vice versa, whereby the heat transfer occurs through a variety of mediums. The heat flow is also from bodies with higher temperatures to those with lower temperatures. The transfer of thermal energy through the wall is shown in **Figure 6.66**, illustrating the same operating principle that is at work in plate heat exchangers. With regard to heat transfer in pipes a differentiation is made in terms
of the direction in which the medium flows, distinguishing between parallel, counter-current and cross-current flow (**Figure 6.67**). The following mediums are suitable heat energy carriers for air heaters:
- Warm water
- Hot water
- Steam
- Gas
- Organic liquids
- Electrical energy

Figure 6.68 illustrates the structure of an **air heater** with warm or hot water. The air heater consists of water chambers, pipes with lamellae (fins) and connections to the heating system. As warm or hot water flows through the pipes, they are surrounded by streams of air that absorb heat energy and simultaneously cool the water. Depending on the heating medium and the desired efficiency as well as make, a variety of lamellae are used for air heaters.

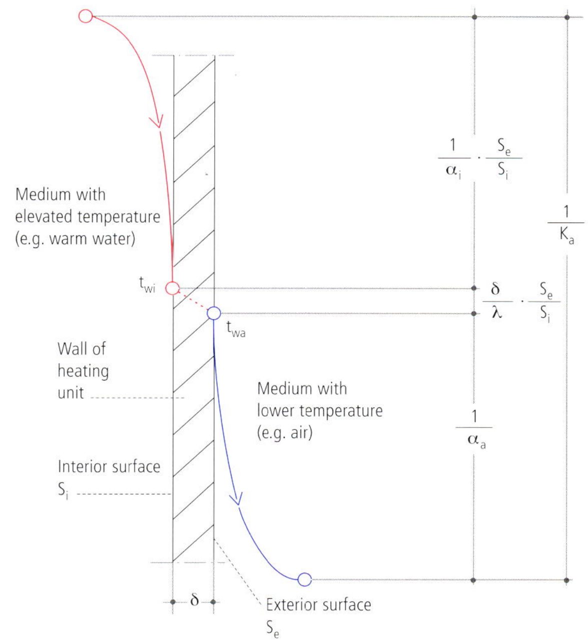

Figure 6.66
Heat transfer through wall

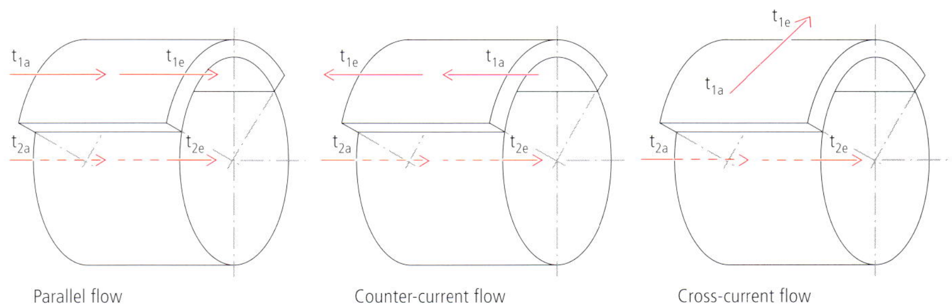

Figure 6.67
Heat transfer at pipe wall section

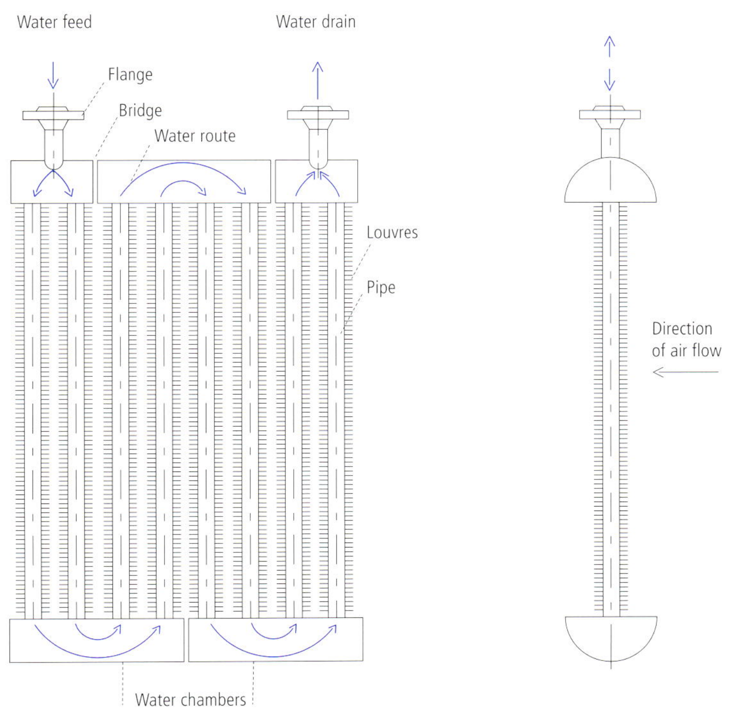

Figure 6.68
Example of tubular air heater in single row array

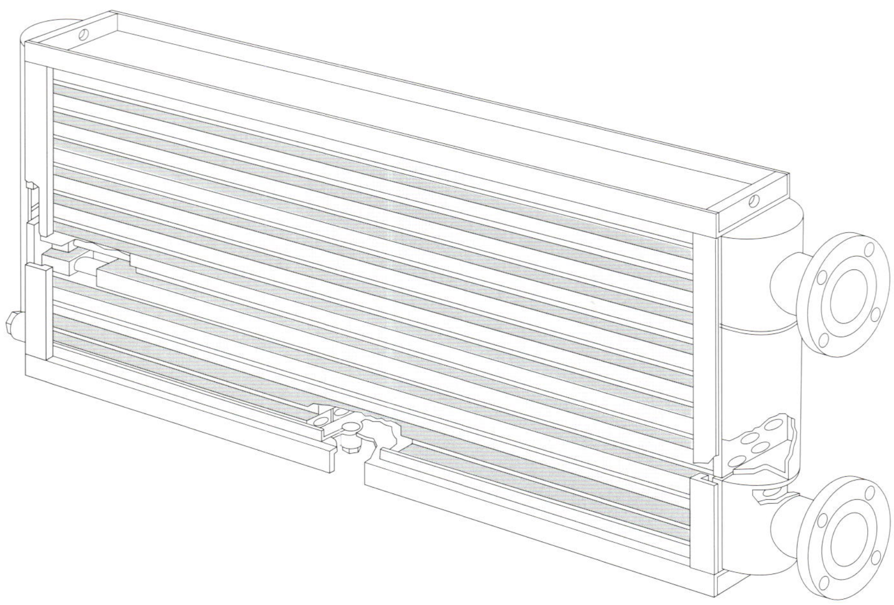

Figure 6.70
Air cooler with horizontal tubes and vertical fins

Air heaters are manufactured of galvanized steel tubes, tin-plated copper or copper with aluminium fins, depending on the condition of the heating medium and the volume of air to be heated. In addition to heating air by means of warm or hot water, vapour etc., electric heaters are also used. Electric air heaters are generally employed for moderate requirements (secondary heater). **Figure 6.69** depicts electric air heaters for duct installation with heating spirals that heat the air passing through them.

Air coolers are used to cool the air that flows through them and to dehumidify it (although this is optional, depending on the cooling medium). Air coolers are principally constructed similarly to air heaters (**Figure 6.70**): turbulent air is transported across the finned tubes, thus cooling and possible emitting water. Air coolers are generally installed in central air conditioning installations or ductwork with horizontal tubes and vertical fins for condensate runoff.

In analogy to the process described for air heaters, coolers operate either on the counter-current or the cross-current principle. The changes in condition vary slightly depending on the construction of the cooler, **Figure 6.71**.

The materials used for coolers are basically the same as those described for air heaters.

Figure 6.69
Electrical air heaters with casing for duct installation, with Cr-Ni resistance wire heating coils, set into ceramic beads and oval pipes

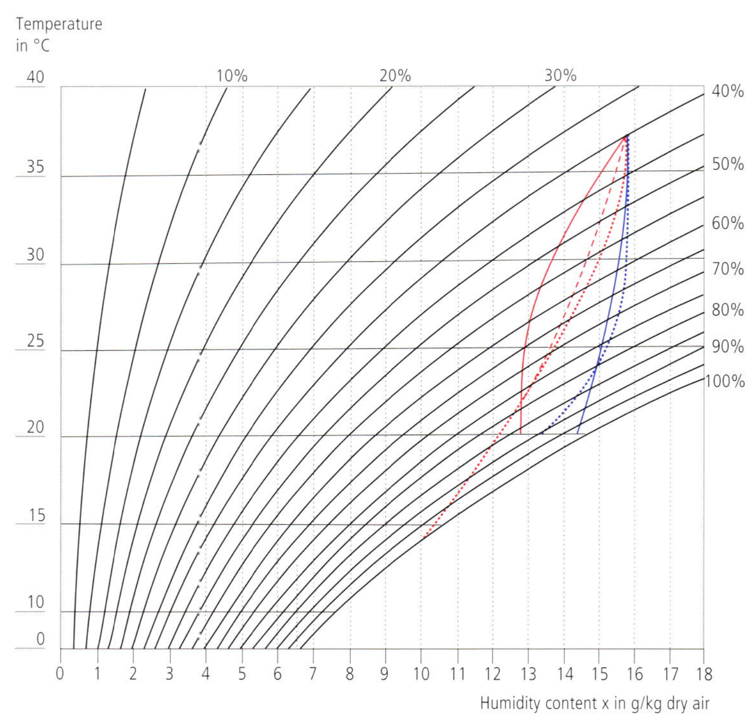

Figure 6.71
Example of cooling curves

- % of relative humidity φ
- Bare tube, parallel current
- Bare tube, cross-current
- Bare tube, counter-current
- Ribbed tube, parallel current
- Ribbed tube, counter-current

6.5.4 Humidifiers

A variety of humidifying aggregates are available for adding humidity to the intake air depending on specific requirements. These are:
- Evaporator humidifiers
- Steam humidifiers
- Tower packing chambers
- Tray evaporators
- Air washers
- High-performance air washers

Evaporator humidifiers are the simplest form of humidification. A water tray is set up inside the casing or in the duct of an AC system. The heating spiral is located in this tray, warming the water in the tray until part of it evaporates and is absorbed by the air flowing past.

Steam humidifiers evolved out of evaporator humidifiers. In this model, saturated steam is usually generated by means of electrical heating. Steam humidifiers consists of a pressure component, in which steam is stored, as well as a water feed and heating cartridges, which heat the supplied water to the required steam temperature. In small systems, stationary steam humidifiers are used, which humidify only the supply air flow for which they are installed. In large systems and for large volumes of steam required for humidification, special steam boilers or steam generating devices are installed. Steam is transported from the boiler or device via piping systems to the humidifying locations and any arising condensate must be returned. Steam humidification is the most hygienic form of humidifying air (prevention of germ or bacteria transfer/ Legionnaire's disease), although it generally leads to high energy and maintenance costs.

Humidification via tower packing chambers or nozzle humidifiers is another form of evaporative humidification. **Figure 6.72** illustrates the principle of a tower packing chamber, **Figure 6.73** the construction of the corresponding device. In this evaporative humidifier, the stream of air passes through a trickle of water, leading to evaporation and hence moisture being absorbed by the air. While this approach to humidification is economical, it is also unhygienic unless carefully and diligently maintained. Moreover, the ultimate moisture level can only be achieved with 70 to 75 % relative humidity. Hence, such humidifiers are only employed in situations where precise control is unimportant.

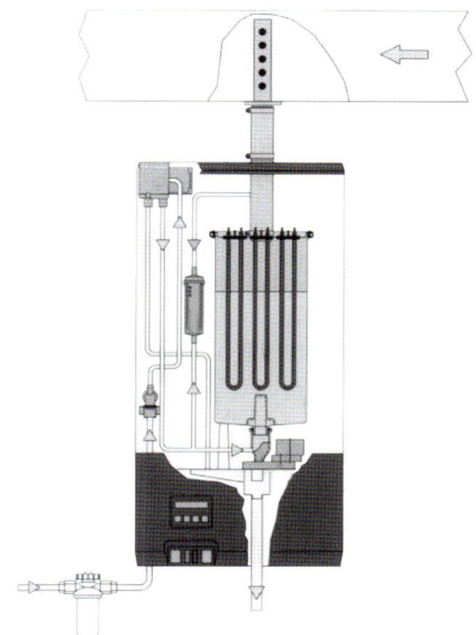

Figure 6.72
Schematic structure of steam humidifier and Condair Esco steam distribution pipe

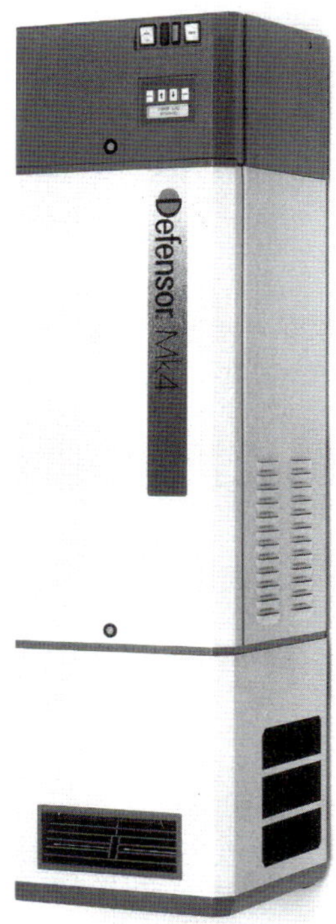

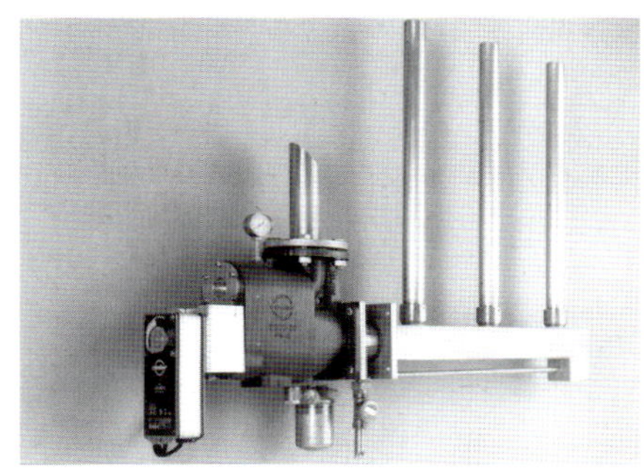

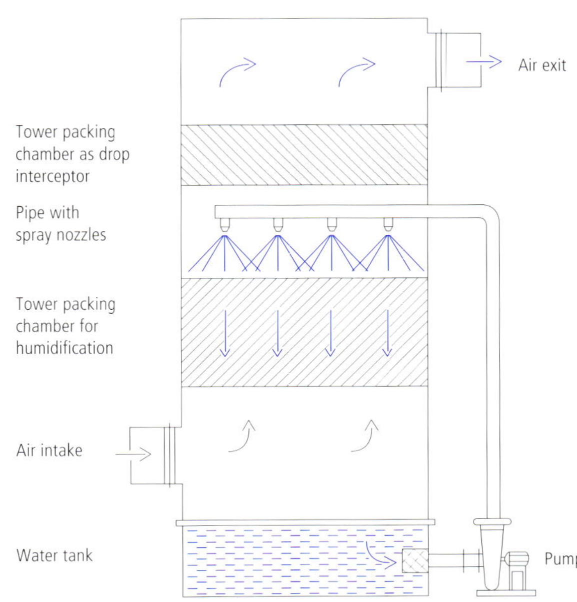

Figure 6.73
Humidifier constructed according to tower packing principle

Air can also be humidified by **spraying a mist of water into the flow of air** – a solution that is, however, rarely employed these days. Tray evaporators offer the simplest solution for applying this method: a stream of water rotating against the tray or disk is dispersed in a fine mist past the edge of the tray. Since scale is deposited in the process of evaporation and mist dispersion, tray humidifiers must be coupled with a calcium interceptor (filter).

In the **ultrasonic mister**, water drips onto a membrane oscillating with ultrasound (> 20 kHz), producing an ultrafine spectrum of droplets that comes out of the mister like steam ("cold steam"). Figure 6.74 contains a diagram of an ultrasonic mister built into an air-conditioner. Since ultrasonic misters still require exorbitant initial investment costs, they are generally only used in smaller units. The droplet size can be set via the oscillating frequency, i.e., the greater the frequency the smaller the droplets (20 kHz = 0.001 to 0.04 mm droplet size).

The most common type of humidification is evaporative humidification in washers. Figure 6.75 depicts an air washer and attached components. Depending on the humidification coefficient, air washers are designed with several rows of nozzles. The water volume at the drop separator not evaporated in the washer is extracted from the washer chamber as the air exits.

High-performance washers are more advanced versions of the air washer. These washers are basically central AC-systems, since they generate an extraordinarily intense blend of air and water via 17 tuyère stocks. This high-performance washer can humidify not only air but also dehumidify when the supply water stream is very cold. High-performance washers can also be employed as oversaturation humidifiers, e.g. in the textile industry or in tobacco processing.

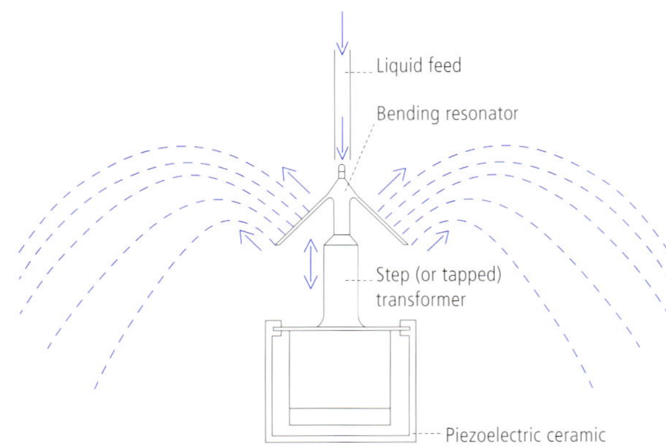

Figure 6.74
Ultrasonic capillary-wave mister with conical resonator

Figure 6.74.1
Ultrasonic mister

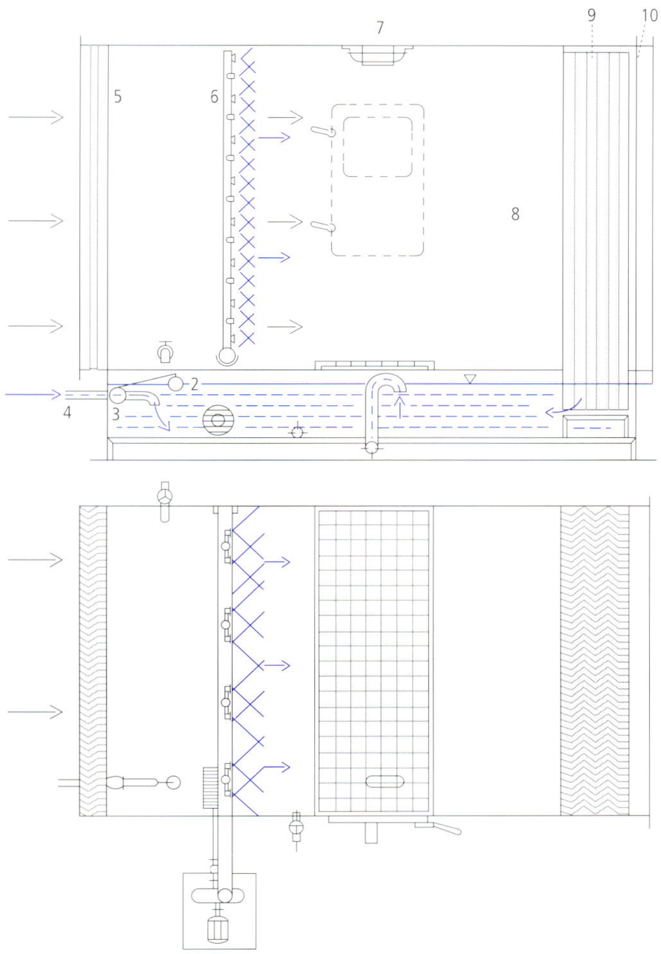

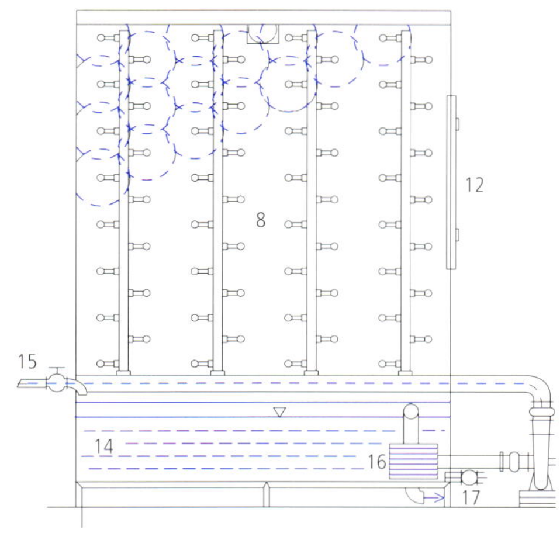

Figure 6.75.1
Air washer with row of nozzles

1 Overflow
2 Floater
3 Floating valve
4 Water feed
5 Rectifier
6 Nozzle attachment with water nozzles
7 Waterproof marine light fixture
8 Nozzle chamber housing
9 Eliminator
10 Packing strips
11 Grille
12 Washer chamber gate with motor
13 Water pump with motor
14 Water tank
15 Feed pipe with hand valve
16 Suction pipe with sieve
17 Emptying with hand valve

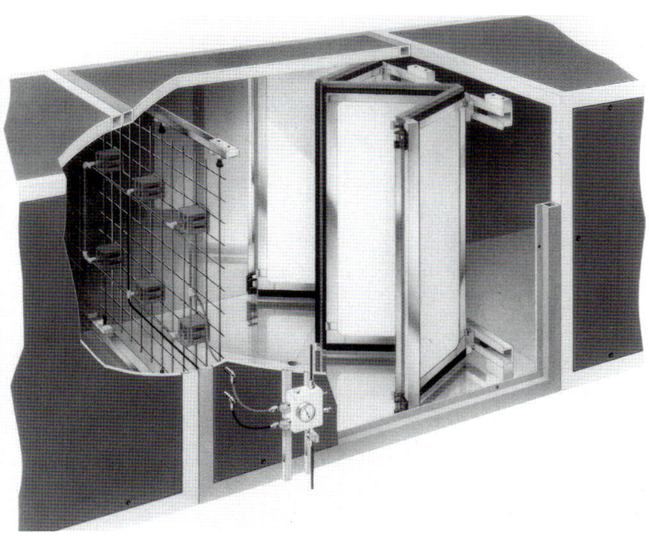

Figure 6.75.2
In the hybrid air humidifier (Condair Dual) an interceptor element is sprayed; the sprayed water evaporates fully from the interceptor

6.5.5 Fans

Fans are fluid flow engines that serve to transport air and gases under pressure of up to 30,000 Pa. A differentiation is made between radial and axial fans as well as cross-current or cross-flow fans. **Figure 6.76** provides an overview of fan types.

Radial fans are differentiated according to feed pressure and blade arrangement.

As to the feed pressure, we differentiate between
- Low pressure fans up to 720 Pa
- Medium pressure fans 720 – 3,000 Pa
- High pressure fans 3,600 – 30,000 Pa

Blade arrangements are divided into cylindrical rotors, which comprise many blades at the circumference of the impeller and others with blade wheels (a few blades, axially extended). Fans with blade wheels are further differentiated into convex and concave curved blades as well as according to application (for example, hot gas fans, explosion-proof fans, transport fans, roof fans etc.).
The principal construction of radial fans is shown in **Figures 6.77/78**. They are usually mounted on a base in combination with oscillation dampers to prevent oscillations from being transferred to the floor (impact sound damping). To transport air to rooms, the fan must overcome a variety of pressures. These are:

P_{st} = static pressure [Pa]
P_d = dynamic pressure [Pa]
The total pressure is $P_t = P_{St} + P_d$ [Pa].

	Design	Diagram	Yield	Pressure coefficient	Application
Axial fans:	Wall fan	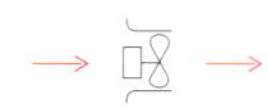	0.10...0.25	0.05...0.10	for window and wall installation
	w/o impeller	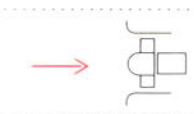	0.15...0.30	0.10...0.30	for low pressures
	w impeller	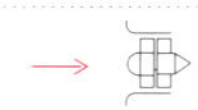	0.30...0.60	0.30...0.60	for higher pressures
	Contrarotation	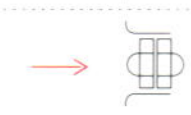	0.20...0.80	1.00...3.00	in special cases
Radial fans:	Backward-bent vane	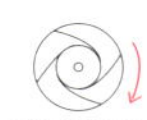	0.20...0.40	0.60...1.00	for high pressures and efficiencies
	Straight vane		0.30...0.60	1.00...2.00	for special applications
	Forward-bent vane		0.40...1.00	2.00...3.00	for low pressures and efficiencies
Cross-flow fans:		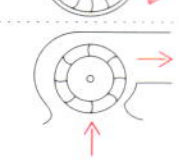	1.00...2.00	2.50...4.00	high pressures and space efficient

Figure 6.76
Overview of fan types, design

Figure 6.77
Radial fan for medium and high pressure range with running wheel

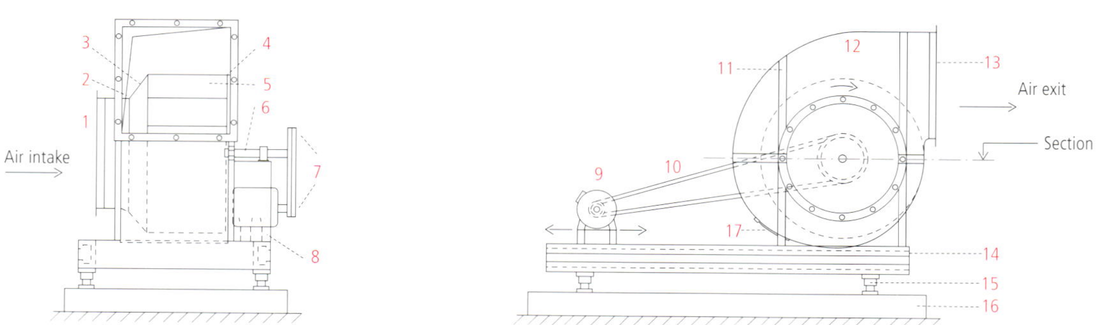

Figure 6.78
Schematic of radial fan with belt drive

1 Suction opening
2 Intake nozzle
3 Cover plate
4 Running wheel
5 Rotor blade
6 Drive shaft
7 V-belt pulley
8 Bearing block
9 Drive
10 V-belt
11 Bracing
12 Housing
13 Pressure connection
14 Base frame
15 Vibration damper
16 Foundation
17 Cleaning port

Figure 6.79 is a diagram indicating the pressure distribution in a ventilating system with static and dynamic pressure on the suction side of the fan, as well as static and dynamic pressure on the pressure side.

Radial fans can be designed to operate as one-directional or two-directional suction fans, with varying fan motors depending on purpose. A differentiation is made between direct drive via the motor shaft, drive via rigid or slip coupling, and belt drive.

Axial fans are those where the air flows through the running wheel in the direction of the axis. Figure 6.80 contains a diagram of an axial fan and a section of the casing, Figure 6.80.1 a frontal view of a fan and its running wheel. The principal components of the axial fan are the hub [root of a blower] with blades, the casing and the motor. The intake nozzles and, in some instances, a guiding wheel in front of or behind the running wheel (similar to driving gear) serve to improve the effectiveness of axial fans. Diffusers at the air outlet, adjusters for the blades and inlet guide vanes (or IGVs) complete the range of applications for axial fans. Smaller axial fan models are also employed as built-in wall and window fans without guide wheel for low pressures (WC venting/extraction fans set into the window of a kitchen etc.). If very high pressures in ventilation systems need to be overcome, counter-current axial fans are installed. Blade adjusters and inlet guide vanes are used to change the volume flow during fan operation. As with radial fans, axial fans are also differentiated according to feed pressures:
- Low-pressure fans 0 – 300 Pa
- Medium-pressure fans 300 – 1,000 Pa
- High-pressure fans above 1,000 Pa

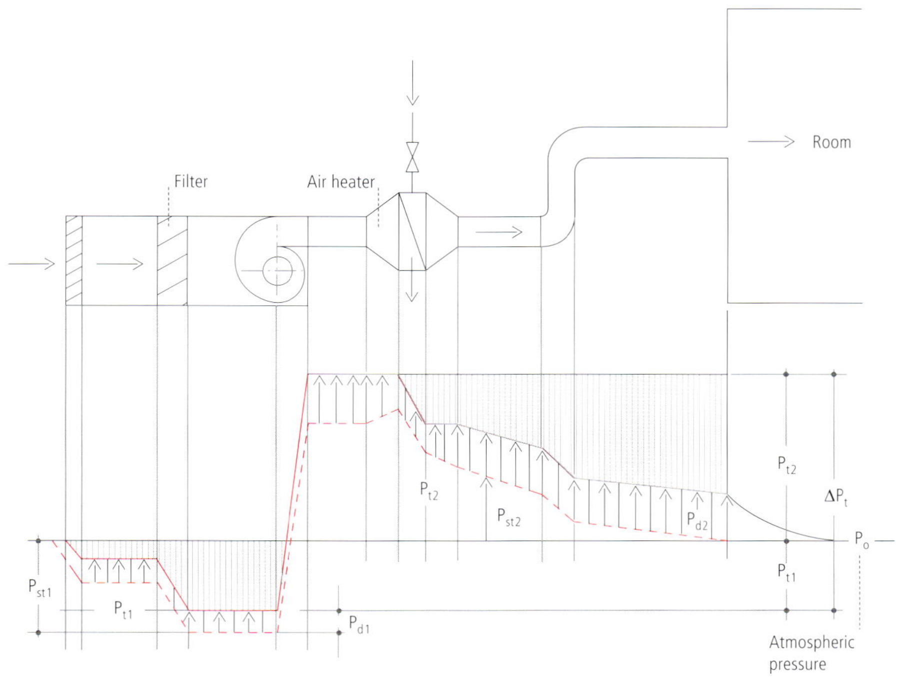

Figure 6.79
Pressure distribution in a ventilation system

Figure 6.80.1
Axial fan with running wheel

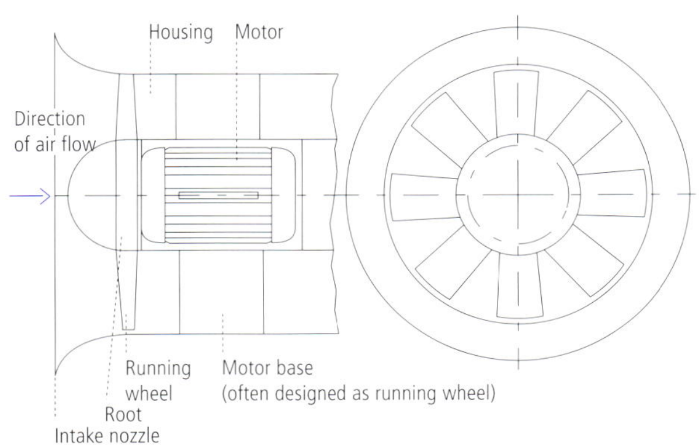

Figure 6.80
Axial fan with housing

Roof fans used to extract stale air from kitchens or WCs (**Figure 6.81**) can accommodate running wheels of the radial or axial type.

The cross-current fan (**Figure 6.82**) is a variation of the radial fan. As the image shows, this fan is equipped with an elongated compressor drum, which both sucks and expels the air at the circumference. By virtue of the air flowing through the blades twice an eddy is created that assumes the function of a conducting device. The running wheel length is proportionate to the volume flow and can be used in trunk devices, fan-coil devices etc. directly in the room due, among other factors, to its low sound impact.

A variety of other special fans exist in addition to the models shown here for a wide range of applications.

6.5.6 Interceptors

In ventilation and especially extraction in industrial operations, different types of interceptors are used in addition to filters. Some of these are:
- Coarse interceptors
- Fine interceptors
- Cyclone dust interceptors etc.

As their name suggests, the principal purpose of interceptors is to intercept coarse dust as well as larger particles (chips, shavings) or foreign matter such as oil or vapours.

6.5.7 Mixing and distributing chambers

Mixing and distributing chambers are used to merge or divide air streams. **Figure 6.83** is a diagram of a mixing chamber with built-in prefilter, in which an outside air stream and a circulation air stream are merged and then conducted into the air-conditioning system once they have passed through the filter.

Mixing and distributing chambers are either standardized components inside air-conditioning systems or special brickwork or concrete modules installed for large, commercial installations.

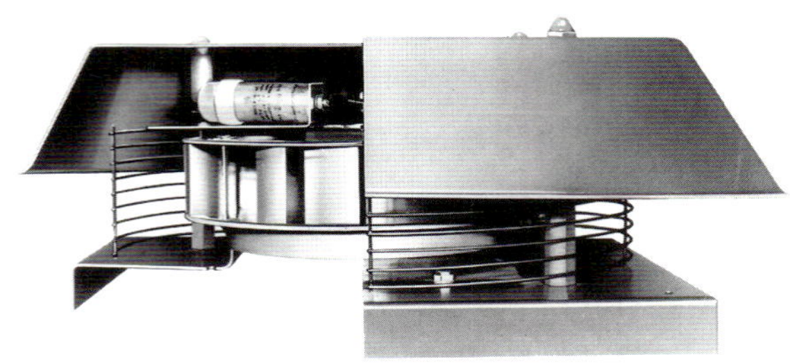

Figure 6.81
Roof fan with running wheel, radial construction

Figure 6.82
Tangential fan, frontal view
(Image: LTG)

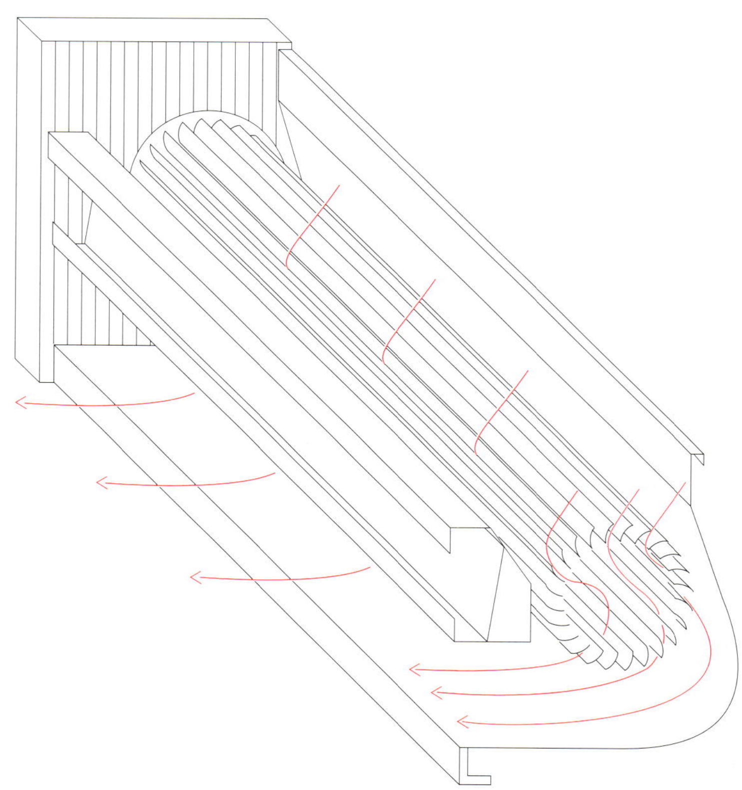

Figure 6.82.1
Tangential fan, schematic

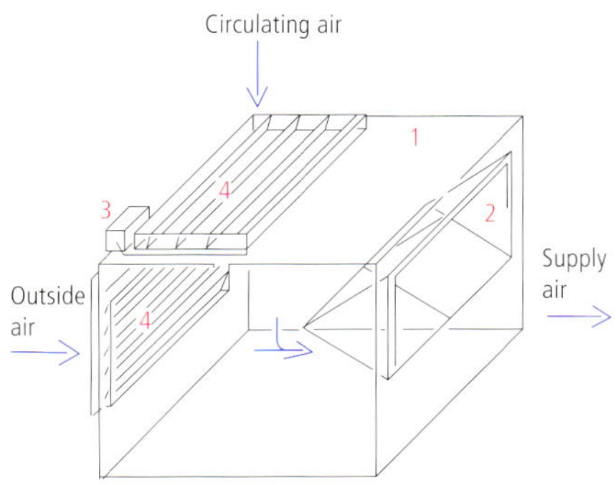

Figure 6.83
Mixing chamber with integrated V-shaped filter

1 Housing chamber
2 V-shaped filter
3 Modulating motor
4 Louvred shutter

6.5.8 Variable air volume box

In VAC systems with variable volume flows, variable air volume boxes are installed on site (close to the interior space) in the ductwork to feed a specific volume of supply air into the room controlled by a thermostatic valve.

In the case of single-duct AC systems (air-only systems, variable volume flow), volume flow regulators transport cold supply air into the room, thereby controlling the cooling energy for cooling load compensation through the air stream.

Figure 6.84 illustrates two volume flow regulators for ducts (rectangular) or pipes (round), respectively, with the corresponding drive and control component.

Figure 6.85 illustrates a volume flow regulator installed inside a mixing chamber (sound-proof box), in which the cold and warm air of a two-duct system is mixed, as well as a section of a volume flow regulator and a diagram of the operating principle (**Figure 6.85.1**).

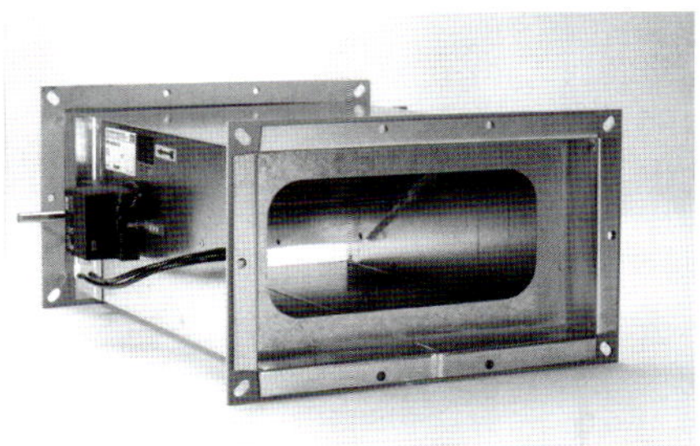

Figure 6.84.1
Volume flow regulator for duct installation (rectangular)

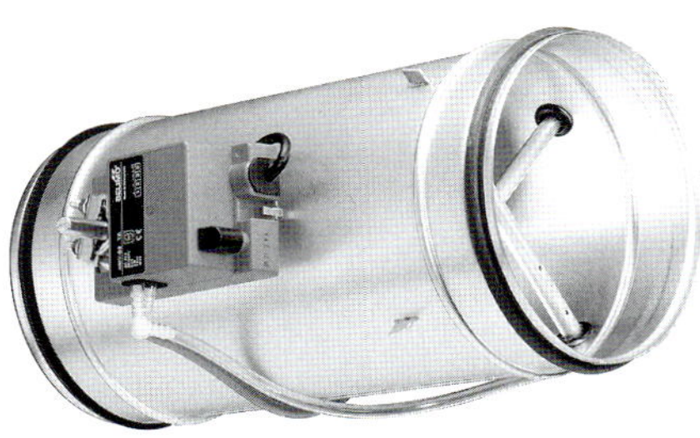

Figure 6.84.2
Volume flow regulator for duct installation (round) for variable volume flow systems, Varycontrol type TVR
(Image: TROX)

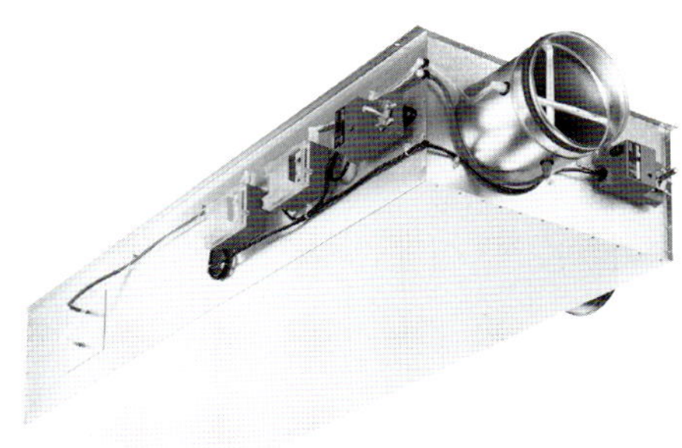

Figure 6.85
Mixing device for variable volume flow systems Varycontrol type TVM
(Image: TROX)

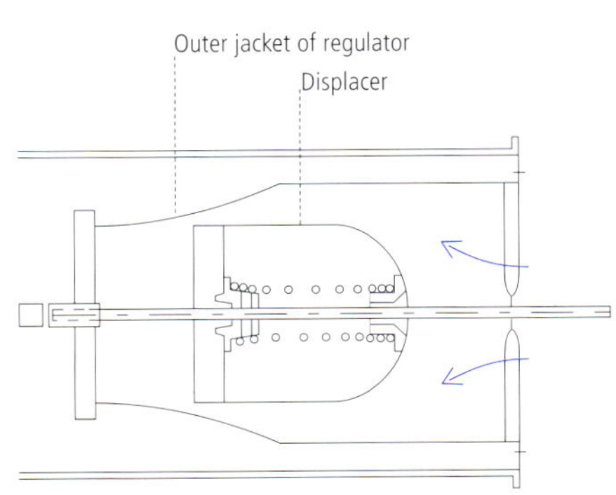

Figure 6.85.1
Section through volume regulator

In two-duct systems, which are, however, relatively rare these days, cold and warm air streams are mixed in situ into a single air stream, whose temperature is that defined for the room supply air and set by the regulator (supply air temperature ranging from warm to cold). **Figure 6.86** provides a section and the functional diagram of the mixing box, while **Figure 6.87** depicts a mixing box with visible constant volume flow regulator and mixing valve for temperature control.

The variable air volume boxes assume the same or a similar function as that of control valves in heating systems (volume flow regulator) or mixing batteries in sanitation (mixing of cold and warm streams of water). In addition to air conditioning, they are a highly essential component to ensure the function of a relevant AC system and are hence of great importance with regard to the system's successful operation and regulating behaviour.

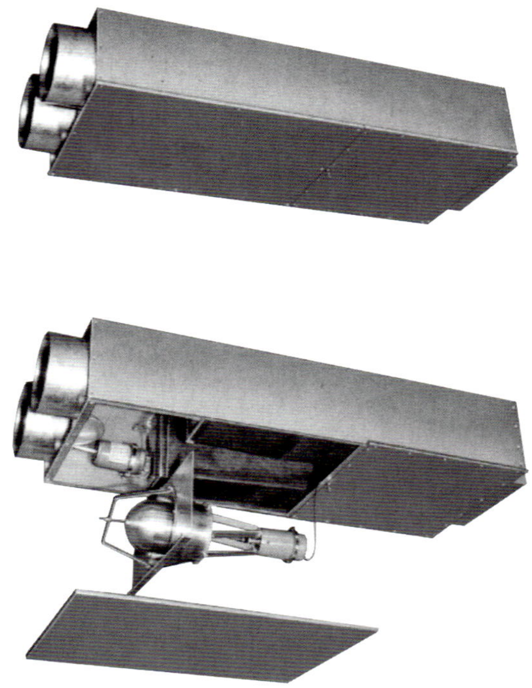

Figure 6.87
Mixing boxes for two-duct system with visible constant volume regulator and mixing valve for temperature control

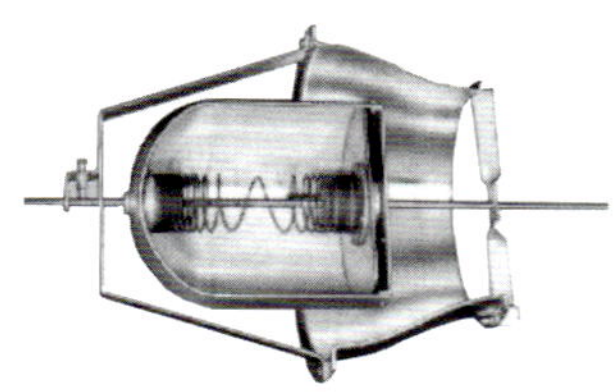

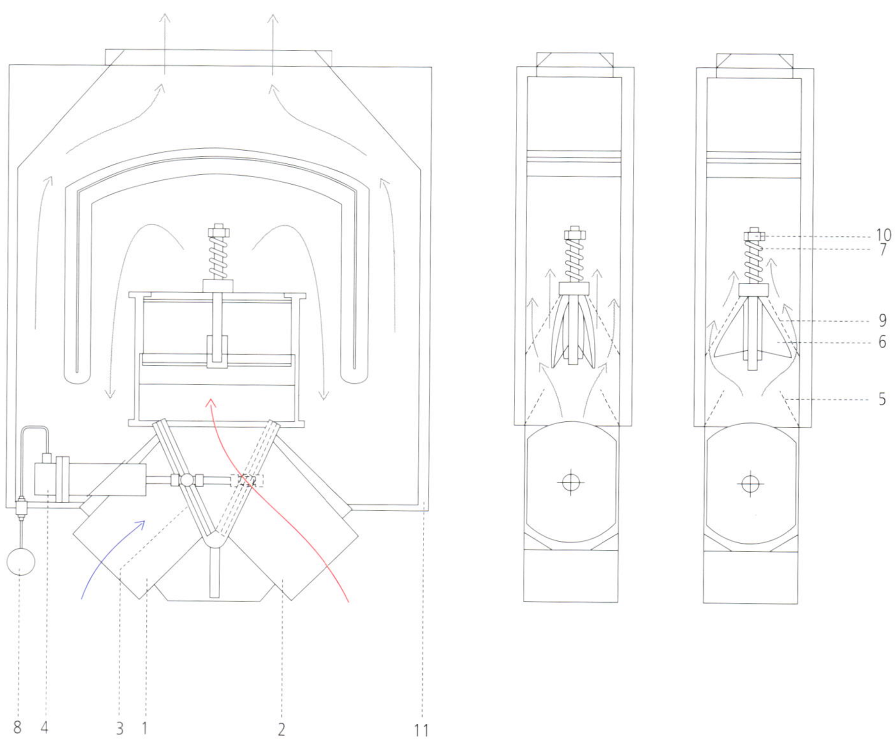

Figure 6.86
Schematic of mixing box

1 Cold air intake fitting
2 Warm air intake fitting
3 Switch-over flap
4 Flap drive
5 Perforated panel
6 Volume control wing
7 Return (reset) spring
8 Thermostat
9 Exit surface
10 Adjusting screw
11 Sound insulation

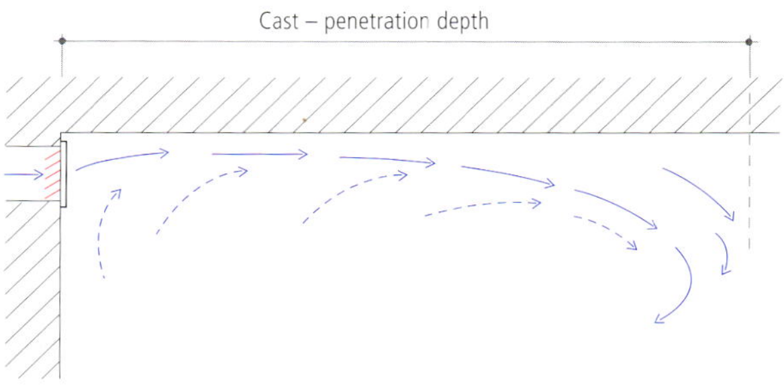

Figure 6.88.1
Air flow resulting from Coanda effect (half or wall streams)

— Primary flow

--- Induced room air
Secondary flow

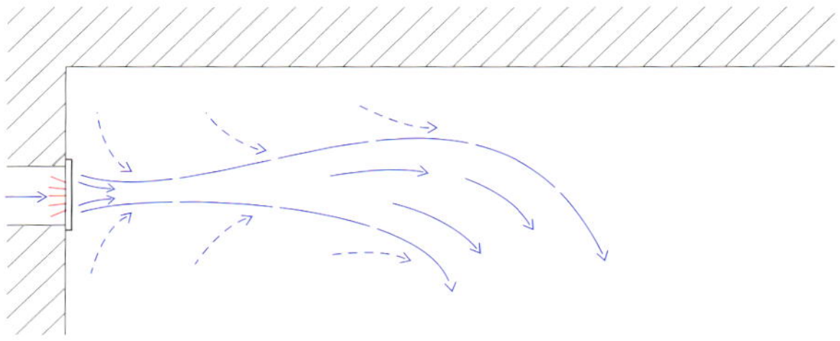

Figure 6.88.2
Free air flow

— Primary flow
--- Secondary flow

Perforated grilles and wire grilles: with and without volume control by means of louvred screen or sliding gate	
Grille with horizontal louvres: with fixed or adjustable louvres, flow can be directed up and down	
Grille with vertical louvres: with fixed or adjustable louvres, flow can be directed right and left	
Grille with vertical and horizontal control: with louvres adjustable individually or together	
Grille with single direction control and volume control: with horizontal louvres for directional control and V-louvres for volume control	
Grille with dual direction control and volume control: with horizontal and vertical louvres for directional control and V-louvres or slide for volume control	
Elongated grille with vertical (or) horizontal louvres: with additional tier of louvres or volume control	
Nozzles: round, rectangular	

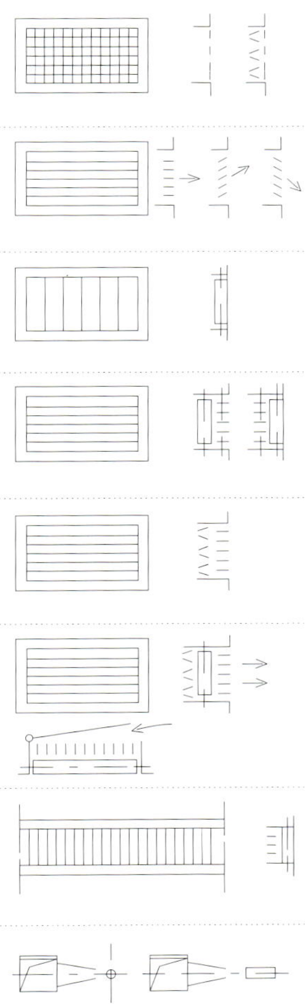

Figure 6.89
Different types of wall grilles

6.5.9 Air distribution outlets

Air distribution outlets comprise:
- Supply air diffusers
- Return air grilles

Air distribution outlets are of great importance to the architect since they are generally visible in the room and thus have a considerable impact on wall and ceiling design. Moreover, they are one of the most important components in any HVAC system, which must be planned with great care to prevent draught or thermal stratification in ventilated rooms. There is a wide variety of air outlets using materials such as steel, aluminium, plastic etc.

Air outlets serve to move air due to a displacement principle (air movement from wall to wall or ceiling to floor, e.g. in clean-room conditions or due to an attenuation effect, whereby air is blown into the room on a tangent or diffusely. In the case of diffuse supply air movement, the supply air is introduced into the room by means of nozzles or spin; when tangential supply air movement is used, flow eddies are formed.

Wall outlets

Wall outlets serve to transport supply air streams near the wall into the room as free stream, half stream or wall stream. If the supply air grille is located directly beneath the ceiling, the air flow is induced only from one side (from bottom to top) and settles at the ceiling. This is referred to as the "Coanda effect" in which the flux of air is literally attached to the ceiling as a result of the one-sided induction and penetrates deeply into the space (penetration depth/cast). **Figure 6.88.1** illustrates the flow distribution with Coanda effect for half streams or wall streams. **Figure 6.88.2** illustrates the flow distribution for a free stream, that is, the air flow can induce on all sides, which reduces the penetration depth (cast).

The air flow shown in the images plots warm or isothermal air streams. The penetration depth of the air flow diminishes when the intake air is cold, since the cold air has the tendency to fall more rapidly as a result of its greater density. The wall outlets (**Figure 6.89**) are manufactured in a variety of designs to respond to a wide range of requirements. These requirements result from both visual and technical concerns. There are many options available on the market, giving the planner a wide selection from which to choose. **Figure 6.90** illustrates several supply air components for wall installation, and **Figure 6.91** an industrial wall outlet in operation.

A

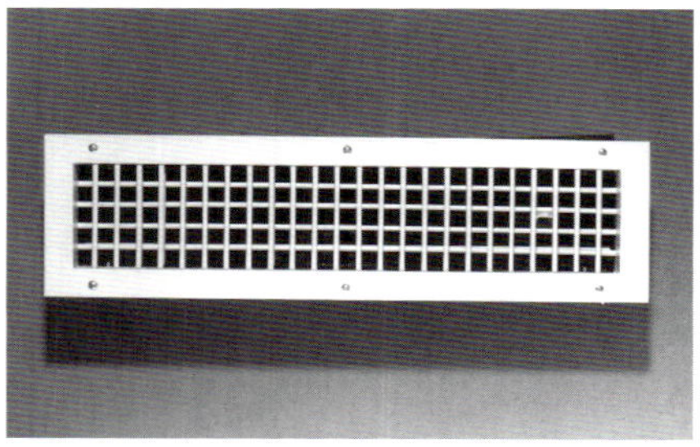
B

C

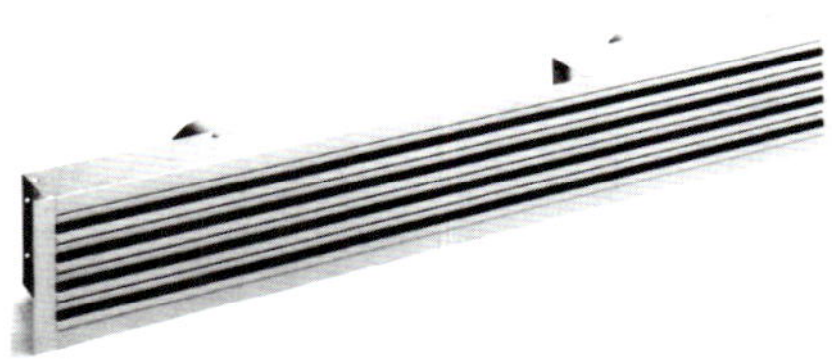
D

E

Figure 6.90
Various air supply elements for wall installation

A – C Supply grille
(Image: TROX)
D Supply/return air combination system for installation in walls, with adjustable nozzles, type LDB
(Image: LTG)
E Wide-cast nozzle, DUK series
(Image: TROX)

Figure 6.91
Industrial air supply for various operating conditions (heat/isothermal/cold air supply)

Supply air is more commonly transported into a room through ceiling outlets than through wall outlets. Again, there are many different models, which distribute the air volumes first along the ceiling (Coanda effect) or allow it to stream freely into the space (spin effect).

Figure 6.92 provides an overview of different supply air diffusers, which can also serve as return air grilles. Ceiling air diffusers can present a louvred, rectangular, square or round image when seen from below to adapt to wide range of design requirements. Figure 6.93 depicts supply air components for ceiling installation currently available on the market.

A

B

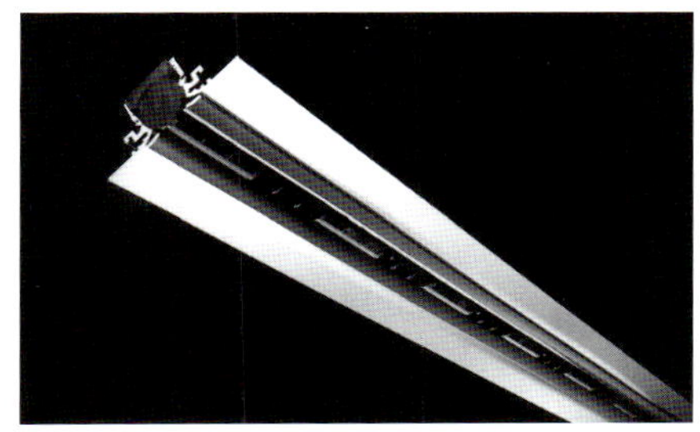

E

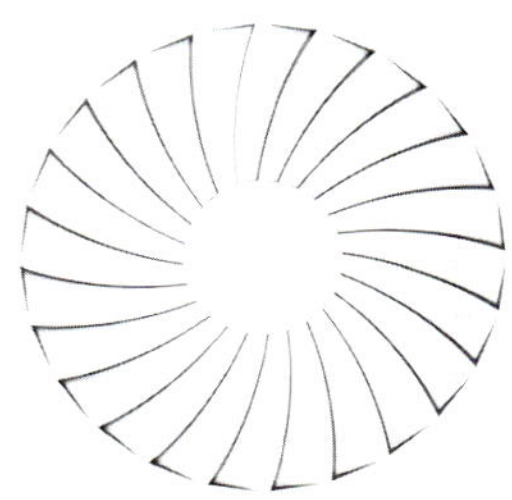

C

F

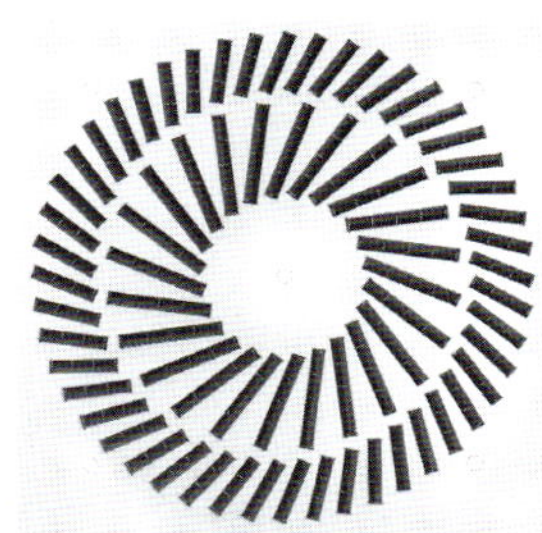

D

Figure 6.93
Various air diffusers for ceiling installation

A Slot outlet, VSD35S
B Spin vent with fixed blades Series RFD
C Spin outlet with fixed blades Series TDF-SilentAIR
D Spin outlet with adjustable blades Series VDW-SilentAIR
E Linear air diffuser type LDB, system clean®
F Swirl diffuser with stackable pyramid plenum box, type DLA 9 P

(Images: TROX A – D, LTG E – F)

Flat air diffusers:
with single flat plate, fixed or adjustable
with several plates

Round, conical air diffusers:
ceiling-mounted (flush), fixed or adjustable, with or without rectifier or volume control
ceiling-mounted (projecting), fixed or adjustable

Square air diffusers:
ceiling-mounted
ceiling-mounted with connecting box, adjustable or fixed

Rectangular air diffusers:
one-sided (also for wall installation)
two-sided and four-sided

Slot outlets:
with or without rectifier or volume control

Perforated plate air outlets:
with plates made from sheeting, plaster etc.

Spin outlets:
in ceiling
above ceiling (also below floor), fixed or adjustable

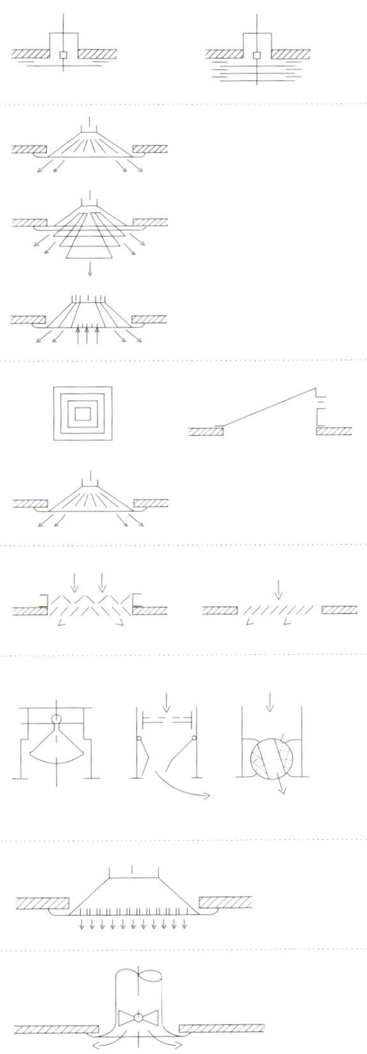

Figure 6.92
Various types of ceiling air diffusers

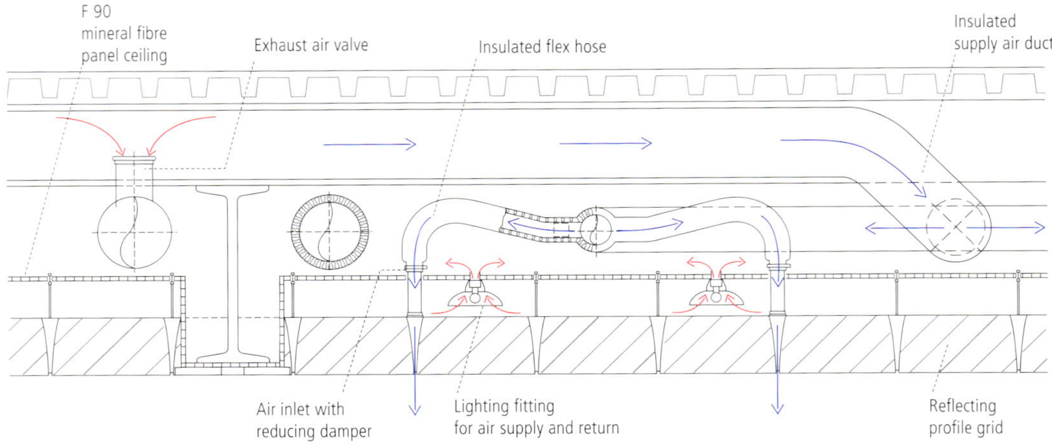

Figure 6.94
Sectional mirror ceiling with air supply through grid louvres (Kiefer)

Aside from supply air diffusers for walls and ceilings, another approach is to create "ventilating ceilings" in the form of perforated or grid ceilings.

Perforated ceilings are ceilings where a large portion of the supply air flows vertically from a pressure space above the ceiling through perforations (representing approx. 1 to 4 % of the total ceiling area) into the space below. Today, perforated panels are generally installed in ceiling components, which allow air to pass through while visually identical panels serve to absorb sound and enclose the space. In contrast to perforated ceilings, grid ceilings make use of the profile grid as an air conducting element and air outlet (**Figure 6.94**).

The sectional reflecting ceiling shown is best suited for large open-plan spaces and takes on the functions of the glare protection grid of a lighting system, the function of diffuse sound dispersion as well as supplying air to the room.

Air movement from bottom to top has become increasingly popular in recent years. Here, fresh air is supplied to the interior near the floor. Again, there are a variety of different system solutions. For fixed seating in theatres or opera houses and similar spaces, fresh air supply via the chair legs is a good alternative (**Figure 6.95**).

In rooms with high thermal loads or others, where supply air cannot be transported into the room from the wall or ceiling area due to other built-in components, air can be introduced through outlets set into the floor. **Figure 6.96** illustrates the principle of a floor outlet.

Figures 6.97.1/2 depict typical air outlets for terraced floor plans (usually employed in lecture halls). The system solution in **Figure 6.97.1** has also been adapted for furnishings in recent years, although this approach met with little success.

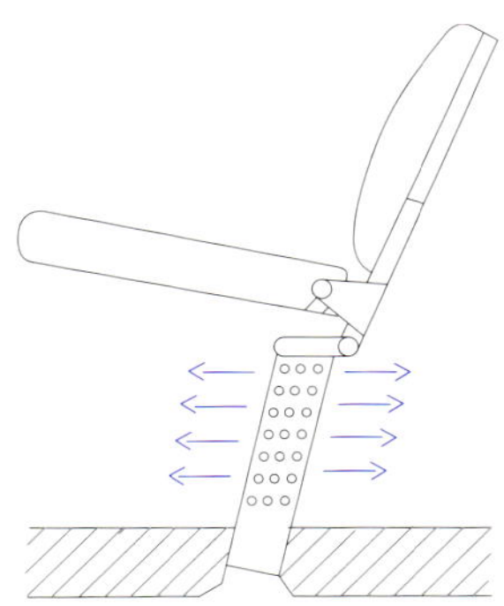

Figure 6.95
Perforated chair leg

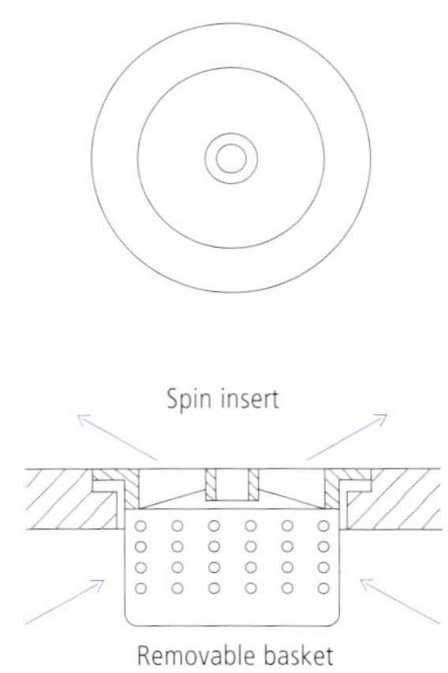

Figure 6.96
Floor vent, top view and section

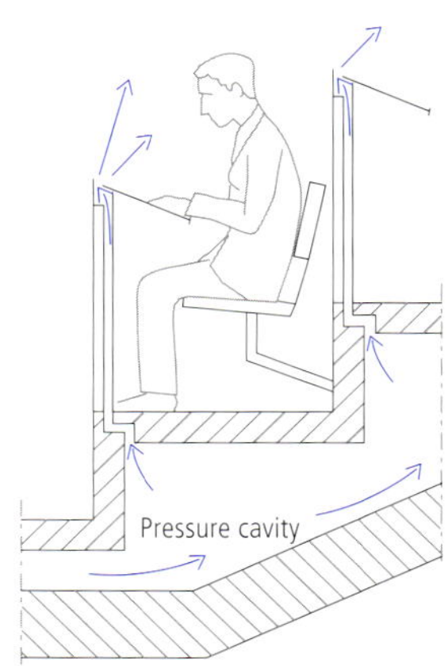

Figure 6.97.1
Schematic of desk ventilation, air supply from pressure cavity into furnishings

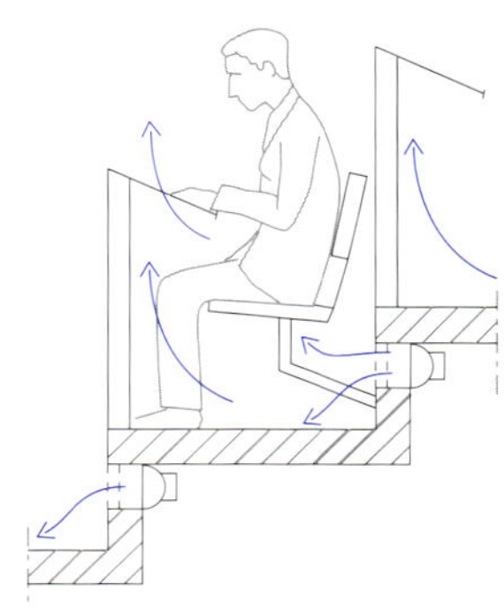

Figure 6.97.2
Air supply via stair vents

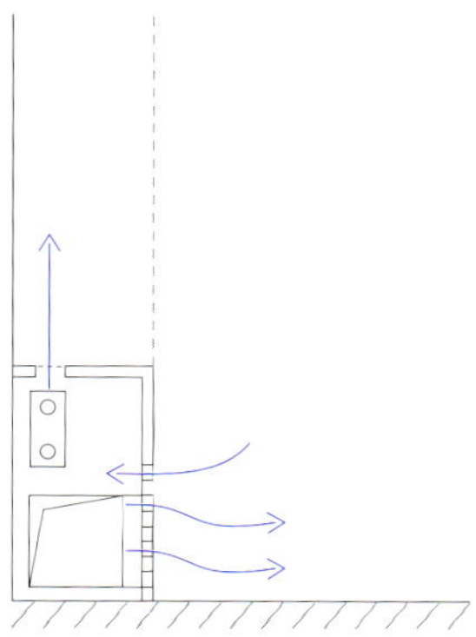

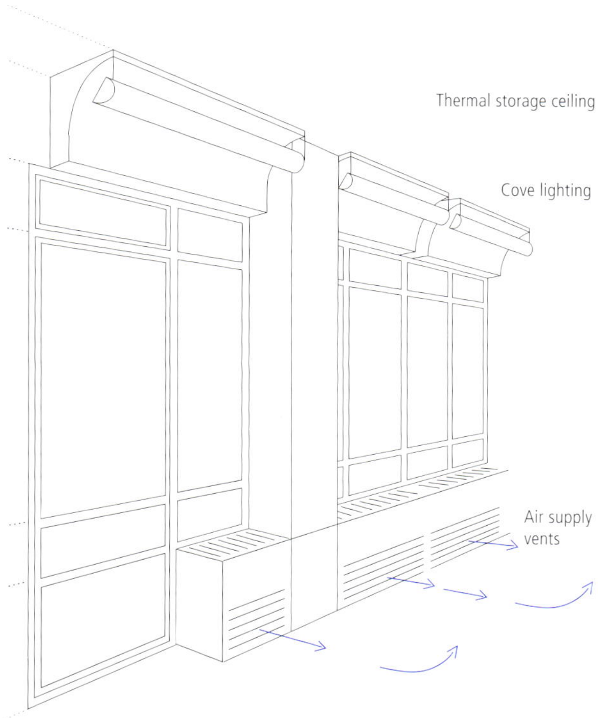

Air supply from the floor up by means of stack effect has, conversely, been highly successful. The principle is laid out in **Figure 6.98.1**. In this approach to ventilation, supply air is blown into the space at very low velocity (< 0.1 m/s) and a minimal temperature of +20 °C. It rises as a result of thermal lift wherever heat sources exhaust the "sea of cold air." This system solution is particularly useful for applications where toxins and odours must be evacuated as cleanly as possible. **Figure 6.98.2** illustrates different forms of stack ventilation outlets.

Many of the above mentioned supply air diffusers can also serve as return air grilles, in which case their appearance is virtually unchanged.

In rare cases, supply air can be introduced into a room through a hollow floor with integrated slits covered by floor to floor carpeting. It is important to ensure that the floor covering is sufficiently permeable to air in order to ensure air penetration.

Figure 6.98.1
Example of fresh air supply by means of stack effect (from bottom to top)

Figure 6.98.2
Example of supply air components for stack ventilation

A Stack ventilation supply unit, series QLK (Image: TROX)

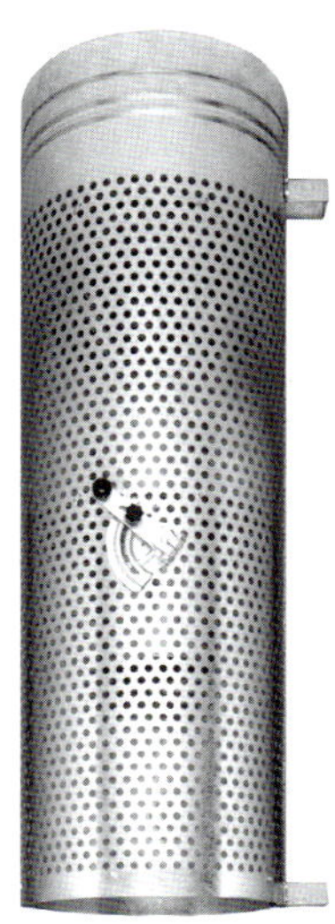

B Round displacement outlet with manual adjustment

C Large stack ventilation air supply through solid wood panelling with micro-perforation, (Image: Nickel, F + E Komponenten)

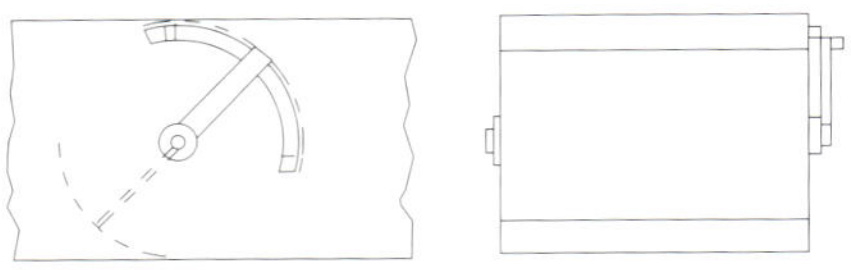

Register (flap) with adjuster

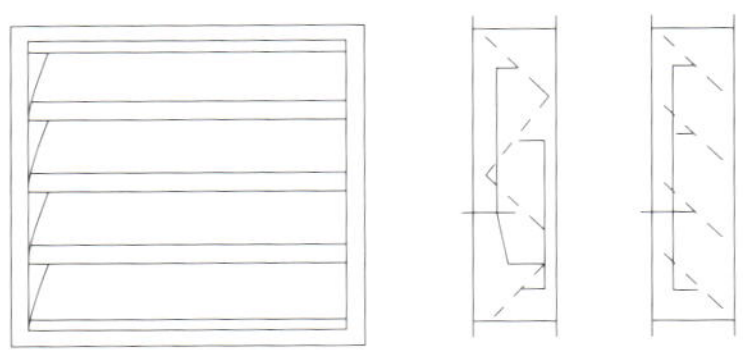

Louvered grille with adjuster
Left: cross-current louvres
Right: parallel louvres

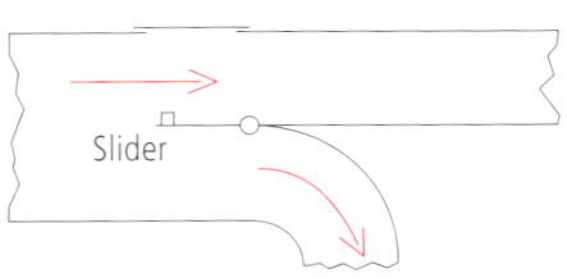

Branch piece with slider

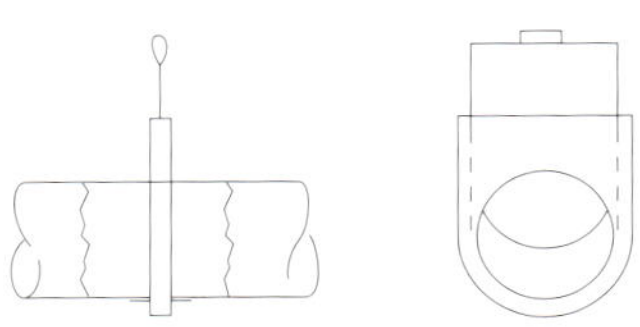

Sliding gate in pipe

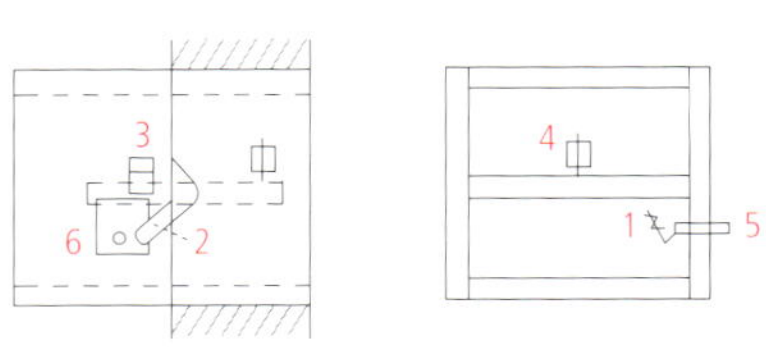

Fire shutter
1 Soldered link
2 Hand lever
3 Engage piece
4 Locking weight
5 Manual release
6 Inspection/maintenance opening

Weatherproofing grilles

Figure 6.99
Registers and grilles for HVAC systems

A

B

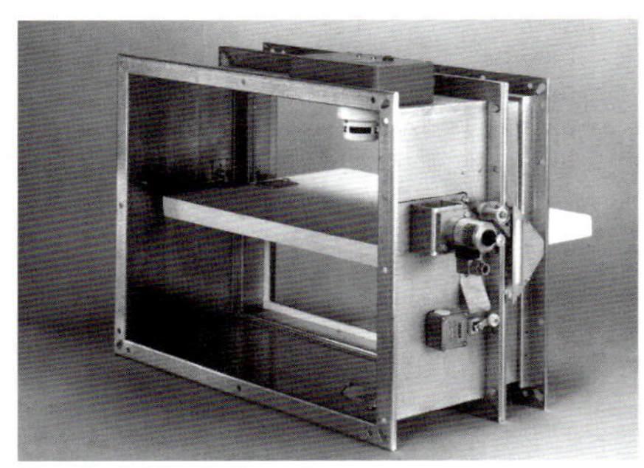

C

E

D

Figure 6.100
Examples of registers for HVAC systems

A Register-regulated secondary heating unit for night heating of supply air in control zone, for installation in constant or variable volume flow systems
B Airtight register AKD for 100 – 300 mm diameter (Image: LTG)
C Fire-protection shutter
D Dual-control register, type ARD (Image: LTG)
E Shut-off and control unit, type ARF (Image: LTG)

6.5.10 Ventilation registers

All ventilation systems require a variety of ventilation registers, as shown in **Figure 6.99**. These are on the one hand controlled registers to regulate air streams (dampers), sluice valves with regulating devices to close or open ventilation systems (internally or externally), devices to regulate diverted air streams, slide (or gate) valves in pipes, fire shutters for wall installation, which separate duct systems in case of fire between fire compartments, and weatherproofing grilles that prevent rainwater from entering into the building with the flow of supply air. **Figure 6.100** depicts several models of the elements mentioned above for duct installation.

6.5.11 Air-conditioning devices / cooling devices / induction units

Central air-conditioning via filters, heaters, coolers, humidifiers and fan-induced air movement is provided by central air-conditioning installations or air-conditioning devices as described below. Walled air-conditioning systems (or concrete encased air conditioning chambers), shown in **Figure 6.101**, are used only for very large air volumes. Since the brickwork and in part concrete, too, allow air to diffuse from the outside into the chamber and vice versa, these types of chambers are usually executed from sandwich rolled steel plates in order to accommodate the individual air conditioning and processing aggregates.

For air volumes of less than 50,000 m^3/h, air-conditioning systems in sheet casings are used. These are delivered to the site ready to install; all that is required is to set them up and connect them to the ductwork. **Figure 6.102-6.104** show different forms of air-conditioning devices in a variety of designs.

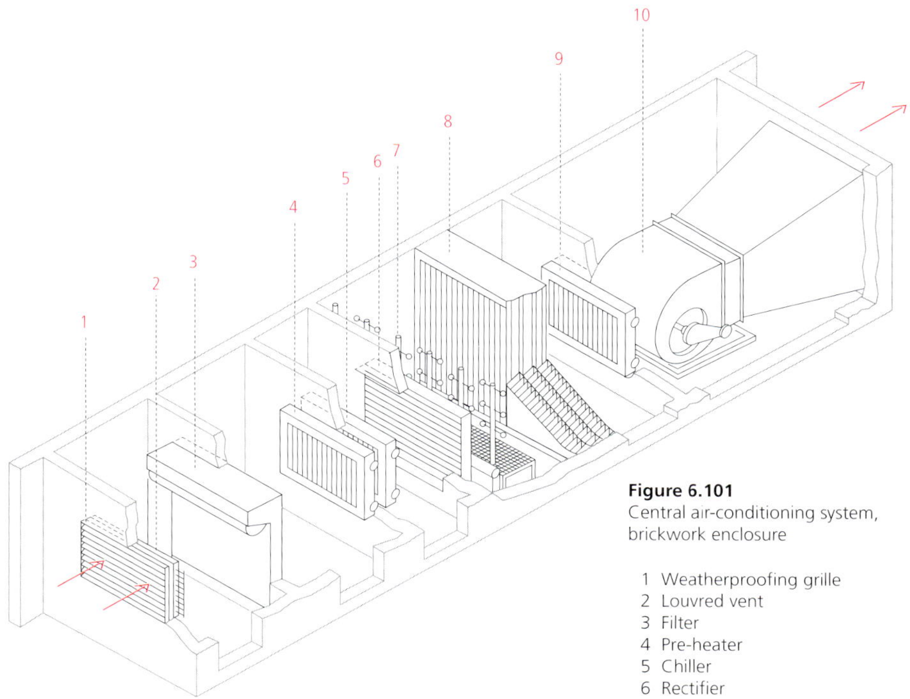

Figure 6.101
Central air-conditioning system, brickwork enclosure

1 Weatherproofing grille
2 Louvred vent
3 Filter
4 Pre-heater
5 Chiller
6 Rectifier
7 Nozzle array
8 Drop interceptor
9 Secondary heater
10 Supply air fan

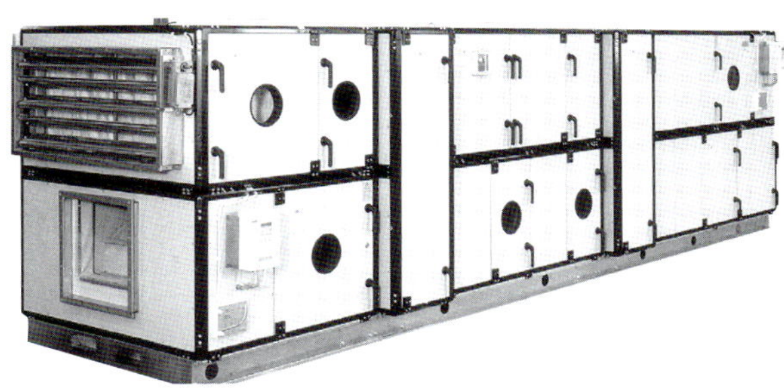

Figure 6.102
Air-condition system with sheet housing, horizontal installation

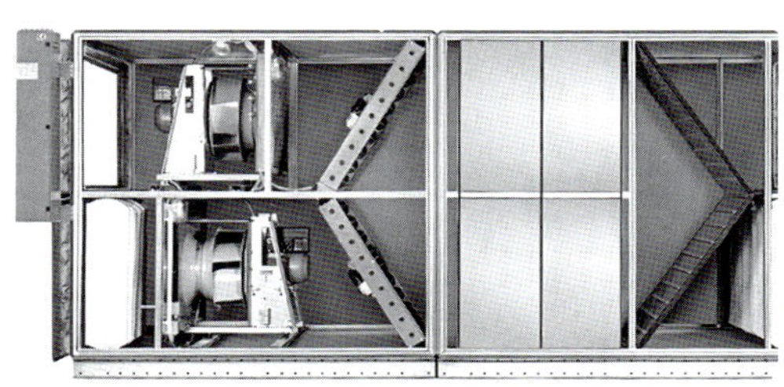

Figure 6.103
Central air-conditioning installation with control and regulation with regenerative heat recovery

Due to the documented high recovery heat quotient of over 90 %, secondary heating of the supply air is usually not necessary with the Resolair® regenerative energy exchanger System if static heating units are installed in the rooms. MENERGA® ventilation systems can be installed in nearly all buildings, e.g. sports centre, residential homes, restaurants, department stores, theatres, cinemas, museums, industrial buildings etc.

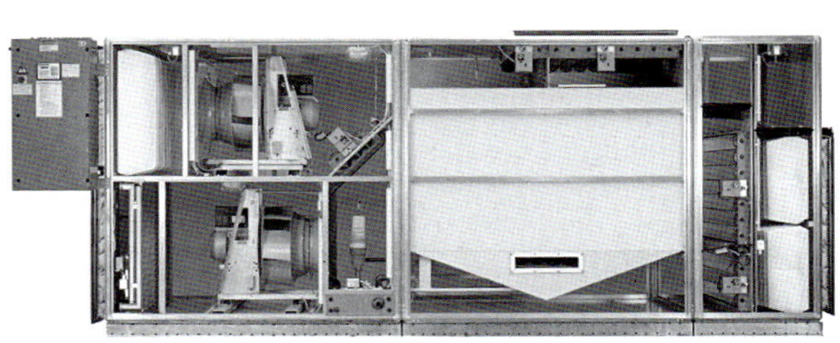

Figure 6.104
Central air-conditioning installation with control and regulation with recuperative heat recovery

The systems based on "adiabatic" evaporative cooling manufactured by MENERGA® offer a solution that minimizes the size of a central refrigerating ma-chine and reduces the energy required for operation. These systems belong to the large Adsolair® group. Outside air can be cooled in summer by humidifying the exhaust air in the heat exchanger. Latent heat transfer yields efficiencies of over 80 %. A combination with an integrated refrigerating machine is particularly energy efficient in situations with very high cooling loads.

In addition to these large devices, there are smaller units for wall installation (room air-conditioners, **Figures 6.105**), which are basically simple cooling and heating devices since they do not add humidity to the supply air. The same is true for room air-conditioners in trunk or cabinet design (**Figure 6.106**): in summer, these devices cool and dehumidify the air in the room and in winter, they provide heat, again without humidification.

Figure 6.107 depicts a cabinet design of a room air-conditioner with built-in cooling machine and ultrasonic humidifier. This type of model is used, for example, in computer rooms where large amounts of circulating air require constant cooling.

Some air-conditioning system must be used in conjunction with secondary treatment in the room itself. Fan-coil systems fall into this category, as do induction air-conditioning systems. **Figure 6.108** illustrates the installation of inductors near the balustrade in combination with a slotted ceiling vent, which accelerates the supply air introduced into the room on a tangent in order to achieve a greater penetration depth (cast).

Figure 6.105.1
Air-conditioning room unit with built-in heat pump

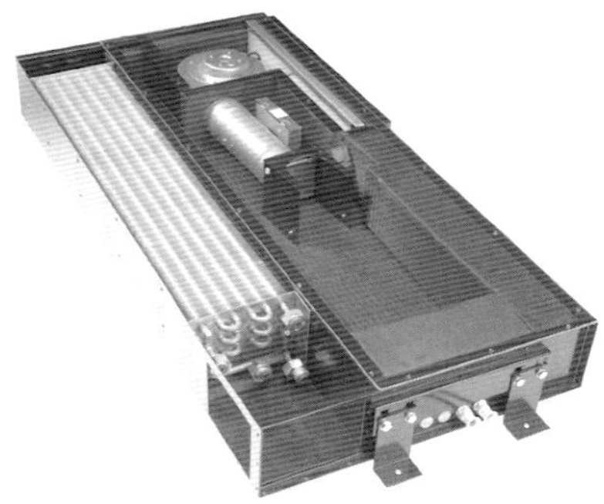

Figure 6.105.2
Fan-coil unit as underfloor unit (Image: Trox)

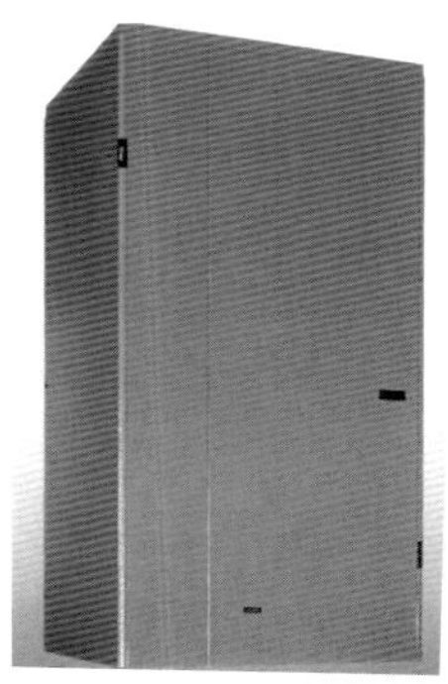

Figure 6.107
Precision air-conditioning unit in cabinet construction with innovative technology

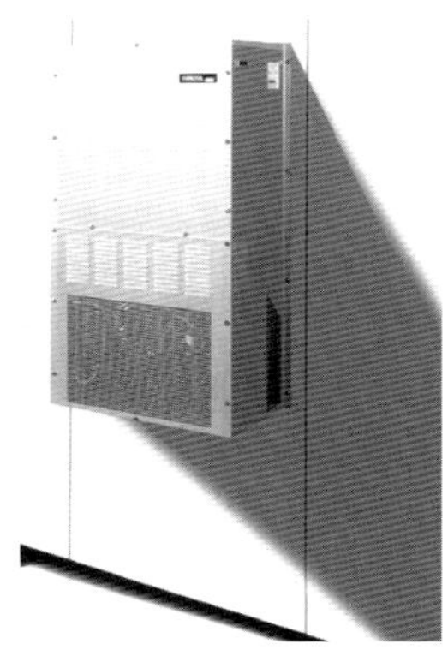

Figure 6.106
Precision air-conditioning unit in container construction, e.g. for mobile radio container

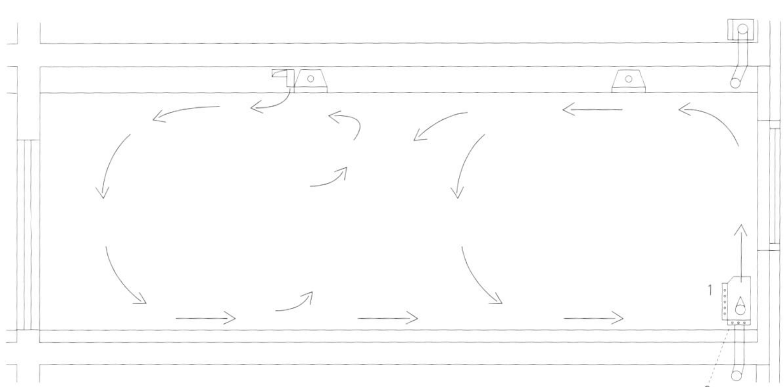

Figure 6.108
Injection technology for optimum room air movement for induction systems and fan-coil systems

1 Cooling component
2 Heating component

The design of an induction unit is illustrated in **Figure 6.109.1**. In this device, supply air (primary air) enters the device on the side and is ejected from the device by nozzles. Primary air thus induces room air, which in turn is transported via coolers or heaters to produce secondary cooling or heating in the room. The blend of primary and secondary air (mix ratio 1:4) is blown into the room on a tangent through grilles located at the window, permeating the room to create hygienic conditions and provide load compensation.

The radiator air-conditioning device shown in **Figure 6.110** is a variation on induction units. It was developed specifically for use in hospitals (wards) and boasts an induction ratio of only 1:1. The face plate of this device is reminiscent of the design used for flat-plate radiators, making it easy to clean.

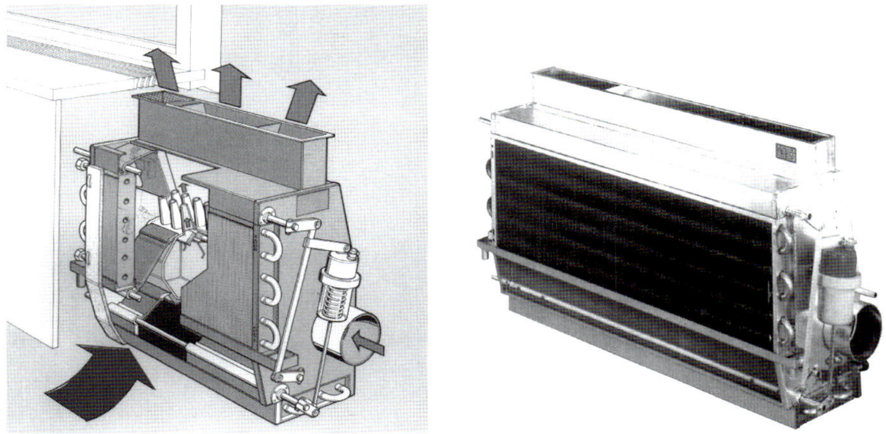

Figure 6.109.1
Four-conductor induction unit for variable volume flow systems, (Type HFW, LTG)

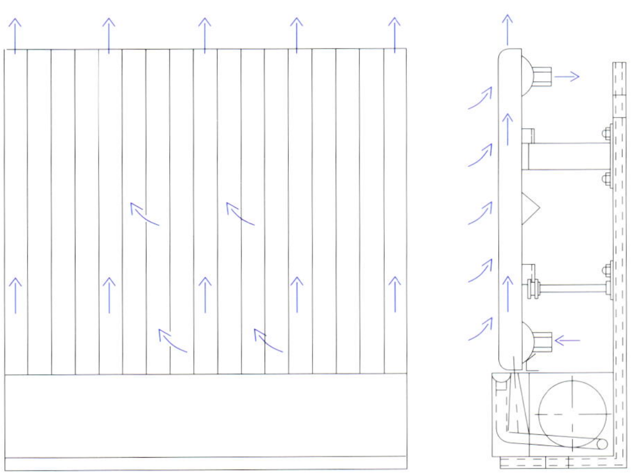

Figure 6.110
Air-conditioning radiator, installation view and section

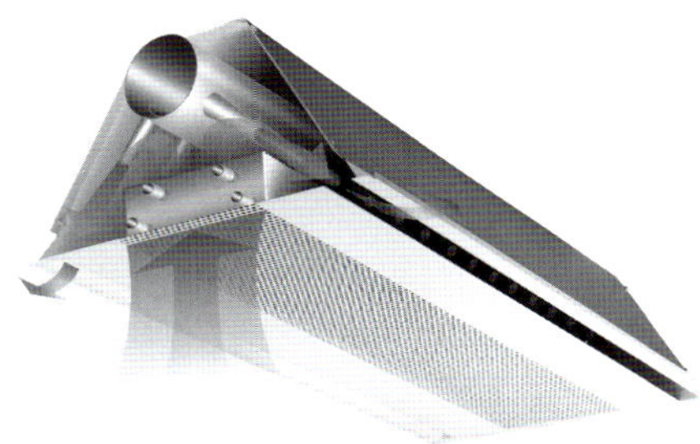

Figure 6.109.2
Ceiling induction outlet (Series DID, TROX)

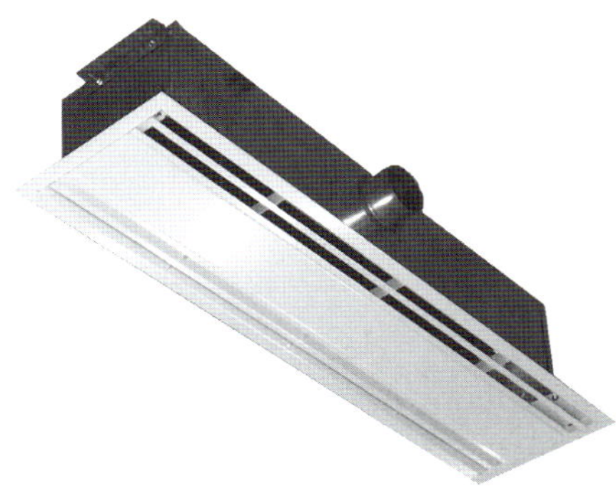

Figure 6.109.3
Ceiling induction unit (Type HDC, LTG)

6.6
Internal air conduction

Internal air conduction is a difficult task concerning primarily:
– Human physiological requirements for warmth
– Production-related technical requirements
– Furnishings and design
– Hygienic demands

In individual cases specific requirements may impose a very definite form of supply air conduction, while other scenarios offer great freedom to choose among a wide variety of options. Hence there is no "one-fits-all" solution, for it is impossible to say that any one of many options of air conduction is the best. Below is an overview of typical internal air conduction models with comments on design limitations.

If one has complete freedom with regard to air conduction and if thermal-physiological, design or other related reasons do not impose a specific approach to air conduction, one should attempt to harness the internal thermal load to the utmost degree in order to diminish air volumes, i.e. air conduction from bottom to top may be preferable to conduction in the other direction.

Figure 6.111 illustrates different options of supplying air to an internal space, differentiated as follows:
– Tangential flow
– Air conduction from ceiling (top-top)
– Air conduction from suspended ceiling (top-top)
– Air conduction by means of jet injection (bottom-top)
– Air conduction from floor area (bottom-top)

Each of these options results in different thermal loads, indicated in **Figure 6.112**. An essential factor is whether the air conduction follows the thermal lift (warm, rising air) or not. If this is not the case, the room is subjected to a 100 % thermal load as a result of the distribution of convective thermal currents. With air conduction from bottom to top, conversely, the thermal load can decrease to 70 %.

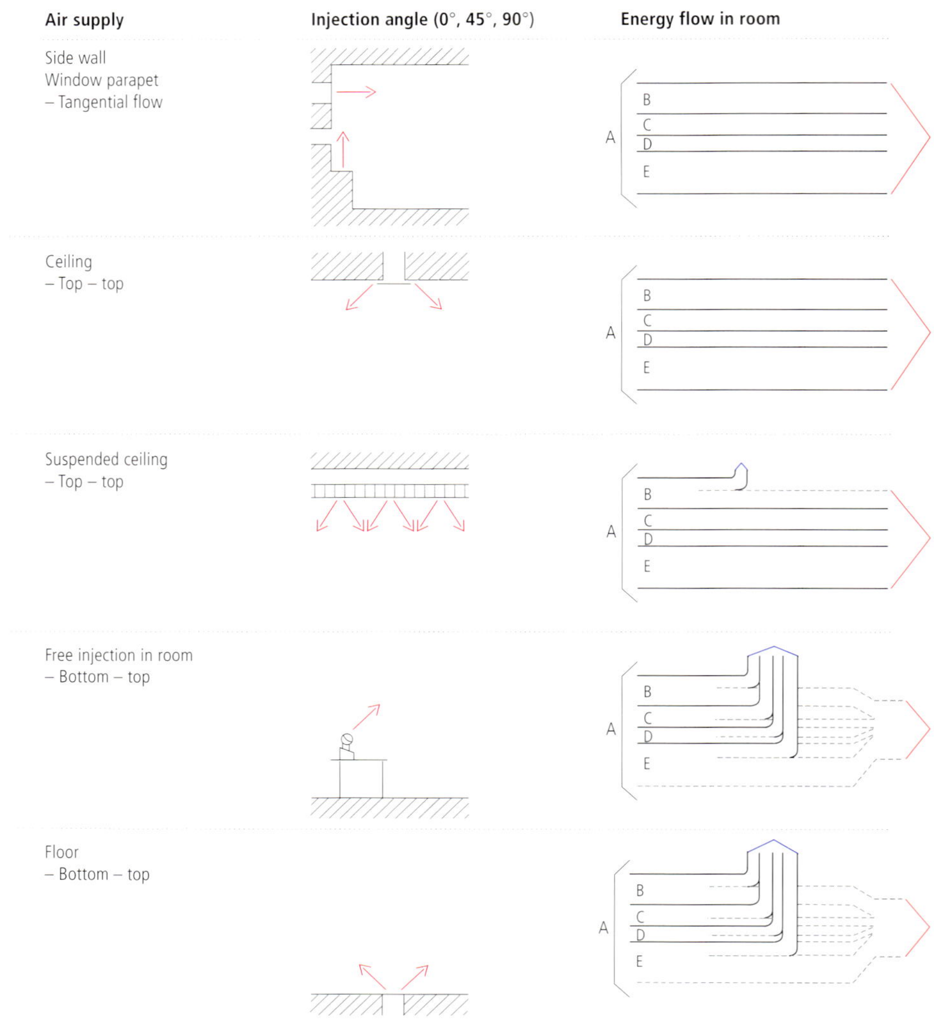

Figure 6.111
Air supply options

A Total load
B Lighting
C Occupants
D Machines
E Facade

Exposes the usable area of the room to load and must be evacuated by HVAC system

Load reduction – joins exhaust air by convection

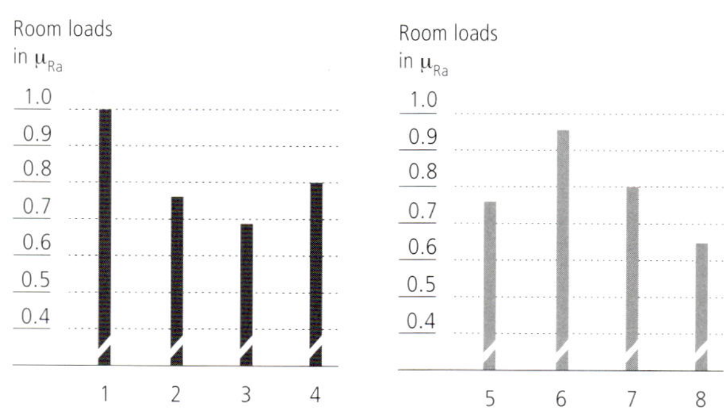

1 Ceiling – ceiling
2 Free jet outlets at floor level
3 Spin outlets at floor level
4 Free jet outlets at floor level and table outlets

5 Load at vertical surrounding surfaces
6 Load at floor level
7 Load at table height
8 Residual heat from air-handling luminaires

Impact of air supply
Load measured at table height, vertical surrounding surfaces as well as residual heat from air-handling luminaires

Impact of heat source locations in room, example: free jet outlets at floor level

Figure 6.112
Thermal loads (mean value)

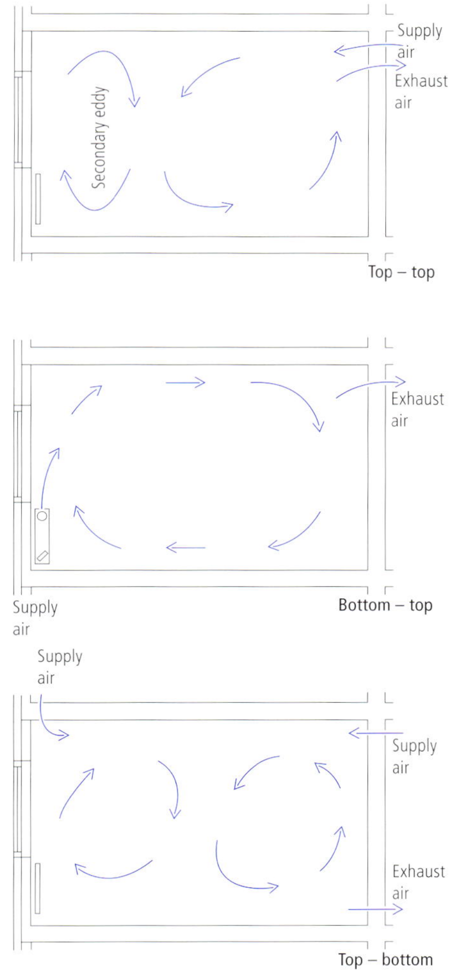

Figure 6.114
Air movement in room, tangential air supply via window sill induction units leading to the formation of an "eddy"

(A to F: sequential documentation photographs/lab images by IKL, Giessen)

Figure 6.113
Air movement through tangential flow

6.6.1 Air conduction with tangential flow

Tangential flows, shown in **Figure 6.113**, result in pronounced eddies in the interior, originating either at the ceiling or near the window sill. To ensure unimpeded return flow in the presence of these eddies, there should be a clearance of approx. 15 cm between all furnishings and room components and the floor near the entry point of supply air. At the same time, the supply air currents should be able to expand freely in the ceiling area up to a height of 0.6 to 1 m, that is, they should not be blocked by cabinetry, wall units or similar built-in features.

Air conduction by tangential flow occurs either via supply air grilles, induction or fan-coil units integrated into the wall (window sill outlets for the latter) or ceiling diffusers, i.e. slotted ceiling outlets. **Figure 6.114** illustrates the course of a typical room air flow (rendered visible with smoke) for tangential air supply via an inductor.

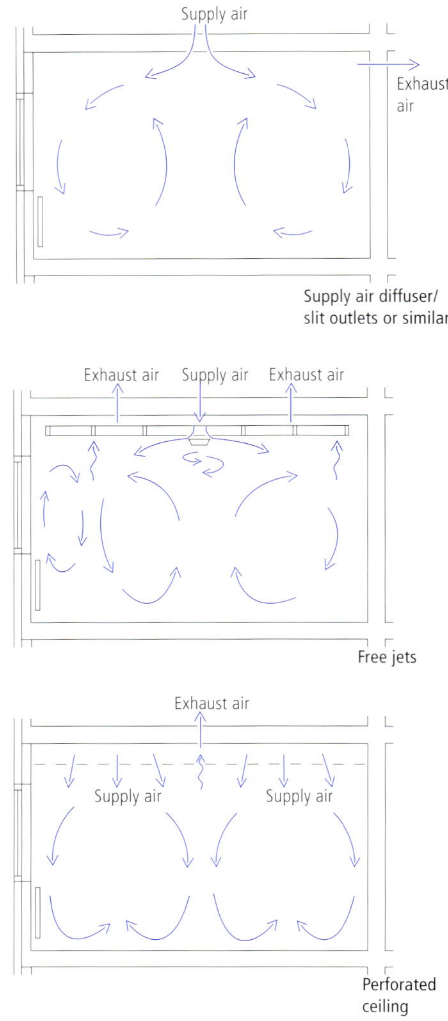

Figure 6.116
Air movement in room with air supplied from top-top

Air supply via jet outlets, smoke injected through an air supply element

(A to F: sequential documentation photographs/lab images by IKL, Giessen)

Figure 6.115
Air movement from top to top for different air supply outlets

6.6.2 Air conduction from top to top

Air conduction from top to top means that the supply air is injected from the ceiling area and exhausted at the ceiling. **Figure 6.115** depicts typical eddies for air conduction from top to top and for different supply air components, such as diffusers or slotted outlets, individual jets or perforated ceilings. When these air supply components are used, the return flow velocity in the floor area is lower than in tangential ventilation, which means that it has no impact on furnishing considerations (floor clearance). However, air conduction from top to top is generally only feasible for a maximum room height of up to 4 m. With greater room heights, this approach becomes problematic since the air currents behave differently depending on the supply air condition (cold or warm). A flow study may therefore be necessary for air conduction from top to top in very large spaces (e.g. halls) in order to determine the correct choice of supply air outlets and the required characteristics.

In air conduction from top to top, the supply air (primary air) permeates the space in an intense manner, inducing room air and thus setting the entire air volume in the room into motion.

Figure 6.116 depicts a room air flow, again rendered visible with smoke, from top to top for supply air influx from a jet.

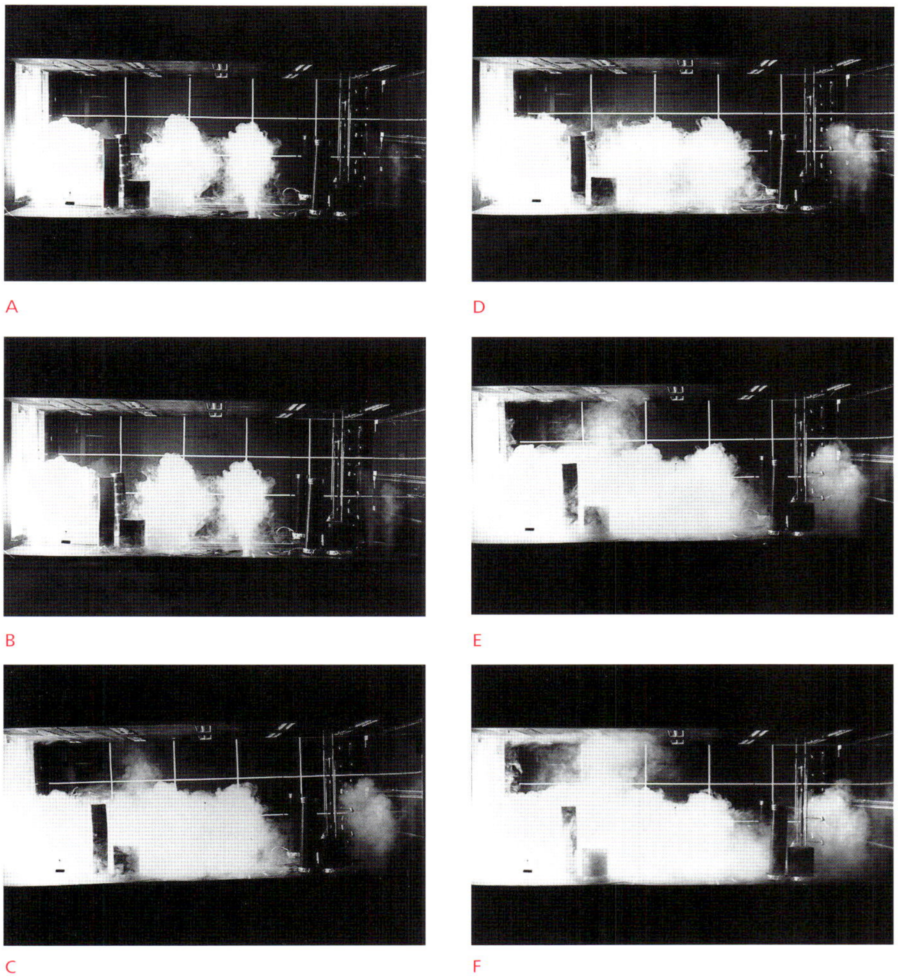

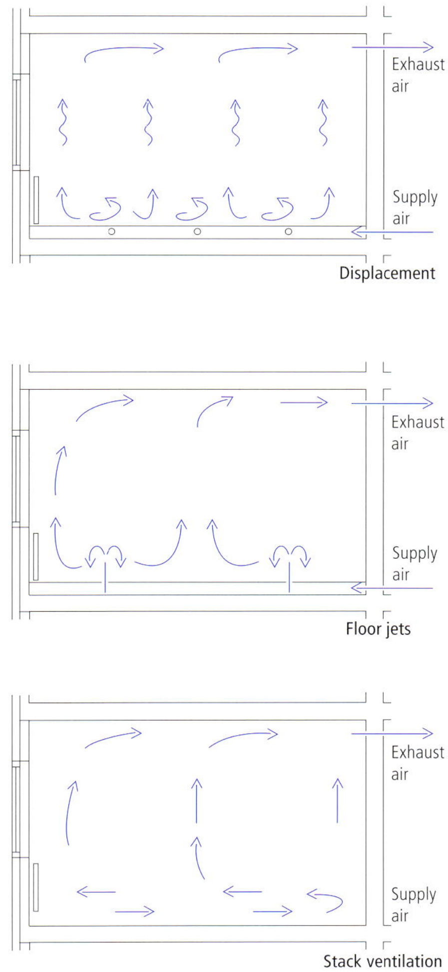

Figure 6.118
Characteristic flow image for air supply through jet outlets at floor level

(A to F: sequential documentation photographs/lab images by IKL, Giessen)

6.6.3 Air conduction from bottom to top

In rooms with high thermal loads (computer labs, heating installations etc.), with great heights (atria across several storeys or similar spaces), or rooms where the interior furnishings can have considerable influence on the flow of air in the space (blocking effects caused by very large furnishings), air conduction from bottom to top should be given primary consideration. When air is conducted from bottom to top, different supply air outlets produce either a displacement flow or a primary air distribution up to a height of 1 m (open jets) or a very gentle, layered flow (stack ventilation). All systems (**Figure 6.117**) share the common characteristic that the air is initially distributed in the floor area as cooling air (air temperature 20 to 21 °C), then siphoned off through heat sources (occupants, machines etc.) and finally rises to the ceiling. **Figure 6.118** depicts the air flow with an open jet for air conduction from bottom to top. The smoke images illustrate the strong impulse with which the air is initially injected into the room and its subsequent spreading across the floor and occupied area.

Figure 6.117
Air movement from bottom to top for different air supply outlets

Figure 6.119 depicts an air flow, visualized with smoke, for stack ventilation. Clearly visible in this image: the siphoning off of the horizontal air flow near the person and the PC and its subsequent rise upward.

6.6.4 Air conduction from top to bottom

(Displacement ventilation with low turbulence)

In rooms where the primary concern is a high degree of cleanliness, the preferred choice is displacement ventilation with low turbulence, where the air flows either from top to bottom or from wall to wall. Displacement ventilation with low turbulence, illustrated in Figure 6.120 (top), is most common in pharmaceutical and electronics environments, where clean-room conditions (classification 1 to 10,000) are vital. Partial displacement ventilation from top to bottom is used, for example, in operating theatres where the operating table is supplied with air of the utmost purity and the contaminated and possibly gaseous air is subsequently exhausted at floor level.

Displacement ventilation with low turbulence can also occur from wall to wall in cases where different degrees of air purity are required in sequential sections within the same space. Figures 6.121.1 – 3 depict laminar air currents (low turbulence, rendered visible with smoke) as well as the turbulence created by an object placed in a laminar air current. The test images with a person demonstrate the impact of flow velocities on optimal displacement ventilation.

Depending on requirements, air conduction in a room can also be effected in a manner entirely different from those described above; other approaches are, however, clearly special or custom solutions.

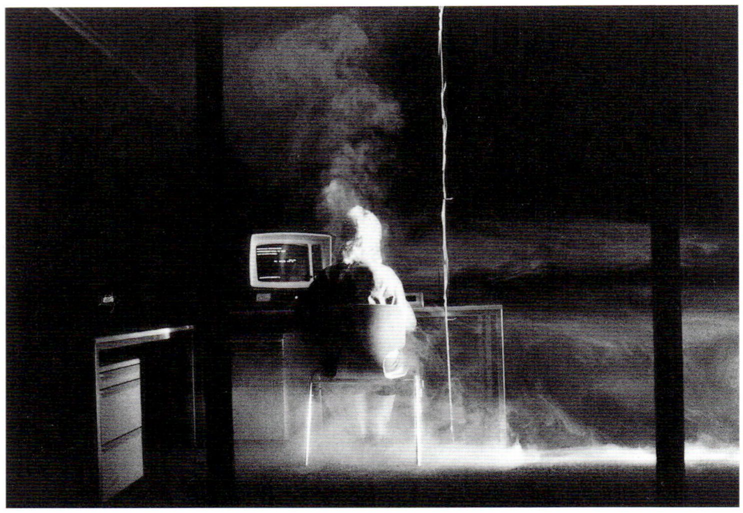

Figure 6.119
Air movement in room during stack ventilation

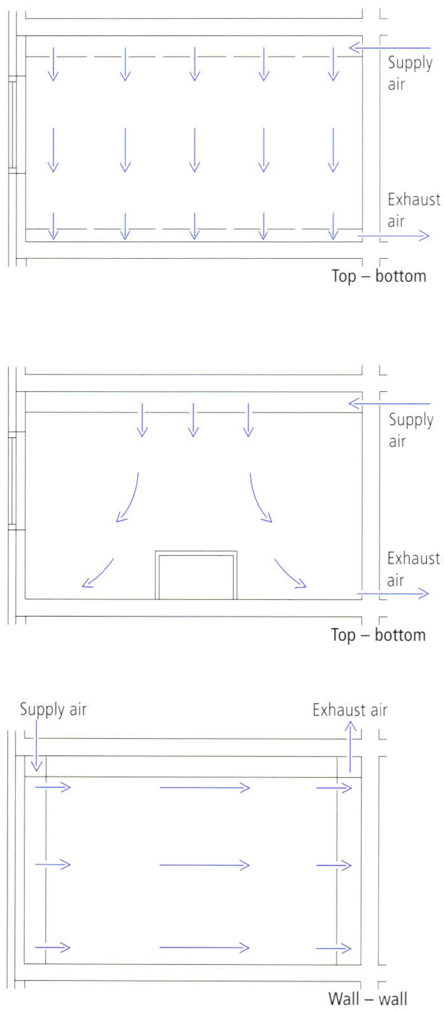

Figure 6.120
Displacement with low turbulence

Figure 6.121.1
Laminar air flow in clean-air room (0.45 m/s)

Figure 6.121.2
Turbulence caused by object in laminar air flow

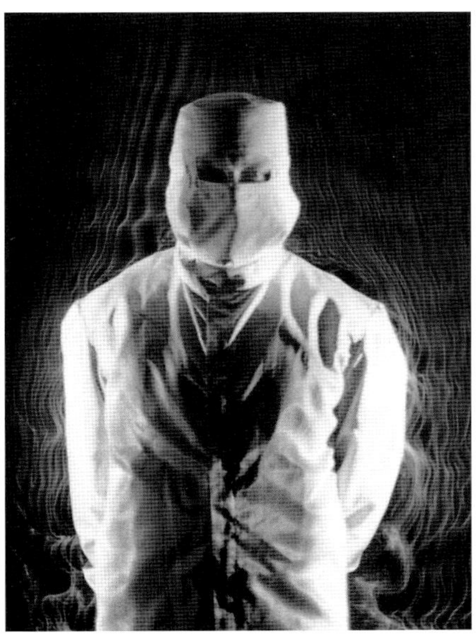

Figure 6.121.3
Flow velocity
0.45 m/s

Flow velocity
0.33 m/s

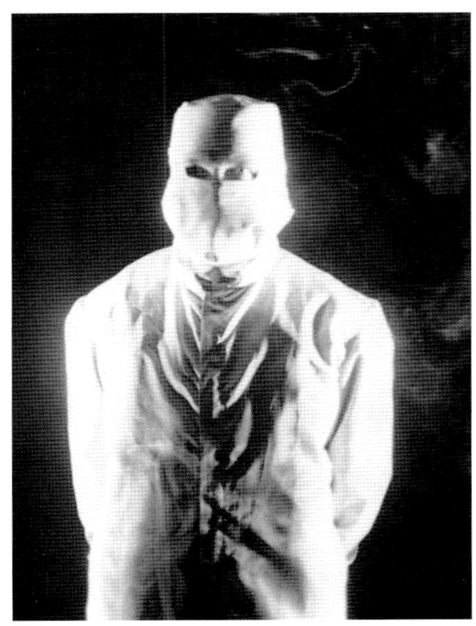

Flow velocity
0.20 m/s

6.7 Ventilation and air-conditioning systems (HVAC systems)

The task of ventilation and air-conditioning systems (HVAC systems) is to evacuate the following loads from rooms:
- Air pollutants (odour, contaminants and ballast)
- Sensible heat loads (cooling loads)
- Latent heat loads (material loads during dehumidification or humidification)

Moreover, these systems serve to create a hygienic and comfortable atmosphere in rooms.

6.7.1 Classification

HVAC-systems are classified according to thermodynamic air treatment functions:
- Ventilation systems (basic configuration)
- Ventilation systems with additional air treatment
- Air-conditioning systems

Table 6.12 lists the classification based on thermodynamic air treatment and ventilating functions. **Figure 6.122** provides an overview of the component design for various thermodynamic air treatment functions.

Beyond ventilation, partial and central air-conditioning systems, HVAC systems also include specialized systems such as industrial extraction, air curtain doors, clean air systems, climatic test chambers and fume dispersion equipment.

HVAC systems can also be subdivided according to the manner by which the energy necessary for load compensation is supplied to the room (by means of air/air and water/water).

This results in the following differentiation:
- All-air systems
- Air and water systems
- All-water systems
- (Direct refrigerant systems)

Table 6.13 and **Figure 6.123** introduce the most common air-conditioning systems. The table provides an overview of applications and additional system parameters and can serve as an initial guide to appropriate system solutions.

Thermodynamic air treatment functions		**HVAC systems and ventilation function**	**w/o ventilation function**
Number	Type	HVAC system	Air circulation system RC
none		Exhaust system EA	Recirculation system RC
one	H C Hu D	Ventilation system OA or W/HVAC	Recirculation system RC
two	HC HHu HD CHu CD HuD	Partial air-conditioning system OA or MI	Recirculation/partial air-conditioning system RC
three	HCHu HCD CHuD HHuD	Partial air-conditioning system OA or MII	Recirculation/partial air-conditioning system RC
four	HCHuD	Air-conditioning system OA or MI	Recirculation/air-conditioning system RC

Example: HVAC – MI = air treatment system with ventilation functions for heating (H), cooling (C), humidifying (Hu) and dehumidifying (D) with fresh air (outside air/OA) and recirculated air (RC)

Table 6.12
Classification and description of HVAC systems after DIN 1946, Part 1

Heating

Cooling
Dehumidifying

Heating
Cooling
Dehumidifying

Heating
Cooling
Dehumidifying

Heating
Humidifying
(Spray humidifier)

Heating
Humidifying
Steam

Heating, cooling, humidifying (spray humidifier), dehumidifying

Cooling, heating, humidifying (spray humidifier), dehumidifying

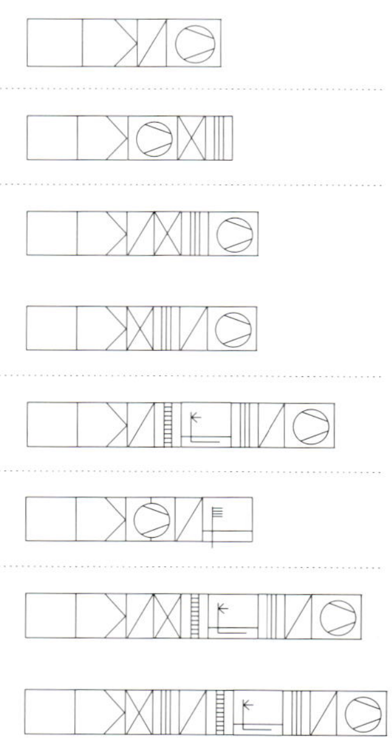

Figure 6.122
System components for thermodynamic air treatment functions (icons after DIN 1946, Part 1)

		A	B	C	D	E	F	G	H	I
Facade	with window sill/parapet	●	●	●	●	●	●	●	●	●
	without window sill/parapet	◍	●	◍	◍	◉	◉	◉	●	◍
	ventilation facade	●	●	●	●	○	◉	◉	●	
	operable windows	●	●	●	●	○	○	○	●	◍
	internal curtains, lintel (opt.)	●	●	●	●	○	○	○	●	●
Ceiling construction	suspended ceiling, smooth	●	●	●	●	●	●	●	●	●
	suspended ceiling with edge joint	◍	●	●	◍	○	○	○	●	●
	suspended modular ceiling	○	●	●	○	◉	◉	◉	●	●
	no suspended ceiling	◉	○	●	◉	●	●	●	●	◉
Room size	open-plan office	●	◍	●	◍	○	○	○	○	●
	group office	●	◍	●	◍	◍	●	●	◍	●
	office for up to 2 occupants	●	○	●	◍	●	●	●	●	●
Room depth	large, above 7m	●	●	◍	●	○	○	○	●	●
	standard	●	◍	●	●	●	●	●	●	●
Room height	high, above 3m	◍	●	●	◍	●	●	◍	●	●
	standard	●	●	◍	●	●	●	●	◍	●
Floor	raised floor	◍	●	◍	◍	●	●	●	●	●
	screed floor	●	◉	●	●	●	●	●	●	●
Lighting fixtures	integrated lighting fixtures	●	●	●	●	●	●	●	●	◍
	mounted lighting fixtures	○	●	●	○	◉	◉	◉	●	◍
	hanging lighting fixtures	○	●	●	◍	○	○	○	●	●
Flexibility	room dividers	◍	◍	◍	◍	●	●	●	◍	●
	furniture dist. and facade dist.	●	◍	●	●	○	○	○	●	●
Heating	integrated into ventilation	○	◍	◉	○	●	●	●	◉	◉
	basic load via heating surface	◍	◍	◍	◍				○	●
	indirect control of heating surfaces	●	○	●	●				●	●
Air vents	below ceiling	◍	◉	◉	◍	◉			◉	○
	in suspended ceiling	●	◉	◉	●	◉			◉	○
	in window sill casing	◉	◉	◍	◍	●	●	●	○	●
	in internal wall	◉	◉	◍	◉	○			●	●
	in floor		●	◍	●	◉			●	◍
Control		●	◍	◍	○	●	●	●	◍	●
High cooling load in room		◍	●	●	○	●	●	◍	●	●
Sound level		◍	◍	●	◍	○	◍	◍	●	●
Comfort	flow velocity	◍	○	●	◍	○	○	○	●	●
	temperature profile/heating	◍	◍	○	◍	◍	◍	◍	◍	◍
	temperature profile/cooling	●	○	○	◍	◍	◍	◍	○	●
	window covering	◍	◍	◍	◍	●	●	●	◍	◍
Contaminant extraction		◍	●	●	◍	○	○	◍	●	◍
Heat extraction		◍	●	●	◍	●	●	◍	●	●
Efficiency	investment costs	◍	○	◍	●	◍	○	◍	◍	◍
	energy costs	●	●	●	○	◍	●	●	◍	●
	maintenance requirements	◍	○	◍	●	○	◍	○	◍	●
	space requirements for sys. installation	○	○	○	○	●	●	●	○	●
	space requirements for ducts/pipes	◍	○	◍	◍	●	●	◍	◍	●

● excellent suitability, very suitable
◍ good suitability, suitable
○ marginal suitability, satisfactory
◉ not suitable

Table 6.13
Overview of most common air treatment systems

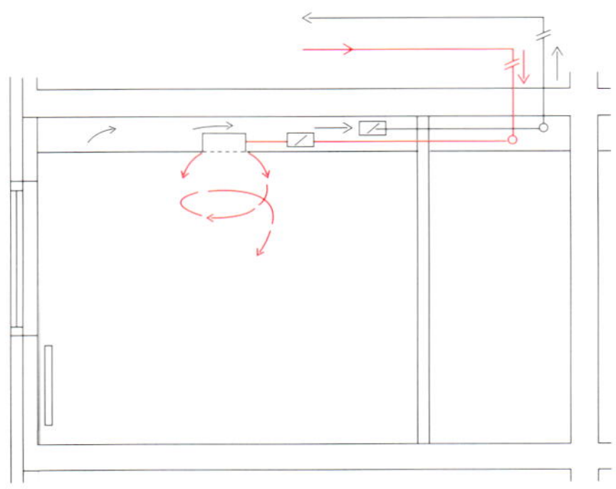

A
Variable volume flow system (VAV)
with static heating –
air supply from top to top

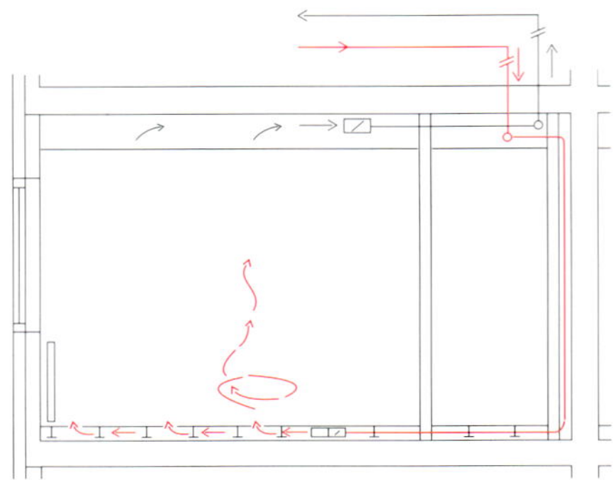

B
Variable volume flow system (VAV)
with static heating –
air supply from bottom to top

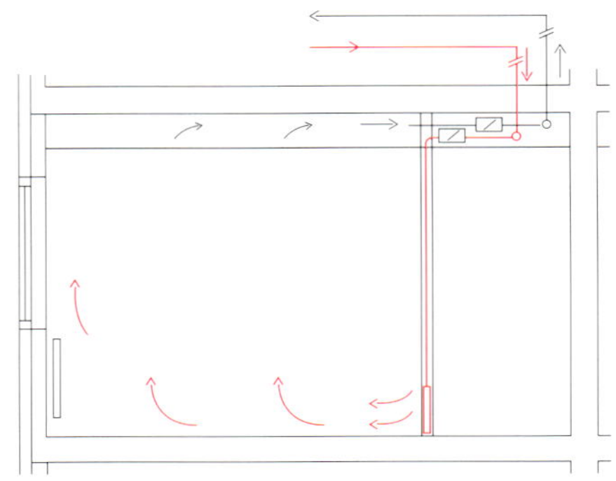

C
Variable volume flow system (VAV)
with displacement and static
heating

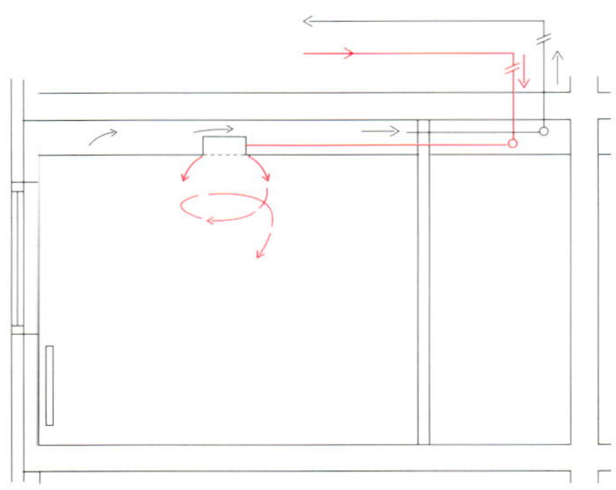

D
Constant volume flow system (CV)
with static heating –
air supply from top to top

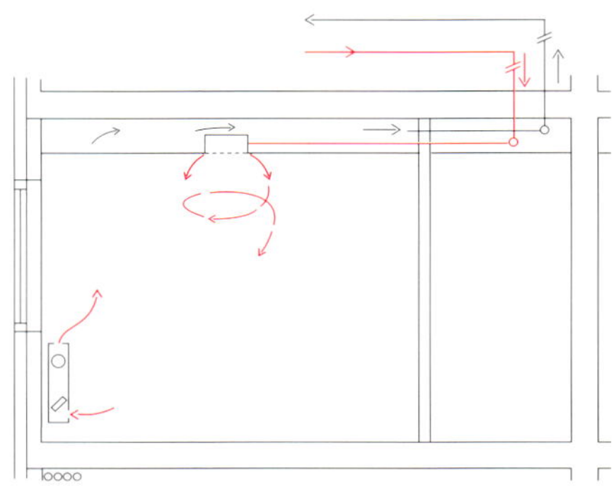

E
Fan-convector system for heating
and cooling (four-conductor)
with central primary air supply

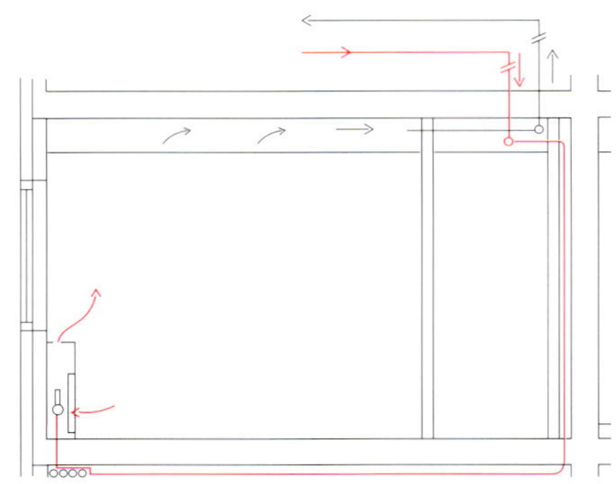

F
Inductionfour-conductor system

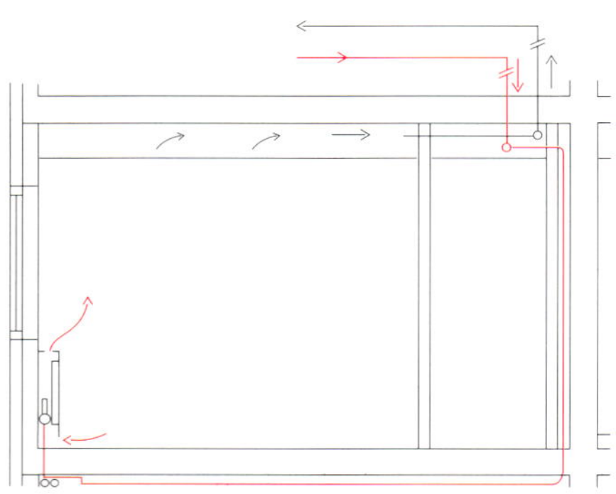

G
Induction-VAV two-conductor system

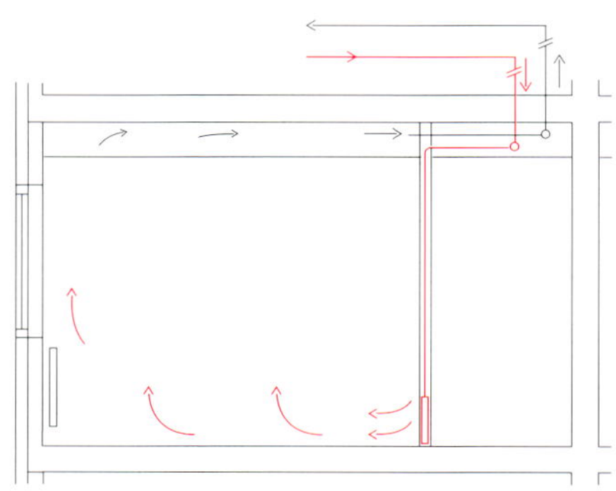

H
Constant volume flow system (CV)
with static heating and displacement

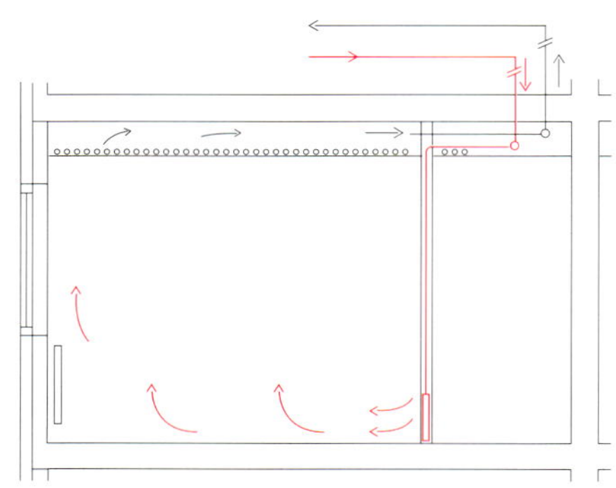

I
Cooling ceiling with central primary air supply,
displacement and static heating

Figure 6.123
Schematic of room-related
installations of different
air-conditioning systems

6.7.2 All-air systems

All-air systems are HVAC systems that transport the energy required for load compensation exclusively by means of air. They comprise ventilation systems, ventilation systems with additional air treatment and air-conditioning systems. All-air systems, in turn, are differentiated into single-duct (supply air) and two-duct (cold and warm air) systems. A further subcategory identifies whether the streams of supply air flow through the supply duct at low velocity (low pressure) or high velocity (high pressure). In English-speaking countries, the latter are commonly referred to as high-velocity systems.

6.7.2.1 Single-duct systems

In single-duct systems air is transported into the rooms via one duct, and the supply air is either treated on site after leaving the central air conditioning plant or not treated at all. The diagram in **Figure 6.124** illustrates the layout of a single-duct system with the central air conditioning plant, the cooling and heating supply, supply air and exhaust air system, as well as the rooms that require air-conditioning. As the diagram shows, the supply air is transported to the rooms without further treatment and injected into them. This approach requires that the rooms have identical or at least very similar load behaviours, making treatment unnecessary.

When rooms have different cooling loads or a different heat requirement or yet, if the heat requirement is the same but occurs at different times, treatment is required within the context of a single-duct system. The diagram in **Figure 6.125** depicts a single-duct system where secondary heaters supply several control zones, that is, where rooms or zones within rooms can be heated to different temperatures. This is a very common system design, since secondary heaters are quite easily integrated into ductwork.

For buildings with smaller rooms or zones with different loads, a multizone single-duct system with mixing valves is the obvious solution. In this system, cold and warm air flows are prepared and mixed centrally; the supply air stream that flows into each room or zone is thus prepared to meet the specific requirements. The diagram in **Figure 6.126** illustrates the layout and design of such a system. This is only useful, however, in situations with a limited number of rooms or zones (max. approx. 10), since the ductwork and the construction of the central installation are too elaborate for other applications.

In buildings where rooms or zones are characterized by very high heat loads and where large volumes of air must circulate and may require cooling on site, multizone single-duct systems with additional ventilators are also an option (**Figure 6.127**).

This system configuration is found in computer processing centres or manufacturing operations (warm operations) and was sometimes employed in high-rises in the past to prepare the air streams for each storey with the help of a subsystem per floor (additional ventilator with additional air treatment).

To improve the regulating faculties of single-duct systems, HVAC systems with variable volumes (VAV) were developed to replace the constant volume flow systems (CV) previously used. In ventilation systems with cooling, and air-conditioning systems with variable volume flow, a volume flow regulator is employed for automatic control (thermostat) in each room or zone and only the volume of supply air is transported into the room (zone) from a cold air duct that is required to maintain the internal climate.

To this end, mixing terminals or VAV boxes are used both on the supply and the return end of the system to decrease or increase the air volume flow into the room.

Figure 6.128 depicts the basic structure of a single-duct system with VAV. In this system solution the room temperature is regulated in winter with the help of the static heating surfaces, blowing supply air into the room at a slightly lower temperature than the room temperature (supply air temperature +18 °C / room temperature +22 °C).

In summer, the regulation occurs by means of volume flow regulators and the function of the heating system is suspended. Today, single-duct systems with VAV are used for a wide range of applications in many buildings. The supply air and air circulation can be either from bottom to top or from top to bottom.

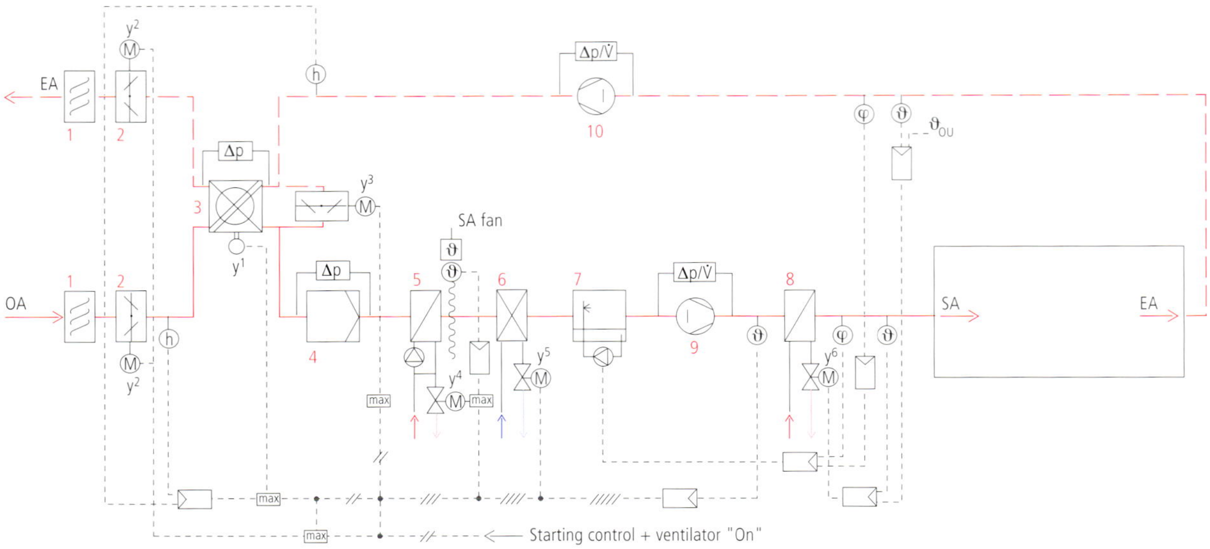

1 Weather protection
2 Louvred vent
3 Mixing chamber
4 Filter
5 Preheater
6 Chiller
7 Washer
8 Secondary heater
9 Supply air fan
10 Exhaust air fan

Figure 6.124
Schematic diagram of single-duct system (CV)

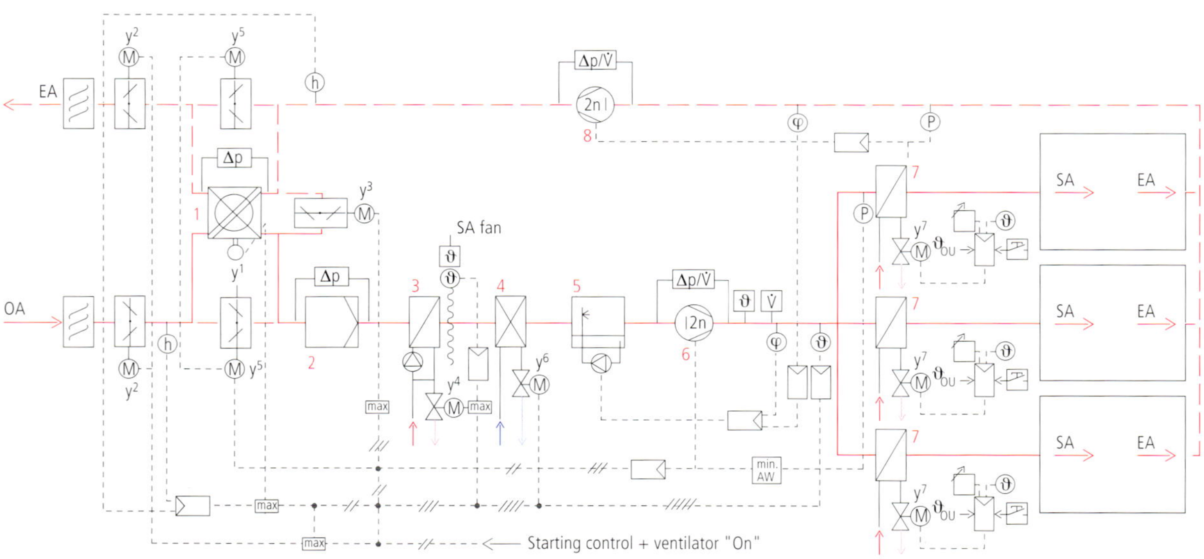

1 Mixing chamber
2 Filter
3 Preheater
4 Chiller
5 Washer
6 Supply air fan
7 Zone secondary heater
8 Exhaust air fan

Figure 6.125
Multizone single-duct systems with secondary heaters (CV)

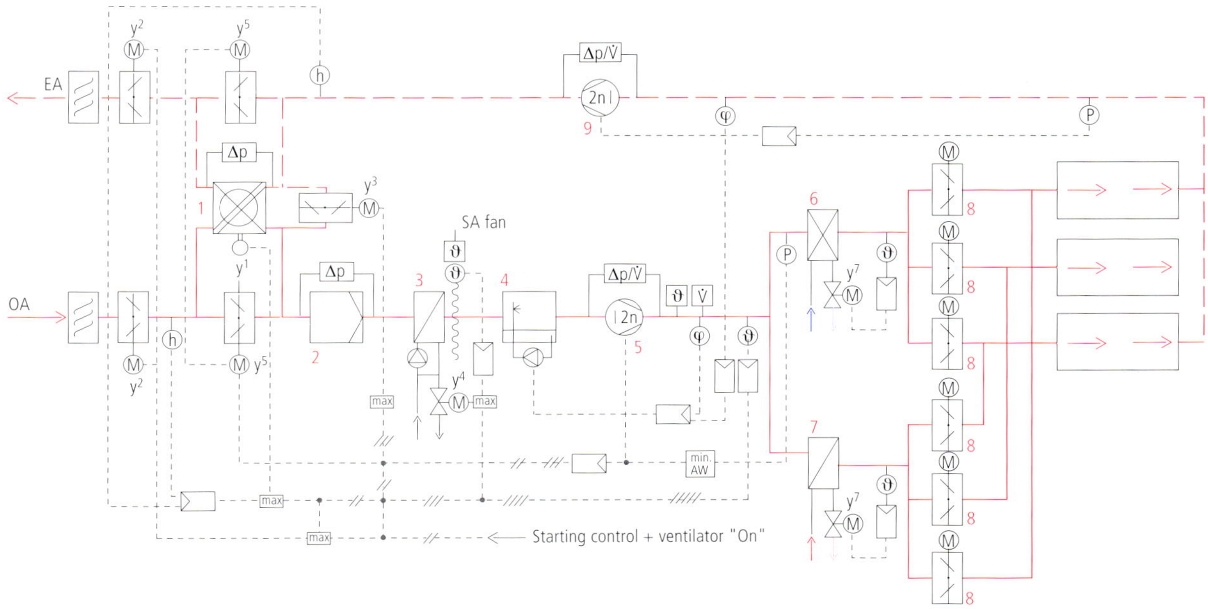

1 Mixing chamber
2 Filter
3 Preheater
4 Washer
5 Supply air fan
6 Chiller
7 Heater
8 Mixing valves
9 Exhaust air fan

Figure 6.126
Multizone single-duct system with mixing valves (CV)

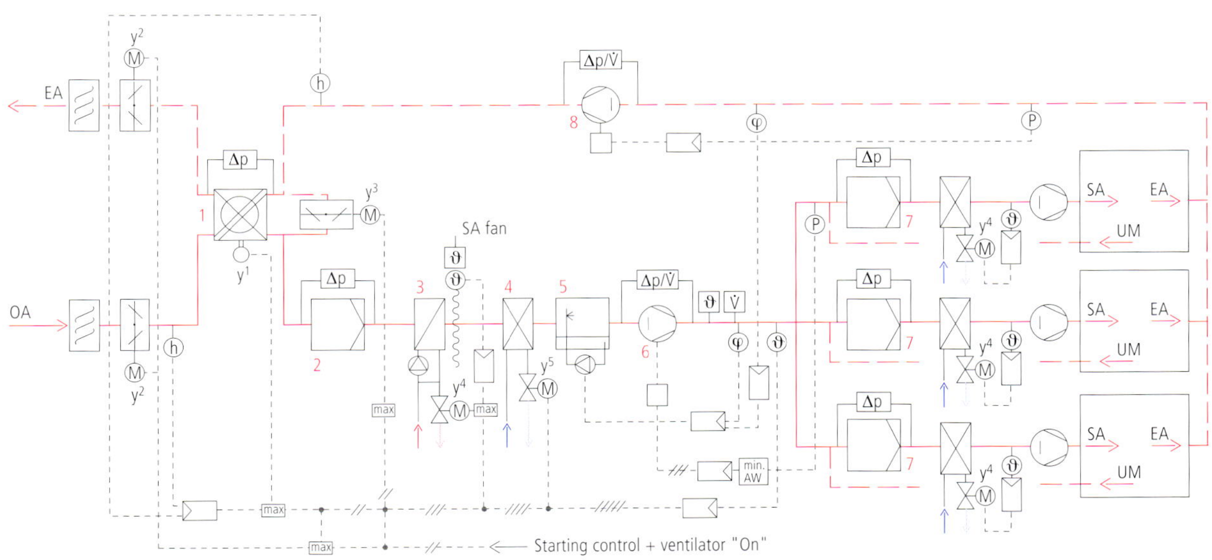

1 Mixing chamber
2 Filter
3 Preheater
4 Chiller
5 Washer
6 Supply air fan
7 Additional device with fa
8 Exhaust air fan

Figure 6.127
Multizone single-duct system with additional fans (CV)

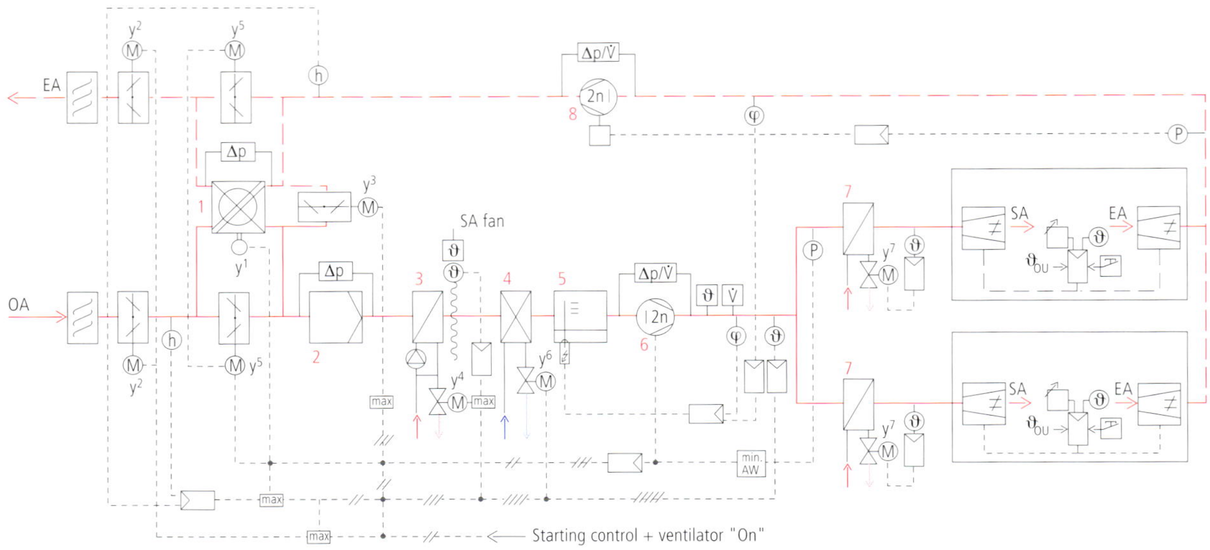

1 Mixing chamber
2 Filter
3 Preheater
4 Chiller
5 Washer
6 Supply air fan
7 Zone secondary heater
8 Exhaust air fan

Figure 6.128
Single-duct system (VV)

6.7.2.2 Two-duct systems

Two-duct systems are ideal if a building features a multitude of regulatory zones due to different loads, and if both temperature and relative humidity must be controlled (**Figure 6.129**). In two-duct systems a cold and a warm air stream are processed and transported through the building in two parallel ducts. At each relevant point within the building (individual rooms or regulatory zones) a mixing box withdraws partial air streams from the warm and cold air duct, mixing them with the supply air to a specific temperature, which allows for sufficient load compensation. This system therefore permits extreme cooling or heating in adjacent rooms. Naturally all other temperatures lying between these extremes are also possible. Since two-duct systems require very elaborate ductwork, they are only employed where compelling arguments exist to do so. Today, this is generally only true in laboratory settings with high requirements, as well as manufacturing and production sites.

6.7.2.3 Low- and high-velocity systems

At the beginning of the 1960s, the US concept of air-conditioning systems for multizone buildings, especially high-rises, was adopted in Europe. American manufacturers introduced high-velocity systems, known as "high-pressure" air-conditioning systems in Europe.

In contrast to low-velocity systems, high-velocity (or pressure) systems are characterized by the elevated air velocities at work in their ducts. In low-velocity systems the mean maximum air velocity is in the region of 6 m/s; in high-velocity systems it is up to 10 m/s and at times even higher. High air velocities produce greater pressure losses in the ductwork, a chain of events that inspired the expression high-pressure systems. However, as detailed calculations showed, pressure recovery in fact means that high-velocity systems do not lead to greater pressure than low-velocity systems and the expression has thus fallen into disfavour in recent years.

6.7.3 Air and water systems

In the previously discussed HVAC systems, the required energy (cooling or heating energy) is supplied to the rooms exclusively through the medium of air. As a consequence, air streams (supply air volumes) arise from the maximum cooling loads and 4- to 6-fold air changes or higher in summer operation. The ductwork must be large enough to transport the corresponding air volume. The resulting demand on space and the wish to achieve greater control inspired efforts as far back as the 1930s in the USA to develop air-conditioning systems that utilize air and water to transport energy into the rooms.

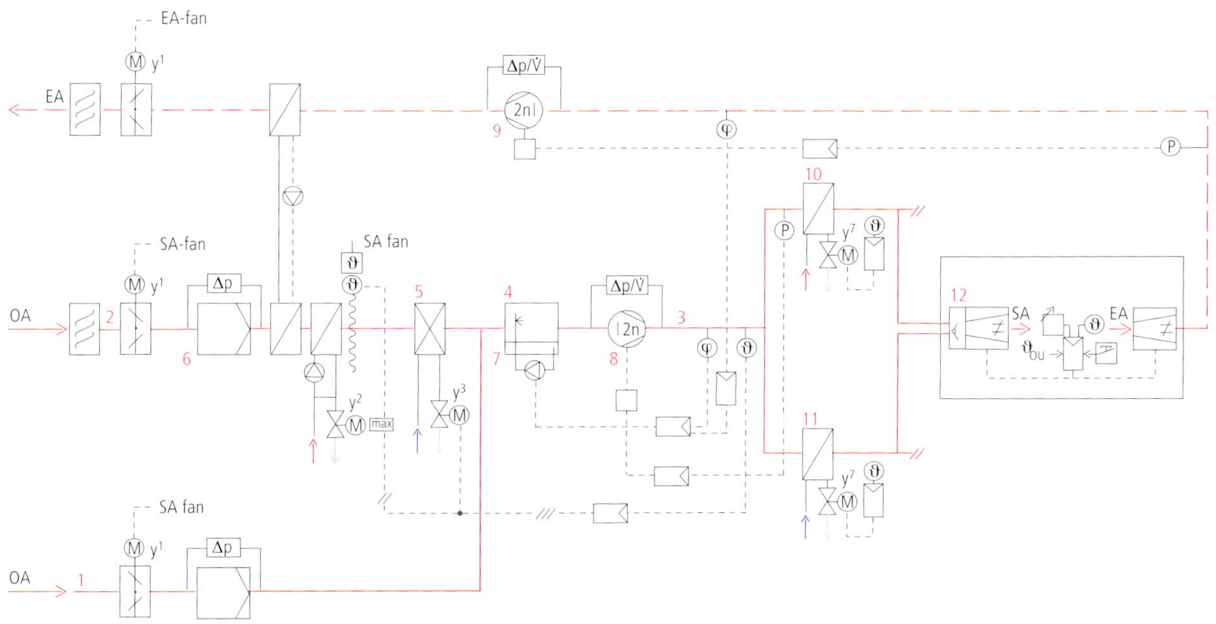

1 Minimum outside air duc
2 Maximum outside air duc
3 Total air duct
4 Minimum air heater
5 Minimum fresh air cooler
6 Filter
7 Washer
8 Supply air fan
9 Exhaust air fan
10 Air heater
11 Air cooler
12 Mixing box

Figure 6.129
Two-duct system with additional minimum outside air duct (CV)

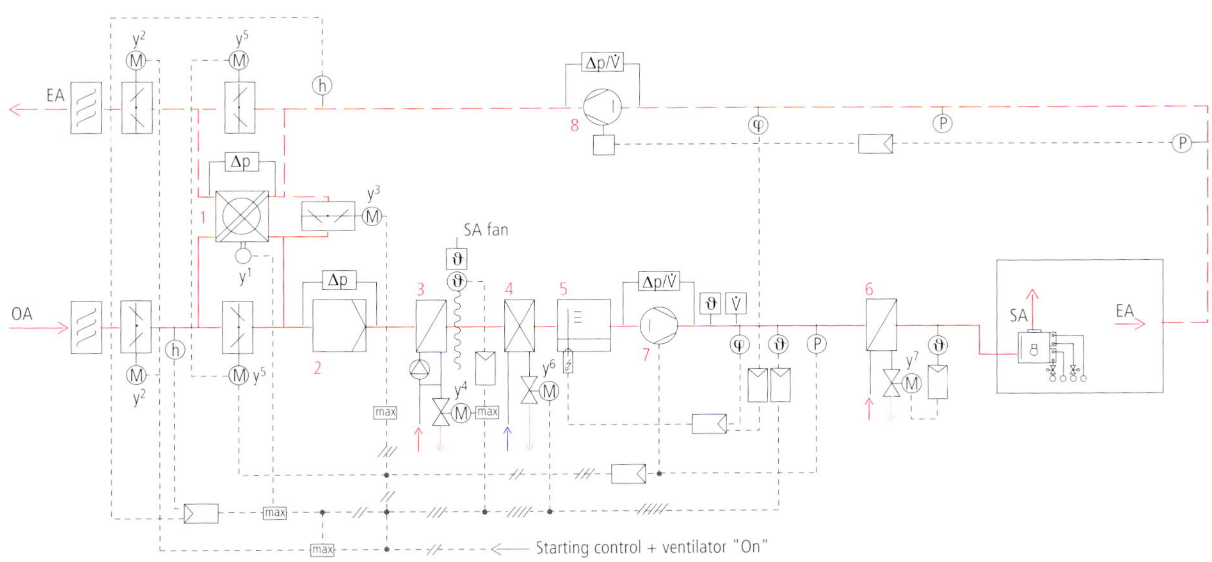

1 Mixing chamber
2 Filter
3 Heater
4 Chiller
5 Washer
6 Secondary heater (optiona
7 Primary (supply air) fan
8 Exhaust air fan

Figure 6.130
Induction air-conditioning system

6.7.3.1 Induction systems

Figure 6.130 illustrates the principle of an induction air-conditioning system. In this system only the volume of air required for a hygienic climate (air change times two) is prepared in a terminal. The supply air (primary air) prepared in the terminal is set to roughly +16 °/80 % rel. humidity in summer and roughly +19 °C/75 % rel. humidity in winter. The primary air is transported through ducts to the rooms and into induction devices, which inject the air through fine jets. The primary air injected by the jets induces room air (secondary air) that is guided across either a cooler or an air heater as it enters the induction device. The result is dry cooling or heating of the secondary air in the room, i.e. the load compensation is achieved at approximately 60 to 70 % by in situ heat exchangers (water). The primary air and the induced secondary air then flows into the room on a tangent through the window sill and permeates the space. The spent air, corresponding to a 2-fold air change, is evacuated and then released outdoors or to utility areas (e.g. garage, storage area or similar) after having been cooled in a heat recovery plant. Induction systems are still in use. In addition to installation below windows, they are also integrated into ceiling areas, supplying air to the room via ceiling jets.

Induction systems are classified into those that supply cooling energy only (2-pipe system/2 water pipes) and those that provide localized heating or cooling (4-pipe system/2 cold water pipes and 2 warm water pipes). Another differentiation characteristic applied to induction systems is whether the controls operate in the water or in the air cycle, that is, whether cooling and heating are driven by a valve (water control) or a slider (air control).

In recent years induction systems have become less popular because they are linked to the disadvantages of tangential flow and because incorrect operation and poor insulation resulted in corrosion in cold water systems (operation with cold water streams that are too cold, followed by condensation and hence corrosion). Nevertheless, these systems continue to offer a viable solution in buildings where minute distinctions between zones and fine-tuned control are required (individual control at each window module) and where space for ductwork is at a premium.

6.7.3.2 Fan-coil systems

Fan-coil systems (fan-heat exchanger systems) operate on a similar principle as multizone single-duct systems with supplementary fans, and are primarily used in hotels. The fan-coil devices consist of a flat box housing a supply air fan, a heat exchanger and sometimes filters. They siphon off the air to be cooled or warmed and blow it back into the room once the processing is completed. Fan-coil systems can either operate purely by circulation or be supplemented from a terminal with a pretreated outside air stream that is drawn off from the room area (in hotel rooms, for example, air siphoned off from bathroom areas).

Similarly to induction systems, fan-coil systems can be mounted below windows or installed above a suspended ceiling through which they supply air to the room via specially designed grilles (typically cooling feature in hotel rooms).

6.7.3.3 Small heat pump systems

When a fan-coil device is equipped with a small refrigerating machine in addition to the fan and the heat exchanger, in principle a small heat pump system has been created. The condenser heat of the fan-coil device is centrally collected via a piping network and transported to the interior areas that require heating energy (heat pump effect). The small heat pump system is equipped with two heat exchangers (condenser/evaporator) capable of alternately heating or cooling the relevant air streams. This system solution did not take hold, however, owing to high maintenance costs.

6.8
Static cooling systems

The term "static cooling" is employed for system solutions where cooling occurs by natural means, i.e. through differences in density or radiation exchange and where cooling energy for load compensation is not necessarily injected into rooms by means of HVAC systems. In Scandinavia a system approach has been pursued for some 20 years which is based on knowledge gained from the operating principles of area heating systems (radiant ceiling panel heating) and which also harnesses the advantage of bringing energy into rooms through the medium of water. The resulting solutions were further developed in Germany and Switzerland and ultimately formed the system group known as static cooling systems.

Static cooling systems are especially advantageous in situations that require individual retrofitting and where the increased demand for cooling energy linked to rising internal cooling loads must be considered. Moreover, natural ventilation ceased to become an option in numerous old buildings due to environmental pollution. These buildings, too, are now faced with the necessity for load compensation in summer.

Therefore static cooling systems fulfil the following functions:
- Reducing the supply air volume to the hygienic minimum
- Diminishing room air velocity and the turbulence quotient (fewer complaints/greater acceptance)
- Reducing the cooling energy consumption
- Harnessing free cooling through water generation in cooling towers
- Reducing shaft and installation area requirements

Static cooling systems include the following:
- Cooling ceilings realized in the form of ceiling systems
- Downdraught cooling in the form of ceiling or stack cooling
- Building component cooling
- Combination of various cooling systems

The principal groups of the cooling systems are presented in **Figure 6.131**. The diagram identifies the working principles, architectural features and manner of producing cooling for each system.

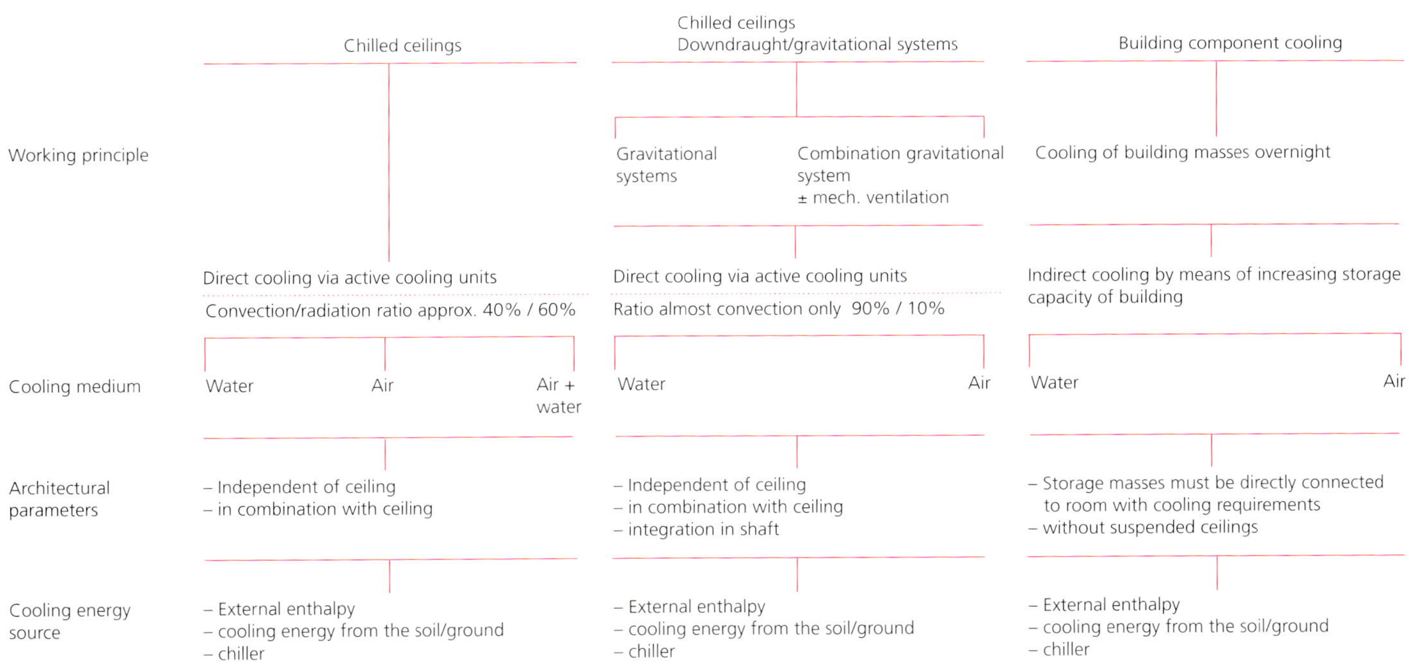

Figure 6.131
Principal groups of "static cooling"

Figures 6.132 – 6.134 illustrate the operating principle of static cooling for varying radiation and convection degrees. It is important to note that cooling ceilings and building component cooling are limited options if:

- Cold water temperatures for cooling purposes lie below the dew point
- Ceiling or wall surface temperatures fall into a range that be uncomfortable for occupants (see section on comfort)
- Design considerations preclude the use of static cooling

With the use of any cooling elements in the room it is vital to avoid these elements dehumidifying the air entering from the outside (natural ventilation), i.e. to prevent vapour formation on the surfaces. The cold water temperatures must be controlled for various systems in a manner that ensures they lie above the relevant dew point, which may lead to a diminished cooling performance. Where cooling ceiling systems are planned for naturally ventilated rooms, one may have to accept that raising the cold water temperatures will not consistently deliver the desired room temperature (cold water temperatures 16 to 21 °C).

Figure 6.132
Cooling ceiling with high radiation component

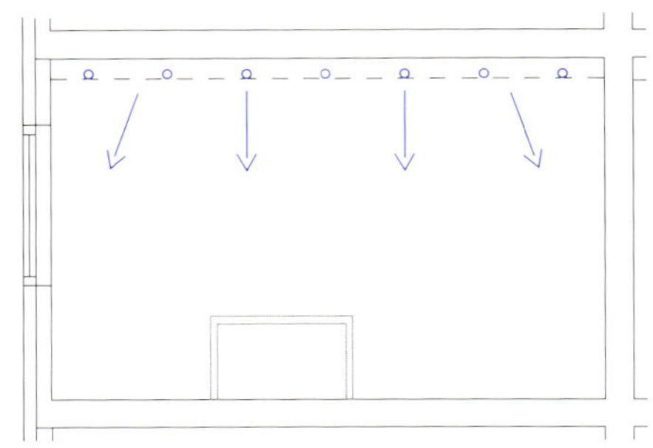

Cooling components in suspended ceiling

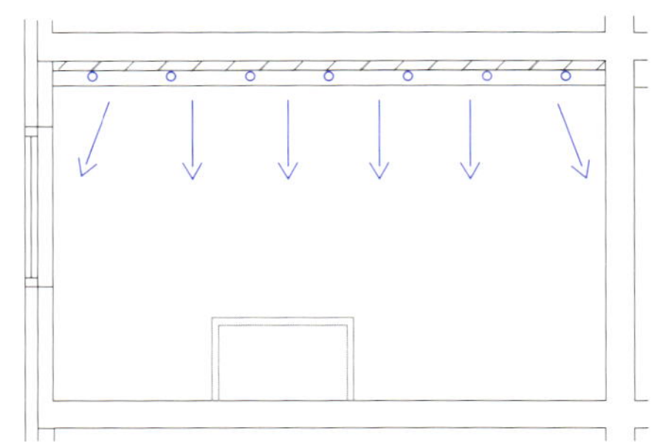

Plaster ceiling with insulation and cooling components

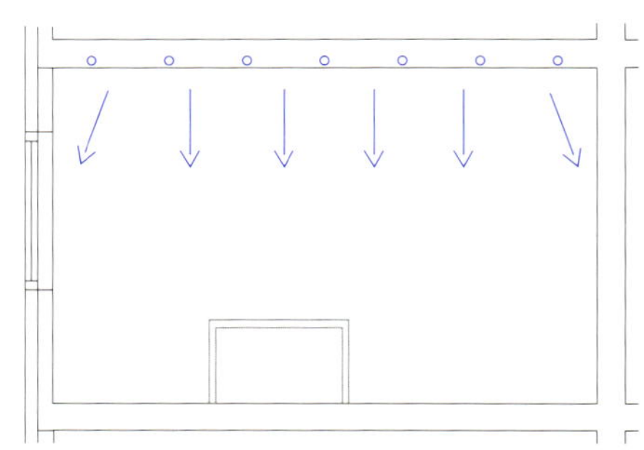

Active building component cooling

Figure 6.133
Cooling ceilings with high convection component

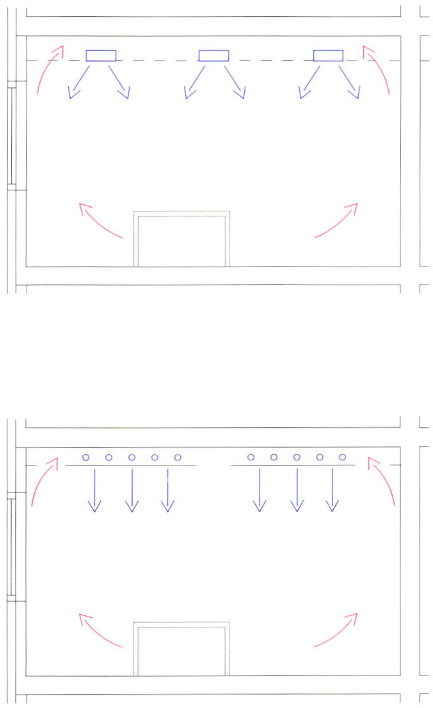

Cooling components in suspended ceiling

Cooling components in suspended ceiling

Figure 6.134
Downdraught/cooling with convection only

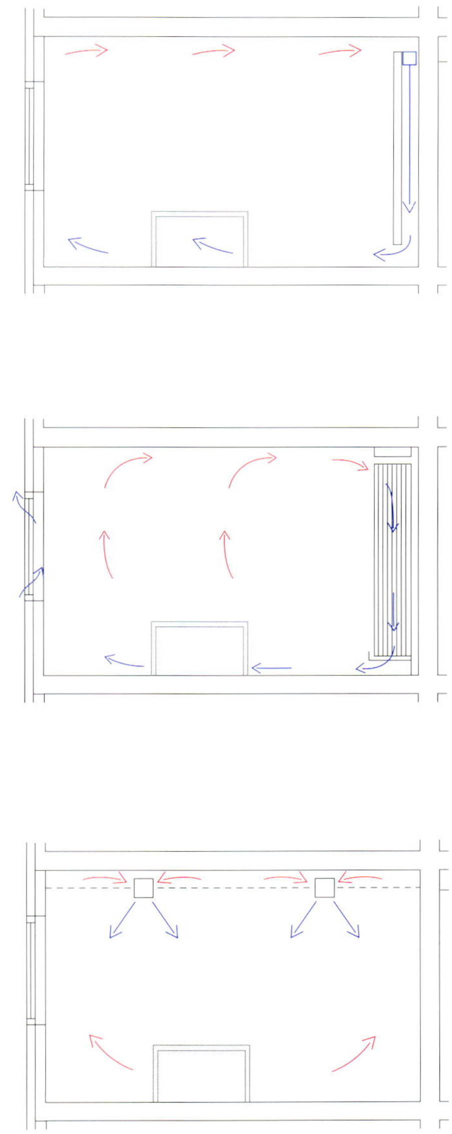

Shaft with cooling component

Water-based cooling body (pipe-/plate radiators)

Cooling beams

6.8.1 Cooling ceiling systems

Cooling ceiling systems are a direct form of cooling by means of active cooling components (ceiling components), whereby 40 % of the thermal loads are eliminated by convection and 60 % by radiation.

The output of a cooling ceiling is the product of the aforementioned parameters for changing cold water temperatures and temperature differences between the cold water and room air (see graph in **Figure 6.135**). As the mean cold water temperature rises, the cooling output decreases; the maximum output is roughly 100 W/m². **Figure 6.136** plots the room temperature in relation to external temperature (maximum absolute humidity 17 g/kg dry air based on 2.5 ac/h by natural ventilation). While the outside temperature rises to 32 °C, the maximum room temperature is capped at 28 °C. **Figure 6.137** depicts the basic structure of a cooling ceiling component and **Figure 6.138.1** a partially assembled cooling ceiling (some panels removed). Other variations on cooling ceiling designs are shown in **Figure 6.138.2**, while **Figure 6.139** presents four executed examples.

Cooling ceilings in combination with HVAC systems achieve maximum efficiency with air movement from bottom to top, which takes advantage of natural convection. Nevertheless, air movement along the ceiling (air supply from top to top) is equally possible and will also yield good results.

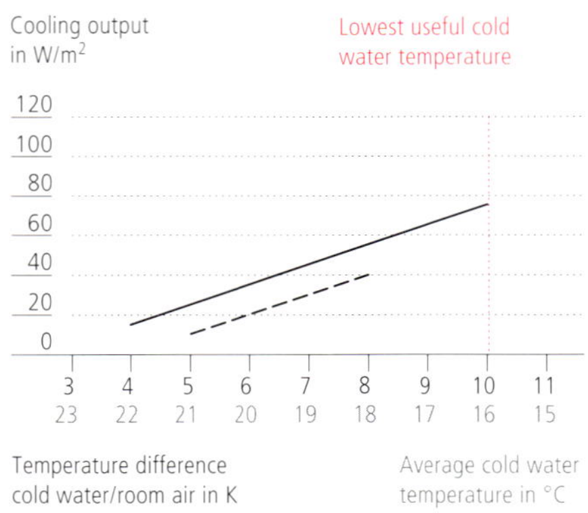

Figure 6.135
Cooling ceiling performance levels for different cold water temperatures

Alpha value approx. 10 W/m² K
Cold water temperature difference 4K
—— Cooling ceiling
- - - Active building component cooling

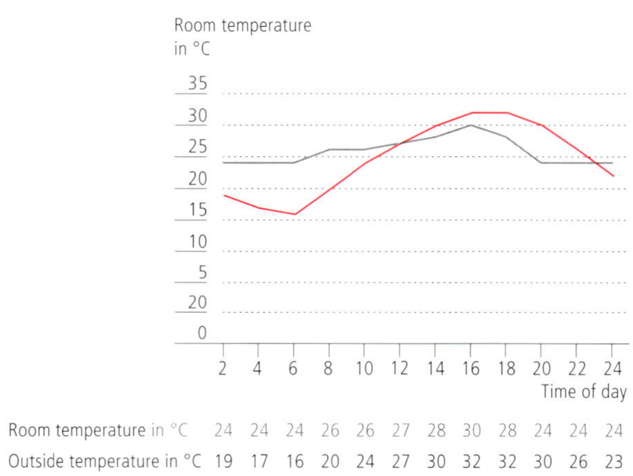

Figure 6.136
Temperature variations (outside air/room air)

Room without air-conditioning system with cooling ceiling, approx. 2.5 ac/h (window ventilation)
—— Room temperature
—— Outside temperature

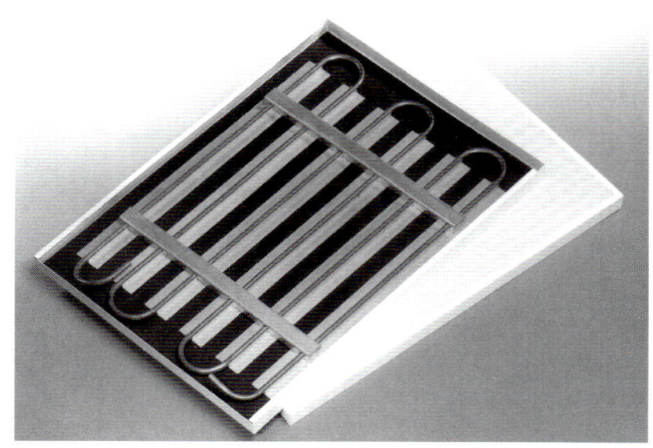

Figure 6.137
Construction of cooling component

Figure 6.138.2
Other cooling ceiling systems

A
Grid cooling ceiling with integrated lighting fixture; the specific cooling output is 164 W/m² (after DIN 4715-1) for a standard grid dimension of 80 mm

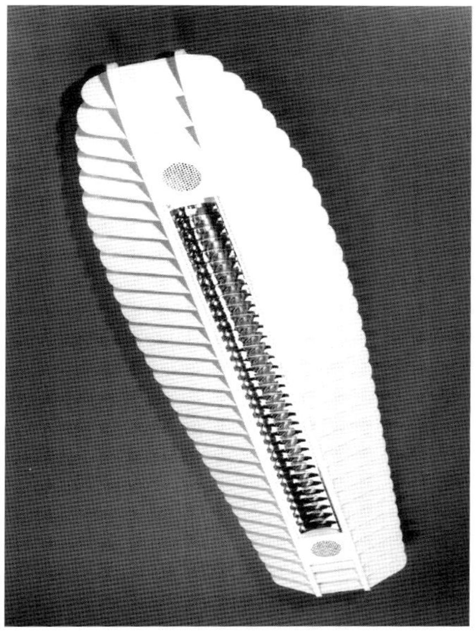

B
Cooling ceiling sail
(Design: Ch. Ingenhoven)

Figure 6.138.1
Cooling ceiling, uncovered

Assembly of a metal cooling ceiling
(Type: SPECTRA)

The panel is held in place with clamps. The thermal contact between aluminium profile and metal panel is ensured by means of the visible magnetic strips
(Image: Zent-Frenger)

A
Teller hall in savings bank

B
Conference room with aluminium cooling ceiling
(Type: ALUP)

C
Ceiling detail in a conference room with cooling ceiling elements above interstitial ceiling

Figure 6.139
Examples of cooling ceilings in buildings

D
Lufthansa, Frankfurt

6.8.2 Downdraught / gravitational cooling

The drawing in **Figure 6.140** represents a section through an office with downdraught cooling in the wall area. In this system, a chiller is installed for each module above a cabinet or wall component. The chiller cools the warm air in the ceiling area sufficiently so that it naturally drops along the wall down to the floor and then flows into the room again from the bottom. A 10-cm-deep installation space (shaft) is constructed across the entire module width in front of the actual corridor walls to guide the movement of the cooling air streams down to the floor. Great care should be taken when using this system with regard to how the cooling air is supplied to the room, since excessive temperature differences between the occupants' head and leg area should be avoided.

Downdraught cooling can also be implemented in the form of supply air injection through a hollow floor and a stack ventilation outlet (**Figure 6.141.1**).

Figure 6.141.2 and the series in **Figures 142.1 – 3** present another option for downdraught or gravity cooling. For this system a cooling component with two convector-like chillers is installed in the ceiling area. These chillers either cool the warm air streams circulating in the ceiling area by means of gravity and vent them through the ceiling, or siphon them through the chillers by means of a primary air system with minimal air volume similar to an induction unit and then return the cooled air to the room. In this solution the cold air drops gradually into the room by virtue of the differences in density between room air and cold air, falling all the way to the floor area.

Finally, **Figure 6.143** presents a cooling system with optional dehumidification by means of an upright tubular radiator (System PS-Comp., Tokyo) used in hot and humid climate zones. The condensation tray installed below the tubular radiator collects any condensate that arises and drains it via a condensate pipe.

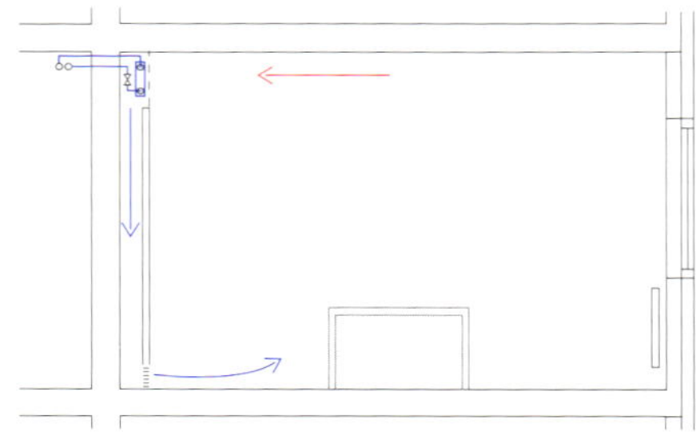

Figure 6.140
Gravitational cooling
Schematic ("static cooling" system, F.H. Schmidt, Berg. Gladbach, Germany)

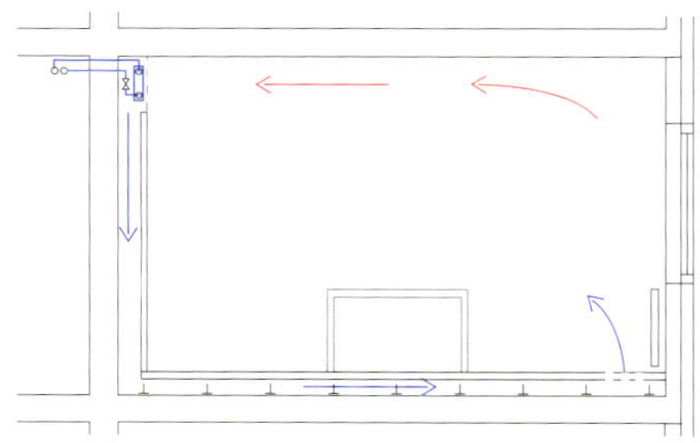

Figure 6.141.1
Gravitational cooling, air supply through hollow floor with stack ventilation
(Image: Gravivent)

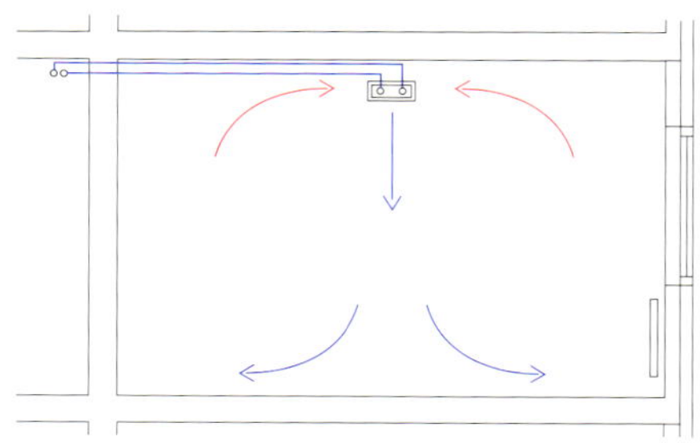

Figure 6.141.2
Gravitational cooling via cooling elements in ceiling area

Figure 6.143
Custom construction, cooling via pipe radiators
Close-up of condensation on surface (left)

Figure 6.142.1
View into cooling ceiling system in a television studio (ceiling cavity)

Figure 6.142.2
Example of gravitational cooling

Downdraught cooling elements during assembly

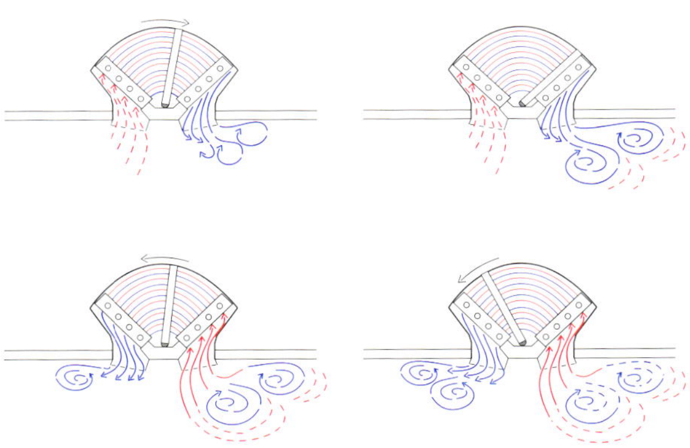

Figure 6.142.3
Example of cool wave® ceiling unit and ceiling unit with integrated light fixture (Images: LTG)

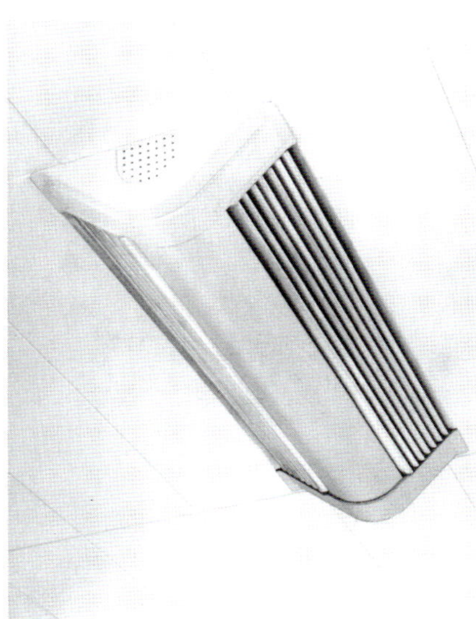

6.8.3 Building component cooling

Building component cooling can operate with water or air to cool thermal storage masses to such a degree that they achieve a large part of the required cooling load compensation while stimulating the radiation exchange between heat-emitting machines, occupants, and ceiling or wall components. System solutions are differentiated into piping systems of varying sizes with specific cooling output, which are either laid in concrete or mounted on top of concrete surfaces. These are shown in **Figure 6.144**. The concept resembles that introduced for floor heating; here plastic capillary tubes (1.2 mm diameter) are filled with cold water to cool ceiling components and thus compensate thermal energy. **Figure 6.145** shows a ceiling installation with capillary tubes prior to plastering. As is the case with cooling ceilings, condensation must be avoided and surface temperatures must remain within the established comfort zone.

Years ago a concept for building component cooling with air was realized in a building project for the European Investment Bank (**Figure 6.146.1**). Cold outside air is conducted through hollow concrete ceilings at night (zone regulated) cooling the concrete masses to 18 °C. The cooling action occurs only when the outside temperatures lie below internal room temperatures (between 7 p.m. and 6 a.m.). The diagrams in **Figure 6.146.2** illustrate the duct layout, the hollow concrete ceiling structure and the operating principle of the system.

The cooling ceiling shown in **Figure 6.144** was further developed in Switzerland by the firm Geilinger, resulting in a system called the "Batiso Ceiling." This solution operates with water-filled pipelines laid directly into the concrete and effecting a gradual or passive cooling of the building component cooling. While of great interest in principle, the system goes hand in hand with a certain level of compromise in terms of controllability (slow-release/formation of control zones). Nevertheless it is gaining more ground especially in combination with stack or natural ventilation, and in response to many clients' demand for a focus on supplementary, gentler cooling systems. All building component cooling systems share the advantage that the cooler surrounding surfaces result in a perceived temperature that is generally described as comfortable.

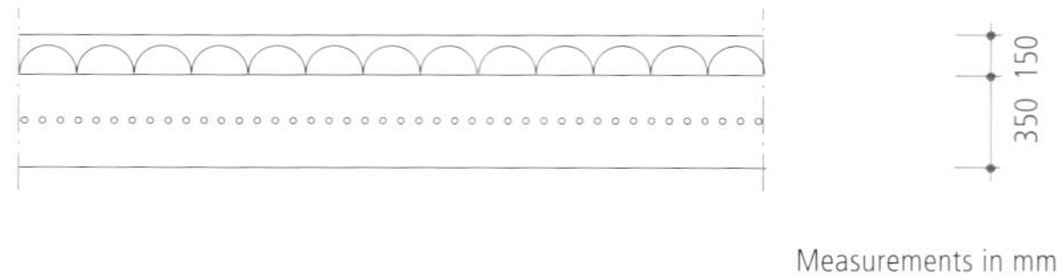

Building component cooling/ heating system in concrete ceilings, floor installation (cooling output approx. 40 W/m²)

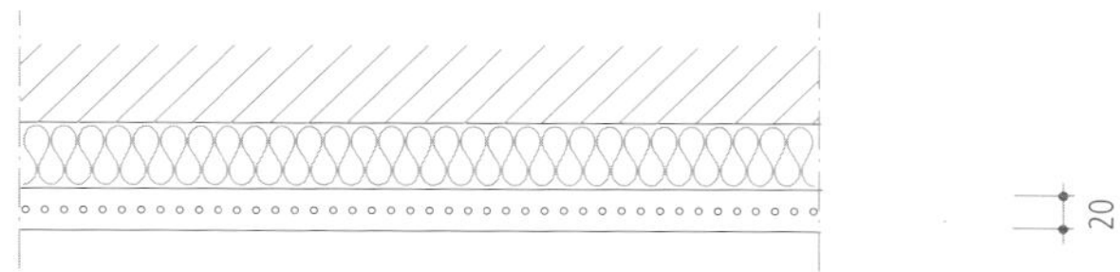

Capillary pipe system (KARO) for heating or cooling, ceiling installation (cooling output up to 70 W/m²)

Figure 6.144
Building component cooling systems

Figure 6.145
Example of ceiling installation

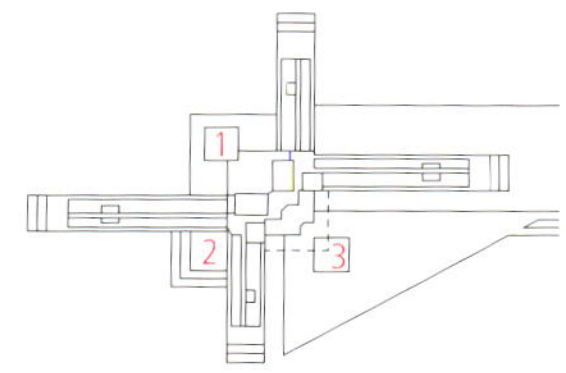

1 Concert halls
2 Restaurants
3 Entrance

Figure 6.146.1
Office building in Luxembourg, high thermal storage construction (Architect: Sir D. Lasdun + Partners, London)

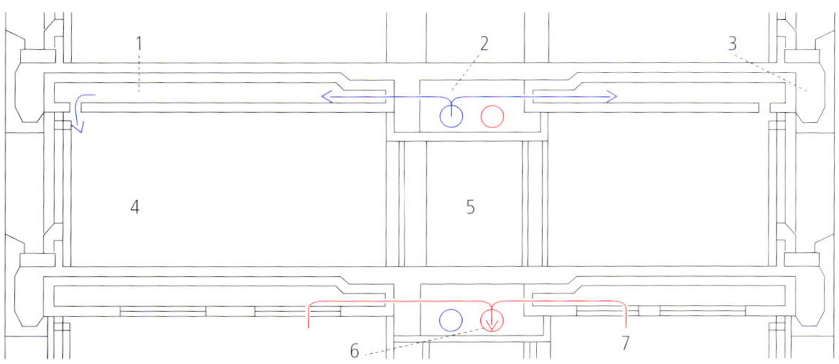

1 Ceiling cavity (prefabricated component) as air ducts
2 Supply air duct
3 In-situ concrete
4 Office
5 Hallway
6 Exhaust air duct
7 Exhaust air through lamp fixture

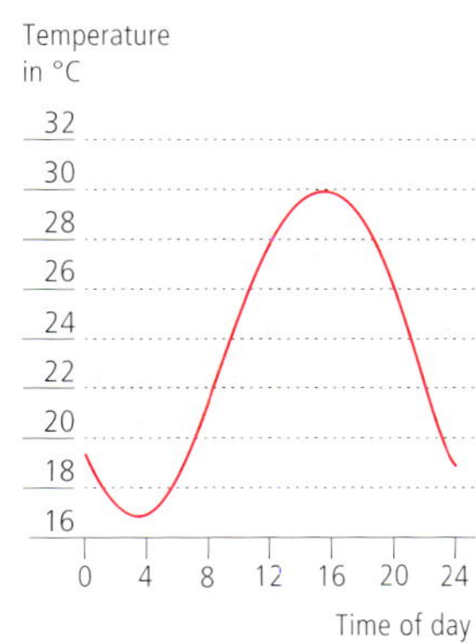

— Outside temperature t_O, sunny day
Range of room temperature R, sunny day

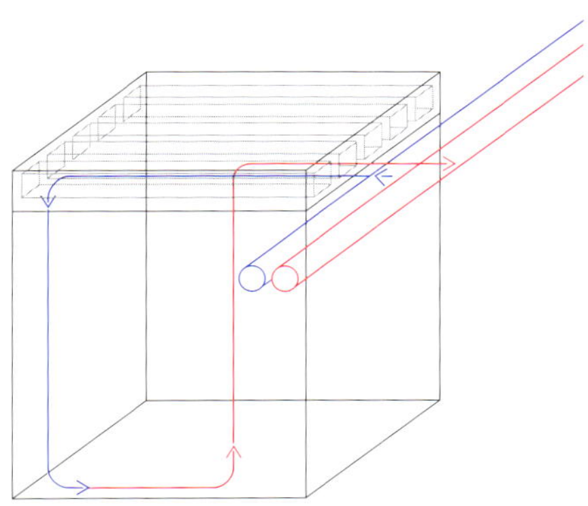

Figure 6.146.2
Diagram of functions and processes during mechanical ventilation to "discharge" the thermal storage ceilings

6.9 Equipment rooms/shafts/ horizontal distributions

6.9.1 Round and rectangular ducts

Round or rectangular ventilation ducts connect supply air terminals (supply air devices) with the areas targeted for supply (inlets).

The duct material should be smooth on the interior, not collect any dust, easy to clean, durable, hygroscopic and non-combustible, corrosion-proof, light and airtight. Depending on the area of application and the air volumes to be transported (especially spent air streams), air ducts consist of:
- Black or galvanized steel plate
- Aluminium plate
- Fireproof plates
- Brickwork and concrete
- Clay pipes
- Plastics or gypsum plates

Galvanized steel plate ducts are used for all general applications; all other ducts are variations for special applications, for example:
- Fireproof ducts
- Moisture-proof ducts
- Acid-proof ducts

In round ducts, air is transported through standardized dimensions with sheets of differing thickness according to the nominal diameter.

The same is true for rectangular ducts, which are also designed with minimum sheet thickness depending on the maximum internal measurement.

In addition to straight ducts, a number of form pieces such as right-angle turns, T-pieces, crosspieces, forks and throat pieces etc. are necessary to execute changes in direction or cross-section.

The diagram in **Figure 6.147** indicating the absolute roughness ε (mm) for different materials is used for the sizing of pipes and ducts:

PVC pipes	0.01
Lock-seamed sheet ducts	0.15
Promabest ducts	0.15
Concrete ducts, smooth	0.5
Rabitz plaster, calendered	1.5 – 2.0
Concrete ducts, rough	1.0 – 3.0
Brickwork ducts, rough	3.0 – 5.0
Flexible pipes	0.2 – 3.0

The diagram also helps to determine the pressure drops per lfm of duct or pipe.

Example:
- Pressure drop in a flexible hose with a d = 100 mm, velocity w = 15 m/s
- Absolute roughness of flexible hose = ε = 0.8 mm
- The pressure drop is 50 Pa/m referenced to the absolute roughness

The pipe friction diagram is primarily used in the preplanning stage to establish rough estimates for pipe and duct sizing. The sizing is based on a design volume flow and assumed velocity. In addition to pressure losses through pipe or duct friction, there are losses caused by individual resistance factors. The individual resistance factors are classified according to resistance coefficients and thus included in the final calculation. Round ducts are manufactured only with set diameters.

Rectangular ducts are generally manufactured for specific applications and are thus available in all possible sizes. The only consideration is that the maximum height to width ratio of the duct should be 1:6 to ensure stability. Round and rectangular ducts are connected with flanges or sliding battens, and the flange connections are used for larger sizes and high-pressure applications.

Aside from rigid ducts or pipes, flexible piping or hose is also used to facilitate installation. Flexible piping and hose is generally available up to a diameter of 400 mm and can also be manufactured with double walls for insulation.

The cross-section of round or rectangular ducts is easy to determine when the volume of air that will flow through them is known. The air volume in m^3/s divided by a maximum velocity in m/s yields the duct cross-section in m^2.

Air-transporting ducts tend to be many times larger than other system installations such as piping, regardless for which application, cable lines and track systems. We can generally assume a multiplier of 10 – 15 by comparison to other supply systems. The area requirements for ducts are particularly problematic in situations where the air change demands for rooms lie between 50 – 100 ac/h (computer labs, pharmaceutical operations, clean-room conditions). There is little choice in these environments but to move the HVAC installations right next to the areas to be supplied into order to prevent enormous construction volumes for extensive duct installations.

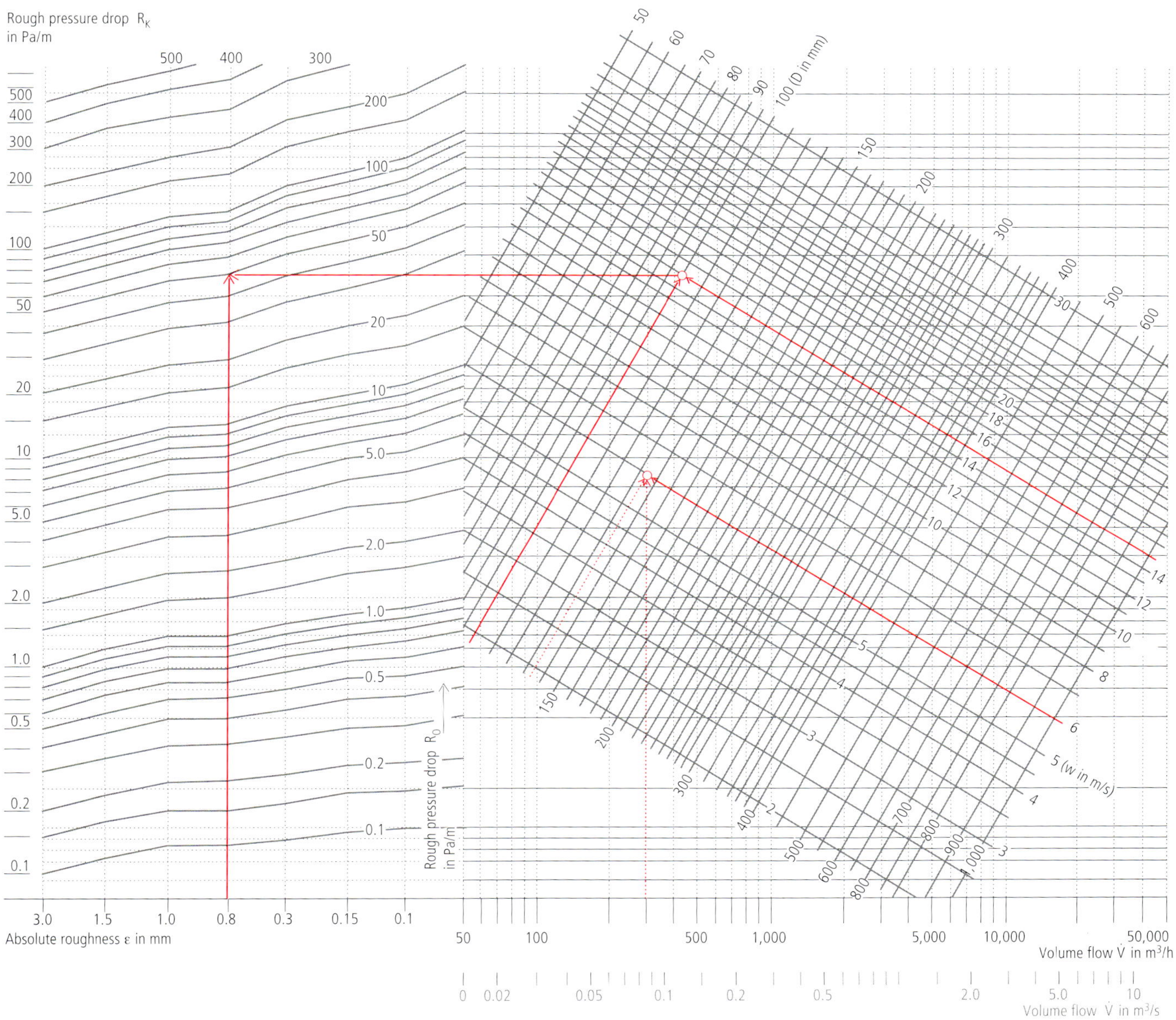

Figure 6.147
Pipe friction diagram for air ducts

6.9.2 Shafts and channels

Vertical distribution trees are installed in shafts (also called chases) and horizontal distributions in channels.

Preferably, shafts and channels should provide the shortest and most direct link between the terminal and the area to be served; central access results in the shortest channels.

In buildings with HVAC systems, the duct dimensions of rectangular or round ducts define the dimension of the total installation and therefore have considerable influence on the area required for shafts.

Small shafts should be operable from the outside without needing to provide man-size access. Large supply ducts, on the other hand, must be accessible for maintenance since they often contain shut-off valves, control devices etc. **Figure 6.148** shows examples of a small and a large supply shaft.

The area requirement for non-accessible and accessible shafts and channels for air-transporting ducts is indicated in **Figures 6.149.1/2**. These diagrams identify the ducts for high- or low-velocity air movement; in practical applications, calculations are generally based on the values of low-velocity systems (single-duct systems). Two-duct systems serving as high-velocity systems have greater area requirements due to a more elaborate duct layout. The anticipated air volumes must be known in order to work with the relevant diagrams. If these are not known, one can work with the diagram shown in **Figure 6.150**: it is based on the thermal transmittance of the pane combination (b-factor) and is principally used for office buildings.

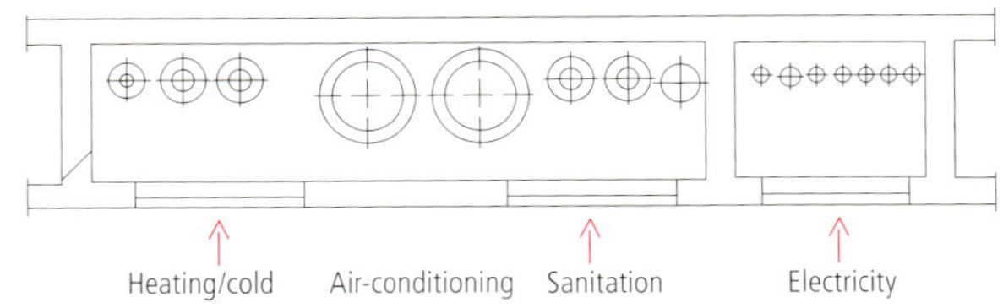

Figure 6.148
Shaft installations

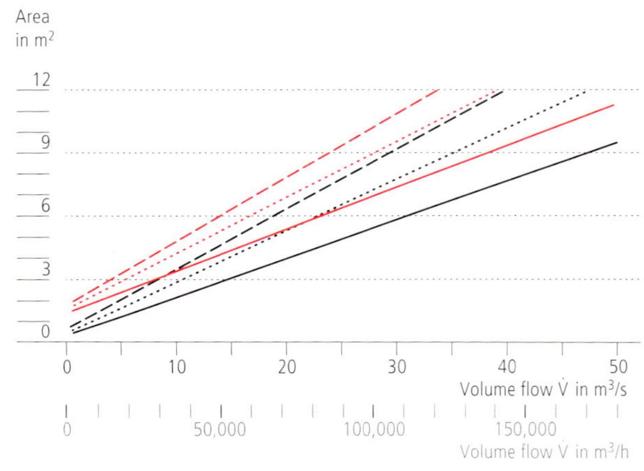

Figure 6.149.1
Area requirement
– Non man-sized (black) and man-sized (red) shafts as well as channels for air-conducting pipes

Two-duct system
Low-velocity system

High-velocity system

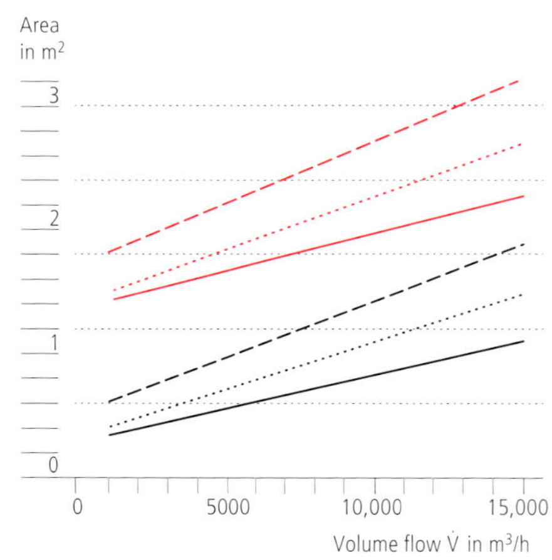

Figure 6.149.2
Area requirement
– Non man-sized (black) and man-sized (red) shafts as well as channels for air-conducting pipes

Two-duct system
Low-velocity system

High-velocity system

Since the space requirement for accessible shafts can be roughly determined with the diagram in **Figure 6.150**, the space requirement in ceiling cavities (plenum) is identified on the premise of the same parameters in **Figure 6.151**. This diagram can also be used for sizing hollow floors targeted for accommodating the installations. **Figure 6.152** provides additional information on duct insulation and anticipated temperature drops.

If features such as fire-protection valves or secondary air-handling devices etc. are installed at the connection points (shaft-ceiling-cavity), they must be installed with maintenance openings or removable floor panels of at least 60 x 60 cm for operation and servicing (in hospitals also for disinfection).

Figure 6.150
Space requirement for man-sized shafts

High-voltage channel (incl. distribution rooms) (1)
Water-bearing pipes (heating/sanitation/cooling) (2)
Air-water system (total shaft area) (3)
All-air (single-duct) system (total shaft area) (4)
All-air (two-duct) system (total shaft area) (5)

(3) + (1) + (2)
(4) + (1) + (2)
(5) + (1) + (2)

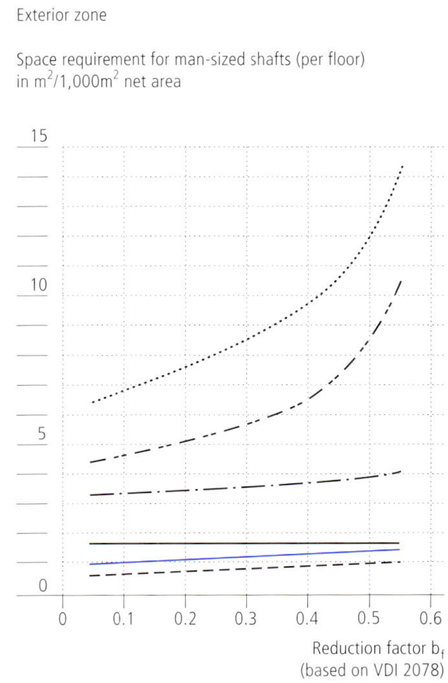

Figure 6.151
Space requirement for ceiling cavities (plenum)/horizontal installation spaces

2-duct system All-air system (with recessed luminaire)
2-duct system All-air system (with mounted luminaire)
Air-water system (with recessed luminaire)
Air-water system (with mounted luminaire)
Heating piping only (with recessed luminaire)

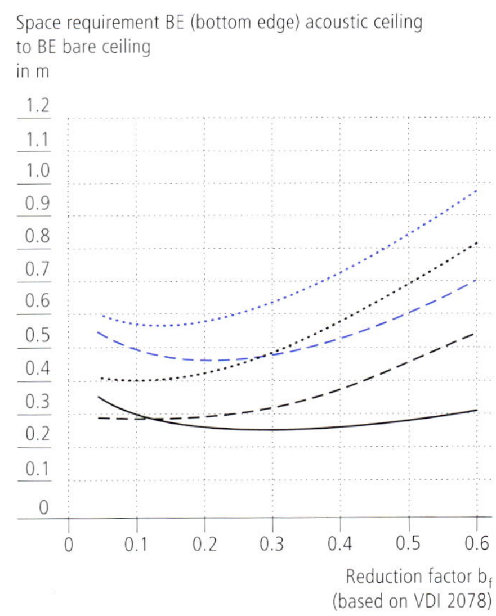

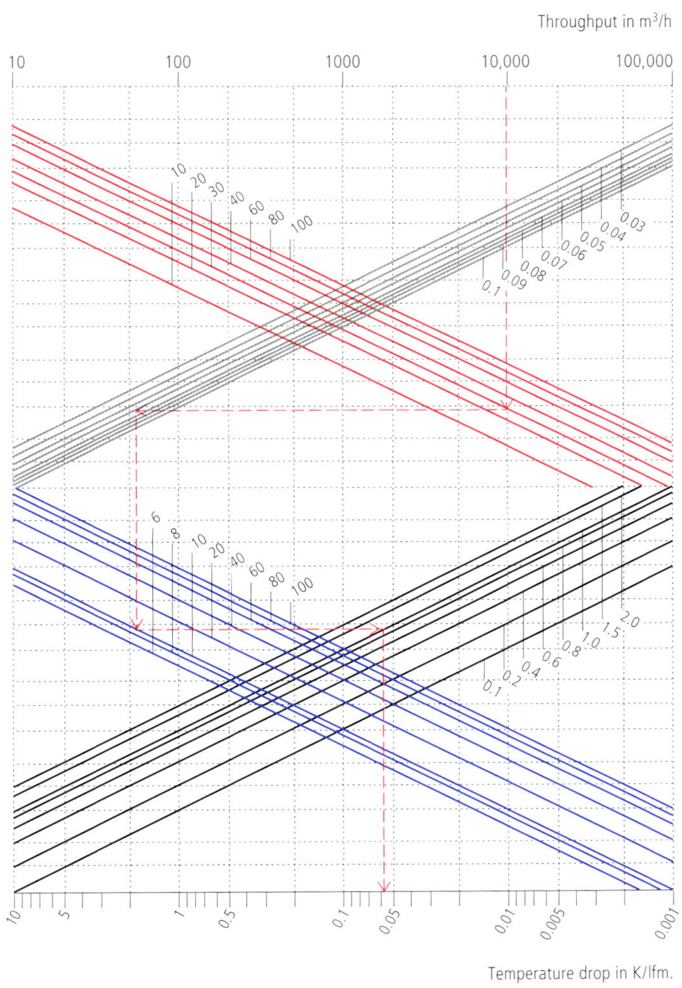

Figure 6.152
Temperature drop in air ducts for insulation with varying thermal conduction coefficients

6.9.3 Equipment rooms

HVAC equipment rooms

The space requirement for HVAC appliances is determined by their number and size. The size results from the volume flow and the number of thermal processing levels as well as sound insulation. The basic plan and height-to-width ratio of the terminal should yield a maximum area use for the placement, operation, maintenance and installation process. Other considerations are the air movement plan, the layout for heating and cooling pipes and cable lines. The graph in **Figure 6.153** illustrates the area requirement and necessary height of central terminals for large system applications. **Figure 6.154** provides the same information for the installation of smaller systems.

Figure 6.155 is useful as a planning tool for sizing equipment rooms for office buildings or buildings slated for similar uses. The graph provided here indicates the area requirements not only for HVAC systems, but also for all other energy supply systems including sanitation (without sprinkler system).

While the air volumes to be processed are the basic parameters for **Figures 6.153** and **6.154**, the sunshade system is the reference point for **Figure 6.155**.

Table 6.14 gives additional information relevant to sizing equipment rooms for HVAC systems.

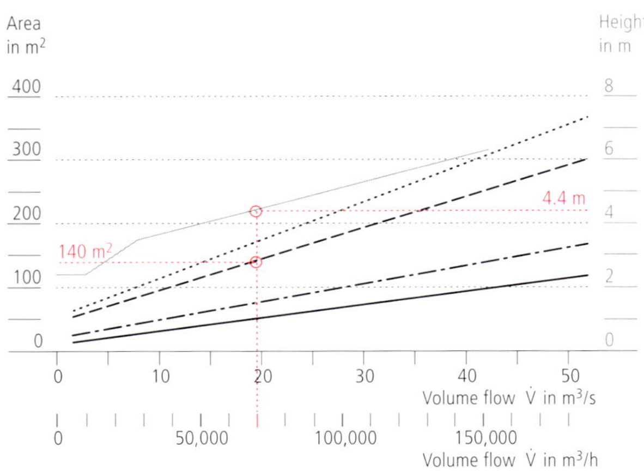

Figure 6.153
Equipment room: area and height parameters for installing individual devices for air flow up to 50 m³/s (180,000 m³/h)

- Room height
- Extract ventilation
- Ventilation (supply and extraction)
- without humidification
- Air-conditioning

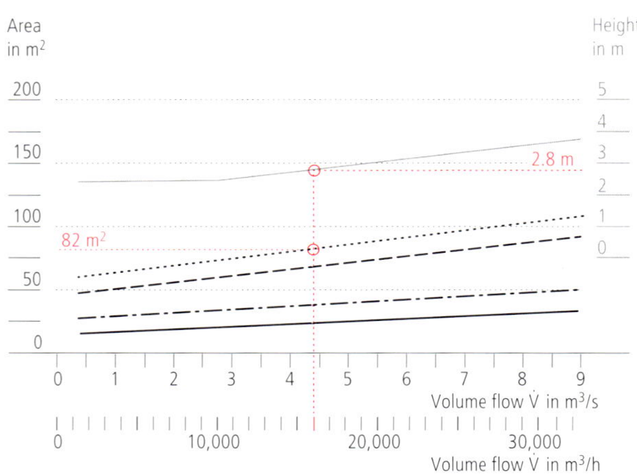

Figure 6.154
Equipment room: area and height parameters for installing individual devices for air flow up to 9 m³/s (30,000 m³/h)

- Room height
- Extract ventilation
- Ventilation (supply and extraction)
- without humidification
- Air-conditioning

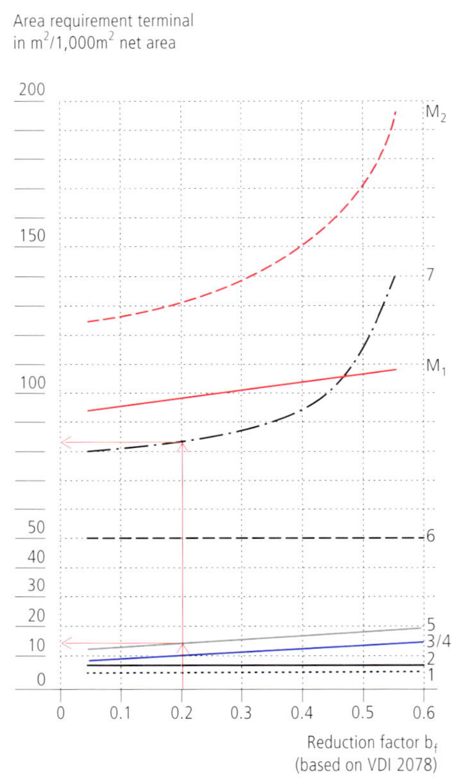

Figure 6.155
Area requirement for equipment rooms

- Sanitation or sprinkler terminal (1)
- Heating terminal (2)
- Cooling terminal (3) and/or recooling plant (4)
- Central power station (5)
- Air-conditioning terminal for air-water system (6)
- Air-conditioning terminal for all-air system (7)
- = (1) + (2) + (3) + (4) + (5) + (6)
- = (1) + (2) + (3) + (4) + (5) + (7)

Space calculation for	**Air output (m³/h)**											
	up to 5,000				5,000 – 10,000				10,000 – 15,000			
	1S	2S	2SA	Minimum height (m)	1S	2S	2SA	Minimum height (m)	1S	2S	2SA	Minimum height (m)
	Area (m²)				Area (m²)				Area (m²)			
Container design:												
Exhaust air	7	12	11	2.50	8	13	16	3.00	8	17	20	3.00
Supply air w/o heating	8	13	16		9	16	17		12	21	23	
Supply air w heating and mixing chamber	9	14	17		12	17	20		18	72	30	
As above, upright	7	12	14		8	13	16		12	21	23	
Combination design for supply and exhaust air:												
Horizontal design	10	17	18	2.50	13	18	21	3.00	17	23	27	3.00
Stacked	9	14	17		12	17	20		14	23	26	
Upright	8	14	16		9	16	17		12	21	23	
Combination design for supply and exhaust air: With cooling, horizontal design (w/o space calculation for chiller)	12	17	20	2.50	16	21	23	3.00	20	29	31	3.00
Air-conditioning systems in container or cabinet form:												
Horizontal	17	23	25	2.50	22	29	30	3.00	30	39	42	3.00
Upright	15	21	22		18	23	26		21	30	33	

1S = 1-step filter
2S = 2-step filter
2SA = 2-step filter and activated carbon filter

Space calculation for	**Area (m²) for air output (m³/h)**						**Room length to width ratio**
	10,000 – 15,000	15,000 – 25,000	25,000 – 35,000	35,000 – 50,000	50,000 – 75,000	75,000 – 100,000	
Humidifying and dehumidifying systems with recirculation fresh air chamber:							
simple filter	26	39	52	72	85	104	1.5 : 1
as above, with add. particulate filter	33	46	58	78	91	110	2.0 : 1
as above, with add. activated carbon filter	39	52	65	85	104	124	2.0 : 1
Humidifying and dehumidifying systems with cooling (w/o space calcul. for chillers:)							
simple filter	33	46	58	78	91	110	1.5 : 1
as above, with add. particulate filter	39	52	65	85	98	117	2.0 : 1
as above, with add. activated carbon filter	46	58	72	91	110	130	2.0 : 1
Air-conditioning systems:							
simple filter	39	52	65	85	98	117	2.6 : 1
as above, with add. particulate filter	46	58	72	91	110	130	2.8 : 1
as above, with add. activated carbon filter	52	65	85	104	124	143	3.0 : 1
Clear room height	3.0 m	3.2 m	3.5 m	3.5 m	3.5 m	4.0 m	

The indicated area requirements also apply to two-duct and high-velocity induction systems, incl. air and filter chambers

Table 6.14
Equipment rooms, HVAC systems (horizontal or upright units)

Figure 6.156
Arrangement of supply and exhaust air equipment rooms

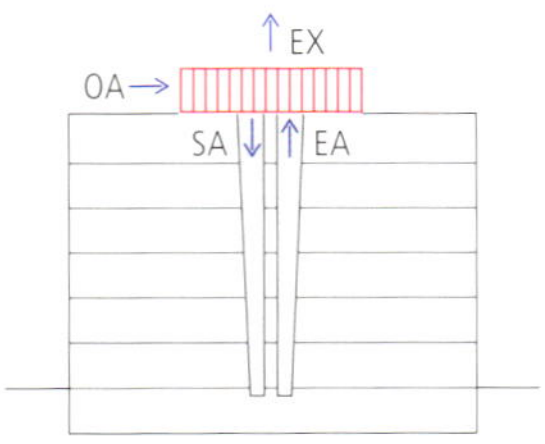

1
Supply and exhaust air terminal on roof
Good solution for outside air intake, escape air venting, energy recovery. Initial installation and maintenance access via vertical paths

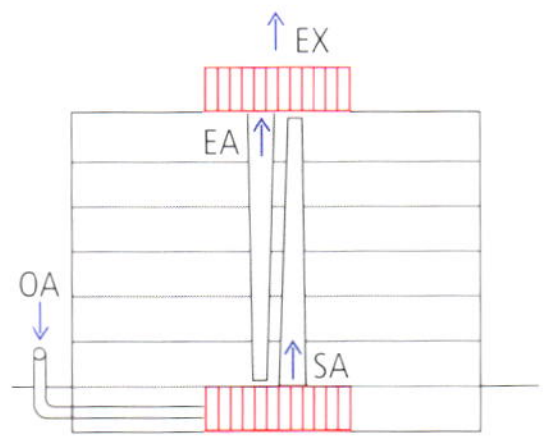

2
Supply terminal in basement, exhaust air terminal on roof
Long paths for outside air intake, good escape air venting, elaborate energy recovery, low space requirements for vertical supply and exhaust air ducts

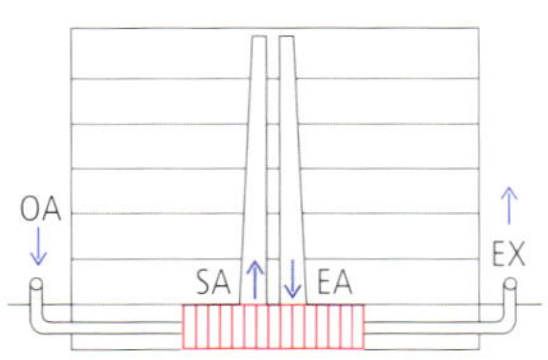

3
Supply and exhaust air terminal in basement
Long paths for outside air intake and escape air venting, good energy recovery options

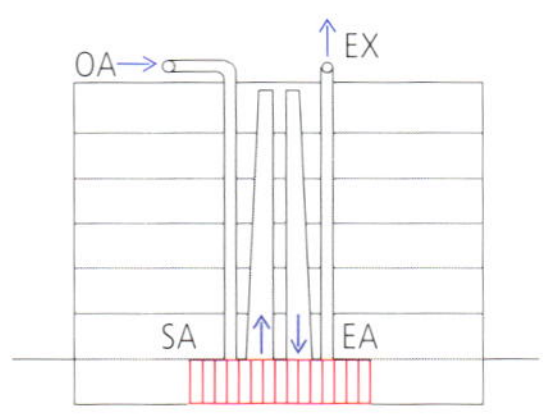

4
Supply air and exhaust air terminal in basement, outside air intake and escape air venting at roof level
Long paths for outside air intake, good energy recovery options

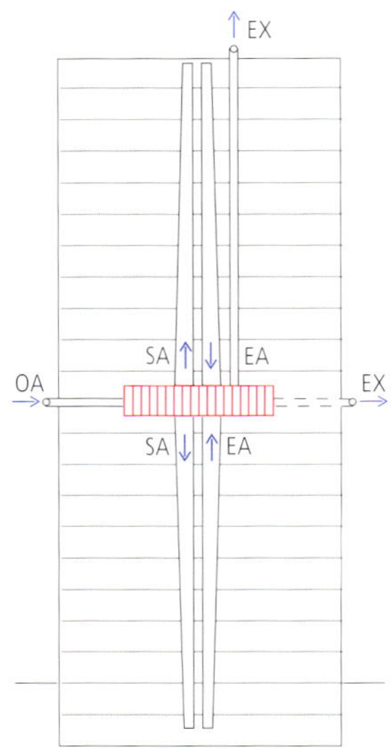

5
Supply and exhaust air terminal in mezzanine
Good location for outside air intake, longer paths for escape air venting via roof, good energy recovery options, good solution for high-rises. Lateral escape air venting possible depending on building use

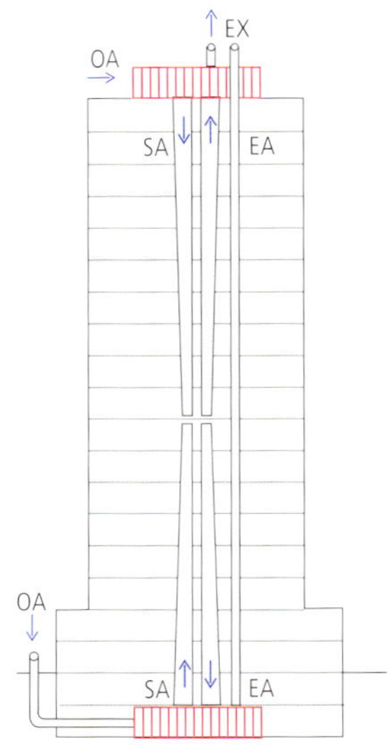

6
Supply and exhaust air terminals in basement and on roof
Frequent arrangement in high-rises with extensive flat areas.
Otherwise as described for Examples 1 and 5

OA Outside air
EX Exhaust air
SA Supply air
EA Escape air
HVAC equipment rooms

HVAC equipment rooms should be as close as possible to the areas they serve and can be placed in a decentralized location to meet this requirement. Thus circulating cooling appliances for computer labs can be installed right next to the lab or even inside it in order to minimize duct routes (shaft areas/floor height to horizontal distribution) and to simultaneously satisfy the specific safety demands. **Figure 6.156** illustrates the impact of various terminal locations on HVAC systems.

This illustration explains the advantages and disadvantages of the different locations. **Figures 6.157/6.159** are designed to convey an impression of possible system dimensions, e.g. air-conditioning systems with necessary control panels or substations for heating and cooling.

Figure 6.157.1/2
Examples of HVAC equipment rooms

Subterminals
(for heating and cooling)
The area requirements for subterminals for heating systems are listed in **Table 6.15**.

The principal parameter for the values in this table is the thermal output of the systems to be supplied; the table provides the required height and area for each system. The area requirement for cooling systems as subterminals is comparable.

Figure 6.158 illustrates the placement of different HVAC terminals on the roof or in the basement including the necessary heating and cooling terminal (basement) and the recovery cooling plant on the roof. Finally, the floor load capacity of system terminals should generally comply with the following guidelines:

HVAC terminals	approx.	5,000 N/m²
Air-conditioning terminals	approx.	10,000 N/m²
Heating terminals	approx.	10,000 N/m²

Pumps and fans with rotating components must be installed in a manner to avoid oscillation and vibration to prevent noise transmission. System terminals should always be placed on floors with an incline towards floor drains, since it is impossible to completely prevent water leakage. Moreover, the floor should be resistant to mechanical loads and slip-proof. Surfaces of ceilings and walls should be as smooth as possible and receive a top coat that is bright and washable. Doors must open in the direction of the escape route; their size depends on the largest component to be transported into the room. Doors should be equipped with baseboards and thresholds and close as tightly as possible.

All wall penetrations for shafts and lines must be executed to fire protection and sound protection codes. System terminals may constitute fire compartments on their own or in combination with the principal shafts and should be treated accordingly. While HVAC terminals do not generally require mechanical ventilation, air-conditioning and heating terminals or heat transfer stations must be fitted with separate ventilation to exhaust possible coolant gases or excessive heat volumes. The temperature in HVAC terminals should not fall below 5 °C or rise above 38 °C to prevent malfunctioning in drives and control systems.

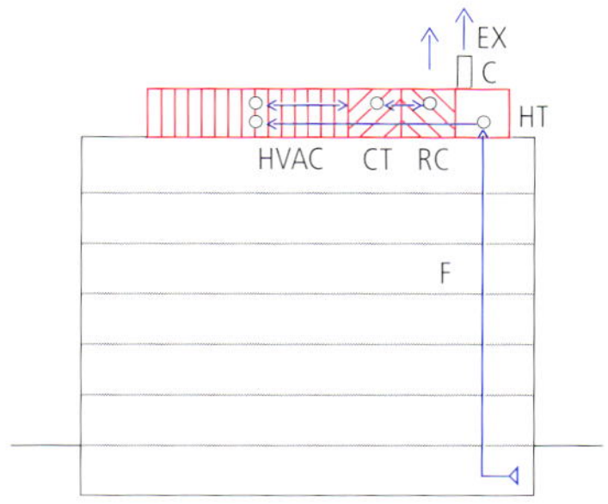

Roof installation of all terminals (HVAC, HT, CT)

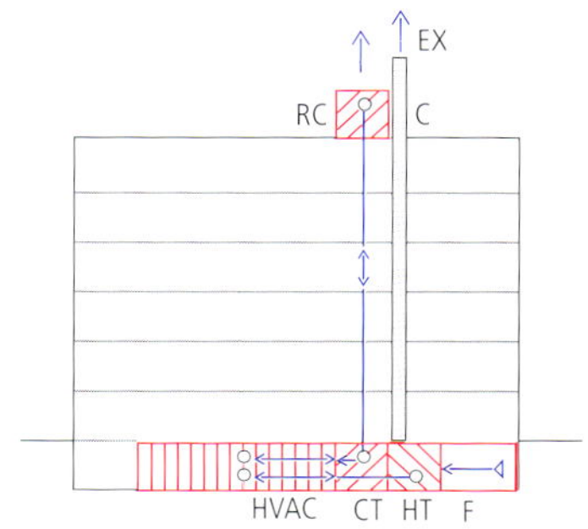

Basement installation of all terminals

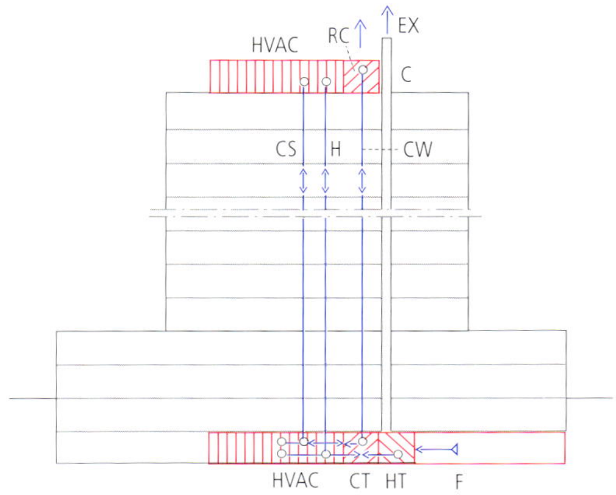

All HVAC terminals on roof and in basement, heating and cooling terminal only in basement and recooling plant on roof

EX	Exhaust air
H	Heating supply
CS	Cooling supply
C	Chimney
F	Fuel supply
HT	Heating terminal
CT	Cooling terminal
HVAC	Heating ventilation and cooling terminal
RC	Recooling
CW	Cooling water

Figure 6.158
Arrangement of heating and cooling terminals

Thermal output MW	**Substation** Area m²	Height in m
to 0.12	to 5	2.4
0.12 to 0.35	5 to 10	2.4
0.35 to 0.95	10 to 20	2.5 (3.0) *)
0.95 to 1.75	20 to 35	2.8 (3.5) *)
1.75 to 3.7	35 to 70	2.8 (3.5) *)
3.7 to 5		2.8 (3.5) *)

*) These room heights are required for upright heat exchangers (steam, water)

Table 6.15
Heating supply (substation), area requirement

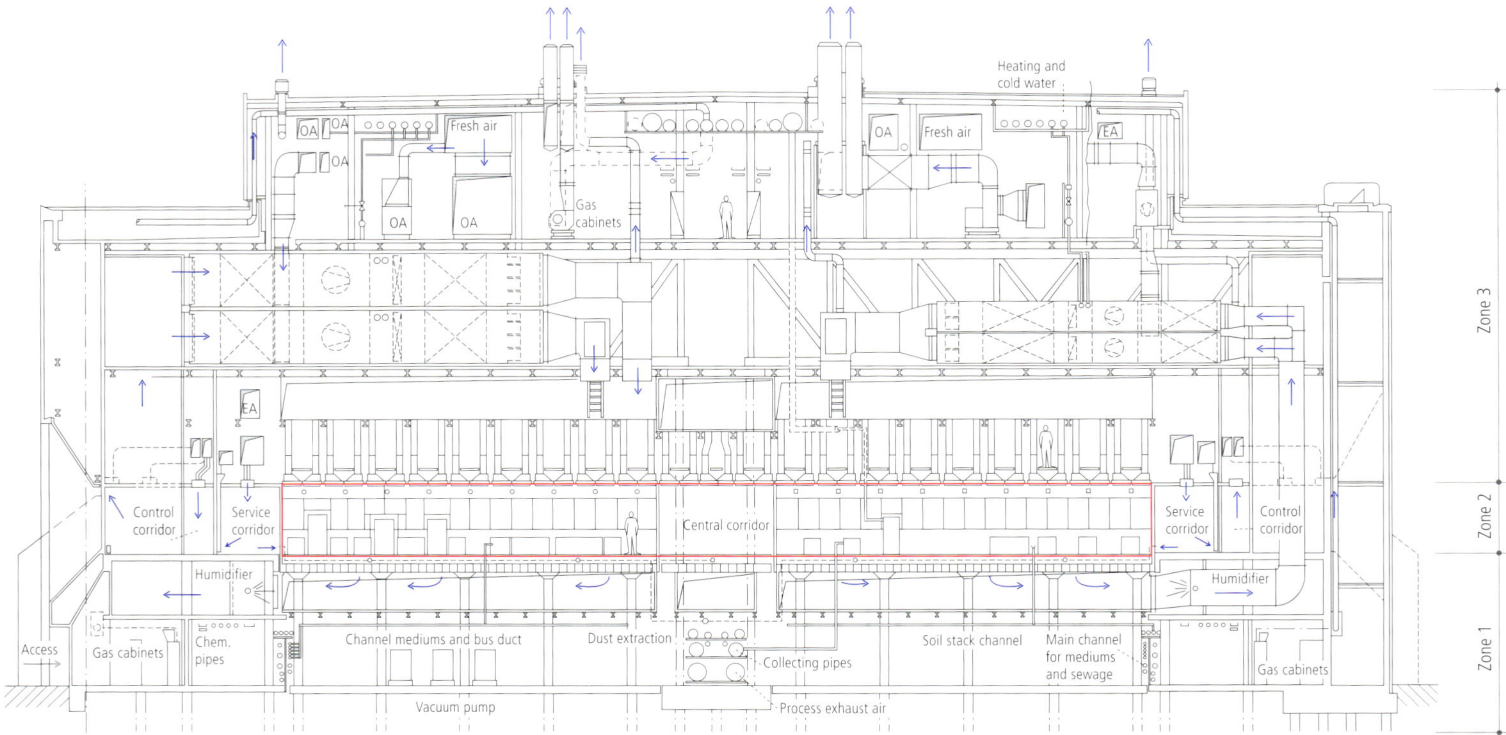

Figure 6.160
Section through building with clean-room conditions (Zone 1 + 3 installation levels, zone 2 usable level)

Figure 6.159 depicts a large terminal during installation with a regenerative heat recovery system mounted on the front.

Figure 6.160 illustrates the importance that air stream processing can have in production processes: this is a cross-section of a building with clean-room conditions, where the production area is located in zone 2 and all other areas are dedicated to systems that serve the production area (mostly HVAC).

The system installations shown in **Figure 6.160** represent an extreme example of technology for a production process. In laboratory buildings, pharmaceutical plants etc., the technical requirements are similarly elaborate.

Figure 6.159
Large-scale air-conditioning plant with heat recovery plant, ready for installation

6.10 Central energy supply

6.10.1 Cooling energy supply

Cooling energy is currently still supplied through electrically powered chillers, because they are cost-effective despite their high rates of energy consumption. Absorption chillers, which utilize waste heat to generate cooling energy (district heat in summer, waste heat from machine installations), offer a viable alternative.

Where soil temperatures range from + 12 °C to + 14 °C, direct cooling through probes in combination with a pile foundation is a feasible option. Generally speaking, the heat exchange produced from probes or horizontal earth coils is insufficient for a tall building, resulting in the need for bivalent cold energy production. In-ground surge tanks can be a useful complement, although these may require drilling deep bore holes. In some cases it is possible to tap into natural aquifers in order to create a seasonal surge tank. Total energy systems, that is, gas-operated combined heat and power plants (CHP) in combination with absorption chillers may be recommendable, as long as they are backed up by standby systems with a sufficient capacity to furnish the required emergency power. Such installations can be very sensible from both an ecological and an economical perspective due to the high rate of primary energy use (approximately 80 – 85 %).

One approach to reducing initial investment costs for chillers and re-cooling plants as well as the connect load and operating costs of electrical systems is the use of ice storage or vacuum-ice-generators, which store a portion of the cold energy produced overnight and release over the course of the following day. **Figure 6.161** illustrates the various technical solutions for the production of cooling energy.

Horizontal earth coils or probes
Direct cooling
Heat pump

Natural aquifers
Cooling or heat pump operation

Water-based cooling
(Desorption)
Cold energy from chiller and
ice storage

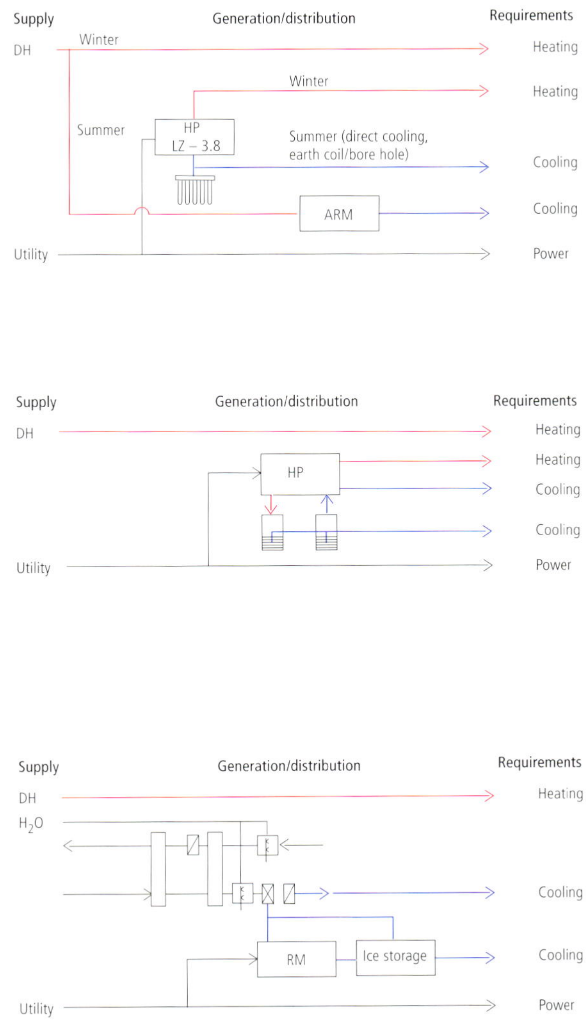

Figure 6.161
Refrigerating plants

6.10.2 Heating energy supply

The most efficient approach to supplying heating energy is the use of district heat, with particular emphasis on heating energy from thermal power stations.

An equally cost-efficient alternative is the use of gas-powered boiler systems based on combustion technology. A combination of heat and power plants and absorption chillers as heat pumps delivers cost-efficient operation, which can offset higher (initial) investment costs.

In-ground surge tanks in combination with heat pumps and earth coils can be integrated into a bivalent structure on a case-by-case basis – an ecological choice where applicable.

Fuel cells – the latest system solution for optimum primary energy use – are still in their infancy in terms of standard use because the investments costs can be prohibitive. However, they should be considered for individual cases.

The active use of solar power through evacuated tube collectors or focal point systems is a feasible solution in combination with architecture designed for this purpose. The inclined surfaces of the building must offer angles that are suited for collector installation.

Figure 6.162 provides a schematic overview of these system solutions.

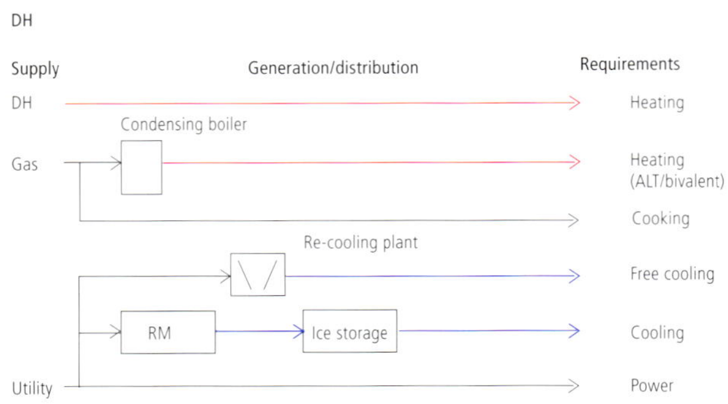

Convential heating energy production from district heat or condensing boiler installation

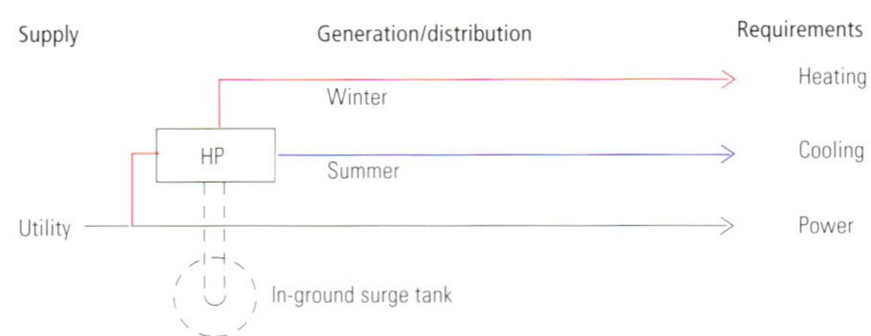

Heating energy production from heat pump with in-ground surge tank

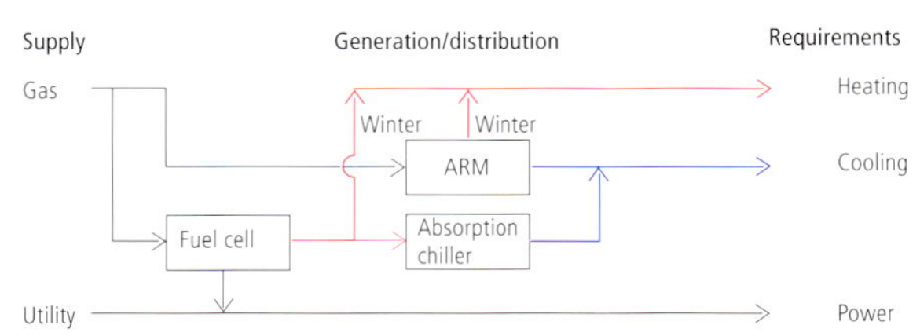

Heating energy production with fuel cell (electrical power) and absorption chiller

Figure 6.162
Heat, power and cooling supply

6.10.3 Electrical power supply

Conventional power supply of buildings from the public utility is still the most efficient option with regard to initial investment costs. Decreasing electricity rates (deregulation) result in lower operating costs. Since high-rises must be equipped with larger stand-by or back-up power supply systems for emergency supply, a combination of power from the CHP and mains input may be the best solution in individual cases. If the waste heat of the CHP can be utilized at the same time, these solutions can be quite cost-efficient. For optimum use of primary fuels (especially gas), one might consider fuel cells for the production of electrical power (and simultaneous production of heating energy). Solar power can be converted through photovoltaic components, although the profitability of this solution rarely matches the positive image it has. Due to the fairly low efficiency quotients, photovoltaic components are still best employed as a complement to electrical power supply.

Figure 6.163.1
Schematic distribution of equipment rooms (ER) with principal building system components, ventilation and air-conditioning (VAC), heating, cooling, fresh water

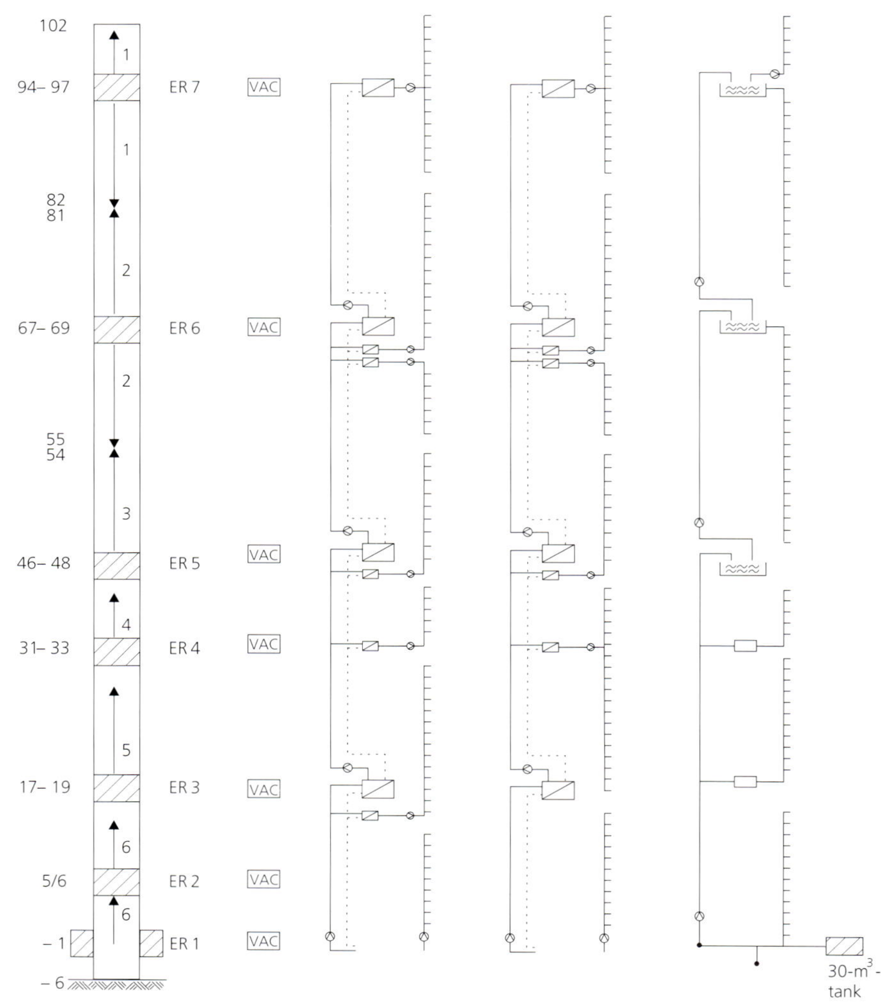

Diagrammatic section of a high-rise, 102 floors   1. VAX   2. Heating   3. Cooling   4. Fresh water

6.11 Power and media distribution

6.11.1 Vertical distribution and horizontal floor distribution

The high-rise illustrated as an example in **Figures 6.163.1** and **6.163.2** contains seven central equipment rooms; the schematic section identifies their location and illustrates sample floors.

As the illustration demonstrates, the equipment room 1 (basement) and the equipment room 6 occupy the greatest amount of space. For the purpose of illustration, these two spaces are fully furnished and are depicted with the full complement of appliances and machines.

The schematic illustrates the location of principal machines and essential characteristics for the following building services:
- Heating
- Cooling
- Water supply/drains
- Fire-extinguishing systems
- High-voltage and low-voltage supply
- Central power supply for the entire complex is provided via central installations located on level -1.

Heating is supplied exclusively from the municipal district heat network, with bilateral feed for safety reasons. This supply is transformed in the equipment rooms 3, 5, 6 and 7 and transported from these stations to the individual floors.

Cooling energy is supplied through large-scale electric heat pumps. To reduce investment and consumption costs, the plan envisions ice-storage systems, which store part of the required cooling energy overnight. Cooling energy on the basis of 6/12 °C or 6/20 °C (cooling ceilings/fan coils) is supplied from a decentralized source at graduated pressure levels, similar to the heating systems.

The central water supply is designed to operate on four pressure levels to avoid excessive network pressures. The relevant transition stations are housed in the equipment rooms 5, 6 and 7. Water supplies for fire-fighting and sprinkler systems are similarly decentralized, scaled to ensure an inexhaustible source of roughly 150 m^3 of water.

The central electrical power is supplied from the municipal medium high voltage grid. At level -1, the entire electrical supply is stepped down from medium high voltage to low voltage. From this point onwards, distribution occurs via low-voltage main distribution cables installed on levels 1, 3, 4, 5, 6 and 7. Power is then distributed from the various low-voltage main distribution cables, which are connected by bus bars, on each individual floor, where distribution cabinets are installed for change-overs and modifications.

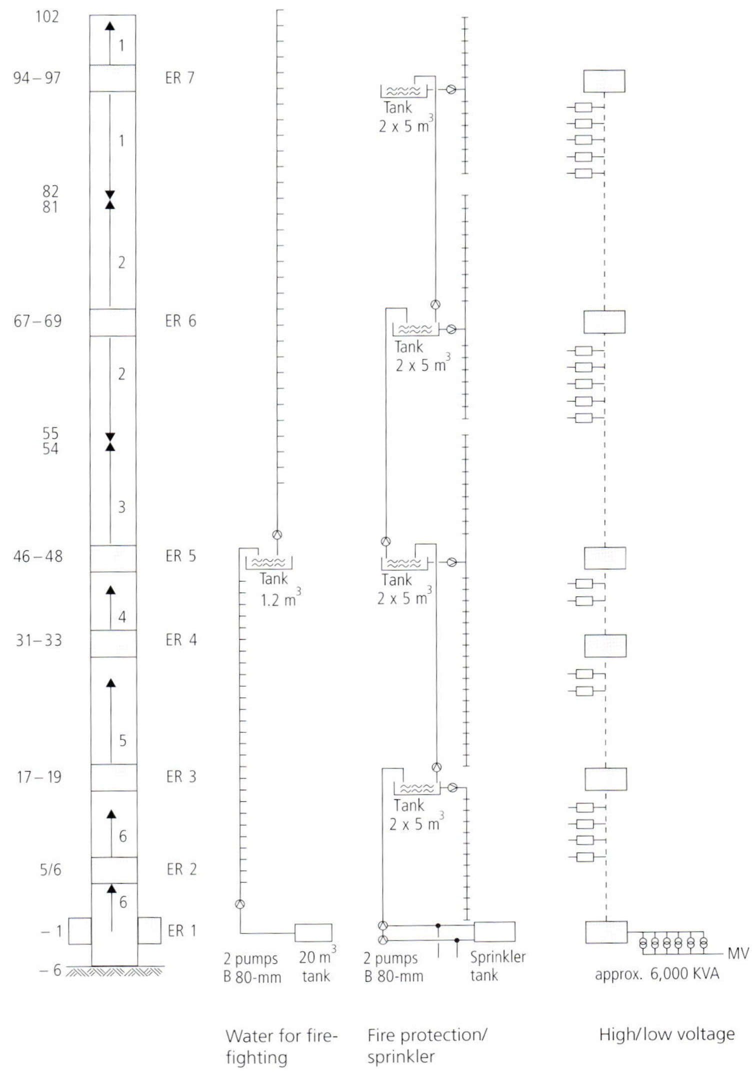

Figure 6.163.2
Schematic distribution of equipment rooms (ER) with principal building system components, water for fire-fighting, fire protection/sprinkler, high-/low-voltage

A broadband network for data and communications technology is designed in an analogous manner.

An emergency power supply system is provided in case of power failure (despite the dual feed). It supplies power to the primary emergency power points (elevators, fire-extinguishing systems, emergency lighting, emergency aggregates, computer processing systems). UPS – uninterruptible power supply – can also be installed (tenant completion). The UPS is an element of the building's "intelligent network".

The building systems are complemented by a central process control system, which supports the building operation and regulates the various operating processes. This central process control system registers all system malfunction messages, prioritizing them in order to satisfy all necessary safety requirements.

6.12 Equipment rooms – centralized versus decentralized

A common topic in all high-rises is the location of the equipment rooms and their integration into the overall design. The most difficult challenge seems to arise with regard to designing an appropriate "head", which should, of course, also house or serve a function.

Another important issue is the decision whether heating and air-conditioning systems should be integrated in a centralized or decentralized fashion, in order to provide separate service for rental areas on individual floors.

The schematic sketch in **Figure 6.165** provides a comparison of the various options, taking into consideration that centralized installations are practical only up to a height of approximately 25 floors in very tall buildings, because any greater height would result in elevated pressure levels, especially in the water systems, which would require installation of heavy-duty industrial fittings.

The space requirement for the necessary equipment rooms and corresponding shaft areas is roughly as follows:

6.12.1 Centralized supply areas

(up to max. 25 upper floors)
- 20 – 25 m^2 per 1000 m^2 usable space for centralized energy supply
- 50 m^2 per 1000 m^2 usable space for central heating and air-conditioning systems
- 4 m^2 per 1000 m^2 usable space for central shaft installations

It is important to remember that centralized systems generally do not occupy valuable usable or lettable areas; instead, they are housed in central equipment rooms in underground levels. However, many underground levels are too small to accommodate equipment rooms and still provide space for other uses, which leads to the construction of additional underground levels.

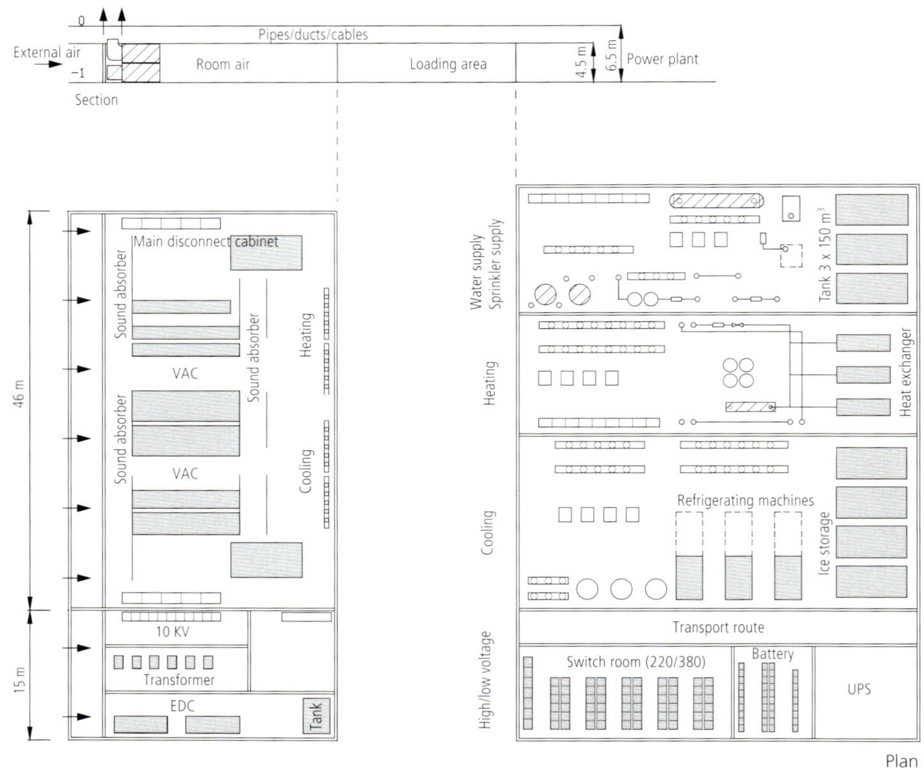

Figure 6.164.1
Equipment room 1 in basement, including central energy supply

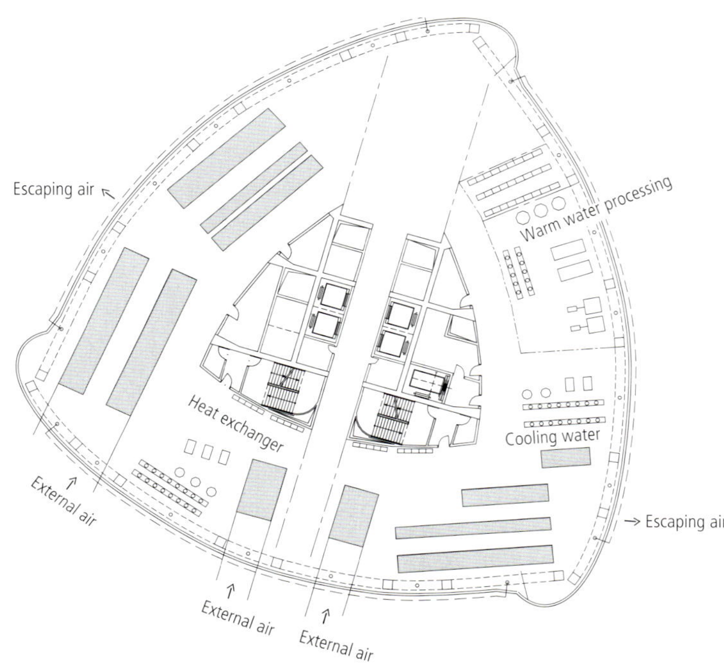

Figure 6.164.2
Plan of equipment room 6, floors 67 – 69

6.12.2 Decentralized supply areas

Decentralized heating and air-conditioning systems (per floor) result in the following space requirements:
- 20 – 25 m^2 per 1000 m^2 usable space for centralized energy supply
- 20 m^2 per 1000 m^2 usable space for central fresh air processing
- approximately 40 m^2 per 1000 m^2 usable space for the installation of decentralized heating and air-conditioning units (processing units installed separately on each floor)
- approximately 1.5 m^2 of shaft area per 1000 m^2 usable space

Case-by-case studies are required for comparative analysis to determine which areas must be constructed at what cost and what the specific value of the lettable areas is. A "core and shell" solution (**Figure 6.165.3**) can be used to add the required installation space for heating and air-conditioning systems to the lettable areas on each floor to avoid losses. High-rises of fifty floors or more should include two-storey transition levels from which the areas lying above or below can be supplied.

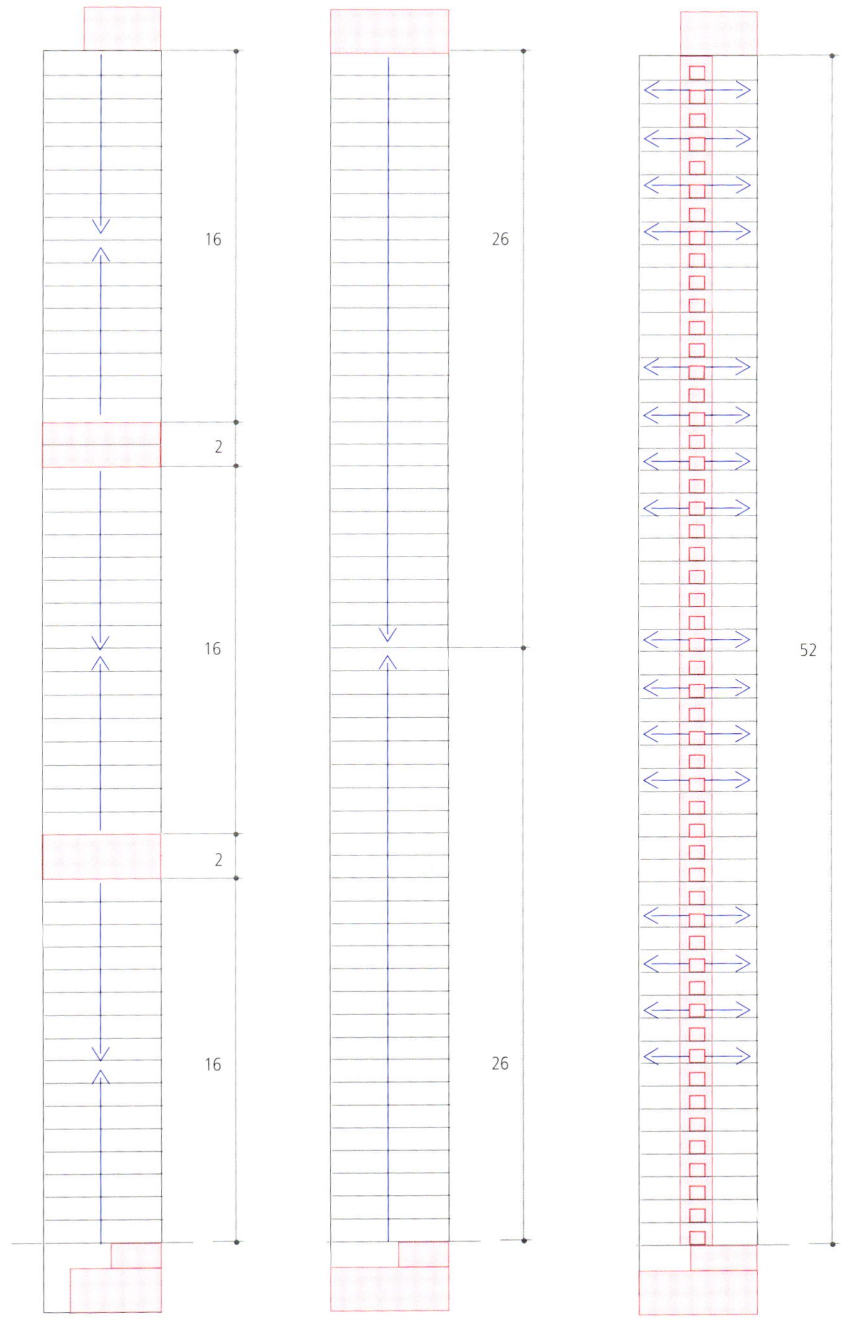

Figure 6.165.1
Central supply
– Basement
– Roof

Figure 6.165.2
Central supply
– Basement
– Intermediate floors
– Roof

Figure 6.165.3
Decentralized supply
(per floor)

Figure 6.165
Diagrammatic sections for possible equipment room locations

6.13 Horizontal distribution

While steel structures generally feature strong girders, because of the sheer depth of each girder, many tall buildings today are constructed with flat (or girderless) slab floors on top of which the entire complement of building services can be installed. This achieves a high degree of flexibility. The complete set of installations (with the exception of air exhaust) located beneath a hollow floor is readily accessible at all times and hence suitable for conversion or retrofitting. **Figure 6.166** illustrates relevant service stations in a section of the building area as well as a schematic section of a service station, while **Figure 6.167** contains a cross-section of a hollow floor with flexible installation.

As a rule of thumb, horizontal distribution of air, media and data communications requires at least 50 – 60 cm hollow installation space per floor in order to satisfy all requirements. This hollow installation space is distributed as follows:

- Installation above ceiling only, approx. 60 cm (shaft connection, max. length 20 – 25 m)
- Installation above floor and above ceiling, approx. 20 cm in ceiling area and approx. 30 cm in floor area (shaft connection as above)

In standard areas, the installation height can be reduced by roughly 20 cm, for example through increased cove height, by employing a comprehensive integrated approach (construction and building systems).

The installation heights listed above for various building systems apply only to rooms with low heat loads (up to approx. 50 W/m^2). In rooms with considerably higher heat loads and higher occupation rates, the space required for a hollow floor to install the required heating and air-conditioning channels can be proportionately greater (e.g. maintenance spaces). The wastewater system also has considerable influence on the installation heights when it is required to accommodate flexible sanitary blocks and when there is considerable distance between shafts (fixed points).

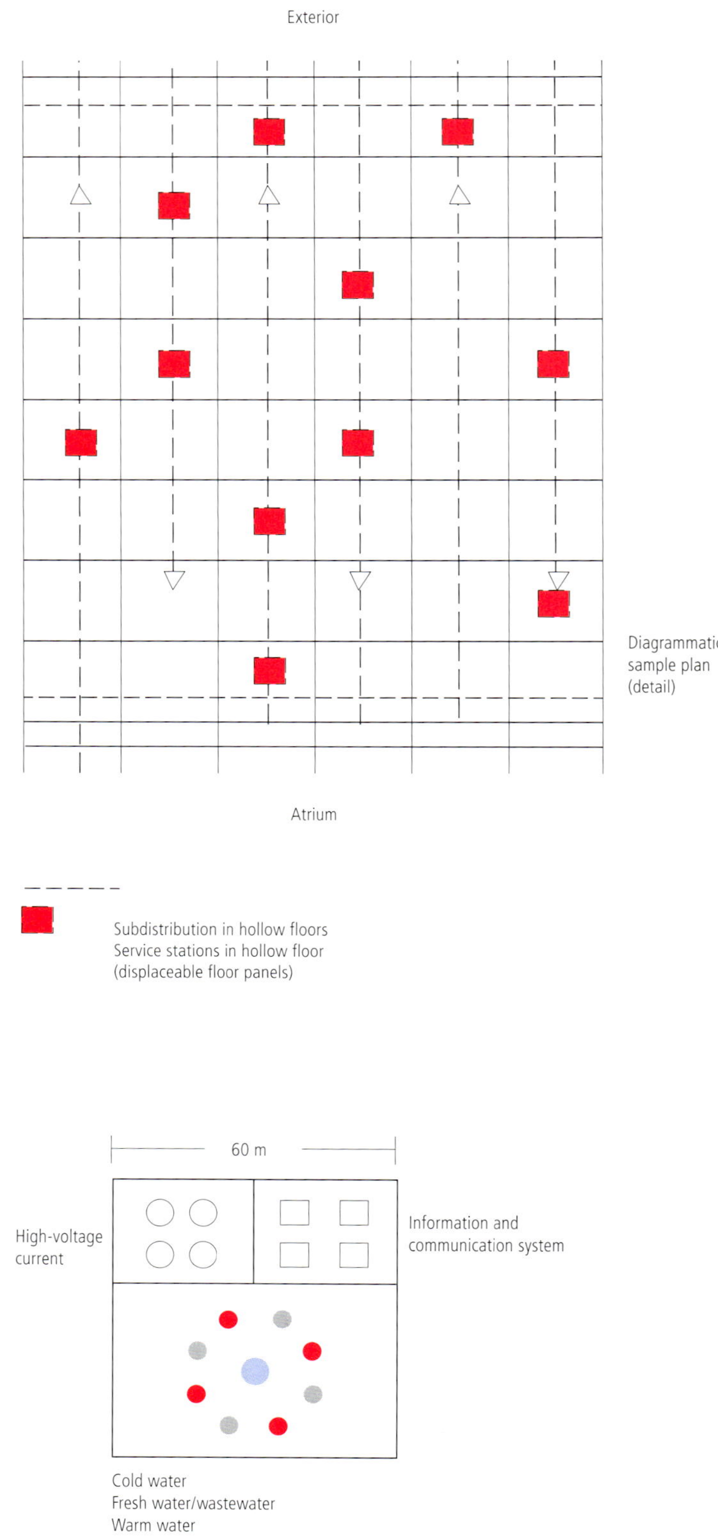

Figure 6.166
Service stations in a high-rise with vertical distribution

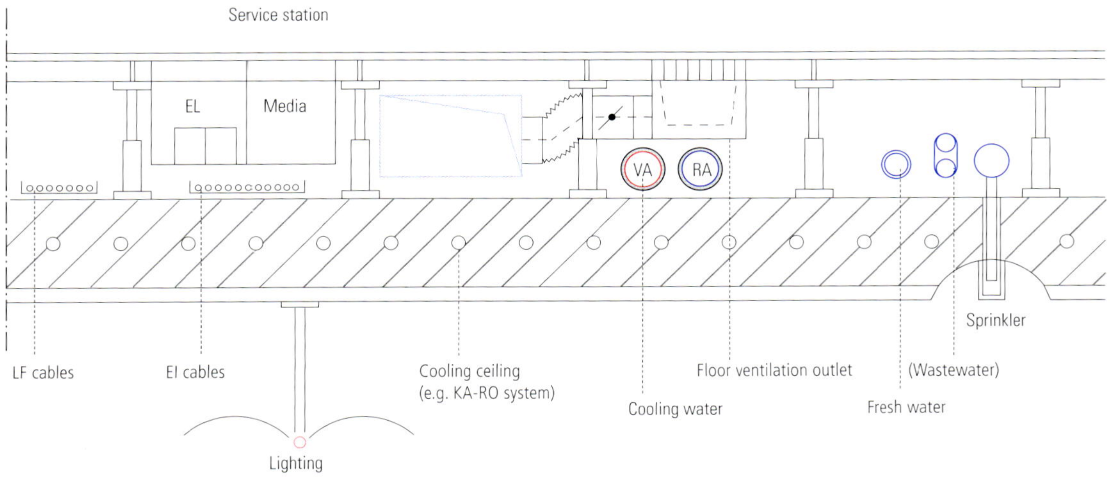

Figure 6.167
Section, flexible hollow-floor installation

7.
Refrigeration and cooling systems

(with Hans-Joachim Kast, Munich)

As well as a medium for heating air, ventilation systems with cooling and air-conditioning and static cooling systems also require a medium for cooling and sometimes dehumidification in summer, or cooling without dehumidification.

In the past, ice, tap or well water was used as a cooling medium. This is no longer feasible for environmental and financial reasons.

Today, the cooling energy required for building systems and to some degree for processing plants is chiefly generated in chillers (refrigerating machines) and recovery cooling plants (static cooling).
The common refrigerants/coolants are:
– Cold water (circulation water 6/12 °C)
– Salt solution (antifreeze-water mix < 0 °C)
– Cooling water (circulation water 18 – 25 °C)

For different building types, the coolant selected and its volume are based on the cooling energy demand (output balance), and on the type of heat exchangers.

A multitude of cooling and refrigerating appliances, ranging from minimum to maximum output, have been developed specifically for air-conditioning.

For major building projects or new urban developments, central refrigerating plants are sometimes installed as central supply systems to serve the surrounding buildings, obviating the need for individual cooling energy production in each building.

It is important to have detailed information on the energy consumption parameters and the control options for the cold energy generators when planning such installations, because cooling fixture regulation (chillers, cooling ceilings etc.) must be compatible with cold generator regulation to ensure efficient and uninterrupted operation.

Thermodynamic cycles in machines or systems, and process cycles in air-conditioning terminals required to achieve a specific cooling effect are roughly classified as follows:

Desired effect	**Process**
Cold generation	Energy supply in the form of electrical or thermal energy to cool a heat carrier to a desired temperature without making use of the waste heat
Heat generation	Energy supply in the form of electrical or thermal energy to raise the temperature of a heat carrier to a desired level
Cold and heat generation	Energy supply in the form of electrical or thermal energy to cool a heat carrier to a desired temperature while making use of the waste heat for heating purposes

The thermodynamic cycles result from different processes based on physical events. The most common are:
– Cold vapour cooling process with cooling agents
– Cold air cooling process with cold air as cooling agent
– Steam-jet cooling process with water vapour as motive agent and water as cooling agent
– Absorption cooling process
– Thermoelectric cooling process supplied by electricity (Peltier components)

The most frequently used machines in cooling and refrigeration technology for HVAC systems are compression chillers and absorption chillers; a detailed discussion of these components follows later in this chapter.

7.1 Output balance sheet (cold energy demand)

7.1.1 Temporal dependence on consumers

In air-conditioning, the sizing of the refrigerating plants broadly depends on the temporal behaviour of the individual consumer demands. Depending on the consumer, the load curve (histogram) may fluctuate greatly both in intensity (= output) and in time. When sizing cold energy generators it is essential to pinpoint the specific day when cumulative consumption peaks. To this end, each consumption point must be analysed in order to determine each individual heat exchanger, and to calculate the interplay of all consumers with regard to simultaneity.

Figure 7.1 presents a histogram for an office air-conditioning system, which is principally defined by the external curve of enthalpy and the plotted cooling load.

Figure 7.2 illustrates the cooling requirements in a commercial kitchen, a consumer with a very specific and limited time profile.

The cooling demand in a computer lab is plotted in Figure 7.3. This demand of this type of consumer remains fairly constant over a 24-hour period. The assumption underlying this histogram is that some personnel must be supplied with fresh air during the day, which increases the cooling demand.

Figure 7.4 provides the summer curve for all three consumers, visualizing the cooling demands in a complex building. The assumption for this simplified rendering was that the maximum demand for all three consumers falls on the same day.

In complex buildings it is generally necessary to examine the entire year in order to determine the true maximum cooling demand. This type of calculation is the ultimate basis for defining the scale of a refrigerating plant.

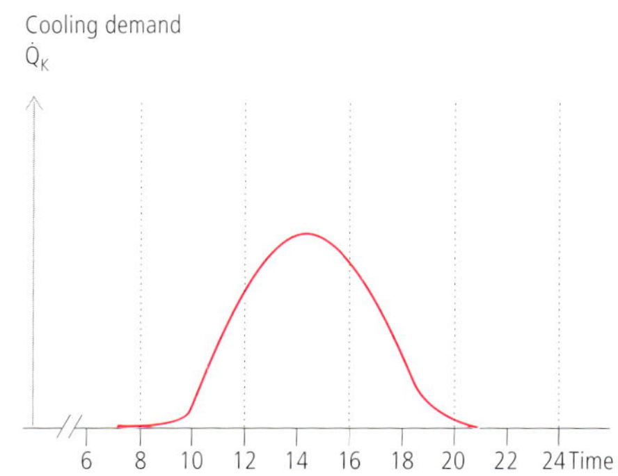

Figure 7.1
Histogram for office air-conditioning (summer day)

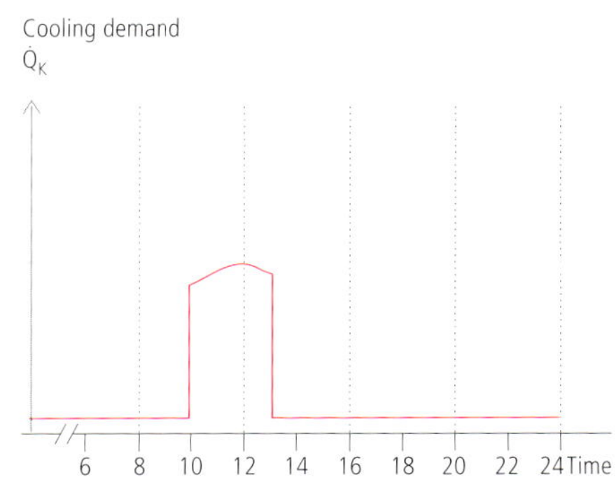

Figure 7.2
Histogram for HVAC installation in a kitchen (summer day)

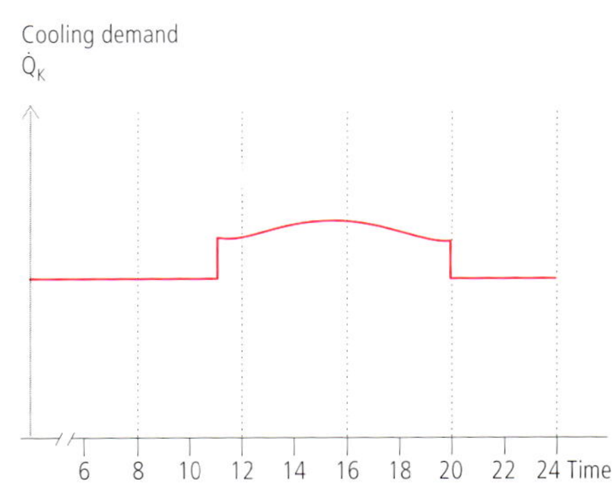

Figure 7.3
Histogram for computer lab (summer day)

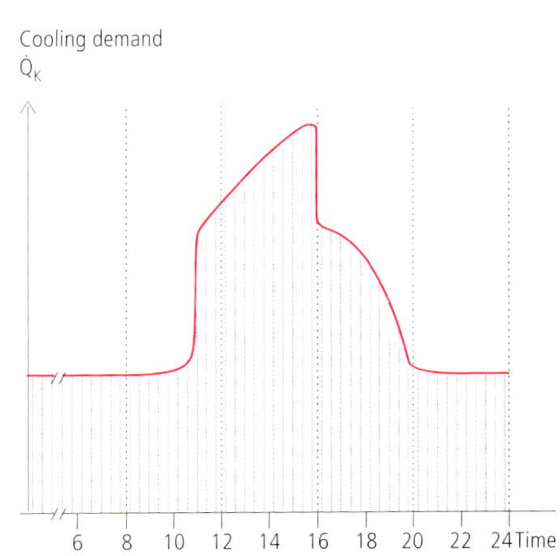

Figure 7.4
Histogram for all consumers (summer day)

7.1.2 Time dependency on power generation

Generally speaking, the refrigerating plant must be able to meet the summary histogram of all consumers. Depending on the type of consumer and the time of year when the maximum demand occurs, time factors may influence the power generation in a refrigerating plant.

The graph in **Figure 7.5** plots the annual curve of the outside temperatures in a reference year. The graph clearly shows that the necessary cold water can be generated in a direct heat exchange with the air according to the required water temperature, while a refrigerating plant is necessary at other times. Depending on the type of consumer, the cold water feed temperatures for air-conditioning vary widely at specific times. For an air-conditioning system with dehumidification the required feed temperatures are generally 6 °C, while feed temperatures of 12 to 15 °C are required for cooling the air in computer labs, or for direct water cooling for large computer terminals. Cold water for use in cooling ceilings should be kept at temperatures of 16 to 20 °C, while cooling water temperatures of 20 °C are sufficient for building component cooling in certain circumstances. The diagram provides an excellent illustration to which degree machine-generated refrigeration is required for the different system configurations and to what degree they can do without.

When a refrigerating plant is used, the average cooling energy output over the course of a day matches the profile shown in **Figure 7.6**. The tiers in the output graph are the result of switching individual output levels on such a plant/machine on or off. According to the steepness of the histogram, one or more output levels may only be required during extremely brief intervals. In some instance the maximum output of a refrigerating plant may be required for only a few hours in the entire year.

The ideal operation of a refrigerating plant is represented in **Figure 7.7**. It demonstrates how a refrigerating plant is operated constantly at its optimum performance level during standard daytime operation and is therefore able to deliver a correspondingly constant performance. At night, the output is transferred to cold storage with a similarly optimized output profile (the output difference illustrated here results from the varying temperature levels).

This operating principle results not only in the most efficient sizing of the refrigerating plant, it also represents additional savings in terms of energy consumption because the plant can be operated at maximum efficiency at all times and because a large portion of the energy production falls into the utility's low-tariff period.

Figure 7.5
Options for free cooling without chillers

1 Cooling and dehumidification
2 Sensitive cooling (data processing)
3 Chilled ceiling
4 Sensitive cooling and building component cooling

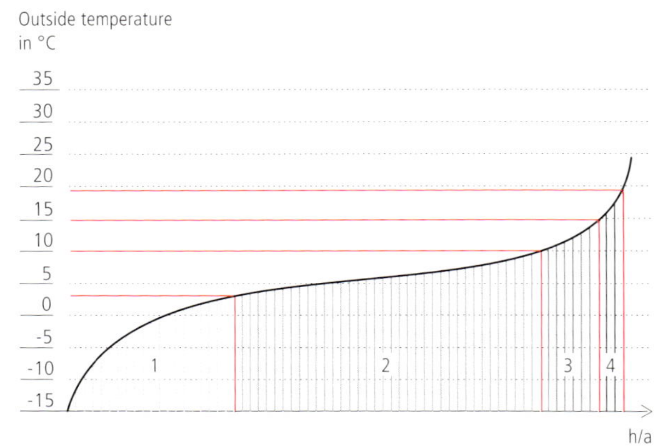

Figure 7.6
Cooling energy output of multi-phase refrigerating plant

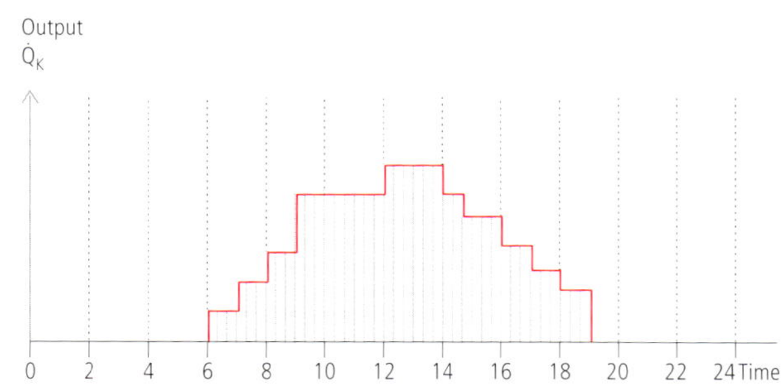

Figure 7.7
Ideal application of refrigerating plant in combination with cold storage

— Q_{st} Stored cooling energy
— Q_{dir} Direct usage of cooling energy

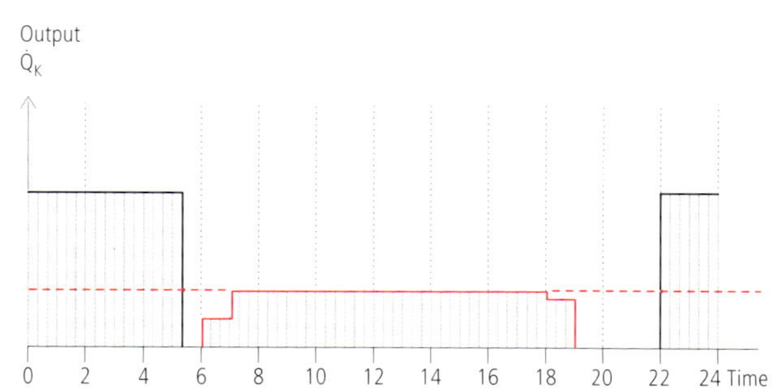

7.1.3 Cumulative curves of consumers and generators

Taking the histogram in **Figure 7.4** as the basis for a refrigeration plant, the resulting operative curve is shown in **Figure 7.8**. The assumption is that two equal basic load power generators (P1 and P2) are combined with a small peak load generator. This configuration has the advantage of putting generators 1 and 2 to relatively efficient and constant use, while calling upon generator 3 only for isolated performance peaks.

When storage technology is used, the curve assumes the profile shown in **Figure 7.9**. Generator 1 covers the constant demand for the computer lab in 24-hour operation, while generator 2 charges a cold storage tank at night and meets some of the additional demand, beyond that of the computer lab, during the day. The storage tank covers the remaining daytime peak load. In contrast to the solution shown in **Figure 7.8**, this translates into a considerable reduction in the connect load.

To summarize, the various time factors and influences must be studied in detail when planning the size and scale of refrigerating plants in order to achieve an optimum solution with regard to investment, operating costs and environmental stress.

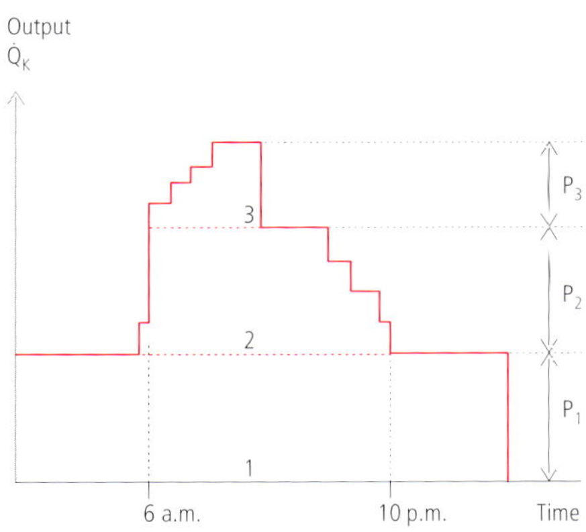

Figure 7.8
Classic approach to meeting consumption demands with several refrigerating plants

— $\dot{Q}_K$
1 Machine 1 with performance P_1
2 Machine 2 with performance P_2
3 Machine 3 with performance P_3

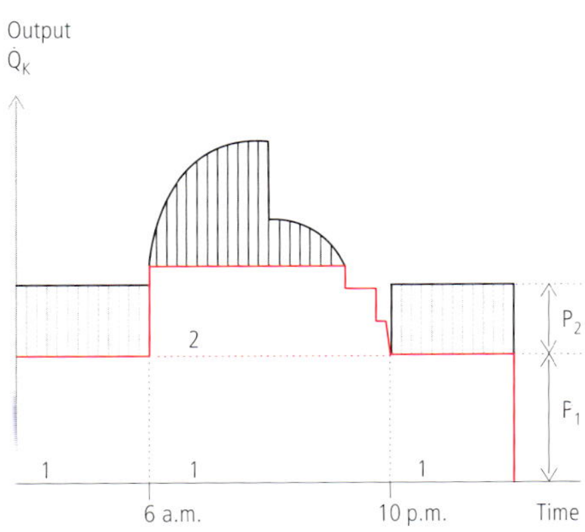

Figure 7.9
Cumulative curve of consumers and generators

Load storage
Discharge storage
1 Machine 1 with performance P_1
2 Machine 2 with performance P_2

7.2 Refrigeration

7.2.1 The refrigeration process

Currently, refrigeration is generally effected by electrically powered refrigerating machines. These operate on the principle that heat is extracted from the medium to be cooled in a thermodynamic cycle by supplying energy; the heat extracted is then transferred to another medium on a correspondingly higher level. Since this thermodynamic cycle is essential for understanding the function of refrigerating systems, it is outlined briefly below.

7.2.1.1 The Carnot cycle

The foundation for the operating principle of refrigerating machines is the Carnot cycle. It is described as an "ideal cycle," as engines that are designed to operate according to the events after Carnot translate a maximum of work into heat and vice versa. The best illustration of the Carnot process is provided in the temperature-entropy diagram (T, S diagram, **Figure 7.10**).

The cycle consists of two isothermal and two isentropic events. Isentropic means reversible and adiabatic. The changes in condition produced by these events are illustrated in **Figure 7.10**. The Carnot process is ideal because the conditional changes from 1 to 2 or 3 to 4 occur on an isentropic line. This means that the change from temperature T_O to temperature T_C would occur by compression and the change from temperature T_C to temperature T_O by expansion without consuming additional energy. Since this is impossible in practice, **Figure 7.11** illustrates the Carnot process in comparison to a real cycle plotted in the h, log p diagram.

Figure 7.10
Carnot cycle

T_C Condensation temperature
T_O Evaporation temperature
$\dot{Q}_C$ Energy output
$\dot{Q}_O$ Supplied energy
P Driving power

Performance of chiller

$$\varepsilon_K = \frac{\dot{Q}_O}{P}$$

Performance of heat pump

$$\varepsilon_W = \frac{\dot{Q}_C}{P}$$

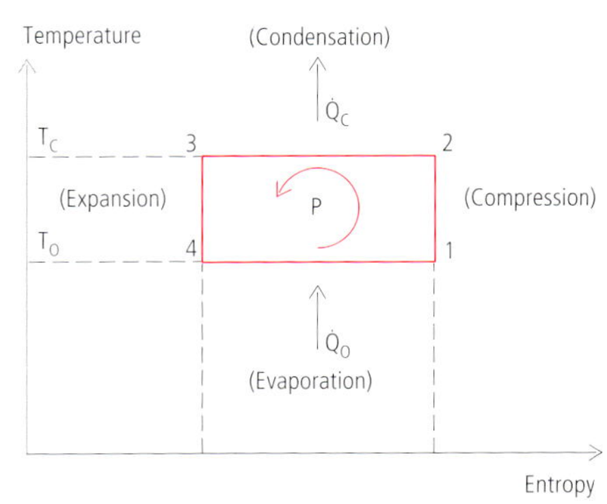

In this diagram the vapour curve of a refrigerating agent is plotted to illustrate the general aggregate conditions of a refrigerating agent during a cycle. This cycle demonstrates how a refrigerating agent is raised from a low temperature to a higher temperature, discharging heat, and drops again to the low temperature level containing a low level of energy, which renders it capable of absorbing more energy.

Figure 7.11
Cold vapour cycle

— Carnot cycle
- - - Actual cycle
A Liquid
B Mixing range (liquid/vapour)
C Vapour
P_C Condensation pressure
P_O Evaporation pressure

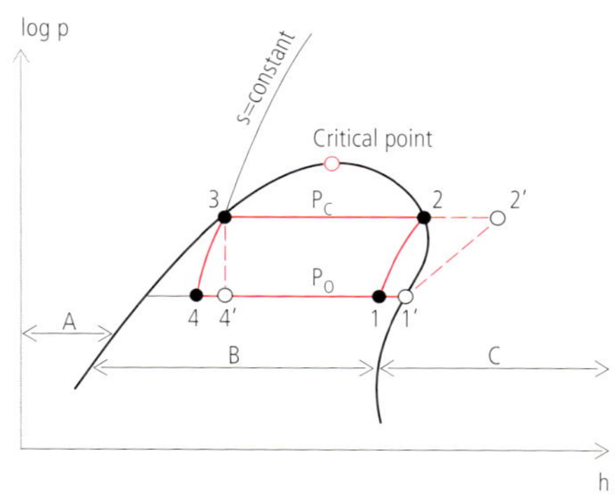

Figure 7.12
Single-phase cold vapour compression process

— Carnot cycle
1–2 Compression
2–3 Condensation
3–4 Expansion (damping)
4–1 Evaporation
P_C Condensation pressure
P_O Evaporation pressure
t_O Evaporation temperature
t_C Condensation temperature

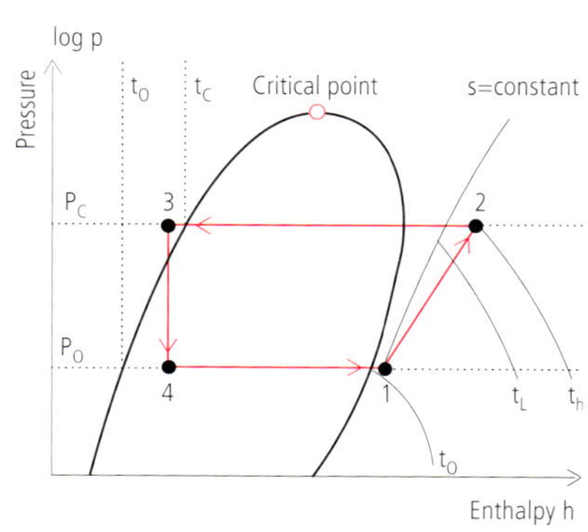

Figure 7.12.1
Schematic of cycle

C Condenser
CP Compressor
D Damper valve
E Evaporator
P_{ELT} Electrical output drive
Temperatures:
t_{co} Compressor/out
t_{ci} Compressor/in
t_{ei} Evaporator/in
t_{eo} Evaporator/out

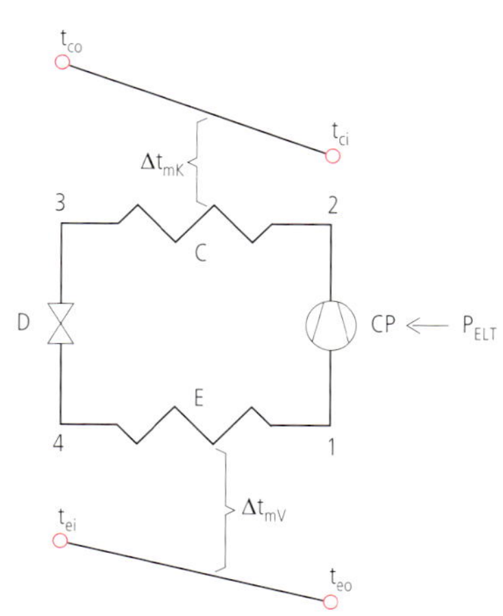

7.2.1.2 The cold vapour machine cycle

Representing over 90 % of all installed systems, the cold vapour compression refrigerating cycle occupies the key role in today's refrigeration and air-conditioning technology. The essential characteristic of this process is the use of refrigerating agents, which evaporate while absorbing a maximum of evaporation heat at a working temperature T_o from a liquid state at the cold end of the cycle, and are liquefied again on the warm end of the cycle under controllable pressures. The process makes use of the physical law that describes the dependency of evaporation/liquefaction temperature on pressure.

Figure 7.12 plots the cycle of a compression refrigerating machine in the h, log-p diagram. This cold steam refrigerating process runs through the following changes in condition:

1 – 2 Density of vapour is increased from pressure p_o to p_c with compression temperature t_h and a correspondingly high enthalpy (energy content);
2 – 3 Transfer of final steam temperature as well as condensation heat via a condenser to an appropriate medium (ambient air or warm water for heating purposes);
3 – 4 Expansion while energy remains constant (h = constant) of p_c to p_o in the expansion organ (accompanied, in part, by evaporation);
4 – 1 Evaporation of refrigerating agent at temperature t_o. The required evaporation heat is with drawn from the medium to be cooled (cold water or sol-water mix).

$(h_1 - h_4)$ is the specific theoretical refrigerating output and $(h_2 - h_1)$ the specific theoretical propulsive output.

The operating behaviour of a refrigerating machine or heat pump system is not only dependent on the thermodynamic cycle. The energy balances for the mass flows on the cold and warm end of the cycle are equally important, together with the resulting temperature differences, as well as the temperature differences resulting from the installed heat exchange surfaces and the attainable thermal transmittance coefficients. There are also various efficiency losses between the absorbed electrical propulsion output of the compressor all the way to the compression work that is actually delivered.

7.2.1.3 The absorption machine cycle

In the absorption refrigerating machine, the vapour produced by the refrigerant in the evaporator is not compressed mechanically as it is in the compression refrigerating machine, but "absorbed" at low evaporative pressure by a solution, hence the name. The refrigerant-enriched solution is raised to a higher liquefying pressure via a pump and then transported to an expulsion device. The refrigerating agent is extracted by means of adding heat, e.g. steam or flue gas heating. An impoverished solution remains and flows back to the absorber via a damper. There it is dispersed through pipes to provide a large surface for the refrigerant steam absorbing and for releasing the emitted solution heat to the cooling water that flows through the pipes. The extracted refrigerating agent is liquefied under pressure p_C by means of heat transfer to the cooling water or cooling air.
After being dampened in the expansion device, it can absorb heat from the medium to be cooled in the evaporator at pressure p_O and at the corresponding evaporation temperature t_O. The refrigerant steam that results from this process flows to the absorber, where it is absorbed by the solution.

Thus, two cycles occur in an absorption refrigeration machine, in contrast to the single cycle in a compression machine. These two cycles are illustrated in **Figure 7.13**, for which a different manner of depiction was chosen to achieve lines for the different solution concentrations. The essential energy supply is in the form of thermal energy in the expulsion device, while the mechanically supplied energy for the solution pump is relatively small. The solution pump, which transports the solvent from the evaporation to the liquefaction pressure, is the only mobile component in the cycle.

The absorption machine cycle with temperature differences is illustrated in **Figure 7.13.1**.

7.2.1.4 Refrigerating agents / refrigerants

Refrigerating agent is the term applied to the substance circulating in refrigerating machines whose changes of state determine the cycle. Essential criteria for these substances are a favourable vapour pressure curve and the smallest possible volumetric refrigerating output in order to keep the circulating refrigerant volume flow and hence the system components as small as possible. The steam pressure curves for different refrigerants are shown in **Figure 7.14**, the volumetric refrigerating output for the same agents in **Figure 7.15**.

The agents must also be chemically stable, and not toxic, explosive or combustible. **Tables 7.1/2** provide lists of common refrigerants in air-conditioning systems as well as all essential parameters relevant to practical application.

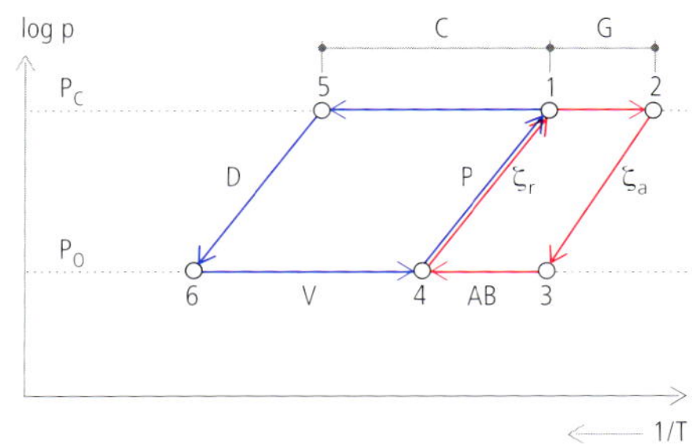

(1–2–3–4–1) Solution cycle
(1–5–6–4–1) Refrigerant cycle
P_C Condensation pressure
P_O Evaporation pressure
C Condenser
V Evaporator
G Generator
AB Absorber
ζ_a Weak solution
ζ_r Rich solution
P Pump
D Damping ζ=1 (refrigerant)

Figure 7.13
Single-phase absorption process in log p, 1/T diagram

Figure 7.13.1 (right)
Schematic of cycle

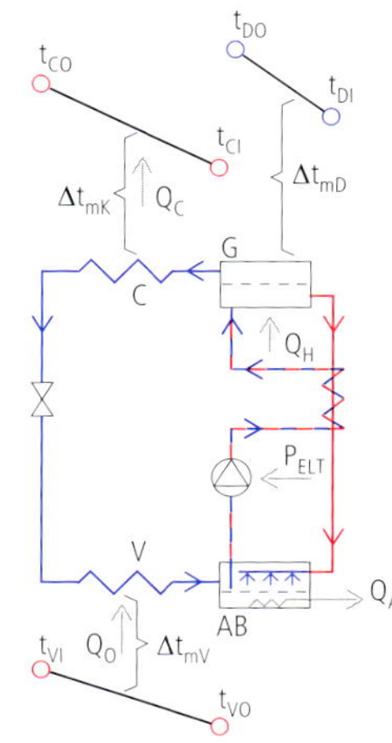

Solution cycle
Refrigerant cycle
G Generator
AB Absorber
C Condenser
V Evaporator
Output:
P_{ELT} Electrical output drive
Q_A Absorber
Q_C Condenser
Q_O Evaporator
Q_H Generator
Temperature:
t_{CI} Condenser/in
t_{CO} Condenser/out
t_{VI} Evaporator/in
t_{VO} Evaporator/out
t_{DI} Generator/in
t_{DO} Generator/out

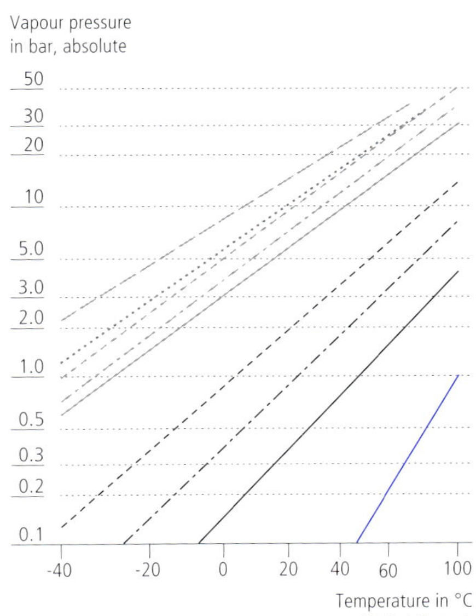

Water
R 113
R 11
R 114
R 12
R 500
R 22
R 502
R 1381

Figure 7.14
Vapour pressure curves of different refrigerants

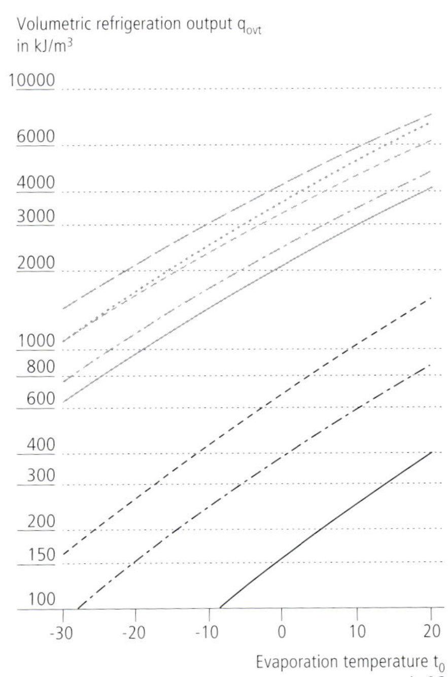

R 113
R 11
R 114
R 12
R 500
R 22/R 502
NH_3
R 1381

Figure 7.15
Theoretical volumetric refrigeration output q_{ovt} of different refrigerants for a temperature of 35 °C in front of expansion valve

Refrigerant	Composition	M	t_c	p_c	NSP*)	Δt_s	t_{26}*)	ODP	HGWP	L	Composition	
		g/mol	°C	bar	°C	K	°C	(R11= 1)			Status	Preferred application
R 744 (carbon dioxide)		44.0	31.1	73.8	- 56.6**)	–	- 10.6	0	0	1	# +	Car/vehicle AC
R 410A (AZ20)	R-32/125 (50/50)	72.6	73.0	49.6	- 52.4	0.2	42.7	0	0.40	(1)	# +	New installations (high operating pressures!)
SUVA9100	R-32/125 (45/55)	75.6	84.1		- 51.0	0.2	43.6	0	0.40	(1)	+	New installations (high operating pressures!)
R 507	R-125/143a (50/50)	98.9	70.9	37.9	- 46.5	0.0	54.8	0	0.96	(1)	# ++	Compact air-conditioning syst. and appliances
R 404A	R-125/143a/134a (44/52/4)	97.6	72.1	37.3	- 46.1	0.7	55.5	0	0.94	(1)	# ++	Compact air-conditioning syst. and appliances
R 407A (KLEA60)	R-32/125a/134a (20/40/40)	90.1	83.0	45.4	- 42.3	6.6	56.5	0	0.49	(1)	# ++	New installations
R 290 (propane)		44.1	96.7	42.5	- 42.1	–	70.3	0	0	3	# ++	R 22 substitute; systems with low fill volume
R 22		86.5	96.1	49.8	- 40.8	–	63.2	0.055	0.37	1	# ++	Commercial, ..., central and split systems
R 407C (AC9000, KLEA66, HX3)	R-32/125/134a (23/25/52)	86.2	86.0	46.5	- 40.0	7.1	59.8	0	0.39	(1)	# ++	Commercial, ..., central and split systems
R 717 (ammonia)		17.0	132.3	113.4	- 33.3	–	59.7	0	0	2b	# ++	Central air-conditioning syst., indirect cooling
RC 270 (cyclopropane)		42.1	125.2	55.8	- 32.9	–	81.6	0	0	3	# +	Systems with low fill volume
ISCEON 49	R-218/134a/600a (9/88/3)	104.0	101.3	41.1	- 31.6	6.9	76.2	0	0.44	(1)	# ++	Car/vehicle AC
R 409A (FX 56)	R-22/124/142b (60/25/15)	97.4	107.0	46.0	- 30.2	8.5	78.2	0.05	0.31	(1)	# ++	Old systems
R 401A (MP 39)	R-22/152a/124 (53/13/34)	94.4	108.0	46.0	- 29.9	6.3	78.1	0.03	0.22	(1)	# ++	Old systems, R 12 applications t_o > -20 °C
R 406A (brines 406A)	R-22/600a/142b (55/4/41)	89.9	114.5	45.8	- 29.8	9.0	85.0	0.06	0.36	(1)	# ++	R 12 substitute, car/vehicle AC
R 134a		102.0	101.2	40.7	- 26.1	–	79.4	0	0.26	1	# ++	Car/vehicle AC, commercial, split systems, ...
R 401C (MP52)	R-22/152a/124 (33/15/52)	101.0	112.7	43.7	- 25.2	6,2	85.1	0.03	0.17	(1)	# ++	R 12 substitute, car/vehicle AC
R 152a		66.0	113.6	44.9	- 25.0	–	84.7	0	0.03	2a	#	Mix component
R 227ea		170.0	101.8	29.3	- 16.5	–	95.7	0	0.60	(1)	# +(+)	Air-conditioning (high temp.), heat pump
R 124		136.5	122.5	36.4	- 13.2	–	104.7	0.03	0.10	1	# +	Air-conditioning (for very high temperatures)
R 600a (isobutane)		58.1	135.0	36.5	- 11.8	–	115.0	0	0	3	# ++	Systems with low fill volume
R 600 (butyraldehyde)		58.1	152.0	38.0	- 0.5	–	129.3	0	0	3	#	Systems with low fill volume
R 123		152.9	183.8	36.7	27.9	–	162.5	0.02	0.02	1	# ++	Air-conditioning, turbo refrigerating machines

°C t_o -60 -50 -40 -30 -20 -10 ±0 10 20

Table 7.1
Refrigerants for air-conditioning and heat pumps

XX (data marked in red) transition refrigerant – permissible until 31.12.1999 (Germany)

XX (data marked in black) long-term use permissible

— Recommended application

M molar mass in g/mol
t_C critical temperature in °C
p_C critical pressure in bar
NSP boiling point *) for atmospheric pressure
Δt_S sliding temperature for 1 bar
t_{26} boiling point *) for 26 bar

\# Commercially available refrigerant
\+ Application in test phase
++ Commercial application

*) Arithmetic mean between boiling point and dew point temperatures for mixed agents
**) Triple point

L: Health hazard classification after E DIN EN 378
L 1 nonflammable
L 2a nonflammable > 3.5 Vol%
L 2b flammable, bottom limit of inflammability > 3.5 Vol%; toxic
L 3 flammable, bottom limit of inflammability < 3.5 Vol%; (low toxicity)
In parentheses: non-binding categories based on E DIN EN 378 criteria

Refrigerant	Composition	M g/mol	t_c °C	P_c bar	NSP*) °C	Δt_s K	t_{26}*) °C	ODP (R11= 1)	HGWP	L	Application conditions: Status	Preferred application
R 410A (AZ20)	R-32/125 (50/50)	72.6	73.0	49.6	- 52.4	0.2	42.7	0	0.40	(1)	# +	New installations, commercial, transport (pressure!)
SUVA9100	R-32/125 (45/55)	75.6	84.1		- 51.0	0.2	43.6	0	0.40	(1)	+	New installations, commercial, transport (pressure!)
R 507 (AZ 50)	R-125/143a (50/50)	98.9	70.9	37.9	- 46.5	0.0	54.8	0	0.96	(1)	# ++	New installations, commercial, transport, food
R 402B (HP81)	R-125/290/22 (38/2/60)	94.7	82.6	44.5	- 46.1	2.5	56.1	0.03	0.52	(1)	# ++	Old and new installation, commercial, ice machines
R 404A (HP62, FX70)	R-125/143a/134a (44/52/4)	97.6	72.1	37.3	- 46.1	0.7	55.5	0	0.94	(1)	# ++	New installations, commercial, transport, food
R 407B (KLEA61)	R-32/125/134a (10/70/20)	102.9	76.0	41.6	- 45.2	4.4	53.5	0	0.70	(1)	# ++	New installations, commercial, transport
HX4	R-32/125/143a/134a (10/33/36/21)	94.5	77.5	40.1	- 44.6	4.1	56.0	0	0.75	(1)	+	New installations, commercial
R 408A (FX10/neu)	R-125/143a/22 (7/46/47)	87.0	83.5	43.4	- 44.1	0.7	58.8	0.023	0.78	(1)	# ++	Old and new installations, commercial
R 407A (KLEA60)	R-32/125/134a (20/40/40)	90.1	83.0	45.4	- 42.3	6.6	56.5	0	0.49	(1)	# ++	New installations, commercial
R 290 (propane)		44.1	96.7	42.5	- 42.1	–	70.3	0	0	3	# ++	Systems with low fill volume
R 22		86.5	96.1	49.8	- 40.8	–	63.2	0.055	0.37	1	# ++	Commercial, industrial ...
R 407C (SUVA9000, KLEA66, HX3)	R-32/125/134a (23/25/52)	86.2	86.0	46.5	- 40.0	7.1	59.8	0	0.39	(1)	# ++	Commercial, industrial ...
R 717 (ammonia)		17.0	132.3	113.4	- 33.3	–	59.7	0	0	2b	# ++	Commercial, industrial, warehouse, ...
RC 270 (cyclopropane)		42.1	125.2	55.8	- 32.9	–	81.6	0	0	3	# +	Systems with low fill volume ...
FX 57	R-22/124/142b (65/25/10)	96.7	105.0	47.0	- 31.8	7.7	75.6	0.05	0.31	(1)	# +	Old installations, refrigerated transportation
R 401B (MP66)	R-22/152a/124 (61/11/28)	92.8	106.1	46.8	- 31.6	5.9	75.5	0.035	0.24	(1)	# ++	Old installations, R 12 applications $t_o < -20$ °C
ISCEON 49	R-218/134a/600a (9/88/3)	104.3	101.3	41.1	- 31.6	6.9	76.2	0	0.44	(1)	# ++	Refrigerated transportation, household, commercial
R 409A (FX56)	R-22/124/142b (60/25/15)	97.4	107.0	46.0	- 30.2	8.5	78.2	0.05	0.31	(1)	# ++	Old installations, commercial
R 401A (MP39)	R-22/152a/124 (53/13/34)	94.4	108.0	46.0	- 29.9	6.3	78.1	0.03	0.22	(1)	# ++	Old installations, R 12 applications $t_o > -20$ °C
R 406A (brines 406A)	R-22/600a/142a (55/4/41)	89.9	114.5	45.8	- 29.8	9.0	85.0	0.06	0.36	(1)	# ++	R 12 substitute, household
R 134a		102.0	101.2	40.7	- 26.1	–	79.4	0	0.26	1	# ++	Household, commercial, air-conditioning, heat pump
R 152a		66.1	113.6	44.9	- 25.0	–	84.7	0	0.03	2a	#	Mix component
R 227ea		170.0	101.8	29.3	- 16.5	–	95.7	0	0.60	(1)	# +(+)	
R 124		136.5	122.5	36.4	- 13.2	–	104.7	0.03	0.10	1	# +	Mix component
R 600a (isobutane)		58.1	135.0	36.5	- 11.8	–	115.0	0	0	3	# ++	Household, systems with low fill volume

°C t_o - 60 - 50 - 40 - 30 - 20 - 10 ± 0 10

Table 7.2
Refrigerants for standard cooling applications

Figure 7.16 gives the h, log-p diagram for the refrigerating agent 134a, plotting actual pressures and temperatures as well as its energy content (enthalpy).

Ammonia (NH_3) can be recommended as a refrigerant to achieve long-term safety and make a positive contribution to the environment. The side effects of this agent, such as burns when inhaling elevated concentrations, are largely avoidable when existing accident spill prevention regulations are applied. Long-term use of this refrigerating agent in industrial process refrigeration has demonstrated its safety when used appropriately.

Figure 7.17 illustrates the combination of ammonia/water as an agent for absorption machines (alternative: water/lithium bromide).

Here are some general comments on the terms CFC and FC: fluorocarbons are organic compounds that contain hydrogen in the molecule in addition to the central carbon atoms. If these hydrogen atoms are fully or partially replaced by the halogens fluoride, chloride, bromide or iodine, one speaks of halogen-fluorocarbons. Those that contain the halogen fluoride play an important role in refrigerating technology as refrigerating agents and HR foam insulating cell gases.

In DIN 8962 the use of chlorofluorocarbons in the refrigerating technology are identified with "R" for "refrigerant" followed by the ciphers relevant to the atomic composition. For applications in other sectors (e.g. aerosols, plastic foams, cleaners and solvents), the common acronym CFC, originally FC, has become standard practice. Since CFCs behave completely differently with regard to the estimated risk they pose to the ozone layer, depending on whether they contain additional hydrogen atoms or no chloride component, the following descriptions are chosen:

CFCs
are chorofluorocarbons, which always contain fluoride and chloride in each molecule but not hydrogen. They are also described as "fully halogenated" hydrogens since in these molecules all hydrogen atoms are replaced by halogens (fluoride, chloride, bromide and iodine). This group of compounds is chemically and thermally stable. It can therefore continue to diffuse unchanged through the troposphere according to the ozone theory and then contribute to ozone depletion in the stratosphere owing to the chloride contained in the molecule.

FCs
are fluorocarbons whose only halogen in the molecule is fluoride. They are also hydrogen-free, that is, fully halogenated and hence similarly chemically and thermally stable. However, since they do not contain chloride, they do not contribute to ozone depletion.

HCFCs
are hydrochlorofluorocarbons. They contain at least one hydrogen atom in each molecule and are therefore described as "partially halogenated." By comparison with the fully halogenated chlorofluorocarbons they possess a lower degree of chemical stability and are therefore subject to considerable decomposition even in the troposphere. Their impact on ozone depletion in the stratosphere is thus insignificant.

HFCs
are hydrofluorocarbons. They also contain at least one hydrogen atom per molecule, the only halogen being fluoride. Since they no longer possess chloride halogen they too, like FCs, do not contribute to ozone depletion.

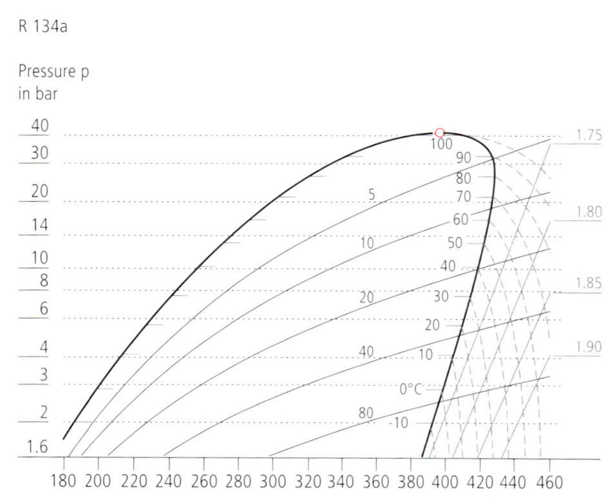

Figure 7.16
h, log-p diagram for 134 a

— Spec. volume, in dm³/kg
— s = entropy
-- Isotherms
○ Critical point
h' = 200 kJ/kg at 0°C

7.2.2 Refrigeration machines

As we have already noted, compression is the most frequently used refrigeration process for generating refrigerating energy.

The operating principle of a refrigerating machine with compressor is illustrated in **Figure 7.18**.

The compressor suctions the refrigerant as cold steam and compresses it (piston compressor/screw-type or worm compressor). As a result the refrigerant is overheated and pressed to the condenser through a pipeline containing an oil interceptor, which separates the oil deposits produced by the compressor. In the condenser the overheated refrigerant steam is condensed because the refrigerant is cooled in the counter current with cooling water or air, which means that heat is extracted from the refrigerant. After it has exited from the condenser, the condensed refrigerant is guided through a dryer, which extracts water and an expansion valve, which reduces the high pressure of the refrigerant to the level of evaporation pressure. Some of the refrigerant already evaporates while this pressure reduction occurs although the heat content remains the same. In the evaporator, the refrigerant absorbs heat from the medium to be cooled (e.g. water), allowing it to evaporate fully.

In the example shown here the evaporator is a tubular heat exchanger through which the water to be cooled (cold water for cooling) flows counter current to the refrigerant. Once the refrigerant has evaporated, it is once again suctioned by the compressor, compressed and the cycle resumes.

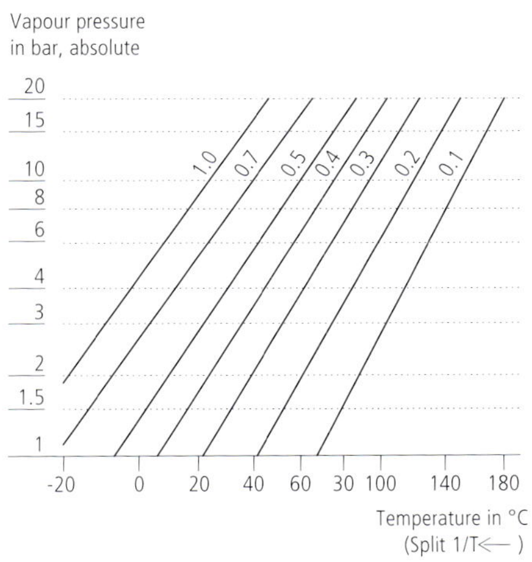

Figure 7.17
Pairing ammonia/water in log p, 1/T-diagram

—— NH_3 concentration

Figure 7.18
Schematic diagram of refrigerating plant with piston compressor and approximate temperature indications during refrigeration process

1 Compressor
2 Oil interceptor
3 Condenser
4 Dryer
5 Expansion valve
6 Evaporator
7 Re-cooling plant
8 Chiller
9 Cold water pump
10 Cooling water pump

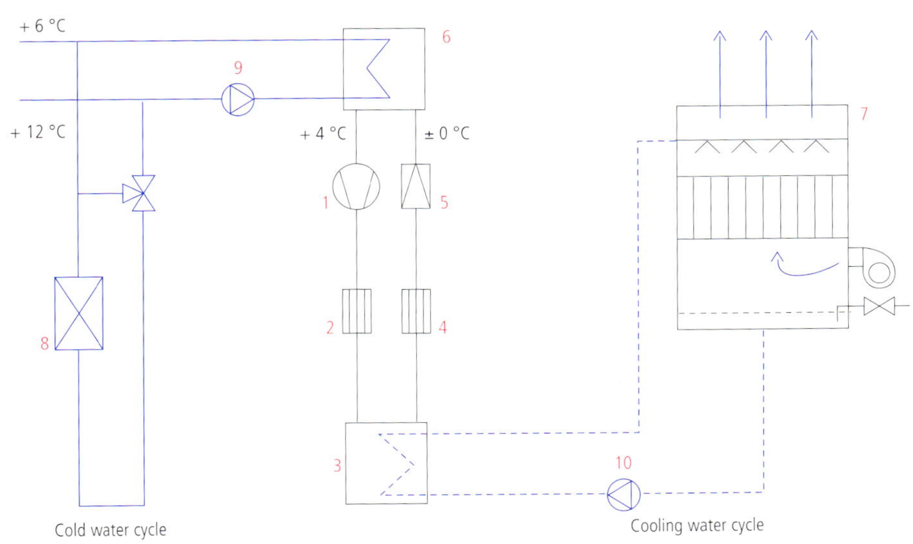

7.2.2.1 Compression refrigerating machines with scroll compressor

As is clearly visible in **Figure 7.19**, the electric motor is housed in the lower half of the compressor and the scroll compressor in the upper half. In the bottom left corner of the image you can see the hook-up for the refrigerant feed (gas) and in the upper compressor housing the refrigerant outlet after compression. The entire refrigerating compressor is hermetically sealed and cooled with a refrigerating agent, i.e. the refrigerant flows around the electric motor to the compressor, where it is, in turn, expelled from the housing.

Two interlocking spirals compress the refrigerant and push it out of the housing into the pipe leading to the condenser. The advantage of the scroll compressor over the piston compressor lies in the fact that it operates exclusively with rotating movements and that constant movement is combined with smooth and quiet operation. Moreover, this configuration does not require valves and the component load behaviour falls into a very favourable range.

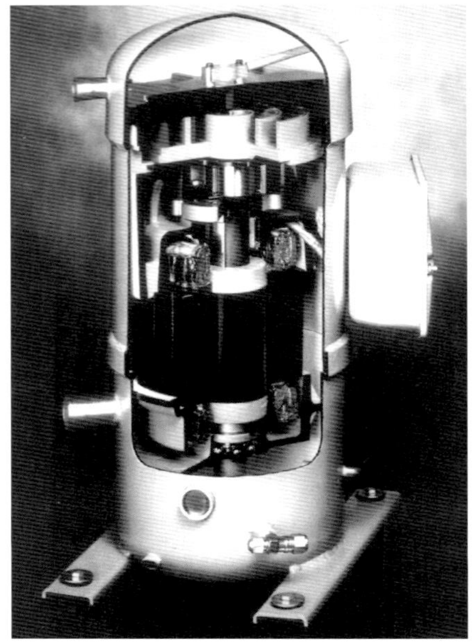

Figure 7.19
Scroll compressor
(Image: Trane)

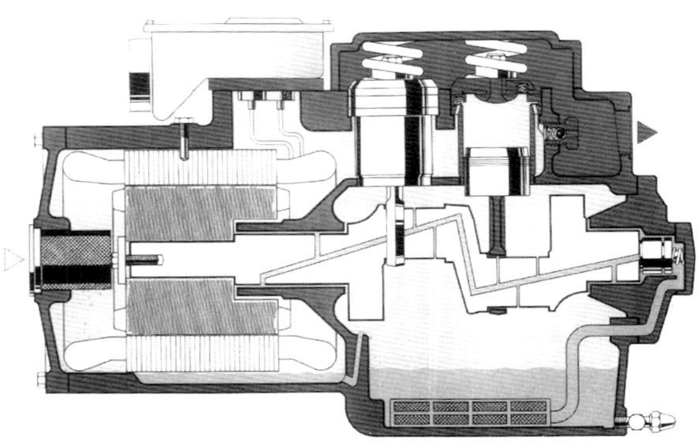

Figure 7.20
Reciprocating piston compressor

Figure 7.21
Reciprocating piston refrigerating plant with separate condenser cycles

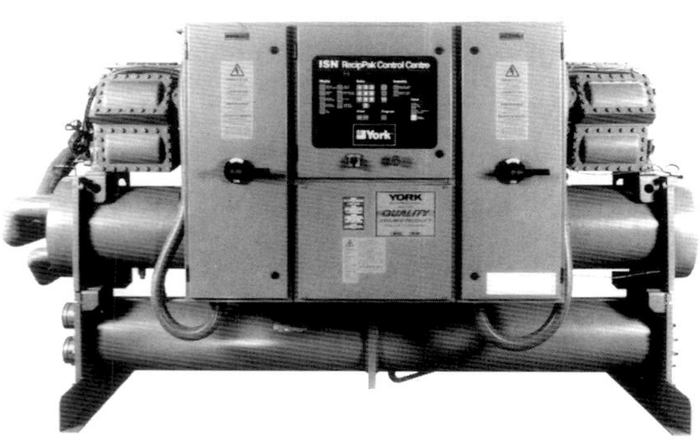

Figure 7.22
Refrigerating plant with piston compressor for cold water processing, YCWZ series, YORK, output range 180 kW – 370 kW
(Image: YORK)

7.2.2.2 Compression refrigerating machines with piston compressor

In a piston or reciprocating piston compressor, the refrigerant is compressed by means of pistons moving back and forth within cylinders and connected to opening and closing valves – a principle that is familiar from car engines. **Figure 7.20** presents a section through a reciprocating piston compressor. Here the refrigerant steam enters the machine on the left side, flows again around the electrical motor, cooling it. In the example shown, the electric motor is linked to a crankshaft that moves two pistons set into the cylinders. The pistons suction the refrigerant steam into the cylinder via an inlet valve and compress it.

Reciprocating piston compressors are the most frequently used compressors in refrigerating machines and we have most long-term data on operation for these compressors. Still, they present some disadvantages compared to worm or scroll compressors: they have a slightly lower efficiency than worm compressors and the machine itself is exposed to greater loads as a result of the to-and-fro movement of the pistons.

Figure 7.21 depicts a reciprocating piston refrigerating machine with the compressor component, the evaporator below it and the separated condensers above it. **Figure 7.22** shows a compact refrigeration machine installed on a floor frame with piston compressors for cold water generation. This machine is typically used for a cooling output of 400 to 450 kW, the output being regulated by activating or deactivating individual piston compressors.

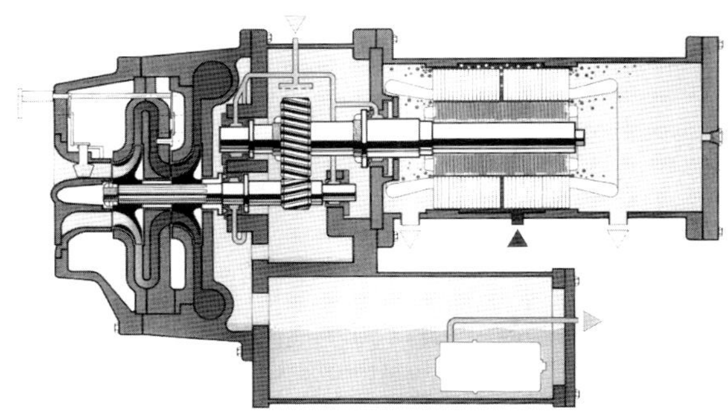

Figure 7.23
Hermetic turbo condenser CCGE series for use with refrigerant R134a
(Image: Trane)

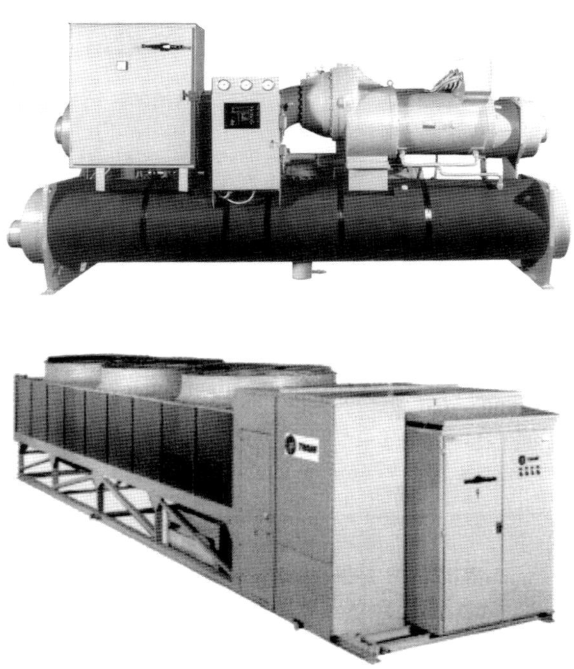

Figure 7.24
Refrigerating plant with hermetic turbo compressor, CVGE series, output range 930 kW – 3,000 kW and compact turbo refrigerating plant, CVAE series, output range 700 kW – 1,200 kW for roof installation
(Image: Trane)

7.2.2.3 Compression refrigerating machine with turbo compressors

For cooling output of more than 450 to 500 kW, it is common to use compression refrigerating machines with turbo compressors (turbo refrigerating machines).

In turbo compressors, the refrigerant is compressed by means of a one- or two-stroke turbine (**Figure 7.23**).

The electric motor of the refrigerating machine powers a turbo compressor via a transmission, whereby the compression of the gas stream in the wheel occurs through acceleration. The kinetic flow energy is translated into a pressure increase in the diffuser.

The turbo compressor shown in **Figure 7.23** is hermetically encapsulated and once again cooled with refrigerants.

The output of turbo compressors is regulated by means of speed control (driven by a frequency transformer), pre-rotation control (adjustable blade in front of wheel entry) or diffuser control (adjustable blades in the diffuser). Another form of output control is achieved by means of hot gas bypass control, with the bypass being formed by means of an overflow (or relief) valve installed between the pressure and the suction side of the turbine.

Turbo compressors are designed to operate with an output range of 300 kW to 30 MW. The images in **Figure 7.24** illustrate a large refrigerating plant with hermetic turbo compressor and a compact model for roof installation, respectively. The evaporator, from which the refrigerant is suctioned off into the turbine (bottom part of image) and the condenser, which is located above the evaporator and connected to the turbine on the pressure side, are clearly visible in the image, as are the connections for the cooling- and cold-water cycles to the evaporator and to the condenser.

The diagram in **Figure 7.25** illustrates the interaction between a turbo refrigerating machine with evaporator and a connected cold water cycle, as well as a condenser and connected cooling cycle.

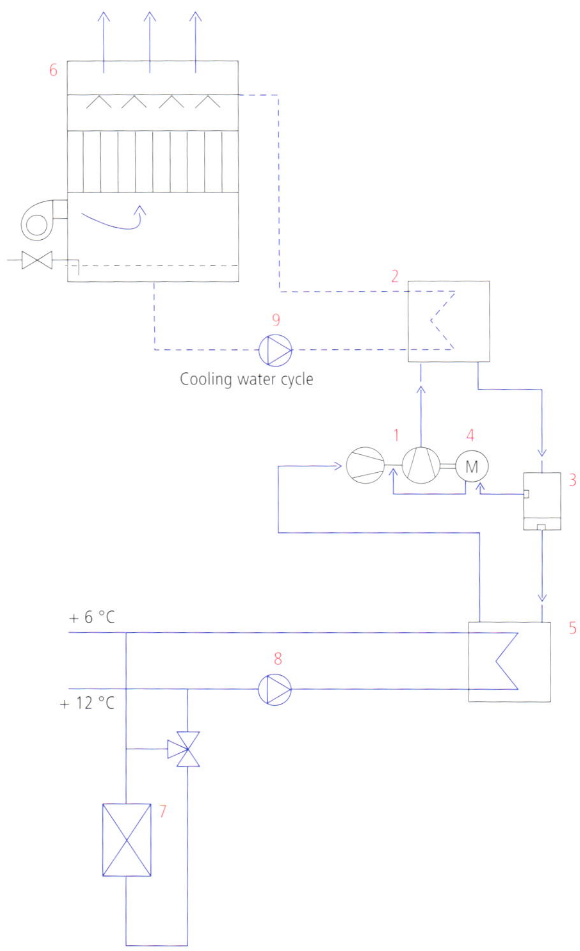

Figure 7.25
Schematic diagram of a refrigerating plant with turbo compressor

1 Two-phase turbo compressor
2 Condenser
3 Economizer
4 Compressor motor
5 Evaporator
6 Re-cooling plant
7 Air chiller
8 Cold water pump
9 Cooling water pump

7.2.2.4 Compression refrigerating machines with screw-type or worm compressors

Figures 7.26.1/2 depict refrigerating machines with condenser and evaporator, and a top-mounted engine with screw-type or worm compressor in a hermetic construction. In the screw-type or worm compressor, compression occurs through interlocking, screw-like drums. The refrigerant is compressed by the diminishing spaces between the drums rotating in the axial direction. The advantage of screw-type or worm compressors is that they produce less heat during the compression process as a result of intensive oil cooling, with the end result that the compression temperature is lower. As with scroll compressors, screw-type compressors also present the additional advantage that in larger installations part of the refrigerant stream is suctioned off to the heat exchanger via an additional suction stub. This additional cooling of the refrigerant makes it possible to achieve an even greater cooling output.

Aside from the compressor types itemized above, rotating piston compressors and sliding vane rotary compressors are also employed to a lesser degree.

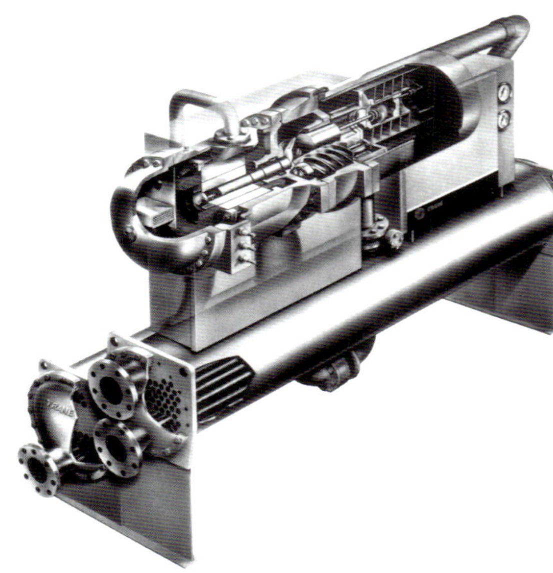

Figure 7.26.1
Cutaway view of compression refrigerating plant with worm compressor
(Image: Trane)

Figure 7.26.2
Compression refrigerating plant with worm compressor,
YS series, output range
330 kW – 1,910 kW
(Image: YORK)

7.2.2.5 Absorption refrigerating machines (chillers)

The absorption refrigeration cycle (**Figure 7.27**) indicates that steam or hot water are required for the operation in the generator. Absorption refrigerating machines are therefore principally used for applications where inexpensive heating energy is available during the cooling period (industrial steam/waste heat from a combined heat and power plant/solar energy etc.).

The absorption refrigerating machine consists of six essential components clearly identified in **Figure 7.27** and visible in the photographs of this type of machine (**Figures 7.28.1/2**).
These components are:
- Absorber (lower part of bottom tank)
- Evaporator (upper part of bottom tank)
- Concentrator (upper tank) (= expeller)
- Condenser (upper tank)
- Solution pump (concentrator/solution pumps)
- Heat exchanger

Unlike the machines described above, absorption refrigerating machines are not equipped with compressors since refrigeration occurs via a thermal process. They offer the tremendous advantage, therefore, of operating with very little vibration and nearly noiselessly. Disadvantages are the large dimensions and the greater recovery output (recovery cooling plants).

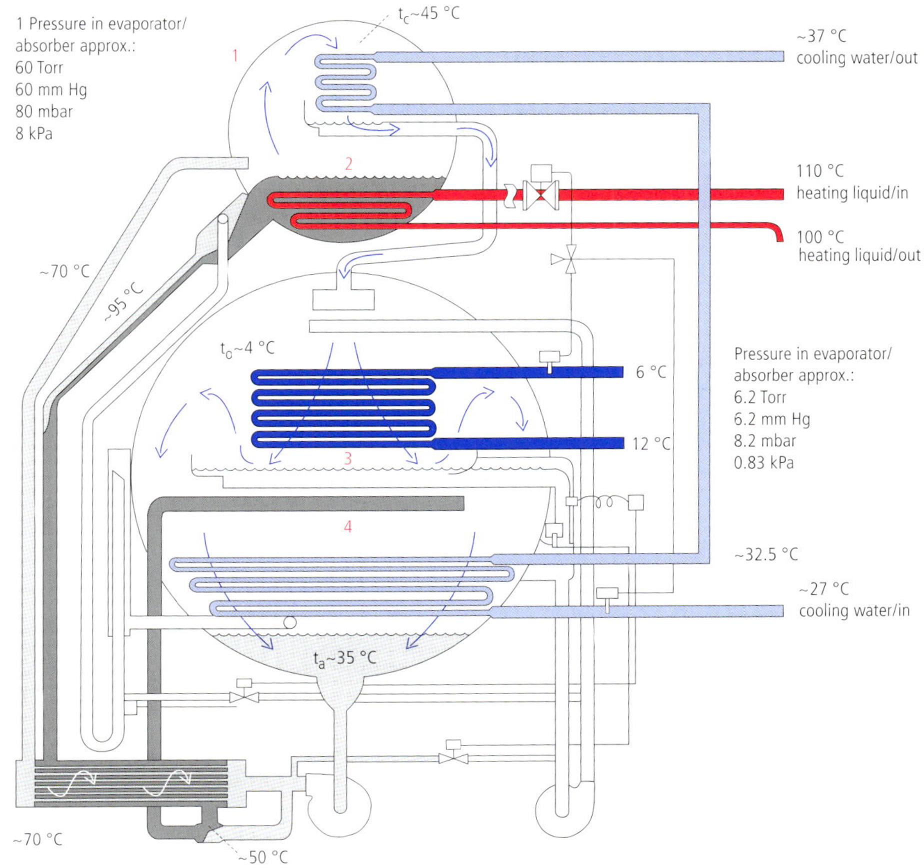

Figure 7.27
Principle of the cycle in a single stage absorption chiller
(Image: YORK)

1 Condenser
2 Generator
3 Evaporator
4 Absorber

Cold water
Cooling water
Concentrated solution (LiBr)
Lithium bromide/water mixture (Li Br)
Steam or hot water
Refrigerant

Figure 7.28.2
Single-phase absorption chiller, ABSC series, Trane, output range 390 kW – 6,000 kW
(Image: Trane)

Figure 7.28.1
Single-phase absorption chiller, YIA series, output range 420 kW – 4,850 kW
(Image: YORK)

If waste heat is not available, direct-fired absorption refrigerating machines can be used (Figure 7.28.3). These machines are equipped with an integrated boiler, which supplies the required expeller output. During periods when the available output is only partially utilized or when there is no refrigerating demand, the thermal output can also be fed into the grid, which can help to replace other heat generators.

Figure 7.28.3
Two-phase, direct-fired absorption chiller,
YPC series, output range
700 kW – 5,300 kW
(Image: YORK)

7.2.2.6 R718 water chiller (water as refrigerant)

Water is the most environmentally friendly and natural refrigerant, and is given the code R718. While R718 has been used as a refrigerant in H_2O-LiBr absorption plants for some time, compression refrigeration with water as refrigerant was not feasible in the past.

Advances in material development in compressor design have made this configuration possible in recent years (compression refrigeration with water as refrigerating agent). A water chiller that operates with water as refrigerant and achieves refrigerating outputs between 0.5 and 1.0 MW depending on the evaporation temperature was developed at the Institut für Luft- und Kältetechnik (ILK or Institute for Air-conditioning and Refrigeration Technology, Dresden, Eng. Heinrich). The principal application lies in air-conditioning, but heat pump applications, especially with the aim of using waste heat, are also possible.

Figure 7.29 illustrates the principle operation of this type of (single-step) R718 water chiller. The relevant cold steam process is plotted in the log p, h diagram in **Figure 7.29.1**. Supply water is expanded to low pressure in the direct evaporator. During this process, some of the water evaporates, thus becoming a refrigerant. The remaining water is cooled to the desired cold water temperature (e.g. 6 °C) and made available to the consumer of cooling or refrigerating energy.

The resulting refrigerant vapour is siphoned off through a connected turbo compressor, compressed to a higher pressure and again liquefied in the condenser as a result of direct contact with the cooling water. The released condensation heat is discharged to the cooling water (e.g. 29 °C).

The innovative direct condensers and direct evaporators, new to refrigeration technology, are possible because the external material streams are identical to water as cold and refrigerating water cycle, thus allowing for heat exchange in direct contact.

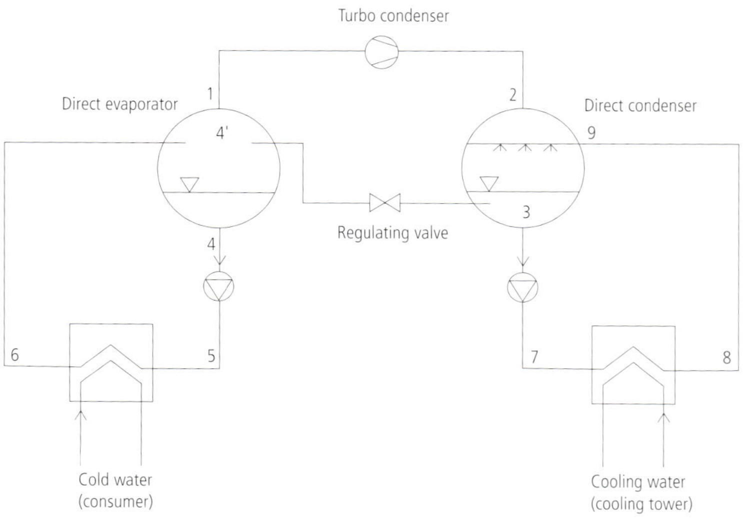

Figure 7.29
R718 water chiller,
dual phase

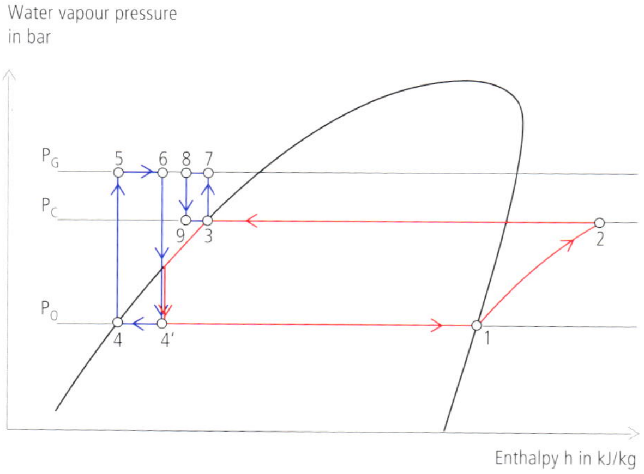

Figure 7.29.1
log p, h diagram of process

1 –2	Compression
2 –3	Liquefaction
3 –4'	Expansion
4'–1	Evaporation
4–5–6–4'	Consumer (cold water)
3–7–8–9	Re-cooling (cooling water)
P_G	Grid pressure
P_C	Condensation pressure
P_O	Evaporation pressure

Figure 7.30
R718 water chiller (800 kW) in test lab at the ILK, Dresden. Evaporator, condenser and two compressor stages are accommodated in the hermetic housing (2 m diameter)
(Aqua Turbo Kältetechnik, Sonneberg)

R718 water chillers achieve higher output levels than conventional refrigerants. Two-stage R718 water chillers, with internal intermediate cooling, can achieve output coefficients of g > 6.

The use of water as a refrigerant in compression refrigeration machines is designed for larger refrigeration outputs from 400 kW and above, particulary in the MW range. Figure 7.30 depicts such a refrigeration machine in the installation phase.

7.2.3 Well cooling

Groundwater stored in the soil can be tapped through wells and used for cooling purposes with a feed temperature of roughly 12 °C. Well cooling was a standard procedure and was, and often still is, used for cooling in industrial engineering processes. For years, however, the use of well water for cooling has diminished since the relevant authorities (e.g. waterworks) no longer grant permission to do so. Thus, well water cooling is primarily used in rural applications, and it is limited even in this area.

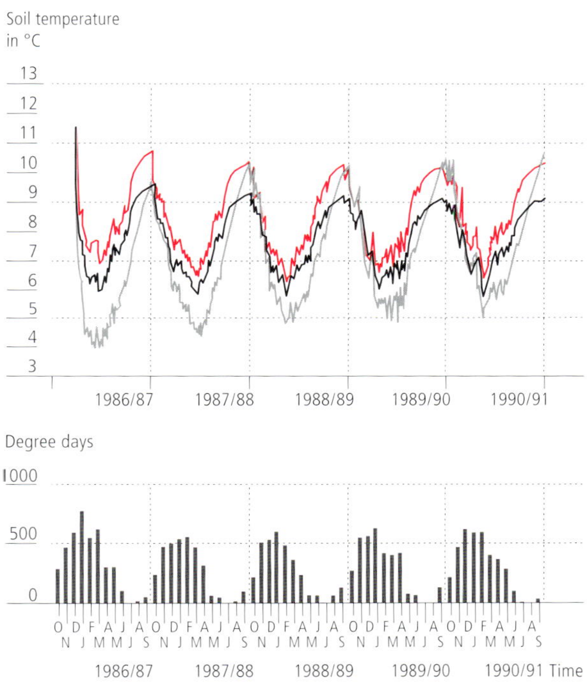

Figure 7.31
Temporal variation in earth temperatures at depths of 5 m, 50 m and 85 m, and at 50 m distance from bore hole in Elgg, Switzerland, over a period of five years of operation
(Degree days 20/12)

85 m
50 m
5 m

7.2.4 Bore holes (Shallow geothermics)

For generating small amounts of refrigeration output, bore holes or earth coils can deliver the necessary cooling energy.

Figure 7.31 records the temporal variation in earth temperatures at depths of 5 m, 50 m and 85 m (bore hole installation in Elgg, near Zurich). As is evident from the graphs, earth temperatures at shallow depths drop by approx. 4 K more than at a depth of 85 m. The temperature fluctuation in the earth depending on heat extraction is also visible (degree days).

When bore holes are positioned, care must be taken to evaluate the temperature interaction between bore holes, because earth heat for more distant regions cannot replenish the extracted heat quickly enough. In **Figure 7.32**, the cone of depression near the bore hole shows the radial temperature distribution around the hole.

Another potential application of bore holes is the direct creation of cold energy (cooling heated water in the soil). Corresponding system solutions can be used for the direct and indirect cooling of building components mentioned above.

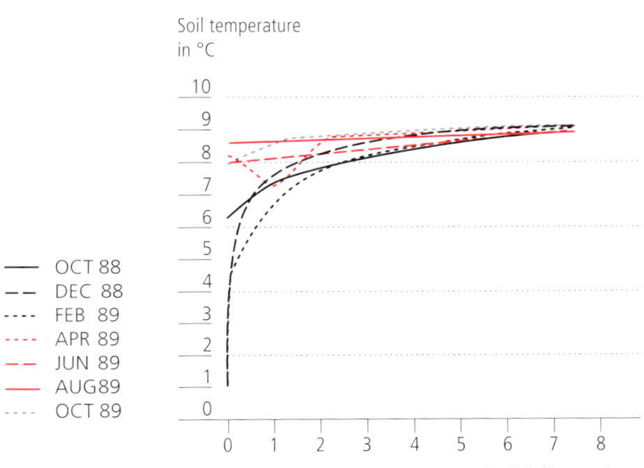

Figure 7.32
Dynamic cone of depression
The graph shows the radial temperature distribution at the installation in Elgg at different times
(after W.J. Eugster et.al.)

It is important to ensure that the supply temperatures of relevant cool cycles are not much lower than + 18 °C and the return temperatures not far above + 22 °C. Utilizing earth cooling means transferring heat emitted from the building to the soil. Therefore, it is necessary to ensure that the annual energy flow is well balanced, i.e. that the heat energy returned to the soil during the warm season is balanced during the cold season. Assuming a spatial heat energy flow of approx. 0.065 W/m^2 in the soil, an average distance of 6 m between bore holes leads to a probe energy yield of approx. 30 W/m of probe lenght. The maximum probe depth should not exceed 100 m of probe lenght; otherwise investment costs are disproportionately high. **Figure 7.33** illustrates the operating principle of the cooling cycle.

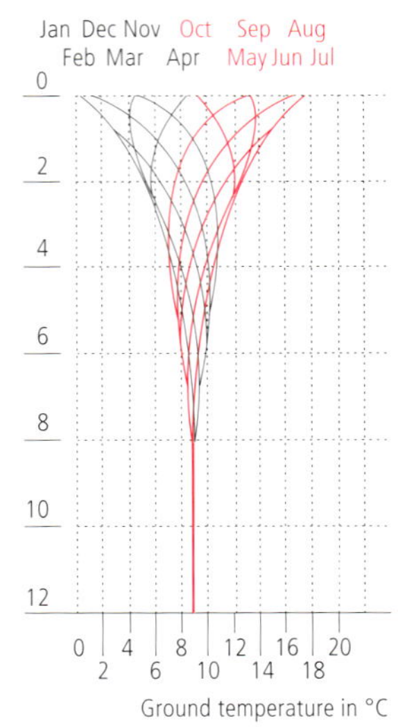

Figure 7.34.1
Ground temperature

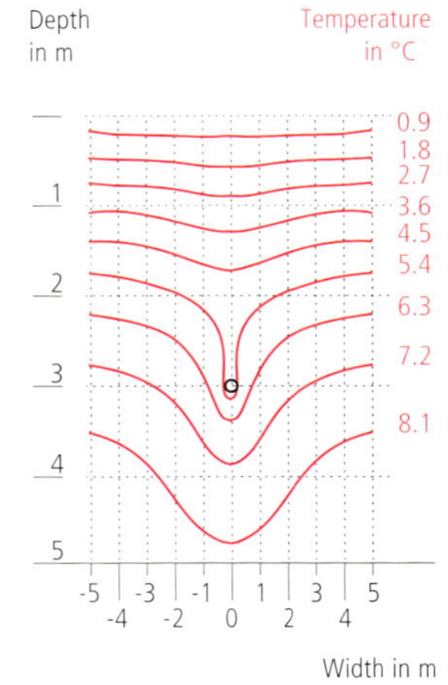

Figure 7.34.2
Cone of depression in ground surrounding earth pipe (Isothermal lines)
Earth surface: 0 °C
Pipe surface: 4.5 °C

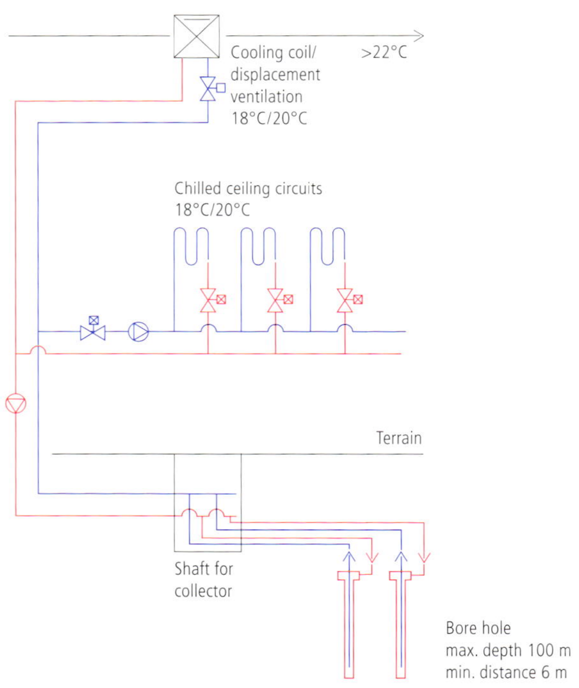

Figure 7.33
Principle of cooling cycle in bore holes

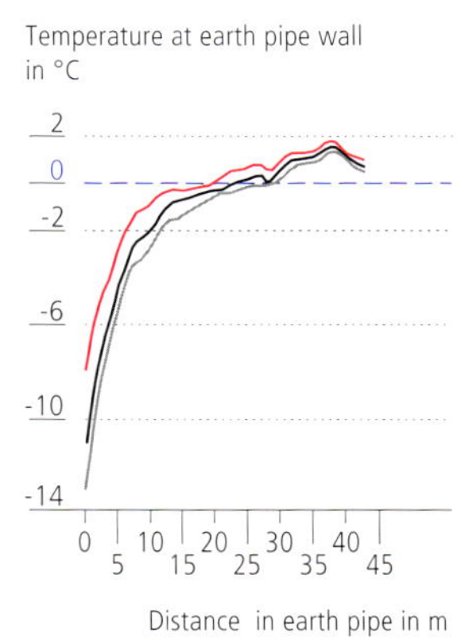

Figure 7.34.3
Temperature in the direction of the earth pipe wall, winter

- Values during max. outside temperature
- Daily average values
- Values during min. outside temperature

Figure 7.34.4
Temperature in the direction of the earth pipe wall, summer

Operation of earth pipe:
9 a.m. to 9 p.m.

- Values during max. outside temperature
- Daily average values
- Values during min. outside temperature

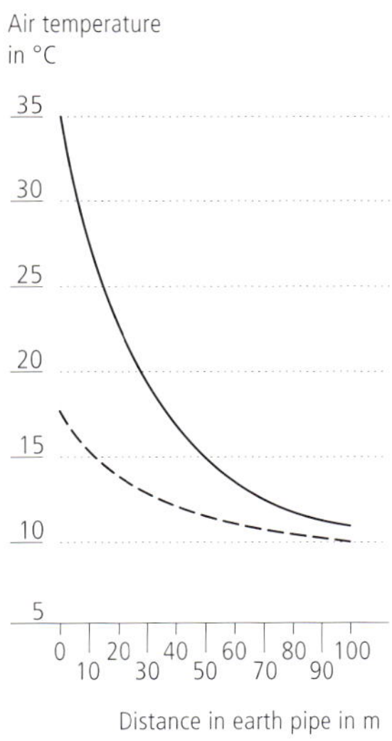

Figure 7.34.5
Air temperature in direction of the earth pipe (summer) at 3 m depth (ø127 mm, V = 140 m³/h) for various outside temperatures

— 35 °C
-- 17.6 °C

7.2.5 Thermal energy in the soil

To cool fresh air in summer before it enters a building and to humidify or heat it in winter, it seems sensible to draw fresh air through pipes in the ground or through a concrete thermal labyrinth below the building.

Thermal tunnel (earth pipeline)

For smaller air volumes, e.g. for induction units in residential buildings, pipes with smaller dimensions (e.g. clay, concrete or plastic pipes) can be installed in the ground. The pipes should be at some distance from the building, because the impact of heat transmission from the building to the pipes would be too great when cooling is required. However, when heating energy is the primary goal, the heat lost through basement walls can naturally be captured in well-ventilated pipelines. In the planning of such installation, the pipe grid or the individual pipes must be buried deep enough in the ground to create either a noticeable heating or cooling effect. The graphs in **Figures 7.34** can serve as a rough guide for sizing. They show the familiar funnel phenomenon of soil temperatures from top to bottom (surface to depth), as well as the temperature curves in the direction of the earth pipe wall in winter and summer. **Figure 7.34.2** shows the cone of depression in the ground surrounding an air-conducting earth pipe at a depth of 3 m, using the material values of a moist clay soil as an index. As indicated for bore holes, buried earth pipes should be spaced at least 8 to 10 m apart to minimize their influence on each other (maximum heat gain for buried pipes (earth heat exchangers) is achieved at a depth of 4 – 6 m, measuring a heat gain of 6 – 6.5 kW/a per m^3 of conducted air volume). In summer, the optimum depth for the earth heat exchanger pipe is approx. 4 – 6 m and the cold energy gain approx. 7 kW/a (assumed pipe length approx. 40 m).

Assuming an installation depth of 3 m, a pipe diameter of 127 mm and an air volume stream of 140 m^3/h (soil as above) produces a steady state condition only after the 100 m mark (**Figure 7.35.5**).

Thermal labyrinth

In conclusion, it is worth noting a research project carried out by the firm Rud. Otto Meyer. A thermal labyrinth, consisting of 140-m-long concrete tunnels, was installed beneath the ground floor of the municipal theatre in Heilbronn. Fresh air is drawn into these tunnels and conducted through the system below the building, where it is heated in winter or cooled in summer. **Figure 7.35.1** shows the thermal labyrinth in the basement during construction (the black arrows in the image indicate the movement of air within the labyrinth). **Figure 7.35** is a cross-section with the schematic of the outside air flow, which reaches the air-conditioning and heating system either directly or after passing through the labyrinth.

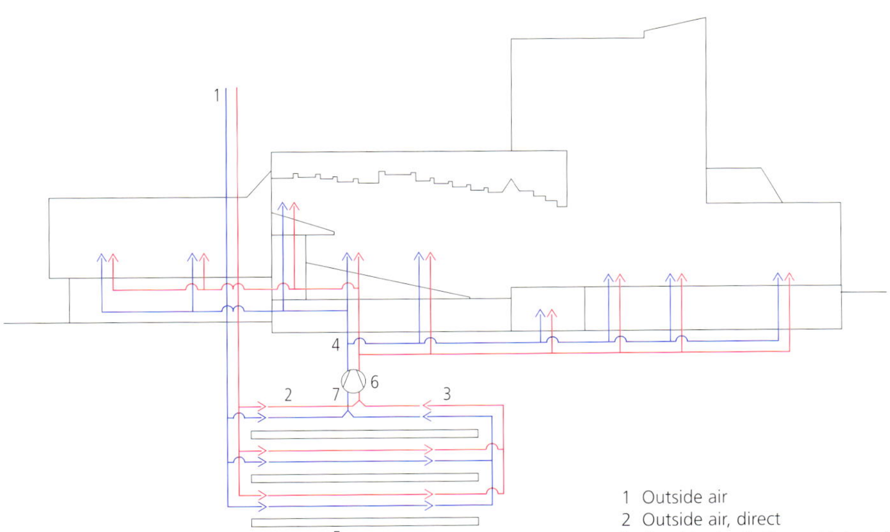

1 Outside air
2 Outside air, direct
3 Outside air, via thermal labyrinth
4 Fresh air to rooms
5 Underground heat exchanger Concrete thermal canal
6 Fan
7 Dampers direct-thermal air

Figure 7.35
Diagram of outside air flow, reaching the air-conditioning system either directly or by way of the labytinth

Figure 7.35.1
Thermal labyrinth in new municipal theatre in Heilbronn, construction arrows indicate air flow)

As the figures above illustrate, fresh air is drawn directly into the building through a ground-level tunnel and conducted through the labyrinth. The ground surface of the concrete tunnel serves as a heat exchanger, which is certainly not ideal for harnessing geothermal energy. In any case, this system yields measurable heating and cooling effects (**Figures 7.36.1/2**). In winter, the outside air heats up in the thermal tunnel by approx. 2 to 4 K, and, in summer, it is cooled by approx. 1to 8 K.

The maximum tunnel transmission rates are shown in **Figure 7.37** in relation to the volume flow, which naturally fluctuates greatly in a theatre building, depending upon use and time frame. The maximum tunnel transmission rate is a result of the daily tunnel transmission rate and the effective transmission surface. When calculating the amortization for this research project, it was found that the additional investment necessary to construct the thermal labyrinth would be paid off over a period of roughly eight to ten years, a finding that argues for further studies in this area and the implementation of similar solutions.

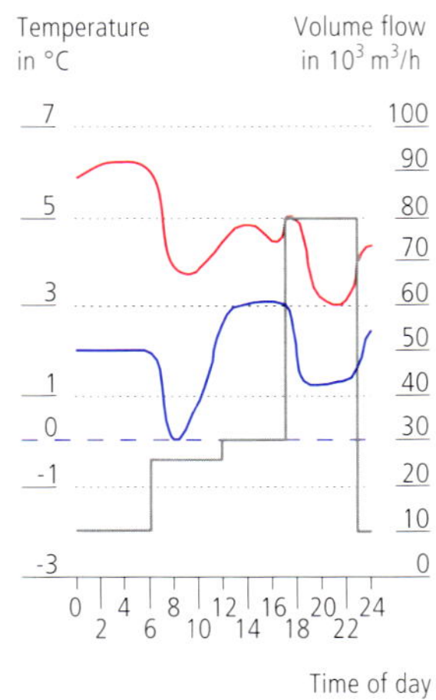

Figure 7.36.1
Recorded temperatures and analysis,
one day in winter 1984

— Start of duct
— End of duct
— Volume flow

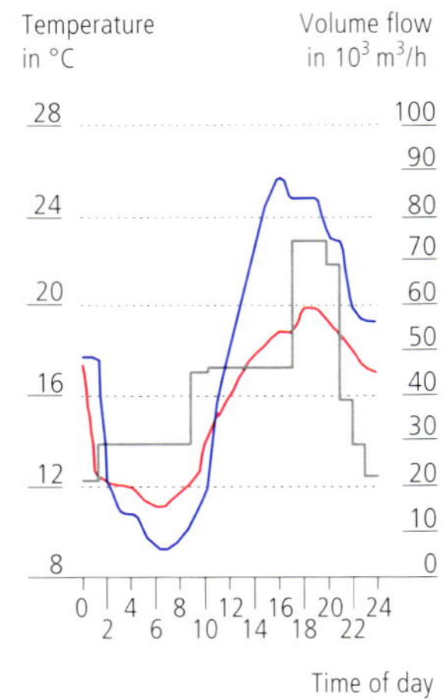

Figure 7.36.2
Recorded temperatures and analysis,
temperature increase and drop,
one day in May 1983

— Air shaft/entrance
— Air shaft/exit
— Volume flow

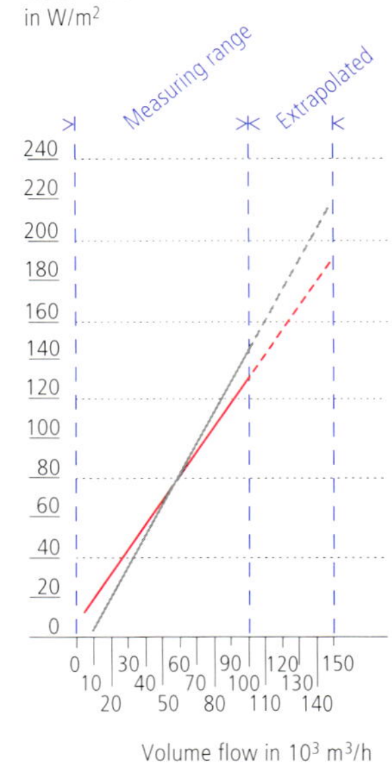

Figure 7.37
Specific maximum tunnel transmission rates
(from representative values)

— Heat
— Cold (August, July)

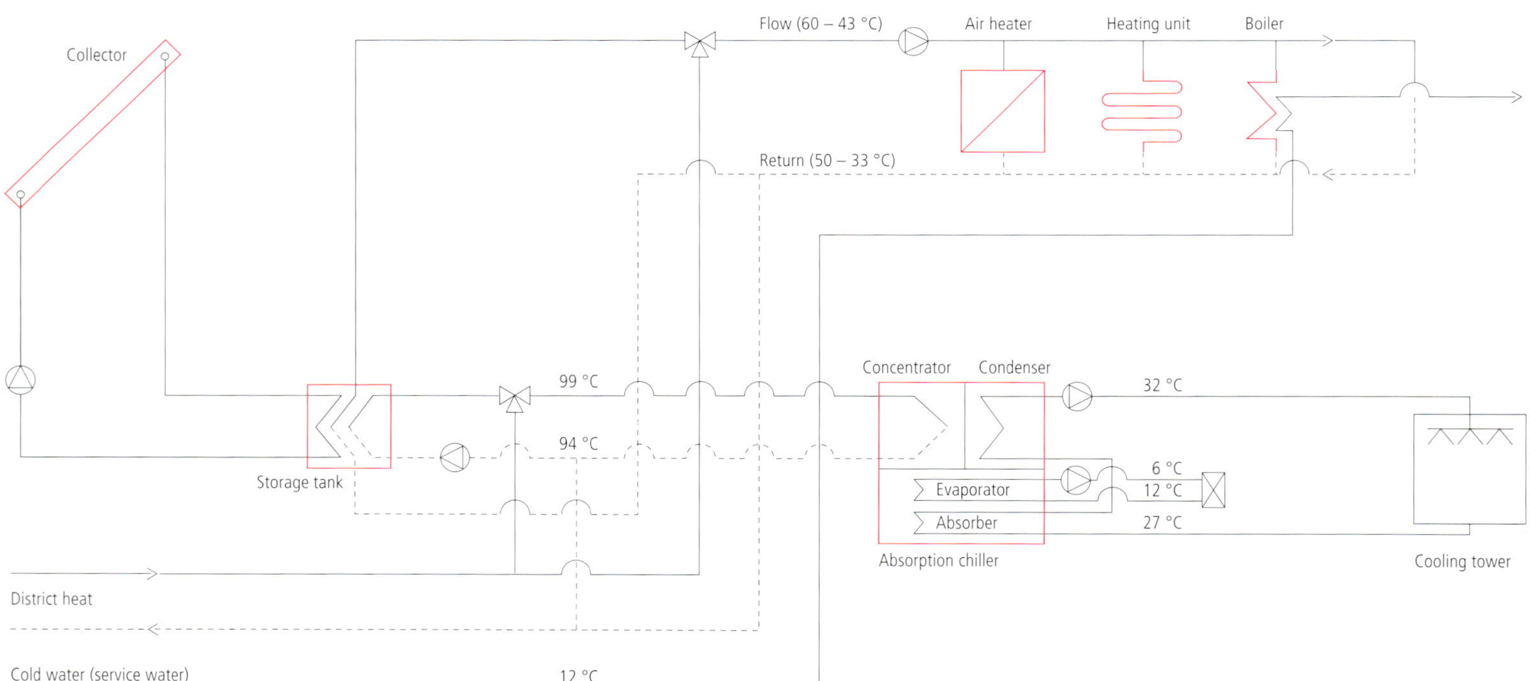

Figure 7.38
Solar energy heating and cooling installation

7.2.6 Generating cooling power with solar energy

Figure 7.38 is a schematic diagram of a system where solar energy delivers both the thermal energy for heating and the thermal energy required to operate a refrigerating plant. Evacuated tube collectors or a focal line system are required for long-term use of this system and in particular for achieving high temperatures and efficiencies (supply temperatures well above 100 °C). The tremendous advantage of the solution presented here is that the periods of high radiation intensity, and hence high cooling loads, coincide with the necessary thermal energy for the absorption-refrigerating plant; in other words, that energy consumption costs will be low.

7.2.7 Cooling by means of sorptive dehumidification

As described, the most frequent method of cold generation for air-conditioning is based on the use of refrigerants, which evaporate in a closed cycle and condense, with the cold end of the process being used for cooling and dehumidifying supply air. This process of cold generation is characterized by a diminished output quotient, especially under summer conditions, since the condenser heat can be evacuated only when its temperature is higher than the outside temperature. At the same time, the evaporation temperature of the process must be kept very low due to the required supply air humidity. The result is a great temperature differential in the refrigerating plant, which yields output quotients (ratio of absorbed to discharged output) in a range of 2 to 3.

The computed output quotients of the refrigerating plant demonstrate a deterioration by roughly 24 % when the outside temperature rises from 25 to 35 °C. Conversely, the quotient could be measurably improved if the evaporation temperatures were raised quite considerably. Sorptive dehumidification, especially, can make a contribution as an alternative to the classic approach to dehumidification by means of depression of the dew point.

One alternative to depression of the dew point lies in the use of hygroscopic materials combined with both solid and liquid sorption systems. In recent years, the most successful from a technical perspective have been sorption generators (Figure 7.39), where the storage mass assumes hygroscopic properties by means of impregnation or coating. For applications aimed at air-conditioning for comfort requirements, one essential characteristic is the temperature necessary for regeneration.

Figure 7.40 compares sorption generators where the hygroscopic material is silica gel in one instance and lithium chloride in the other. Dehumidification was factored in as a function of the regeneration temperatures for nearly identical parameters. The evidence shows that storage material impregnated with lithium bromide is more advantageous, especially at low regeneration temperatures, achieving dehumidification of 2.8 g/kg even at regeneration temperatures of roughly 50 °C.

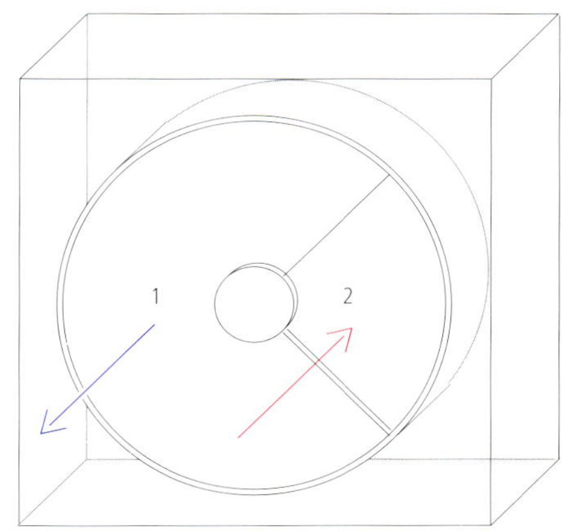

Figure 7.39
Sorption regenerator

1 Exhaust air, regeneration of sorption regenerator
2 Supply air, dehumidification of outside air

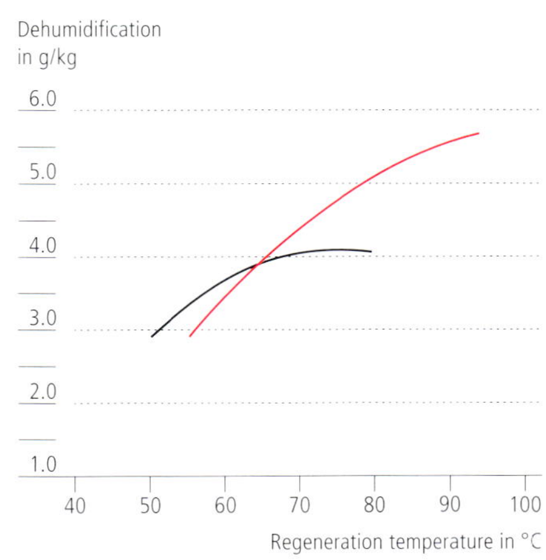

Figure 7.40
Dehumidification related to regeneration temperature

LiCl-cellulose rotor
W_{Aa} = 2.5 m/s, n = 21.8 1/h
Silica gel rotor
W_{Aa} = 2.9 m/s
Inlet conditions
$t_E \simeq 28$ °C, x_E = 10 g/kg

Figure 7.41 illustrates the integration of sorptive dehumidification and **Figure 7.41.1** the operating principle of the process. Outside air (1) is dehumidified as it passes through the sorption generator. As evaporation and hydration heat is released, the temperature rises until it reaches the condition (2). The subsequent heat recovery lowers the temperature relative to the heat recovery quotient of the heat recovery system. The temperature decrease occurs along a line of constant moisture content until it reaches condition (3). The subsequent dry cooling in the supply air makes it possible to achieve the supply air point (4). The cooling load in the room is evacuated from (4) to (5). Once the cooling load has been evacuated, the heat recovery from (6) to (7) and another temperature increase from (7) to (8) is carried out. This temperature increase of up to 44 °C is necessary for the continuous regeneration of the sorption generator. The energy generated is the actual driving energy of the process, with the exception of fans and pumps. The water volume absorbed during the dehumidification phase is transported out of the sorption generator by the warmed air resulting in condition (9).

Integrating the sorption generator, which can be operated with a higher rotation as an enthalpy recovery plant in winter, and combining it with the heat recovery device achieves a clear improvement in the energy efficiency in the winter months than comparable conventional air-conditioning systems.

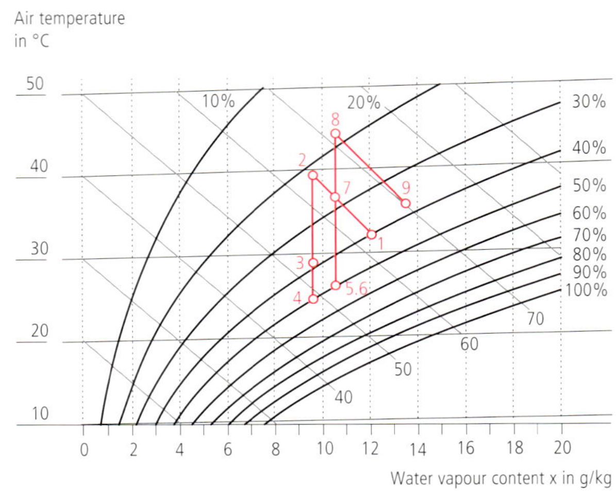

Figure 7.41.1
Percentages of sorptive dehumidification, h, x diagram

1-2 Sorptive drying of outside air
2-3 Heat recovery (supply air)
3-4 Dry cooling of supply air
4-5 Absorption of cooling load in room
6-7 Heat recovery (exhaust air)
7-8 Heating to regeneration temperature level
8-9 Regeneration of sorption regenerator

Figure 7.41
Integrating sorptive dehumidification into air-conditioning supply for a room with chilled ceiling and stack ventilation

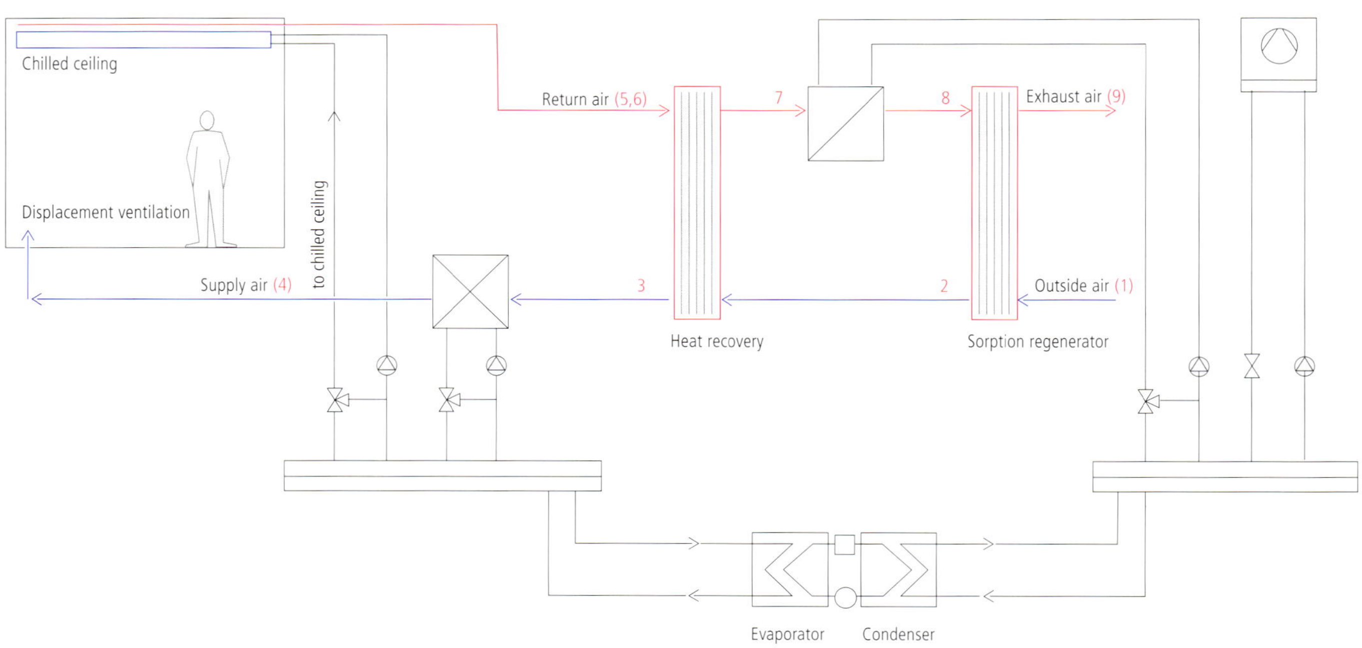

7.2.8 Desorption process

The desorption process or air cooling via adsorptive dehumidification and adiabatic cooling is a system designed to generate cooling energy using heating energy. **Figures 7.42.1/2** show the system layout, indicating temperatures and humidities for a specific test case and relevant graphs in the chart.

Air cooling by means of adsorptive dehumidification and adiabatic cooling does not require a cooling plant in the usual sense, when dehumidification is not required. Cooling, e.g. with a gas-motor-driven heat pump, becomes necessary when the outside air is to be cooled and dehumidified. The cooling system can operate without H-CFCs, ammonia or brine cycles. Instead, it functions by spraying cold water on to the supply and return air, thus creating adiabatic changes that simultaneously cool and humidify. The outside air is initially dehumidified because the packing of the heated drying rotor (hygroscopic silica gel) contains very little moisture and is, therefore, able to absorb water vapour from the outside airflow. Next, the outside air is cooled in a heat recovery unit. Further adiabatic cooling occurs by spraying cold water onto the outside airflow, which then evaporates and lowers the supply air temperature while increasing the moisture content. Similar processes can be implemented, e.g. when excess heat from a manufacturing process is to be used, or when thermal energy is created in collector installations, since it can be assumed that the greatest cooling energy is needed when high outside temperatures coincide with high heat gains from solar radiation.

Figure 7.42.2
Test curve of air for sorptive dehumidification and adiabatic cooling

Fresh air
1 Outside air condition
2 Air condition after dehumidification
3 Air condition after heat exchange
4 Air condition after adiabatic cooling
Return air
5 Room air condition
6 after adiabatic cooling
7 after heat exchange
8 after additional heat gain
9 after sorption generator

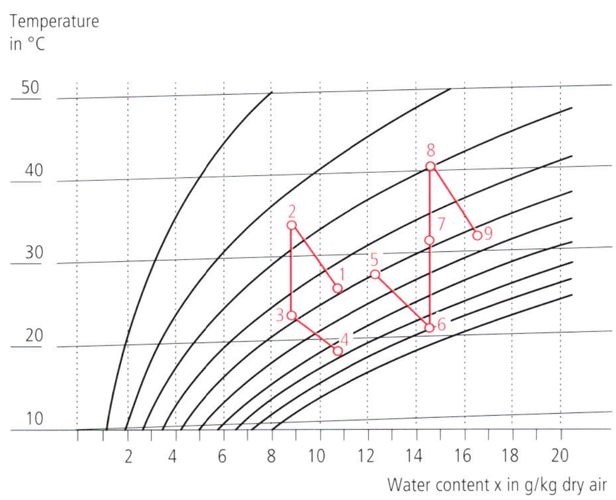

Figure 7.42.1
Air cooling by means of absorptive dehumidification and adiabatic cooling with different heat sources, integrated into a total-energy system

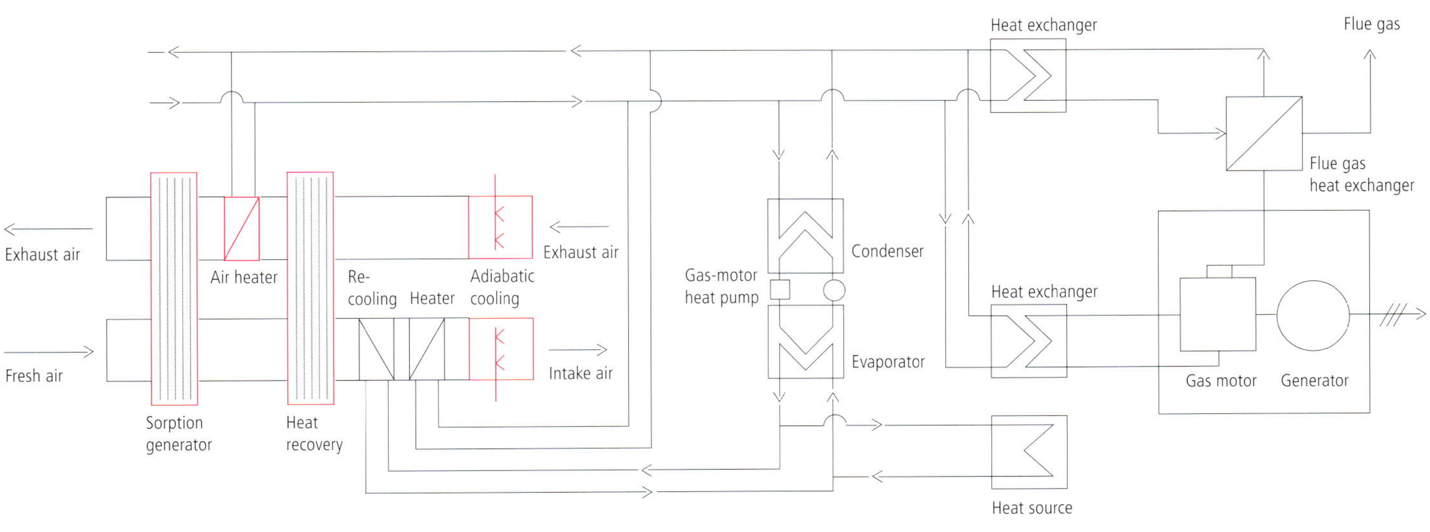

An alternative sorption technique was developed at the Centre for Applied Energy Research (ZAE) in Bavaria, where dehumidification occurs by means of a liquid sorption system (**Figure 7.43.1**). In this system the outside air is dehumidified with a highly concentrated brine solution in a cooled absorption process (1) (**Figure 7.43.2**). The dehumidified (and pre-cooled) air is then adiabatically cooled (sprinkling) and conditioned for injection into the building (2).

The dehumidification process is re-cooled (4) in an integrated cycle system with an indirect evaporative cooler. The air humidity is absorbed by the brine solution, which is therefore less concentrated after the dehumidification process has been completed.

This weak brine solution is intercepted and collected. A thermal regeneration process at temperatures of 60 °C and above is employed to expel the absorbed water from the brine solution, increasing the concentration of the solution and making it once again available for the sorption process. The brine solutions can be stored for any length of time; hence, they are a viable alternative for long-term storage of the concentrated energy required for dehumidification. This fact and the low regeneration temperatures suggest combining brine solutions with solar collectors for the regeneration process since storing the energy in the brine solution can help to match the available solar heat to the cooling demand of the building.

Figure 7.43.1
Schematic representation of air-conditioning by means of sorption and saline solution

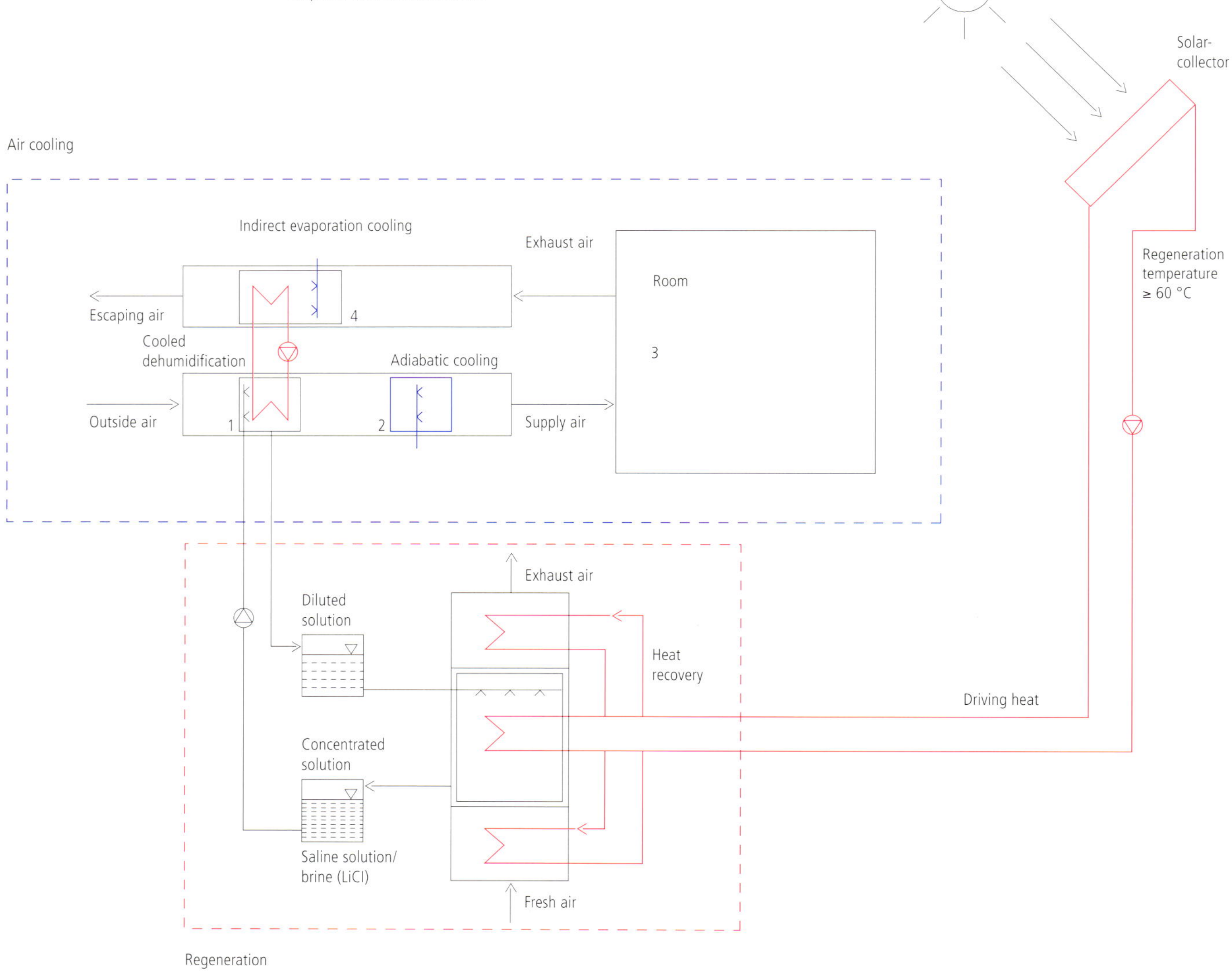

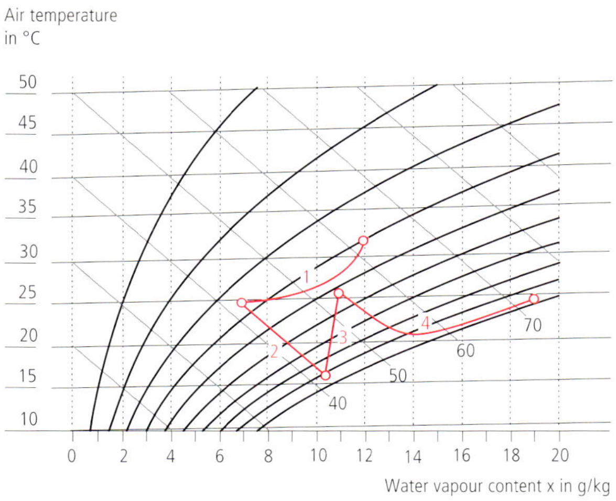

Figure 7.43.2
h, x diagram of air-conditioning by means of sorption and saline solution

1 Cooled dehumidification
2 Adiabatic cooling of supply air
3 Air-conditioning
4 Indirect evaporation cooling

7.2.9 Other processes

Other processes, albeit rarely used, exist in addition to the system solutions discussed above. Some of these are:
– Steam jet refrigerating process
– Thermoelectric refrigerating energy generation
– Hybrid dehumidification systems etc.

Steam jet refrigerating process

In this process, water, and in the simplest scenario the water cycle itself, is used as a refrigerant and water vapour as the driving energy. Steam exits from one or more jets, suctioning steam from the evaporator. The velocity of the mix steam is then reduced in the diffuser and converted into a pressure that corresponds to the condensation pressure of 40 – 50 mbar. The condensate emitted by the condenser is transported in part to the boiler and in part to secondary condensers. The water flowing back from the cold water cycle is dispersed in the evaporator and re-cooled to the exit temperature by means of evaporating a partial volume. Since the process occurs in a high vacuum, multiphase ejectors are needed to unload the cycle. **Figure 7.44** illustrates the operating principle and the layout of a steam jet refrigerating plant.

Figure 7.44
Steam jet refrigerating plant

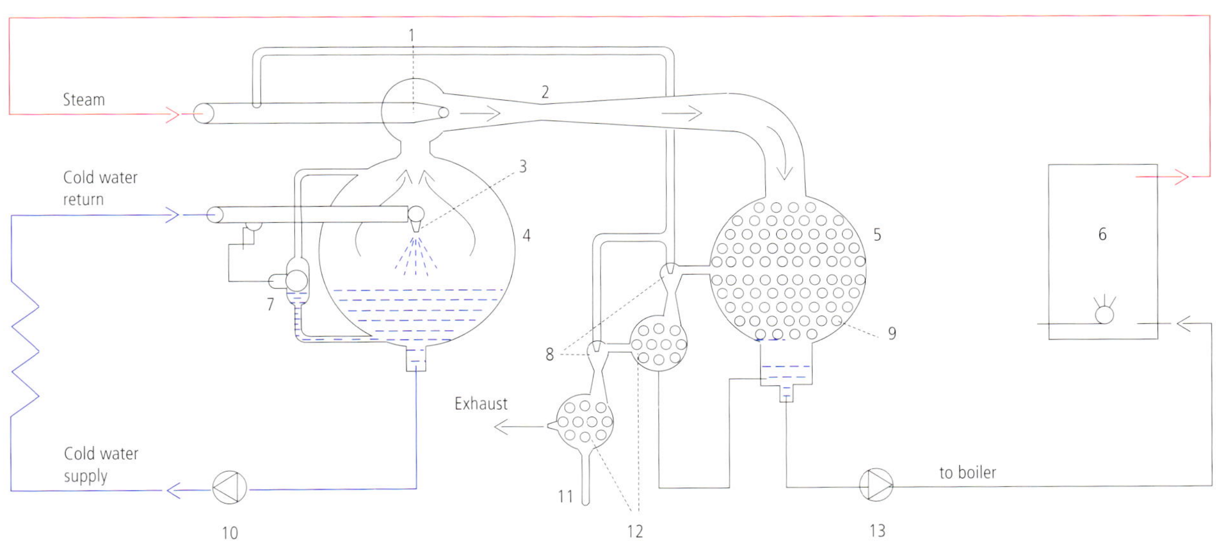

1 Nozzle
2 Diffuser
3 Spray nozzle
4 Evaporator
5 Condenser
6 Boiler
7 Float valve
8 Ejectors
9 Water pipes
10 Pump
11 Outflow
12 Auxiliary condensers
13 Condensate pump

Thermoelectric generation of cooling energy

The system of thermoelectric generation of cooling energy is based on feeding a circuit consisting of two different metallic conductors with direct current voltage. One contact point cools and the other is heated. The contact points also exchange their temperature behaviour if the poles are changed. As conductors, metals develop a moderate thermal force. The limitations of applying this effect for cooling energy generation have been in part overcome in recent years with the successful use of semiconductors. P- and n-type semiconductor materials are connected to a Peltier element via copper bridges (**Figure 7.45**). Sequencing such elements creates a block-like Peltier battery constructed so that all cold copper bridges form the heat absorbing and all warm bridges the heat releasing side of the block. The refrigerating or heat pump devices manufactured from these blocks with an output absorption rate of up to approx. 1 kW are still slightly more expensive to operate than traditional compression and absorption machines, since the Carnot quotient reaches no more than 10 – 20 %. The advantage of the system is that it requires neither moveable parts nor liquids.

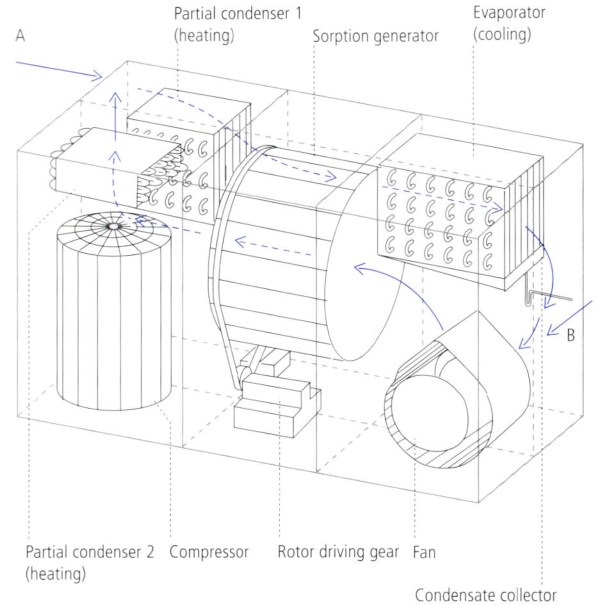

Figure 7.46
Hybrid dehumidifier
(ILK Dresden)

A Room air (partial volume flow 1)
B Room air (partial volume flow 2)

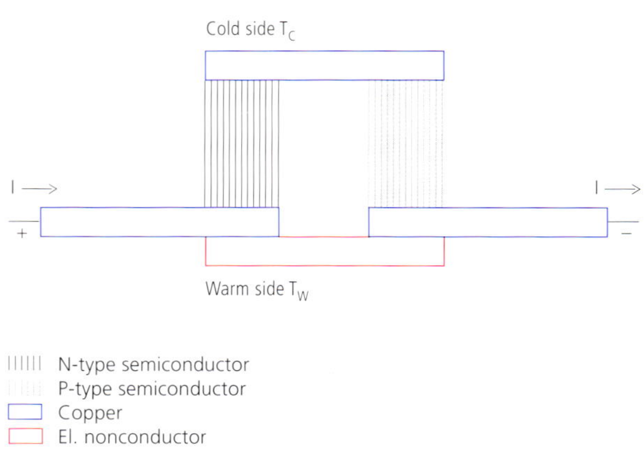

Figure 7.45
Schematic diagram of Peltier element
The thermal output on the warm side is
$P_H = P_C + U \cdot I$ (W)
where
P_C Cooling output on cold side in W
U Voltage in V
I Intensity of current in A
$U = L * \Delta t$
L Seebeck voltage
Δt Temperature difference between warm and cold side

Hybrid dehumidification (with sorption wheel)

The hybrid dehumidifier harnesses the advantages of the condensation principle and the sorption process, operating on the same principle as a heat pump. The characteristic functional components of both dehumidification processes are combined:
- Heat pump
- Sorption dehumidifier

The regeneration airflow is heated in the partial condenser 1, absorbing humidity in the regeneration sector and cooling and dehumidifying it in the evaporator. The second partial airflow is added to the mix before it reaches the fan and pressed through the dry sector of the sorption wheel. During this process, moisture is transferred to the wheel. The dehumidified total airflow absorbs the residual heat from the condenser 2 and is then blown to the outside. Typical applications for such devices are rooms or buildings where humidity levels must be kept low, i.e. which require constant dehumidification. **Figure 7.46** illustrates a hybrid dehumidifier with sorption wheel and heat pump.

7.3
Heat pumps

When refrigerating machines are simultaneously utilized for heat generation, they are called heat pumps. Heat pumps can be motor-driven with electricity, gas or oil as energy source, or via steam turbines (large installations). The examples below are only a small selection from a large spectrum; the descriptions are further limited to those where heat exchange occurs water to water, water to air and air to water, since these are the most common, while heat exchange via solar installations, absorber roofs, the soil etc. is still relatively rare and limited to small building objects.

7.3.1 Heat pumps, water to water exchange

The following paragraphs introduce several heat pumps with simplified cycles, giving examples of their refrigerating output, absorption, thermal output, output quotients and measurements, and operating weight.

The diagrams in **Figure 7.47** illustrate a water-to-water heat pump installed in series with piston compressors in a semi-hermetic construction. The drawing clearly identifies the drive engine with piston compressor as well as a tubular flow cooler (insulated and the control panel containing all switching, regulating and safety devices necessary for operation). The diagram illustrates the refrigerating cycle, beginning with the compressor to the condenser, the heat exchange by means of process or warm water, the return flow of the refrigerant to the evaporator, where evaporation heat is either siphoned off from a cold water appliance or heat is withdrawn from an energy source (surface water/ground water).

The performance figure (or heating rate) in **Table 7.3** indicates the ratio of the heating output emitted in the condenser to the compressor output. As the table shows, the performance figures lie in the range of 4 to 4.5.

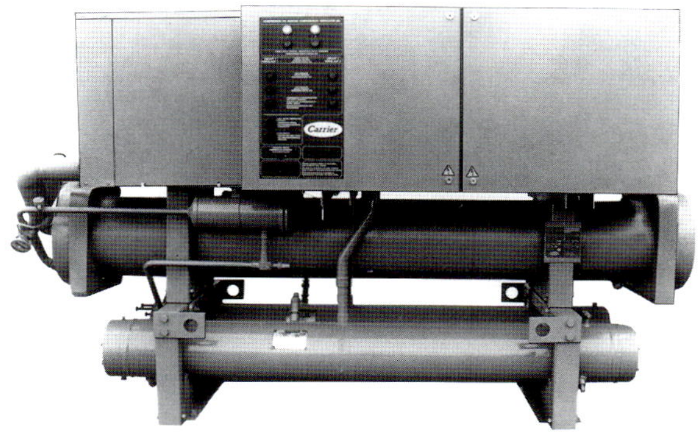

Schematic (1 condenser)

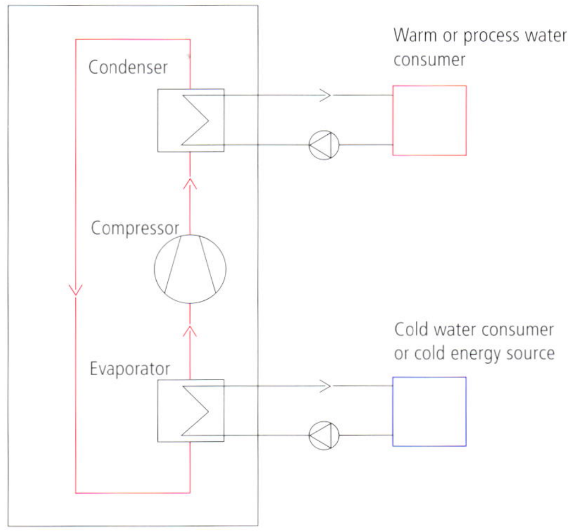

Schematic (2 condensers)

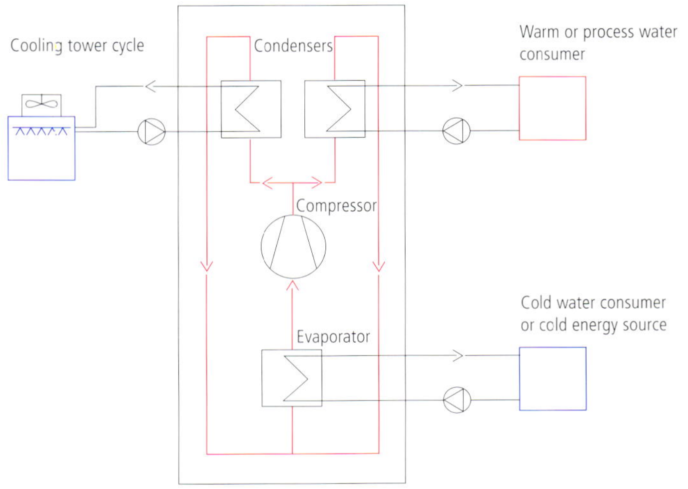

Figure 7.47
Heat pump and schematic of water-water heat pumps with one or two condensers

Type	Cooling output (kW)	Absorption (kW)	Heating output (kW)	Output coefficient	Dimensions (mm) L	W	H	Operating weight (kg)
Machines with one cold cycle:	44.0	14.1	58.1	4.1	1,870	710	1,170	550
	56.6	17.9	74.5	4.1	1,870	710	1,170	570
	69.6	21.1	90.7	4.3	2,090	710	1,170	620
	84.0	26.4	110.4	4.2	2,160	790	1,220	720
	103.0	30.9	133.9	4.3	2,160	830	1,220	850
	134.3	39.0	173.3	4.4	2,230	1,010	1,420	1,300
	170.1	47.3	217.4	4.6	2,400	1,010	1,420	1,480
	199.3	59.5	258.8	4.3	2,420	1,010	1,420	1,510
	266.5	75.3	341.8	4.5	2,850	1,070	1,520	1,980
Machines with two cold cycles:	139.2	42.2	181.4	4.3	2,410	1,020	1,300	1,380
	168.0	52.7	220.7	4.2	2,480	1,040	1,300	1,470
	205.6	61.7	267.3	4.3	2,670	1,040	1,320	1,850
	265.1	77.7	342.8	4.4	2,850	1,170	1,550	2,640
	339.9	94.6	434.5	4.6	2,850	1,170	1,550	2,810
	393.4	119.0	512.4	4.3	2,880	1,170	1,550	2,950
	474.0	135.2	609.2	4.5	4,045	1,370	1,690	3,510
	547.4	150.2	697.6	4.6	4,045	1,370	1,690	3,730

Table 7.3
Output data of a water-water heat pump

Data based on
- Cold water 12/7 °C,
- Warm water 35/40 °C.

Machines with heat exchangers in nominal size

7.3.2 Heat pumps, water to air exchange

If the condenser heat is emitted directly into the air via a finned exchanger, we refer to water to air exchange.

Figure 7.48 depicts a water to air heat pump in compact design, which generates warm air up to 45 °C. This cabinet houses a fully thermal compressor with regulator, an air-cooled louvred tubular condenser, a thermostatic evaporator, and two axial fans. The entire cabinet is insulated and soundproofed. It is a compact model that is virtually ready for operation upon delivery. **Table 7.4** once again provides information on several performance data, to give an idea of the parameters, measurements and operating weight.

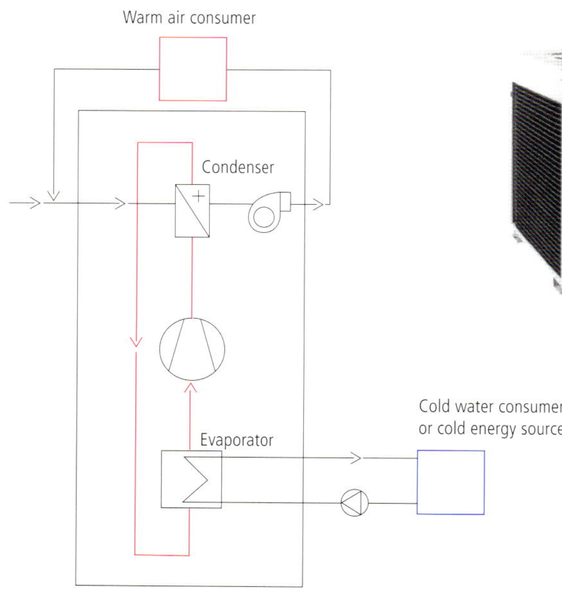

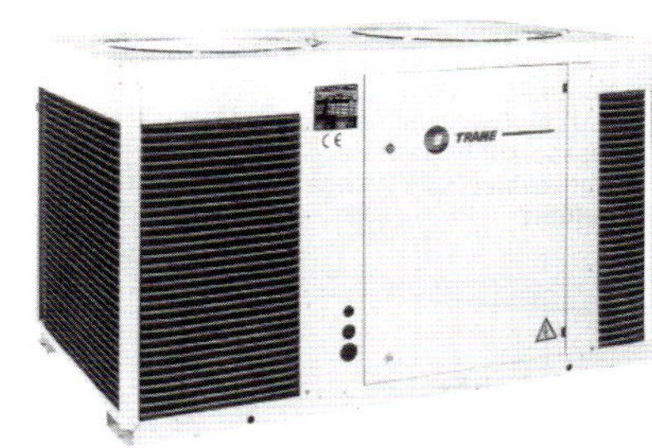

Figure 7.48
Water-air heat pump, CGA/CXA series and schematic of heat pump (Image: Trane)

Model		CGA/CXA 075	CGA/CXA 100	CGA/CXA 125	CGA/CXA 150	CGA/CXA 200	CGA/CXA 250
Cooling output CGA	kW	14.0	18.0	22.5	28.0	36.0	45.0
Current consumption, cooling	kW	5.7	7.4	8.9	11.3	14.7	17.9
Heating output CXA	kW	15.1	19.8	23.7	30.2	39.6	47.4
Current consumption, heating	kW	5.7	7.2	8.9	11.3	14.8	18.0
Air volume	m³/h	7,700	9,000	10,600	15,400	18,000	21,240
Width x depth	mm	1,600 x 950	1,600 x 950	1,260 x 1.050	1,260 x 1.050	1,800 x 950	2,200 x 1,500
Height	mm	1,050	1,050	1,050	1,050	1,050	1,050
Weight CGA/CXA	kg	241/260	241/260	250/268	407/440	410/445	463/505

Table 7.4
Output data of a water-air heat pump for R134a refrigerants

7.4 Cooling towers, re-cooling systems

If the condenser heat of a chiller is not used for heating purposes, it must be rejected to the atmosphere.

In the past, the standard approach was to cool condensers with water drawn from communal grids, wells or open waters – a method that is no longer feasible for ecological and economical reasons. Water-cooled condensers are therefore linked to a re-cooling plant (or cooling tower) executed either in an open or enclosed design. In addition there are re-cooling plants designed specifically for static cooling.

7.4.1 Open re-cooling plants

Open re-cooling plants (water re-cooling plants) operate on the principle that the water originating from the condenser is collected and sprinkled into air, cooling the outside air to a greater or lesser degree depending on its temperature and humidity. The water then returns to the condenser and the cycle begins anew. A minute percentage of the water is cooled when sensible heat is discharged to the cooler air, the greater percentage is, however, caused by the evaporation of a small amount of the water. The diagram in **Figure 7.49** illustrates the cycle in an open re-cooling system. The cooling water is pumped through the condenser and flows to the water inlet of the re-cooling plant. Here, as mentioned, the cooling water is sprayed into the air and drips into a water-collecting basin across a trickling installation in the opposite direction to the air flow generated by a fan. The re-cooling plant is supplied with potable water to replace the evaporated water volumes and to compensate for clarifying water. The potable water supply is regulated with the help of a floating valve control.

Figure 7.50 illustrates the system once more in detail with an open re-cooling plant installed in the open air. To the left, the drawing clearly shows the fan generating the necessary air flow in the opposite direction to the water flow. The re-cooling plant consists of the following basic components:
- Upper drop interceptor (moisture eliminator)
- "Spray tree" nozzles for water distribution
- Trickling installation for spraying the water to be cooled into the counter-current air
- Lower part of housing with air louvres
- Lower (cold) water collecting basin with floating valve, suction sieve and water outlet connecting pipe
- Tap and overflow

The re-cooling plant is installed on a vibration reducing foundation block or on base strips, with the radial ventilator for air circulation including the motor swing-attached to the housing of the plant.

For large plants that serve a wide range of consumers it may be best to operate with a storage tank, in which condensator water (32 to 35 °C) from different condenser is collected in a preliminary stage. The re-cooling cycle is then connected to this intermediate storage. Similarly, the individual condensers draw the water volumes required for cooling from the cold storage (roughly 25 to 28 °C).

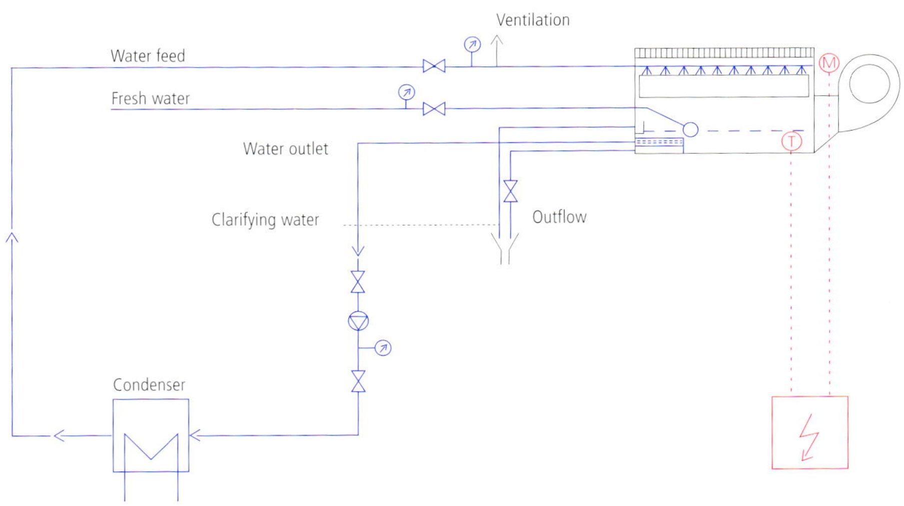

Figure 7.49
Schematic of cycle in open re-cooling plant

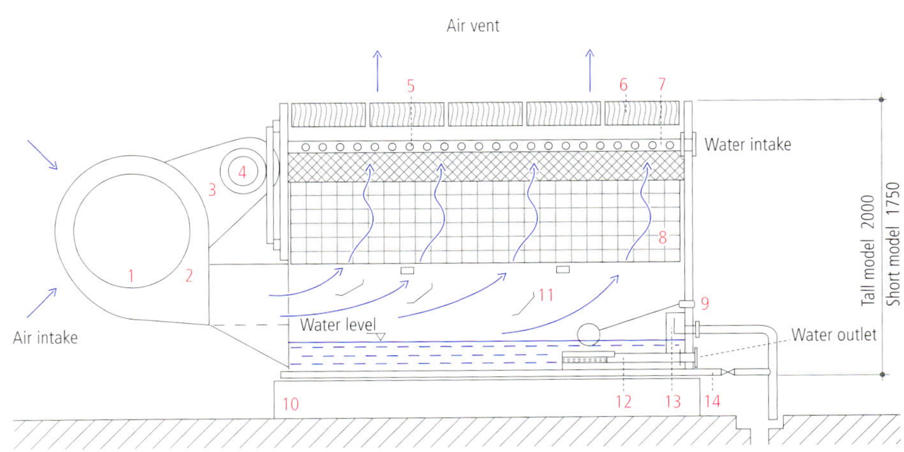

1 Suction filter
2 Radial ventilator
3 Belt screen
4 Motor
5 Spray nozzles
6 Drop interceptor
7 "Spray tree" nozzle
8 Trickling installation
9 Floating valve
10 Base strip
11 Air louvre
12 Suction sieve
13 Overflow siphon
14 Tap (outflow)

Figure 7.50
Open re-cooling plant

Since fans can be a considerable source of noise, they should be designed either to suction or blow at the top (**Figure 7.51**), or be fitted with sound-proofing in conventional installations.

Figure 7.52 depicts open re-cooling plants (cooling water temperatures 26/32 °C) in situ.

Open re-cooling plants achieve excellent efficiencies, demonstrated in relatively high performance rates referenced to the required installation area. One disadvantage is the effect of water vapour in cool seasons as a result of the strong evaporation.

The circulating water (= the water to be cooled) generally requires purification (filtering) in order to prevent carrying the dispersed matter (dust, dirt etc.) that accumulates from spraying it into the air from penetrating into the piping (risk of scale formation, corrosion).

It is also important to bleed roughly 10 % of the circulating water out of the cycle at regular intervals to avoid increased pollution of the water as a result of particles that cannot be intercepted by the filter (risk of corrosion). The hourly, maximum fresh water supply is 6 kg/h for each kW of cooling performance; it replaces evaporated water, sprinkling losses and filtered-out water. Precipitation from water droplets carried away from the cooling tower cycle onto metal facades, for example, may lead to spotting and corrosion due to the softness of the water.

7.4.2 Closed circuit re-cooling plants

Closed circuit re-cooling plants were developed to overcome the disadvantages of open re-cooling plants (pollution and water vapour). **Figure 7.53.1** is a diagram of a closed circuit re-cooling plant, while **Figures 7.53.2/3/4** depict installed re-cooling plants. The diagram illustrates a second heat exchanger cycle linked to the cooling water cycle; this second cycle makes use of the condenser output of the chiller.
In a closed circuit re-cooling plant the cycle of the condenser is described as the primary cycle and that of the re-cooling plant as the secondary cycle.

The primary cycle consists of a condenser with primary pump through which a mix of antifreeze and water flows, a membrane expansion valve and a tubular heat exchanger.

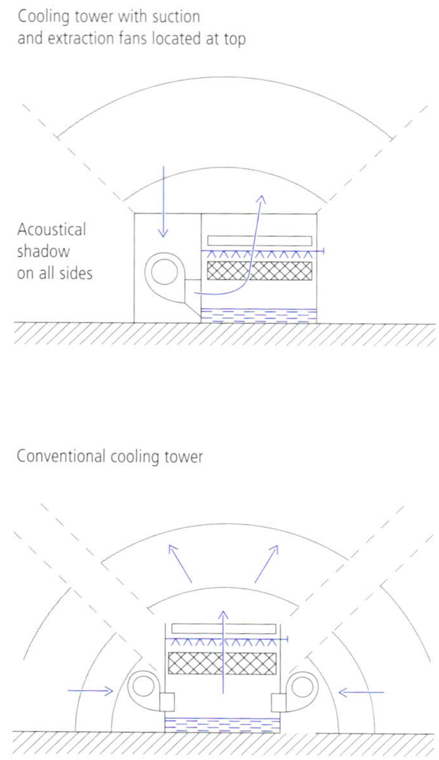

Figure 7.51
Sound insulation in re-cooling plants

Figure 7.52.1
Cooling tower, open re-cooling plant
(Image: Grohl)

Figure 7.52.2
TXV cross-current cooling tower with axial fans (suction principle) with sound dampers
(Baltimore Aircoil)

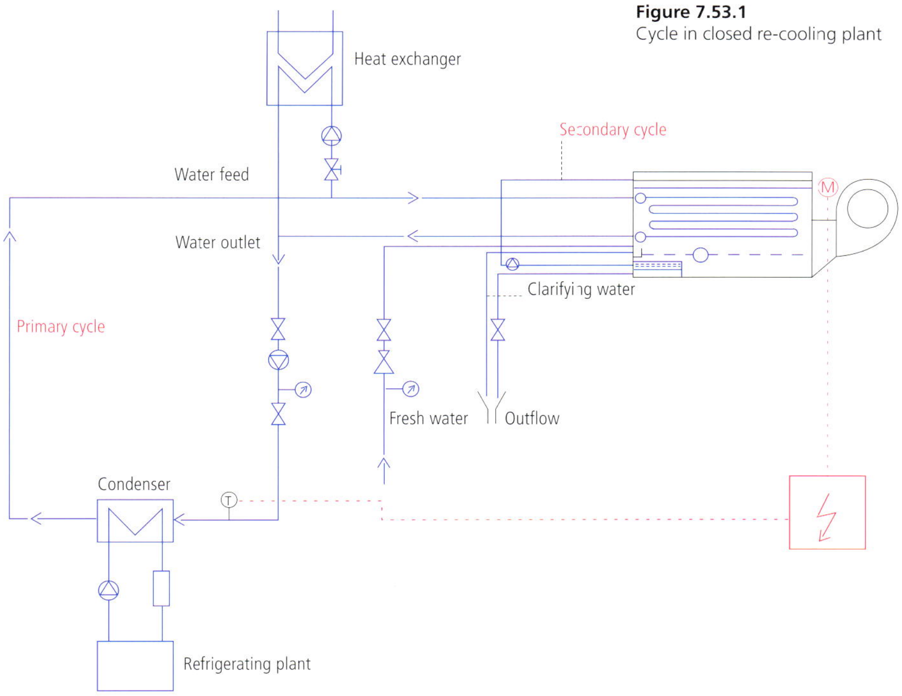

Figure 7.53.1
Cycle in closed re-cooling plant

In summer and the transitional seasons, when there is no risk of water vapour forming, the coiled piping is sprinkled with water from a secondary cycle to achieve a greater cooling effect and better heat transfer at the coils. Because water is sprinkled from the secondary cycle, the water droplets carried along by the air flow must be intercepted in the upper area of the re-cooling plant (drip interceptor). The sprinkled water is then collected in a water-collecting basin (secondary basin) located at the bottom of the installation. An electric basin heater prevents the water from freezing in the transitional season (night frost). The potable water supply pipe, and if necessary the primary system piping, must also be fitted with secondary heating. In winter, when there is a risk of fog formation, the sprinkler system is turned off since relatively cold air flows circulate against the primary water streams during this period, resulting in better heat exchange in the counter-current principle.

Figure 7.54 portrays roof installations of air-cooled condensers with axial fans. These are direct re-cooled condensers, which have already been explained in the context of heat pumps.

Figure 7.53.2 (above)
Evaporation cooler with closed cycle, series VXI, in counter-current principle with pressure radial ventilators and suction and pressure sound dampers
(Baltimore Aircoil)

Figure 7.53.4 (right)
Evaporation cooler with closed cycle, series VFL; image documents sound ranging by means of enveloping surface method
(Baltimore Aircoil)

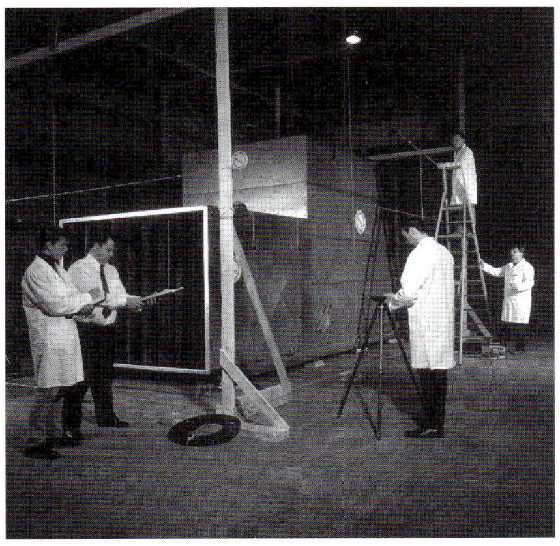

Figure 7.53.3
Evaporation liquefier, series 1500; combined cross-, counter-current design with suction axial ventilators
(Baltimore Aircoil)

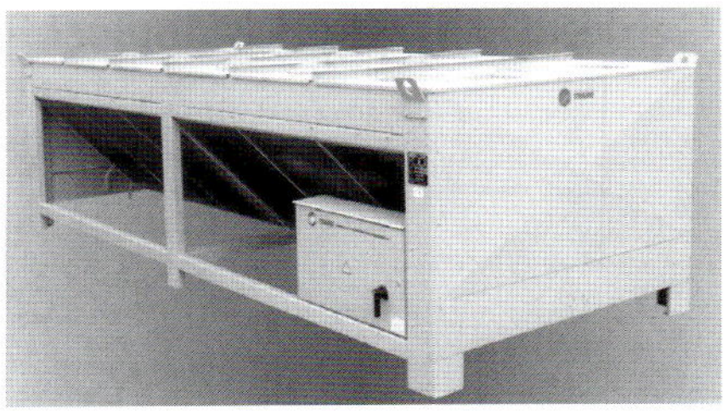

Figure 7.54
Re-cooling plants, air-cooled condensers
(Series RTVCA, Trane)

7.4.3 Re-cooling plants for static cooling

Static cooling (often referred to as winter cooling) occurs when cooling is required in cold seasons due to high heat gains in rooms. During these periods it seems sensible to switch chillers off in order to conserve energy and to cool directly or indirectly with outside air. Static cooling can be achieved by transporting the cooling water through a chiller linked to the HVAC systems (outside air circulation) or through cooling towers, where cooling performance is enhanced by evaporation induced by trickling water across the chiller. Static cooling can be used when outside temperatures are at 8 to 10 °C and below.

Figure 7.55 is a diagram of static cooling for a building.

Figure 7.56 illustrates an integrated system with static cooling and ice-storage technology. Static cooling is particularly advantageous when the cooling water temperatures required for the cooling process are relatively high, as is the case, for example, in induction air-conditioning systems, cooling ceilings, circulation chillers without dehumidification. It should be used whenever possible since it leads to considerable savings in energy costs and greatly improves the lifecycle of the chiller.

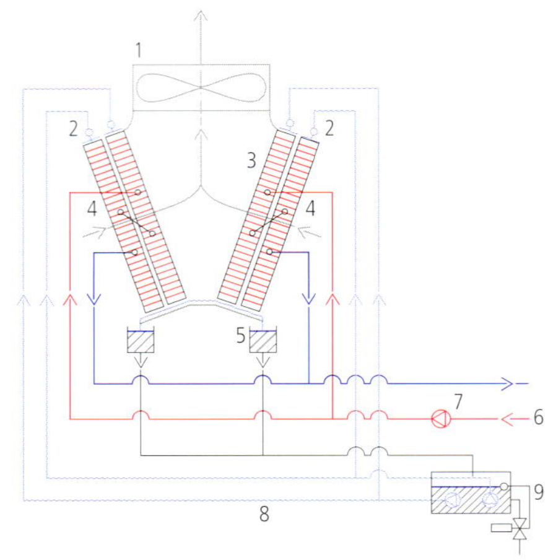

Figure 7.55
Free cooling (hybrid cooling)
Air is drawn by the fans through the slanted chiller louvres, which are sprinkled with water from above. The water evaporates in the air flow, the evaporation heat is withdrawn from the chiller, whose medium – usually a water / glycol blend – is thus cooled.
(Figure Yäggi – Günther)

Optimum configuration of angle, and air and water volumes, can avoid leakage as well as drainage of sprinkling water.

1 Fans
2 Sprinkling water
3 Condenser units
4 Air
5 Water collection
6 Heat source
7 Primary cooling cycle
8 Secondary cooling cycle (Sprinkling water)
9 Water container with filter and automatic refill

Figure 7.56
Chilled ceiling with mechanical as well as free cooling (additional ice-storage installation is possible)

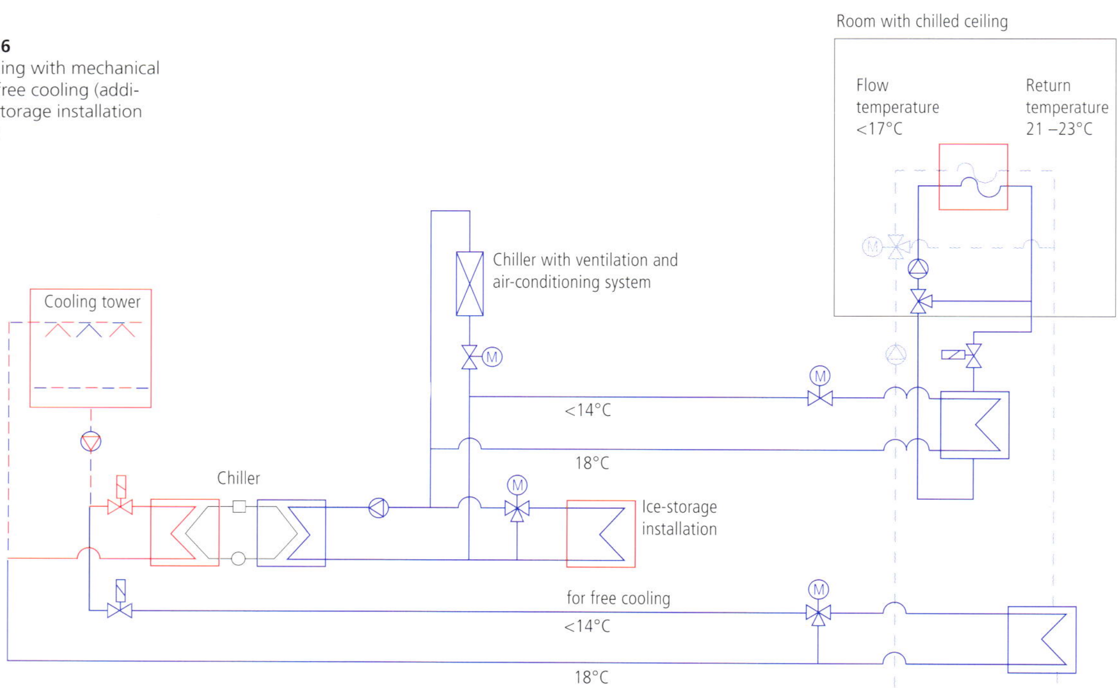

7.5 Ice-storage systems

The advantages of ice-storage systems over systems without storage have already been indicated in Figures 7.7 and 7.9. In principle, coldness can be stored in the form of cold water, although the low temperature differences between storage water (≤ 0 °C) and supply water (+ 6 °C) mean that the storage tanks must be very large. Even if the return temperature of the cold water were raised to + 15 °C, the effective mean temperature difference would be no more than $\Delta t = 10$ K. Combined with the specific heat of water, the storage capacity would be $Q = 11.6$ kWh/m^3.

To avoid having to construct such large and heavy water storage tanks, ice-storage tanks are being used instead. In addition to the temperature difference, the melting heat of water (93 kWh/m^3) is used to increase the specific storage capacity. Current technology focuses on three types, whose principal differences lie in the heat transfer for the formation of ice, or the manner by which the ice is melted.

7.5.1 System options

The most traditional technology is positive transition. The system is composed of an insulated galvanized steel or concrete tank. The tank contains a series of metal plates, copper pipes or corrosion-proof iron pipes (**Figure 7.57**). These plates or pipes comprise the evaporator of the chiller: they are filled with a liquid refrigerant that evaporates at a temperature below 0 °C.

Positive transfer

The outer surface of the plates or pipes serves as the carrier for the ice layer that is formed from the water stored in the tank. The water of the secondary cycle flows through the tank and is chilled as it comes into direct contact with the ice. The water in the container is set into motion by compressed air or an agitator to ensure that transparent or crystal ice forms evenly across all surfaces. If the consumer cycle is a pressure system, the tank must be hydraulically separated from the system by a heat exchanger. In this type of system, shown in **Figure 7.58**, the heat transfer is described as "positive" because the direction of the heat flow is identical for the charge and the discharge phases.

Alternating, internal transfer

Alternating, internal heat transfer is illustrated in **Figure 7.59**. This system consists of an insulated water tank with walls of synthetic resin or polyethylene. The interior is equipped with a system of thin pipes, generally plastic, laid in evenly spaced coils or spirals.

A water/antifreeze mix flows through these pipes and is cooled to below 0 °C with a chiller when storage is required. The pipe exterior serves as the carrier surface for the ice layer that forms from the water contained in the tank. In contrast to the previously described system, the water in the tank is not linked to the heat carrier cycle, but serves exclusively as a storage medium. Here, too, the water/antifreeze mix first flows through the pipes during the discharge phase and then supply cold energy to the consumer. The heat transfer process in this system is described as "alternating, internal."

Alternating, external transfer

The third system illustrated in **Figure 7.60** features alternating, external heat transfer. This type of system consists of an insulation storage tank constructed in either steel or concrete. The tank interior is packed with many vessels that contain water. These vessels constitute the ice storage. A mix of water and antifreeze circulates between the vessels and is periodically cooled to below zero by a chiller. According to whether the temperature of the heat carrier is negative or positive, the water contained inside the vessels eighter freezes or thaws. This heat carrier provides direct supply to the heat exchanger of the connected appliances. This system is the only one suited to integration into a closed pressure cycle of cold generation and distribution. Its advantage is that the water/antifreeze mix is then distributed across the entire cooling system.

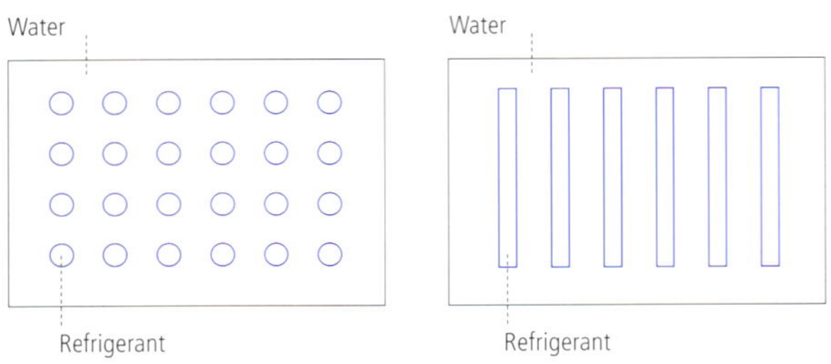

Figure 7.57
Heat exchanger for ice storage

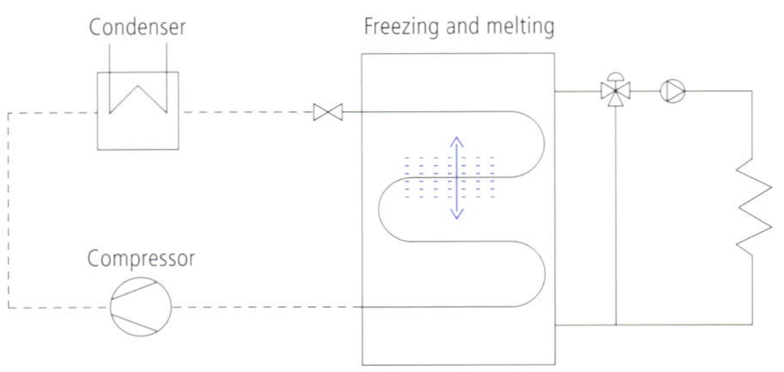

Figure 7.58
Principle of positive transfer

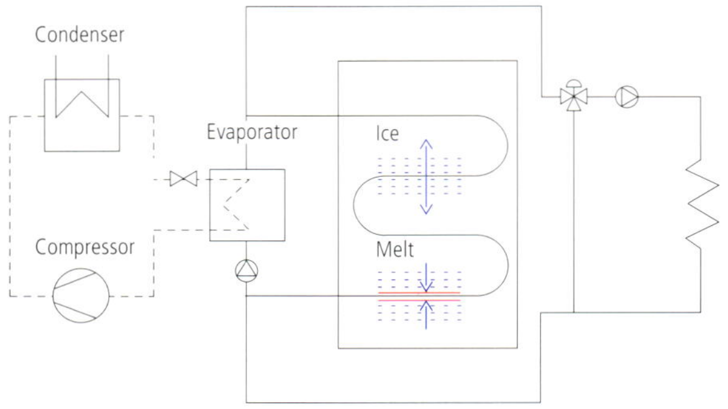

Figure 7.59
Principle of alternating transfer

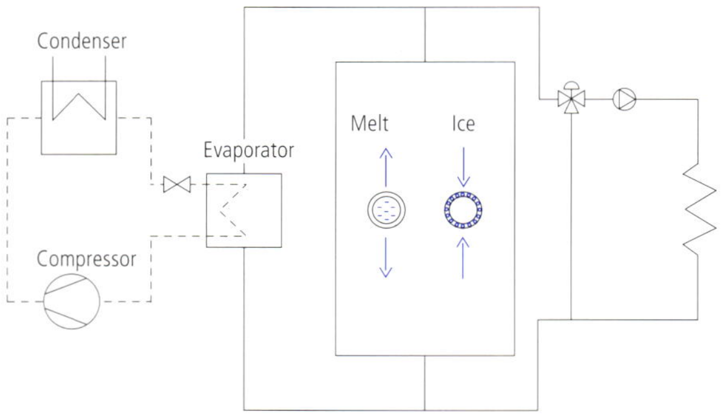

Figure 7.60
Principle of external alternating transfer

	Air-cooled compressor/ condenser	**Air-cooled condenser** (split system)	**Cooling tower**	**Spray condensation** (evaporative cooling)
Air consumption m³/h kW cooling output	300 – 500	300 – 500	80 – 120	80 – 120
Water consumption l/h kW cooling output	0	0	2	2
Connect load for refrigerating plant – kW/kW cooling output Assuming: - t° PKW: +6 °C - t° outside air: dry +32 °C humid +21 °C				
Compressor	0.25	0.25	0.23	0.20
Fan	0.06	0.06	0.03	0.03
Cooling water pump	–	–	0.03	–
Spray water pump	–	–	–	0.01
Total connect load	0.31	0.31	0.29	0.24
Distance between refrigerating plant and condenser	0	Average	High	Average
Installation effort	Low	High	Average	High
Operation and maintenance	Low	Low	Elaborate	Average

Table 7.5
Energy consumption of condensing systems

	Air-cooled compressor	**Cooling tower**	**Spray condensation** (evaporative cooling)
Total connect load in kW/kW Cooling output	0.26	0.28	0.23

Table 7.6
Energy consumption of condensing systems in combination with thermal storage

7.5.2 Planning parameters

Using ice storage has an impact on the sizing of the chillers and the re-cooling plant or cooling tower.

Table 7.5 lists the energy consumption of a cooling plant in relation to different re-cooling systems. The data show that evaporation re-cooling plants are the most efficient in terms of energy consumption based on the lower condensation temperature. The air-cooled condenser, on the other hand, offers the simplest solution overall.

The data in **Table 7.6** are based on a daily output of 10 K and a corresponding simultaneous drop in condensation temperature. In terms of storage, the data indicate that the air-cooled condenser is as efficient from an energy perspective as a "wet" cooling tower. Influences of this type should be considered when making a decision with regard to ice-storage systems.

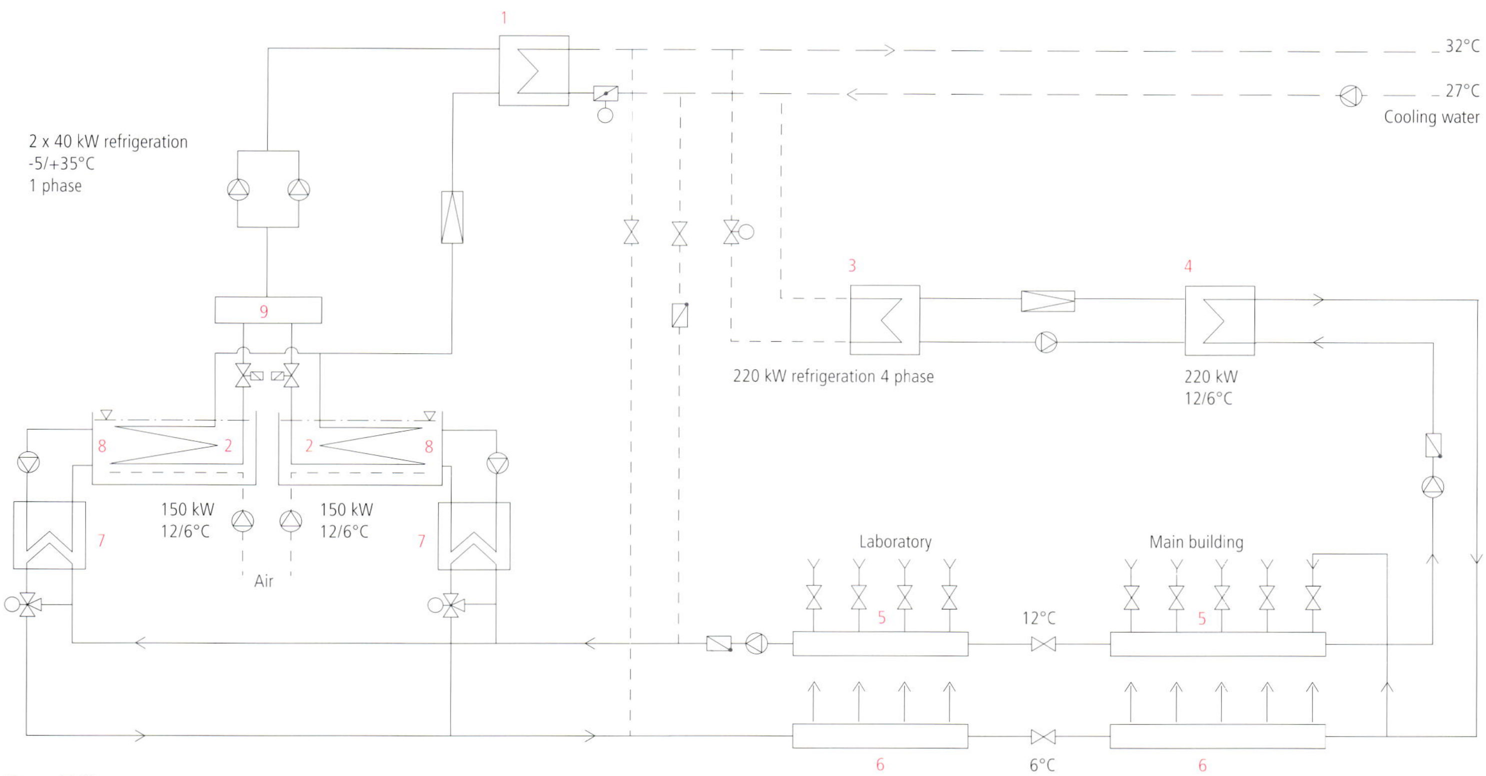

Figure 7.61
Schematic of ice storage with refrigeration and integration into cold water generation for building (right)

1 Condenser/ice-storage-system
2 Evaporator/ice-storage-system
3 Condenser
4 Evaporator
5 Cold water collector
6 Cold water distribution
7 Heat exchanger
8 Ice-storage tank
9 Refrigerant collector

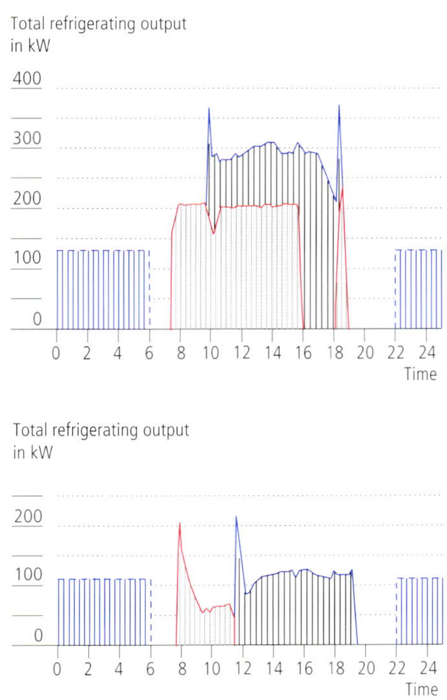

Figure 7.62
Refrigeration curves for average i.e. low load of refrigeration supply plants

Refrigerating plant
Ice-storage discharge
Ice-storage load

7.5.3 Operating parameters

Figure 7.61 is a system diagram with two ice-storage facilities operating on the principle of positive transfer. This installation was tested and analysed for a longer period of time to gather information on the exploitation of the storage capacity. Developing a control parameter was an essential criterion; this control mechanism made it possible to unload the ice storage in the most efficient manner to achieve optimum exploitation of the storage coupled with minimum operating periods of the chiller during the effective period. **Figure 7.62** plots the chiller output on a day with average demand and a day with low demand. The graphs clearly show that the ice storage tank is the ideal tool for transferring the generation of cold energy to the night.

Aside from taking advantage of off-peak rates for night power, this system solution also allows the chillers to run evenly at maximum performance without being subject to consumer specific fluctuations. **Figure 7.63.1** shows an open ice-storage tank and **Figure 7.63.2** a compact model of an ice-storage tank.

Figure 7.63.1
Close-up of ice formation on evaporator pipes

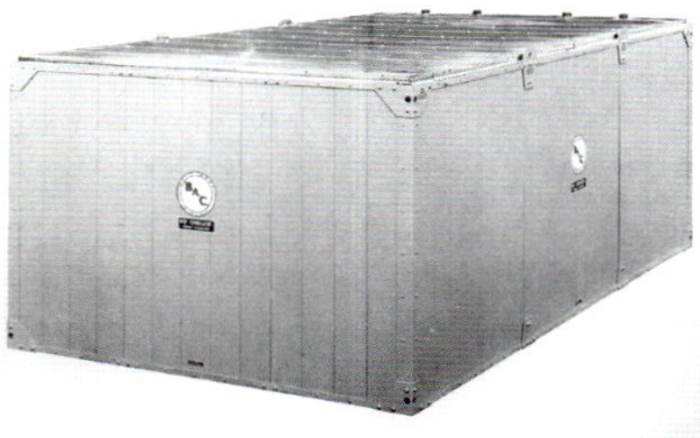

Figure 7.63.2
Modular ice storage, type ICE Chiller® TSU-M
(Image: Baltimore Aircoil)

7.6 Cooling energy distribution (From central plant to consumer)

Cooling energy is not distributed in the same manner as heating energy. Instead, it is planned on a case-by-case basis according to demand and available space. Hence there are no rigid categories such as top and bottom distribution, although these may also occur in cooling systems.

7.6.1 Chilled water distribution

Cooling energy, that is cold water ranging in temperature from 6 to 15 °C and soles (brines) in the 0 °C range, are distributed in pipes that are differentiated into supply, or feed, and return, or back flow pipes. **Figure 7.64** illustrates a very simplified cold water distribution grid with a turbo chiller and two absorption chillers. Three separate supply pipes run from these chillers to a cold water distributor that, in turn, supplies cold water to the individual consumers. The scale of the distribution pipes is based on:
- Number of chillers or cooling appliances
- Differences in cold water supply and return temperatures
- Differences in grid pressures

As **Figure 7.64** illustrates, the individual coolers are supplied from the distributor and the cold water is transported to the collector after heat exchange with warm air.

From the collector, the water stream is first conducted across a metering orifice, which releases direct feed from the distributor into the return flow in the event of unstable pressure conditions. The primary pumps transport the return water to the individual chillers, where it is processed anew.

In smaller, closed systems (direct cooling for a computer lab) distribution in the conventional sense can be omitted. Here, the chiller can be designed to function simultaneously as a direct evaporator, which means that refrigerant flows directly through the pipes to the evaporators, where it is injected.

Figure 7.65 shows a typical smaller distribution centre with heating and cooling distributors. These are virtually indistinguishable to the layperson. The expert can readily identify cold or warm water pipes on the basis of the insulation.

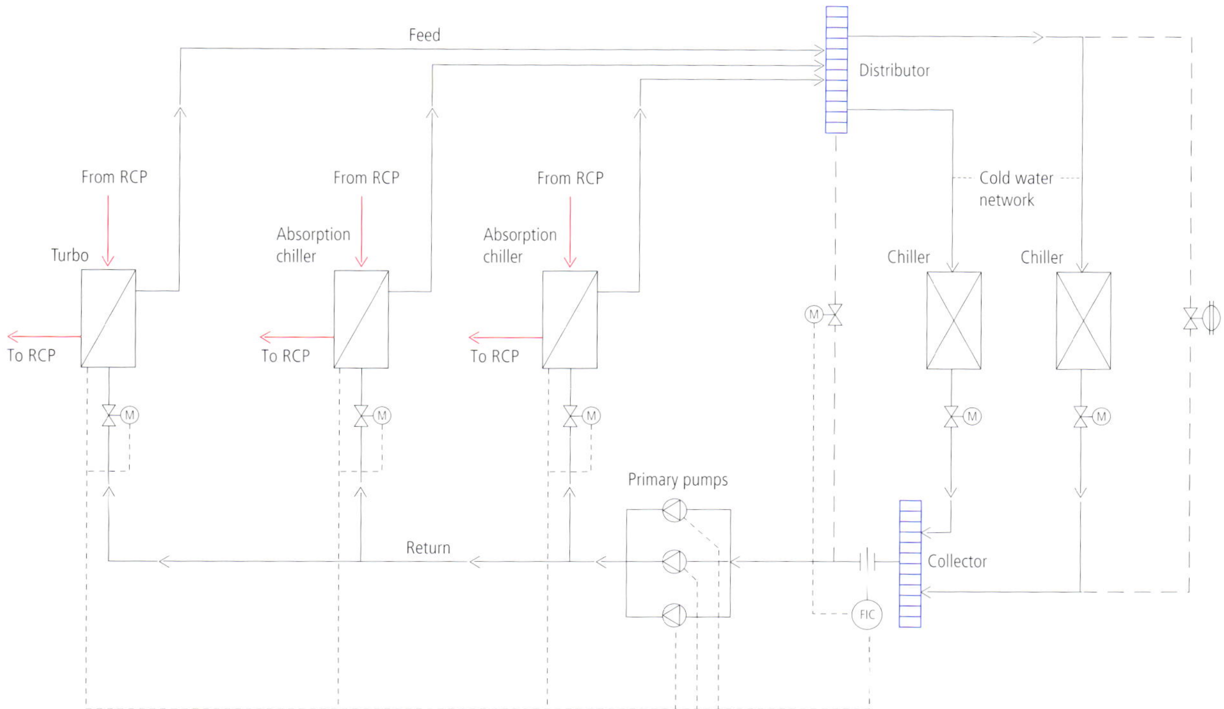

Figure 7.64
Simplified chilled water distribution grid

Figure 7.65
Distribution, heating and refrigeration distribution

For sizing the pipelines it is best to use a friction diagram, much like the process used for estimating pipelines for heating systems. The diagram provides data on the pressure gradient based on set conditions, which are used to determine the pump pressure. The total pressure loss, which must be overcome by a pump, is the result of pipe friction losses and losses cause by individual resistance factors, similar to the conditions described for duct systems. **Figure 7.66** is such a diagram for water.

This diagram was drafted for a rough steel pipe. The pressure drop in a pipe of a diameter and a given roughness is a product of the resistance coefficient. Plastic and copper pipes are much smoother, which translates into less friction loss. It is therefore essential to refer to the appropriate pipe friction diagram for the material of the pipe.

Cold water piping grids generally exhibit a temperature range of 3 to 8 K, with supply temperatures of 5 to 6 °C or 14 to 20 °C, depending on whether the pipes are linked to chillers for cooling with dehumidification or coolers and cooling components for cooling without dehumidification.

The difference between supply and return should be as large as possible in order to keep the circulating water volumes to a minimum. At the same time, it is important not to size pipes too small, which would minimize pressure losses, since the energy consumption rates of electrical power, especially by pumps, are quite considerable, particularly for large refrigerating and cooling water systems.

Variable volume flows are generally considered in the planning for extensive cold water pipe grids, similar to the approach described for ventilation systems, to ensure that only those cold water volumes are circulated that are truly necessary for the cooling energy distribution.

7.6.2 Cooling water grids

Cooling water grids are laid out similarly to chilled water grids, i.e. they consist mostly of seamless steel pipes that connect the condensers of the chillers to the re-cooling plant. It is standard practice to install one re-cooling plant for each chiller to create a direct link between condenser and re-cooling plant and back again. Intermediate storage tanks are used in rare occasions, which means that several chillers operate in connection with one re-cooling plant unit. Since cooling water grids frequently cover a large area, the pipelines should be generously sized, calculating on the basis of an average cooling water temperature of 30 °C (for re-cooling plants) or 45 to 50 °C (for heat pumps) and in conjunction with the appropriate pipe friction diagrams.

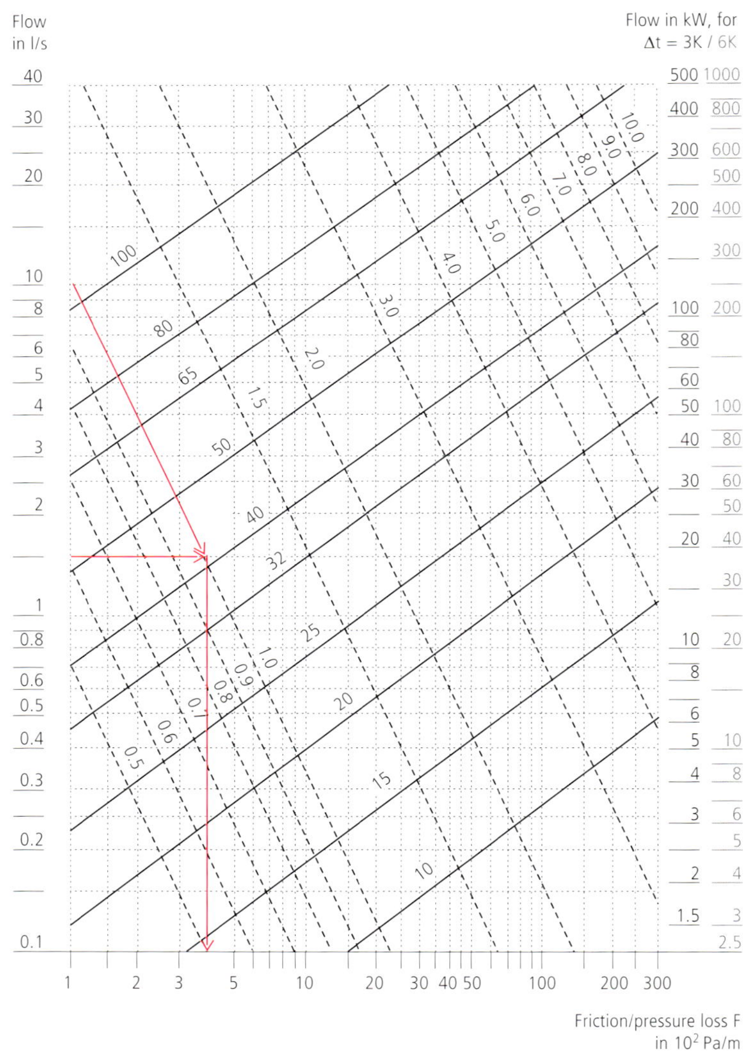

Figure 7.66
Flow chart for friction loss in black threaded pipes, for cold water at 10 °C

— Pipe diameter (standard diameter)
--- Velocity in m/s

7.6.3 Piping / insulation

Materials

The standard materials for cooling and refrigeration systems are:
- Seamlessly drawn copper pipes
- Seamless steel pipes
- Seamless soft steel pipes

Steel pipes are joined by welding, bolts or flanges, whereas copper pipes are soldered.

In selecting materials for cooling systems, it is important to ensure that they are compatible, that is, to ensure that there is no risk of corrosion due to electrochemical processes ("electrochemical attack"). Thus, only galvanized piping should be used for cold water pipes and cooling water systems in cooling installations, since most tubular heat exchangers and components in chillers, and sometimes in re-cooling plants, are made of copper, which means that there is a risk of local corrosion.

Corrosion

All unintended changes in the original metallic condition emanating from the surface of a metal caused by chemical or electrochemical reactions are called corrosion. The common characteristic of all types of corrosion is that metals form bonds with gases like air, O_2, H_2O (steam) and CO_2 at the surface (dry corrosion) or migrate to the electrolyte due to electric potentials in the ion when an electrolyte is present (wet corrosion). The result is that the metals or alloys are no longer able to fulfil their original function. Corrosion damage in cold water and cooling water pipes can be caused by a variety of factors. These are:
- Local corrosion from inside to outside
- Surface corrosion on the inside
- Surface corrosion on the outside
- Copper abrasion in copper pipes when flow velocities are too high (> 1.2 – 1.5 m/s)

External corrosion on black steel pipes is common when condensation is allowed to form as a result of insufficient damp-proof insulation.

Internal corrosion may also occur as a result of electrochemical reactions when various metals are used if a galvanic element is created through direct contact between metals and the metals come into contact with a liquid that contains dissolved salts. This liquid then acts as an electric conductor (current). The cold water, and especially the cooling water, may therefore require pretreatment to avoid corrosion damage.
The best corrosion protection is to use a suitable material such as copper, copper-coated bronze or chrome-nickel steel. Corrosion and furring, as well as microbial growth (e.g. algae, bacteria) are important issues for open systems, in contrast to closed systems (cold water cycle). Evaporation in the re-cooling plant creates a constant loss of water, which means the system is constantly replenished with potable water and hence a continuous feed of salts and hardening constituents. Desalination may be required, although any such measure should be discussed with the manufacturer.

Insulation (cold/warm)

While the return pipes (connecting pipes between re-cooling plant and condenser) do not require insulation, insulation of cold water pipes is an important general issue to prevent corrosion on the outside.
Insulation for cold water pipes (especially those transporting water below 10 °C) should have the following properties:
- Low conductivity
- High durability
- Good form stability
- Low vapour diffusion
- Sufficient temperature resistance
- Non-combustibility
- Good workability

The following materials are appropriate for insulation:
- Glass fibre and mineral wool (may not be appropriate in some situations)
- Cellular or foamed glass with closed cells and excellent form stability
- Perlite insulation
- Aluminium foil
- Expanded foam insulation (conditional use)

To prevent vapour from settling on pipes and fittings when temperatures fall below the dew point, vapour barriers must be used to raise the diffusion resistance on the heated side. The building materials, insulating materials and barrier materials suitable for use as vapour barriers and their vapour-resistance values are listed in **Table 7.7**. The higher the vapour-resistance values, the more resistant to vapour the relevant building is, or insulation material to an equally thick layer of air.

Table 7.7
Parameters for water vapour and diffusion resistance coefficients

	Vapour diffusion resistance coefficients
Materials:	
Plaster, mortar	10 – 35
Cast-in-place concrete	70 – 150
Prefabricated concrete comp. also expanded concrete with closed structure	70 – 150
Lightweight concrete, plasterboard	5 – 15
Asbestos (fibrous) cement panels	20 – 50
Brickwork with solid and cored bricks	5 – 10
Clinker bricks	50 – 100
Sandlime brick	5 – 25
Glass mosaic or cleaving clinker	100 – 300
Wood	40
Plywood	50 – 400
Hard fibre board	70
Chipboard (particle board)	50 – 100
Insulating materials:	
Wood wool lightweight building boards	2 – 5
Corkboard	5 – 10
Porous fibre board	5
Fibre insulating material	1 – 3
Polystyrene particle foam depending on pipe diameter	20 – 100
HR polystyrene extrusion foam	80 – 300
Rigid polyurethane foam	30 – 100
Rigid polyvinyl chloride (PVC) foam	150 – 300
Rigid phenolic foam	30 – 50
Foam glass	Basically vapour-proof
Waterproofing layers:	
Uncoated bitumen roofing felt	2,000 – 3,000
Roofing felt	15,000 – 100,000
Polyethylene foil	100,000
Aluminium foil, thickness ≥ 0.05 mm	Basically vapour-proof

Table 7.8
Difference $\Delta\vartheta_{Dew}$ in K between air and surface temperature when condensate begins to form

Air temperature ϑ_L in °C	$\Delta\vartheta_{Dew}$ **% for relative air humidity of** 30	35	40	45	50	55	60	65	70	75	80	85	90	95
– 20	–	10.4	9.1	8.0	7.0	6.0	5.2	4.5	3.7	2.9	2.3	1.7	1.1	0.5
– 15	12.3	10.8	9.6	8.3	7.3	6.4	5.4	4.6	3.8	3.1	2.5	1.8	1.2	0.6
– 10	12.9	11.3	9.9	8.7	7.6	6.6	5.7	4.8	3.9	3.2	2.5	1.8	1.2	0.6
– 5	13.4	11.7	10.3	9.0	7.9	6.8	5.8	5.0	4.1	3.3	2.6	1.9	1.2	0.6
± 0	13.9	12.2	10.7	9.3	8.1	7.1	6.0	5.1	4.2	3.5	2.7	1.9	1.3	0.7
2	14.3	12.6	11.0	9.7	8.5	7.4	6.4	5.4	4.6	3.8	3.0	2.2	1.5	0.7
4	14.7	13.0	11.4	10.1	8.9	7.7	6.7	5.8	4.9	4.0	3.1	2.3	1.5	0.7
6	15.1	13.4	11.8	10.4	9.2	8.1	7.0	6.1	5.1	4.1	3.2	2.3	1.5	0.7
8	15.6	13.8	12.2	10.8	9.6	8.4	7.3	6.2	5.1	4.2	3.2	2.3	1.5	0.8
10	16.0	14.2	12.6	11.2	10.0	8.6	7.4	6.3	5.2	4.2	3.3	2.4	1.6	0.8
12	16.5	14.6	13.0	11.6	10.1	8.8	7.5	6.3	5.3	4.3	3.3	2.4	1.6	0.8
14	16.9	15.1	13.4	11.7	10.3	8.9	7.6	6.5	5.4	4.3	3.4	2.5	1.6	0.8
16	17.4	15.5	13.6	11.9	10.4	9.0	7.8	6.6	5.4	4.4	3.5	2.5	1.7	0.8
18	17.8	15.7	13.8	12.1	10.6	9.2	7.9	6.7	5.6	4.5	3.5	2.6	1.7	0.8
20	18.1	15.9	14.0	12.3	10.7	9.3	8.0	6.8	5.6	4.6	3.6	2.6	1.7	0.8
22	18.4	16.1	14.2	12.5	10.9	9.5	8.1	6.9	5.7	4.7	3.6	2.6	1.7	0.8
24	18.6	16.4	14.4	12.6	11.1	9.6	8.2	7.0	5.8	4.7	3.7	2.7	1.8	0.8
26	18.9	16.6	14.7	12.8	11.2	9.7	8.4	7.1	5.9	4.8	3.7	2.7	1.8	0.9
28	19.2	16.9	14.9	13.0	11.4	9.9	8.5	7.2	6.0	4.9	3.8	2.8	1.8	0.9
30	19.5	17.1	15.1	13.2	11.6	10.1	8.6	7.3	6.1	5.0	3.8	2.8	1.8	0.9
35	20.2	17.7	15.7	13.7	12.0	10.4	9.0	7.6	6.3	5.1	4.0	2.9	1.9	0.9
40	20.9	18.4	16.1	14.2	12.4	10.8	9.3	7.9	6.5	5.3	4.1	3.0	2.0	1.0
45	21.6	19.0	16.7	14.7	12.8	11.2	9.6	8.1	6.8	5.5	4.3	3.1	2.1	1.0
50	22.3	19.7	17.3	15.2	13.3	11.6	9.9	8.4	7.0	5.7	4.4	3.2	2.1	1.0

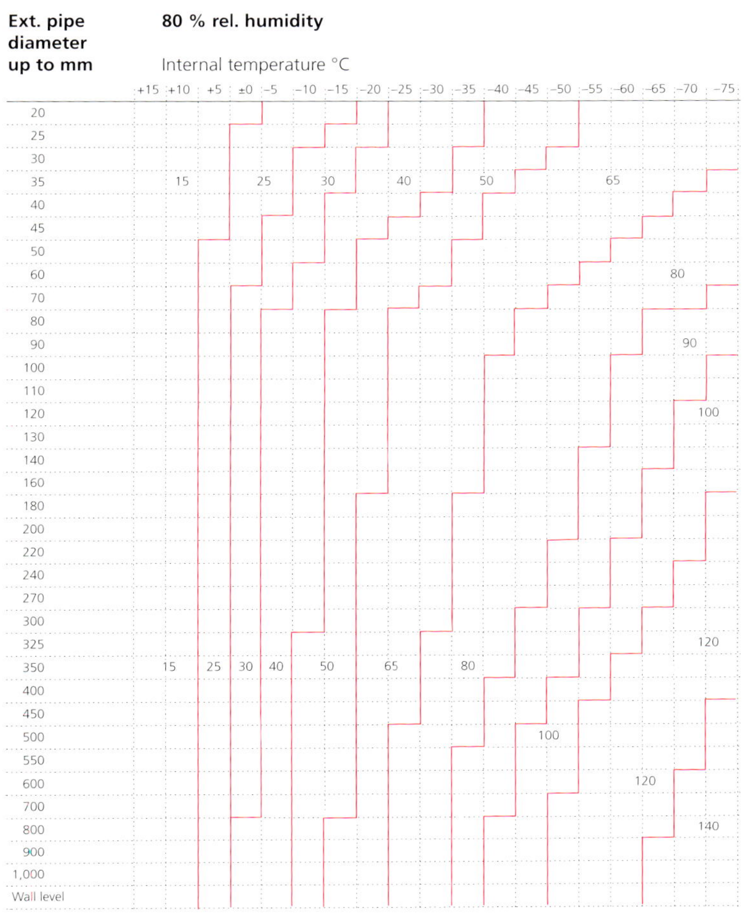

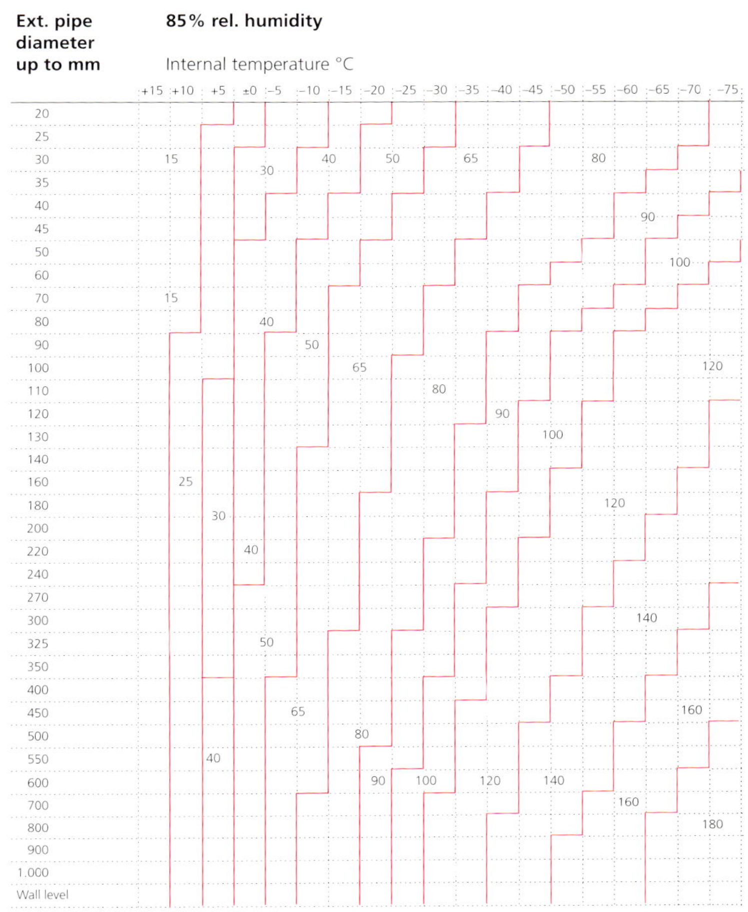

Table 7.9
Required insulation in mm to prevent condensation on plastic foam insulation for cold water pipes for 80 and 85 % rel. air humidity, respectively, ambient temperature + 20 °C and motionless air
Thermal conductivity
λ = 0.04 W/m K for
qm = 10 °C and
λ = 0.033 W/m K for
qm = 100 °C (below)

Condensation occurs when the air near the pipe surface falls below the dew point. **Table 7.8** lists the temperature differences between air and surface temperature at the point when condensation begins to form. **Table 7.9** indicates the relationship between external pipe diameter, internal pipe temperature and vapour barrier thickness for cellular foam insulation on cold water pipes. The table highlights the interactions between specific parameters: as the table shows, the thickness of the vapour barrier must increase as relative humidity increases in order to prevent condensation.

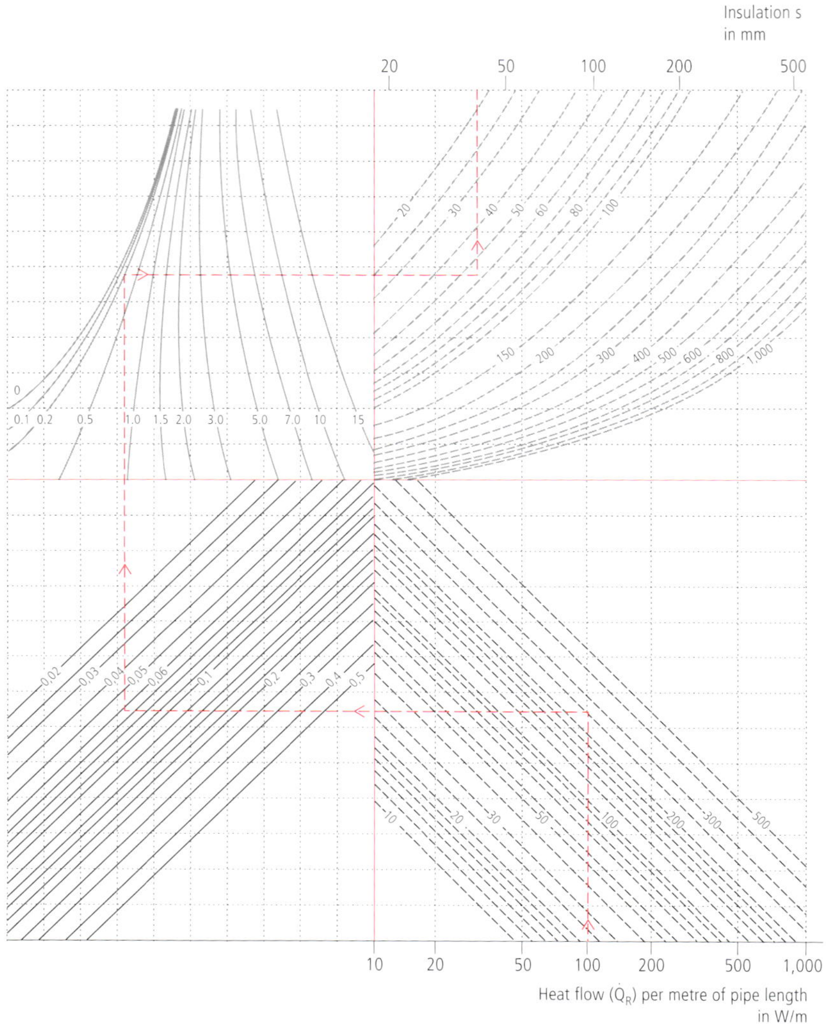

The diagram in **Figure 7.67** is used to determine the correct thickness of the vapour barrier referenced to the allowable heat flow per metre of pipe length, the conductivity of the insulation, the temperature differences between medium and air and the G-factor, which expresses heat transmission at the pipe diameter.

Figure 7.68 illustrates the considerable, and necessary, effort put into insulation for cold water pipes. Clearly evident in the image, the evaporators and condensers of refrigerating machines, as well as the pipes leading off from them, are enclosed in a hard insulating casing of galvanized steel sheeting, which covers the actual vapour barrier. The heat insulation around the evaporators is also visible (black insulating sleeve).

Figure 7.67
Insulation layers for pipes according to thermal flow per m of pipe length, temperature differences between medium and air, thermal conductivity, as well as

$$G = \frac{2\lambda}{\alpha_a d_i}$$

and the pipe diameter

—— $\vartheta_M - \vartheta_L$ in K
- - - λ in W/mK
—— G
- - - d_i in mm

Figure 7.68
Refrigeration plant with cold-water piping (insulated)

7.7. Refrigeration plants/shafts/horizontal distributions

Refrigeration plants are rooms that house refrigerating machines, expansion tanks, cold water and cooling water distributors, as well as the main pumps. There should also be sufficient space for control panels and regulators. The size of the rooms and the re-cooling plants is indicated in **Table 7.10**.

It is important that sufficient space is reserved in front of and behind the refrigerating machines (from a longitudinal perspective) for the pipe connections and for pulling out the pipe registers (condenser/evaporator). The ground floor of a building would be the ideal location for such an engine room, making it easy to bring the machines into the building or to replace components at a later date. However, since this is rarely feasible in practice, most refrigerating machines are installed in the basement, whereas re-cooling plants are usually installed on the roof. The installation rooms should be sufficiently ventilated, depending on the refrigerant used, to prevent refrigerant leaks posing a health hazard.

The air from a refrigeration room must be evacuated at foot level, since refrigerants are heavier than air and settle near the floor. The temperature in these installation rooms should not exceed 40 °C, and the heat emitted by drive engines must be evacuated.

The machine room and the areas leading to it must be readily accessible since refrigerating machine components are very heavy and difficult to transport. Refrigerating machines and main pumps are installed on non-oscillating platforms, the machine room itself must have excellent sound insulation because sound levels of such machines may reach up to 130 db(A). Re-cooling plants should not be installed in basements, although rare exceptions are possible, because this is inevitably linked to exorbitant costs due to the space required, as well as extensive sound insulation.

Refrigerating output	Equipment room			Re-cooling plant, open design			
('W)	Area m^2	Room height m	Note	Area m^2	Room height m	Total weight t)*	Note
12,000 – 120,000	10 – 20	3.50	Equipment rooms in basement whenever possible. For an output of up to 120,000 W, the chiller should be placed on top of the HVAC installation in the equipment room.	20	2.80	0.4 – 0.8	Areas apply to re-cooling towers with casing. Roof installation wherever possible.
120,000 – 350,000	20 – 40	3.00		25	3.20	0.8 – 1.8	
350,000 – 700,000	40 – 60	3.00		50	3.40	1.8 – 4.0	
700,000 – 1,120,000	60 – 80	3.50		60	3.40	4.0 – 6.0	
1,120,000 – 1,750,000	85	3.50		90	4.00	6.0 – 9.0	
1,750,000 – 2,350,000	100	4.00		130	4.00	9.0 – 11.0	

)* including water and piping

k Note: The area dimensions should be increased by 20% for closed re-cooling plants

Table 7.10
Refrigerating plant dimensions

The volume flow required to mechanically ventilate an engine room housing a refrigerating machine is calculated as follows

$$\dot{V} = 50 \sqrt[3]{W^2}$$

Where:
W = Filing weight [kg]
$\dot{V}$ = Volume flow [m^3/h].

For piston compressors, the estimated filling weight ranges from 0.25 to 0.45 kg/kW and for turbo compressors from 0.40 to 0.70 kg/kW.

Figure 7.69 depicts a smaller plant with worm compressors for a malthouse, **Figure 7.70** a large plant (9 MW output) for a television station.

Table 7.10 (previous page) indicates the space requirements for refrigerating machines and re-cooling plants, including distributors, collectors and other central components. The necessary minimum heights and total loads are also listed.

The shaft area required for refrigerating and cold-water pipes can be calculated to occupy roughly 4 % of the available engine room area.

Installation heights for horizontal distributions in engine rooms range from 0.3 to 1.5 m according to the pipe dimensions and number. On standard floors, the heights for cooling water pipes range from 0.1 to 0.2 m, with the exception of industrial applications.

Figure 7.69
Refrigerating plant, worm compressors

Figure 7.70
Refrigerating plant, cold water turbo installation

1. Building performance

2. Comfort

3. Integrated approaches to planning

4. Heating systems

5. Sanitary and fire-extinguishing systems

6. Ventilation and air-conditioning technology

7. Refrigerating and cooling systems

8. Power plants

9. Illumination

10. Daylight technology

11. Low-voltage current systems

12. Conveying systems

8.
Power plants

(with Dieter Jakob, Frankfurt)

Reliable electricity supply and electrical installations have an enormous impact on the quality of buildings. The construction and operation of all building systems and installations is virtually impossible without electrical energy. Electricity supply and relevant electrical installations are therefore a vital topic.

The advantages of using electrical energy are:
- Transformability into other forms of energy, such as light, heat, mechanical forces
- Universal applicability
- Low transfer losses

The disadvantages of working with electrical energy are:
- Accidents
- Fire risk
- Risk of explosion
- Electromagnetic radiation (e.g. transmitters, microwaves, x-rays)
- Electrosmog
- Strong magnetic force in case of short circuit

Still, the advantages clearly outweigh the disadvantages despite the dangers associated with electrical energy, making it indispensable in households, technical installations and especially for building systems. Specific safety measures must be taken to counteract the dangers associated with electrical installations, and these are set forth in standardized guidelines and regulations.

8.1
Electrical energy demand (performance)

In order to determine the electrical energy demand, we must begin by cataloguing all areas in a building where power is consumed. There are several principal consumption points and a multitude of small consumption points, for which a rough estimate is generally undertaken at the outset of a project. (For the sake of simplicity, these consumption points are referred to as "consumers" in the following sections.) The principal consumers must be defined early on to provide a basis for determining the type of supply and to engage in tariff or rate negotiations with the utility company (or electricity board (UK)).

Permits are often necessary for large consumers or for consumers that may impact the electricity grid. Local standards and guidelines vary widely. Several principal consumers are presented in the sections that follow.

8.1.1 Motors

Buildings, especially air-conditioned buildings, often contain motors as large consumers. The most common types are:
- Small motors for AC and DC power (up to a maximum of 1 kW)
- Alternating-current motors driven by a wide variety of connection methods
- Commutator motors for AC and DC power
- Polyphase-synchronous motors
- Asynchronous motors

In HVAC systems, the most common motors are:
- Asynchronous motors (variable voltage, variable frequency, controller motors (VVVF), which allow continuous speed control, the most common solution today)
- Slip ring motors (allow switching to several default speed levels)
- Pole changing motors (allow switching to two or at most three default speed levels)

Motors are furthermore classified according to:
- Type/design (A, B, V)
- System of protection (IP11 to IP33)
- Rotor type
- Insulation class (A to H)
- Free shaft end (1 or 2)
- Ventilation type
- Nominal operation type etc.

All large motors must be tested to see if they can or should be used as medium/high voltage motors in order to save copper cables or trays (investment costs) wherever possible (large motors in chillers/large fan units).

8.1.2 Receptacle consumers

These are all consumers supplied from receptacles or convenience outlets. This category includes office machines, small kitchen appliances, small cleaning appliances and a multitude of other appliances and equipment for daily use. Receptacle consumers generally represent only a small fraction of the total power consumption in a large building project.

8.1.3 Elevators and hoists

The motors for elevators and hoists generally have a connect load of 3 to 30 kW and are rarely a major factor in the total energy balance. However, since they are started with great frequency and "draw" high starting currents, they should be taken into consideration and may create what is referred to as an "unquiet" grid. For the purpose of the energy balance, only the consumption rate during continuous operation is considered, unless the elevator and hoist installations represent an essential part of the total building consumption, for example in apartment buildings, in which case the starting phase must be factored in as well. When sizing the supply, it is important to take the starting current and the resulting drop in current into consideration.

8.1.4 Lighting systems

In addition to large motors and kitchen appliances (especially in commercial kitchens), lighting systems are major power consumers. It is important to consider, however, that lighting systems should be divided into those for high-quality and others for low-quality areas, and the corresponding differences in the connect loads. A connect load of 8 to 12 W/m^2 can be assumed for the purpose of a rough calculation. The targeted connect load for modern, energy-efficient lighting is 6 – 10 W/m^2.

8.1.5 Kitchen appliances

Commercial kitchens are often the principal individual consumers in building objects (with the exception of computer equipment) and play a considerable role in the load estimate. Therefore, the energy demand for large and commercial kitchens should be determined at the outset (with the help of kitchen consultants) and factored into the load estimate. (For kitchens with electric heating, the connect load for roughly 750 diners is approximately 400 kW; for 3,000 diners it rises to roughly 1,200 kW; interpolation is possible).

The trend towards fast preparation with precooked foodstuffs has had a considerable influence on the increase in short-term spikes, amounting to easily twice the load required for traditional kitchens.
The chief contributing factors are superfast induction elements and regenerators. To limit these peak loads, load optimization measures are undertaken for commercial kitchens: these measures may include cyclical on- and off-switching of the supply to different appliances.

8.1.6 Load estimating and simultaneity factor

Table 8.1 contains a sample load estimate for an office building complete with installation output, simultaneity factor and the anticipated, simultaneous demand on the grid.

This load estimate reflects a "moment in time" and can only serve as a rough initial guide towards designing the transformer installation. It is marginally useful for rate negotiations, and of no use with regard to defining a complete energy supply system because it gives no indication as to the daily and annual consumption patterns.

	P_i	F_S	P_S
Service/equipment floor			
Elevator 1	27.0 kVA	0.5	13.5 kVA
Elevator 2	23.0 kVA	0.5	11.5 kVA
HVAC	120.0 kVA	0.9	108.0 kVA
4th floor			
Distribution terminal	60.0 kVA	0.8	48.0 kVA
3rd floor			
Distribution terminal	60.0 kVA	0.8	48.0 kVA
Kitchen	380.0 kVA	0.7	266.0 kVA
2nd floor			
Distribution terminal	60.0 kVA	0.8	48.0 kVA
Computing	400.0 kVA	0.9	360.0 kVA
1st floor			
Distribution terminal	60.0 kVA	0.8	48.0 kVA
Ground floor			
Distribution terminal	80.0 kVA	0.8	64.0 kVA
Basement			
Distribution terminal	25.0 kVA	0.8	20.0 kVA
Heating	160.0 kVA	0.9	144.0 kVA
Sanitation (central equipment)	20.0 kVA	0.8	16.0 kVA
HVAC	300.0 kVA	0.8	240.0 kVA
Refrigeration	300.0 kVA	0.9	270.0 kVA
Standby power plant	630.0 kVA	1.0	630.0 kVA
	2,705.0 kVA		2,235.0 kVA

Table 8.1
Sample load estimate

P_i Installed load
F_S Simultaneity factor
P_S Anticipated simultaneous load drawn off from grid

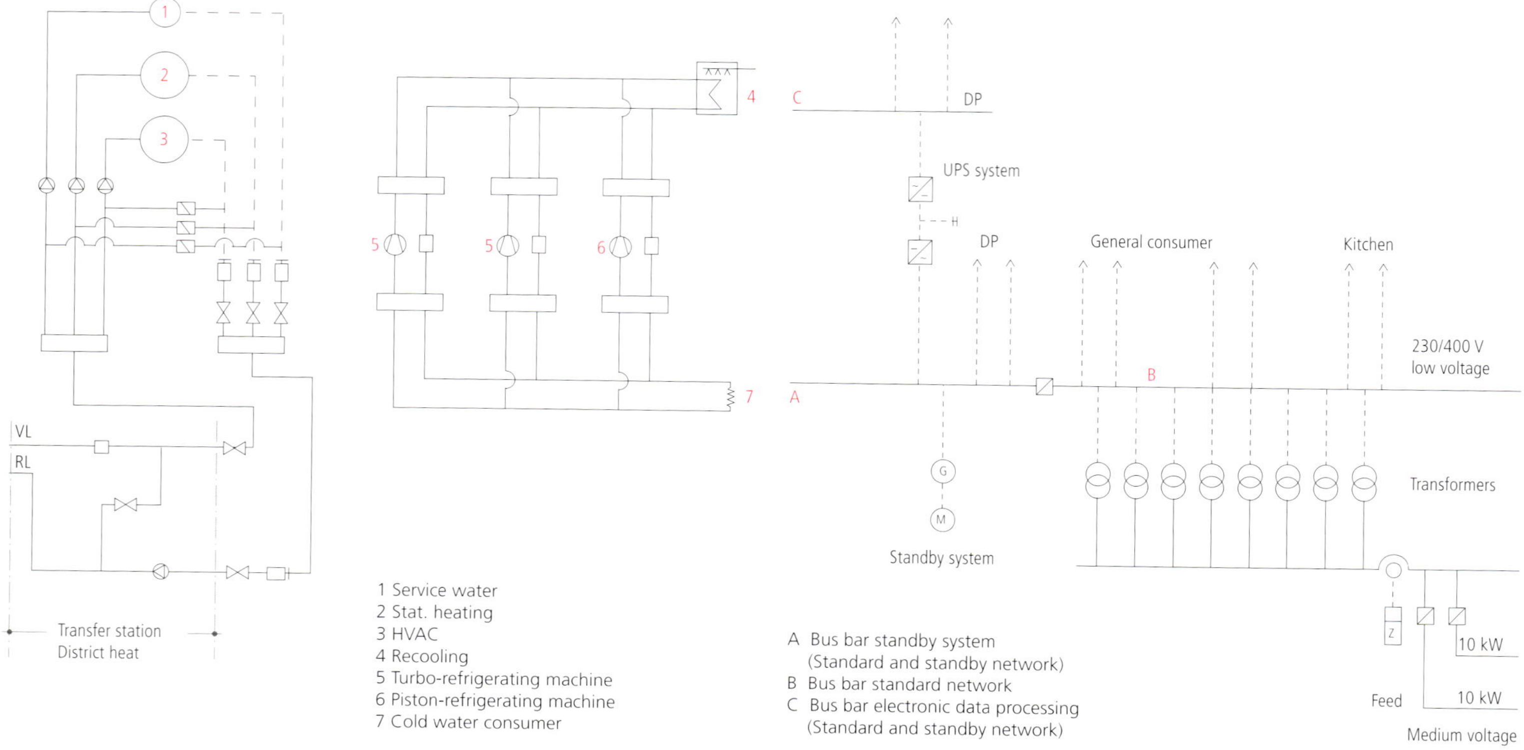

Figure 8.1
Electrical supply scheme with statistical UPS system and corresponding heating and cooling scheme (conventional solution)

If the aim is to develop a total energy supply concept and to enter into detailed rate negotiations with the utility, a far more precise document on the energy demand is required. **Figure 8.1** illustrates a conventional heating and cooling scheme for a fairly large office complex (heat supply via district heating/cold energy production via electricity) as well as the electrical supply scheme with subsidiary installations (Uninterruptible Power Supply (UPS)/standby power equipment). **Figure 8.2.1** indicates the electrical energy demand for the same building (in kW) at the anticipated peak time. As the graph clearly illustrates, the operations in the building are plotted hour by hour based on the assumed operations performed by each individual consumer, especially the large consumers. The end result is a typical consumption pattern, which can be assumed to remain more or less constant during other months. When this is in doubt, similar graphs can be drafted for other months, especially for seasons when the building does not require electrical energy for cooling. **Figure 8.2.2** indicates the anticipated annual demand in kWh, a useful and necessary piece of information for developing a good energy concept and for detailed rate negotiations.

System	Office building	Industrial	Hospitals	Residential
Lighting	0.80 – 0.90	0.95	0.70 – 0.90	0.70 – 0.90
Outlets	0.20 – 0.40	0.10	0.10 – 0.20	0.10 – 0.25
HVAC	0.70 – 0.90	0.80 – 1.00		
Heating	0.80 – 1.00	0.80 – 1.00	0.80 – 1.00	0.80 – 1.00
Refrigeration	0.80 – 1.00	0.90 – 1.00	0.90 – 1.00	0.90 – 1.00
Kitchens	0.60 – 0.75	0.60 – 0.75	0.60 – 0.60	0.60 – 0.80
Elevators	0.50 – 0.70	0.30 – 0.90	0.80 – 1.00	0.60 – 0.70
Hoisting equipment		0.80 – 0.90		
Other	0.30 – 0.40	0.35 – 0.45	0.60 – 0.85	0.30 – 0.60
Total	0.7 – 0.8	0.3 – 0.7	0.6 – 0.8	0.3 – 0.6

Table 8.2
Approximate values for simultaneity factors of consumers

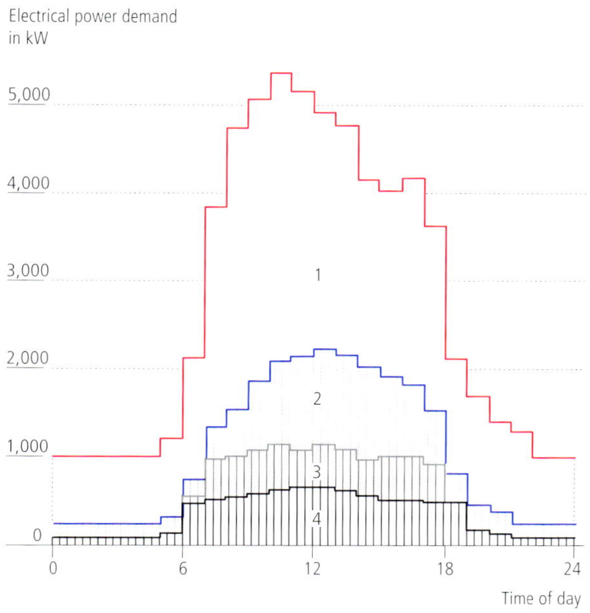

Figure 8.2.1
Electrical power demand per hour in kW, July, hot summer

1 Computers, kitchen appliances, lighting, elevators and conveyors, communications and low voltage, office equipment and terminals, pumps
2 Refrigerating machines, special core areas and general office areas
3 Ventilation, HVAC systems, general office areas
4 Ventilation, HVAC, special core areas

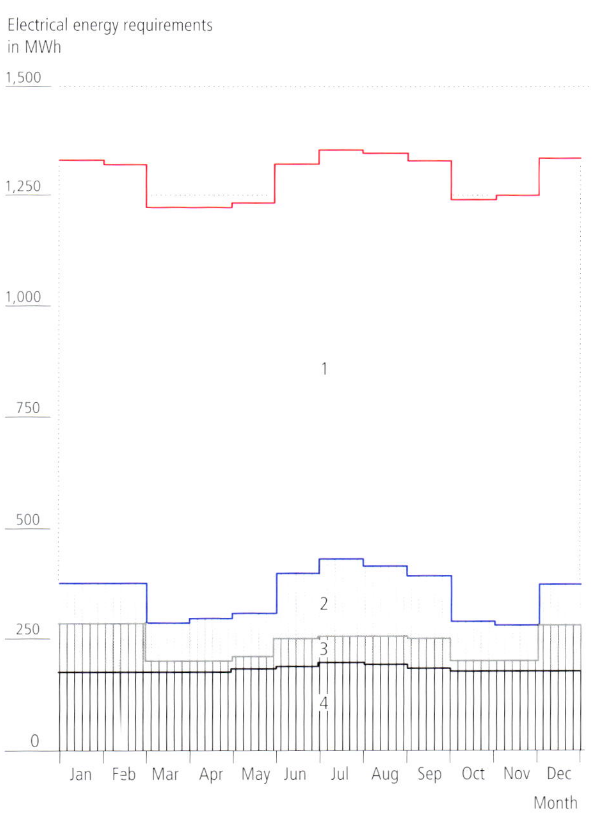

Figure 8.2.2
Annual progression of electrical energy requirements

Simultaneity factors are essential for drafting a load estimate. These factors result from the fact that all electrical consumers are never in use at the same time in a building and that the sum of consumer connect loads is therefore not an accurate guide. The simultaneity factor indicates the simultaneous load on the grid in relation to the sum of all connect loads. The larger and more varied the number of consumers, the smaller the simultaneity factor (< 1). **Table 8.2** indicates some approximate values for simultaneity factors of consumers in different building types.

The demand values given below can serve as a rough initial estimate of typical and specific connect loads:

Office buildings	50 – 100 W/m²
Hospitals	120 – 50 W/m²
Laboratories (incl. animal cages)	100 – 350 W/m²
Housing	4 – 6 kW/unit
Retail (department stores, etc.)	50 – 60 W/m²
Computer labs (etc.)	400 – 800 W/m²
Commercial kitchens	400 – 700 W/D (D = diner)

It is vital to properly design systems for large consumers such as lighting, large motors, transformers etc. from the very beginning because each transformation of electrical energy also results in some energy losses. A design that is either too generous or falls far short of the actual demand will compromise the efficacy of the transformation. Inductive consumers (transformers, motors, inductance coils) are especially important in this regard. To form the magnetic field, they require inductive reactance, which has no effective or real output. The current heat losses caused by inductive reactance should be reduced as much as possible with the help of compensatory installations to minimize both the individual circuit and the grid circuit. Most utilities impose a surcharge on large consumers, who fail to cap the inductive reactance they produce. The boundary values of the maximum permissible output factor, for which no inductive reactance fee is charged, is specified by the utility in the electricity supply contracts. In principle, inductive reactance output can be limited by precise sizing of electrical motors and transformers and by using low-loss ballast devices in lighting systems.

8.2 General information

8.2.1 Electricity generation

Electricity generation is differentiated by generator type into thermal and hydroelectric power stations. In Central Europe, most electrical energy is produced in thermal power stations. In Switzerland, on the other hand, 57 % of electrical energy is produced in river and storage (or stored-water) power stations, 2 % in thermal power stations, 39 % in nuclear power stations and 2 % in alternative power stations.

Thermal power stations are chiefly designed as steam, gas and diesel power stations, driven by steam and gas turbines as well as steam piston and gas engines. Steam piston engines provide only some of the required generating power; the bulk is generated with gas and steam turbines. Water turbines are another option, frequently used in Switzerland, which has sufficient utilization options for hydroelectric power. The largest steam turbines yield up to 600 MW in power stations, while the largest water turbines have an electric output of up to roughly 150 MW.

The above-mentioned machines are used to drive generators that produce electricity. Induction in the generators (also in dynamos) is also harnessed. Electricity is generated by rotating an electric conductor system in magnetic fields (DC generator) or, vice versa, rotating a magnetic field inside a conductor system (AC generator, internal pole generator) and tapping the voltage that is induced at the conductor ends. The fixed component is called a stator and the moving component is called an armature (rotor). **Figure 8.3** illustrates the process of induction when a wire coil is rotated in a magnetic field.

If a rectangular wire coil rotates in a homogenous magnetic field, its margins cut magnetic force lines. When the coil begins its movement from a vertical position, as shown in **Figure 8.3**, the number of cut force lines is nil. After one rotation from A to B, the number of cut force lines is still very low, increasing only as the curve continues, for example, from C to D and reaching its peak at the rotation through the magnetic axis (north-south pole). The coil has described a rotation of roughly 90°, and the voltage augments from zero to a peak value. If the coil ends are linked to two coil rings and if the brushes sitting on top of the coil rings are connected with a closure wire, an induction current runs through the wire, which also reaches its peak after a total rotation of 90°.

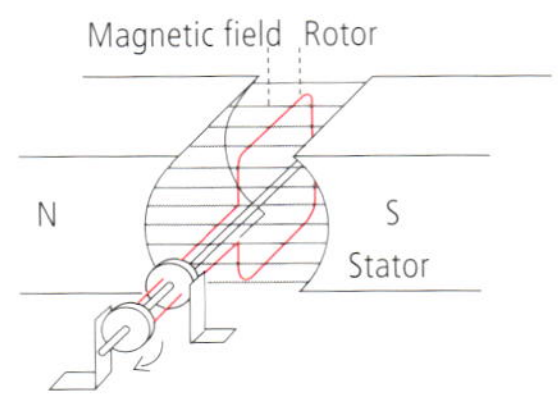

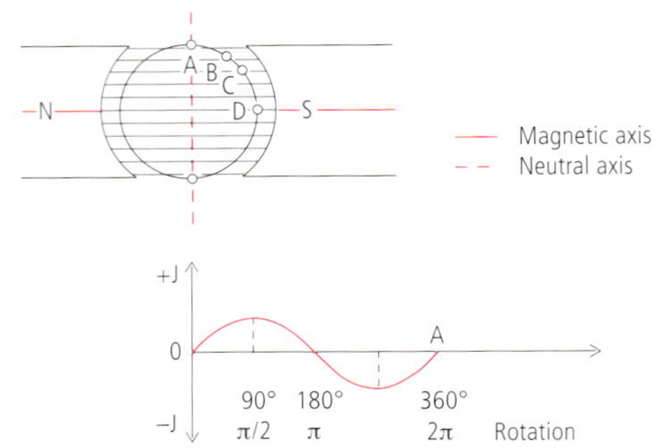

Figure 8.3
Induction current in a wire coil

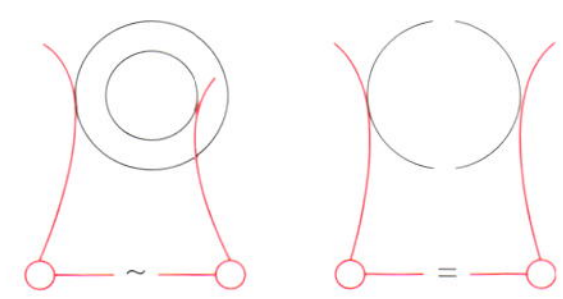

Figure 8.4
Alternating current generator with internal poles

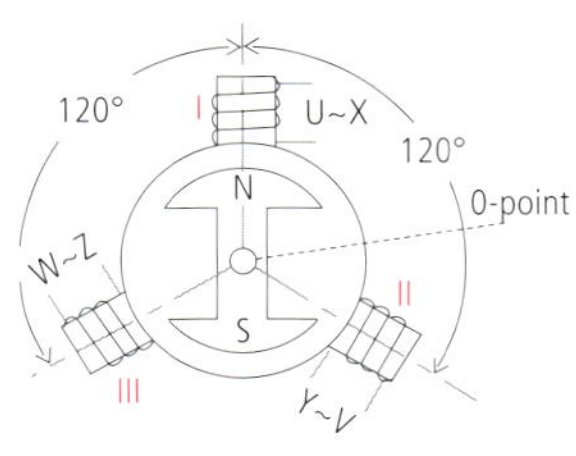

Schematic of rotary current generator

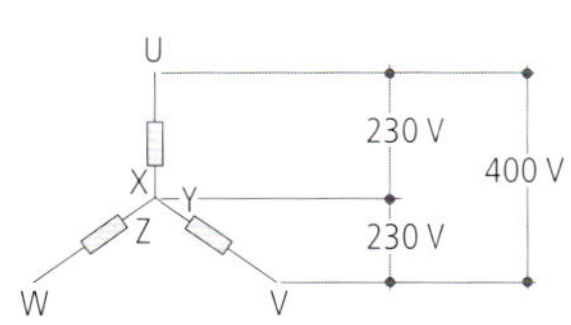

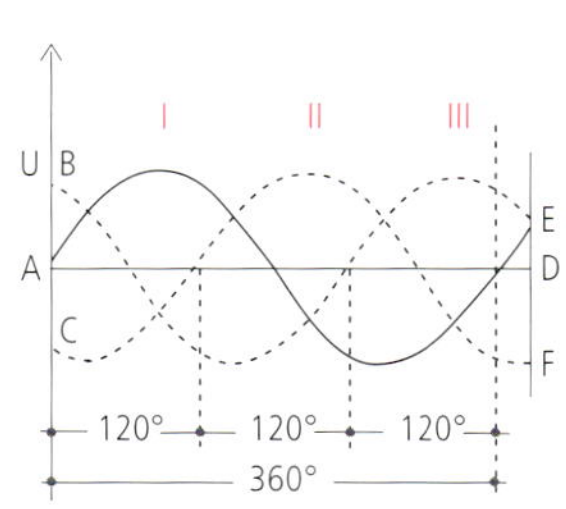

Voltage in relation to neutral conductor

Figure 8.5
Schematic diagram of rotary or three-phase current

If the coil continues to rotate, the intensity or strength of the current and the voltage decrease once again to nil after a total rotation of 180°, and peak again after a total rotation to 270°, etc. Each time the coil passes through the neutral zone, the current changes direction, and this is why it is referred to as alternating current (AC). In intensity and direction, alternating current produces a sine wave.

If the coil ends are not connected to coil rings, but to two collector segments, as shown in **Figure 8.4**, the collector segments are transferred to the other brush every time the coil passes through the neutral zone. An induction current is tapped. While the intensity alternates between nil and a peak value, the direction remains unchanged (= direct current or DC).

A three-phase alternating current is produced when three separate coil systems are housed in a motor casing.

The diagram in **Figure 8.5** illustrates a rotary current generator or three-phase alternator and the phase displacement in rotary current. The induction voltages display three sinus curves, displaced by 120 ° and the course the voltage takes in one of the three coil systems. Rotary current can be tapped at the L1/L2/L3 (U, V, W) terminals. If a neutral conductor is carried along from the zero point, a phase voltage is present between the neutral conductor and the phase conductor, which is 1.73 (= $\sqrt{3}$) times smaller than the so-called interlinked or phase-to-phase voltage. This means that two different voltages are available.

8.2.2 Electrical power supply

From the power station, where the current is generated, the power is distributed via high-voltage (110 to 400 kV) overhead or open lines across an integrated network. The large power stations are linked by the integrated network. In large cities or major industrial sites, transformers are used to step the voltage down from high voltage to medium/high voltage (1 to 30 kV) and to feed it into the local supply grid (utility grid). Within the cities, the current is distributed at the aforementioned medium/high voltage level to additional transformers, which are either installed in the buildings of large consumers or set up as stand-alone utility stations outside of buildings. These transformers effect yet another change in voltage from medium/high to low (ranging from 50 to 1,000 V), in Central Europe mostly to 230/400 V.

Very low voltages (below 50 V) are then generated on-site by additional step-down transformers, and are chiefly used for supplying electricity to signalling and communications systems.

Figure 8.6 contains a diagram of the principal electricity supply outside of buildings, while **Figure 8.7** illustrates the supply grid in Europe.

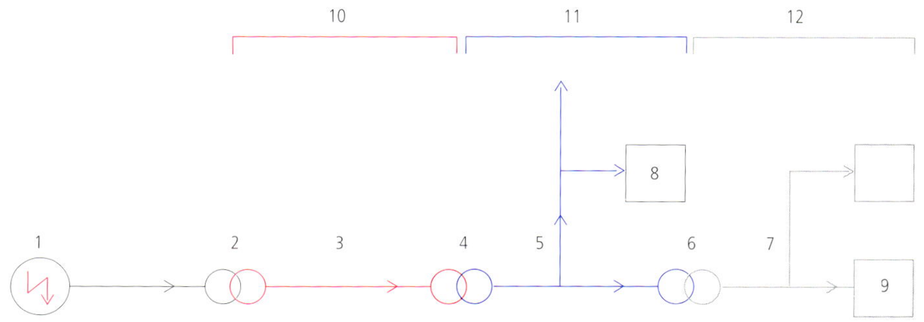

Figure 8.6
Principal electrical power supply

1 Power station
2 Transformers
3 High-voltage lines
4 Distribution substation
5 Medium/high voltage lines
6 Transformer stations
7 Low-voltage lines
8 Medium/high voltage consumers
9 Low-voltage consumers
10 High voltage 110 – 380 kV
11 Medium/high voltage 1 – 30 kV
12 Low voltage 50 – 1000 V

Figure 8.7
Supply grid in Europe

Power connections (including 380 kV lines under construction)

- 380 kV
- 220 kV
- HVDC
- Transformer stations or power plants

8.2.3 Rates

Rates are divided into those for low voltage and those for medium/high voltage consumers. Low-voltage consumers are directly hooked up to the 400 V grid and have an approximate maximum connect load of 200 kVA, which is determined by the utility.

Consumers with demands that exceed the capacity of the low-voltage grid are supplied with medium/ high voltage in the range from 1 to 30 kV, generally between 10 to 20 kV. Rates are subject to market pricing under the revised European laws on deregulation (free market for various local suppliers). The utilities generally offer a wide variety of different rates. Consumers supplied from the medium/high voltage grid are generally offered special contracts, which take their specific consumption patterns into consideration. The rates are generally calculated according to the following parameters:

Work-dependent costs, whereby the cost (price per kWh) for "work" (kWh) tapped from the grid can fluctuate depending on the time of day (off-peak rates).

Performance-dependent costs, which are payable for the supply of a specific output. Hence, performance dependent costs can partially offset or defer the connection costs for a new hook-up.

The most common calculation is the **incentive rate**, which staggers the rate according to the output that is used, and where a monthly or annual peak is measured according to a quarter- or half-hourly average. Some utilities have introduced adjustments for consumption-dependent parameters such as usage duration or percentage of total maximum into their rate calculations.

The common factor in all rates is that the average cost is lowest when the consumption peak falls outside the utility peak load (utility peak load generally between 7 a.m. and 9 a.m. or 5 p.m. and 7 p.m.). Most utilities offer special off-peak (night) rates (10 p.m. to 6 a.m.), which are commonly factored into plans for an energy concept. The **usage duration** in hours is another important parameter: it is the product of the usage duration of the tapped work (kWh) divided by the peak load (kW) tapped during that period (in analogy to total load hours for heating and cooling systems).

In summer, all energy costs may be considerably cheaper (up to 50 %) than in winter (higher demand).

Rate negotiations are feasible only for very large consumers. Medium-sized and small consumers usually have only two options to reduce their electricity bill. They can:
- Select the most advantageous rate on offer
- Adapt their usage to existing rate structures

With deregulation, however, all this may change in the future; Austrian consumers, for example, might be able to purchase their electricity from Spain.

To select the correct rate, it is essential to have detailed knowledge of the peak demand patterns (in terms of time and absolute peak) and of the usage hours in order to determine advantageous electricity rates in collaboration with the utility. There are several options of adapting the tapping pattern to the existing rates, and these should be taken into consideration even when systems are in the planning stage. The following measures yield the best results:
- Switching all non-essential consumers off during the peak load period (load decrease programme)
- Shifting the connection time of short-term consumers to a low-load period (monitoring with central systems for large installations/bus systems in residential applications).

By subjecting the operational processes to a detailed analysis and adapting usage behaviour accordingly, electricity costs can be reduced by roughly 15 %.

8.2.4 Terms

For a better understanding of the following sections, the most common terms are briefly explained:

Voltage (V) in volts = difference between two potentials
Current (I) in amperes = flux of free electrons from negative to positive potential
Resistance (R) in ohm = physical characteristic of the conducting material (indicates how many free electrons are available in the current flux)

Ohm's Law:
Voltage = Current I · Resistance R
(Or: the voltage is the product of current and resistance)

$$V = I \cdot R$$

The current I is decisive for the sizing of electrical systems. The effective output P for pure ohmian loads is:

Direct current: $P = V \cdot I$

Alternating current: $P = V \cdot I \cdot \cos\varphi$

Rotary current: $P = \sqrt{3}\, V \cdot I \cdot \cos\varphi$

To transfer more output, for example, one can increase either the voltage (V) or the current (I). Increased voltage requires more insulation for high-voltage lines, while increased current leads to higher cross-sectional efficiency.

Direct current = flux of current in one direction (from negative to positive potential)
Alternating current = flux of current in alternating directions (number of changes per second is frequency in hertz)
Rotary current = three alternating current cycles, displaced by 120° each
Frequency (f) in hertz (Hz) = number of directional changes (full oscillation in alternating current) per second

Power (P) in watts (W) = effective output or work in a specified time period ($P = I \cdot V \cdot \cos\varphi$)
Work (W) in watt hours (Wh) = power · time ($W = P \cdot t$)
Performance efficiency (P) in watt (W) = power available in the form of energy (drive power, lighting, etc.)
Idle efficiency (Q) in volt-amperes (VA) = output required for an induction consumer to form the magnetic field
Apparent efficiency (A) in volt-amperes (VA) = output exacting a load on the electric conductor without being immediately effective.

The interactions between performance, idle efficiency and apparent efficiency are shown in **Figure 8.8.1**. The performance efficiency (P) indicates how much energy must be made available, for example, to provide power to run a motor. The unit for power is BTU per hour, watt, kilowatt or megawatt. The idle efficiency, illustrated in **Figure 8.8.2**, can be understood as a kind of energy storage. Devices constantly absorb output in their storage and release it to the grid.

Induction consumers (coils) are so-called accumulators, while capacitating consumers are so-called voltage regulators. In application, there are neither purely induction nor purely capacitating consumers.

They influence each other and are loaded with an ohmian component. This creates fluctuating energy fluxes in the supply grid, which cannot be transformed into working energy. All electrical systems are exposed to an additional load for this reason, which translates into further losses in addition to the increased sizing for cables, switches and grids; the losses increase exponentially to the transferred current. In order to keep them as low as possible, the output factor cosine φ is given with 0.85 to 0.95 and should be maintained. Reactance causes a phase difference between V and I, the cosine of which is called pf, see **Figure 8.8.1**

Compensation = measure to improve the power factor (cosine φ), thus diminishing transfer losses. **Figure 8.9** illustrates the relationship and shows that the apparent efficiency has decreased due to the improved cosine φ as a result of the compensation and is approaching the performance or working efficiency.

Starting current = current that flows when the consumer is switched on (started). It can be many times greater than the current that flows in the operating state, see **Figure 8.10**.

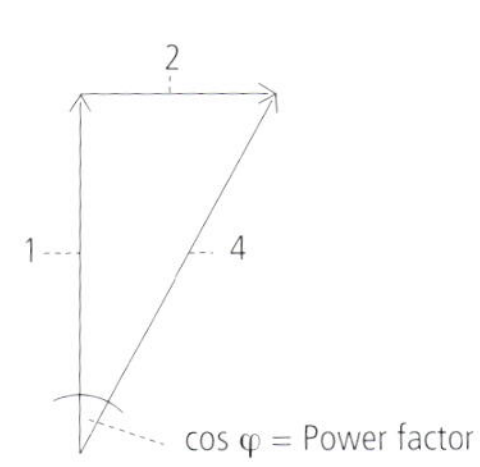

Figure 8.8.1
Performance, idle efficiency and apparent efficiency

1 Performance efficiency P (watts)
2 Inductive idle efficiency Q (VA)
3 Compensation (capacitive idle efficiency)
4 Apparent efficiency A (VA)

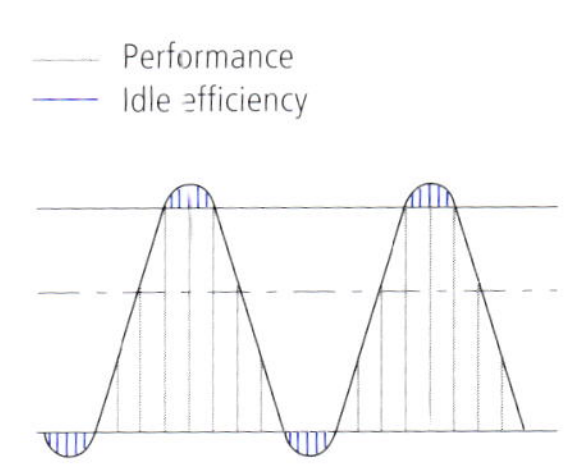

Figure 8.8.2
Idle efficiency

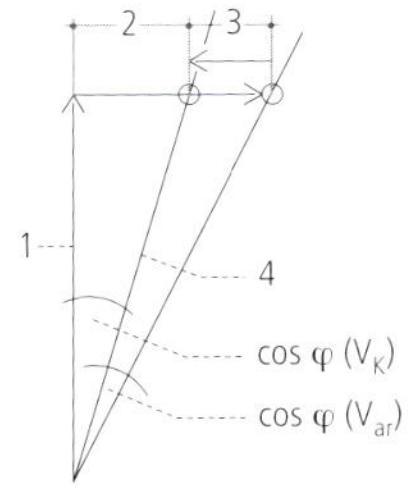

Figure 8.9
Reduction of idle efficiency

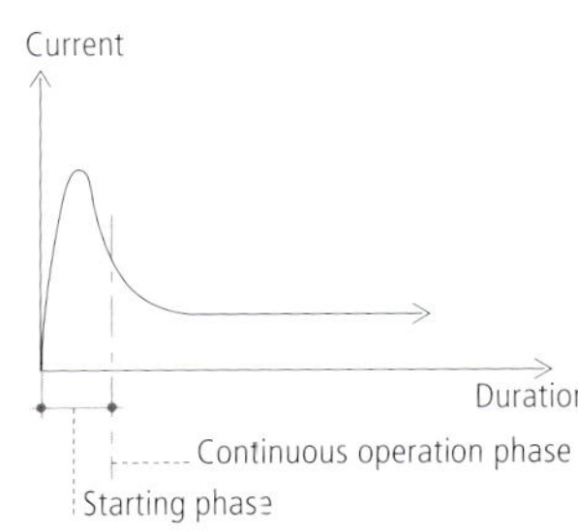

Figure 8.10
Starting current of a motor

8.3 Power feed

The power feed in the building is in either the low voltage or the medium/high voltage range, depending on the supply grid and especially on the power demand of the connected object. The various power supply utilities generally define the installation conditions.

8.3.1 Low-voltage feed

Low-voltage feed is generally via a cable provided by the utility, and connected to the house line from the outside.

8.3.1.1 Main disconnect

The main disconnect is usually set up in a box, which should be readily accessible to the utility workers. For multifamily homes or large housing blocks, the main disconnect is installed in an equipment room that is accessible at all times, whereas smaller building objects, such as single-family homes, are equipped with a box on an exterior wall as shown in **Figure 8.11**. The diagram in **Figure 8.12** illustrates a main disconnect cabinet installed on the boundary of the property to provide service for several smaller buildings.

For larger building projects, e.g. commercial and industrial buildings, hotels and restaurants etc., the main disconnect is located in a feed section (usually one to two) of the low-voltage grid. These installation configurations are subject to regulations specified by the energy supplier.

The illustration also indicates the specific conditions that apply to a main disconnect installed in a cabinet. If the main disconnect is not installed outside of the building or on an exterior wall of the building, it is set up in a special equipment or meter room (multifamily homes).

Niche installation

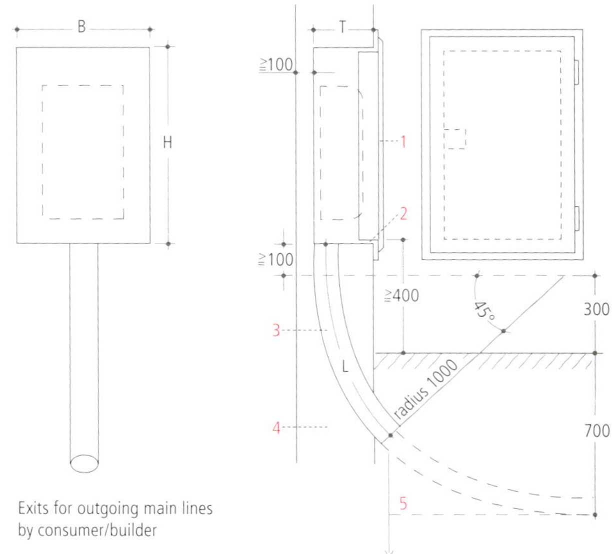

Figure 8.11
Example of main disconnect for low-voltage supply

L Curve length of pipe (700 mm)

1 Door
2 L-bar frame (50/50/5 mm)
3 Pipe (80 – 100 mm)
4 Exterior or garden wall
5 Cable trench bottom

Internal niche measurements

Main disconnect size	Minimum measurements in mm H	W	D
1x3 NH00 2x3 NH00 100 A	800	600	200
1x3 NH2 400 A	1200	600	240

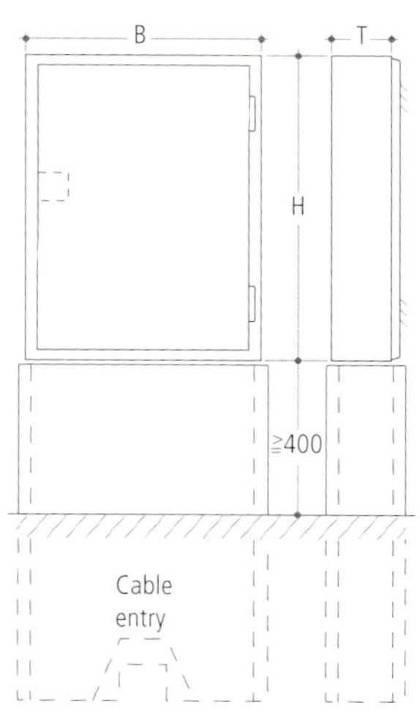

Figure 8.12
Main disconnect box

Measurements of protective housing

Main disconnect size	Minimum measurements in mm H	W	D
1x3 NH00 2x3 NH00 100 A	800	600	210
1x3 NH2 400 A	1200	800	270

8.3.1.2 Meter room

When a meter room is set up for a larger building project (e.g. a multistorey apartment building), the first step is to install the incoming service feeder. Special cable ducts must be laid in the external wall of the building for cables and piping, and the sizing of the ducts is determined by the cable sizing defined by the utility. **Figure 8.13** illustrates the installation of a cable duct in an exterior wall.

Buried connections to the meter room must adhere to the minimum measurements indicated in **Figure 8.14** (depth below site surface).

The utility generally determines how the incoming service feeder is installed up to the point where it is hooked up to the building service switch; all relevant decisions must be taken in conjunction with the utility, and this is one of the services the utility provides. Incoming service feeders and boxes must never be laid or installed in rooms exposed to fire or explosion risks. This restriction also includes damp rooms (including garages) or rooms, where temperatures may be too high.

Figure 8.15 features an example of a meter room with ground-fault circuit interrupter (GFCI).

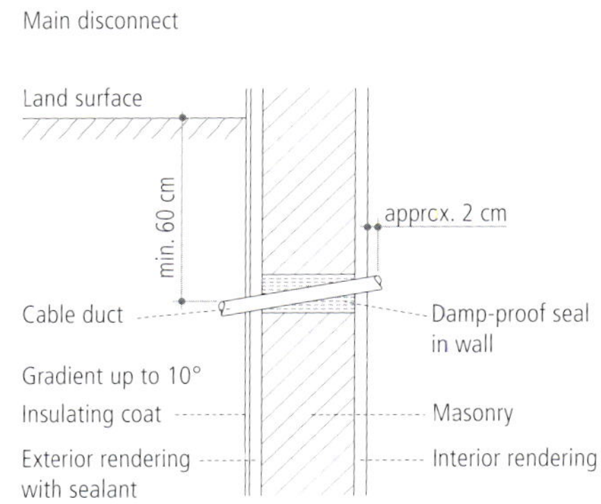

Figure 8.13
Example of cable duct installation in exterior wall

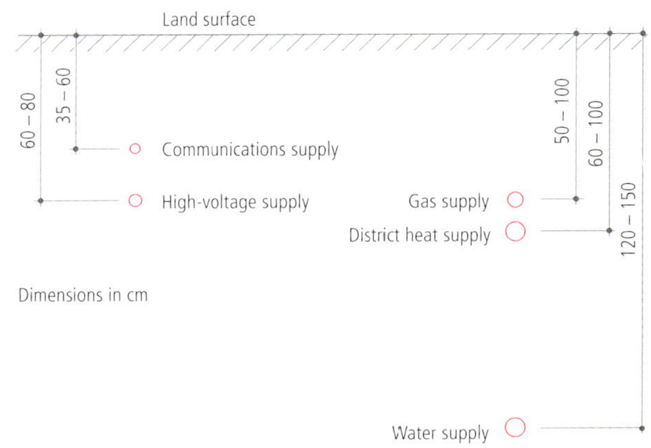

Figure 8.14
Depth below site surface for buried connections (according to DIN 18012)

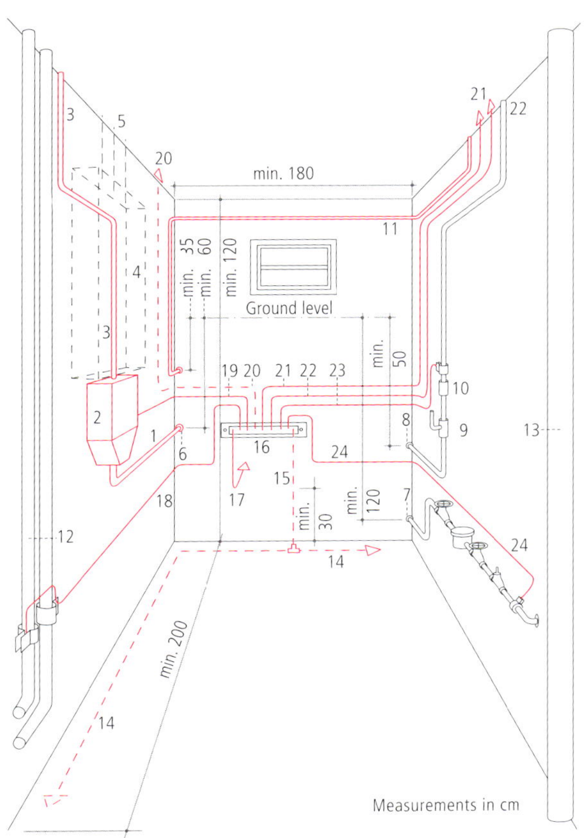

Figure 8.15
Example of meter room (according to DIN 18012) with ground-fault circuit interrupter (GFCI)

1 Incoming service feeder for high voltage
2 High-voltage main disconnect box
3 High-voltage main line
4 Meter (optional)
5 High-voltage lines from meter to circuit distribution
6 Cable duct
7 Main disconnect for water with meter
8 Main disconnect for gas
9 Gas main shut-off
10 Insulator
11 Main disconnect for communications line
12 Heating pipes
13 Sewage pipe
14 Foundation ground
15 Connection lug for foundation ground
16 Equipotential bus bar with GFCI
17 Lightning protection system
18 Heating pipes
19 Connection line for protective measure in TV net
20 Grounded conductors for protective measure in TT net
21 Telecom system
22 Antenna system
23 Gas pipes
24 Water pipes

Meter rooms must be adjacent to an exterior wall, through which the feed lines are laid. Depending on the situation, they may be located in the basement (or first basement level in large buildings) or on the ground floor. All doors to equipment rooms must open in the direction of the escape route. In smaller buildings, meter rooms may also house the supply services for power, telecommunications, water, gas and district heat.

Universal guidelines specify that every building must be equipped with a ground connection installed on top of a ground-fault circuit interrupter.

The GFCI illustrated in **Figure 8.15** links the connections of other system components such as heating pipes, connection lines for grounding, grounded conductors, telecom system, antenna system and gas- or water-conducting pipes, and acts as a compensator. This compensation is necessary to ensure that life-threatening fault or residual currents in conduits and pipes are diverted into the ground.

The ground connection also serves as a ground for lighting protection, antenna and telecom systems, thereby ensuring equipotential compensation.

The ground connection generally consists of galvanized strip steel with a cross-section of 100 mm² (cross-section preferably 30 x 3.5 mm, also 40 x 2.5 mm). It must be embedded in a corrosion-proof manner in the concrete foundations of the building (encased by concrete on all sides).

Figure 8.16 illustrates the construction of a ground connection. The rebars of the footing steel reinforcement can be used provided certain criteria with regard to sizing and connection are fulfilled.

Figure 8.17
Main supply system

Main supply line according to DIN 18 015, Part 1 Curve A
Distribution line NYM 4x10 mm
Trip line or conduit (Ø 29)

1 House connection box
2 Meter cubicles according to DIN 43 870
3 Circuit distributors

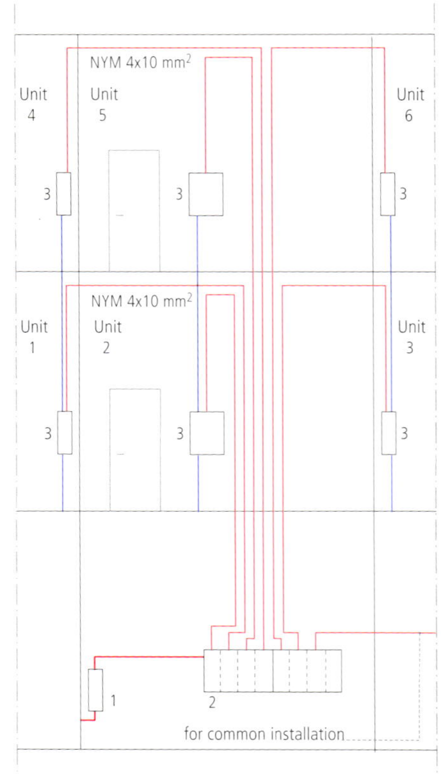

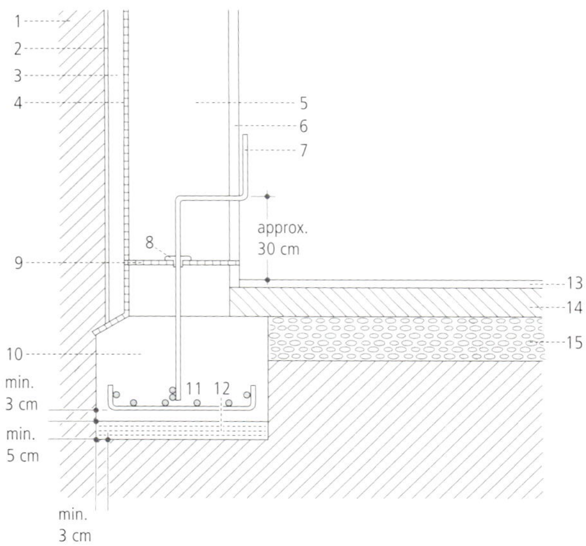

Figure 8.16
Construction of a ground connection

1 Soil
2 Protective layer
3 Polystyrene HR foam
4 Bitumen sheeting
5 Masonry
6 Interior rendering
7 Connection lug, exposed end approx. 150 cm
8 Bituminous seal
9 Insulating board
10 Reinforced concrete foundation
11 Foundation ground
12 Granular subbase
13 Screed
14 Concrete
15 Ash layer

8.3.1.3 Measuring devices (meters)

The main supply line leads from the house connection box (junction box) to the measuring devices (main switchboard). When several consumers are connected to the same line, the meters are installed in a separate meter room. When there are only a few meters, the main equipment room also houses the meters (Figure 8.17).

For large housing projects or similar building projects, there may be several meter rooms in addition to the equipment room. Like the latter, meter rooms must be accessible at all times to the utility workers and should never be assigned to any additional uses. The room must be constructed to specific minimum dimensions in cases where it houses a great number of meters. These dimensions are shown in Figure 8.18. Generally speaking, it is prohibited to install gas, water, heating or sewage pipes or lifting and hoisting installations in these rooms. The number of measuring points is dependent on the number of units to be counted (e.g. apartments), and in some cases on reserved points for additional units. They are defined differently by each utility.

Figure 8.18
Minimum dimensions for meter rooms

1 Min. 500 mm escape route width according to VDE 0100 Part 729
2 Max. 750 mm door width on meter cubicles
3 Min. 1500 mm meter room width for single-sided equipment installation
4 Min. 1750 mm meter room width for two-sided equipment installation
5 When narrower doors are used on cubicles (500 – 250 mm), the meter room width can be reduced accordingly

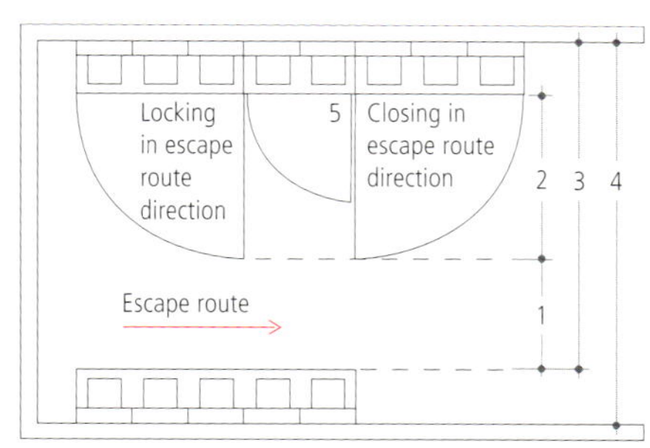

1 Min. 700 mm corridor width according to VDE 0100 Part 729
2 Min. 900 mm meter room width for single-sided equipment installation
3 Min. 1100 mm meter room width for two-sided equipment installation

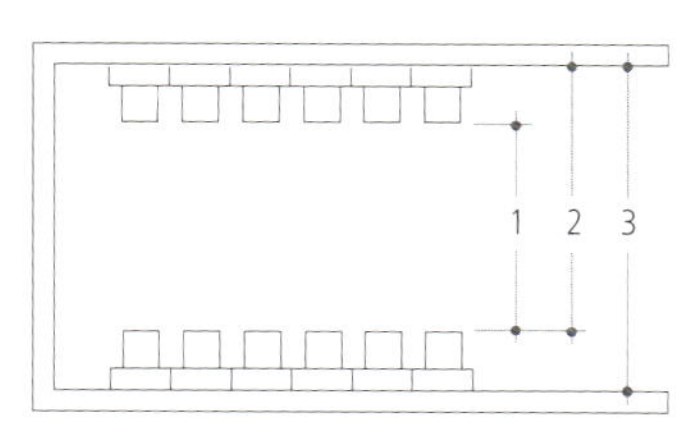

Gas, water, heating and sewage pipes are prohibited in meter rooms with meter boards

8.3.2 Medium-voltage feed

For larger building projects with a connect load of over 240 kVA, power is supplied according to a different set of criteria from those described above. In developed low-voltage networks (in cities) the indicated criteria value may be considerably higher. The network structure is the decisive parameter.

8.3.2.1 Medium-voltage plant

If electricity is supplied from the medium-voltage network, the consumers are either connected to a medium voltage ring or to one or several conductors from the substation. Very large building projects are sometimes equipped with several medium-voltage feeds, if losses in transmission by low-voltage cable or busbar systems (231/400 V) are too high. Thus many industrial projects feature several medium voltage stations at heavy load points, and the energy is distributed from these plants at the low-voltage level along the shortest possible route. The number of medium-voltage stations is generally determined according to the following criteria:
- Investment for current transmission including switchboard, cables and conduits to consumer
- Construction costs for installing the relevant systems
- Running costs resulting for transmission losses of the electric energy

To begin with, medium-voltage electric energy is supplied via the medium-voltage station.

The medium-voltage station houses different cells dedicated to supply, transmission and measuring.

Figure 8.19 illustrates the layout of a medium-voltage station switchgear, where the individual cells are arranged in a single row. They can also be arranged in two facing rows with a minimum distance of 1.2 m between rows. **Figure 8.20** depicts a typical medium-voltage station in an office complex. Rooms for such stations should be located in an area where access is provided at all times via either external stairs or internal stairwells, as the same access criteria apply as previously described for house connections.

Responsibilities for the station are divided between the utility and the client. The utility is responsible for the feed cells and the measuring devices. The client is responsible for the main switch, as well as switching and safety installations for the transformers.

Figure 8.19
Layout of medium-voltage switchgear with branch feed

1 Feeder cell
2 Transfer cell
3 Meter cell
4.1 – 4.3 Discharge cells to transformer 1/2/3

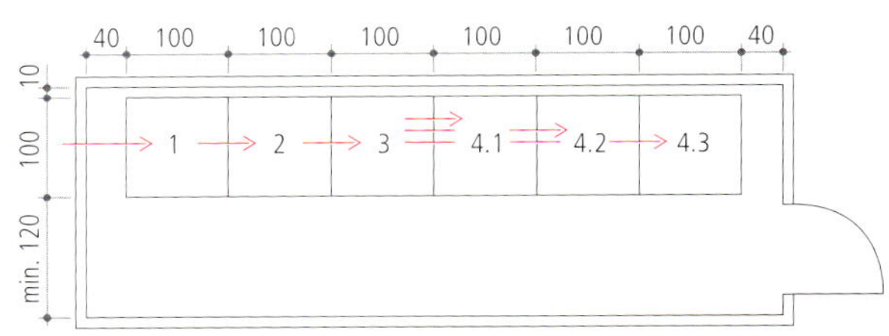

Measurements in cm

Figure 8.20
Medium voltage switchgear in office complex

8.3.2.2 Transformers

Once electric energy has been fed into the building via cables or busbars, transferred and measured, the current is transformed from the medium-voltage level, generally 10 to 20 kV, and stepped down to the low-voltage level (0.4 kV). This is achieved by transformers, which are selected according to the following criteria:
- Nominal voltage
- Nominal power, capacity (kVA)
- Connection symbol
- Type
- Insulating medium
- Insulation class

The **nominal voltage** is determined by the voltage of the network to be supplied and by the voltage required by the consumers (e.g. 400/230 V).

The **nominal output** is determined by the output of the consumers (load) connected to the transformer.

The **connection symbol** expresses the purpose for which the transformer is used (an alphanumeric sequence that indicates the switching of the windings and the location of the windings in relation to each other). The most common connection symbol for domestic users with residential building systems and lighting systems is Dy5, since these transformers are suitable for grounding and for many other protective measures, and because the neutral point can be exposed to continuous nominal voltage.

The **type** (construction) is largely determined by the type of cooling and by the conditions for idle and short circuit losses (universal standards apply).

With regard to **cooling**, transformers are either liquid-filled (the liquid acting as the coolant) or dry (air-cooled).

Clophene was the standard coolant for liquid-filled transformers in the past. Today, in recognition of the new environmental standards, liquid-filled transformers use oil (oil-insulated transformers). These transformers present a fire hazard indoors because of potential oil leaks and therefore require elaborate constructional measures (a separate cell for each transformer).

Oil-insulated transformers

Oil-insulated transformers are generally used in substations, large industrial operations and the network of the utility. **Figure 8.21** depicts an oil-insulated transformer; the cooling fins are clearly visible on the outside, as are the expansion valve and the protective measures.

The transformer rests on an undercarriage with wheels for easy transport. The connection poles of the high voltage cables with ceramic insulators can be seen on the top cover plate. The principle of the transformer is based on the arrangement of two coils (windings) on a common iron core.

A voltage impressed on the primary coil (high voltage) produces a magnetic current in the iron core which in turn produces an induction voltage of the same frequency in the winding of the secondary coil (low voltage).

Oil-insulated transformers with an output of up to 630 kVA must be equipped with an oil drip pan capable of catching the liquid content (0.7 m^3) of the transformer (**Figure 8.22**). When several transformers are set up in a row, a single drip pan is usually sufficient for all. For transformers with capacities ranging from 800 to 2,500 kVA, a pit with a total catch volume of 2 m^3 must be installed.

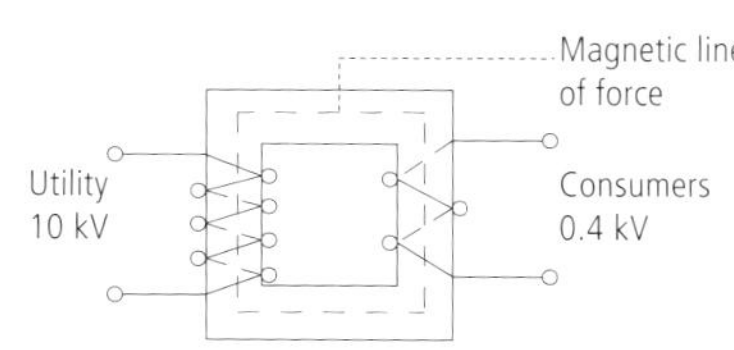

Figure 8.21
Oil-insulated transformer

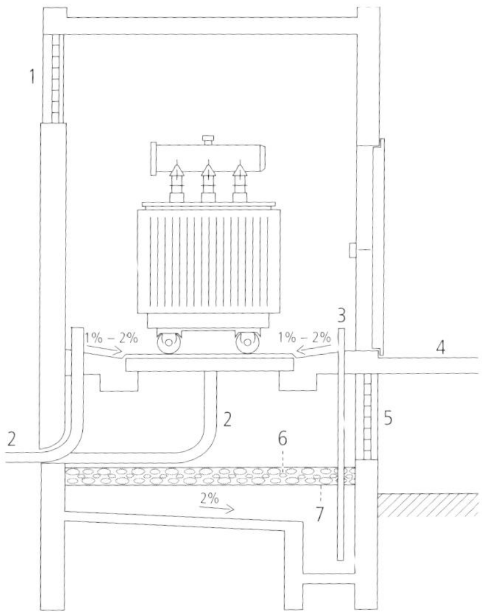

Figure 8.22
Interior installation of a liquid-filled transformer

1 Exhaust opening with guard grille
2 Cable duct
3 Coupler for pump
4 Ramp
5 Air inlet opening with guard grille
6 Gravel or crushed stone layer
7 Galvanized flat rolled steel grating

d = 15 x Ø duct

Dry-type transformers (moulded resin transformers)
Dry type transformers are far easier to install and offer the advantage of almost unrestricted choice of location as long as a sufficient volume of cooling air can be supplied.

Moulded resin transformers, see **Figure 8.23**, consist of single coils (aluminium or copper windings) cast into a high-insulating artificial resin. As the image shows, the individual coils are mounted onto a sectional steel yoke.

Dry-type transformers require less space than comparable oil-insulated transformers and there are not special constructional measures. **Figure 8.24** depicts an installation with several dry-type transformers in a single room. The compact shape and absence of fire hazard make it possible to install these transformers in a protective housing, which can be set up in the equipment room. This is an ideal approach to creating distribution centres (**Figure 8.25**).

Figure 8.23
Dry-type transformer

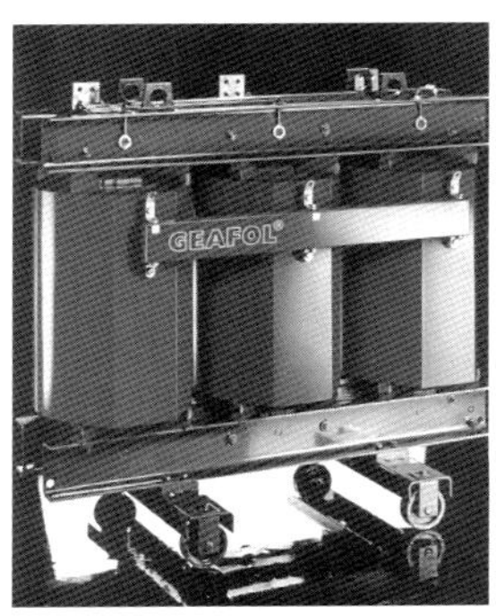

Figure 8.24
Dry-type transformers in common transformer room

Figure 8.25
Dry-type transformer in protective housing

Transformer type	Capacity in kVA	Transformer size (approx.) L/W/H in m	Space requirements L/W/H in m	Weight in kg
Dry-type	250	1.50/0.70/1.28	an additional 2m on 3 sides	1,300
	400	1.55/0.80/1.50		1,700
	630	1.70/0.80/1.60		2,250
	1,000	1.85/1.00/1.80		3,290
	1,600	2.10/1.00/2.10		4,940
	2,000	2.25/1.30/2.15		5,940
	2,500	2.30/1.30/2.40		6,810

Oil-insulated transformers 12/0.4 kV 24/0.4 kV	Transformer measurements L/W/H in m	Maintenance access space (all sides)	Transformer cell/ measurements L/W/H in m	Weight in kg
250 kVA	1.35/0.85/1.70	0.70	2.75/2.25/2.30	1,140
315 kVA	1.50/0.90/1.70	0.70	2.90/2.45/2.30	1,410
400 kVA	1.60/0.95/1.85	0.70	3.00/2.50/2.30	1,550
500 kVA	1.75/0.85/1.85	0.75	3.25/2.35/2.30	1,790
630 kVA	1.80/0.95/2.00	0.75	3.30/2.45/2.50	2,090
800 kVA	1.90/1.10/2.15	0.80	3.50/2.70/2.50	2,630
1,000 kVA	1.85/1.15/2.30	0.80	3.45/2.75/2.70	2,970
1,250 kVA	1.95/1.25/2.40	0.85	3.65/2.95/2.80	3,330
1,600 kVA	2.10/1.25/2.55	0.85	3.80/2.95/2.90	4,140
2,000 kVA	2.10/1.30/2.60	0.90	3.90/3.10/2.90	5,210
2,500 kVA	2.20/1.30/2.85	0.90	4.00/3.10/3.20	6,060

Table 8.3
Overview of transformer types and space requirements

Table 8.3 provides an overview of typical transformers and concomitant space requirements. It is important to consider that water-bearing pipelines must never be laid through transformer rooms or vaults. The surfaces of these rooms should be smooth and finished in a dust-bonding top coating.

The heat generated by transformers (10 % of the capacity) is disposed of by means of natural or mechanical ventilation.

When natural ventilation is used, transformers must be installed against an external wall, and ventilation trenches must be provided on the outside for transporting the transformer into the building and also to allow fresh air to flow in and exhaust air to flow out. Transformers should be installed on double floors for ease of installation of large cables with their required bending radius (double floor height 0.8 m). Low-voltage cables leading to the switchboard plants can also be laid in this double floor.

Future replacement of the transformers and the grounding system (connection to ground-fault circuit interrupter) should be given special consideration when planning a transformer station.

8.4 Power distribution

Once the electrical energy has been stepped down to the voltage required in the building, it is distributed via switchboard, i.e. the energy is fed through cables or bus ducts from the transformer to the low voltage plant, which can be fed by several transformers.

8.4.1 Low-voltage power switchboard

The low-voltage power switchboard accommodates circuit breakers and safety fuses, which are allocated to the individual supply areas and thus create the link between the collecting bus bar and the supply areas. All other electrical components are arranged in cabinets (**Figure 8.26**), with relevant cabinets set up in a separate room.

The dimensions of switchboards are governed by strict regulations for electrical equipment rooms. This is especially true for the placement of safety locks and escape routes, as well as ventilation and smoke extraction for these areas. Since low-voltage power switchboards in cabinet design are enclosed on all sides, they can be installed in generally accessible areas, although this is not recommended in larger buildings for safety reasons. It is best to install the cabinets in a separate room to prevent inexpert tampering with the switchboards.

The number of units depends on the size and number of consumers and the necessity for sub-distributions. Otherwise the same conditions described for transformer rooms apply to low-voltage power switchboards. They, too, should be located against an external wall whenever possible to achieve natural ventilation. If this is not feasible, mechanical ventilation and smoke extraction must be provided.

Figure 8.26
Low-voltage power switchboard (cells in row formation)

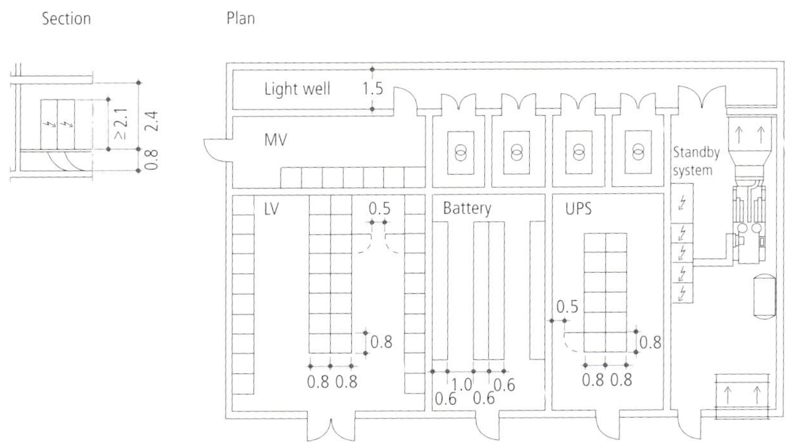

Measurements in m

Figure 8.27
Example of switchboard
(plan and section)

Figure 8.27 illustrates the layout of a switchboard with the principal dimensions for devices, the required space for circulation and installation and the escape routes. This example is designed for a total connect load of roughly 2,500 kVA. The average ceiling load is 10 kN/m².

Figure 8.28 contains a typical wiring diagram with medium/high voltage power switchboard, transformers and low-voltage power switchboard.

In this example the feed from the network is provided via two feed cells installed by the utility. The diagram also identifies the transfer and metering on the medium/high voltage busduct, linked to the individual transformers.

The transformers are followed by the circuit breakers for the low voltage power switchboard and the low voltage collecting bus bar. The latter runs through the individual switchboard cells, distributing energy across the relevant switches and safety fuses, cables or busducts to the consumers.

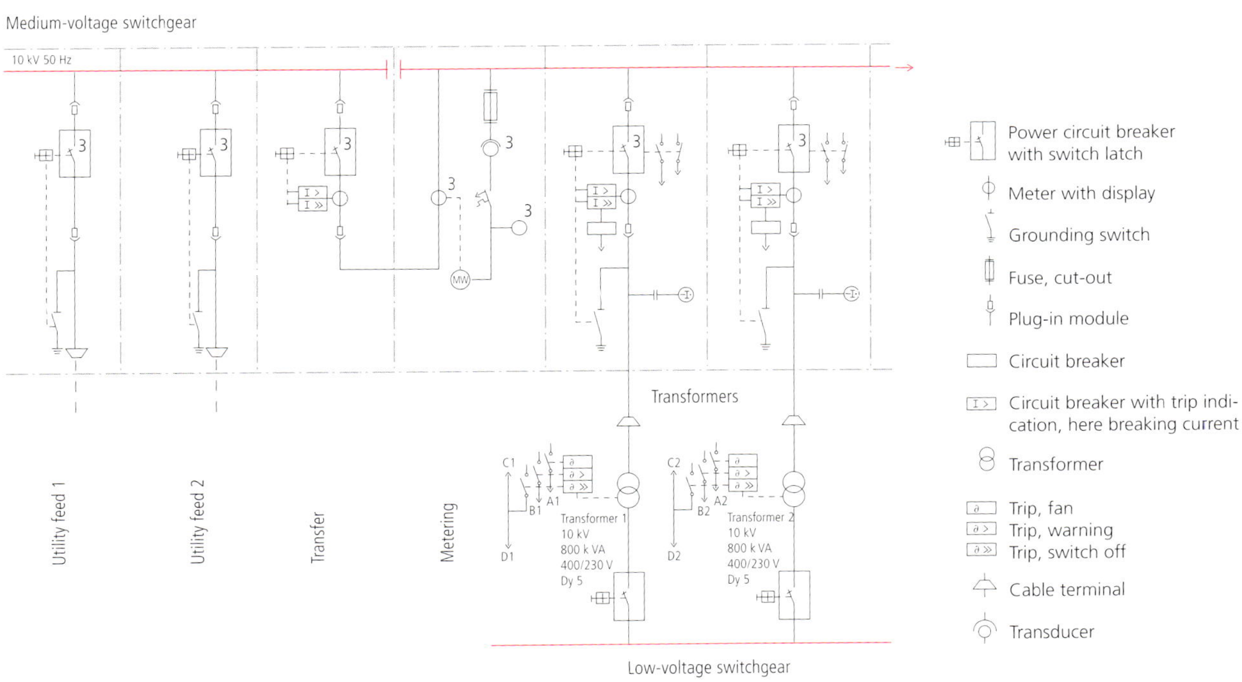

Figure 7.28
Simplified diagram of utility feed input, transformation and distribution in a building

As previously noted, it may be sensible to supply energy on the medium/high voltage level to large consumers. This helps to avoid high operating currents and the concomitant losses associated with energy supply on the low-voltage level. Depending on the structure, a differentiation is made between centralized and decentralized systems, as illustrated in **Figure 8.29**.

Centralized systems preceded by transformers should be located as close as possible to the largest consumer in the building (load stress; next to the chiller, ventilation installation or beneath the kitchen or DP centre). The installation paths and available work space for installation should be considered.

Decentralized systems are generally used in industrial operations, where they are installed directly next to a large consumption point, as shown in **Figure 8.30**.

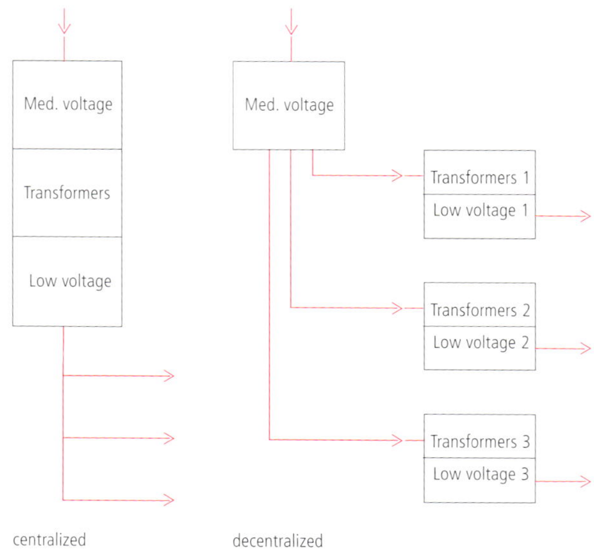

Figure 8.29
Centralized and decentralized switchgear systems

8.4.2 Supply systems

Depending on building type and consumption, as well as on the scale and requirements for supply safety, main current supply systems take on a variety of forms. Electrical energy is supplied either via cables or raceway systems. **Figure 8.31** provides a comparison of supply systems for buildings with the corresponding advantages and disadvantages and supply safety ratings.

For smaller residential objects, tree and group arrangements are the standard installation forms (**Figure 8.17**), while larger objects demand a high degree of supply safety and are therefore often supplied from ring installations. Double supply is typically used for operations with high safety requirements and high demands (several major loads); this is a standard solution for large computer centres, industrial facilities and similar buildings. The decision as to whether cables or bars are preferable is often based on the need for rigid versus flexible installations. Rigid installations are generally realized with cables, while flexible installations are generally provided with a bar system. **Table 8.4** lists the conditions and requirements for rigid and flexible installations.

Figure 8.30
Heavy-load system in industrial setting

Table 8.4
Fixed and flexible installation options

Conditions	Fixed cable systems	Flexible raceway systems
Precise device layout in planning phase	Required	Not required
Supply from distribution terminal	Possible with bottom-guided cable systems	Adaptable on site at connection point
Overhead conveyors Crane runways	Possible with bottom-guided cable systems	Limited options
Short circuit and selective behaviour	Must be reconfigured after each modification. Impedances cannot be determined with accuracy as a result of different cable types	Can be determined with accuracy in planning phase. Low impedances through type-tested systems
Installation effort	High installation effort as a result of radial distribution across conduits etc.	Basic installation only, few fastening points (approx. 2 m) required. 2 man assembly
Use of simultaneity factor	Only acceptable upon distribution	On site across entire system
Fire loads	High	Low
Retrofitting Modifications	VDE requires that power is shut down (night and weekend work), installation associated with high material expenditure, dust accumulation, chipping or chiselling work etc.	Possible within collectively agreed working hours for live systems. No power interruption necessary
Voltage drop en route to consumer	Cannot be defined with accuracy for retrofit installations	Low and accurately definable
Revisions Maintenance	At central point in switchgear	Raceways generally beyond arm's reach, scaffolding, ladders etc. required
Apparent power compensation	Centrally, via self-regulating systems	On site for apparent power relay
System utility in case of manufacturing process changes or assignment to new uses	Cable installations only marginally usable	Usable
Operator liability	High risk, especially due to complexity following modifications	Low risk due to type-tested systems

Figure 8.31
Comparison of supply systems

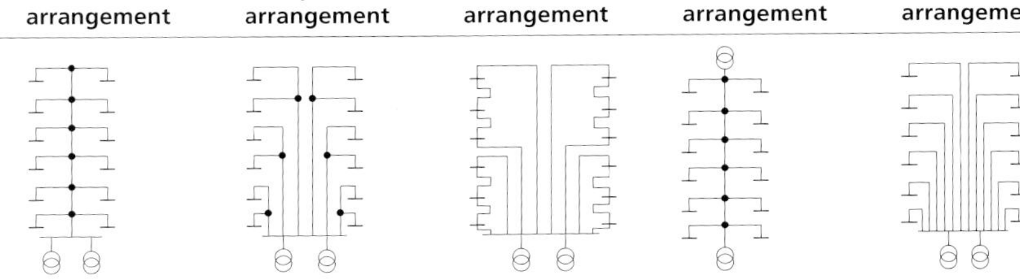

	Tree arrangement	Group arrangement	Ring arrangement	Dual arrangement	Individual arrangement
Riser	Predominantly collecting bus bar system Fewer cables	Collecting bus bar or cable	Collecting bus bar or cable	Predominantly collecting bus bars, fewer cables	Cable
Advantages	– Clear layout – Small low-voltage main distribution – Sizing based on mean load of all distribution terminals	– Only partial loss of power in case of disturbance – Smaller conductor cross-sections	– Good security of supply – Smaller conductor cross-section	– Advantages as with tree arrangement, however, better security of supply	– Power outage only in one distribution in case of disturbance
Disadvantages	– Total loss of power in case of disturbance – Difficult assembly for cables due to large cross-sections	– Larger low-voltage main distribution – Load compensation only within distribution groups	– Somewhat more comprehensible for operating personnel	– Maintenance and operation are somewhat more difficult	– Large low-voltage main distribution – Large shafts – No compensation possible for load shifts
Security of supply	Low	Average	High	Average	High

8.4.2.1 Cable systems

Cables generally consist of several copper cores and an additional insulating layer. Black and brown cores serve as outer conductors (phase conductors), green/yellow cores as protective conductors, and light blue cores as neutral conductors.

Installations for smaller buildings are generally executed by laying cables from central distribution boxes (inside the apartment) to the individual load points (switches, sockets, outlets) as shown in **Figure 8.32**.

For reasons of safety and accessibility, cables and lines should always be laid parallel to the edges of the room. This makes it possible to reconstruct the cable paths at a later point (standard regulations and guidelines, e.g. DIN 18015, Part 1/3, must be considered). Hidden cables should always be laid in specific zones to avoid damage to electricity cables or lines during subsequent work on walls or ceilings. **Figure 8.33** and **8.34** illustrate the relevant installation zones in residential building.

For larger building projects with major equipment and high connect loads, cables are laid on cable trays (or raceways).

Cable trays are generally manufactured from galvanized steel and designed for subsequent additions and expansion. **Figure 8.35** illustrates standard cable trays including all special pieces such as bends, junctions and T-junctions.

The space required for cable tray installations is frequently underestimated and **Figure 8.36** is included to illustrate the considerable space required for large cables or a great number of cables (in this example, cables are installed inside a shaft). Floor installations are another option as shown in **Figure 8.37**, which illustrates the fixed or rigid installation for a commercial application. This solution calls for special cable trenches, which accommodate the underfloor cables.

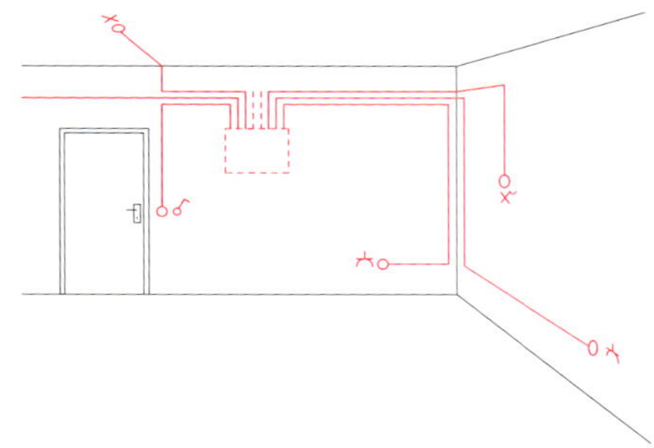

Figure 8.32
Installation with central distribution boxes (diagram)

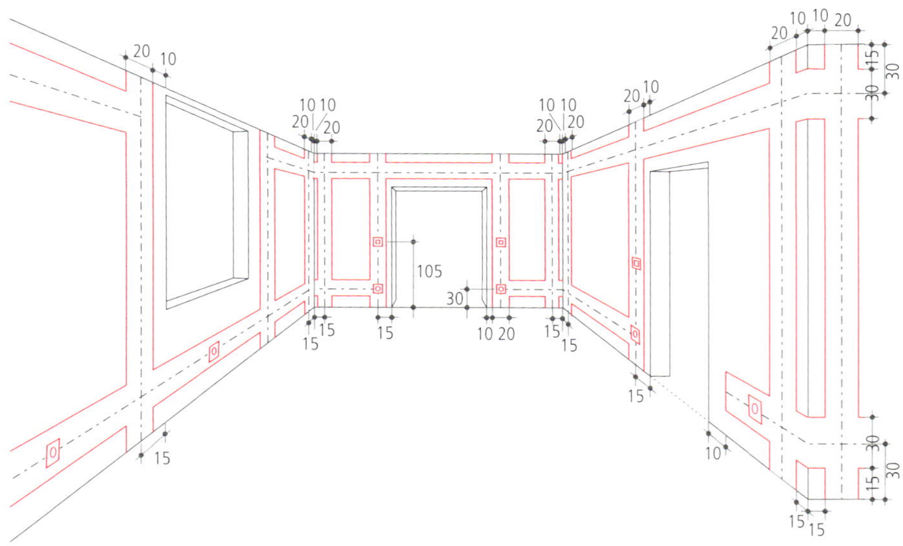

Figure 8.33
Installation zones and preferred measurements for apartment rooms in cm

Installation zones
Preferred measurements for electricity lines
Preferred height for switches
Preferred height for outlets

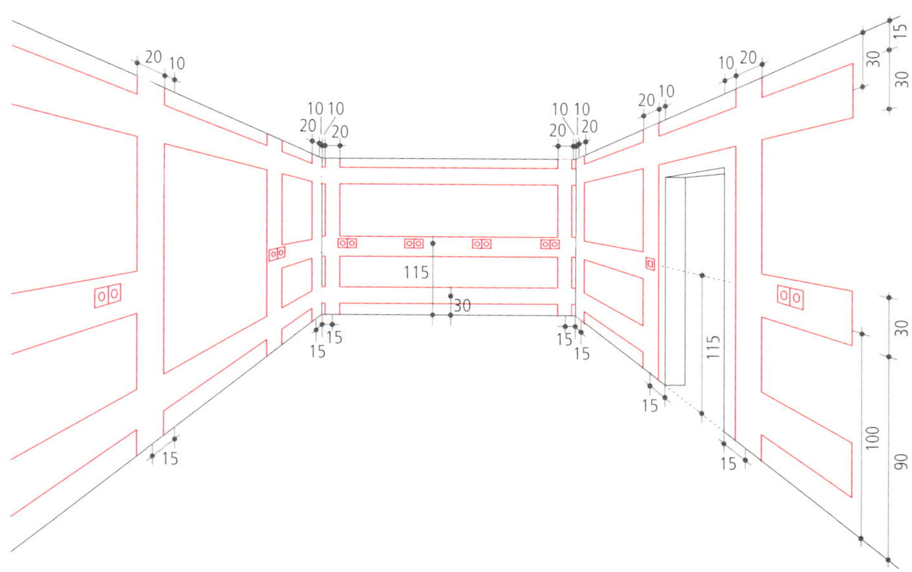

Figure 8.34
Installation zones and preferred measurements for kitchens, utility and similar rooms (according to DIN 18015, Part 3)

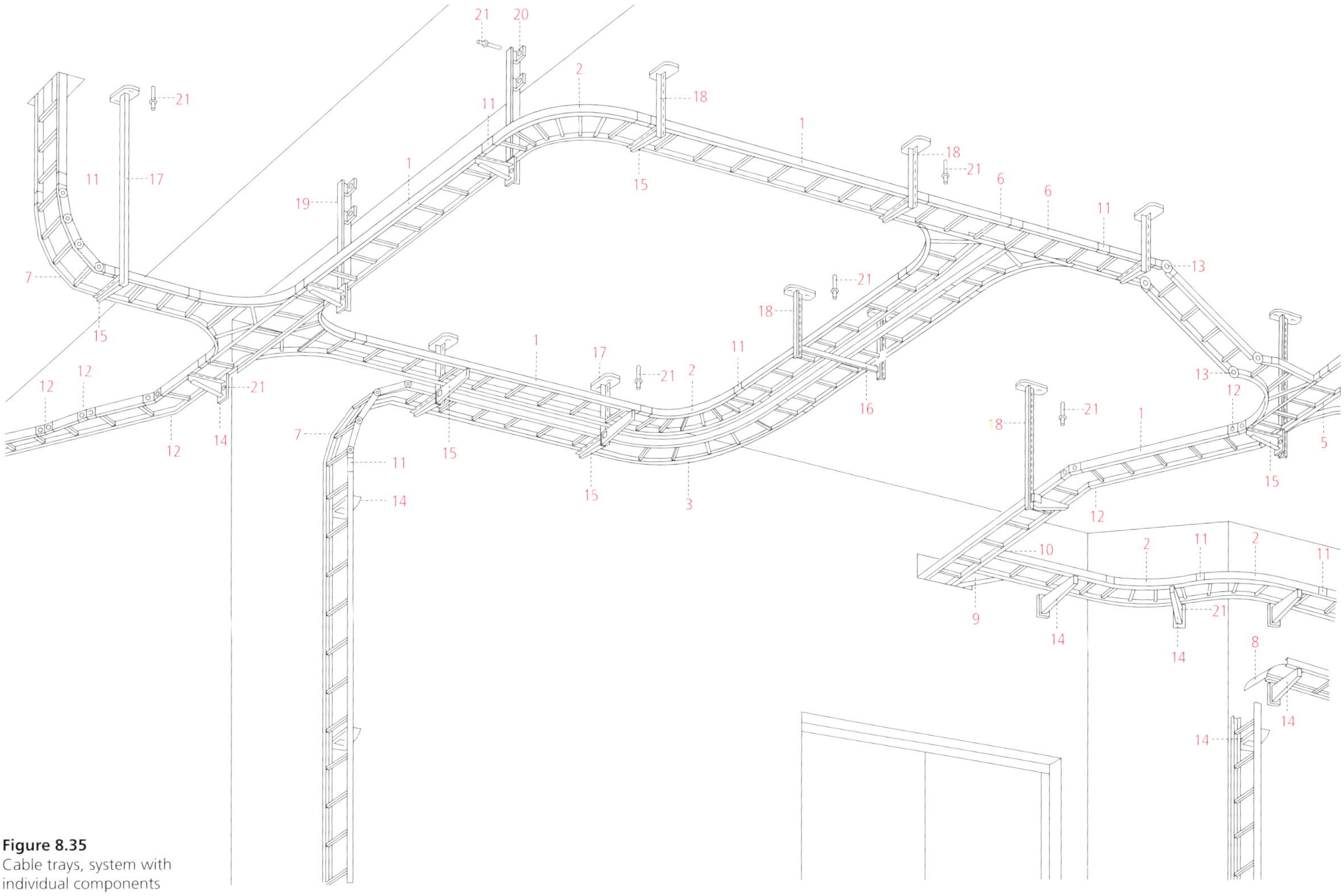

Figure 8.35
Cable trays, system with individual components

1 Cable tray
2 Bend, inside
3 Bend, outside
4 Junction
5 Junction
6 T-junction
7 Vertical bend
8 Junction plate
9 Corner plate
10 Connecting piece
11 Connection at joint
12 Angle connection
13 Joint
14 Wall bracket
15 Stud bracket
16 Cross tie-bar
17 Suspending strut
18 Suspending strut
19 T-strut
20 Angled anchoring bracket
21 Steel straddling dowel

Figure 8.36
Cable trays for extensive cabling networks (view into shaft)

8.4.2.2 Busways (busduct/cablebus)

Busways are used to transmit high currents for reasons of safety and to provide flexibility. Given the same copper cross-section, busways can be exposed to higher currents than comparable cable systems. Energy can be transferred in the system at specific points prefabricated to factory specifications. Other advantages of busways are their very low fire hazard due to few combustible components and the lower space requirements for the installation. The higher investment costs for short transmission distances are a disadvantage.

Figure 8.38 also depicts the electrical supply system for an industrial application, in this example via a cablebus. A comparison with **Figure 8.37** reveals, even at first glance, that the installation is much less elaborate and much easier to read. **Figure 8.39** depicts a cablebus in the corridor of a large building. Like cable trays, the cablebus is installed in suitable structural components (here suspended from the ceiling).

8.4.3 Wiring dimensions for small building objects

For smaller building objects, especially residential homes, lines must be laid to supply the connect loads for electrical appliances listed in **Table 8.5**. A preliminary sizing of the cross-section can be undertaken on the basis of the connect load (alternating current or rotary i.e. three-phase current) and depending on the maximum allowable cable length from the counter listed in **Table 8.6**. The table indicates the maximum connect loads for alternating or rotary i.e. three-phase current (kVA) as well as the maximum fuse protection of the current (ampères). For example, taking a maximum load of 3.5 kVA (AC) in the table and following a line straight up in the table, we can see that a cross-section of 1.5 mm^2 corresponds to a maximum cable length of 35 m. Conversely, we can read the table to determine the maximum connect load based on the cable cross-sections and the maximum lengths.

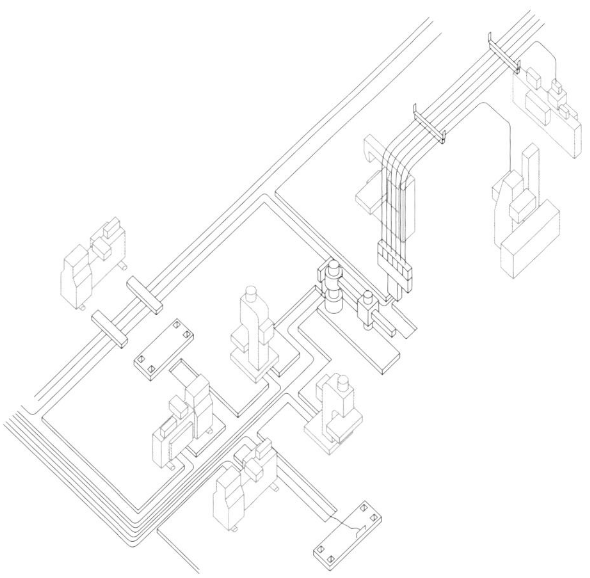

Figure 8.37
Fixed, underfloor cable installation

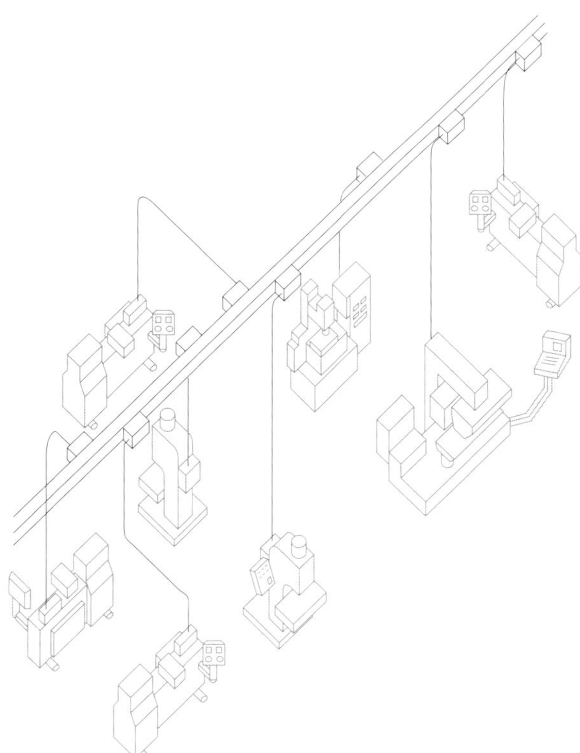

Figure 8.38
Flexible busway installation

Figure 8.39
Ceiling-mounted busway system

Electric appliance		Connect load in kW Alternating current	Rotary current
Electric stove		8.0…14.0	
Cool top cooking platform			6.0… 8.5
Built-in oven			2.5… 5.0
Microwave oven		1.0… 2.0	
Grill unit		0.8… 3.3	
Toaster/warmer plate		0.9… 1.7	
Foodmixer/juicer/blender		0.2	
Pressure cooker/waffle iron		1.0… 2.0	
Coffee maker		0.7… 1.2	
Deep fryer		1.6… 2.0	
Range hood		0.3	
Water heater for boiling water	3 l/ 5 l	2.0	
Warm water tank	5 l/10 l/15 l	2.0	
Warm water tank	15 l/30 l		4.0
Warm water tank	50 l – 150		6.0
Flow water tank	30 l – 120 l		21.0
Instantaneous water heater			18.0/21.0/24.0
Upright electric water tank	200 l – 1000 l		2.0…18.0
Iron		1.0	
Ironing machine		2.1… 3.3	
Tumbler		0.4	
Wash combination		3.2	
Washing machine		3.3	7.5
Dryer		3.3	
Hair dryer		0.8	
Hand dryer		2.1	
Towel dryer		0.6	
Air humidifier		0.1	
Infrared radiator/sun bed		0.2… 2.2	
Solarium		2.8	4.0
Sauna		3.5	4.5…18.0
Bathroom radiator		1.0… 2.0	
Refrigerator, freezer		0.2	
Fridge/freezer combination		0.3	
Dishwasher		3.5	4.5
Rinsing/flushing installation		3.5	5.0
Vacuum		1.0	
Beating vacuum cleaner		0.6	
Shoe polisher		0.2	
Floor polisher		0.5	

Table 8.5
Connect loads of electronic appliances

S in mm²	Cable lengths for rotary current (half lengths for alternating current) in m													
120									451	361	282	225	201	180
95								446	357	285	223	178	159	143
70							417	327	263	210	164	131	117	
50						376	298	235	188	150	117			
35					376	263	209	164	131	105				
25				376	268	188	149	117	94					
16			300	240	172	120	95	75						
10		235	188	150	107	75	60							
6	225	141	113	90	64	45								
4	150	94	75	60	43									
2.5	94	57	47	38										
1.5	56	35	28											
LSS	10	16	20	25	35	50	63	80	100	125	160	200	(224)	250 A
3 ~	6.6	11	13	16	23	33	41	53	66	82	105	132	(147)	165 kVA
1 ~	2.2	3.5	4.4	5.5	7.7	11	14	18	22	28	35	44	(49)	55 kVA

Group 1 (one or more single-conductor cables in conduits)

Group 2 (Multi-conductor cables, e.g. non-metallic sheathed wires and cables, tape conduits, flexible conduits)

Table 8.6
Max. allowable cable lengths from meter (3 % voltage drop, sizing example)

LSS Fuse protection
3 ~ Max. connect load, rotary current
1 ~ Max. connect load, alternating current

Apartment floor area in m²	Number of circuits depending on finishing standards *) A Average standard	B High standard
up to 45	2	3
45 to 55	3	4
55 to 75	4	6
75 to 100	5	7
greater than 100	6	8

*) Additional circuits must be provided for basement and utility rooms allocated to individual apartments. A separate circuit is required for convenience outlets in recreation or workshop areas in residential buildings.

Table 8.7
Number of circuits for convenience outlets and lighting in apartments
(DIN 18015, Part 2)

8.4.4 Secondary distribution

Distribution terminals are required to avoid having to lay a separate cable from the power station to each individual consumption point or group of consumers in a building. This would be far too complex and inefficient. The required power is therefore delivered through cables with the largest possible cross-section from the low-voltage main distribution or the counter to the consumption point. It is only at this point that the energy is divided (distributed) in smaller cross-sections, whereby the necessary safety precautions are implemented for the individual consumption groups or appliances.

The scale of the secondary distribution in a building is determined by the circuit demands, that is, the number of appliances and devices connected to it. Table 8.7 provides an overview of the number of circuits for average and top-grade interior appointments based on the usable area.

Special terminals, which are usually built into the wall, are used to distribute electricity within a residential space. Prior to wiring such a terminal, it is necessary to create a wiring diagram for the secondary distribution to illustrate the number of appliances and outlets and the maximum fuse protection, usually indicated in the documentation provided by the manufacturers of the relevant appliances. **Figure 8.40** contains a wiring diagram for secondary distribution in an apartment; **Figure 8.41** depicts a typical apartment circuit panel.

In large building objects, the secondary distribution is larger in scale and generally housed in cabinets that include all required fuse protection and switching elements. **Figure 8.42** shows a typical floor-to-floor distribution terminal in cabinet form and **Figure 8.43** the wiring inside the cabinet.

In industrial facilities, secondary distributions can also be designed and installed in a way to expose (that is, to make visible) all fuse protection and switching elements, encapsulating the entire floor-to-floor distribution against dust and water (**Figure 8.44**).

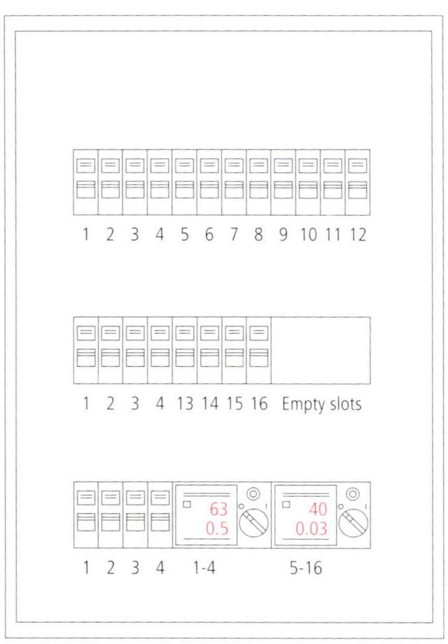

Figure 8.41
Circuit panel

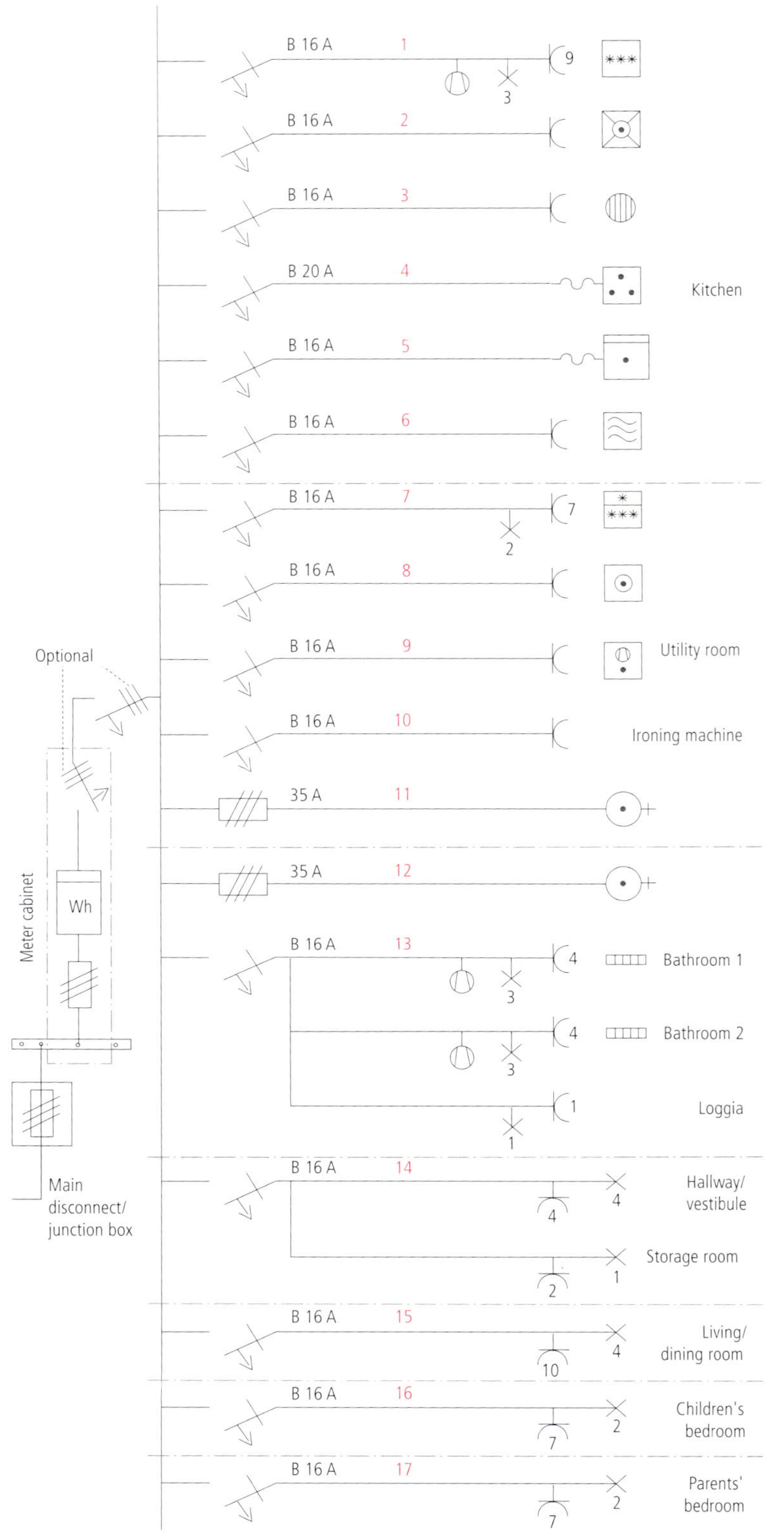

Figure 8.40
Wiring diagram for secondary distribution in an apartment (for symbol definitions, see Chapter 8.4.5)

Figure 8.42
Floor distribution terminal

Figure 8.44
Subdistribution in an industrial setting

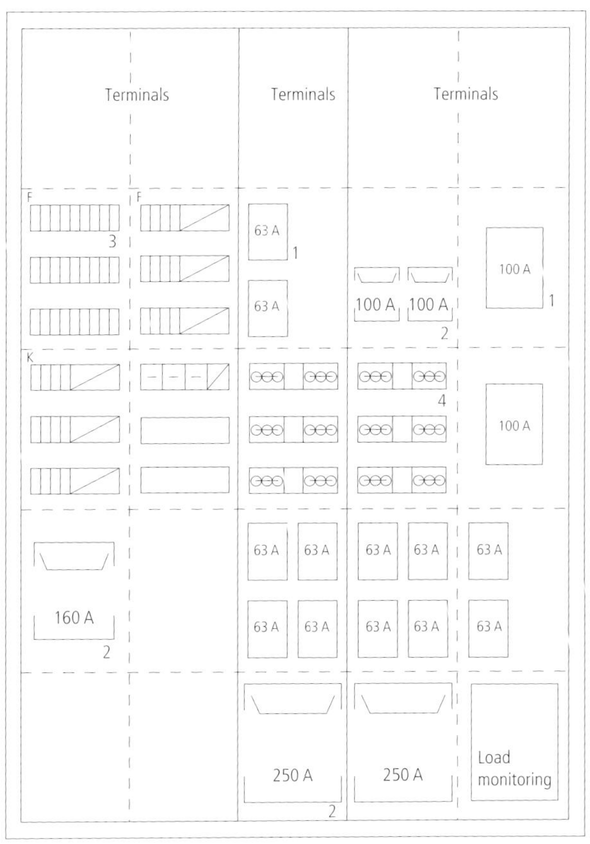

Figure 8.43
Wiring inside floor distribution terminal

1 Circuit breakers
2 Low-voltage fuse isolator
3 Automatic cutout, line safety switch
4 Cabinet fuse

8.4.5 Wiring

The power is supplied from the distribution terminal to the individual consumers on one floor according to what is called a circuit or wiring diagram.

Within an individual floor electricity is generally distributed in wiring, although trunking (UK) or raceways (USA) are used in rare cases (especially industrial applications or large computing centres).

Wiring is installed as:
- Exposed wiring
- Concealed wiring
- Conduit wiring

In exposed wiring, light plastic-sheathed cables are installed on top of the wall with cable clamps, primarily in areas where concealed wiring is not an option due to the required system of protection (garages/basements).

Concealed wiring is installed beneath the plaster. Installation must be executed horizontally and vertically according to a wiring plan. Diagonal wiring is only permitted on ceilings. This type of wiring is not accessible for maintenance or modification.

The external elements of an installation such as switches, sockets, and junction boxes make it possible to retrace the approximate location of concealed wiring even after the installation has been completed. (**Figure 8.33** already identified some of the installation zones and preferred dimensions for apartments, with the exception of kitchens.)

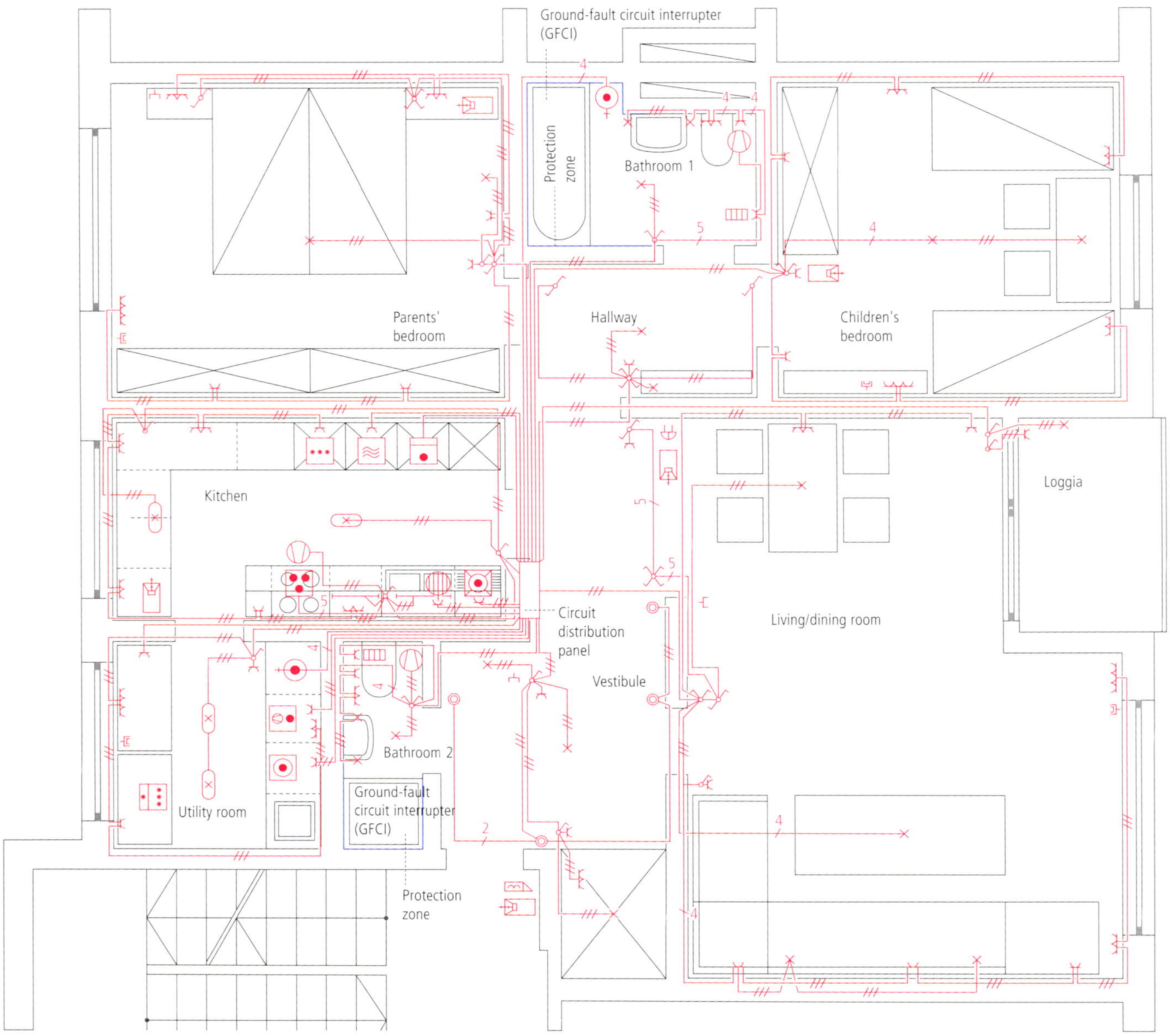

Figure 8.45
Example of wiring diagram
for an apartment

For conduit wiring, installation conduits (or hollow ducts) are laid into mortised channels in preparation for the electrical wires that are drawn through later. Installation conduits can also be laid onto the formwork or cast into concrete. With this type of wiring installation, access to wires and cables for maintenance and modification is possible at any time.

In residential objects, electricity is distributed to the consumers from the branch circuit via concealed wires, as shown in **Figure 8.45**. Residential installations can be executed with or without junction boxes.

When junction boxes are used, they are installed at every branching point, as shown in **Figure 8.46**. These junction boxes are generally located in the top third of the wall (30 cm below the ceiling) or in the ceiling.

When junction boxes are not used (appliance connectors only) (**Figure 8.47**), switch boxes with additional distribution space are employed to connect wires inside these boxes.

The advantage of this approach is that the installation can be checked at any time by simply removing the electric equipment (switch or outlet) without having to open the branch boxes (risk of damaging the surfaces).

While power is therefore generally distributed through concealed wiring in smaller applications, a different approach to floor distribution must be employed for the distribution of larger cable cross-sections and volumes. This is especially true for areas with high branch circuit demands (e.g. laboratories, offices, manufacturing sites, hospitals).

Since all participants in the planning of a construction project (architect, structural engineer, building services engineers and others, including the client) should be able to determine the location of outlets, switches, plugs, junction boxes etc., the following pages provide an overview of common circuit and wiring symbols (**Table 8.8**).

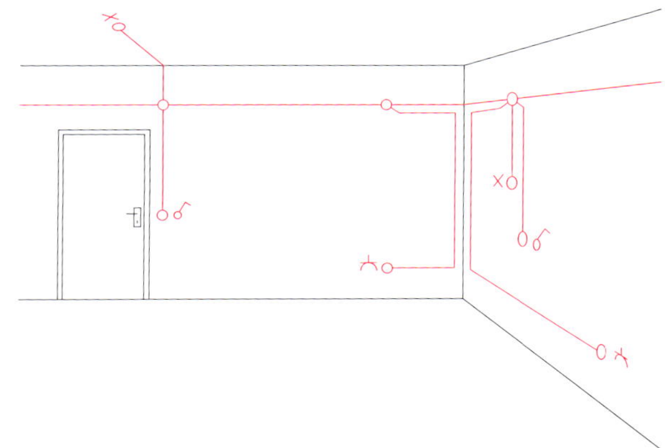

Figure 8.46
Installation with junction boxes (schematic)

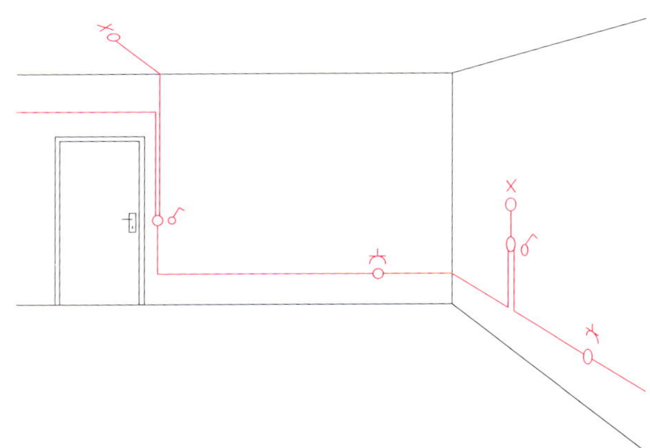

Figure 8.47
Installation with appliance connectors (schematic)

Symbol	Description
	Conductor, wire, cable
	Conductor, movable
	Conductor, shielded
	Buried conductor, buried wire
	Overhead line, open line
	Cable conduit or duct, wiring conduit
m	Exposed conductor
m	Conductor in plaster
m	Concealed conductor
	Conductor or cable, not connected
3x1,5Cu	Cable with three copper conductors, 1.5 mm^2
3N~50Hz 400V	3-phase four-wire network with three external conductors and one neutral
	Conductor in cable, example with three conductors
	Conduit with three conductors
3	Conduit with three conductors, abbreviated symbol
	Protective conductor (PC)
	Neutral conductor with protective function (PCN)
	Neutral conduction (N), middle (or third) conductor (M)
	Three conductors, one neutral conductor, one protective conductor
	Conduit, leading up
	Conduit, leading down
	Conduit, leading up and down
	Connection of conductors
	Conductor junction (form 1)
	Conductor junction (form 2)
	Connection (e.g. terminal) (circle may also be solid)
	Connection box, junction box
	Distribution box, general
	Socket, general Empty socket, general
	Branch socket
	Loop-through socket
	Main disconnect, general, shown with line
	Distribution terminal, shown with 5 connections
	Frame or boundary line
	Protective ground
	Primary cell, primary element, accumulator
220/8V	Transformer with 2 windings
	Rectifier
	Frequency changer
U const.	Voltage stabilizer
	Fuse, general
D II 10 A	Screw-in fuse, example: 10 A, Type D, three-polar
00 25 A	Low-voltage/high-voltage fuse (LH), example: 25 A, size 00
	Fuse isolator switch
	Fuse switch
10 A	Switch, example: 10 A, three-pole
4	Error voltage circuit breaker, four-pole
	Circuit breaker
3	Motor-protection, three-pole, with thermal and magnetic trip, example: single-pole
	Emergency switch
	Switch, general
	Switch with control light
	Cutout, one-pole Switch 1/1
	Cutout, two-pole Switch 1/2
	Multicircuit switch, one-pole Switch 5/1
	Double-throw switch, one-pole Switch 6/1
	Intermediate switch Switch 7/1
	Switch with pull cord
t	Time switch, one-pole
	Sensor switch
	Sensor switch with light
	Impulsing switch, time pulse relay
	Proximity sensor
	Contact sensor
	Proximity switch (cutout or disconnecting switch)
	Contact switch (intermediate switch)
	Dimmer
	Socket, general
	Safety plug socket
3/N/PE	Safety plug socket, example shown for rotary current, five-pole
	Safety plug socket, disconnectible
	Safety plug socket with locked switch
3	Safety plug socket, shown as three-pin plug
	Optional symbol
	Socket with isolating transformer, e.g. for electric razor
	Telephone jack
	Antenna jack
Wh	Electric meter, watt-hour meter

Time switch, economizer
Time relay
Flasher relay, shown with a flash frequency of 5/min
Audio frequency remote control relay
Luminaire, general
Luminaire outlet, shown with line
Exposed luminaire outlet, shown with line leading off to left
Luminaire with switch
Luminaire with adjustable brightness
Safety luminaire, permanent connection
Safety luminaire, emergency lamp on separate circuit
Safety luminaire with integrated emergency power supply
Searchlight, general
Spotlight
Floodlight
Luminaire with additional, permanently connected safety luminaire
Luminaire with additional, permanently connected safety luminaire
Luminaire for discharge lamp, general
Luminaire for fluorescent lamp, general
Luminaire with 3 fluorescent lamps
Luminaire with 5 fluorescent lamps
Ballast device for discharge lamps
Starter for fluorescent lamp
Electronic appliance, general
Kitchen appliance
Electric stove, general
Microwave oven
Oven (conventional)
Hotplate
Deep fryer
Hot water tank
Instantaneous water heater
Hot water appliance, shown with line
Infrared grill
Washing machine
Dryer
Dishwasher
Hand dryer, hair dryer
Heating element
Thermal storage heater/radiator
Infrared radiator
Fan
Air-conditioner
Refrigerator, deep-freeze Number of stars, see DIN 8950, Part 2
Freezer Number of stars, see DIN 8950, Part 2
Motor, general
Generator
Converter
Star delta voltage, Y-voltage
Telephone, general
Telephone, intraoffice
Telephone, exchange line
Telephone, long-distance line
Telephone for two or more extensions
Intercom system, e.g. front door or gate intercom
Duplex system, e.g. front door or gate duplex system
Loudspeaker, general
Microphone, general
Loudspeaker/microphone
Switching centre, general
Alarm clock, bell
Buzzer
Gong, one-stroke bell
Horn, hooter
Siren
Signal lamp, general
Door opener
Time recorder
Fire push-button manual alarm
Temperature alarm
Key-operated switch Guard alarm
Vibration detector
Pass lock for switching routes in safety systems
Smoke detector, automatic, light-dependent principle
Fire alarm, automatic
Dimmer switch
Antenna, general
Amplifier, general; triangle points in direction of amplification

Table 8.8
Wiring diagram symbols and definitions

8.4.6 Bus system

With rising demands in recent years, cabling has become increasingly complex and confusing, even in residential applications. There are an increasing number of functions to be fulfilled by a multitude of appliances and devices. The functional demands in a modern home are summarized in **Figure 8.48**. This overview demonstrates that higher demands on the flexibility and comfort delivered by the electrical installations, coupled with a desire to reduce energy consumption, has necessitated a novel approach to building systems in this area, which will revolutionize electrical installations in buildings – in short, the evolution of bus technology.

The technology is based on a common European concept, the European Installation Bus (or EIB). Many manufacturers have joined the European Installation Bus Association (EIBA).

The member firms of EIBA ensure that bus-compatible products are available in the marketplace. This makes it possible to run appliances and devices from different manufacturers on and through one and the same EIB system.

In conventional electrical installations, each function requires a dedicated line and each control system a separate network. By contrast, the "instabus" EIB, for example, makes it possible to control, monitor and record all operational functions and processes via a joint line. Power is transported directly to the consumer. Reducing the number of lines is only one advantage of this approach. Other advantages include: installation in a building is far easier to execute, expand and modify. The bus system is easily adapted to changes, such as new uses or new room divisions, simply by reassigning (that is, establishing new parameters for) the bus specifications without having to install any new lines. This reassignment of parameters is achieved, for example, with the help of a PC connected to an "instabus" EIB system. The same computer software required for initial start-up, the EIB Tool Software, is employed to implement the necessary modifications.

With corresponding interfaces, the "instabus" EIB can also be connected to the control rooms of other systems for building automation systems (or BAS) or to a public telephone network (e.g. ISDN). This means that "instabus" EIB is equally efficient in single-family houses, hotels, schools, banks, office buildings or complex purpose buildings.

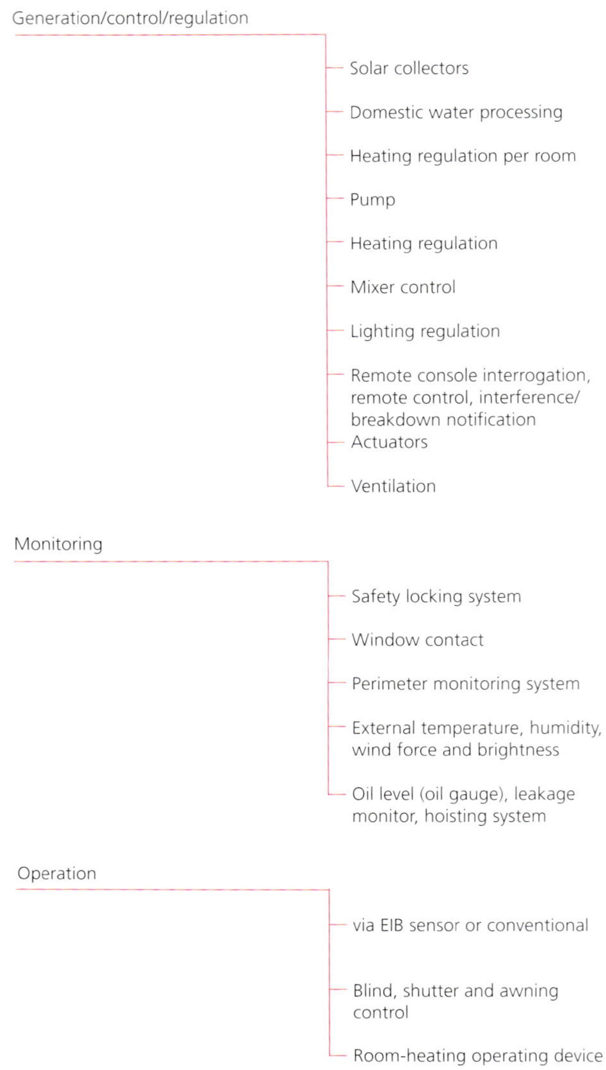

Figure 8.48
Functional demands in modern residential home

Transmission technology
The "instabus" EIB is a decentralized, event-controlled bus system with serial data transmission employed to control, monitor and record building system functions. **Figure 8.49** contains a diagram of the functional principle.

All users connected to the system can exchange data via a shared transmission path. Data transmission is serial and is executed according to precisely defined rules (bus protocol). The information to be transmitted is "packaged" into a telegram and transported by a sensor (pick-up) via the bus to one or more actors (command recipients). Each recipient confirms receipt of the telegram after the transmission has been successfully executed. If this confirmation is not relayed, the transmission is repeated up to three times. Should confirmation still be outstanding after these repeats, the transmission process is cancelled and the error recorded in the sender's memory.

The transmission process in the "instabus" EIB is not physically separated because the supply voltage (DC 24 V) for the bus subscribers is transmitted at the same time. The telegrams are modulated upon this DC voltage; a logical zero (0) represents a transmitted impulse, and a logical one (1) is interpreted as no impulse being present.

The individual data of the telegrams are asynchronously transmitted. Transmission is synchronized by means of start and stop bits.

Access to the bus as a shared physical communication medium for asynchronous transmission must be clearly defined. In the case of the "instabus" EIB, the CSMA/CA (or: carrier sense multiple access with collision detection) process is used. As the name suggests, the process ensures collision-free bus access without diminishing bus data throughput.

The transmission medium can also be a separately installed twin-core copper cable, the existing 230 V cable or a wireless connection.

For the "instabus", for example, a twin-core copper cable is ideal in new construction projects because it keeps the overall cabling effort to a minimum, reduces the fire load and helps to create a flexible high-performance network. The cable is laid in all areas where data will be recorded, controlled and processed. The control line is a closed system that is completely secure from external interference. Only the power cable leads to the consumers.

Bus-signal transmission via the power cable is an interesting alternative, however, for renovation or conversion projects in older buildings. The only modification is that all switches and relays have to be exchanged to intelligent EIB components. This solution saves the inconvenience of having to install new lines.

All technical systems in the home, such as automated shutters, retractable blinds, lighting, appliances such as washing machines, dish washers, refrigerators, instantaneous water heaters, as well as telephone and television cables are connected to the "instabus" (**Figure 8.50**).

Since the "instabus" is a decentralized system in which the intelligence is distributed across actors and sensors, any malfunction of an individual component does not affect the operation of the system as a whole and all connected devices and appliance.

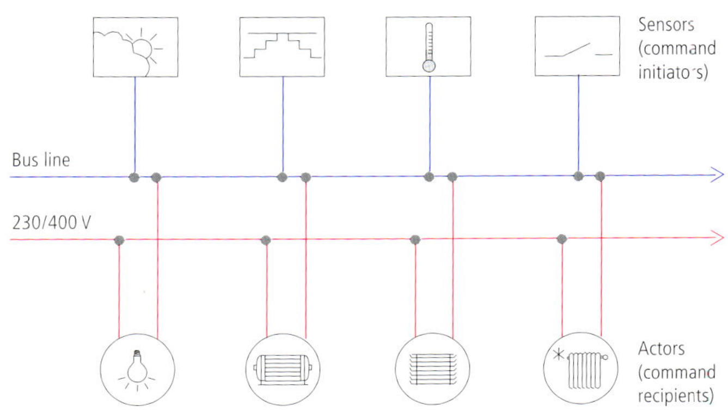

Figure 8.49
Functional diagram of bus system

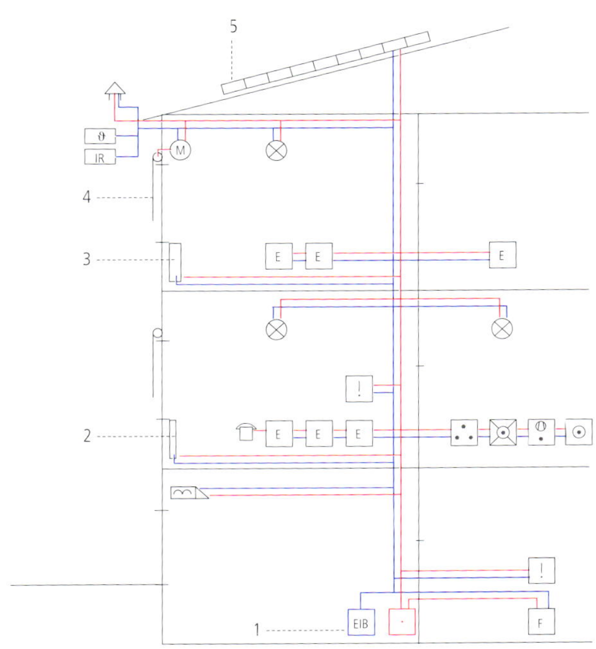

Figure 8.50
Building with bus system (Instabus EIB)

— Primary bus
— 220 V power line
1 Instabus EIB, modem
2 Heating unit
3 Convector
4 Shutter
5 Photovoltaic element
F Furnace

Sensors are operating instruments such as switches (for example, for lights) or sensing elements for brightness, humidity and temperature. Actors are operating recipients that convert commands into actions, for example, light relays, switchgears for luminaires, motors or blinds, and fan drives for heating elements etc. **Figure 8.51** illustrates a selection of sensors and actors connected to a PC (HomeAssistant).

Like any other device, the central unit (PC) is also connected to the "instabus". The HomeAssistant is linked via a standard interface (the so-called Com-interface) in the PC and the communications connector (bus connector with interface). This enables the HomeAssistant to interact actively and passively with the bus operation.

As a bus-connected component, the HomeAssistant analyses the data transmitted via the bus system and displays them on the monitor. It thus keeps track, for example, of open windows and doors, closed shutters and whether the stove and washing machine are in use. It also records any error messages transmitted from other devices connected to the "instabus". By analysing the transmitted data, the HomeAssistant is able to identify the source of the message and the type of the malfunction.

The HomeAssistant is also an interface to public networks and therefore requires a link to the public telephone network. If occupants want to use the HomeAssistant for watching television, it must also be connected to an antenna socket and a compatible TV card installed in the PC.

The HomeAssistant can also activate operations in the bus system. This means that all actions, which are normally initiated at switches etc. in the house, can also be activated on the screen. The HomeAssistant can also be used to allocate specific functions to the "instabus" devices.

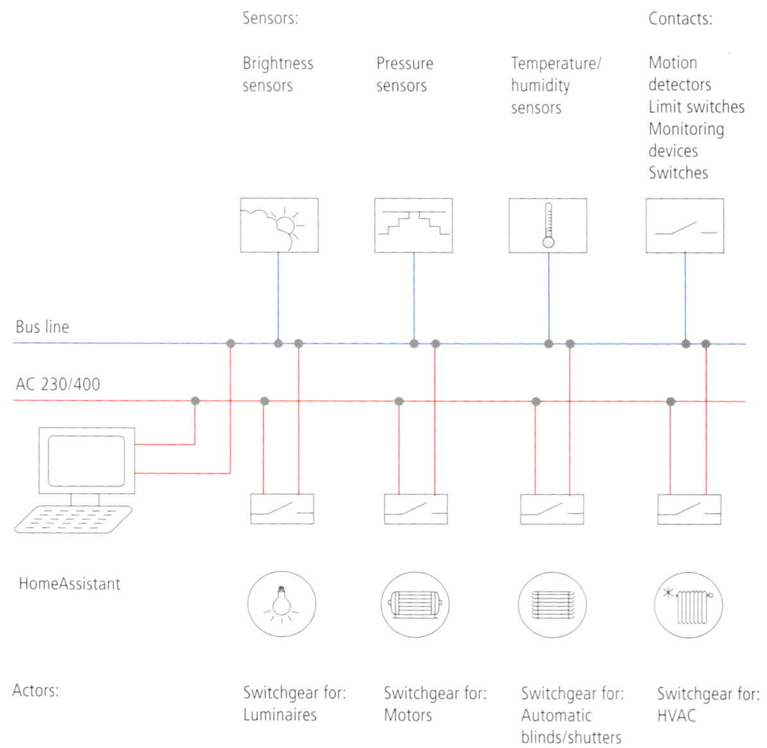

Figure 8.51
Expanded bus system (schematic)

Thus, home occupants use the HomeAssistant to set the room temperature for the heating system. They define which switch operates which lamp or shutter. Once these allocations have been completed, all decentralized components continue to operate independently.

The HomeAssistant is thus a comfortable operating unit within the decentralized "instabus" system:
- It delivers simple and understandable information on the bus devices at one or several locations and
- provides user-friendly operation, configuration and control of bus components at one or more locations.

In this new type of installation the standard operating components such as thermostats and switches not only assume additional functions, but are also somewhat different in appearance (see **Figure 8.52**).

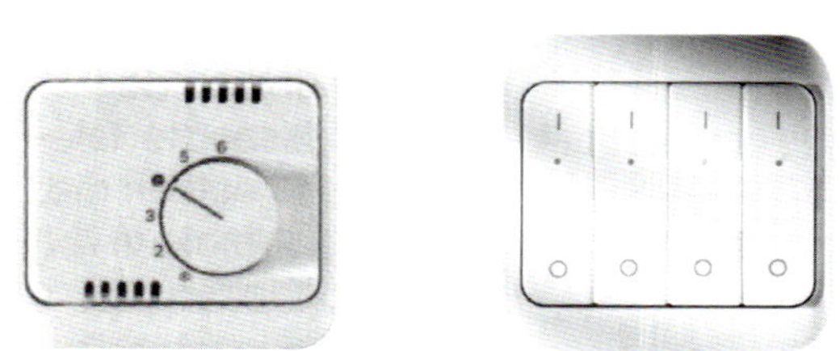

Controller

Push-button switch (4 buttons)

Figure 8.52
Instabus EIB switches

8.5 Floor installation

8.5.1 Underfloor electrical installations

The different subgroups of underfloor electrical installations are listed in **Table 8.9**.

Underfloor raceways are commonly used for administration, industrial and similar types of buildings characterized by heavy concentrations of electronic equipment and large room depths. The types of installation system illustrated in **Figure 8.53** offer appropriate solutions for these buildings.

The illustrated systems include:
- Underfloor installations with ceiling penetrations
- Underfloor systems with built-in units that rise above the floor
- Underfloor systems with screed-covering conduits and built-in units
- Floor-mounted installations

Table 8.9
Overview of different underfloor systems

Underfloor installation			
Underfloor electrical installation systems		**High-rise and finishing-dependent systems**	
Closed systems	Open systems	High-rise	Finishing
Closed conduit for connections, screed-covered SC + UC + UCC	Open feed conduit, flush with screed *)	Cellular steel ceiling SC + UC + UCC	Raised floor SC + UC + UCC
Closed conduit for connections, flush with screed SC	Open conduit for connections, flush with screed SC + UC	Precast concrete ceiling SC + UCC	Screed hollow floor SC + UCC
In-concrete conduit: – flush with unfinished floor and screed covering SC + UC + UCC – in neutral ceiling zone SC + UC		Ceiling penetrations SC + UC + UCC	

Connections:
SC Surface connection (above floor units)
UC Underfloor connection (floor level units)
UCC Underfloor connection only via height-variable units (conduit outlets)
*) no connection possible

Figure 8.53
Overview of different underfloor systems

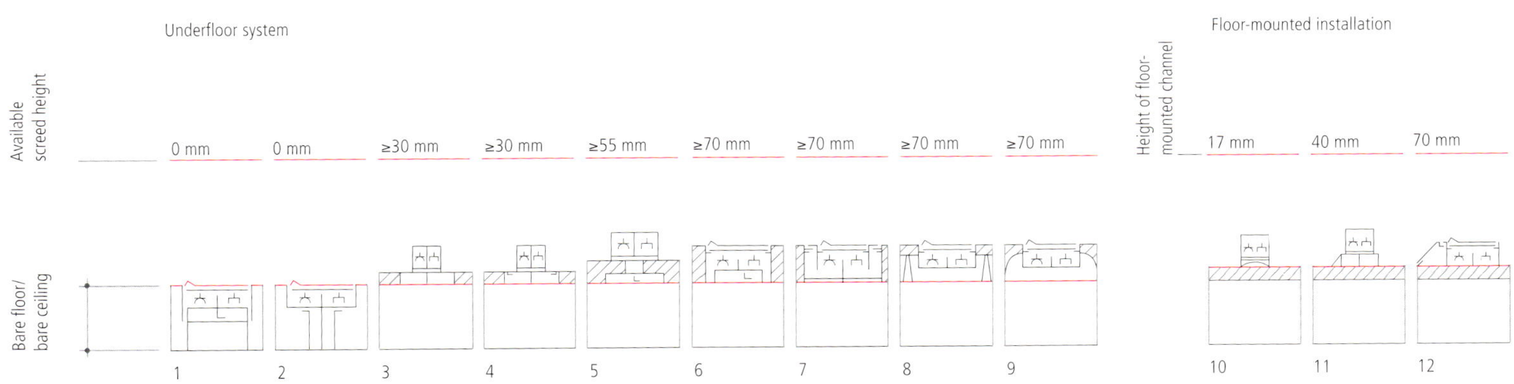

1 In-concrete channel in neutral ceiling zone with surface-mounted underfloor socket and built-in units that are flush with the floor
2 Ceiling penetration with surface-mounted underfloor socket and built-in units that are flush with the floor
3 Channel flush with screed with built-in units rising above floor level
4 Open channel for connections, flush with screed with built-in units rising above floor level
5 Screed-covered channel with built-in units that are flush with floor
6 Open channel for connections, flush with screed with built-in units that are flush with the floor
7 Raised floor with built-in units that are flush with the floor
8 Raised floor with built-in units that are flush with the floor
9 Hollow floor with built-in units that are flush with the floor
10 Floor-mounted sectional installation with built-in units above floor level
11 Floor-mounted installation channel with built-in units above floor level
12 Floor-mounted installation channel with built-in units above floor level

The illustration indicates the required screed depth for each system and also depicts the subsequent appearance of the electrical units. **Figures 8.54/55** show different built-in elements (flush with the floor and floor-mounted) commonly used today for different requirements, although elements that rise above the floor are only employed in areas where it is necessary to do so for safety precautions (e.g. in damp rooms etc.).

Figure 8.56 illustrates installation options for ceiling penetrations. This approach is rare today for reasons of fire safety and is only viable for smaller systems with less concentrated cabling (> 5 x 5 m).

Figures 8.57/58 illustrate how installation units can be mounted on a cellular steel ceiling construction.

Figure 8.54
Built-in unit flush with floor

Figure 8.55
Various models of built-in units above floor level

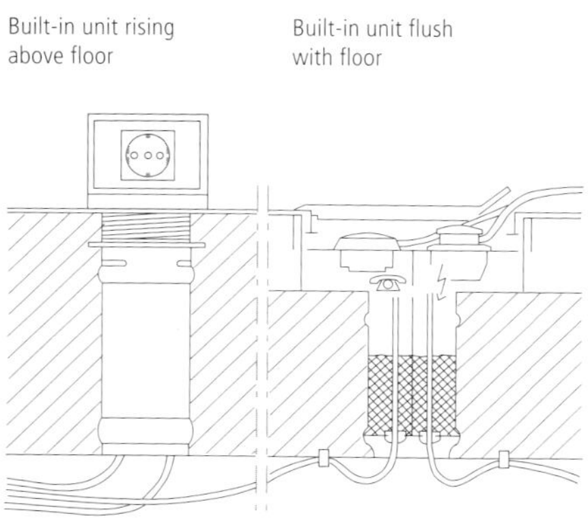

Figure 8.56
Installation options with ceiling penetrations – floor-mounted and underfloor connection systems

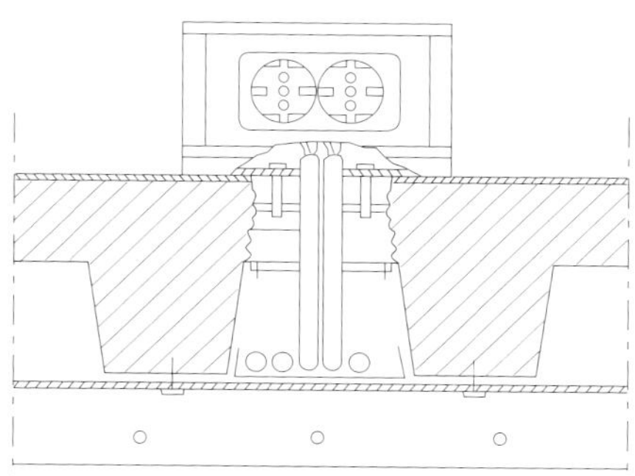

Figure 8.57
Built-in unit above floor, connected and operational, mounted on cellular steel ceiling

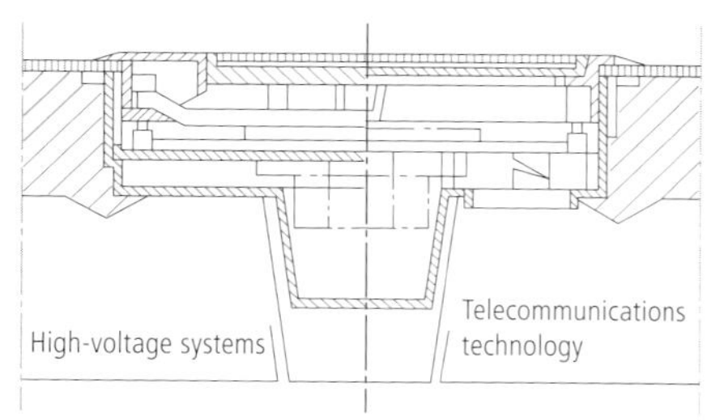

Figure 8.58
Built-in unit, flush with floor

Figure 8.59 illustrates a unit that is flush with the floor level used with an in-concrete channel (channel built into the neutral zone of the ceiling), which is sometimes still used.

Depending on the building configuration and the total floor area on each level, the electrical installation is carried out via one or several riser shafts. The floor distribution systems are also located in these shafts.

As **Figure 8.60** illustrates, the wiring is laid from these points onwards across main distribution channels or connecting channels, and the corresponding inserts are integrated or mounted depending on requirements.

The dimension of the installation channels is based on the overall line-up and installation density and is generally defined by the electrical engineer after consultation with the client.

Figure 8.61 depicts the completion of the roughing-in stage prior to laying the screed; the formwork for channels and wall outlets is clearly visible.

The construction of a hollow floor with form slabs or form sheeting is a good option in instances with high wiring density and greater requirements for flexibility (**Figure 8.62**). This type of floor is delivered at depths ranging from 8 to 15 cm (gross floor depth) and allows a higher density of installation than channel systems.

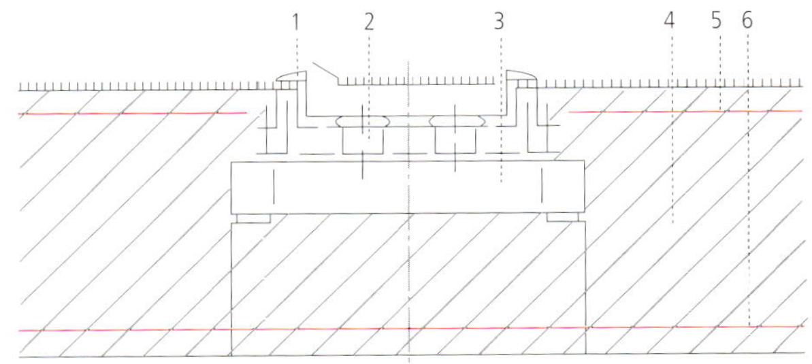

Figure 8.59
Built-in unit, flush with floor with in-concrete channel

1 Protective carpet frame with hinged cover
2 Installation device
3 In-concrete underfloor socket
4 Bare ceiling reinforcement (top)
5 Bare ceiling
6 Bare ceiling reinforcement (bottom)

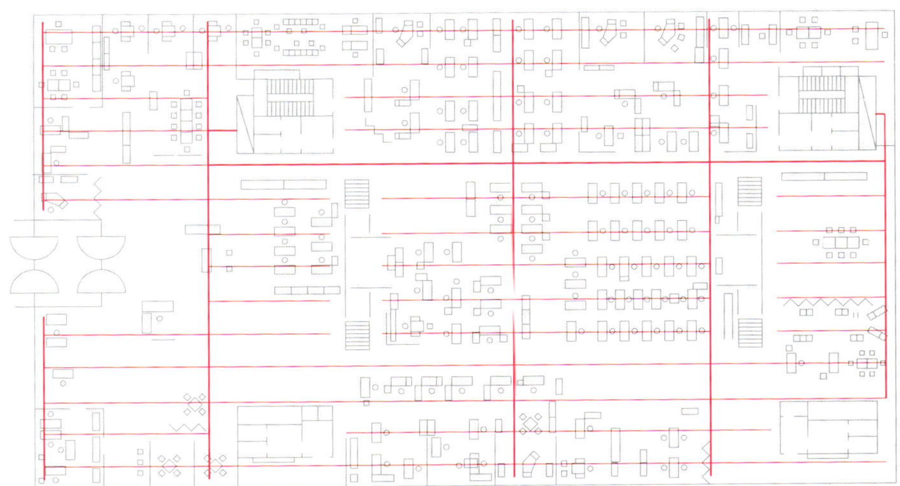

Figure 8.60
Electrical installations in building

Riser with floor distribution

Underfloor channels:
Main channel
Connecting channel

Figure 8.61
Closed, screed-covered channels mounted on top of bare ceiling; shown with underfloor sockets arranged on a grid in preparation for the installation of built-in units that are flush with the floor

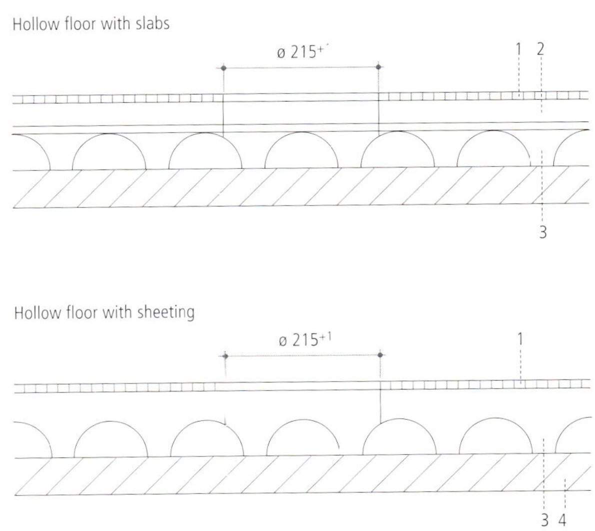

Figure 8.62
Differences in hollow floor construction (diagram)

1 Floor covering
2 Floating screed
3 Sheeting or slabs
4 Concrete floor

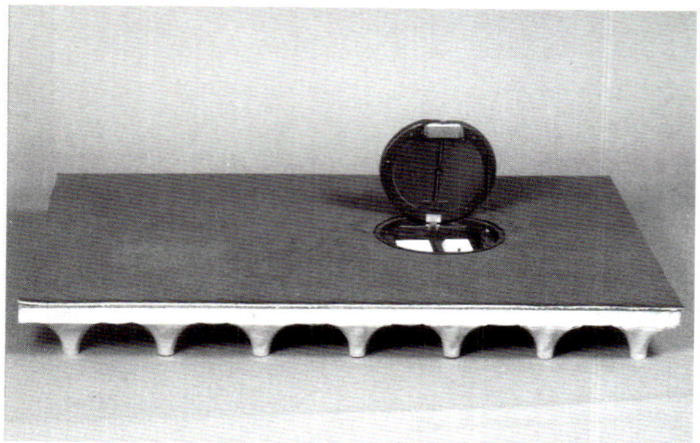

Figure 8.63
Finish hollow floor element

Practical experience has shown that hollow floors generally allow no more than three to four subsequent installations before reaching a density level that makes further installations impossible. Hence, the use of these floors is limited for building projects where frequent future modifications are likely, and especially where major upgrades or changes to the electrical installations will be required in the long term and the gross floor depth is no more than 10 – 12 cm.

Hollow floors are always fitted with units that terminate flush with the floor level (**Figure 8.63**). The cost of having this system installed is considerably higher by comparison to a floor channel systems, unless the latter are installed with an aperture width of less than 2.5/2.5 m.

A raised floor is a good choice for an installation that offers maximum flexibility and accessibility at all times. **Figure 8.64** illustrates the basic structure of a raised floor with connection options; in addition to the electrical installation, this type of floor can also accommodate the ducts for heating, ventilation and air conditioning systems with air supply from above or below.

Raised floors are primarily used for applications where each floor must accommodate a high concentration of electrical equipment, or for computer mainframe spaces. **Figure 8.65** illustrates a typical arrangement of underfloor distribution for raised floor installations with high-voltage cabling, telecommunications and data transmission technology. Raised floors are generally built to a height of 15 cm (electrical installations only) and 60 cm (electrical and HVAC installations). Although the initial investment costs are very high, these floors offer almost unlimited flexibility and accessibility. A combination of raised floors in the corridors and floor channel systems in individual offices is frequently employed in axially planned buildings to achieve a higher degree of flexibility in comparison with installing a floor channel system alone.

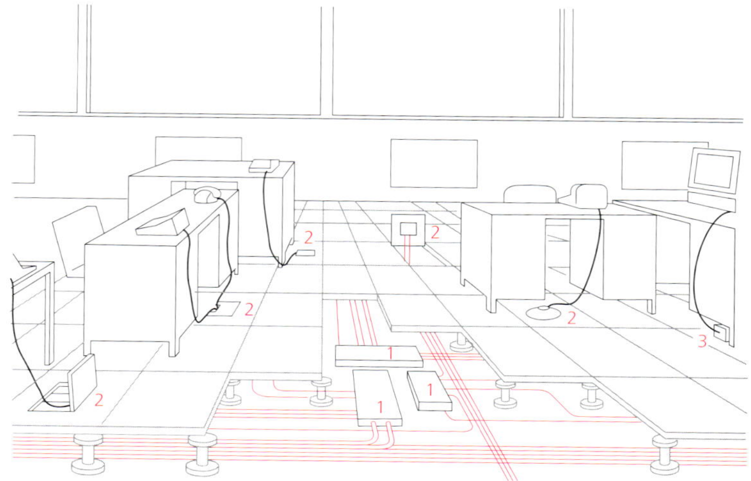

Figure 8.64
Connection options with raised floor installation (diagrammatic sketch)

1 Underfloor distributor for plug-in high-voltage, telecommunications and data transmission connections
2 Floor level connection with inserts for 1 – 6 devices
3 Floor-mounted connection box for 1 – 8 devices

Figure 8.65
Arrangement of underfloor distributor in raised floor installation

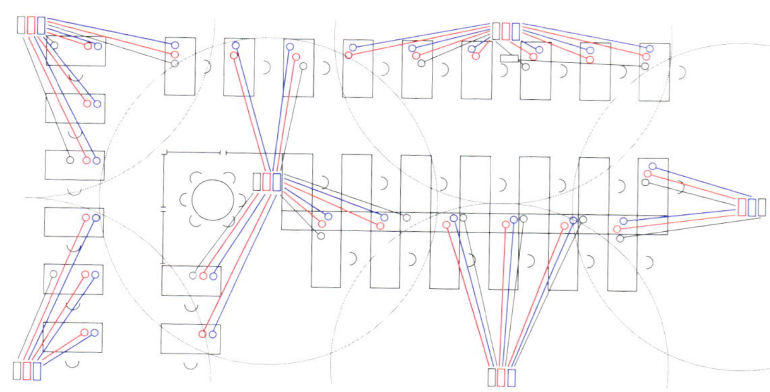

Connections at workstation with floor level installation units:
- ○ High-voltage connection
- ○ Telecommunications connection
- ○ Data transmission connection

Flexible lines between undefloor distributors and floor level installation units:
- — High-voltage connection
- — Telecommunications connection
- — Data transmission connection

Distribution triangulation station consisting of a high-voltage, a telecommunications and a data transmission underfloor distributor:
- ▯ High-voltage systems
- ▯ Telecommunications systems
- ▯ Data transmission systems

Figure 8.66
Installation channel
(nominal size 173)
with attached data channel

8.5.2 Windowsill installation

Windowsill installations are commonly used in offices, laboratories, hospital wards and similar spaces unless the concentration of electrical equipment is too high. Figure 8.66 illustrates a typical windowsill channel for use in front of a radiator, for example, or an induction device (top and bottom grille as air vent/air supply), and Figure 8.67 contains an installation diagram of the individual elements of this type of windowsill channel system. Windowsill channels are primarily used in individual rooms where the density of installation is fairly moderate and where housings are required at any rate for the heating units or air-conditioning units on the walls below the window.

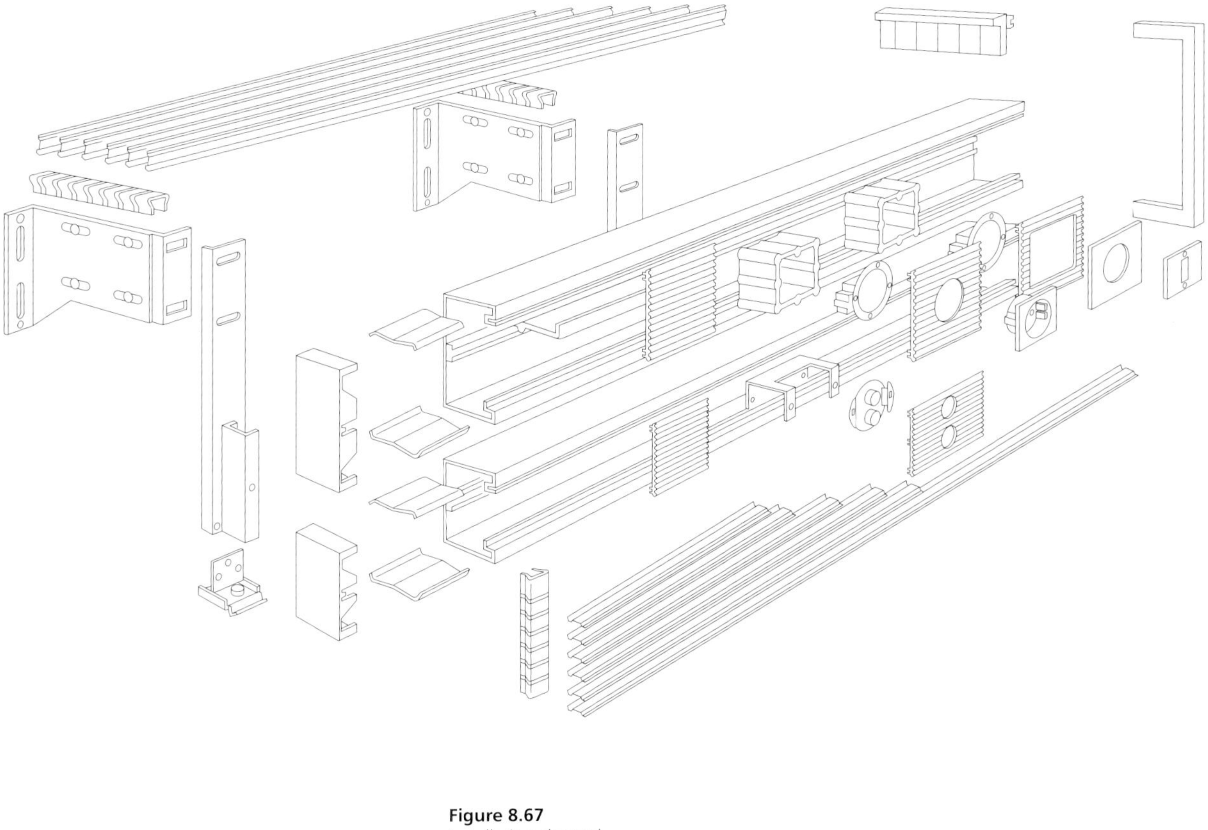

Figure 8.67
Installation channel –
system overview

The windowsill channels are complemented by special wall or mounted installation channels. In laboratories, open suspended media channels are suspended above the worktables, supplying the laboratory tables with all necessary media, as well as power and light. Figure 8.68 portrays this type of media channel with electrical connectors and plug-in couplings for gaseous media.

Figure 8.69 illustrates a floor-mounted installation channel with various installation options. Floor-mounted installation channels are commonly employed for renovation projects and can sometimes be combined with baseboard heating. This increases the shaft height by roughly 10 cm in the windowsill area.

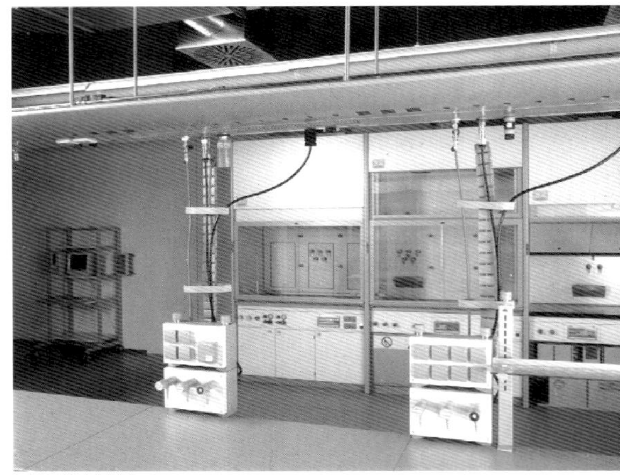

Figure 8.68
Suspended media channel for power, light and communications media connections

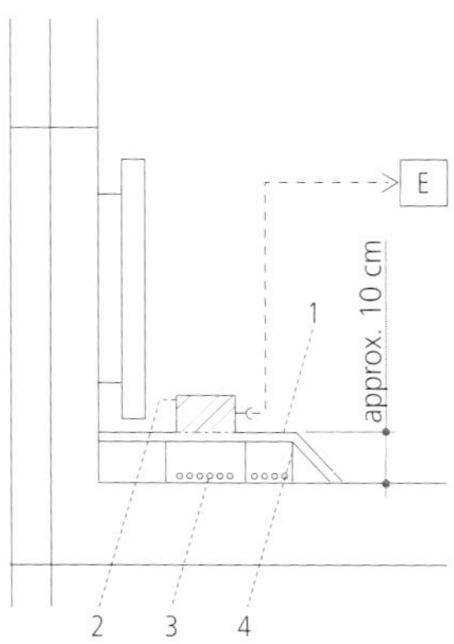

Structure of a unit

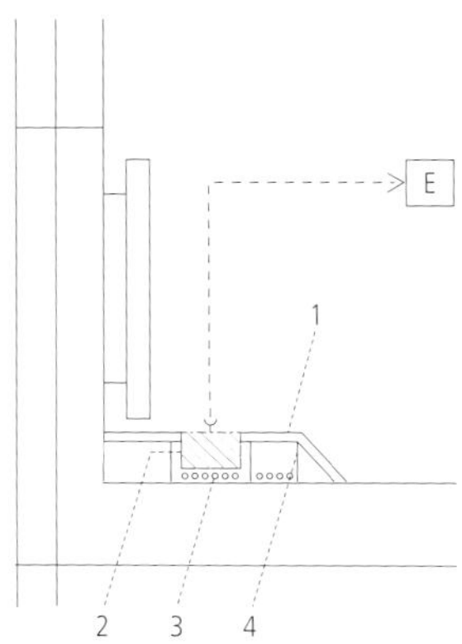

Integration of a floor-level unit

Figure 8.69
Installation diagram of floor-mounted installation channel

1 Channel cover
2 Built-in unit, e.g. socket
3 Cables
4 Channel partition

8.5.3 Floor-to-ceiling electrical / communication raceway poles

The distribution of high- and low-voltage cables via raceway poles is very economical and flexible installation. Figure 8.70 illustrates how the poles are connected to the distribution system in the ceiling. In Central Europe, raceway poles are primarily used in workshops or similar work environments; in North America they are widely used in office buildings to supply several workstations from a single pole, as shown in Figure 8.71 / 72. Unfortunately, raceway poles are not very aesthetic and this limits their use to some degree. In all other aspects, however, they are extremely practical and offer complete installation flexibility because the poles can be set up and connected in any location within the room. Cables are drawn into the poles from the ceiling. Suspended supply units (pendulum suspended conduits) or similar units can also be used instead of poles in spaces where the cables are drawn down from the ceiling.

8.5.4 Wall installations

Wall installations are widely used for one-time installation (for example, in residential buildings). This is also a common installation type in combination with cell-type construction. In this instance, vertical raceways in the door areas are a sensible solution.

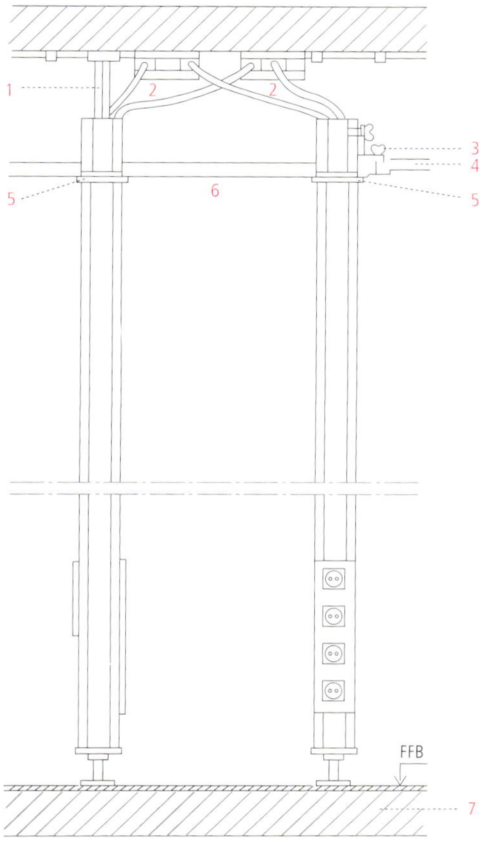

Figure 8.70
Connection of raceway poles

1 Crown piece
2 Ceiling connection socket
3 Clamp fastening
4 Ceiling rail (builder/ client installed)
5 Cover plate
6 Suspended ceiling
7 Bare ceiling

Figure 8.71
Placement of ceiling connection sockets in raceway pole systems (diagram)

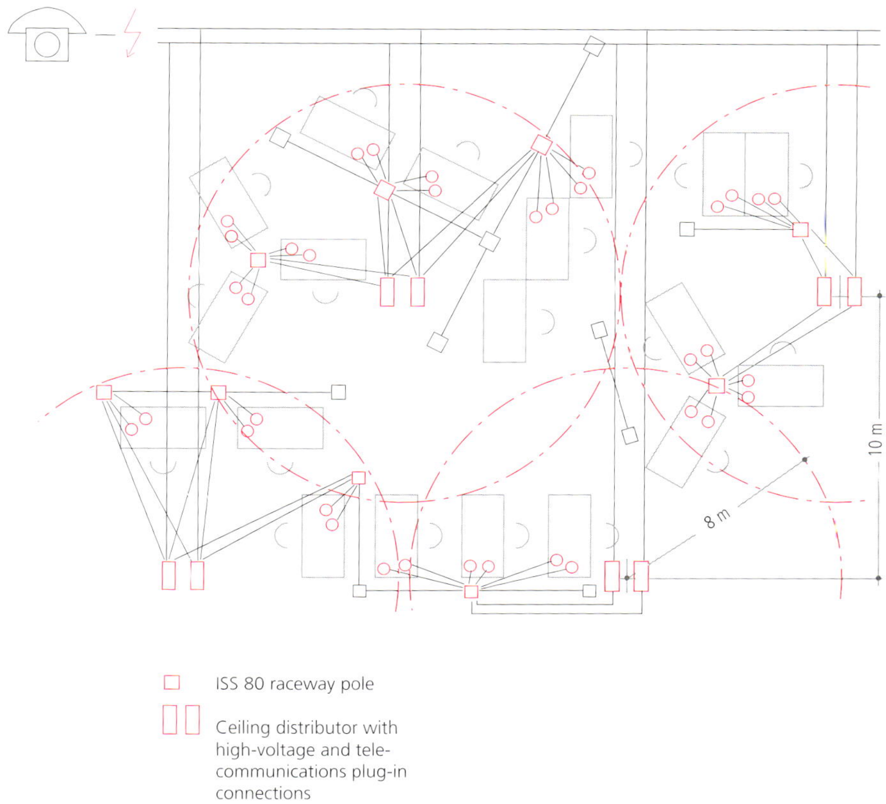

Figure 8.72
Installation cable as vertical supply column

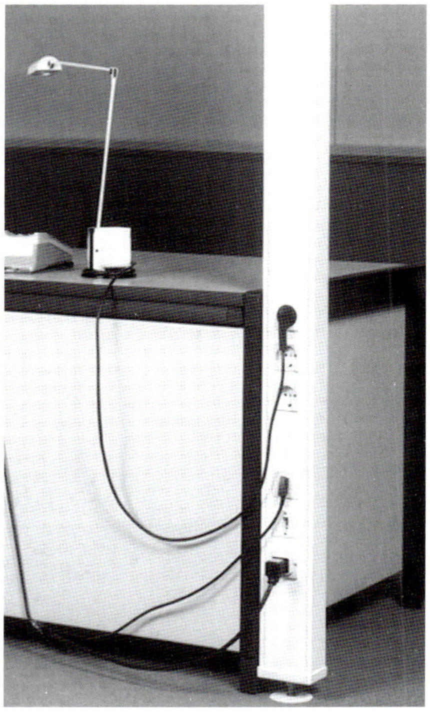

8.6
Protective measures and protective systems

While electrical power offers many advantages, it can also be a source of danger, and calls for protective measures. The guidelines and regional regulations set forth in DIN 57100 and VDE 0100 (Association of German Electrical Engineers), for example, contain detailed descriptions of these protective measures for people and electrical installations.

People are at risk of receiving an electric shock, while the systems themselves may be irretrievably damaged through thermal and mechanical loads resulting from an error.

Risk to persons
Dangerous body currents can be categorized according to their effects as follows:

Physical effects
– Protein coagulation
– Internal burns
– Current burns at the current contact point

Physiological effects
– (Cardiac) fibrillation (alternating current)
– Asphyxia/apnea
– Muscle contractions

Chemical effect
– Breakdown of cell fluid (direct current)

The severity of the effects of electric shock is influenced by various factors. Some of these are:
– Type of current (D.C. or A.C.)
– Duration of exposure
– Frequency
– Intensity or strength of current
– Physical condition of person exposed to shock

Protection of persons
Direct contact with low-voltage systems (400 V/230 V) can be fatal. Nevertheless, fatal accidents involving electricity are relatively rare, and most cases are caused by disregard for the safety regulations and technical guidelines. Protective measures have therefore been defined for each safety category, described in **Tables 8.10/8.11**. The safety features listed in the first category are divided into three areas, each capable of preventing accidental exposure and injury. All safety features are designed for coordinated use in order to provide multiple layers of protection.

– Prevention of direct contact
– Protection in case of direct contact
– Protection in case of indirect contact

IP	Contact and foreign body protection
0.	No protection
1.	Protection against penetration of solid foreign bodies greater than 50 mm Ø
2.	Protection against penetration of solid foreign bodies greater than 12 mm Ø Protection against touching
3.	Protection against penetration of solid foreign bodies greater than 2.5 mm Ø Protection against contact with tools
4.	Protection against penetration of solid foreign bodies above 1 mm Ø Protection against contact with wires and fine tools
5.	Dust-protected; while dust penetration cannot be fully prevented, it is prevented from settling or accumulating at sensitive points. Complete contact protection
6.	Dust-proof; no dust is allowed to penetrate

Table 8.10
Safety category designations

IP	Protection against water
.0	No protection
.1	Protection against dripping water (falling directly from above)
.2	Protection against dripping water (falling down at a 15° angle)
.3	Protection against spray water, rain protection
4.	Protection against spray water from all directions
.5	Protection against hose spray from all directions (hoseproof)
.6	Flood protection
.7	Dip protection
.8 --m	Submersion protection

Table 8.11
Safety categories for electrical installations
(SEV 1053, 1992)

Prevention of direct contact
Direct contact is prevented by both organizational and technical means.

The organizational means are chiefly focussed on training personnel, since they must work directly with electrical systems and are therefore subject to the highest degree of accident risk.

Protection in case of direct contact
Live electrical equipment, which can pose a danger to people, must be protected against accidental direct contact. Necessary openings in the devices are in contrast with the protection requirements. Recommendations developed by the I.E.C. (International Electrotechnical Commission) define protection against foreign bodies (X = first code) and water (Y = second code). The safety category of devices and systems is indicated with IP XY.

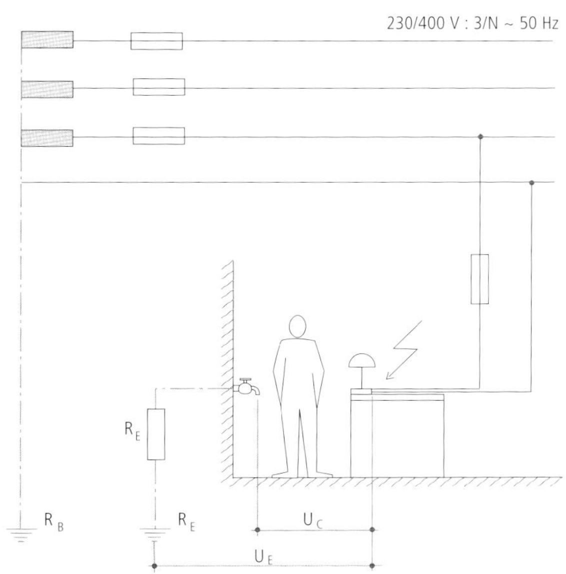

Figure 8.73
Error voltage (U_E) and contact current (U_C) with ground

Protection in case of indirect contact
An example of indirect contact is when a housing that is not designed for a live current becomes charged through some malfunction. Any contact with the housing would pose a risk of electric shock. The protective measure ensures that the circuit (or fuse) is tripped when such errors occur. The voltage, which is present in the housing only in case of malfunction, is referred to as error voltage and occurs between the metal housing and the ground. The housing of devices are grounded via a conductive connection (differential between potentials = voltage). The ground (on which the person is standing) has an equal potential when the protector ground is used, that is, it prevents error voltage. For example, if a live conductor in a device comes into contact with its metal housing, the error voltage is conducted away from the person via the ground fault protector (**Figure 8.73**). The conductive connection between the device and the ground creates an error voltage that is powerful enough to trip the protective system (fuse), preventing an accident. The fuse action provides notification of the error. This type of system is called TN-S net.

In the aforementioned system, the neutral point of a transformer is also grounded (**Figure 8.74**) to conduct overcurrent to the ground. In special environments such as operating theatres, for example, the neutral point is not grounded (insulated neutral point) because an automatic interrupt can be very disadvantageous for patients. Instead, errors are signalled by special fault detectors.

Figure 8.74
Wiring diagram for transfomer

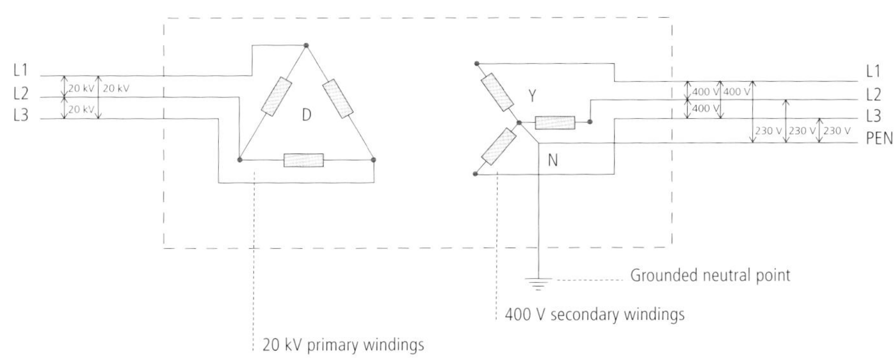

The protection systems for low-voltage power supply systems are designated by three letters, for example in accordance with DIN/VDE:

The first letter indicates the neutral point treatment:
I = Insulated neutral point
T = Grounded neutral point

The second letter indicates how error voltage is conducted from a component:
N = Conduction via neutral conductor
T = Conduction via ground (earth)

The third letter indicates the arrangement of the neutral and the ground conductor:
S = Neutral (N) and ground are separated
C = Neutral and ground are combined in one conductor

The most common system in buildings is the TN system, which is itself differentiated into three variations:

TN-C system
Neutral and ground conductor are unified
TN-S system
A separate ground conductor is laid from the neutral point of the transformer
TN-C-S system
The first part of the system is laid out as a TN-C system, the second part as a TN-S-system.

The TN-C-S is the most commonly used system in building technology (**Figure 8.75**). (A separate ground conductor is required for conductor diameters of less than 10 mm^2/TN-S system.) In other words, a TN-S system must exist at the very least from the subdistributors onwards.

TN systems are easily installed and protection by means of overcurrent protection devices has proven efficient and safe. In households in and above all in high-risk areas, an additional error voltage switch or GFC (ground-fault circuit interrupter) is indicated, which immediately trips the current in response to very minor error voltages (30 mA).

Potential equilization is achieved by connecting all conductive components. Potential equalizers are used to create a conductive connection, for example in a building, between water pipes, heating and gas pipes, the lighting ground, the ground fault, steel structures, elevator tracks etc. and the combined neutral and ground conductor. Connecting this combined conductor to the main potential equalizer prevents excessive contact currents in case of a short circuit or body contact.

Rooms where risks are especially high, for example bathrooms, are equipped with additional potential equalization by linking all conductive metal components to the ground.

Despite the aforementioned ease of installation, the TN system has some characteristics that can be hazardous if the installation is inexpertly done or does not adhere to the regulations. General rules stipulate that the fault currents must be tripped within a timeframe of 0.2 seconds for circuits of up to 35 ampere and within 5 seconds for all other circuits. Hence the cross-section areas of conductors and protective devices must be properly coordinated.

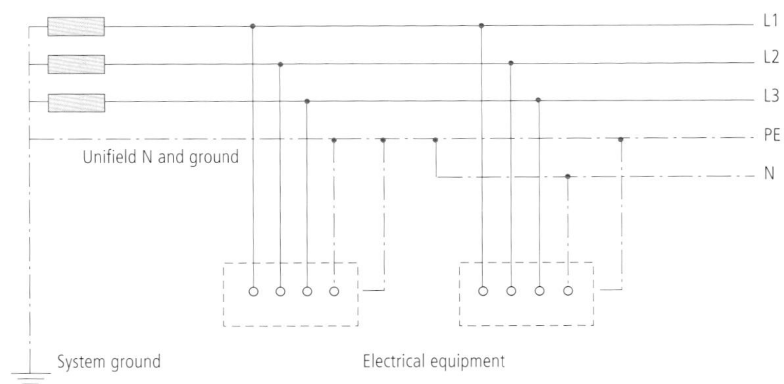

Figure 8.75
Power networks in building systems
(Safety systems)

8.7 Installation equipment

Installation equipment comprises all switches, sockets and connector boxes, which are selected according to application site (dry or damp rooms), function and design considerations.

8.7.1 Switches

A differentiation is made between standard and flat switches, their various functions and installation modes (concealed or exposed) and application areas (dry or damp rooms). **Figure 8.76** illustrates some standard switches, a selection from the wide range of products currently in use.

8.7.2 Sockets/connector boxes

Sockets are used to plug in electrical appliances. They are differentiated according to connection conditions, room situation (dry/damp) and design characteristics. **Figure 8.77** illustrates a small selection of the many sockets available for a variety of uses and conditions.

Connector boxes are subject to the same criteria described above and are also available in a wide range of models. **Figure 8.78** illustrates commonly used connector boxes.

A

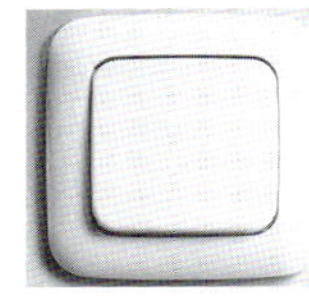
B

C

D

E

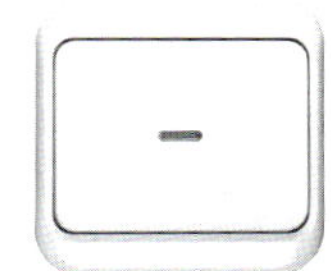
F

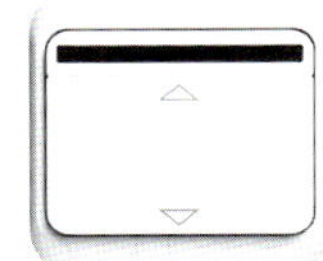
G

Figure 8.76
Switches
A Moisture-proof switch, surface type
B Rocker-type switch, standard
C Flat (rocker-type) switch
D Economizer (time switch) for flush
E Flat switch with stainless cover
F Rocker-type switch with large toggle surface and light
G Dimmer switch, manual and remote control and infrared remote control device

Figure 8.77
Sockets
– Schuco® socket
– 3-phase current socket
– Moisture-proof socket
– Telephone jack

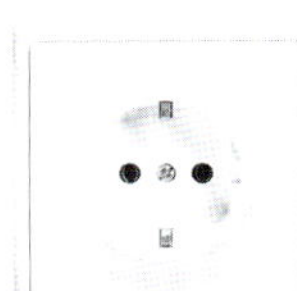
Alternating current (single-phase)

Rotary (3-phase) current

Moisture-proof current

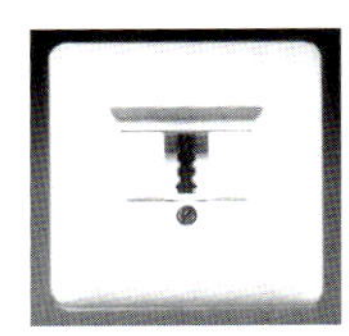
Telephone jack

Figure 8.78
Connector boxes

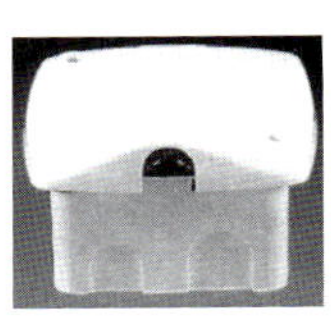

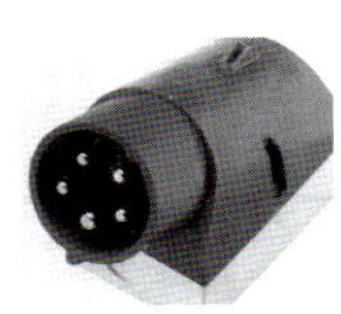

8.8 Installations for special voltage requirements

Electrical transformers are used wherever there are special voltage requirements (laboratories, computer processing, machine units).

8.8.1 Motor generator sets

Special voltages, especially constant voltage, are usually produced with the help of motor generator sets. In this configuration, an electric motor drives a generator that in turn generates the direct voltage, transferring it via a distributor to the direct voltage mains or supplying individual consumers. Motor generator sets move large masses. The prefabricated module units are therefore mounted on bases and pedestals to avoid the transmission of impact sound and to distribute concentrated loads.
Figure 8.79 shows the front and interior view of a standard transformer unit.

The frequency is transformed (50 Hz to 400 Hz) by frequency transformers. One portion of the energy is transmitted in a pure transformation process from the stator of an asynchronous or induction generator to the rotor, and another portion, which corresponds to the difference between the output and the input frequency, is transmitted mechanically via a rotor. The output frequency increases when the rotor is driven counter to the rotary field through a coupled motor.

Like other transformer units, frequency transformers are set up close to the consumers as units that are ready for use. In electronic data processing areas, frequency transformers are now generally installed as static plants with no special requirements for the installation area. However, it is best to install such units on double or raised floors to improve the conditions for installation.

As mentioned, plants with rotating masses require pedestals, which are not required for static plants. The transformation produces a high level of set noise and relevant equipment rooms may need to be soundproofed. Transformer units generate heat due to energy losses (roughly 10 % of the rated power). This heat must be eliminated from the equipment rooms through sufficient ventilation. Alternatively, the equipment room can be cooled to a maximum operating temperature of 40 °C using a return air device. When frequency transformers are integrated into a security concept (data processing plant), the same parameters apply to associated return air cooling devices and to the cold water production.

If frequencies in the range of radio waves are generated by frequency transformers, the relevant equipment rooms must be appropriately screened and all electrical lines must be equipped with filters.

Figure 8.79
Exterior and interior view of transformer unit

8.8.2 Uninterruptible power supply (UPS installations)

Certain work environments rely heavily on computing centres and all their secondary systems at all times – this is especially true for banks, insurance companies and computerized production plants – even in case of total power failure. A UPS plant must be installed in these cases to improve the availability of supply voltage.

Redundant power generating plants are equipped with standby passive reserve units or additional active UPS blocks that follow in the partial-load range of parallel operation.

The need for constant supply voltage even in the case of power failure is, however, not the only indication of redundant parallel operating plants. They should also be employed where there are slight deviations between the mains and the consumer frequencies; the relevant information is supplied by the manufacturers (EDP equipment). **Figure 8.80** contains a block diagram of a static UPS plant with a manual bypass switch and an electronic shift mechanism. The UPS plant is composed of a rectifier, a downstream current rectifier and a battery-operated intermediate circuit. The consumers are connected to the current rectifier. Operation of the UPS plant is initiated via the battery in case of deviation in the input voltage or the nominal values; in other words, the energy required to maintain operation is drawn from the battery.

Like transformer units, UPS plants (**Figure 8.81**) are also delivered as complete cabinet units and installed on raised floors. UPS units are ventilated with air supplied through the raised floor and cooled through air exhaust in the ceiling area to guarantee compatible room temperatures.

As **Figure 8.81** shows, a UPS plant can occupy a fair amount of space, and this must be considered in buildings where such plants are required from the very beginning.

Attention should be given in particular to the following aspects:
- Overall efficiency
- Quality of output voltage
- Lifecycle of accumulators in DC. intermediate circuit
- Low current regeneration

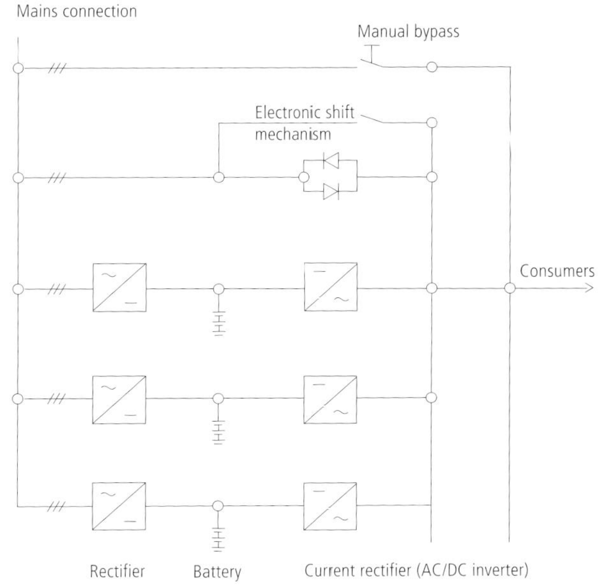

Figure 8.80
Block diagram of a static UPS plant with manual bypass and electronic shift mechanism

Figure 8.81
UPS plant
(Uninterrupted Power Supply)

8.9
Domestic power supply

As the name suggest, domestic power supply plants supply power that is not drawn from the mains supply to a building or complex of buildings. It is important to differentiate between systems that are operated primarily in case of power failure, where standby plants deliver the necessary power, and other systems that generally supply a portion of the electrical power and whose waste heat is continuously utilized.
The latter are combined heat and power plants (CHP).

8.9.1 Standby power plants

Standby power plants – also referred to as emergency power diesel plants or emergency power aggregates – consist of an internal combustion engine with a follow-up generator, mounted on a joint base (**Figure 8.82**). Emergency power diesel plants must be installed in separate rooms because they generate high noise levels and produce high volumes of thermal energy. The following factors must be taken into consideration with regard to the location and dimensions of these rooms:

- Combustion air and ventilation air supply
- Air dissipation of radiant heat produced by the engine and the generator
- Evacuation of cooling air or waste heat with water (water-cooled engines)
- Waste gas evacuation through roof
- Coordination with corresponding low-voltage switch plants
- Safe cable runs etc.

The dimension of the equipment room must allow for a minimum clearance of 1.5 m around the aggregate. The space must also accommodate additional components such as switch plants, fuel storage, starter battery etc. It is vitally important to allow for additional space for sound absorbers near the air supply and air ventilation, which can sometimes be quite considerable.

When planning an equipment room, it is important to decide in advance how the equipment will be installed (flexible or static) to determine the design of the base; the cabling and any other special measures should also be laid out in advance.

In smaller applications, the emergency power diesel aggregates and the switch plant are installed in one room; for larger systems, these components must be set up in two separate rooms.

Since emergency power diesel plants create fairly strong vibrations, which could lead to impact sound transmission, special attention must be paid to the placement of the machine and the separation of the unit from the building. This is true for both roof-installed emergency power diesel plants and basement-installed plants. A separate base (elastic rubber or spring bedding) is generally only useful or necessary for emergency power diesel plants with a performance range over 250 kVA. For heavy-duty emergency power capacities, the emergency power plant is spread across several engines for parallel operation.

Emergency power plants can run on oil from biomass, diesel oil or gas as primary energy carriers, although gas offers the advantage of obviating the need for stocking fuel.

The engines are cooled either by indirect water cooling (engine cooler, untreated water demand approx. 35 to 50 l/kWh) or water cooling that is directly linked to a water network (in combination, for example, with a recooling plant, in which case a cooling-water storage tank is required).

For air-cooled engines, there are requirements for the air volumes necessary to evacuate the radiant heat, generator waste heat and the combustion air needed for various allowable temperature differences.

Figure 8.82
Emergency power diesel plant (emergency power aggregate, 135 kVA with 1500 rpm) with fan cooling attached to motor

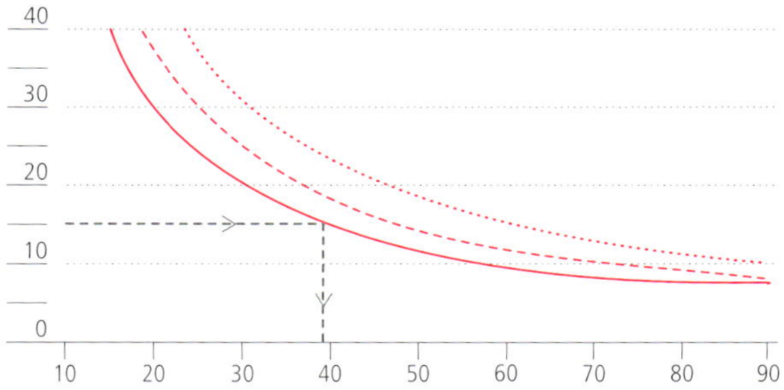

Figure 8.83
Air volumes required for room ventilation

····· below 80 kVA
- - - 80 to 120 kVA
—— above 120 kVA

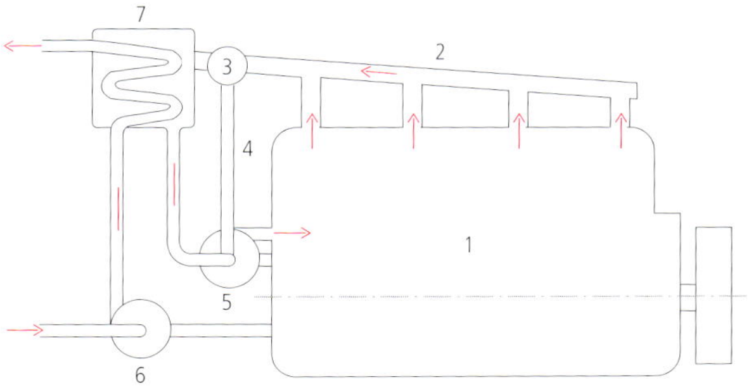

Figure 8.84
Diagram of indirect water cooling for an emergency power plant

1 Diesel motor
2 Mains
3 Thermostat
4 Warm up pipe
5 Circulating pump
6 Pipe water pump
7 Chiller

Figure 8.85
500 kVA emergency power diesel aggregates (TRHS518 V16) with 1500 rpm. The demonstration installation of this plant in the Siemens research facility in Erlangen, Germany, also includes insulation for the waste gas pipes.

The air volumes plotted in **Figure 8.83** are shown in relation to performance. Significant air volumes are required for room ventilation for large emergency power diesel plants, resulting in large structural air intake shafts and escaping air ducts, which must always be installed to rise above the roof. If shafts and ducts of this kind are not feasible, the air volumes can be reduced to roughly 10 %, that is, the essential portion of the combustion air. This system design must, however, be combined with water cooling (see diagram in **Figure 8.84**).

Generally speaking, the standby power system must be properly cooled to prevent overheating of the engine. Some examples of cooling air volumes for engines are listed below:

95 m^3/h kW for a room temperature of 30 °C
115 m^3/h kW for a room temperature of 37 °C
140 m^3/h kW for a room temperature of 45 °C

The operating temperatures of the engine range from 85 to 95 °C. Temperatures in equipment rooms for standby power systems should not exceed 40 °C to ensure that generator function is guaranteed at all times and to provide appropriate room temperatures for maintenance staff.

The vent pipes should be well insulated to reduce excessive heat emissions; the same is true for water carrying pipe systems (**Figure 8.85**).

To reduce exhaust gas noise, chamber sound dampers or sound absorbers must be built into the vent pipes, whereby sound absorbers achieve almost complete sound damping without impeding the flow rate. These absorbers are used whenever annoyance caused by noise is to be avoided at all costs for surrounding buildings. The same considerations apply to outside air intake and exhaust air evacuation.

While fuel storage is not necessary for gas-powered engines, it is required for diesel-powered engines. Fuel is generally stored in one tank for the daily supply, supplemented by an additional larger storage tank (**Figure 8.86**). The tank for the daily supply should store sufficient diesel fuel for eight to ten hours of operation; the sizing is estimated as follows:

Engine performance (kW) x 3 = tank capacity in litres

The larger storage tank can be sized according to requirements and may hold several days worth of fuel.

Figure 8.89 (plan and section, (see **Chapter 8.9.2.1**) illustrates the typical layout of a standby power plant with air-cooled engine. The drawing does not include the required intake air and exhaust air sound absorber (length 1.5 to 2 m). The standby diesel system shown in **Figure 8.89** represents a low-capacity plant.

Figure 8.86
A 250 l tank is installed below the 300 l fuel tank for a 400 kVA aggregate. Temperature-controlled indirect cooling is provided through the hydraulic pipes.

8.9.1.1 Consumption areas entitled to standby power supply

The entitlement to standby power supply is generally based on the building type (e.g. high-rise), its use (hospitals, public assembly rooms) and other construction codes for special building systems (large sprinkler systems, fire services elevators etc.). The following consumption areas are generally entitled to standby power supply:
- Emergency lighting
- DP systems
- Air-conditioning for DP systems
- Refrigerating systems for DP systems/food storage/refrigerated goods
- Elevator systems
- Essential machines for production processes

In case of doubt, the precise demand for standby power should be established in consultation with the client or user and with the relevant authorities.

8.9.1.2 Sizing the standby power system

The standby power system is designed exclusively for the aforementioned consumption areas entitled to standby power (summation of all connect loads) taking the corresponding reactive factor into account in addition to the sum of the apparent power. Sizing is also dependent on the following factors:
- The largest consumer supplied by the standby power system (including starting current, where applicable)
- The base load to which other loads are modulated
- The network into which the system is designed to feed (230/400 V or other)
- The installation height of the aggregate
- Special installation conditions (elevated ambient temperatures, humidity, contamination)

8.9.2 Combined heat and power plant (CHP)

Combined heat and power plants are basically the same aggregates as standby power plants. In contrast to them, however, CHPs not only provide back-up power in case of power failure, but also continuously cover a portion of the regular power supply. Hence the same parameters apply in terms of system layout and primary energy supply, which can be gas or oil as described for standby systems. CHPs can also be equipped with engines, which can be converted from gas to oil at any time, resulting in additional redundancy.

The outmost care must be taken in sizing a CHP plant both in consideration of the high initial investment and the special maintenance required once the plant is in operation.

8.9.2.1 Sizing a CHP plant

CHP plants are sized in consideration of two principal factors:
- Thermal efficiency requirements
- Power requirements (base load/standby)

Numerous economic efficiencies studies have demonstrated that CHP plants should cover only a percentage of the total power requirements (25 to 35 %); this is generally sufficient to meet the thermal efficiency requirements.

Figures 8.2.1 and **8.2.2** (see **Chapter 8.1**) serve as an example of sizing a CHP plant. The graphs plot the power requirements in an office building over the course of a day, and the annual power requirements, respectively. **Figure 8.87.1** (left) provides yet another overview of typical diurnal power requirements, and **Figure 8.87.2** (below) the annual requirements.

Figure 8.87.1
Power requirements and power supply in kW, hot summer's day in July

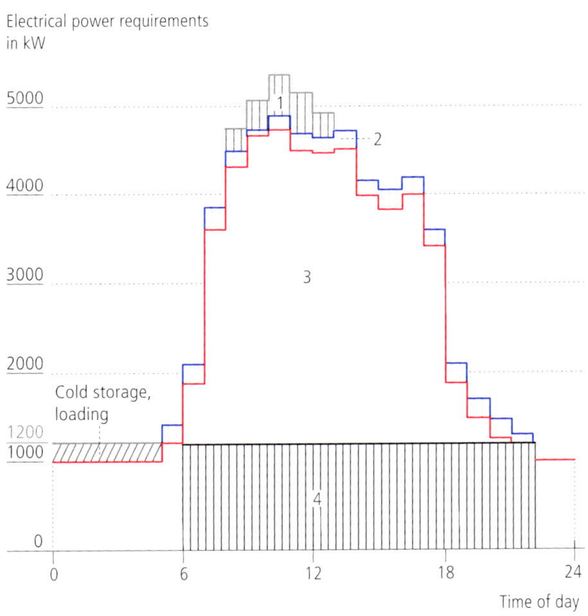

1 1. Power savings: Kitchen appliances partially operated with CHP-HT waste heat
2 2. Power savings: Cold energy partially produced by means of CHP-HT-LT waste heat (cold storage unloading)
3 Utility power max. 3490 kW
4 Power savings: (Utility power)

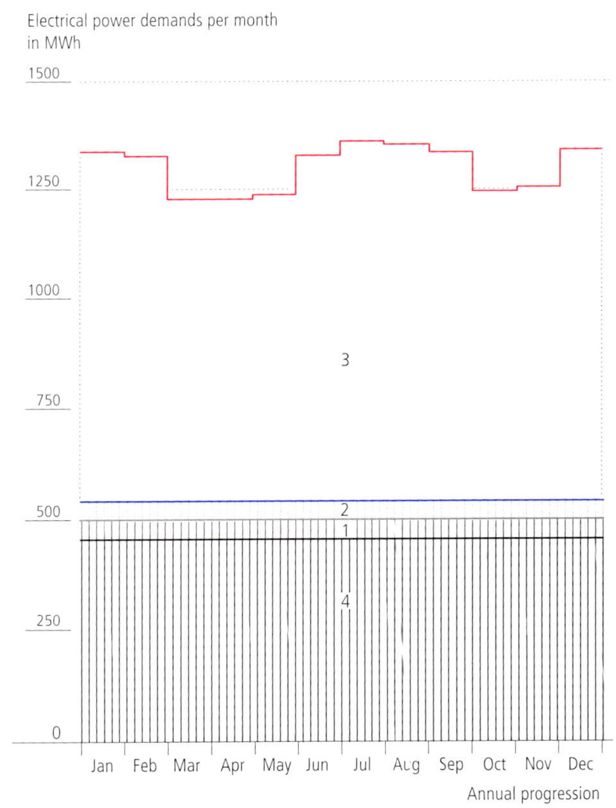

Figure 8.87.2
Annual power requirements and supply

1 1. Demand reduction: Kitchen appliances partially supplied from CHP-HT waste heat
2 2. Demand reduction: Cold energy partially produced by means of CHP-HT-LT waste heat (cold storage unloading)
3 Utility power
4 CHP power

The optimum size of the CHP system is calculated on the basis of an in-depth analysis of the power demands (power and thermal energy) for systems of different capacities and a comparison of investment costs, operating costs and annual system operating periods (no less than 4,000 h/a), as well as the total annual costs. In the example shown in **Figure 8.87.1**, the power demand for a large office complex is 1,200 kW with the CHP system, whereby the graph clearly demonstrates that the CHP is operated exclusively during peak rate hours (6 a.m. to 10 p.m.), whereas the off-peak demands are covered entirely by the public or mains power supply since the use of the CHP system would be unprofitable during those hours. However, the CHP plant not only covers a percentage of the power demand, it also contributes indirectly towards a reduction of this demand. The power demand of 1,200 kW shown in **Figure 8.87.1** is supplied by a generator.

The indirect reduction (reduction of power demand) is achieved, for example, by operating kitchen appliances with high-temperature waste heat instead of electrical power. Similarly, refrigeration is partially provided by absorption refrigerating machines instead of electricity-powered refrigerating machines. This reduces the electrical power demand and optimizes the voltage drop.

Figure 8.87.2 plots the annual power demand and the supply from the CHP and the public utility, respectively. As the graph demonstrates, 42 % of the annual demand is covered by the CHP plant and 58 % by the mains supply. Given these conditions, CHP systems can be amortized over a relatively short period. The system layout and its differences to a conventional system is shown in **Figures 8.88.1/2**, illustrating different operating conditions.

Figure 8.88.1
Heating/cooling diagram, hot summer's day in July

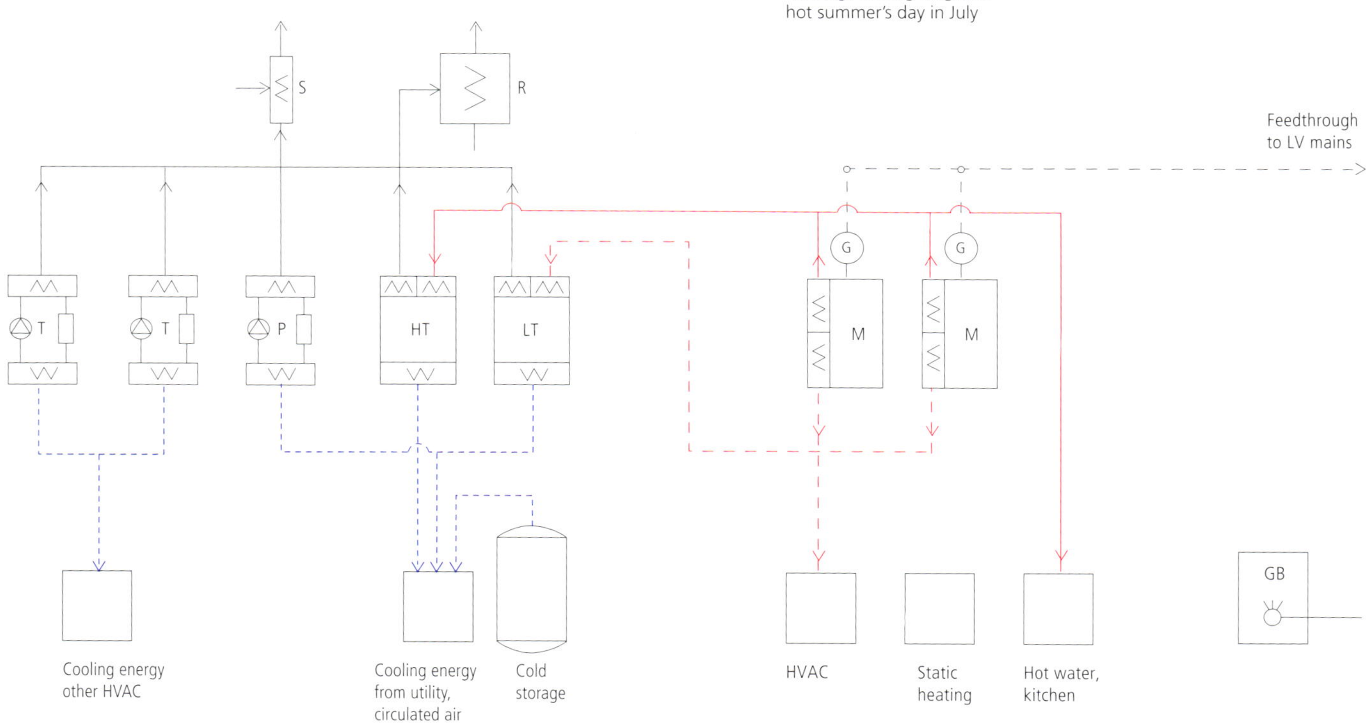

– – Cold energy
– – Low-temperature heat
—— CHP waste heat
—— Recooling
– – Feedthrough

HT High-temperature absorption refrigerating machine
LT Low-temperature absorption refrigerating machine
HVAC Heating, ventilation and air-conditioning
T Turbo refrigerating machine
P Piston compressor refrigerating machine

GB Gas boiler
G Generator
R Recooling plant
S Service water warm up
M Diesel/gas motor

In summer, the entire supply of 1,200 kW (electrical power) from three CHP modules (with a fourth as standby module) is fed into the main low-voltage distribution. At the same time, the waste gas heat is transferred through a waste gas heat exchanger to a high temperature absorber for cold energy production and the motor and oil waste heat to a low-temperature absorber, again for cold energy production. Whenever VAC systems require thermal energy during this period, it is supplied at the same time. The absorption refrigerating machines, which require very little energy in contrast to piston and turbo refrigerating machines, produce a considerable amount of the cold energy required for a computing centre. Should the cold energy supplied by the absorbers be insufficient, additional cold energy is fed into the system from a reciprocating piston compressor machine.

Thermal energy, which cannot be supplied to the consumers from the high-temperature absorbers and other refrigerating machines, is carried off via a recooling plant. To reduce power demands in the kitchen, some of the high-temperature heat (higher than 100 °C) is supplied to the kitchen appliances.

In winter (daytime operation/**Figure 8.88.2**), cold energy for power production by means of CHP systems is supplied exclusively from the waste heat of the waste gas heat exchanger (indirectly via the high-temperature absorber) and if necessary from a cold storage, which is loaded overnight. During this period hot water is supplied to the kitchen from the waste gas heat exchanger. Low-temperature thermal energy from the engine and oil cooling and waste heat from the high-temperature absorber (heat pump system) are supplied to the HVAC heat consumers, supplemented by a boiler system, which supplies additional heat volumes that may be required to heat the building.

Figure 8.88.2
Heating/cooling diagram, cold winter day in January

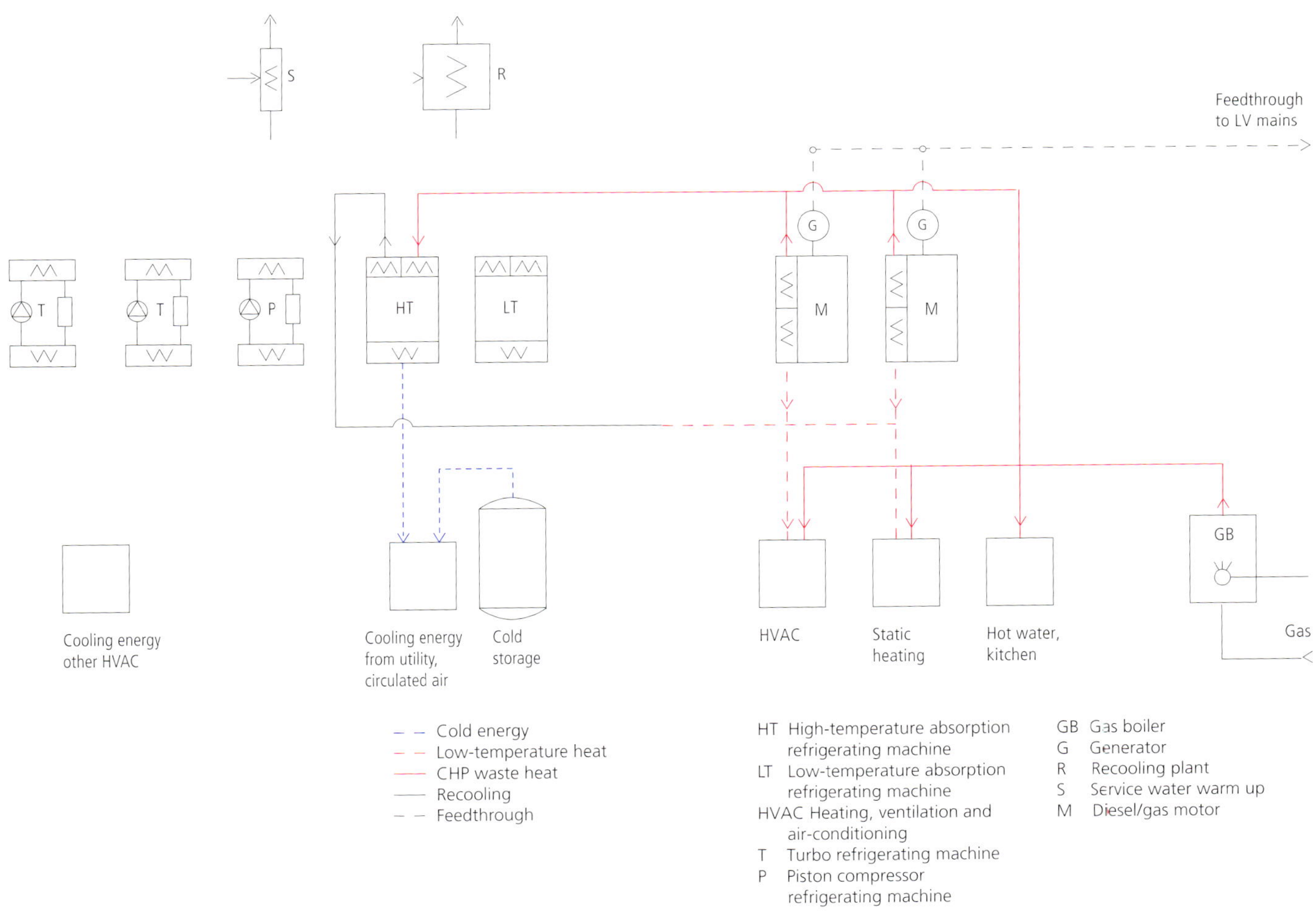

The simplified diagram of a CHP plant illustrates the complex composition of a wide variety of technical applications and optimized utilisation of the primary energy (in this case, gas). **Figure 8.89** illustrates a possible layout for a standby power system that is equally applicable to a CHP system.

CHP system design is a complex matter that involves extensive experience and complex calculations. In general, however, the amortization rates for CHP systems are quite reasonable (amortization in less than seven years) for a building or complex with continuous high power demands and where energy from waste heat can be harnessed (for example, to produce cold energy). The simultaneous utilization of electrical and thermal energy is an essential prerequisite for the use of CHP systems. If this prerequisite is not given, the cost of a CHP system is unlikely to be amortized.

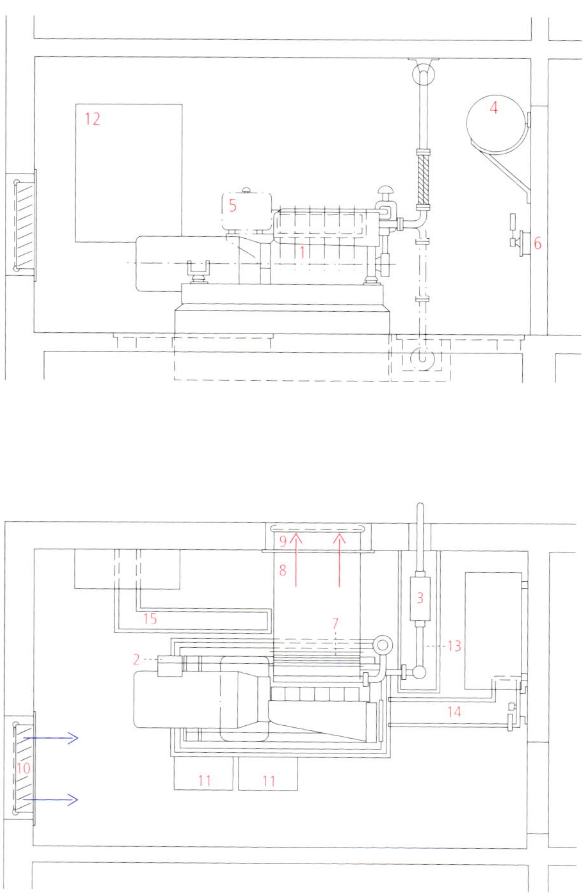

Figure 8.89
Layout example for aggregates with air-cooled motors

1 Diesel generator aggregate
2 Terminal block
3 Exhaust muffler
4 Fuel tank
5 Fuel tank (attached)
6 Semi-rotary fuel hand pump
7 Elastic intermediate piece
8 Exhaust air duct
9 Exhaust air grille
10 Fresh air grille
11 Battery
12 Switchgear cabinet
13 Exhaust duct
14 Duct for fuel pipes
15 Cable duct

8.9.3 Battery power systems

For short periods, buildings can also draw their power from battery systems, which are charged either from the mains or from standby power systems.

Stationary batteries as energy storage systems thus ensure continuous power supply to all essential consumers in case of power interruption or total power failure. While the battery voltage must be transformed by frequency converters into AC voltage for three-phase current consumers, DC consumers can draw power directly from the battery, which is charged by an autonomous battery charger. Correct charging has a strong influence on the battery's lifecycle. It is vital, therefore, to maintain the correct charging current and voltage.

Charged batteries from which no loads are drawn off will automatically self-discharge by means of internal chemical processes. A float charge voltage is applied to continuously supply the battery with a low charging current to maintain its start-up condition.

In change-over circuits (**Figure 8.90**), consumers are supplied from the three-phase (low-voltage) network via a power rectifier. In case of power failure, the consumers are switched over to the battery, drawing their power from it until the system voltage is restored (**Figure 8.90**, bottom).

In contrast to change-over circuits, parallel circuits means that the battery and the consumer are connected in parallel and supplied via a joint battery charger. In case of power failure, power continues to be supplied to the consumers from the battery without interruption. This arrangement is necessary for systems that cannot tolerate power interruptions (computing centres); lead- or nickel-cadmium batteries are used for these installations. The batteries are composed of cells, consisting of a cell box, positive plates, negative plates and the electrolyte (**Figure 8.91**). As distinction is made between the following lead accumulators (or batteries):

- Batteries with Planté-type or large surface plates, power adapter and high-voltage capacitance
- Block batteries with steel plates and grid plates or stationary grid plates

Large-surface plate batteries have a longer life expectancy and slightly better characteristics for high-voltage discharge, while block batteries are less expensive. Nickel-cadmium batteries (NiCd batteries) are classified as medium-load and normal-load (or off-peak load) batteries (discharge time: one hour and more). In addition, there are high-voltage load batteries with a discharge time of less than one hour. The technical data of batteries are listed in **Table 8.12**.

Nickel-cadmium batteries are used in environments with ambient temperatures of -40 to -10 °C, lead-cadmium batteries for temperatures of -10 to +55 °C. Therefore, lead batteries with a lifetime of 12 to 20 years are the only viable option for standard applications in buildings. The cell number and type is determined by the nominal voltage, discharge time and discharge current. The battery capacity must be chosen to ensure that a specific current can be drawn off within a predetermined discharge period without falling short of the permissible voltage limits. The capacity should be sufficiently high to allow for the aging of the batteries and possible system expansions.

		Lead batteries		Nickel-cadmium batteries	
		Block	Planté-type	VSM	VSX
Nominal voltage	V/cell	2.00	2.00	1.20	1.20
Float charge voltage	V/cell	2.23 ± 1%	2.23 ± 1%	1.40 ± 1%	1.40 ± 1%
Continuous charge voltage					
Floating electromotive					
force (buffer voltage)	V/cell	2.40	2.40	1.55	1.55
Quick charge voltage	V/cell	2.35...2.40	2.35...2.40	1.50...1.55	1.50...1.55
Discharge voltage (10 h...$^1/_2$ h)	V/cell	1.80...1.72	1.80...1.72	1.10...0.80	1.10...1.05
Nominal capacity	Ah	C_{10}	K_{10}	K_5	K_5
Capacity as percentage	10 h	100%	100%	103%	103%
of nominal capacity	5 h	90%	90%	100%	100%
for different discharge	3 h	81%	82%	96%	99%
times	1 h	63%	64%	83%	95%
	$^1/_2$ h	50%	52%	80%	90%
Max. allowable charging currents					
at begin of gassing	A/100 Ah	8.00	8.00	10	10
at closing	A/100 Ah	4.00	4.00	5	5
Load factor		1.10	1.10	1.40	1.20
Energy efficiency factor (approx.)		0.75	0.77	0.60	0.75
Electrolyte		Diluted sulphuric acid		Diluted potash lye	
Density (approx.)	g/cm³	1.24	1.22	1.17...1.20	1.17...1.20

Nominal voltage = number of cells · nominal voltage of a cell (V)
Capacity = current that can be drawn from a battery (Ah) depending on: discharge voltage and time as well as closing voltage, density and temperature of electrolyte, condition of battery
Nominal capacity = capacity indicated by manufacturer (for defined discharge time and nominal discharge)

Table 8.12
Batteries – technical data (right)

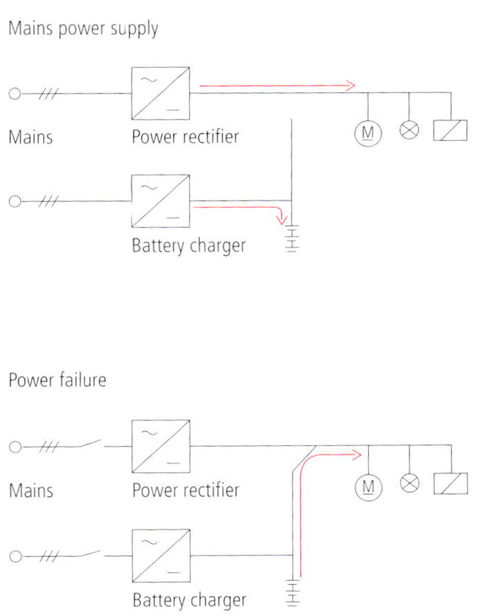

Figure 8.90
Change-over circuit

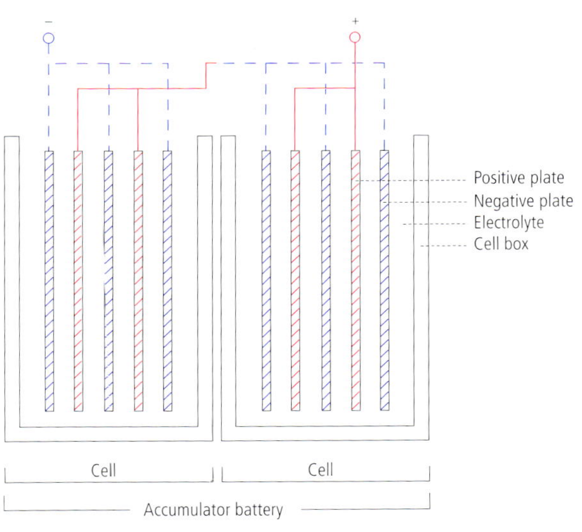

Figure 8.91
Battery structure

Battery systems with nominal voltages up to 250 V are housed in separate battery rooms, subject to the codes established for all electrical equipment rooms. The regulations pertaining to damp rooms must be satisfied when installing electrical systems and other components in battery rooms. All doors to battery rooms must open outwards; signs prohibiting open flames and smoking must be clearly posted on all doors.

Depending on the size of the installation, batteries are mounted in cabinets, on floor pads or individually on insulators. Battery and battery charger can also be installed in a common cabinet, provided the charge load does not exceed 2 kW. Since gases are released when the batteries are charging, discharging and at rest, provisions must be made to dilute the gaseous mixture sufficiently through natural or mechanical ventilation to eliminate any explosion hazard. The necessary air volume for battery room ventilation is calculated according to the number of cells and the hydrogen-generating current (ampere). The ventilation requirements are calculated as follows:

$V = 55 \cdot n \cdot I$ (litre per hour)

Where:
n = number of cells
I = current in amperes
(caused as a result of hydrogen generation)
55= factor for required air volume

It is important to ensure that outside air enters at floor level (for both natural and mechanical ventilation) and is exhausted at ceiling height on the opposite side of the space (cross-ventilation, air movement bottom-top). Negative pressure should be minimized in battery rooms to prevent acid-laden air from escaping to the outside or into adjacent rooms.

Figure 8.92
Battery room

Owing to the presence of acid-laden air, the room surfaces should be tiled or at the very least receive a treatment that prevents the acid from penetrating into walls, ceilings or floors. All motors for air supply and air exhaust components must be explosion-proof (temperature class G1) and the battery may be charged only if perfect cross-ventilation is ensured. **Figure 8.92** depicts a battery room.

8.9.4 Photovoltaics

Photovoltaics is a process where light is transformed into electricity based on the photovoltaic effect that describes the interaction between light and the basic material of the solar cells. In other words, electrical charge carriers are released and collected by the metallic contacts of the solar cell, generating direct voltage. Depending on their structure and basic material, solar cells achieve efficiency coefficients of up to 20 %, utilizing the incident global radiation on a building. Photovoltaic elements contain different types of cells, which have different efficiency coefficients. These are typically:
- Monocrystalline silicon cells, efficiency coefficient 14 %, direct voltage approx. 0.48 V, direct current approximately 2.9 A
- Polycrystalline silicon cells, efficiency coefficient 12 %, direct voltage approx. 0.46 V, direct current approx. 2.7 A
- Amorphous solar cells – silicon plated onto support substance (opaque module), efficiency coefficient 5 %, direct voltage approx. 63 V, direct current approx. 0.43 A
- Amorphous solar cells – semi-transparent, efficiency coefficient approx. 4 %, direct voltage approx. 63 V, direct current approx. 0.37 A

The outer appearance of the solar cells is shown in **Figure 8.93**. The corresponding solar modules are generally embedded between two glass panes and supertransparent film to protect them from the elements (**Figure 8.94**).

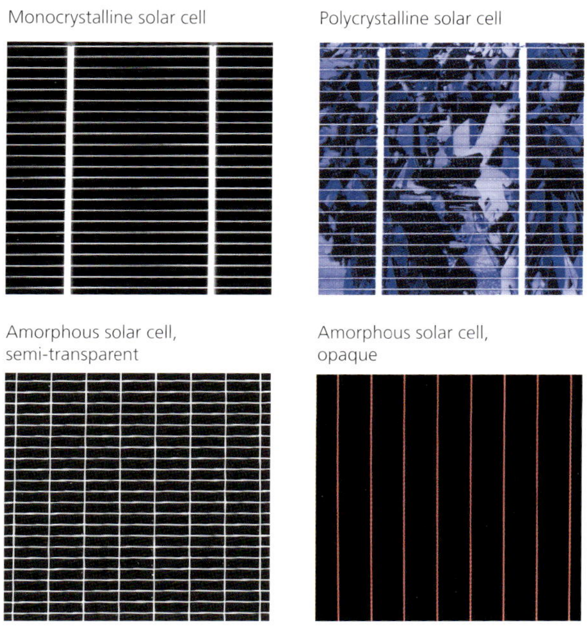

Figure 8.93
Photovoltaic modules

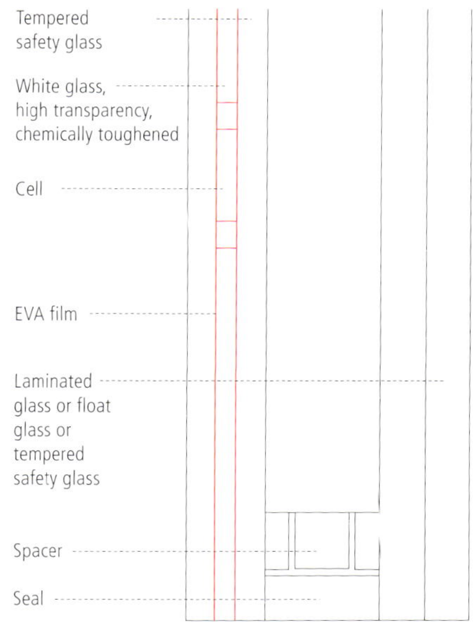

Figure 8.94
Insulating glass in module, structural diagram

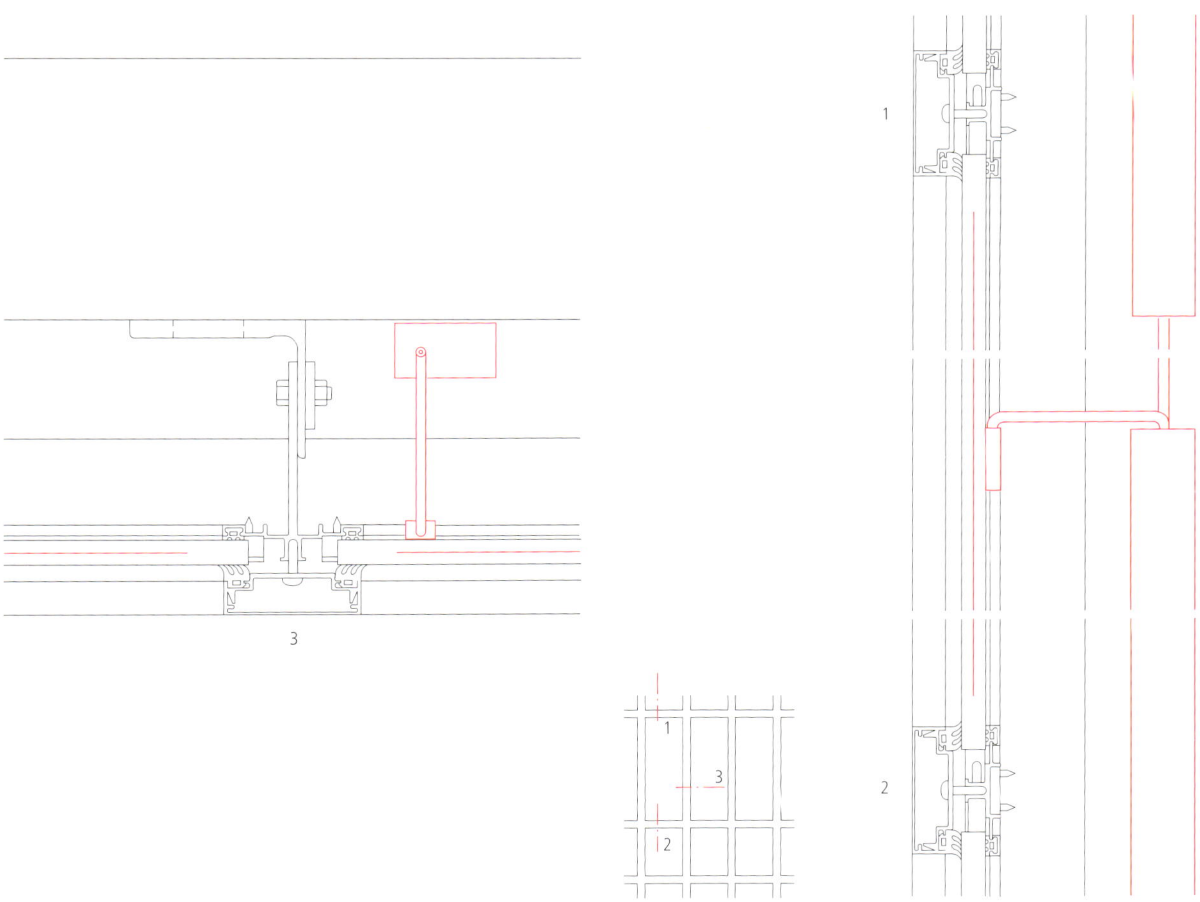

Figure 8.95.1
Facade section, cold facade system CW 80 with "opaque" amorphous modules

Crystalline modules (monocrystalline and polycrystalline cells) have high efficiency coefficients, depending on the basic material, and offer many design options. However, they are much more expensive than amorphous modules, with efficiency coefficients of approx. 5 %. Furthermore, the latter have the advantage of a considerably higher initial voltage, leading to simpler system designs. The use of photovoltaic elements (photovoltaic generators) makes sense only for the sections of a building that receive direct sunshine. This section of the building facades or roofs gives off the greatest amount of energy, so that even a small solar gain due to diffuse insolation with a very low efficiency coefficient is noticeable. Photovoltaic elements should utilize the direct and the diffuse radiation as much as possible. Hence, the active surfaces should face southeast or southwest. **Figure 8.95.1** shows an example of a cold facade section with "opaque" modules (System CW 80, Schünco International).

The wiring diagram in **Figure 8.95.2** illustrates how the electrically gained energy is collected and transported to the electrical energy units of a house via a transformer. Since photovoltaic systems generate electric power of approx. 1.0 – 1.5 W with full solar radiation (>800 W/m²) and since approx. 100 kW/a of electrical energy can be gained per square metre, they are at this point not yet economically competitive (long amortization periods). Despite the sometimes excessive investment costs, it seems increasingly sensible and necessary to make greater use of photovoltaic systems in the future, both to advance the technology and to protect the environment. Current investment costs of approx. 767 EUR/m² module surface generate annual savings of approx. 38 EUR/m² module surface, resulting in an amortization rate of roughly 20 years.

Figure 8.95.2
Wiring diagram of a photovoltaic installation

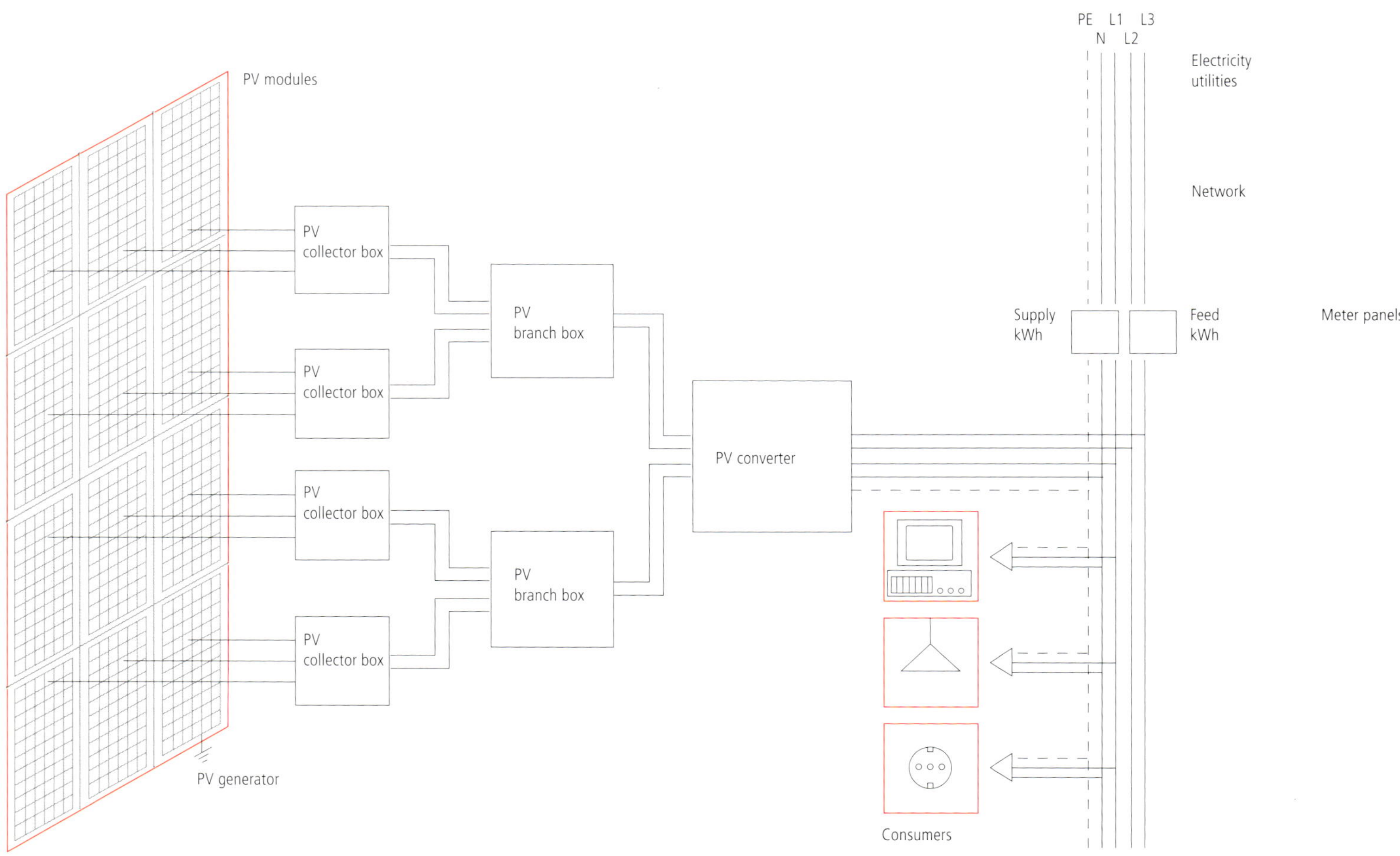

8.10 Lightning protection systems

These systems protect buildings against lightning strokes by catching and conducting lightning voltage to the ground.

External lightning protection systems consist of:
- Air terminals
- Down conductors
- Ground terminals

The principle of external lightning protection in combination with internal potential equalization for connection to the internal lightning protection is illustrated in **Figure 8.96**.

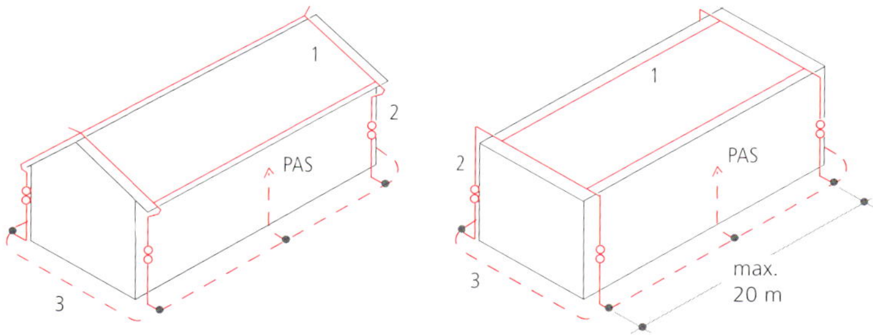

Figure 8.96
Principle of external lightning protection with link to potential equalization (or compensation) and connection to internal lightning protection

1 Air terminal
2 Down conductor
3 Ground terminal

Table 8.13
Number of down conductors depending on building form

Circumference of outer roof edge	**Number of down conductors** Symmetrical building	Asymmetrical building	Gable roof up to a max. length of width of 12 m
... 20 m	1	1	1
21... 49 m	2	2	2
50... 69 m	4	3	2
70... 89 m	4	4	4
90...109 m	6	5	4
110...129 m	6	6	6
130...149 m	8	7	6

The air terminal comprises all metal building components (wires and rods) that are installed on top of, above or next to the structure, and it serves as a striking point for the lightning.

Generally speaking, a protective mesh (mesh size 10 m · 20 m) is installed on the roof, regardless of the building height. For buildings up to a total height of 20 m, lightning lines or rods can be installed, each creating a 45° protection zone. In shape, the protection zone created by lightning lines is similar to a tent or prism with semicircular edges, while lightning rods or masts create a cone-shaped protection zone.

The lightning is conducted away from the structure across a conductive path between the air terminal or arrester and the ground terminal. One down conductor must be installed per 20 m of roof area measured along the outer edge.

The ground terminal conducts the lightning voltage safely into the soil. It generally consists of a foundation ground around the building or a surface ground (or individual ground in special cases).

Table 8.13 contains a list of the number of down conductors required for different building types and roof dimensions (measured along the outer edge).

There are three principal types of external lightning protection, differentiated according to the type of air terminals used. These are:
- Cage or "mesh" systems
- Shield wire systems with protective zone
- Mast systems with protective zone

Figures 8.97.1 illustrate various "mesh" systems, clearly marking the 20 m distance between down conductors.

Figure 8.97.1
"Mesh" air terminal

Mesh/air terminal/mast

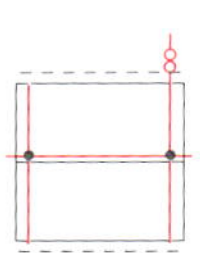

Circumference < 20 m :
1 down conductor sufficient

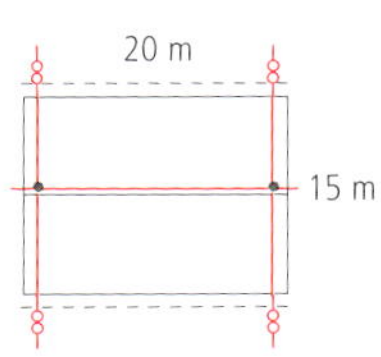

Number of down conductors:
70 m : 20 m = 3.5
that is, 4 down conductors

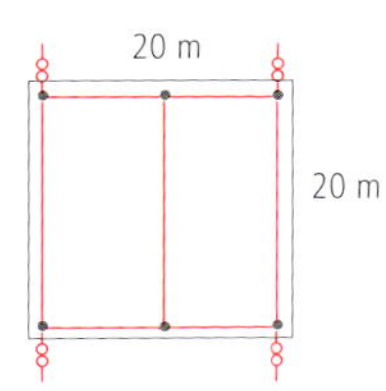

Number of down conductors:
80 m : 20 m = 4
that is, 4 down conductors

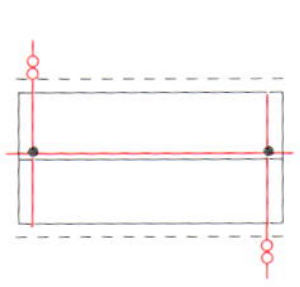

Number of down conductors:
60 m : 20 m = 3
width ≥ 12 m: 3 - 1 = 2
that is, 2 down conductors

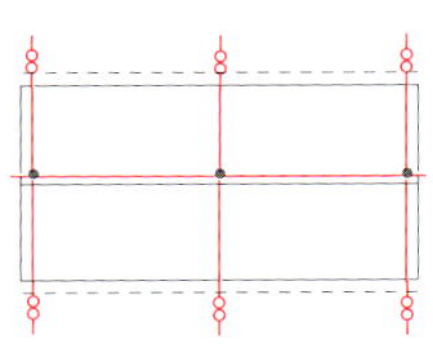

Number of down conductors:
92 m : 20 m = 4. 6
that is, 5 down conductors
Symmetrical building: 5 + 1 = 6
that is, 6 down conductors

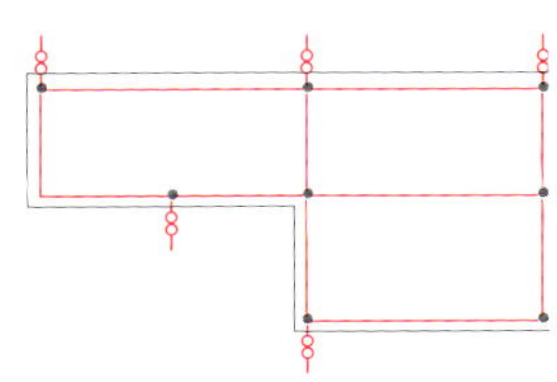

Number of down conductors:
120 m : 20 m = 6
that is, 6 down conductors

Figure 8.97.2 illustrates systems that rise above the building and **Figure 8.98**, systems with masts that are installed either on top of the building or next to the structure. The maximum mast height above ground is 30 m; masts higher than 30 m are no longer effective.

Figure 8.99 illustrates the lightning protection system on the roof of a very tall building, identifying the necessary intermediate loops (1/2). Care must be taken to maintain a distance of 40 m between the intermediate loops and between each intermediate loop and external down conductors. This factor helps to establish the required number of intermediate loops and external down conductors. If intermediate loops (or internal down conductors) are not possible, the number of external down conductors must be increased accordingly, taking care, however, to maintain a minimum distance of 10 m between external down conductors.

Metal building components outside of the building or on the building skin (elevator tracks, fire escapes) can be used as down conductors. In steel skeleton structures, the skeleton itself can be used as a down conductor, while the rebars in reinforced concrete structures can serve a similar purpose, as long as there is good conduction between the components and the ground. If this is not the case, separate down conductors must either be integrated into the reinforced concrete or mounted onto the structure externally. On metal facades, continuous vertical and conductive metal sections (sectional components of the substructure) can be used. Here, too, it is important to a conducting transition and, if necessary, to establish a connection to the individual components of the metal facade. Metal downspouts can also be used as down conductors, provided the joints are welded or connected with welded or riveted joints. They must be connected to the ground terminal. Metal installation inside the building must never be used as down conductors, although their base must be connected to the potential equalizer or the ground terminal.

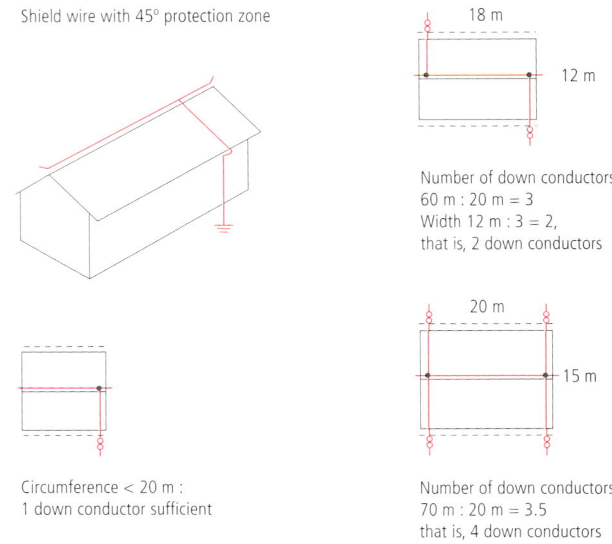

Figure 8.97.2
Shield wire system

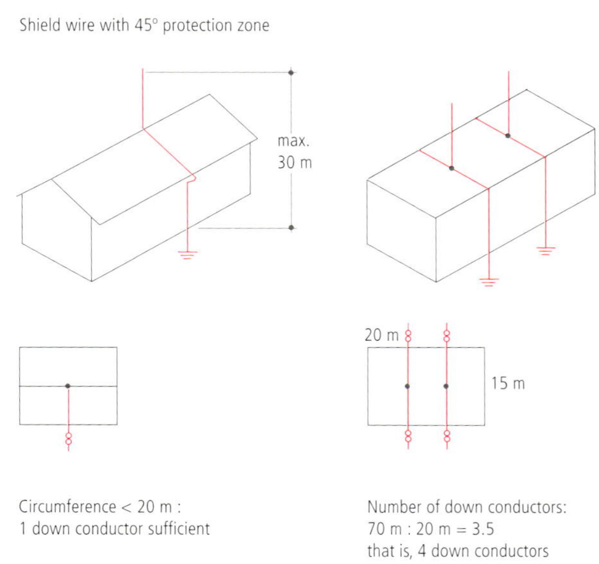

Figure 8.98
Mast system

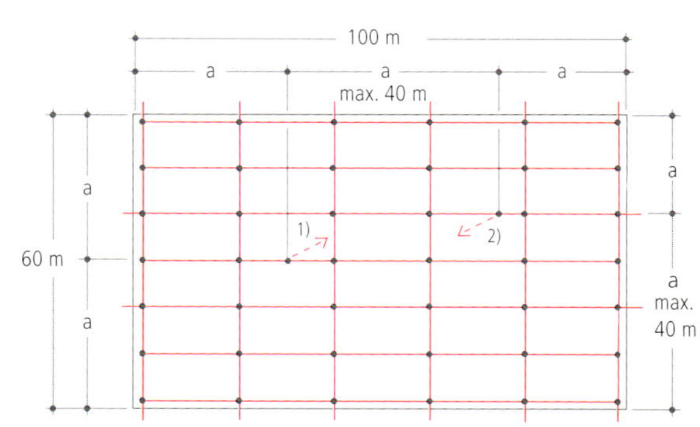

Figure 8.99
"Mesh" system on large surfaces

1) and 2): additional internal down conductors (a ≤ 40 m)

8.11 Substations/shafts/distribution

Substations are rooms in which large electrical equipment is housed, usually on the ground floor or in the basement of a building. There are some exceptions where medium-voltage switchgear and the transformers for the low-voltage distribution are housed separately on a different floor (an additional basement level, ground floor, upper floor or attic floor). Water pipes are generally prohibited in electrical equipment rooms, which are moreover maintained as dust-free environments. Mechanical or natural ventilation must be provided for the waste heat from switchgear, transformers and main distribution. Care must be taken to provide installation space for the various components (transformers, standby power systems) to allow for any subsequent installations or replacements. Electrical equipment room space requirements for rooms without standby systems or UPS are roughly 6 m^2 per 1,000 usable area. The room should be located as close as possible to the heavy load areas within the building (that is, areas with high power demands). **Table 8.14** gives detailed information on space requirements for medium-voltage switchgear and transformers. The size of the low-voltage distribution is roughly equal to that of the transformer installation (approx. 50 % of the values given in **Table 8.14**; see **Tables 8.15/8.16**). Floors in electrical equipment rooms should carry loads of 1,000 kN/m^2; this is the value used for the sizing calculation, allowing leeway for large area loads and localized loads. The space requirements for standby power systems are listed in **Table 8.17** in relation to the aggregate capacity. Large storage tanks are not included in these calculations.

The space requirements for UPS systems vary widely: in high-tech buildings with advanced electronic equipment, these dimensions may represent 100 – 200 % of the values given in **Table 8.14**.

Space requirements for medium-voltage switchgear and transformers	**Area** in m^2	**Comments**
Maximum capacity up to 100 kVA Secondary measurement, load-break switch with fuses and transformer in one cell	8	Minimum room height for medium voltage switchgear rooms 10 kV : 3.0 m 20 kV : 3.5 m 30 kV : 5.0 m
Maximum capacity up to 100 kVA Joint utility and DBP sector, secondary (medium or low voltage) measurement, consisting of: 1 cable feeder cell, if necessary with measurement set, 1 transformer end cell with 1 transformer	15	The room heights are in reference to open switchgear installations. When circuit breakers are included, it is necessary to check whether the room height is sufficient
Maximum capacity up to 100 kVA Separate utility and DBP sector, secondary (medium or low voltage) measurement, consisting of: 1 cable feeder cell, if necessary with measurement set, 1 transformer end cell with 1 transformer	15	The height of the transformer containers must be determined in relation to ventilation cross-sections and transformer losses
Maximum capacity above 100 kVA up to 500 kVA Secondary measurement and 1 transformer Separate utility and DBP sector, secondary (medium or low voltage) measurement, consisting of: 1 or 2 cable feeder cells, if necessary with measurement set, 1 transformer end cell and 1 transformer cell, for one cable feeder cell for two cable feeder cells	 17 25	Medium voltage switchgear should be located at ground level and as close to the peak load consumption point as possible The area requirements for medium-voltage systems apply only to row configurations up to 20 kV
Maximum capacity above 100 kVA up to 500 kVA Medium voltage measurement and 1 transformer Separate utility and DBP sector, consisting of: 1 or 2 cable feeder cells, 1 measuring cell, 1 transformer end cell and 1 transformer cell, for one cable feeder cell for two cable feeder cells	 23 25	
Maximum capacity above 500 kVA up to 1260 kVA Medium voltage measurement and 2 transformers Separate utility and DBP sector, consisting of: 2 cable feeder cells, 1 measuring cell, 1 transfer cell, 2 transformer end cells and 2 transformer cells	50 (65) *)	
Maximum capacity above 1260 kVA Medium voltage measurement and 3 transformers Separate utility and DBP sector, consisting of: 2 cable feeder cells, 1 measuring cell, 1 transfer cell, 3 transformer end cells and 3 transformer cells **)	70 (80) *)	*) (for row 30) **) If necessary, add one cable feeder cells for direct feed

Table 8.14
Space requirements for medium-voltage switchgear and transformers

Table 8.15
Low-voltage supply systems

Space requirements for low-voltage switchgear	Area in m^2	Comments
		Low-voltage switchgear should be installed at the peak load consumption point Minimum room height for low-voltage switchgear rooms
Cabinet switchgear configuration, per field including operating aisle	2.5	2.5 m
Free-standing field with rear access, per field including operating aisle at front and back	3.0	2.5 m

Table 8.16
Space requirements for risers and floor distribution

Space requirements for floor distribution terminals	Area in m^2	Comments
Insulating-encapsulated or steel plate distribution terminals for surface mounting or concealed installation	3 – 8	2.20 m minimum room height 1.20 m minimum room depth Distribution terminals should be installed at the peak load consumption point

Table 8.17
Space requirements for emergency power systems (fixed)

Capacity kVA	Capacity m^2	Room height m (on top of concrete topping)	Cubic room volume m^3	Fuel storage volumes l	Comments
					The room dimensions are applicable for aggregates with air-cooled diesel motors up to 150 kVA and water-cooled diesel motors ranging from 12 to 4,000 kVA
63	35	3.1	115	700	The room dimensions also allow space for the required emergency power switchgear panels (with the exception of low-voltage distribution)
160	50	3.4	170	2,000	The length-to-width ratio ranges from 1.5:1 to 1.9:1
250	65	3.6	235	3,000	Adequate installation and assembly openings must be provided
400	80	4.1	330	5,000	Larger rooms may be required for higher capacities due to ventilation requirements (heat radiating from the motors)
					Water-cooled systems, which are also fan cooled, require an additional cubic room volume of 10% for the cooling system
					A new concept for emergency power systems is currently in development. One of the issues the new concept will address is the installation of coolers on footings. The indicated area requirements will increase by roughly 5 m^2. This nullifies the 10% increase in cubic volume for coolers
					The fuel storage corresponds to roughly 40 hours of operating time

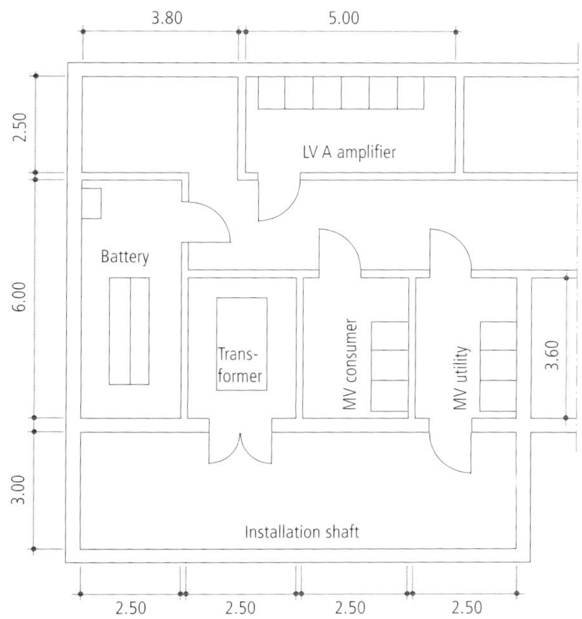

Figure 8.100
Electrical equipment room with total connect load of roughly 500 kW

Shafts for vertical cable and tray distribution occupy roughly 2 % of the equipment area or an estimated 1 m² per 1,000 m² usable area. In addition, roughly 3 – 5 m² distribution space must be provided per 5,000 m² supply area.

Horizontal distribution has already been described in **Chapter 8.5**. For ceiling installations, the space requirements for the ceiling cavity can be calculated in conjunction with the necessary requirements for lighting systems (approx. 20 – 30 cm).

In conclusion, **Figures 8.100 and 8.101** offer a schematic overview of electrical equipment rooms with all essential aggregates, room dimensions and locations, house connections, installation paths and corridors (escape routes).

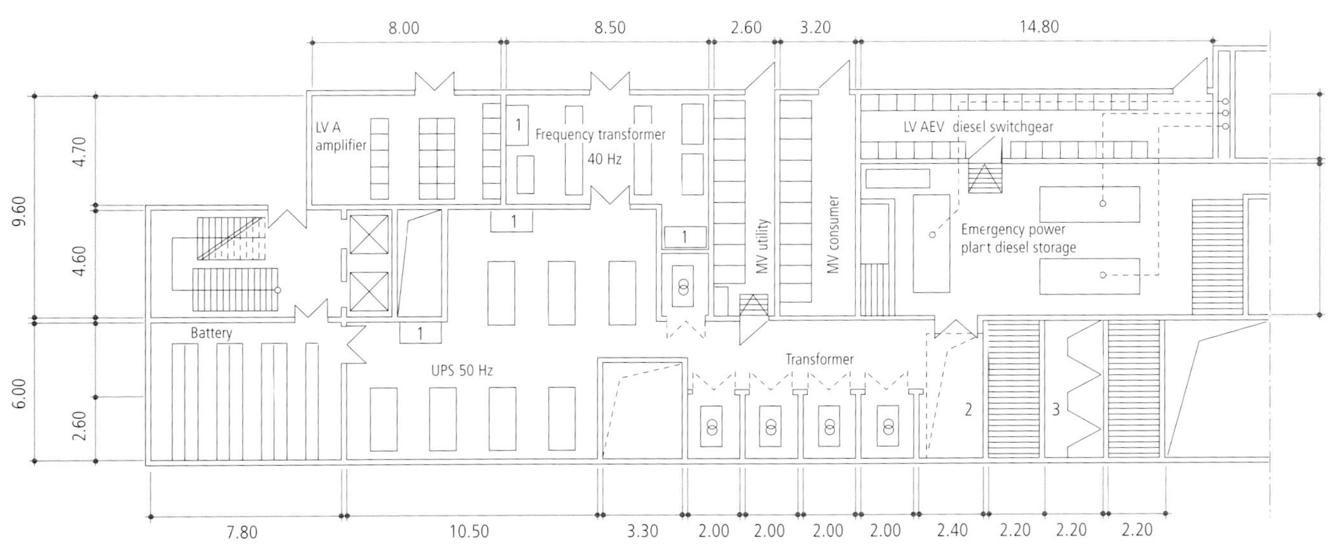

Figure 8.101
Electrical equipment room, roughly 3000 kW with battery/UPS/emergency power plant

1 Recirculating air cooling apparatus
2 Installation shaft
3 Cooling apparatus

1. Building performance
2. Comfort
3. Integrated planning models
4. Heating systems
5. Sanitation and fire protection
6. Ventilation and air-conditioning technology
7. Refrigerating and cooling systems
8. Power plants

9. Lighting

10. Daylighting
11. Low-voltage systems
12. Transportation systems

9. Lighting

9.1 The human eye

Optical stimuli and images are perceived by the human eye and processed in the brain. In this section the focus is on the influence of light on human comfort and the influence of glare on task performance. The working of the human eye is also briefly described.

Figure 9.1 provides a vertical section of the human eye. In the process of seeing, an image of an object is produced by the lens on the retina. The retina is a membrane at the back of the eye that is sensitive to light. The image is focused on the retina by the lens changing shape.

The aperture of the pupil is influenced by the illuminance with which the image is reproduced on the retina. The photoreceptors (light-sensitive cells) of the retina consist of rod cells and cone cells. The cone cells enable daylight vision. Only a narrow wavelength band of the total spectrum emitted by the sun stimulates the retina, thus producing a sensation of light (400 to 700 nm). The cone cells are characterized by a relative spectral receptivity, plotted in the "V (λ)-curve" in **Figure 9.2**. They allow colour vision and are concentrated near the central direction of vision. This yields a high degree of focus.

The rod cells are much more sensitive than the cone cells and provide night vision. They do not enable colour vision and yield only a relatively moderate degree of focus.

The ability of the human eye to change perception and sensitivity over a wide range is called adaptation. Although the eye can adapt rapidly to certain levels of luminance, it is incapable of simultaneously evaluating luminances in the entire possible range; instead it is able only to do so on a scale of 1:200 depending on the adaptation condition. The eye is stimulated by luminance not illuminance. Stable perception occurs when the surrounding luminance and the interior field luminance are in harmony. The contrast sensitivity of the human eye is another function that depends on luminance and not illuminance.

In addition to sensitivity (receptivity), focus is another important eye function that allows individual perception of contours that lie side by side in the range of vision. Focus increases as the contrast be-tween object and surroundings increases. Light focused on an image in a dark room creates a very high contrast, thereby increasing the focus. Excessive contrast is experienced as glare and the visual capacity of the eye is diminished both by physiological and psychological (discomfort) glare. However, the higher the level of illumination in a room, the smaller the required contrast in order to achieve a certain degree of focus.

Figure 9.1
Cross-section of the human eye

1 Cornea
2 Pupil
3 Lens
4 Iris
5 Vitreous humor
6 Ciliary body
7 Central fovea
8 Blind spot
9 Sclera
10 Choroid
11 Retina
12 Optic nerve

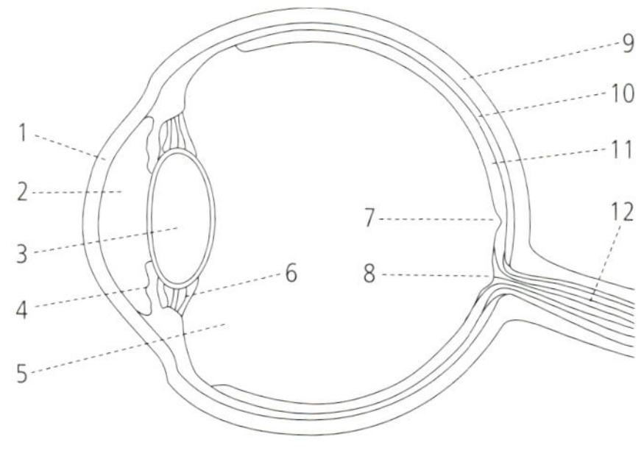

Figure 9.2
V (λ) curve

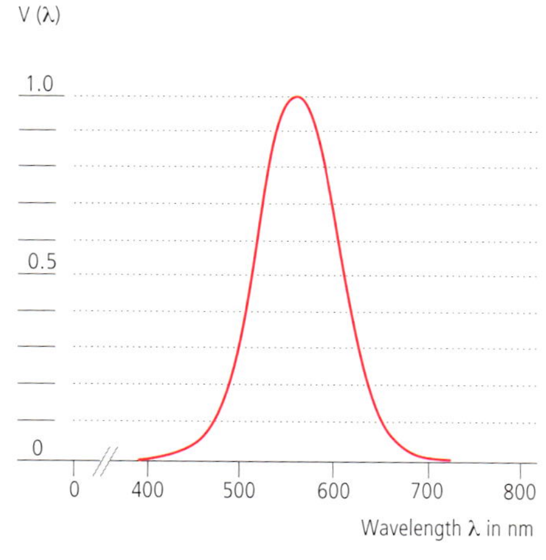

9.2 Lighting effectiveness criteria

9.2.1 General criteria

Light has played a major role for its symbolic content and quality in all civilizations. Ancient building masters considered the sun as a source of light in the design and construction of buildings, and in this sense they were also lighting planners. Working with daylight became too complex a field for architects alone only in the wake of technological advances and the advent of a wide range of lamps and luminaires.

Light is a creative tool, perhaps the most effective of all, for we perceive our environment through our eyes and our perception, in turn, is directly influenced by how objects are lit.

Lighting is used to facilitate work, to enhance colour, brightness and structure, to enlarge rooms, to promote intimacy, to attract attention, to soothe and to stimulate, to view and to enjoy.

As the most important information carrier, light supports our sensory impressions perceived by the eye, with 80 % of all information taken in by vision. Human beings experience their surroundings as a result of the interplay of light, material and space.

Light is therefore a relevant environmental factor with considerable influence on the vision and comfort of human beings. For a specific visual task the quality of a lighting installation is determined by the interplay of a variety of parameters, such as:
- Illuminance
- Luminance distribution
- Brightness ratio
- Visual environment
- Interplay of artificial light and daylight
- Light colour and colour rendering
- Shading and light direction
- Simultaneity
- Glare control

Good lighting technology creates an harmonious environment for the relevant task, providing the necessary prerequisites for seeing and recognition as well as accents for zones dedicated to relaxation.

Light planning is also planning the visual environment of human beings. Its principal aim is to create the conditions for perception that enable efficient working, safe orientation, comfort and aesthetics. Different lighting systems and lamps that cast different shades of light can be used to create a unique ambience in a room, capable of delivering all the parameters we have described above.

Figure 9.3 depicts a room with varying ambience created by changes in lighting and light hues.

When planning the lighting for any room, it is important to begin by studying the intended ambience or the range of different types of ambience required to meet all the demands of the illuminated room. Lamps, luminaires etc. are merely tools for creating such an ambience. Measured values provide little information about the quality of light; they merely define the technical parameters for the system.

Figure 9.3
Different mood settings in a room as a result of various lighting options

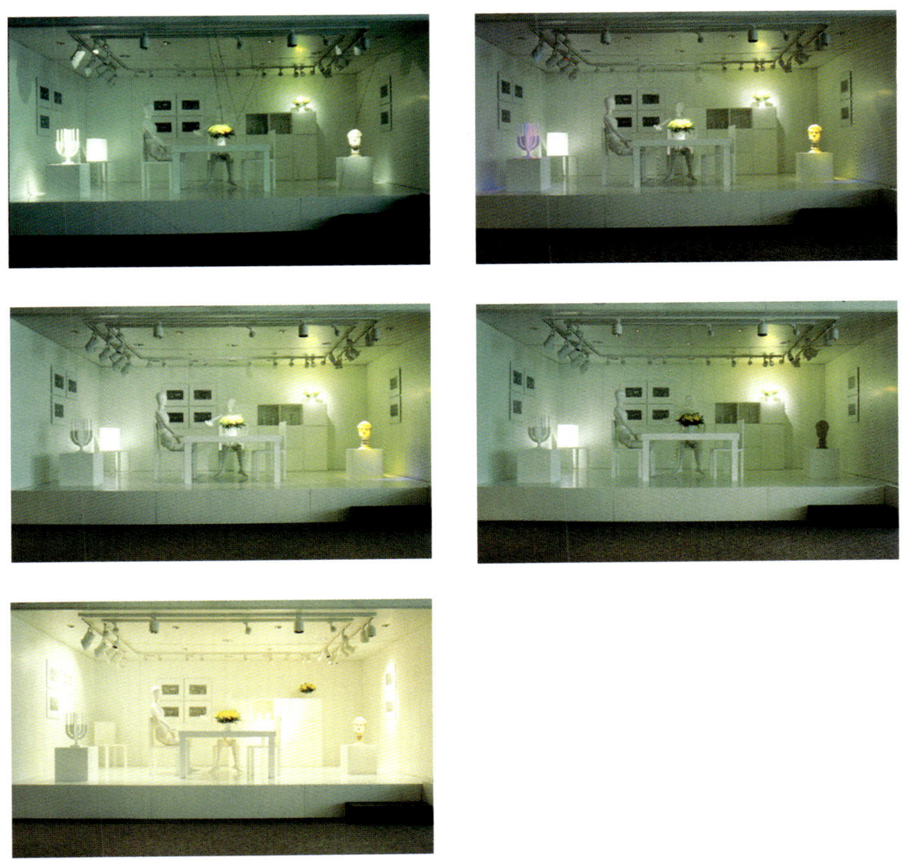

9.2.2 Terms – symbols – dimensions

Glossary
A comprehensive glossary of lighting terms and definitions is contained in the CIE publication No. 17 (International Lighting Vocabulary); the terms used below are broadly synonymous with the CIE list.

Absorptance (formerly absorption factor)
Ratio of the absorbed radiant energy or luminous flux to the incident flux i.e. incident luminous flux.

Absorption
Conversion of radiant energy to a different form of energy by interaction with matter.

Accommodation
Ability of the eye to adjust focus on an object at any distance to the eye to obtain visual acuity.

Adaptation
1. The process by which visual organ adapts to the luminances and colours in its visual range.
2. The final state of this process.

Ballast
Device used with discharge lamps for stabilizing the current in the discharge.

Note 1:
A differentiation is made between ohmic, inductive and capacitive ballast devices, as well as combinations of any of these and electronic ballast devices.

Note 2:
The ballast can also effect the ignition of a lamp, either on its own or in combination with a starting device. The starting device is sometimes built into the ballast device.

Black body (Planckian radiator)
Thermal radiator that fully absorbs all incident radiation, regardless of wavelength, direction of incidence or polarization.

This radiator has, for any wavelength, the maximum spectral concentration of radiant exitance at a given temperature.

Blended-light lamp
Lamp containing, in one bulb, a high-pressure mercury vapour discharge and an incandescent lamp filament connected in series.

The bulb may be diffusing or coated with a fluorescent material.

Brightness
The attribute of a visual sensation according to which an area appears to emit more or less light.

Note:
By definition, brightness is also an attribute of colour.

In British recommendations the term "brightness" is now reserved to describe the brightness of an object (or surface etc.) colour; "luminosity" should be used in all other instances.

Chromaticity diagram
Plane diagram showing the result of mixtures of colour stimuli, each chromaticity being represented unambiguously by a single point on the diagram.

Colour rendering
The effect of an illuminant on the colour appearance of objects illuminated by it, in conscious or subconscious comparison to the colour appearance of the same objects under a reference luminant, i.e. in general: the relationship between the colour rendered and the original colour.

Colour rendering index (R_a) of a light source
Measure of the degree to which the colour appearances of objects illuminated by the source conform to those of the same objects illuminated by a reference illuminant for specified conditions.

Colour stimulus (chromatic stimulus)
The physically defined radiant beam penetrating the eye and producing a sensation of colour.

Colour temperature
Temperature of the black body that emits radiation of the same chromaticity as the radiation considered.
Unit: Kelvin (K).

Contrast
Subjective assessment of the difference in appearance of two parts of a field of view seen simultaneously or successively (simultaneous contrast, successive contrast).

Correlated colour temperature
The temperature of the Planckian radiator whose perceived colour most closely resembles that of a given stimulus at the same brightness in a uniform colour table.
Unit: Kelvin (K)

Cosine law of incidence
The law that states that illuminance at a point on a plane is proportional to the cosine of the angle of light incidence (the angle between the direction of incident light and the normal to the plane).

$$E = \frac{I}{d^2} \cos \alpha$$

Diffused lighting
Lighting in which the light on the working plane or on an object is not incident predominantly from a particular direction.

Diffuse reflection
Diffusion by reflection in which, on the macroscopic scale, there is no regular reflection.

Diffuse transmission
Transmission in which, on the macroscopic scale, there is no regular transmission.

Direct lighting
Lighting by means of luminaires with a light distribution such that 90 to 100 % of the emitted luminous flux reaches the working plane directly, assuming that this plane is unbounded.

Direct flux on a surface
The luminous flux received by the surface direct from the luminaires of an installation.

Discharge lamp
Lamp in which the light is produced, directly or indirectly, by an electric discharge through a gas, a metal vapour or a mixture of several gases and vapours.

Discomfort glare
Type of glare that causes a sensation of discomfort without necessarily diminishing the capacity of the visual function.

Divergence of a beam (one-tenth peak divergence)
The angular extent in a specified plane in which the luminous intensity drops to half (1/2) (one peak divergence) or 1/10 (one-tenth peak divergence) of the maximum.

Drip-proof luminaire
Luminaire so constructed that, when mounted in its designed position, it is capable of withstanding drops of water falling in a substantially vertical direction.

Dust-proof luminaire
Luminaire so constructed that dust of a specified nature and fineness cannot enter it when it is used in a dust-laden atmosphere.

Dust-protected luminaire
Luminaire so constructed that it can withstand a certain amount of dust of a specified nature and fineness penetrating it when used in a dust-laden atmosphere without having an adverse effect on its safe operation.

Emergency lighting
Lighting provided for use when the supply to the normal lighting fails.

Even or uniform lighting
Lighting by means of luminaires with a light distribution such that 40 to 60 % of the emitted luminous flux reaches the working plane in a direct manner, assuming that this plane is unbounded.

Fluorescence
Photoluminescence that persists for an extremely short time after excitation.

Fluorescent lamp
Low-pressure discharge lamp in which the ultraviolet radiation is transformed into light by means of a fluorescent material on the inner surface of the outer bulb.

Frequency (f, v)
The frequency indicates the number of times a periodic event recurs during a specific unit of time.
Unit: Hertz (Hz): 1 Hz = 1 s^{-1}

Gas discharge
Conduction of electricity in gases and vapours due to movement of ions produced by collisions between electrons and gas molecules. The discharge of electromagnetic radiation effected by this process is harnessed in almost all areas of lighting technology.

Gas discharge lamp
Lamp in which the light is produced by an electric discharge through a gas (e.g. xenon, neon, helium, nitrogen or carbon dioxide lamps).

Note:
The expression "neon tube" is frequently used incorrectly as a generalization of other tubular discharge lamps.

General lighting
Uniform lighting of an area without provision for special zone requirements within the space.

Glare
Condition of vision in which there is discomfort or a reduction in visual ability due to an unsuitable distribution or range of luminance or to extreme contrasts of space or time.

Glow discharge
Discharge of electricity in which the secondary emission from the cathode (due to ion bombardment) is much higher than the thermal emission.

Halogen lamp
Gas-filled lamp containing a certain proportion of halogens.

Hue
Attribute of a visual sensation according to which an area appears to be similar to one of the perceived colours, such as blue, green, yellow, red and violet.

Illuminance (E) at a point on a surface
Quotient of luminous flux incident on an element or surface containing the point, and the area of that element.
Unit: lux (lx).

Incandescent (electric) lamp
Lamp in which light is produced by means of an element heated to incandescence by the passage of an electric current.

Increased safety luminaire (formerly flame-proof luminaire)
Enclosed luminaire that satisfies the appropriate regulations for use in situations where there is risk of explosion.

Indirect flux on a surface
The luminous flux received by the surface from a lighting installation after reflection from other surfaces.

Indirect lighting
Lighting by means of luminaires with a light distribution such that only 0 to 10 % of the emitted luminous flux reaches the working plane direct, assuming that this plane is unbounded.

Infrared radiation (IR)
[Optical] radiation for which the wavelengths are longer than those for visible radiation and smaller than approximately 1 mm.

Inverse square law
The law that states that the illuminance at a point on a plane perpendicular to the line joining the point and a source is inversely proportional to the square distance between the source and the plane:

$$E = \frac{I}{d^2}$$

Iso-intensity curve (Iso-candela curve)
Curve traced on an imaginary sphere with the source at its centre and joining all the points corresponding to those directions in which the luminous intensity is the same, or a plane projection of this curve.

Isoluminance curve
Locus of points on a surface where the luminance is the same, for given positions of the observer and of the given source(s) in relation to the surface.

Isolux curve
Locus of points on a surface where the illuminance has the same value.

Jet-proof luminaire
Luminaire constructed to withstand a direct jet of water from any direction.

Lamellae louvre
A louvre whose principal main shielding elements are in the form of straight strips of opaque or translucent material.

Lamp
Fixture for an artificial light source.

Light
Radiation capable of causing a visual sensation, i.e. visible radiation.

Light stimulus
Visible radiation entering the eye and producing a sensation of light.

Lighting with directed light
Type of lighting in which the working plane or an object is lit predominantly from one direction.

Localized lighting
Localized lighting in addition to general lighting, to provide sufficient illumination for a specific visual task at a workstation.

Louvre
Screen or series of lamellae louvres made of translucent or opaque components, whose geometry is designed to prevent lamps from being directly visible over a given angle.

Luminaire
Apparatus that distributes, filters or transforms the light given by lamps and which includes all the items necessary for fixing and protecting these lamps and for connecting them to the supply circuit.

Luminaire efficiency
The ratio of the luminous flux emitted by the luminaire to be measured under specific practical conditions and the sum of luminous flux emitted by the individual lamps inside the luminaire.

Luminance (L)
in a given direction, at a given point of a luminous or illuminated surface:

The quotient of the luminous flux transmitted by an elementary beam passing through the given point and propagating in the solid angle containing the direction, and the product of the solid angle, the area of a section of that beam containing the given point, and the angle between the normal to that section and the direction of the beam.
Unit: candela per square metre (cd/m^2).

Luminance contrast (C_L)
The relative difference in luminance between two parts of a visual field. Where the size of the two parts differs greatly, the contrast is defined as:

$$C_L = \frac{L_2 - L_1}{L_1}$$

where:
L_1 = luminance of the smallest part
L_2 = luminance of the greatest part

Where the size of the two parts is equal, the contrast is defined as:

$$C_L = \frac{L_2 - L_1}{L_2 + L_1} \text{ or } \frac{L_2}{L_1}$$

where:
L_1 = luminance of one part
L_2 = luminance of the other part

Luminance distribution curve
Curve representing the luminance of a luminaire in a vertical plane as a function of the angle from the nadir.

Luminous efficacy (of a source)
Quotient of the luminous flux emitted and the power consumed.
Unit: lumen per watt (lm/W).

Luminance factor (b)
at a point, of a non-self-radiating body, in a given direction, under specific conditions of illumination:

Ratio of the luminance of the body to that of a perfect reflecting or transmitting diffuser identically illuminated.

Luminous flux Φ
The output of light emitted by a luminous surface or received by an illuminated surface. It corresponds to quantity derived from radiant flux by evaluating the radiation according to its action upon the CIE standard photometric observer.
Unit: lumen (lm).

Luminous intensity (I)
of a source in a given direction:
Quotient of the luminous flux leaving the source, propagated in an element of solid angle containing the given direction and the element of solid angle.
Unit: candela (cd).

Note:
The luminous intensity of luminaires is normally given either in a luminous intensity diagram or in an isocandela diagram.

Luminous intensity diagram (Luminous intensity table)
Luminous intensity shown in the form of a polar diagram or table, in terms of candela per 1,000 lumens of luminous flux emitted by a lamp.

The diagram for non-symmetrical light distributions should give the light distribution of a luminaire in at least two planes:
In a vertical plane through the longitudinal axis of the luminaire and in a plane at right angles to that axis.

Note:
The luminous intensity diagram (luminous intensity table) can be used to provide a rough idea of the light distribution of the luminaire, to calculate the illuminance value beneath the luminaire and to calculate the luminance distribution of the luminaire.

Luminous intensity distribution curve of a lamp or luminaire
The curve, normally with polar coordinates, which plots the distribution of the luminous intensities on a plane bisecting the source as a function of the direction.

Maintenance factor
Ratio of the average illuminance on the working plant after a specific period of use of a lighting installation to the average illuminance obtained under the same conditions for a new installation.

Mercury vapour lamp
Discharge lamp in which light is chiefly produced in a mercury vapour.

Metal halide lamp
Discharge lamp in which the major portion of light is produced by the radiation from a mixture of a metallic vapour (e.g. mercury) and the products of the dissociation of halides (e.g. halides of thallium, indium and sodium).

Metal vapour lamp
Discharge lamp, e.g. the "metal halide lamp", the "mercury vapour lamp" and the "sodium vapour lamp", in which the light is chiefly produced in a metallic vapour.

Mixed reflection
Partly regular and partly diffuse reflection.

Monochromatic radiation
Electromagnetic radiation characterized by a single frequency or wavelength.
In practice, radiation of a very small range of frequencies, i.e. range of wavelengths that can be described by stating a single frequency or wavelength.

Operating efficiency of a luminaire
The ratio of the luminous flux emitted by the luminaire to be measured under specific practical conditions and the sum of luminous flux emitted by the individual lamps when the latter are operated outside of the luminaire under default conditions.

Pendant luminaire
Luminaire provided with a cord, chain, tube etc. for suspension from a ceiling or wall bracket.

Physiological glare
Type of glare, which results in diminished visual perception without necessarily causing discomfort.

Predominantly direct lighting
Type of lighting with luminaires, which emit 60 to 90 % of luminous flux directly onto a working plane assumed to be unbounded.

Radiant efficiency (η_e) of a source of radiation
Ratio of the radiant flux emitted to the power consumed.

Radiant energy (Q_e, Q)
Energy that is emitted, transferred or received in the form of radiation.
Unit: joule (J).

Radiant power (Φ_e, Φ)
Power that is emitted, transferred or received in the form of radiation.
Unit: watt (W).

Radiation
- Emission or transfer of energy in the form of electromagnetic waves or particles.
- These electromagnetic waves or particles.

Rainproof luminaire
Luminaire constructed to prevent the penetration of rain and for outdoor use.

Reference surface
Surface on which recommended illumance quotients are based or on which they are measured.

Reflectance (formerly reflection factor)
Ratio of the reflected radiant or luminous flux to the incident radiant or luminous flux.

Reflection
Return of radiation by a surface.

Reflector lamp
Lamp in which part of the bulb is coated with a reflecting material, either diffuse or specular.

Refraction
Change in the direction of propagation of radiation determined by change in the velocity of propagation in passing through an optically heterogeneous medium, or in passing from one medium to another.

Regular (or specular) reflection
Reflection, without diffusion, in accordance with the laws of optical reflection.

Room index
Parameter representing the geometry of a room, used in calculating the utilization factor or the utilance.

Note 1:
Unless otherwise indicated, the room index is given by the formula

$$\frac{l \cdot b}{h\,(l + b)}$$

where:
l is the length of the room, b is the width and h is the distance of the luminaires above the working plane.

Note 2:
In British practice, the "ceiling cavity index" is calculated from the same formula except that h represents the distance between the ceiling to the luminaires.

Safety lighting
Lighting provided to ensure sufficient light and hence safe access to exits when the general or standard lighting fails.

Saturation
Attribute of visual sensation that permits a judgement to be made of the proportion of pure chromatic colour in the total sensation.

Screen
That part of a luminaire designed to prevent direct visual contact with lamps over a given range of angles.

Sodium vapour lamp
Discharge lamp in which the light is chiefly produced in a sodium vapour.

Solid angle (Ω)
The cone whose point is located at the centre of a sphere and terminated by a calotte of the spherical surface, whose area is numerically equal to the square of the radius.
Unit: steradian (sr).

Spectral distribution
The spectral concentration of a radiation measured in terms of the relative spectral distribution of an arbitrary photometric quantity (luminous flux, luminous intensity etc.).

Spectral luminous efficiency V (λ) for monochromatic radiation at wavelength λ:

Ratio of radiant flux at wavelength (λ_m) to the radiant flux at wavelength λ so that both types of radiation give the same relative sensitivity under specific photometric conditions.

(λ_m) must be selected to ensure that the maximum of this ratio equals 1.

Specular reflector
That part of a luminaire designed to reflect the luminous flux of the lamps in required directions by means of specular reflection.

Starter
Device for starting a discharge lamp (in particular a fluorescent lamp) that provides for the necessary preheating of the electrodes and/or causes a voltage surge in combination with the ballast.

Starting device
Device that provides the conditions required for starting a discharge lamp.

Suspension height
Height of luminaires above reference surface.

Thermal radiation
Radiation emitted by the thermal agitation of the particles of matter (atoms, molecules, ions).

Transmission
Passage of radiation through a medium without change of frequency.

Transmittance (formerly transmission factor)
Ratio of the transmitted radiant or luminous flux to the incident radiant or flux.

Ultraviolet radiation (UV radiation)
Radiation where the wavelengths of the monochromatic components are shorter than those for visible radiation and longer than 100 nm.

Uniform-chromaticity-scale (UCS) diagram;
UCS diagram
A two-dimensional diagram in which the coordinates are defined so that equal distances represent as closely as possible equal steps of colour discrimination of colour stimuli of the same luminance throughout the diagram.

Uniformity ratio of illuminance
The ratio of illuminance on a given plane expressed as either
- the ratio of the minimum to the maximum illuminance
- the ratio of the minimum to the average illuminance

Note:
In some countries, the reciprocal of these ratios is used, which is greater than 1.

Upper (lower) hemispherical luminous flux of a source
The luminous flux emitted above (below) a horizontal plane passing through the source.

Utilance
Ratio of the utilized flux to the luminous flux emitted by the luminaires (see also utilization factor).

Utilization factor
Ratio of the utilized flux to the luminous flux emitted by lamps.

Utilized flux
Luminous flux received on the reference surface under consideration.

Note:
Unless otherwise indicated, the reference surface is the working plane.

Visual field
The total angular extent of the space in which an object can be perceived when the eye(s) regard(s) an object directly ahead. The yield can be monocular or binocular.

Watertight luminaire
Luminaire constructed to withstand the penetration of water when immersed to a specified depth, but not intended for permanent use under water.

Note:
The term "submersible luminaire" applies to a luminaire constructed to withstand submersion in water to a specified depth indefinitely.

Wavelength (λ)
Distance in direction of propagation of a periodic wave between two successive points at which the phase is the same (at the same time).
Unit: metre (m).

Working plane
Reference surface defined as the plane on which work is usually done.

Note:
For interior lighting, unless otherwise specified, this plane is assumed to be horizontal to the floor at a height of 0.85 m and bounded by the walls of the room.

Zoning for a specific reference surface within the room is permitted.

Table 9.1 gives additional symbols, formulae and units for photometric and radiation parameters.

Parameter	Symbol	Equation	Unit	Abbre-viation
Light volume	Q	$Q = \int \Phi\, dt$	Lumen/hour Lumen/second	lm h lm s
Luminous flux	Φ	$\Phi = dQ/dt$	Lumen	lm
Illuminance	E	$E = d\Phi/dA$	Lux	lx
Luminous intensity	I	$I = d\Phi/d\Omega$	Candela	cd
Luminance	L	$L = dI/dA \cos \alpha$	Candela/m^2	cd/m^2
Photometric radiant equivalent	K	$K = \Phi/\Phi_e$	Lumen per Watt	lm/W
Radiant energy, radiant volume	Q		Joule, Watt second, Kilowatt hour	J/s, Ws kWh
Radiant power	Φ	$\Phi = dQ/dt$	Watt, Joule per second	W J/s
Radiant intensity	E	$E = d\Phi/dA$	Watt/m^2	W/m^2
Radiance	I	$I = d\Phi/d\Omega$	Watt per steradian (sr)	W/sr
Spectral energy	L	$L = dI/dA \cos \alpha$	Watt per steradian (sr) and m^2	W/sr m^2
Absorptance		$a = d\Phi_a/d\Phi_o$		
Reflectance		$P = d\Phi_r/d\Phi_o$		
Transmittance		$t = d\Phi_t/d\Phi_o$		

A = reference surface
Ω = solid angle, into which the luminous flux of a light source is radiated
α = angle between direction of perception and standard surface in relation to observed surface

Ω = solid angle, into which the luminous flux of a light source is radiated
α = angle between direction of perception and standard surface in relation to observed surface

Table 9.1
Equations, symbols, relationships and units for photometric and radiant parameters

9.2.3 Perception

Relaxation, increased performance and concentration on the task at hand are possible only when the perceptive faculties in the human brain can operate without interference. Incorrect distribution of luminance in a room, glare, poor colour rendering, and inappropriate interior design all inhibit perception. On the other hand, when perception is in no way inhibited, the basic senses of the eye, such vision, speed and contrast sensitivity, are optimized. Optimum working conditions can be achieved by harmonizing the luminance ratio at the workstation (interior field luminance) and the luminance ratio in the working environment (surrounding luminance).

In traffic areas, recognizability of the human face and signage is a criterion of appropriate lighting.

At workstations where vision is especially important, for example, office workstations with PCs, contrast sensitivity should be the principal criterion for creating a stable range of perception.

Figure 9.4 plots contrast sensitivity in dependence on internal luminance. As the graph clearly shows, fluctuations in contrast sensitivity are markedly higher in the range of low interior field luminance than in the range of high interior field luminance. Hence the interior field luminance at the work place should be defined so as to satisfy as many demands as possible and to ensure perception with consideration of the specific task within a stable range of contrast sensitivity (see example).

As **Figure 9.4** shows, the fluctuation of contrast sensitivity is drastically reduced in the indicated range (only 7 %). The reduction of illuminance to 300 lux (standard) for an office with daylight oriented workstations, introduced primarily for energy efficiency, is fully compatible with the functions and perception of the human eye.

Figure 9.5 identifies the range of stable and unstable perception in dependence on surrounding (L_s in cd/m^2) and interior field luminance (L_i in cd/m^2). The graph shows that surrounding luminance should lie between 1/10 and 2/3 of interior field luminance; moreover, it should never exceed the interior field luminance. This is difficult to achieve, however, with luminaires that are part of the surroundings (source of surrounding luminance) and near windows.

Windows without glare protection can yield a luminance of up to several 1,000 cd/m^2. At workstations in private offices, high luminance near the window is less problematic than in large open-plan offices. Moreover, there are other factors against lowering the surrounding luminance to the proportionate range of interior field luminance in the design of windows (for example, reducing the transmittance of light, visual contact from the inside out and the outside in etc.).

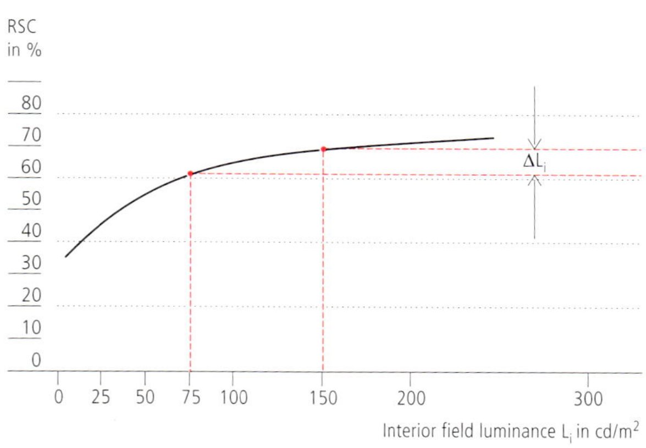

Figure 9.4
Contrast sensitivity (% of RSC) in dependence of interior field luminance (L_i in cd/m^2)

ΔL_i approx. 75 cd/m^2 / ΔRCS approx. 7 %

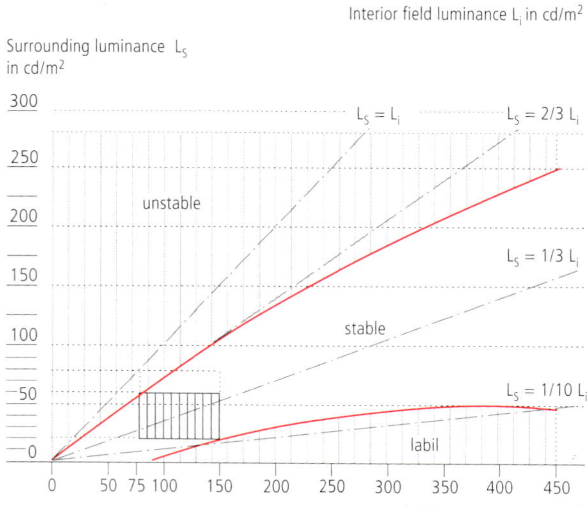

Figure 9.5
Stable and unstable range of perception for surrounding luminance L_s in dependence on interior field luminance L_i

9.2.4 Illuminance – luminance

The illuminance E is the quotient of luminous flux incident on a surface and the area of that element. Illuminance (lux, lx) is, as **Figure 9.6** illustrates, imperceptible by the human eye and therefore only a theoretical parameter. Light is perceptible only when it is cast onto material and reflected by it. The luminance L in cd/m^2 is the light evaluated by the human eye. With diffuse reflecting materials, it is dependent on the reflectance of the material (ρ).

The eye is stimulated by luminance and not by illuminance.
The relationship between luminance and illuminance is expressed as

$$L = E \cdot \frac{\rho}{\pi}$$

L = luminance (cd/m^2)
E = illuminance (lux)
ρ = reflection factor (rho)
π = constant 3.14.

Several additional terms are also used in practice:
Mean illuminance E is the arithmetic value of the illuminance in a room or zone.

Standard maintained illuminance E_n is the nominal value of the mean illuminance in the room or in a zone within the room dedicated to a specific task, and for which a lighting system has to be designed.

Cylindrical illuminance E_c is the mathematical mean of the vertical illuminance factors E_v at a specific point.

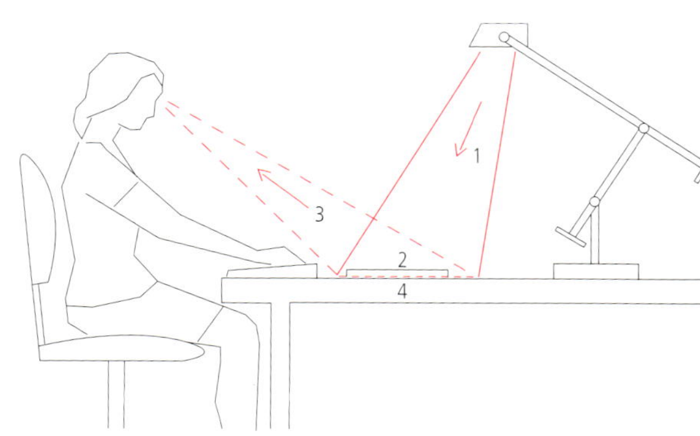

Figure 9.6
Relationship between luminous flux, illuminance and luminance (The human eye is stimulated by luminance, not illuminance)

1 Luminous flux (lumen)
2 Illuminance (lux)
3 Luminance (candela)
4 Reflectance

9.2.5 Glare
9.2.5.1 Direct and indirect glare

Glare is understood as visual discomfort caused by excessive luminance, i.e. excessive differences in brightness. It is important to remember that the human eye is capable of perception in a wide range (**Figure 9.7**) and that this perception is described as the visual range. In particular, the human eye is notable for its ability to adapt to different lighting conditions. Thus we can see our environment in sunlight and in moonlight, although the brightness of the two differs by a factor of 105. The performance of the human eye spans an even greater range – a faint, glowing star in the nocturnal sky is still perceived, although the illuminance that reaches the eye is no more than 10 – 12 lux. This adaptability is achieved only to a very small degree by the pupil, which translates incident light at an approximate ratio of 1:16. The principal effort necessary for adaptation is performed by the retina.

Although vision is possible across a wide range of luminance, the boundaries for contrast perception are far more narrowly defined for each individual lighting scenario. This is due to the fact that the human eye cannot evaluate the full range of visible luminance at the same time; instead it adapts to one specific, narrowly defined partial range, within which it is capable of differentiated perception. Objects whose luminance is too high for a specific state of adaptation, create glare and appear unfocussed and overly bright. Objects whose luminance is too low appear unfocussed and too dark. While the eye is able to adapt to new luminance conditions, it can only do so by selecting one new, equally limited partial range. The process of adaptation takes some time, and adaptation to brighter situations is more rapid than adaptation to dark conditions.

With regard to direct and indirect glare, a differentiation is made between:
- Physiological glare, which leads to reduced visibility, contrast sensitivity and shape recognition
- Psychological glare, evaluated solely on the basis of the visual discomfort it causes

Exposure to physiological glare in a room over a longer period of time leads to fatigue, which diminishes the ability to concentrate and interferes with performance. Direct glare is chiefly caused by luminaires and lamps, incident sunlight or the sky. Reflections of light at the workstation or from work materials that interfere with the task are called reflective glare and lead to diminished conditions for vision and contrast.

The following measures can be employed to limit glare:
- Covering the bottom of the luminaire or lamp
- Blending the direct and indirect components in lighting systems
- Indirect radiation via metallic ceiling components or reflectors

Reflective glare, resulting in the kind of interference illustrated in **Figures 9.8/9.9**, can largely be avoided by adjusting the placement of luminaires in relation to the workstation and by using matt, non-reflective surfaces at the workstation (e.g. keyboards, desk surface, paper etc.).

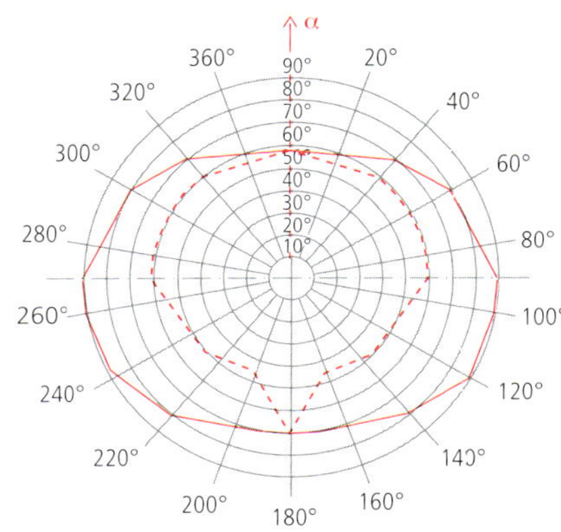

Figure 9.7
Field of vision of the human eye (binocular vision)

—— One-eyed
- - - Binocular
α Angle of vision

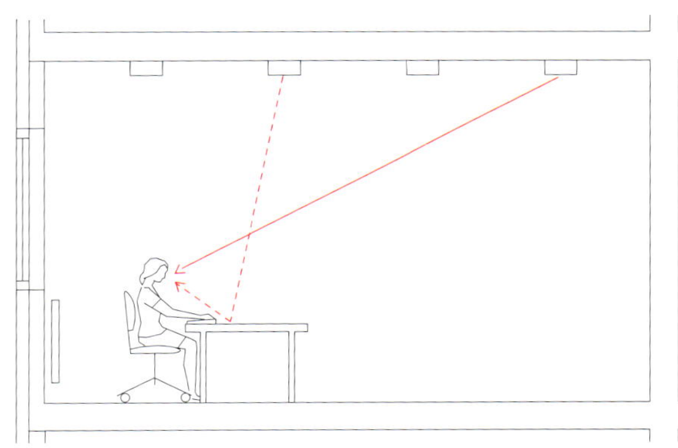

Figure 9.8
Direct glare/indirect glare interferes with visual performance and comfort

Figure 9.9
Reflections interfere with contrast vision

9.2.5.2 Daylight glare

As mentioned, interiors should contain only limited areas where luminance is higher than 400 to 500 cd/m². The exterior and peripheral zones near windows are often characterized by a luminance that far exceeds desirable levels for the interior. For this reason, luminance-reducing elements are frequently needed to limit glare at the window, unless other external features naturally contribute to reducing glare.

The luminance of an overcast sky can reach values of several thousand cd/m² (up to roughly 6,000 cd/m²). Even bright walls on other buildings that lie within the field of vision can interfere with perception as a result of their excessive luminance (up to roughly 70,000 cd/m²). When designing workstation layouts it is therefore essential to take measures to limit luminance differences causing increased brightness ratios as the result of the placement of window openings and the built environment.

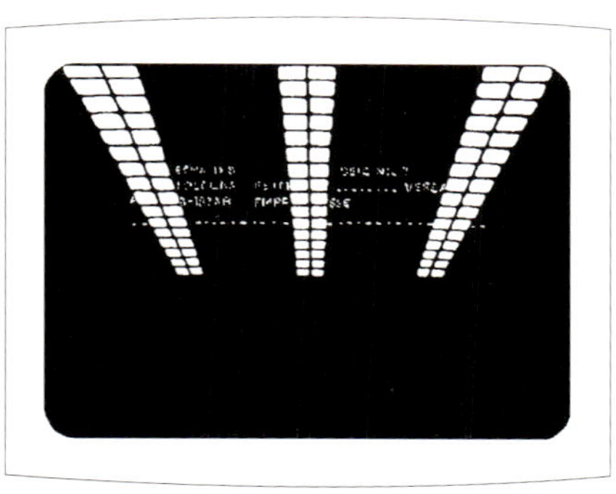

Figure 9.10
White overhead lighting strips reflected on screen

Figure 9.11
Preventing undesired reflections on screens

9.2.5.3 Glare near monitors

Glare is a frequent occurrence on or near monitors as a result of overlapping luminous surfaces nearby, such as luminaires and windows.

Monitors have been greatly improved in recent years with regard to surface contrast, and a luminance of 300 to 400 cd/m² in the immediate surroundings, which are reflected in the monitor, no longer creates too much discomfort. When the luminance is even greater in the vicinity of the monitor, there may be overlapping reflections and the clarity of what is on the screen may be greatly reduced (**Figure 9.10**). Monitors with a dark background and bright characters are more susceptible to this effect than those with a bright background and dark characters.

Reflective glare originating from keyboards and glossy areas on the monitors is rare for modern monitors and screens, since manufacturers have addressed this problem and are producing matt keyboards. Horizontal monitors and screens may well be introduced in the near future and this will add yet another factor to consider.

Visibility of screens and monitors may be complicated by luminaires and high luminance near the window, as well as by very bright walls or very bright clothing, on which artificial or daylight falls.

Incorrect placement of a monitor workstation (**Figure 9.11**) can be as great a factor contributing to fatigue and diminished concentration as poor lighting. While visibility increases with higher illuminance in general, it is reduced at the monitor (**Figure 9.12**).

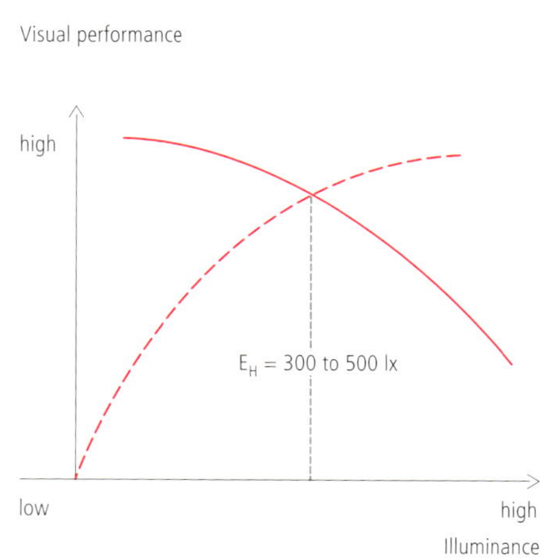

Figure 9.12
Deteriorating visual performance as a result of increased illuminance

— Screen
--- Document

9.2.6 Hue – colour temperature

Hue defines the colour perception of light. The hues relevant to lighting are divided into three groups:

Warm white (ww)
Colour temperature below 3,300 K
This hue is produced by incandescent lamps, halogen lamps, fluorescent lamps and discharge lamps with very high red content.

Neutral white (nw)
Colour temperature 3,300 to 5,000 K
This colour temperature, predominantly 4,000 K, is produced by fluorescent lamps used as secondary lighting for daylighting and gas discharge lamps (in interiors chiefly by metal halide lamps).

Daylight white (dw)
Colour temperature above 5,000 K
This hue is chiefly produced by special fluorescent lamps and gas discharge lamps.

In daylight, a low colour temperature is experienced as comfortable at dawn and dusk and a high colour temperature as equally comfortable during the day when illuminance is very high. The relationship between colour temperature and light spectrum is illustrated in **Figure 9.13**.

The relationship is also evident in artificial lighting conditions. When illuminance is low, incandescent lamps and halogen lamps with a colour temperature below 3,000 K are experienced as comfortable. In offices or exhibition areas, where illuminance is higher, whiter hues with a higher blue component and a colour temperature of approx. 4,000 K are preferred. Fluorescent lamps and discharge lamps with daylight white hues are the best choice for very high artificial illuminance. **Figure 9.14** indicates the comfort range in dependence on illuminance and colour temperature (after Kruithoff).

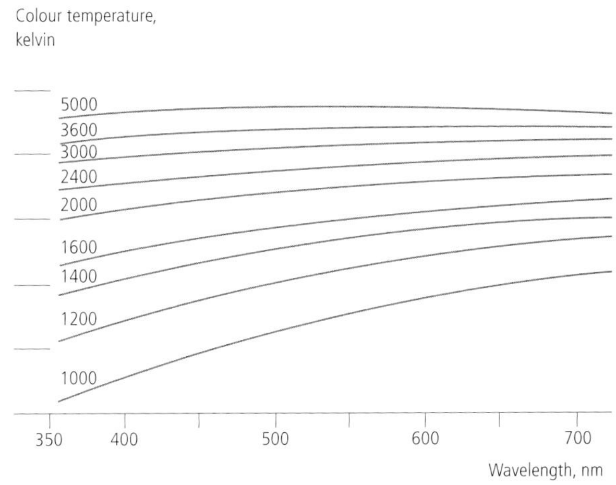

Figure 9.13
The spectral composition of the radiation emitted by a standardized projector (black projector). As the graph illustrates, long wavelengths, that is the yellow-red spectrum, are dominant for a colour temperature of 1,000 K.

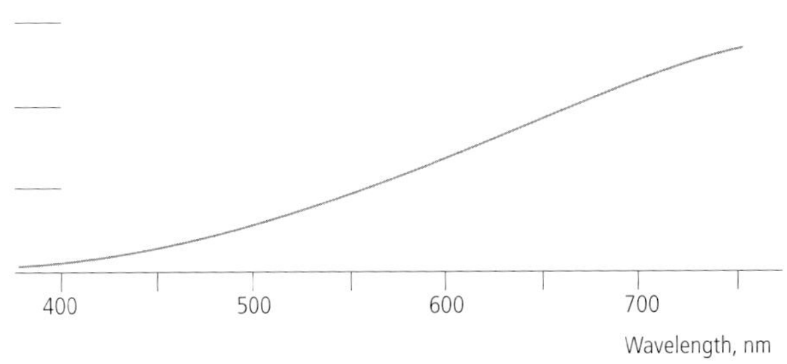

Spectral composition of light emitted by an incandescent lamp (2,700 K)

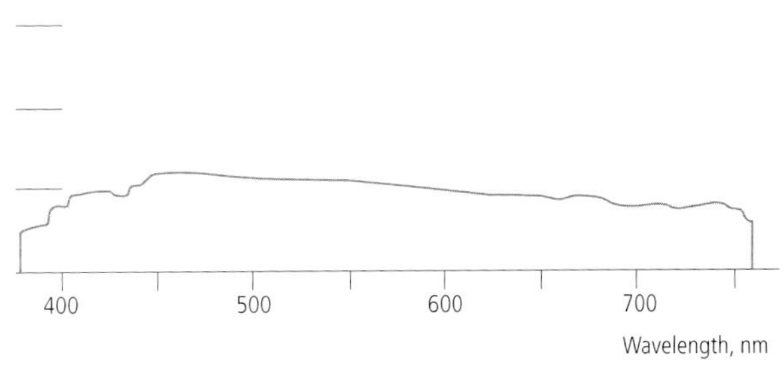

Balanced daylight spectrum for 5,000 K

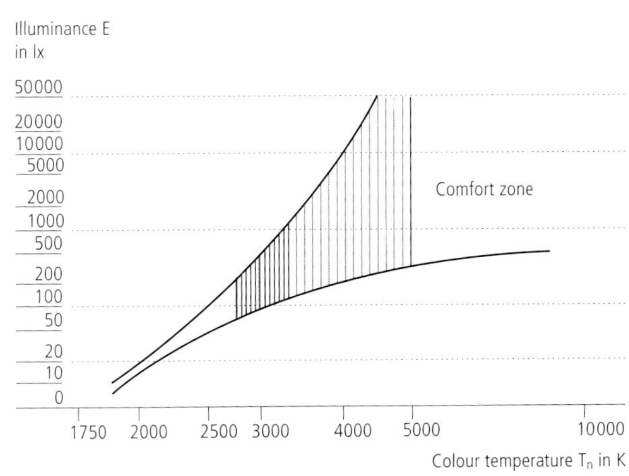

Figure 9.14
Comfort zone according to Kruithoff, colour temperature T_n in dependence on illuminance

— Incandescent bulb
— Halogen bulb
||||||| Fluorescent tube ww
|||||| Fluorescent tube nw

Incandescent lamps

High pressure sodium lamps
White son

Fluorescent lamps, warm white
"DE LUXE"

Metal halide lamps
HGI-WDL

Figure 9.15
Examples of colour rendering resulting from different light sources with corresponding spectrums
(Regent, Basel)

9.2.7 Light spectrum – colour rendering

For artificial lighting, the requirements with regard to colour rendering are dependent on the specific task. Since lamps emit different light spectrums they are also characterized by different colour rendering properties. Lamps with continuous spectrums generally achieve better colour rendering than lamps, whose spectrums feature peaks in the visible range.

The colour rendering properties of specific lamps are expressed in the colour rendering index R_a and by special colour rendering values in indices.

According to specific lighting tasks, the index for colour rendition properties in conjunction is based on the following categories:

R_a types (colour rendering index)

1 A	R_a 100 – 90
1 B	R_a 89 – 90
2 A	R_a 79 – 70
2 B	R_a 69 – 60
3	R_a 59 – 40
4	R_a < 40

Identical colours have an R_a = 100. Colour shifts are expressed in values < 100 in different classifications (types).

The indices are evaluated by comparing then with different test colours. These test colours represent only a small selection of all existing colours; as **Figure 9.15** demonstrates one should always select the colour of a material by exposing it to the light that will illuminate it.

Lamps with continuous spectrum are preferable to lamps with non-continuous spectrum because colour nuances are more evident in a continuous spectrum.

Classifying lamps in accordance with standardized colour rendering properties has led to the development of lamps with non-continuous spectrums and high light efficiencies, which are often poorly equipped to provide good rendering of certain hues. These non-continuous spectrums may also have a negative impact on human comfort (see light spectrums in **Figure 9.15**).

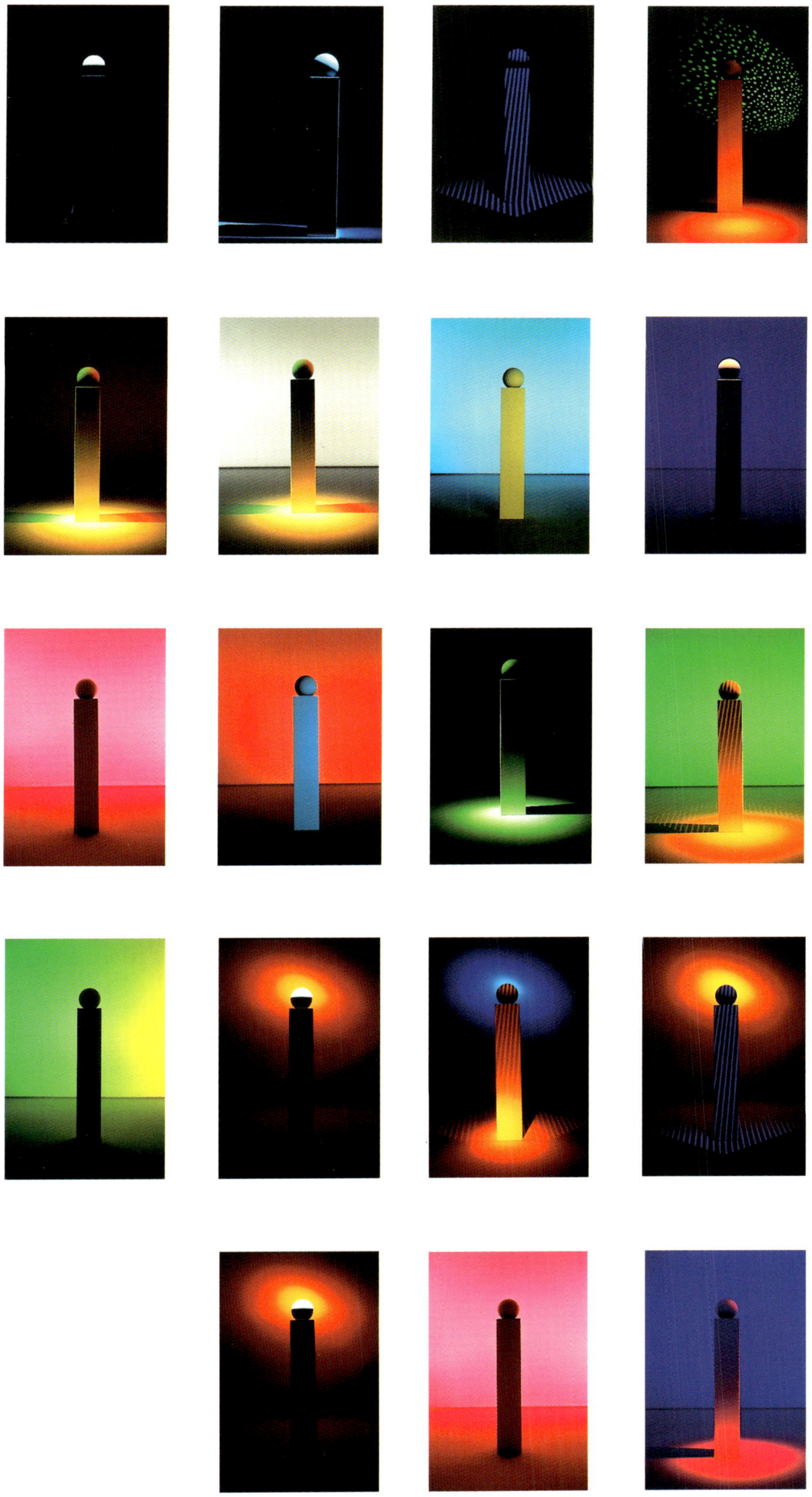

9.2.8 Light direction and shadowing

Sufficient shadowing increases the perception of illuminated bodies and surface structures by enhancing their plasticity. Lighting systems that are too diffuse with predominantly indirect light cast hardly any shadows, which can be experienced as monotonous and even unpleasant. A balanced ratio of direct and indirect lighting contributes to creating light and shadow of an intensity that is perceived as agreeable. **Figure 9.16** demonstrates these relationships on a single object lit with a wide variety of lighting options, creating various moods.

Depending on the task and the ambience, interior lighting can be flat (few shadows) or plastic (more shadows). Excessive shadowing can be avoided with the appropriate placement of several luminaires, in (considering the radiant property of the luminaires) and through reflective surrounding surfaces.

Shadowing can be evaluated by means of the ratio of the cylindrical illuminance (E_c) to the horizontal illuminance (E_h). The ratio of E_c to E_h should not fall below 0.3 if harsh shadows are to be avoided.

Figure 9.16
Creating a variety of moods by using different types of lighting, colours and shadow effects (Erco)

Recommendations	Nominal illuminance E_n (lux)	Colour temperature groups	Colour rendering index	Glare index
Office:				
Reception, telephone switchboard, simple tasks	300	nw, ww	1, 2, 3	2
General tasks, data processing	500	nw	1, 2, 3	1, 2
Drawing (drafting), punched-card processing	1,000	nw	1, 2, 3	1
Open-plan offices	1,000	nw	1, 2, 3	1
Meeting rooms	500	nw, ww	1, 2, 3	2
Retail/sales areas:				
Salesrooms	300 – 500	nw, ww	1, 2, 3	2
Department stores	500 – 700	nw	1, 2, 3	2
Self-service areas	750 – 1,000	nw	1, 2, 3	2
Shop windows, display cases	> 1,000	dw, nw, ww	1, 2	–
Exhibitions, museums, libraries:				
General exhibition rooms, collections, stacks, reading room	300	nw, ww	1, 2	1, 2
Reading stations	500	nw, ww	1, 2	1, 2
Apartments:				
Utility rooms	100	nw, ww	1, 2, 3	–
Laundry room	200	nw, ww	1, 2, 3	–
Kitchen	300	nw, ww	1, 2, 3	–
Bathroom	300	nw, ww	1, 2, 3	–
Reading, writing, manual tasks	500	nw, ww	1, 2, 3	–
Hospitality:				
Entrance, reception	300	nw, ww	1, 2, 3	–
Kitchen, laundry, linen	500	nw, ww	1, 2, 3	2
Restaurant, dining rooms	200	nw, ww	1, 2, 3	–
Self-service areas	500	nw, ww	1, 2, 3	2
Buffet, office	500	nw, ww	1, 2, 3	2
Reading, writing, manual tasks	500	nw, ww	1, 2, 3	1, 2
Bathroom	300	nw, ww	1, 2	–
Meeting rooms	500	nw, ww	1, 2, 3	2
Theatres, concert halls, cinemas:				
Entrance, lobby, cloakroom	200	nw, ww	1, 2	–
Box office	300	nw, ww	1, 2	–
Rehearsal rooms, dressing rooms,	200	nw, ww	1, 2	2
auditorium during intermission	up to 200	nw, ww	1, 2	–
Music stands	500	nw, ww	1, 2, 3	–
Hospitals, clinics:				
Waiting rooms	300	nw, ww	2	2
Utility rooms	500	nw, ww	2	2
Examination and treatment rooms	1,000	dw, nw	1, 2	1
Lab, surgery preparation	1,000	dw, nw	1, 2	2
Operating theatre	1,000	dw, nw	1	1
Operating area (e.g. table)	10,000	dw, nw	1	1
Patient rooms, general lighting	100	nw	2	1
Schools:				
Classrooms, day school	300	nw, ww	2, 3	1, 2
Classrooms, night school	500	nw, ww	2, 3	1, 2
Auditoriums, chemistry and physics labs	500	nw, ww	2, 3	1, 2
Teachers' rooms, offices	500	nw, ww	2, 3	2
Circulation areas, hallways, stairs	100	–	2, 3	–
Changing, washing and shower facilities	100	–	2, 3	–
Gyms, pools	300	dw, nw, ww	2, 3	2
Steel factories, machines, appliances:				
Machine and assembly work				
rough	300	nw, ww	2, 3	3
medium	500	nw, ww	2, 3	2
fine	750	dw, nw	1, 2, 3	2
precision	1,000	dw, nw	1, 2, 3	1, 2
Adjusting, testing, calibrating	1,000	dw, nw	1, 2, 3	1, 2
Electronic products:				
General tasks	300	nw, ww	2, 3	2
Assembly tasks				
rough	300	nw, ww	2, 3	2, 3
medium	500	nw, ww	2, 3	2, 3
fine	750	dw, nw	1, 2, 3	2
precision	1,000	dw, nw	1, 2, 3	1, 2
Adjusting, testing, calibrating	1,000	dw, nw	1, 2, 3	1, 2
Instruments, clocks:				
Machine and assembly work				
fine	1,000	dw, nw	1, 2, 3	1
precision	1,500 +	dw, nw	1, 2, 3	1
Adjusting, testing, calibrating	1,500 +	dw, nw	1, 2, 3	1
Surface treatment:				
Sandblasting	200	–	–	–
Cleansing and pickling baths	200	–	–	2, 3
Electroplating, hot galvanizing, painting, polishing	300	nw, ww	2, 3	2
Colour mixing, fine polishing, laboratory	1,000	dw, nw	1, 2	1, 2

Table 9.2
Recommended values for the design of lighting systems

9.2.9 Other lighting criteria

Other criteria may come into play when lighting systems are planned for rooms with specific requirements or demands for a unique ambience. The principal criterion in mood-setting rooms is the interplay of hues in dependence on the surrounding materials and lighting zones.

In carpentry shops, metal workshops and work environments with rapidly rotating parts, the prevention of stroboscopic effects can be another important factor. Stroboscopic effects are the result of fluctuations in luminous flux produced by the alternating flux in observing moving parts. They can impede vision or even create deceptive images (a rapidly turning saw blade may appear to be standing still).

In some rooms, generous amounts of daylight are essential; for specific uses, emitting certain spectrums is very important, e.g. for degerminating (or bactericidal) lamps, plant-growing lamps etc.

The following tables give recommendations for illuminance, hue, colour rendering and glare protection as a planning guide (source: Regent, Basel). The information in **Table 9.2** is organized by room functions; **Table 9.3** (Osram lamps) provides additional information on the selection of lamps for a wide variety of room types.

Table 9.3
Application example of fluorescent lamps with various colour temperatures

Application		**Full spectrum lamps**	**dw daylight white**		**nw neutral white**				**ww warm white**			**Special colour temperatures**	
		965	860	950	640	840	940	740	630	830	930	827	(76)
Office and administration:	Offices, corridors	●				●				●			
	Meeting rooms, boardrooms									●		●	
Industry, crafts and trades:	Electronics					●							
	Textile manufacturing	●	●	●		●	●						
	Woodworking	●	●	●		●							
	Steel and rolling mills				●				●				
	Graphic trades, laboratory	●	●	●		●							
	Colour testing	●		●			●						
	Warehouse, shipping and receiving					●							
School and teaching rooms:	Auditoriums, classrooms, kindergartens	●				●				●		●	
	Library, reading room	●				●				●		●	
Sales rooms:	Groceries, general		●			●				●		●	
	Baked goods											●	
	Refrigerated goods (counters and chests)		●										
	Cheese, fruit, vegetables											●	
	Fish											●	
	Meat, sausages, cold cuts												●
	Textiles, leather goods	●	●	●			●			●	●	●	
	Furniture, rugs									●	●	●	
	Sporting goods, toys, paper and stationery					●	●			●	●		
	Photography, clocks, watches, jewellery	●	●	●						●	●		
	Cosmetics, hair salon										●	●	
	Flowers	●		●							●	●	
	Department stores, supermarkets	●				●	●			●	●	●	
Entertainment and hospitality areas:	Restaurants, pubs, hotels											●	
	Theatres, concert halls, lobbies, museums											●	
Event areas:	Exhibition and trade fair halls					●				●			
	Sports and multipurpose facilities					●							
	Galleries		●	●							●	●	
Clinic and doctor's office:	Diagnosis and treatment	●	●	●									
	Patient rooms, waiting rooms	●									●	●	
Apartment:	Living rooms										●	●	
	Kitchen, bathroom, recreation, basement					●	●				●	●	
Outdoor illumination:	Streets, paths, pedestrian zones							●	●				

9.3 Theoretical luminance model

9.3.1 General information

As mentioned in the section on perception, the following sensory faculties of the human eye are optimized by taking the specific characteristics of human visual perception into consideration:
- Visual performance
- Speed of perception
- Contrast sensitivity

While dysfunctional perception caused by glare from lamps, windows and reflections is compensated for by the human brain, it does cause fatigue, stress and increased absenteeism.

If the reflectance of room materials and surface structures is known, and the radiant characteristics of a lamp are equally known, the luminance in a room can be modelled in advance to verify whether the combination of material and light will create an ambience that promotes optimized visual performance.

9.3.2 Surface structure and reflectance

When light falls onto a surface, some of the radiation is returned (reflected) by the surface. The remaining incident light is absorbed. Reflectance is the ratio of the reflected radiant flux to the incident radiant flux. **Table 9.4** indicates the reflectance for different surfaces, complemented by the relevant luminance values in **Table 9.5**.

In transparent and partially transparent materials (glass pane), both the outer and the inner glass surface reflect a fraction of the incident light. The more acute the angle of incidence, the more light is reflected.

Surface	Reflectance
Concrete	
light	0.5
dark	0.2
Granite, gneiss	0.2
White marble	0.6
Cement clinker	
light	0.5
dark	0.2
Bricks	
yellow	0.3 – 0.4
red	0.2 – 0.3
Wall panels, glazed	
white	0.8
light colour	0.6
dark colour	0.2 – 0.4
black	0.2
"Perfecta" panels, unfinished	0.2 – 0.3
Asbestos cement (eternit)	
light	0.35
dark	0.2
Lime plaster	
rough	0.5 – 0.6
smooth	0.4
White gypsum	
new	0.8
old	0.6
Ceiling panel, gypsum	
light	0.6 – 0.8
white, perforated	0.5 – 0.7
Mineral fibre slabs, light	0.6
Floor slabs, plastic	
light grey	0.4
dark grey	0.2
Textile floor covering	
light	0.3 – 0.5
medium	0.2
dark	0.1
Windows (double glazed)	
Standard glass	0.2
with white louvred stores	0.5
Antisun glass	0.25
Wood	
Birch, light maple, untreated	0.6
Fir, untreated, new	0.5
Fir, untreated, old	0.3
Wood, stained	
light	0.4
dark	0.15
Coatings	
Ceiling white, new	0.85
Ceiling white, light	0.7 – 0.8
Dispersion paint white	0.8
Aluminium bronze	0.7
Oil and synthetic resin paint	
white	0.85
light buff	0.65
light yellow	0.6
yellow	0.4
dark yellow	0.15
light blue	0.4
blue	0.2
dark blue	0.1
light green	0.7
green	0.5
dark green	0.2
light red	0.3
red	0.2
dark red	0.1
beige	0.4
light brown	0.2
dark brown	0.1
light grey	0.6
grey	0.3
dark grey	0.15
black	0.03 – 0.1

Table 9.4
Reflectance values for different surfaces

to Table 9.4
The figures under reflectance indicate recommended values for p_1, p_2 and p_3 for standard white fluorescent light of a 4,200 to 4,500 K colour temperature.

Deluxe lamps yield slightly higher reflectance coefficients for red colours and slightly lower reflectance coefficients for green colours.

Warm-white standard lamps amplify the values for the yellow to light red range and reduce the values for green to blue. Rough surfaces generally result in lower reflectance coefficients than smooth surfaces.

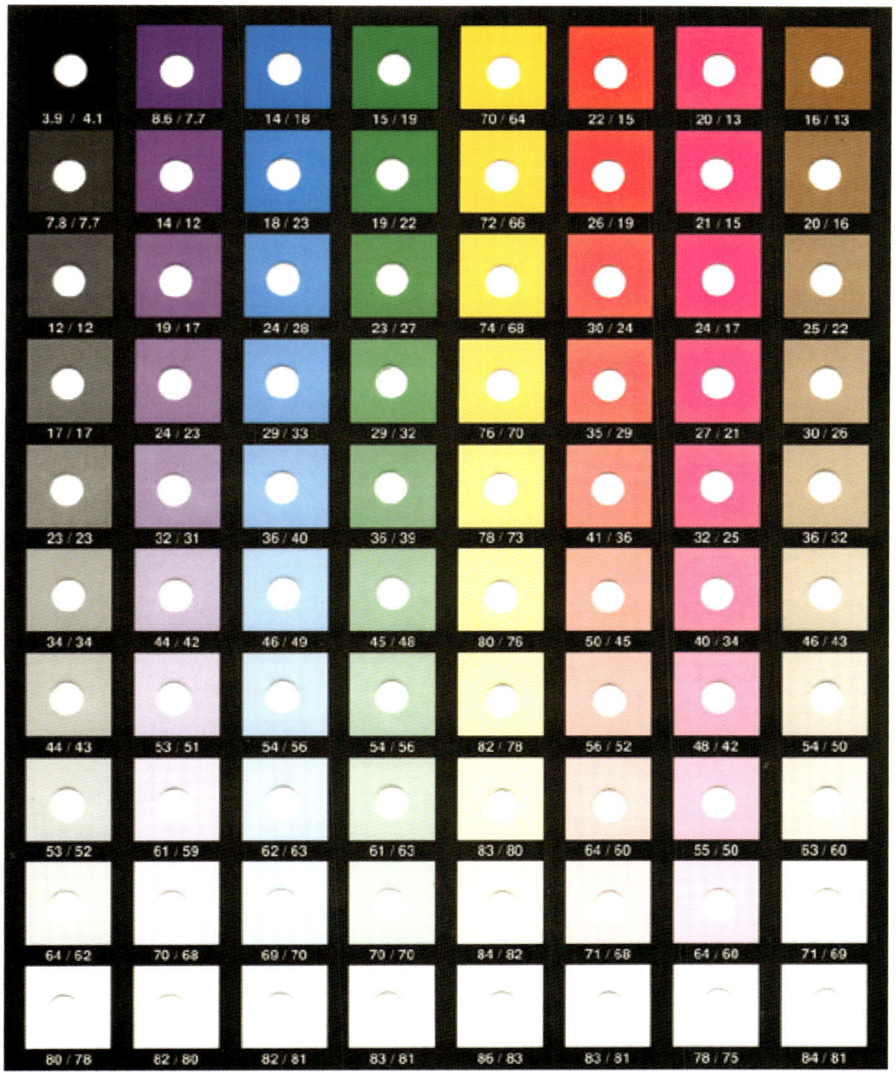

Table 9.5
Table of reflection values, colour table used to determine average luminance factors β 45/0 (%)

Values on left side:
valid for light emitted by incandescent bulbs, 2,454 K, $R_a \simeq 100$ (standard light type A), applicable for so-called warm hues (warm white)

Values on right side:
valid for light emitted by fluorescent tubes, 6,500 K, $R_a \simeq 92$, applicable for so-called cold hues (daylight white)

The following types of reflection are differentiated (**Figure 9.17**):
- Specular reflection (from mirrors, aluminium, silver, glass etc.)
- Diffuse reflection (from matt materials)
- Predominantly diffuse reflection (from high-gloss paint, velvet and high-gloss surfaces)
- Predominantly specular reflection (from matt aluminium)

Many materials cause specular and diffuse reflections simultaneously in different directions. If diffuse material is coated in a clear varnish, the material acts mostly in a diffuse manner when it is perpendicular to incident radiation. However, when the incident radiation falls on the material at a shallow angle, its effect is predominantly that of a mirror.

The indicated reflectance for materials generally refers to diffuse reflection. Specular reflection must be indicated in conjunction with the relevant angle of incidence.

Figure 9.17
Schematic illustration of different types of reflection

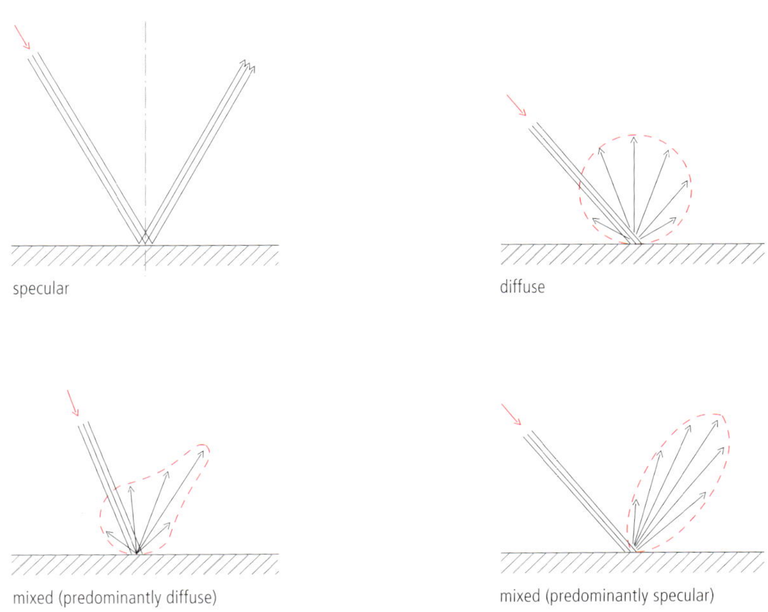

9.3.3 Colour

Colour rendering and colour rendering indices have already been discussed. Two systems have been devised to evaluate colours.

9.3.3.1 The CIE system

The CIE colour triangle gives a mathematical value for any colour with the help of two hue coordinates (**Figure 9.18**). All lamp manufacturers indicate these coordinates for each lamp they produce.

The disadvantage of the colour triangle is that identical distances on the colour scale do not correspond to identical intervals on the x/y axis. The uniform chromaticity scale (or UCS diagram) was developed for this reason.

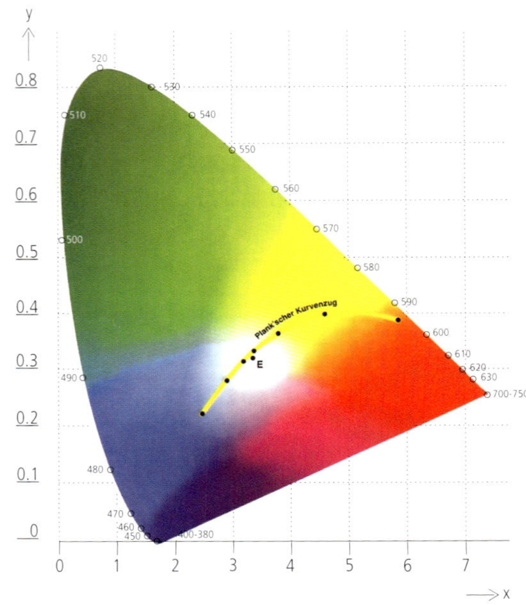

Figure 9.18
The CIE colour triangle

9.3.3.2 The Munsell system

The Munsell system is ideally suited to identifying colour in daylight conditions. It evaluates hue, brightness (value or greyness) and saturation (chroma). The system offers a scale for each of these parameters. When all these are collated, the result is a book of colour tables. Brightness is indicated by means of a grey scale from 0 (black) to 10 (white). Saturation is classified into 16 different levels.

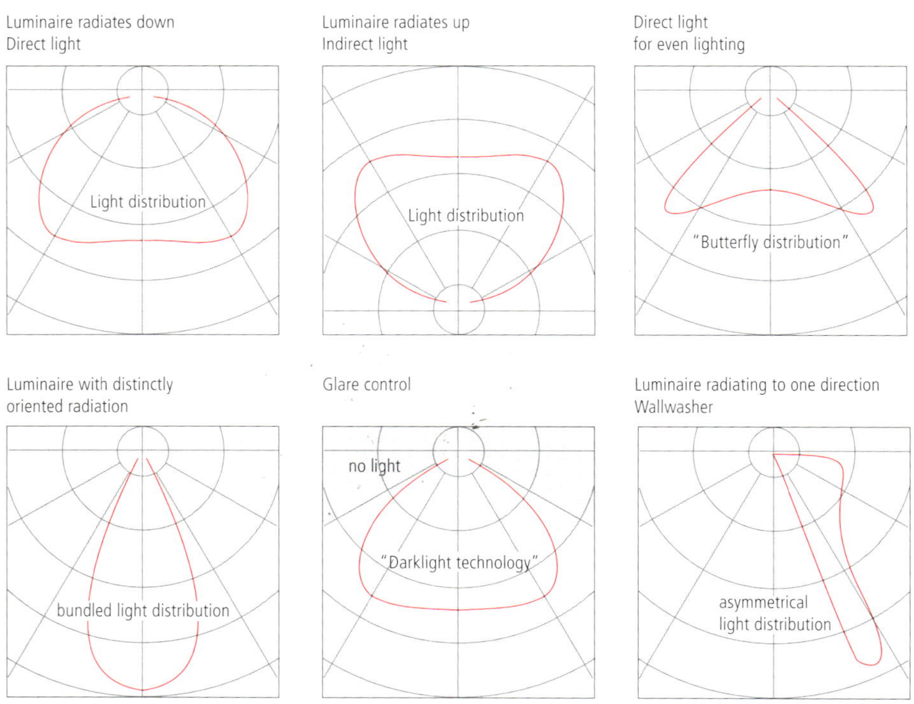

Figure 9.19
Light distribution curves of different luminaires

9.3.4 Intensity distribution characteristics

The intensity distribution characteristic of a luminaire indicates the luminous intensity it emits for a given solid angle. It is plotted in a luminous intensity (cd) curve (CDC), which generally provides longitudinal (90 to 270° plane) and crosswise (0 to 180° plane) curves in relation to the luminaire. Examples of different LIDs for luminaires are illustrated in the section on specific luminaires (see pages 408/409). **Figure 9.19** presents the principal radiant characteristics of luminaires.

Based on the locus of a luminaire in a room, one can determine how much light falls onto the working plane, the walls and the ceilings, and the angles of incidence on materials. This makes it possible to judge the effect of materials.

A point source in a frosted glass globe at the centre of a room will emit roughly equal amounts of light in all directions and hence illuminate all surfaces evenly. A halogen spotlight with a narrow-beam reflector at the same locus will perhaps light only a small image on the wall and direct all its energy towards that point. Hence, different luminaires at the same locus but with differing intensity distribution characteristics can create distinctly different effects (see also **Figure 9.16**).

In the example that follows, different effects have been designed for office rooms: for office luminaires with uniform intensity distribution characteristics, luminaires with uniform and diffuse characteristics and luminaires with exclusively diffuse intensity distribution characteristics.

The illuminance on the working plane will decrease as the diffuse intensity characteristic of the luminaire increases, although the connect load stays constant. The luminance of walls and the ceiling is considerably higher when diffuse lighting systems are used. The intensity distribution characteristic therefore determines not only the brightness of an object or a surface in a room, but also its effect.

Direct lighting systems create a relatively high degree of contrast, which adds plasticity to the objects in the room.

Exclusively diffuse lighting systems radiate energy diffusely into the room, comparable to the diffuse radiation in nature under an overcast sky. Shadows are either very subtle or altogether absent and objects can appear uninteresting without cast shadows and light reflections due to the lack in plasticity.

The aim should be to select an intensity distribution that is appropriate for the room and the visual task, which will properly light the work area and tools (interior field) and add the appropriate measure of brightness to the room (surroundings).

9.3.5 Example of modelling theoretical luminance in an office space

If the reflectance of the materials in the room is known, the resulting luminance can be calculated in advance. To do so, it is necessary to understand the intensity distribution characteristic. A luminaire with uniform (direct) intensity distribution will strongly illuminate the lower half of an office space, while a luminaire with diffuse intensity distribution will cast light on the ceiling.

The following paragraphs present calculations of the illuminance and luminance for different lighting systems and illustrations of the:
- Illuminance on the working plane
- Distribution of luminance on the ceiling
- Distribution of luminance on a side wall

The following lighting systems were used:
- Direct lighting with downlights (**Figure 9.20**)
- Indirect (diffuse) lighting (**Figure 9.21**)
- Lighting with pendant luminaires (**Figure 9.22**)
- Indirect (diffuse) lighting on a chilled ceiling combined with direct lighting with downlights (**Figure 9.23**)

Even in these abstract images, the figures give a sense of the effect in the room across the ceiling, the walls and the working plane.

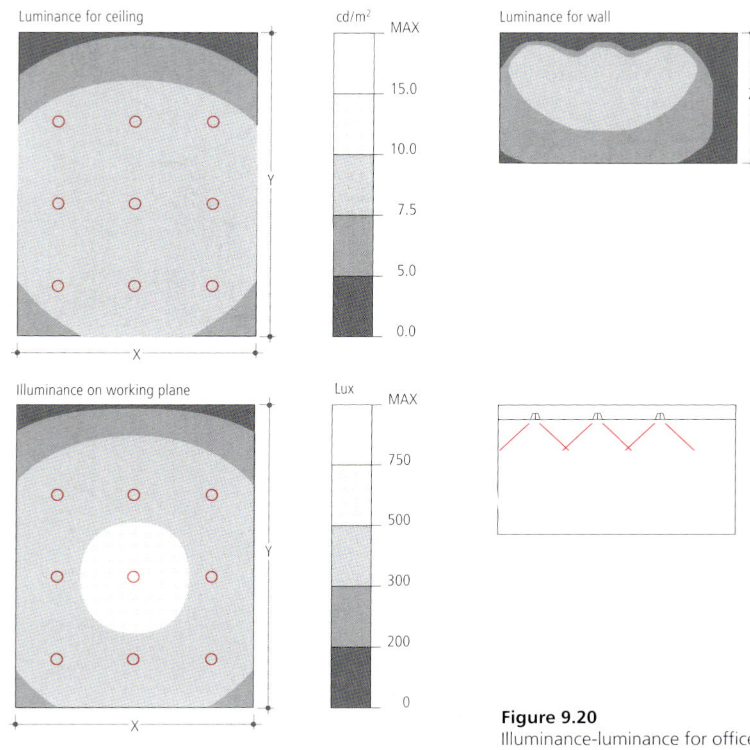

Figure 9.20
Illuminance-luminance for office with direct lighting provided by downlights

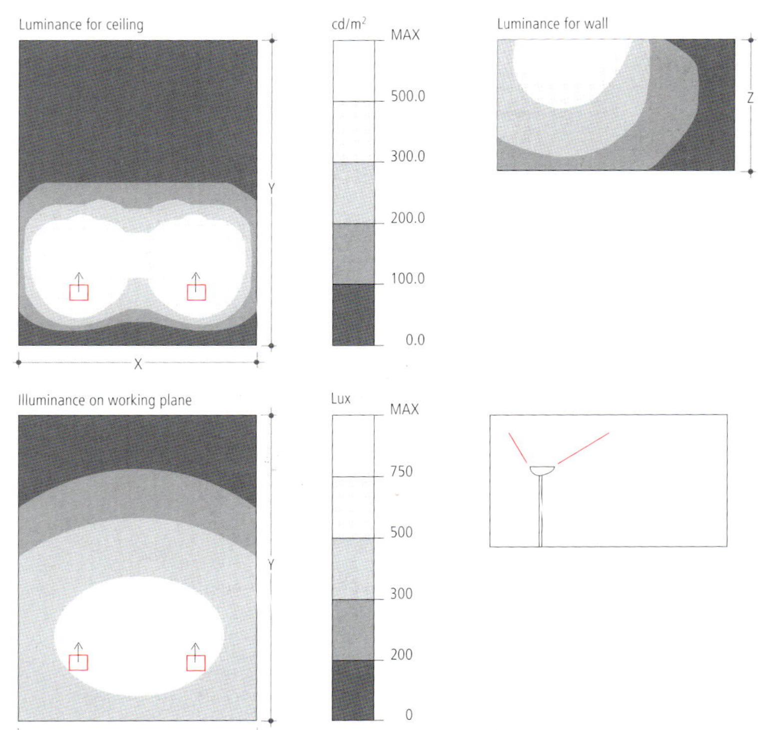

Figure 9.21
Illuminance-luminance for office with indirect lighting

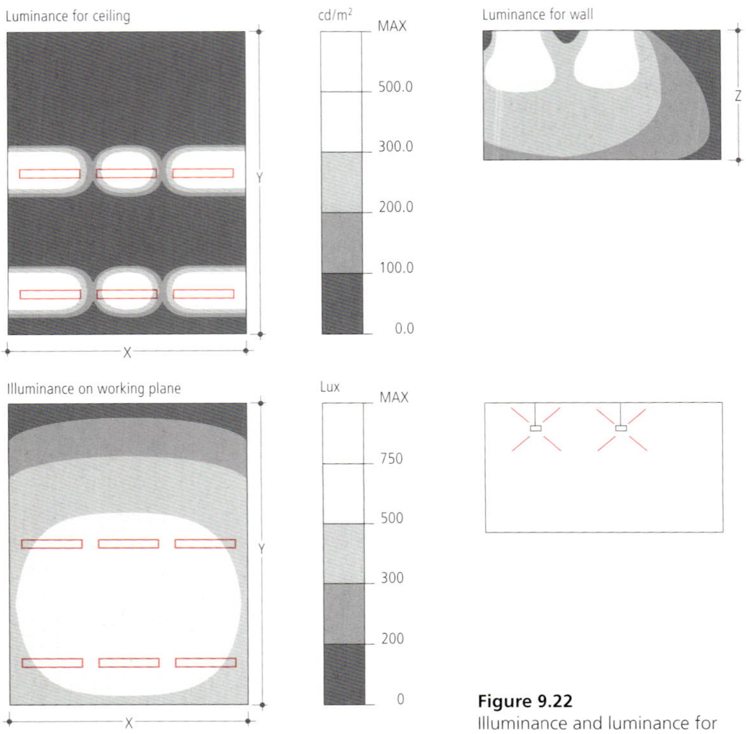

Figure 9.22
Illuminance and luminance for direct/indirect lighting provided by suspended luminaires

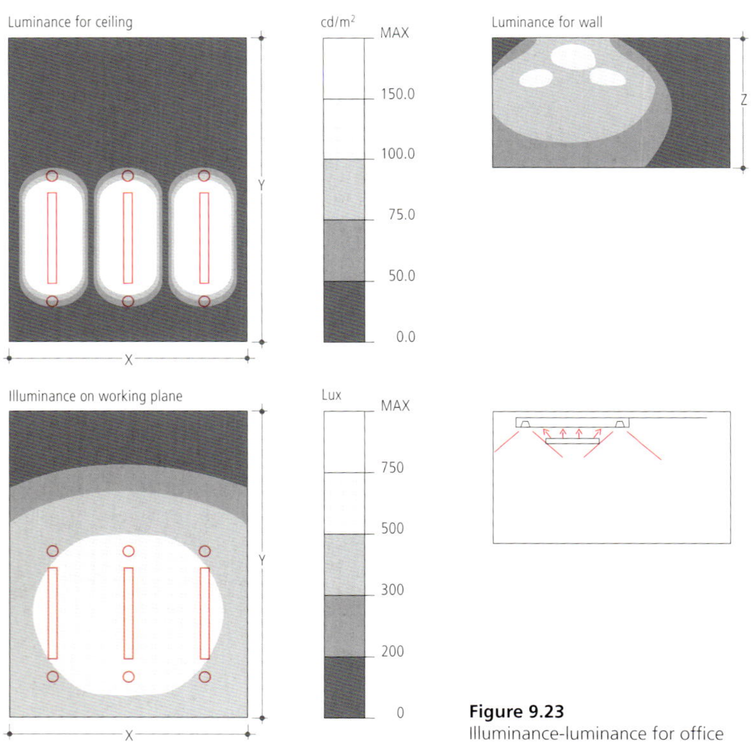

Figure 9.23
Illuminance-luminance for office with indirect light on chilled ceiling and direct lighting provided by downlights

9.4 Light sources (Lamps)

9.4.1 Overview

Electric light sources (lamps) transform electrical energy into light. In addition to light energy, this transformation always produces a greater or lesser amount of thermal energy, since on average, each lamp transforms only roughly 20 % of the supplied electrical energy into light and 80 % into heat.

Lamps are divided into three principal groups:

Incandescent lamps (incandescent luminous radiator) where fixed bodies heated to high temperatures radiate the light (e.g. incandescent lamps, halogen lamps and similar). The average lifespan of incandescent lamps ranges between 1,000 to 3,000 hours and also depends on the degree to which the lamp can transfer the thermal energy it produces directly to its surroundings.

Discharge lamps
where light is produced by means of the electric discharge in a gaseous medium (fluorescent, high-pressure gas discharge lamps etc.). Discharge lamps require conventional or electronic ballast devices both as starters and for operation. The average lifespan of discharge lamps ranges between 4,000 to 12,000 hours under the aforementioned conditions.

Induction lamps
where the metal-vapour-gas mixture forms the induction loop of a high-frequency generator. The atoms charged by the exchange of energy produce UV radiation, which is transformed into visible light with the help of luminous matter. The average lifespan of induction lamps is roughly 60,000 hours.

The ability of a light source to transform electrical energy into light is called "luminous efficacy".
The luminous efficacy is therefore the luminous flux, divided by the connect load (including that required by the ballast device).
Its units are given in lumen/watt (lm/W).
Figure 9.24 provides an overview of typical lamps.

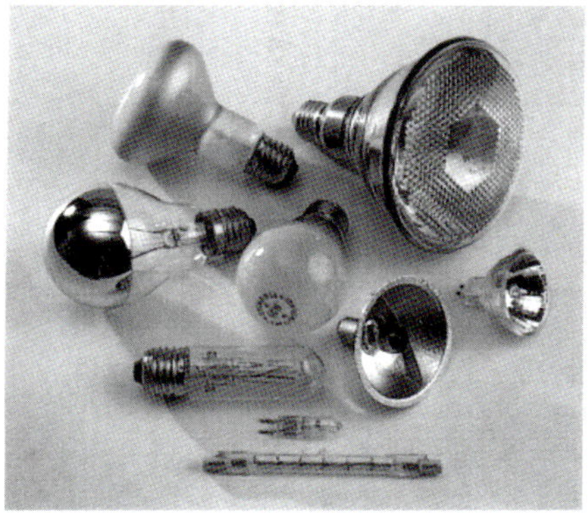

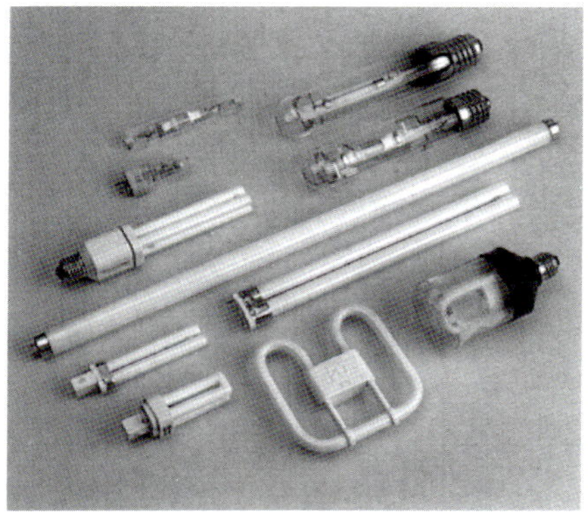

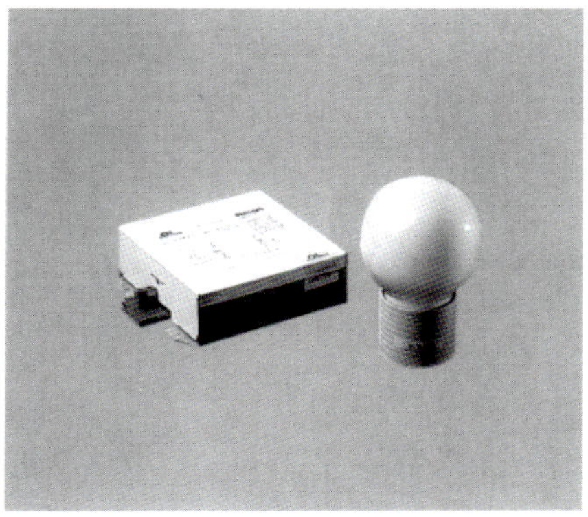

Figure 9.24
Overview of different types of bulbs

9.4.2 Lamps/luminous efficacy

Over the past twenty years, the industry has focused on developing more efficient lamps, that is, on boosting the luminous efficacy of lamps (lumen/watt; a roughly 100 % efficiency increase since 1950).

The increase in luminous efficacy was chiefly made possible with the introduction of lamps with continuous spectrums (see colour rendering characteristics). While the daylight spectrum contains all wavelengths of visible light more or less to the same degree, lamps have uneven spectrums, and may contain only a few narrow wavelength bands.

The luminous efficacy of the most common light sources is shown in **Figure 9.25**, which illustrates the degree of variance in these values. While incandescent light sources (incandescent lamps, halogen lamps etc.) achieve a range of approximately 20 lumen/watt, the luminous efficacy of other lamps is much higher. Luminous efficacy plays an important role in terms of energy consumption and economic efficiency, since lighting systems are one of the principal sources of energy consumption.

Figure 9.25
Luminous efficiency of common light sources

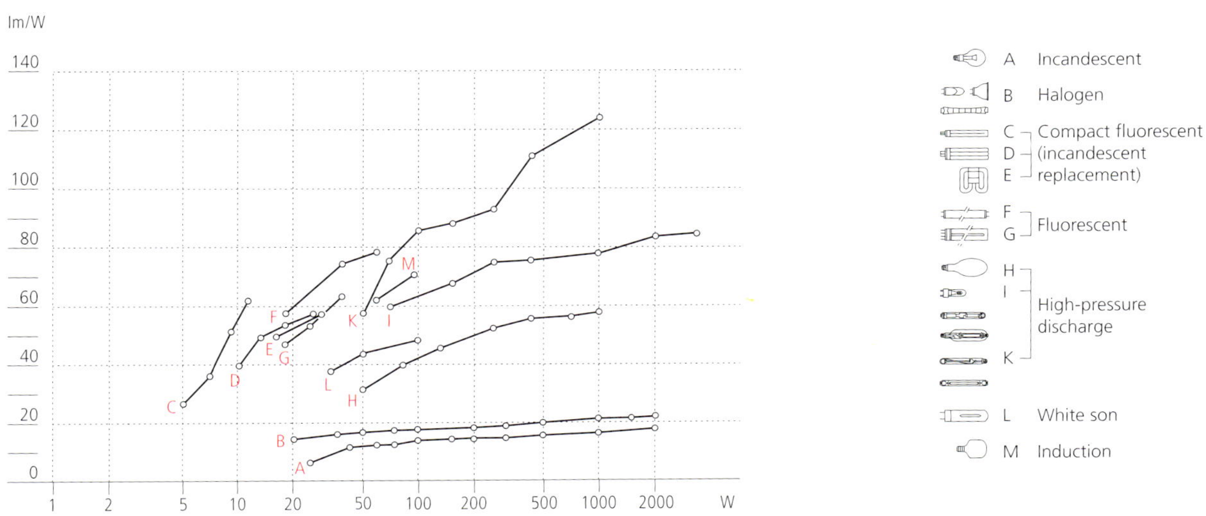

9.4.3 Lamps – colour rendering indeces

The colour rendering properties (R_a) (Or: Color Rendering Index 'CRI', US) of lamps are evaluated on the basis of 14 test colours, indicated in **Figure 9.26**. A lamp is evaluated as 1 A (based on the test colours) if its deviation from a continuous daylight spectrum is no more than 10 %.

The colour rendering index is:

1, 1 A	R_a 90 – 100
1 B	R_a 80 – 89
2, 2 A	R_a 70 – 79
2 B	R_a 60 – 69
3	R_a 40 – 59
4	R_a 20 – 39

For each lamp, the test must establish the range where the test colours are not achieved, i.e. where they fall outside of the 90 – 100 range. This is true for many lamps, and the selection of lamps with regard to the surfaces they should illuminate must therefore be made with great care because colour changes of material surfaces can occur when the analogous colour is missing in the spectrum of the particular lamp. It is important to note here that the palette of all possible colours is naturally far greater than the 14 test colours and that lamps classified as 1 A may produce colour deviations compared to fluorescent lamps with a continuous spectrum and compared to daylight. **Figure 9.27** shows the spectrums of lamps in comparison with the daylight spectrum (D65 standard/after Osram). This illustration clearly demonstrates that there are a number of lamps, which do not have a continuous spectrum, and whose white light is simply a mixture of a few colour bands.

Table 9.6 provides a rough evaluation of lamps, categorized according to the colour rendering indices R_a for fluorescent lamps.

Table 9.7 contains data on a variety of lamp types, including connect load, luminous flux and colour rendering index.

Colour rendering index (R_a)		**Light colour dw daylight white** above 5,000 K	**Light colour nw neutral white** above 4,000 K	**Light colour ww warm white** above 3,300 K
Level 1 very good	1 A R_a 90 – 100	950 daylight POWERSTAR HQI/D	940 bright white	930 warm hue incandescent lamp halogen lamp
	1 B R_a 80 – 89	860 daylight	840 bright white POWERSTAR HQ/NDL	830 warm hue 827 POWERSTAR HQ/NDL
Level 2 good	2 A R_a 70 – 79	10 daylight	740 universal white	Combination light lamps HWL-R DE LUXE
	2 B R_a 60 – 69		640 bright white POWERSTAR HQ/N HWL combination light lamps	Mercury discharge (or arc) lamps High-pressure discharge lamps HQL SUPER DE LUXE Sodium vapour (discharge) lamps VALOX® NAV DE LUXE
Level 3 less good	R_a 40 – 59		Mercury discharge (or arc) and high-pressure discharge lamps HQL	630 warm hue Mercury discharge (or arc) lamps High-pressure discharge lamps HQL DE LUXE
Level 4	R_a 20 – 39			VIALOX® NAV STANDARD VIALOX® NAV SUPER

Table 9.6
Lamp categories based on colour rendering indices R_a for fluorescent lamps

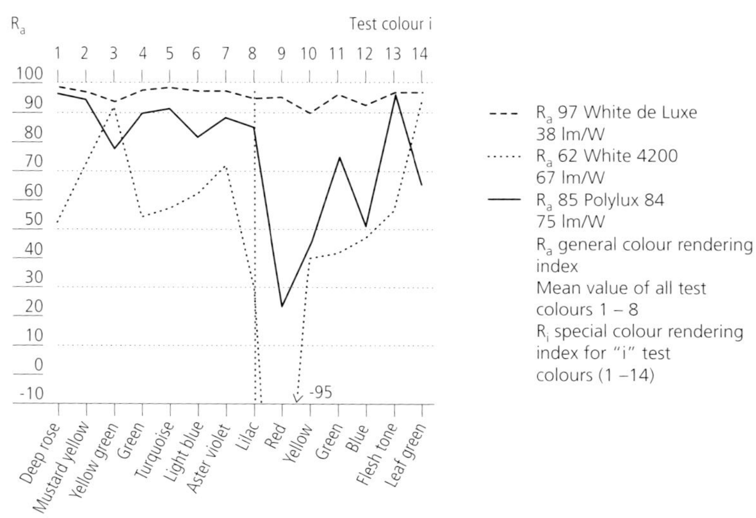

Figure 9.26
Unique colour rendering indices (R_a) of fluorescent lamps

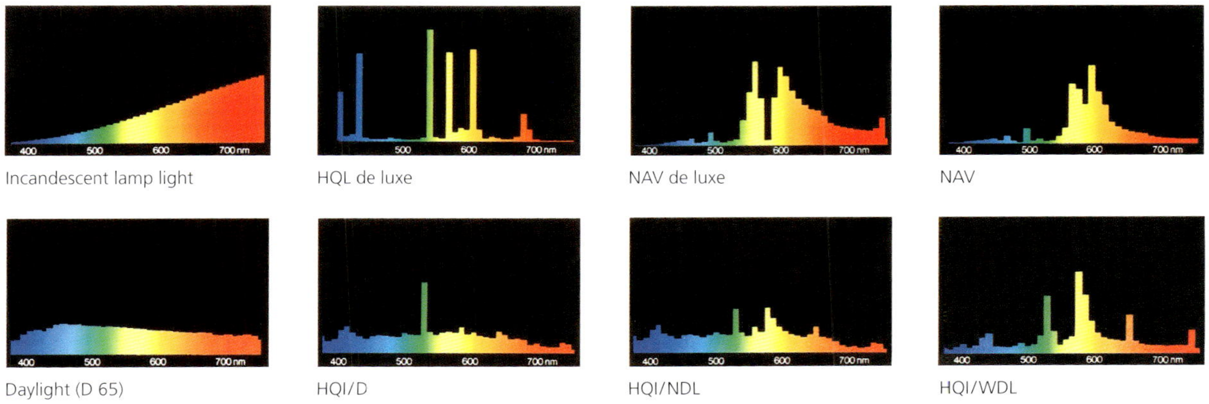

Figure 9.27
High-pressure discharge lamp spectrums (+ incandescent lamps) in comparison with daylight (D 65 standard)

Table 9.7
Light sources (Regent, Basel)

▬ not permissible
▭ permissible

Note: The values listed in the table for light or luminous flux and R_a factors should be checked against the values given in the product information provided by individual manufacturers.

Type	P	P	Base	Ø	L	Opening angle / I	Light flux	lm/W	R_a	Position
Type	W	total W	culot	mm	mm	A	lm			position
Halogen reflector lamps with protective glass										
12.50HLCK14	50	12	GX5, 3	50	45	14°	900	18	100	P360
12.50HLCK27	50	12	GX5, 3	50	45	27°	900	18	100	P360
12.50HLCK40	50	12	GX5, 3	50	45	40°	900	18	100	P360
Sodium vapour high-pressure discharge lamps, ellipsoid shape with phosphorus										
NAHL 50	50	60	E27	72	156	0.76	3,300	55	20	P360
NAHL 70	70	83	E27	72	156	1.00	5,800	70	20	P360
NAHL100	100	116	E40	77	186	1.20	9,500	82	20	P360
NAHL150	150	165	E40	92	227	1.80	13,500	82	20	P360
NAHL250	250	274	E40	92	227	3.00	25,000	91	20	P360
NAHL400	400	432	E40	122	292	4.45	47,000	108	20	P360
Sodium vapour high-pressure discharge lamps, tubular, clear										
NAHT 50	50	60	E27	38	159	0.76	4,000	66	20	P360
NAHT 70	70	83	E27	38	159	1.00	6,500	73	20	P360
NAHT100	100	116	E40	48	211	1.20	10,000	86	20	P360
NAHT150	150	165	E40	48	211	1.80	14,000	85	20	P360
NAHT250	250	274	E40	48	257	3.00	27,000	98	20	P360
NAHT400	400	432	E40	48	283	4.45	47,000	108	20	P360
Sodium vapour high-pressure discharge lamps, tubular, base at both ends, clear										
NAHTS 70	70	84	R7s	20	114	1.00	7,000	84	20	P45
NAHTS250	250	275	Fc2	23	206	3.00	25,500	92	20	P45
NAHTS400	400	450	Fc2	23	206	4.40	48,000	106	20	P45
Mercury discharge (or arc) lamps, ellipsoid shape with phosphorus										
HGL 50	50	60	E27	56	126	0.60	1,800	30	50	P360
HGL 80	80	90	E27	72	156	0.80	3,700	41	50	P360
HGL125	125	138	E27	77	177	1.15	6,300	45	49	P360
HGL250	250	272	E40	91	227	2.13	13,000	47	46	P360
HGL400	400	428	E40	122	292	3.25	22,000	51	44	P360
Halogen metal vapour lamps, base on one side, clear										
HGIT 70	75	88	G12	25	84	1.00	5,200	49	85	P360
HGIT 150	150	170	G12	25	84	1.80	12,000	70	85	P360
HGIT 250	250	275	E40	46	220	3.00	19,000	69	90	H150
HGIT 400	400	440	E40	46	285	4.00	33,000	75	90	P360
HGIT1000	1,000	1,050	E40	76	340	9.50	80,000	76	93	P 60
HGIT2000	2,000	2,080	E40	100	430	8.80	190,000	91	60	H150
Halogen metal vapour lamps, base at both ends, clear										
HGITS 70	75	88	R7s	10	114	1.00	5,500	64	80	P45
HGITS150	150	170	R7s	23	132	1.80	11,250	66	85	P45
HGITS250	250	275	Fc2	25	163	3.00	20,000	72	93	P45
HGITS400	360	385	Fc2	31	206	3.50	25,000	65	90	P45
Induction lamps, lifespan 60,000 hours										
QL55	55	55		85	150		3,500	65	80	P360
QL85	85	85		110	192		5,500	65	80	P360

P360 P60 P45 P20 P4 H150 H110 H30 HS45 HS30

9.5 Luminaires

The industry offers a wide variety of luminaires for all the aforementioned lamps. The luminaires differ greatly in form as well as in terms of lighting quality. Aesthetics and ambient qualities are more important for luminaires in home environments than economic efficiency and glare criteria.

On the other hand, glare criteria and economic efficiency play an important role in luminaires for work environments.

9.5.1 Luminaires for incandescent lamps

Luminaires for incandescent lamps are most often used for residential environments. Incandescent downlights are well known for lighting areas, although these are increasingly replaced by halogen downlights and downlights for compact fluorescent lamps. Reflectors for incandescent lamps are also very common. **Figure 9.28** depicts the home of H. Seidler illuminated by daylight (top). The change in the overall ambience at night with incandescent lamp downlights is evident in the second photograph (bottom).

Figures 9.29/9.30 depict typical luminaires for incandescent lamps.

Figure 9.28
H. Seidler's residence shown in daytime and nocturnal light ambience

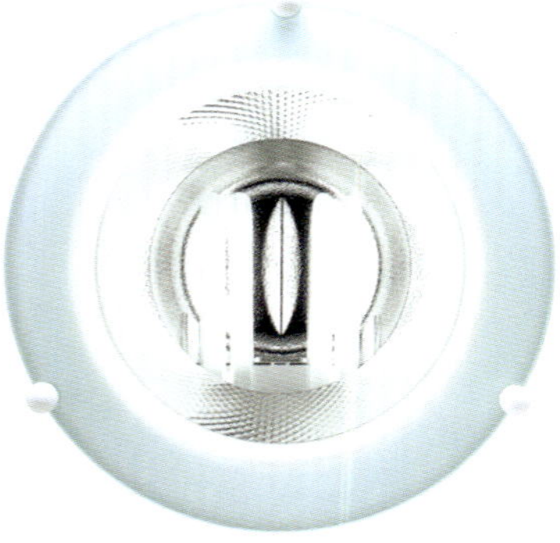

Figure 9.29
Downlights with compact fluorescent lamp, model "Panos", Design by Sottsass Associati, (Image: Zumtobel Staff)

Figure 9.30
Wall luminaire with compact fluorescent lamp model "Rhapsody", (Image: Zumtobel Staff)

9.5.2 Luminaires for halogen lamps

Many luminaires have been developed in recent years to fit the small size of halogen lamps, and this in turn has inspired the manufacturers of halogen lamps to design an even wider range of models.

Luminaires for halogen lamps are especially common for applications where bundled light is a desirable effect. Consequently, a wide range of very small reflectors and track lighting with many different beam angles have been developed.

Downlights for halogen lamps with a diameter of 5 to 10 cm are also useable for many applications (**Figures 9.31/9.32**), although they require transformers.

For luminaires with lenses, which bundle or disperse beams, small halogen lamps with a relatively good focal point are an excellent solution. A multitude of special systems for halogen lamps have been developed for this purpose.

Figure 9.33 depicts a wall luminaire for halogen reflectors, which emits bundled light upwards and downwards. The intensity distribution is adjusted with the help of a lens. **Figure 9.34** depicts a lens reflector for a 100 W halogen lamp, **Figure 9.35** a floodlight for use with high-pressure lamps and **Figure 9.36** various floodlights in combination with a track system.

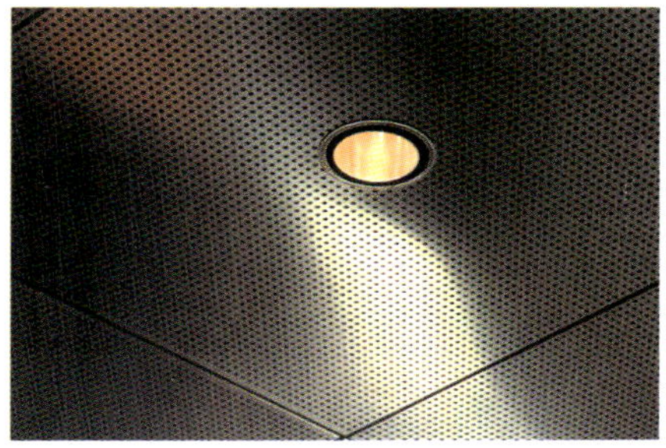

Figure 9.31
Built-in HAL downlight

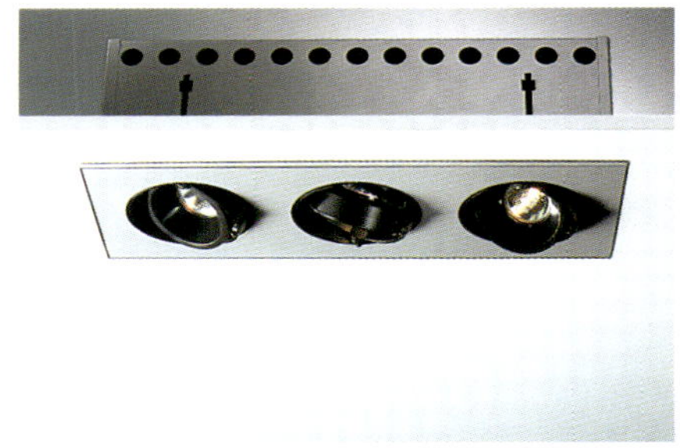

Figure 9.32
Downlight with HAL "cold light reflector" luminaires integrated into the ceiling (downlight module with Gimbal directional reflectors, ERCO)

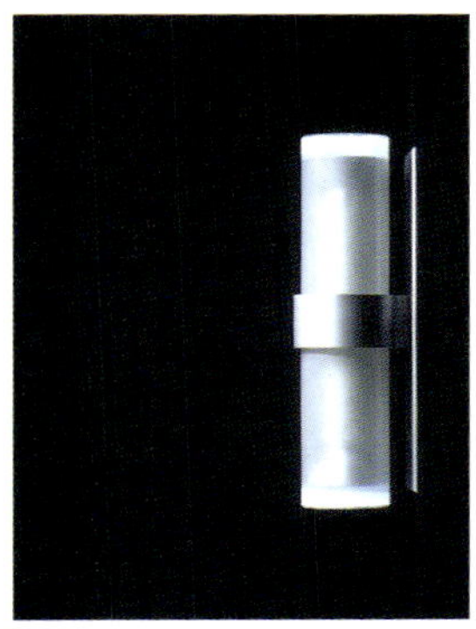

Figure 9.33
Wall luminaire for HAL reflectors
(Image: Adelmann)

Figure 9.34
Lens chandelier with HAL reflector
(Image: Spectral)

Figure 9.35
Floodlight for high-pressure lamps
(Image: BEGA)

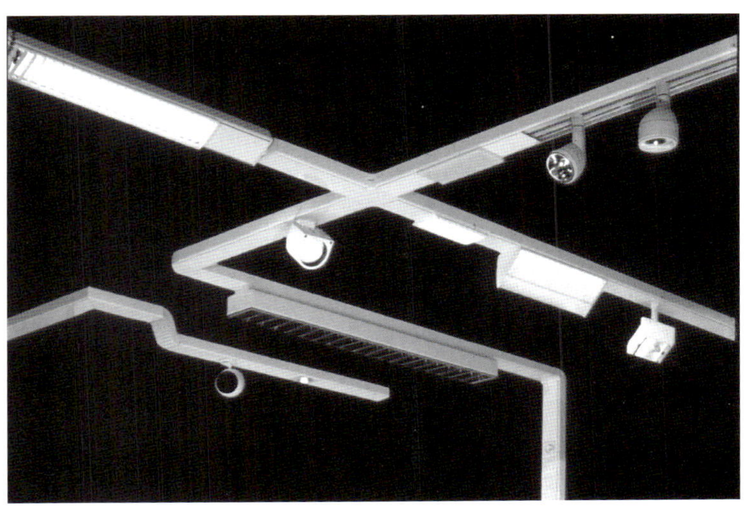

Figure 9.36
Various floodlights (spotlight) for track lighting

9.5.3 Luminaires for compact fluorescent lamps

Compact fluorescent lamps are particularly popular for use in downlights as an alternative to incandescent lamps. Square luminaires are also frequently combined with compact fluorescent lamps.

Signage lamps and emergency lighting units are another area where compact fluorescent lamps are frequently used for their long lifespan and higher luminous efficacy in comparison with both incandescent and halogen lamps.

Exterior lighting is increasingly realized with compact fluorescent lamps. It should however be noted that the luminous flux of compact fluorescent lamps can fall severely when external temperatures are low, depending on the construction of the luminaire.

A section of a downlight for compact fluorescent lamps and different installation models are illustrated in **Figures 9.37.1/2**.

Figure 9.38 demonstrates the effect of direct light cast by ceiling washers in a room: the parabolic patterns are clearly visible on the wall.

Figure 9.39 shows an example in a forecourt.

Figure 9.40 captures an area in the new entrance to the Louvre, lit with compact fluorescent downlights.

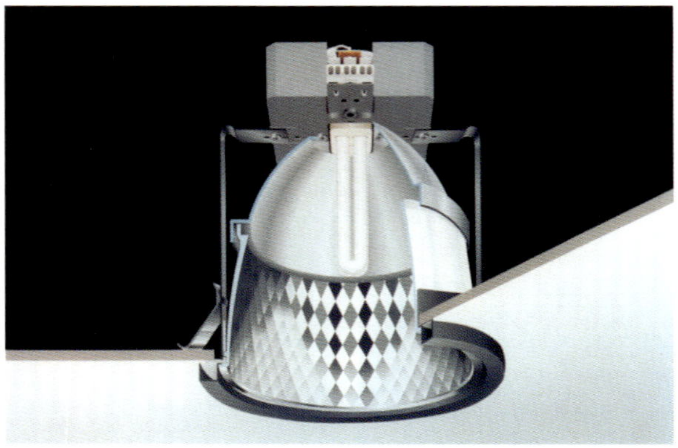

Figure 9.37.1
Downlight for compact fluorescent lamp

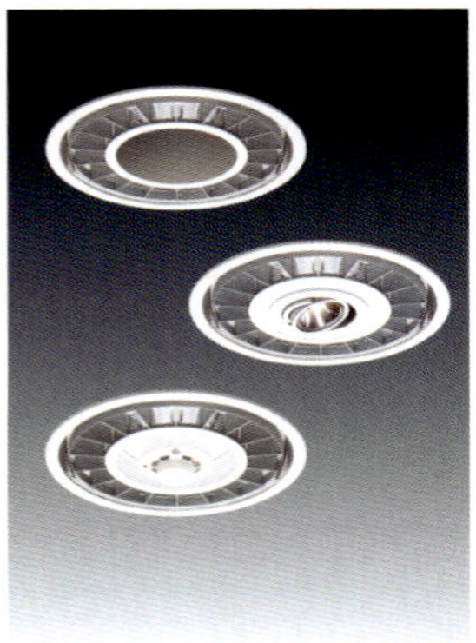

Figure 9.37.2
Recessed or mounted luminaires for compact fluorescent lamps with integrated video camera, cardanic reflector or smoke detector
(Circlet, WILA)

Figure 9.38
Room lighting with ceiling washers for compact fluorescent lamp

Figure 9.39
Forecourt lighting with downlights
(Image: Sill)

Figure 9.40
Downlights, entrance to the Louvre (Paris)

9.5.4 Luminaires for fluorescent lamps

Luminaires for fluorescent lamps play an important role in industrial applications because of their high degree of efficiency. A multitude of luminaires have been developed for fluorescent lamps, which differ greatly according to application and comfort requirements.

The development of luminaire diffusers over the past twenty years has greatly increased the visual comfort these luminaires provide. The luminance of luminaire diffusers for fluorescent lamps is very low in the cut-out range with values below 200 cd/m^2. The directional effect of the mirrors allows for a wide range of radiation angles and achieves high output efficiencies.

The following images illustrate the structural principal and luminous intensity diagram of luminaires with fluorescent lamps.

Figure 9.41 illustrates possible intensity distribution characteristics of luminaire with luminaire diffusers. Installation options for luminaires in ceiling systems are shown in **Figure 9.42**.

Figures 9.43 to **9.49** depict various integrated ceiling luminaires with their principal technical data (luminous intensity diagrams, LID) and the construction of the luminaires themselves.

Figure 9.41
Luminaires for fluorescent lamps: light reflection

wide beam

narrow angle

narrow angle, asymmetrical

Recessed luminaires, universally adaptable, for ceilings with hidden tracks

Recessed luminaires, universally adaptable, for ceilings with visible tracks

Recessed luminaires, univarsally adaptable for ceilings with saw-cut installation ports

Recessed luminaires for ceiling panelling, available for modules 100, 150 and 200

Recessed luminaires for renovation and upgrades where existing ceiling systems are preserved

Figure 9.42
Luminaires for fluorescent lamps: recessed installations

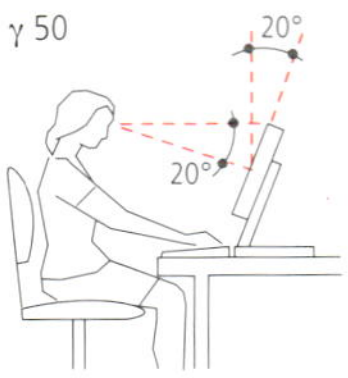

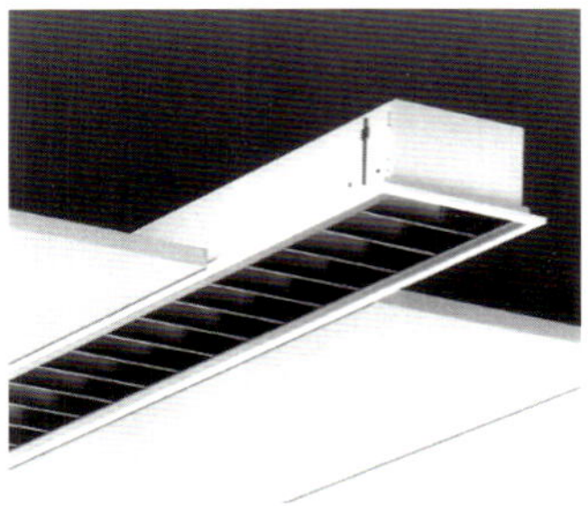

Distribution of luminous intensity

According to DIN 5032, Part 4, reference plane $C_0 - C_{180}$

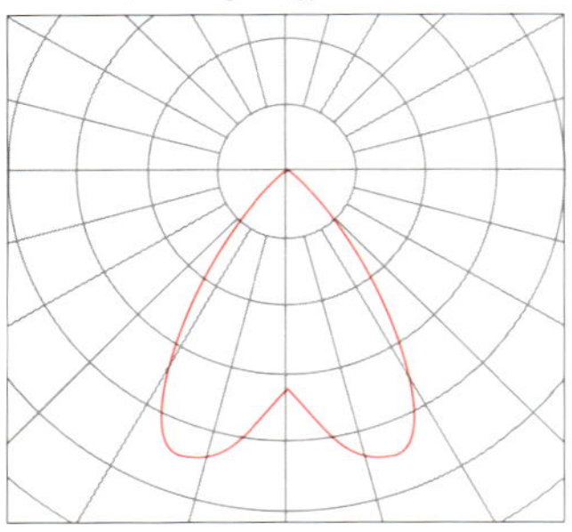

Efficiency $\eta_{LB} = 60\%$

Utilization factor % of η_{LB}

ρ	Ceiling	0.8	0.8	0.7	0.7	0.5	0.5	0.5	0.3	0
	Walls	0.5	0.3	0.5	0.3	0.5	0.3	0.3	0.3	0
	Floor	0.3	0.1	0.2	0.1	0.3	0.3	0.1	0.1	0
Room index k	0.60	40	34	38	34	38	35	34	34	31
	0.80	45	39	43	39	44	40	39	38	35
	1.00	50	44	48	43	48	45	43	43	40
	1.25	56	49	53	49	53	50	48	48	45
	1.50	59	51	55	51	56	53	50	50	47
	2.00	63	55	59	54	59	57	54	53	51
	3.00	68	59	63	58	63	62	57	57	55
	5.00	72	61	65	60	66	65	59	58	56

Figure 9.43
Mirror grid reflector 50°

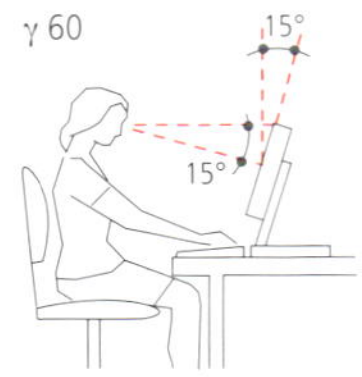

Distribution of luminous intensity

According to DIN 5032, Part 4, reference plane $C_0 - C_{180}$

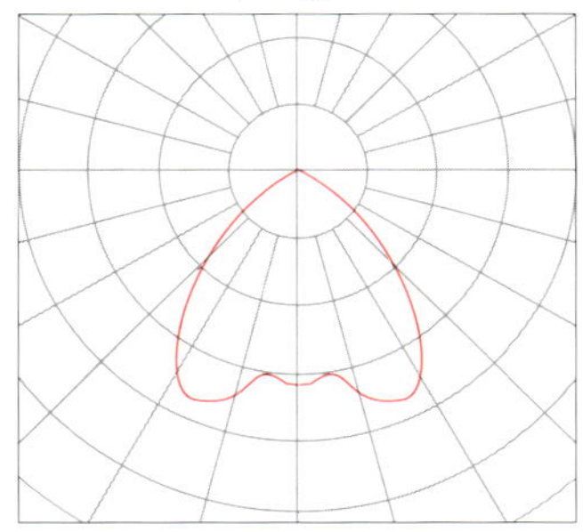

Efficiency $\eta_{LB} = 68\%$

Utilization factor % of η_{LB}

ρ	Ceiling	0.8	0.8	0.7	0.7	0.5	0.5	0.5	0.3	0
	Walls	0.5	0.3	0.5	0.3	0.5	0.3	0.3	0.3	0
	Floor	0.3	0.1	0.2	0.1	0.3	0.3	0.1	0.1	0
Room index k	0.60	38	32	37	32	37	32	31	31	27
	0.80	47	40	45	40	45	40	39	39	35
	1.00	53	45	51	45	51	46	45	44	40
	1.25	60	51	57	51	57	53	50	50	46
	1.50	64	55	60	55	61	57	54	53	50
	2.00	70	59	65	59	65	62	58	57	54
	3.00	77	65	70	65	71	68	64	63	60
	5.00	81	68	74	68	74	73	67	66	63

Figure 9.44
Mirror grid reflector 60°

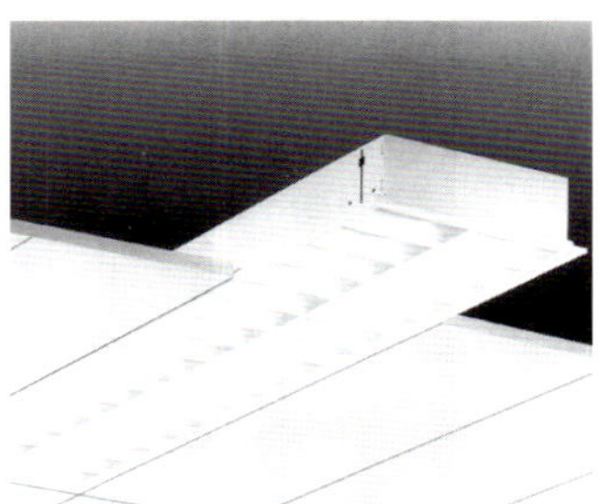

Distribution of luminous intensity

According to DIN 5032, Part 4, reference plane $C_0 - C_{180}$

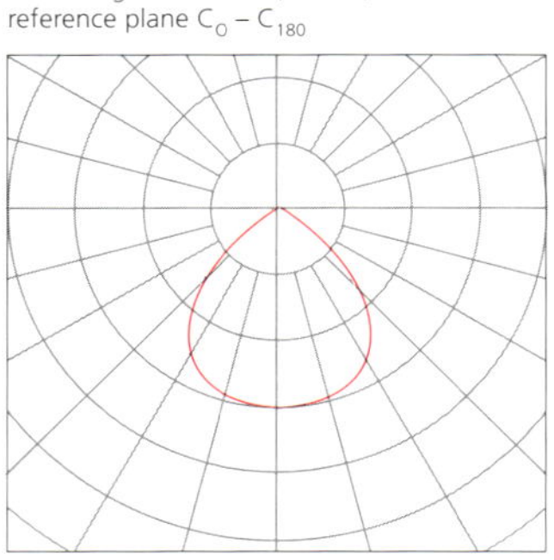

Efficiency $\eta_{LB} = 62\%$

Utilization factor % of η_{LB}

ρ	Ceiling	0.8	0.8	0.7	0.7	0.5	0.5	0.5	0.3	0
	Walls	0.5	0.3	0.5	0.3	0.5	0.3	0.3	0.3	0
	Floor	0.3	0.1	0.2	0.1	0.3	0.3	0.1	0.1	0
Room index k	0.60	34	28	33	28	33	28	28	27	24
	0.80	41	35	40	34	39	35	34	34	30
	1.00	47	39	44	39	44	40	39	38	34
	1.25	53	45	50	44	50	46	44	43	40
	1.50	57	48	53	48	53	50	47	47	43
	2.00	62	52	57	52	57	54	51	50	47
	3.00	68	58	62	57	63	60	56	55	53
	5.00	73	61	66	61	66	65	59	58	56

Figure 9.45
White large grid luminaire

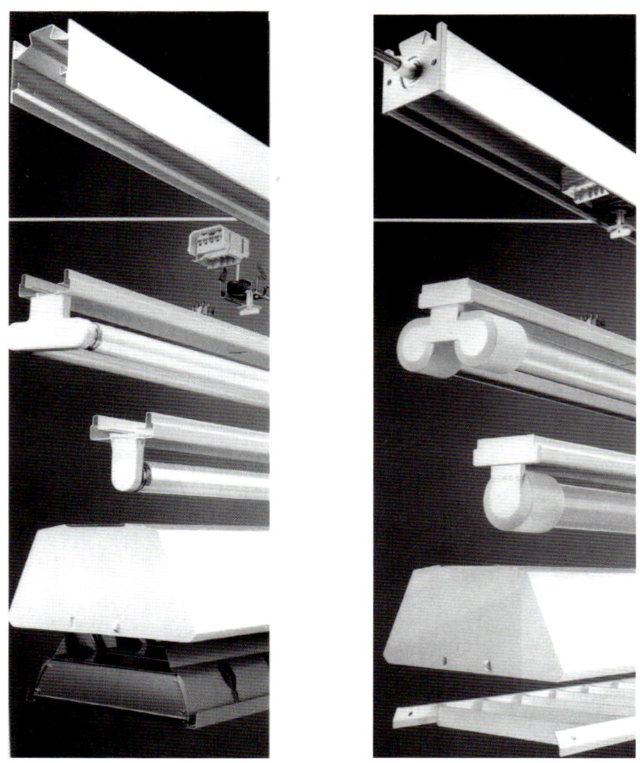

Figure 9.46
Luminaire construction (mounting rail/fluorescent lamps/reflector housing/grid component)

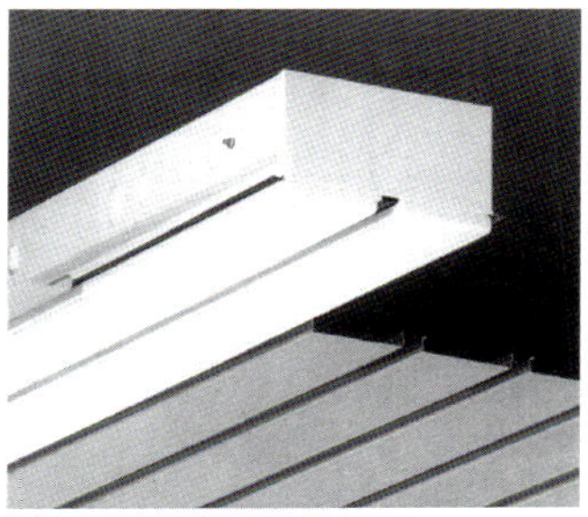

Distribution of luminous intensity

According to DIN 5032, Part 4, reference plane $C_0 - C_{180}$

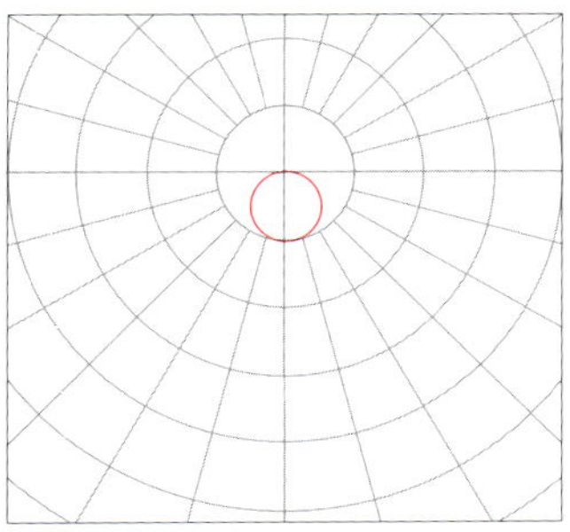

Efficiency $\eta_{LB} = 35\%$

Utilization factor % of η_{LB}

	Ceiling	0.8	0.8	0.7	0.7	0.5	0.5	0.5	0.3	0
ρ	Walls	0.5	0.3	0.5	0.3	0.5	0.3	0.3	0.3	0
	Floor	0.3	0.1	0.2	0.1	0.3	0.3	0.1	0.1	0
Room index k	0.60	15	12	15	12	14	12	11	11	9
	0.80	19	15	18	15	18	15	14	14	11
	1.00	22	18	21	17	20	17	17	16	14
	1.25	25	20	23	20	23	20	20	19	16
	1.50	28	22	26	22	25	23	21	21	18
	2.00	31	25	28	25	28	25	24	23	21
	3.00	35	29	32	28	31	29	27	26	24
	5.00	38	32	34	31	34	33	30	29	27

Figure 9.47
Recessed panel luminaires

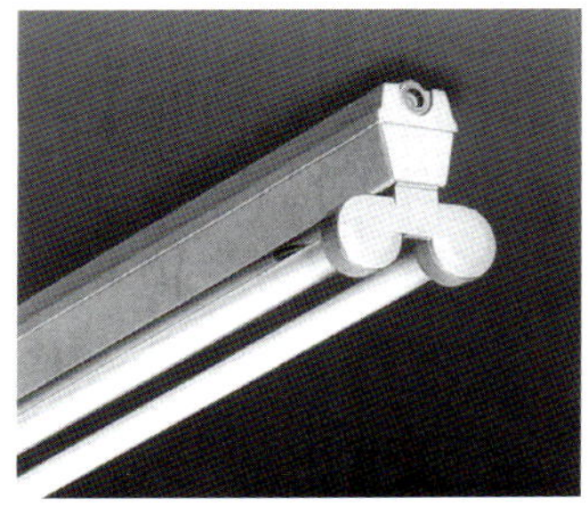

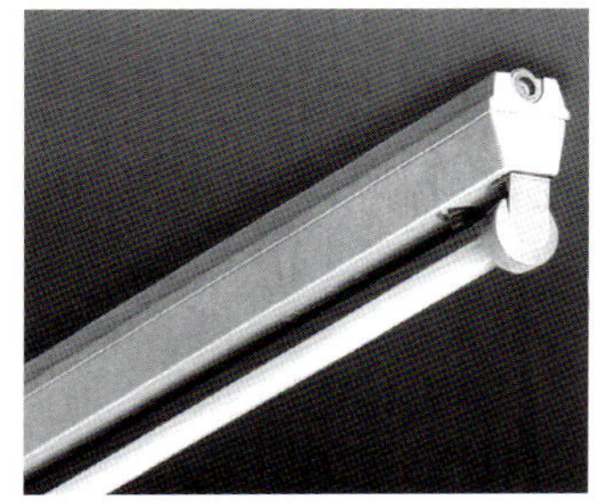

Distribution of luminous intensity

According to DIN 5032, Part 4, reference plane $C_0 - C_{180}$

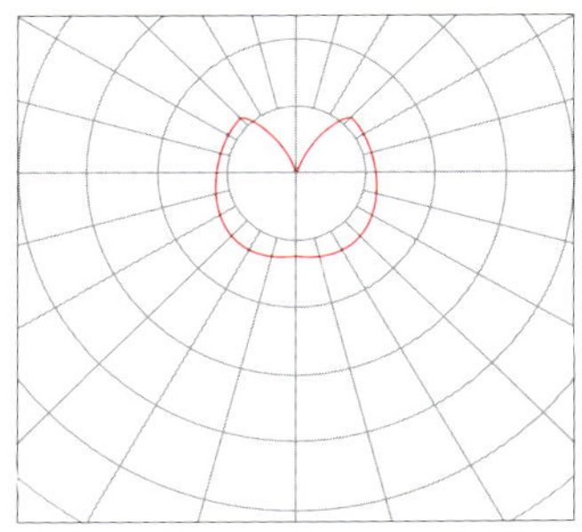

Efficiency $\eta_{LB} = 93\%$

Utilization factor % of η_{LB}

	Ceiling	0.8	0.8	0.7	0.7	0.5	0.5	0.5	0.3	0
ρ	Walls	0.5	0.3	0.5	0.3	0.5	0.3	0.3	0.3	0
	Floor	0.3	0.1	0.2	0.1	0.3	0.3	0.1	0.1	0
Room index k	0.60	34	25	31	24	28	22	21	19	11
	0.80	43	33	39	31	35	28	27	24	15
	1.00	50	39	45	36	40	33	32	28	18
	1.25	58	45	51	43	46	39	38	33	22
	1.50	63	50	56	47	50	44	42	36	25
	2.00	71	57	63	54	56	50	47	41	29
	3.00	82	66	71	62	64	59	55	48	36
	5.00	91	74	79	70	72	68	62	55	41

Figure 9.48
Striplight

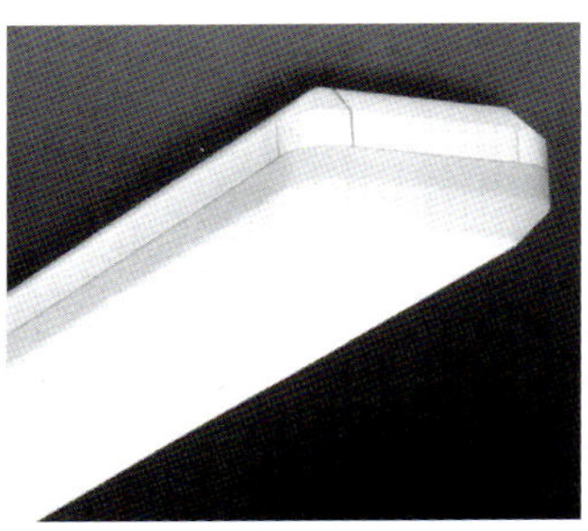

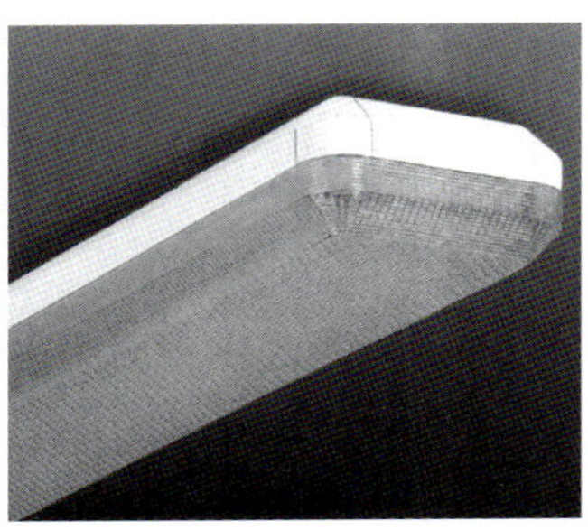

Distribution of luminous intensity

According to DIN 5032, Part 4, reference plane $C_0 - C_{180}$

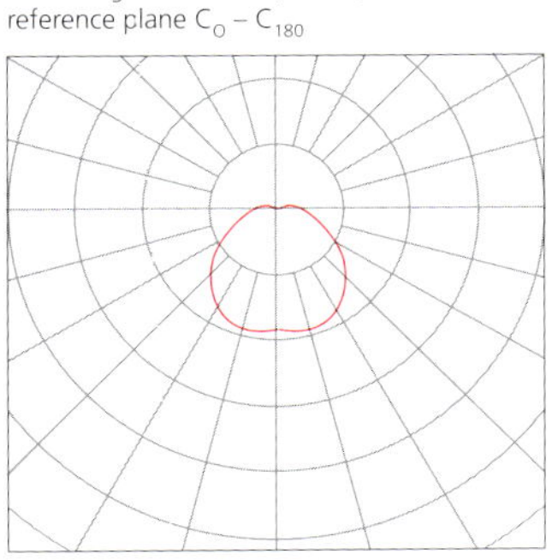

Efficiency $\eta_{LB} = 59\%$

Utilization factor % of η_{LB}

	Ceiling	0.8	0.8	0.7	0.7	0.5	0.5	0.5	0.3	0
ρ	Walls	0.5	0.3	0.5	0.3	0.5	0.3	0.3	0.3	0
	Floor	0.3	0.1	0.2	0.1	0.3	0.3	0.1	0.1	0
Room index k	0.60	27	21	26	21	25	21	20	20	16
	0.80	33	27	32	26	31	26	25	25	20
	1.00	38	31	36	30	35	30	29	28	24
	1.25	44	36	41	35	40	35	34	33	28
	1.50	48	39	44	38	43	39	37	36	31
	2.00	53	43	48	43	47	44	41	40	35
	3.00	59	49	54	48	53	50	47	45	41
	5.00	65	54	58	53	57	55	51	49	45

Figure 9.49.1
Prismatic wraparound luminaire

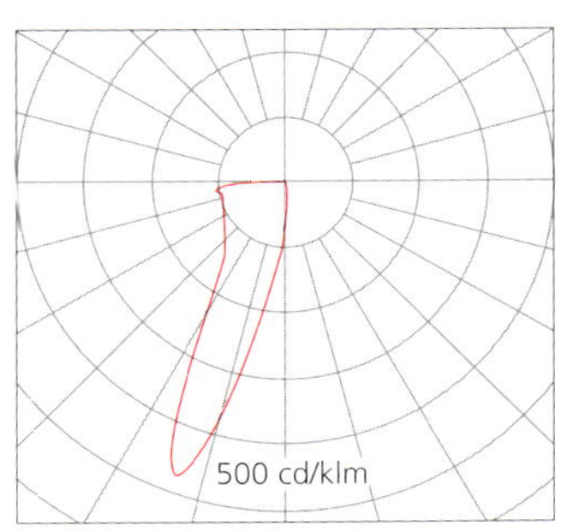

Figure 9.49.2
Recessed wall floodlight

9.5.5 Luminaires for discharge lamps

As mentioned above, luminaires for discharge lamps are used for large halls and athletics facilities, and for floodlighting. In interiors, discharge lamps have not taken hold because of their high luminous efficiency per lamp. The introduction of HQT lamps with a performance of 35, 75 and 150 W has opened the field for using discharge lamps in interiors with low ceilings.

Figure 9.50 shows a hall luminaire for discharge lamps. In interiors, halogen metal vapour lamps are used in floor uplights. **Figure 9.51** shows such a luminaire for indirect interior lighting. **Figure 9.52** illustrates the effect of this luminaire in the room.

9.5.6 Luminaires for safety lighting

Safety luminaires provide minimal lighting in case of power failure, especially along escape routes. Illuminated signage along these routes is also included in most building projects.

Figure 9.53 illustrates the function of safety lighting

Figure 9.52
Indirect luminaire for discharge lamp

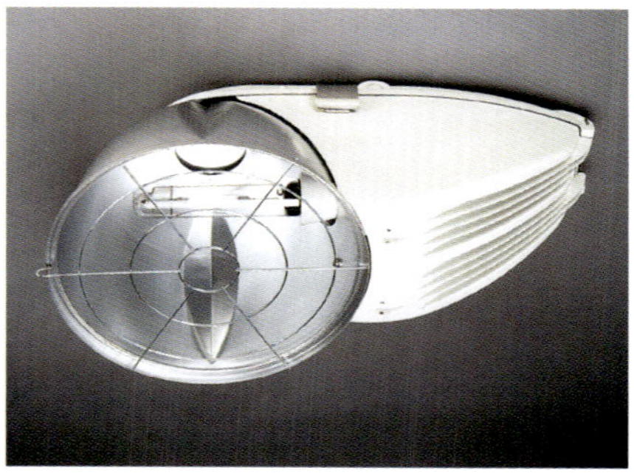

Figure 9.50
Hall reflector luminaire for high-pressure discharge lamp with ventilated "Radflekt" aluminium reflector for horizontal lamp-burner case (Model: "Radbay", Thorn Licht GmbH, Dortmund)

Figure 9.51
Floor luminaire (indirect light) for discharge lamp
(Image: Zumtobel Staff)

Figure 9.53
Emergency lighting function

9.5.7 Air-handling luminaires

Examples of air-handling large area luminaires are discussed in the following paragraphs.

Figure 9.54 shows luminous ceiling units with integrated supply air slot diffusers in an office. As the image of the ceiling illustrates only some of the luminaires are equipped with slot diffusers because each has a range of 3 to 4 m (horizontal penetration depth) and are only required at 8-m intervals (for large open-plan areas). For smaller rooms, the range should be scaled to one supply air unit per luminaire.

Figure 9.55 illustrates a special lighting design for offices with cove lighting and integrated air-handling slots (supply air diffuser and grille air return). Half of the unit visible in the image is dedicated to serve as an air supply unit (50 %) and the other half as a suction vent to evacuate stale air from the room (50 %).

The schematic in Figure 9.55 depicts the air movement in the room: it enters on a tangent and exits on the side adjacent to the corridor (partly via cabinet/closet units, partly via the exhaust section in the ventilation rail of the luminaire).

Figure 9.56 depicts a typical ceiling diffuser, which distributes air in four directions, in combination with a square luminaire. This ventilation unit can transport large volumes of air into the room, while at the same time being fully compatible in terms of design and materials with the luminaire and the ceiling.

Figure 9.57 shows a point vent, Vari Drall (LTG) in combination with strip lighting. In the example shown, the vent is located at the intersection of two lighting strips.

These examples are only a small selection of the many options available today.

Figure 9.54
Luminous ceiling units with supply air slot diffusers

Figure 9.55
Cove with lighting and air-handling functions

Interior view of a wall unit (room divider) with supply air slot diffusers (tangential air intake)

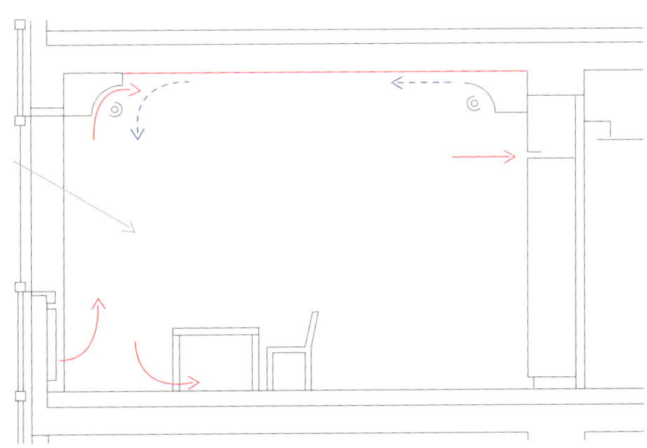

Winter operation, schematic diagram of ventilation functions

Figure 9.56
Ceiling luminaire with diffusing unit

Figure 9.57
Point vent (Vari Drall) in combination with strip lighting

Luminaires with integrated circular air supply slots, as shown in **Figure 9.58**, are a particularly successful example of combining light and air-handling functions.

The round luminaires with integrated jet nozzles shown in **Figure 9.59** are different in design but equally successful as an integrated concept. The fresh air nozzle (with perforated louvre) is clearly visible in the illustration; the fitting on the left side is used to vent air directly through the luminaire and also provide a direct exhaust outlet for the heat generated by the luminaire without adding to the (thermal) load in the room.

A mirror grid ceiling (**Figure 9.60**) is a room-enclosing element that fulfils the functions of a reflector, fresh air supply and diffuse sound absorption. **Figure 9.60** shows a mirror grid ceiling with integrated jet air outlet in an open-plan area.

Figure 9.58
Round luminaire (HAL) with air supply slot

Figure 9.60
Mirror grid ceiling with integrated point vent

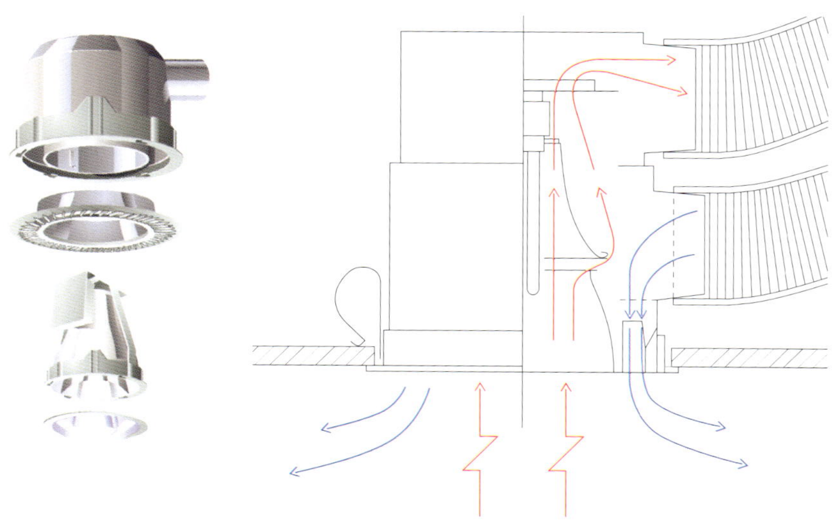

Figure 9.59
Recessed luminaire with air-conditioning function for ventilation and exhaust (WILA)

The following examples feature objects where architects have collaborated with lighting designers and air-conditioning engineers to meet specific design requirements. **Figure 9.61** shows a ceiling design in the form of an oversized reflector for a teller hall, where the standard reflector has been transformed into larger than normal, ad-justable individual elements; narrow vents have been integrated into the supporting structure to transport the required supply air into the space.

With indirect lighting via matt, mirrored aluminium ceilings or matt chrome ceilings, illumination (light source) is separate from the ceiling (with the exception of emergency lighting or spots) and linear vents can be easily integrated (**Figure 9.62**).

Figure 9.63 shows large, pillared luminaires in a hotel lobby. The luminaires are supported on a solid pedestal that can also accommodate an air vent to inject supply air into the area by means of displacement ventilation. In the example shown, the air vent is not integrated into the luminaire itself, but one could easily envision how such a solution might be implemented.

Figure 9.61
Sculpted aluminium ceiling with integrated luminaires and vent

Figure 9.63
Pillared luminaires with option for integrated air vent for displacement ventilation

Figure 9.62
Mirrored ceiling with integrated air vents and indirect lighting

9.5.8 Lighting system and chilled ceiling panels

Chilled ceiling panels have been used for several years to reduce the air volumes in air-conditioned buildings. Chilled ceilings can compensate for much of the thermal load in the room with cold water, that is, chilled ceiling components in the form of panel or flat ceilings provide some of the cooling load compensation. Warm air rising is cooled at the ceiling filled with cold water and gradually flows back down into the room (gravity principle). To achieve hygienic air conditions, chilled ceilings are generally combined with air conduction from bottom to top. Since various ceiling components can be used for chilled ceilings, luminaires are easily integrated into these systems. **Figure 9.64** shows a partially opened ceiling, which serves both as a luminaire and a cooling component. In this example, the bottom ceiling components (visible ceiling components) serve as a support structure for the cooling elements and the luminaires are suspended between the cooling elements. It is rare that an entire office area is equipped with cooling elements and round or square luminaires as well as custom designed ceiling elements (**Figure 9.65**) are often used.

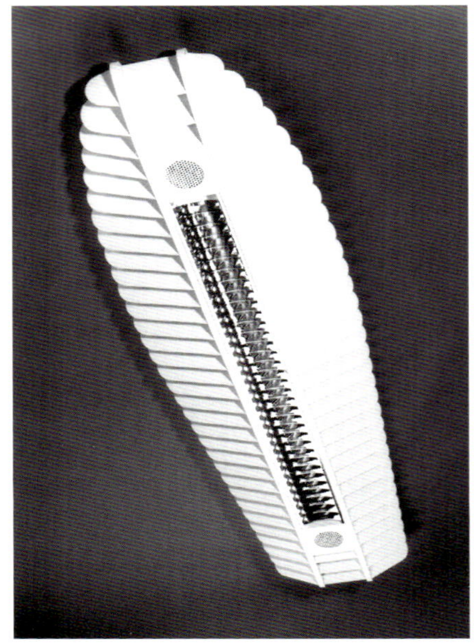

Figure 9.65
Designer ceiling-cooling-acoustic fixture with integrated luminaires

Figure 9.64
Chilled ceiling panels with recessed luminaires

9.6 Economy

The direct costs of a lighting system are composed of investment costs for:
– Luminaires
– Lamps
– Installation and connection

Luminaires are supplied with electrical energy; the following costs must therefore be taken into consideration:
– Supply line costs
– Subdistribution costs
– Riser costs
– Power supply and transformer costs etc.

Since luminaires emit heat, the following costs are also relevant for the lighting system in rooms with HVAC systems for the connect load of the total lighting installation:
– Proportional costs for ventilation and air-conditioning
– Proportional costs for heating
– Proportional costs for the refrigeration machine
– Proportional costs arising from ductwork according to number and height of storeys

For large high-rise projects, all these cost projections are relevant when comparing lighting systems. Selecting a system with low connect loads saves costs in terms of luminaires and lamps, and translates into savings for energy consumption and for the HVAC systems.

Economic analysis alsos take the service of capital for interest and capital expenditure write-off into consideration.

Operating costs
Operating costs include the costs for the power consumption of the luminaire and the necessary ballast:

– Lamp replacement costs and installation, with the lifecycle of the lamp and the annual usage playing a decisive role
– Cleaning costs for the luminaires
– Maintenance and repair costs for the luminaires

As with the investments, it is important to factor in the indirect operating costs, such as the proportional maintenance costs of the electrical system and heat loss through transformers and conduits.

For air-conditioned buildings, the power consumed by fans and chillers may be a multiple of the power consumed by the luminaire itself.

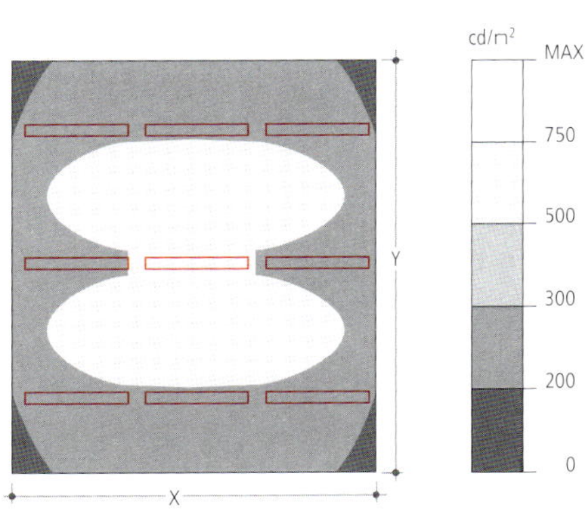

Figure 9.66
Office space with new L 36 W luminaires based on computer calculation

Luminous intensity distribution at working plane (0.80 cm)

Number	Luminaire type	Lph	Lamp(s)/(lm)	C/V factor
9	ZUMTOBEL RAB A 1/36W (VVG) TST	1.871	T26 1/3W/3.450	0.95/0.80

Summary of all calculation results (RPkt: 10, 10, 10/1)

Mean values	Illuminance Total	Direct	Indirect	Luminance Total
Ceiling	78	0	78	17.4
Walls	150	82	68	23.9
Working plane	**460**	**390**	**70**	51.2 (Rho = 35%)

Specific connect load: 17.0 W/qm/460 Lx = > 3.7 W/qm/100 Lx

9.6.1 Calculation method

During the initial planning, the general practice is to simply establish the illuminance of a lighting system in a room and to verify compliance with governing standards.

Two methods are employed:
– Efficiency method
– Computer-aided method

The efficiency method takes the dimensions of the room, the reflectances (reflecting factors) for ceiling, walls and working plane as well as the intensity distribution characteristics and the luminaire efficiency into consideration.

A computer program was used in the following example to calculate a solution for a 5-m-deep and 4.37-m-wide office, to draft a lighting proposal with L 36 W fluorescent lamps and luminaire diffusers. The result (**Figure 9.66**) pro-poses nine L 36 W luminaires for the room, whereby the specific connect load for the room is assumed as being 17 W/m^2.

Luminous intensity distribution curve

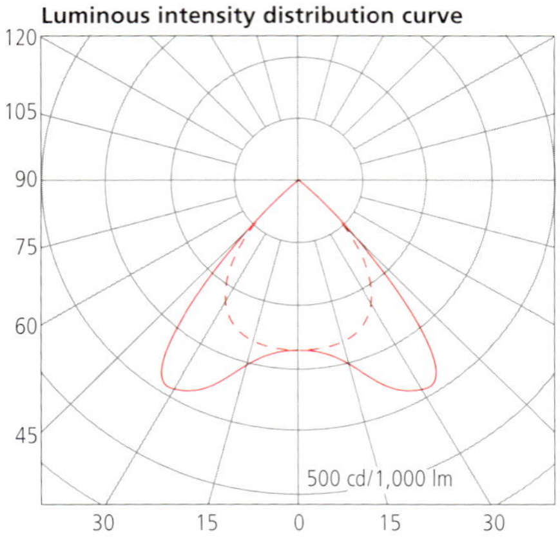

∢	C – 0	C – 90	C – 180
0	267	267	267
5	269	267	269
10	276	267	276
15	296	263	296
20	332	254	332
25	360	240	360
30	377	222	377
35	359	193	359
40	251	146	251
45	69	72	69
50	2	2	2
55	0	0	0
60	0	0	0
65	0	0	0
70	0	0	0
75	0	0	0
80	0	0	0
85	0	0	0
95	0	0	0
105	0	0	0
115	0	0	0
125	0	0	0
135	0	0	0
145	0	0	0
155	0	0	0
165	0	0	0
175	0	0	0
180	0	0	0

Utilization factor

Correction factors	Operating efficiency	0.59
18W : – –	LITG classification	Λ50
36W : 1.00	BZ classification	BZ2/1/BZ3/2/BZ4
58W : 0.98	UTE classification	0.59Λ

Reflectance	Ceiling	80	80	80	80	50	50	50	30	0
	Walls	50	50	30	30	50	30	30	30	0
	Floor	30	10	30	10	30	30	10	10	0
Room index k	0.60	37	35	32	31	36	32	31	31	28
	0.80	43	40	39	37	41	38	36	36	33
	1.00	48	44	43	41	45	42	40	40	37
	1.25	51	47	47	44	48	45	43	42	39
	1.50	54	49	49	45	50	47	44	44	41
	2.00	56	50	52	47	52	49	46	45	42
	2.50	58	52	54	49	54	50	47	46	43
	3.00	61	53	56	50	55	52	49	48	45
	4.00	64	56	61	53	59	56	52	51	48
	5.00	66	56	62	54	60	57	53	52	49

Table 9.8
Lighting proposals for office space based on efficiency method

The required number of luminaires was calculated for the same room with the efficiency method; the result is shown in **Table 9.8**.

The first determination was to propose L 36 W large area diffusing luminaires. The calculation indicates that 7.9 luminaires are needed, with the geometry of the space calling for nine L 36 W luminaires.

Both methods calculate the number of required luminaires on the basis of the mean illuminance in the room.

The example can also be executed for L 58 W luminaires (see example 1.2 (Option 1) in **Table 9.8**). The calculated number of luminaires is rounded off to a total of six.

If the required illuminance is known for the zone relevant to the visual task, (zone illustrated in **Figure 9.67**), the number of luminaires suggested according to the efficiency method is revealed as exceeding actual requirements. Only six L 36 W luminaires, or four L 58 W luminaires, are required for the sample room when the calculation is executed for the actual area where the proposed illuminance is required.

Project			EXAMPLE	OPTION 1	OPTION 2
			1.1. L36W	1.2. L58W	1.1. L36W Zonal calculation
Room			OFFICE SPACE		
Measurements: width	a	m	4.37		Zonal 3.5
length	b	m	5.00		Zonal 3.5
area	A = a · b	m²	2.,87		Zonal 12.25
height	h_R	m	2.85		
luminaire height above working plane (for ceiling luminaires $h = h_R - 0.85$)		m	2.00		
Room index $k = \frac{a \cdot b}{h\,(a+b)}$			1.45		
Reflectance ρ for ceiling/walls/working plane			0.8/0.5/0.3		
Room use/visual task					
Standard illuminance according to DIN 5035 I based on design plan (calculated)		lx	500		
Light colour group according to DIN 5035 I based on selected lamp			/21		
Colour rendition index classification according to DIN 5035 I based on selected lamp			A		
Glare control classification according to DIN 5035 I, based on curve, luminaire permissible up to			/ / lx	lx	lx
Luminaire type					
Fastening/mounting type					
Lamp type			L36W/21	L58W/21	L36W/21
Nominal power Φ		lm	3.450	5.400	3.450
Conversion factor f			1.00	0.98	1.00
Application factor f_1			0.95	0.95	0.95
Table value η^1_B			0.53	0.53	0.53
Utilization factor $\eta^1_B = \eta_B \cdot f \cdot f_1$			0.50	0.50	0.50
Number of lamps $n = \frac{1.25 \cdot E \cdot A}{\Phi \cdot \eta_B} = \frac{1.25 \cdot E \cdot a \cdot b}{\Phi \cdot \eta_B \cdot f \cdot f_1}$			7.9	5.0	4.4
Number of luminaires based on calculation (number of lamps n rounded off)			8 → 9	5 → 6	4
AFTER COMPUTER TESTING			6	4	MATCHES COMPUTER

9.7
Lighting systems in practice

9.7.1 Office buildings
9.7.1.1 Office rooms

Office rooms should have an illuminance between 300 and 750 lx to achieve interior field luminance of 75 to 150 cd/m². This illuminance poses few problems for standard office activities and for visual performance at PC workstations. The following solutions are employed in contemporary office lighting:
- Luminaire with direct lighting characteristics (mounted, integrated into ceiling or pendant) (**Figures 9.68**)
- Suspended luminaire (pendant) with direct and indirect lighting characteristics (**Figures 9.69**)
- Compact fluorescent downlights, integrated into the ceiling
- Direct-lighting ceiling systems with additional desk luminaire (**Figure 9.70**)
- Indirect lighting systems
- Combination of direct and indirect systems, **Figure 9.71**
- Individually designed systems in combination with chilled ceilings

A connect load of 8 to 15 W/m² can be assumed for direct systems, and 10 to 25 W/m² for indirect systems.

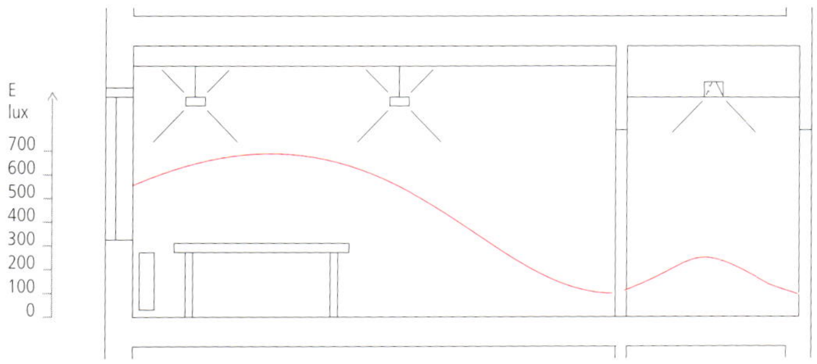

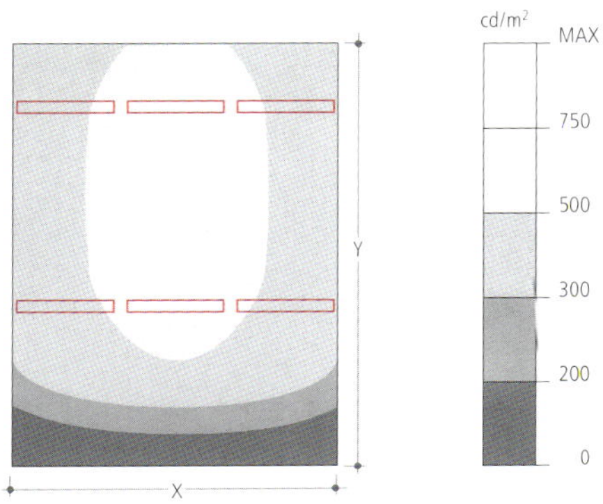

Figure 9.67
Luminous intensity distribution, zone lighting for office space; luminous intensity distribution, zone lighting on floor grid

Y Area length: 5,00 m
X Area width: 4,38 m

Figure 9.68.1
Office with direct lighting from mirror grid luminaires

Figure 9.68.2
Lighting system with direct distribution
"Mellow Light IV" model,
Design by Studio & Partners
(Image: Zumtobel Staff)

Figure 9.68.3
Universally usable pendant or mounted grid luminaire with precisely adjusted indirect light output
(Model: "Challenge", Thorn Licht GmbH, Arnsberg)

Figure 9.70
Office with downlights and individual luminaires at workstation

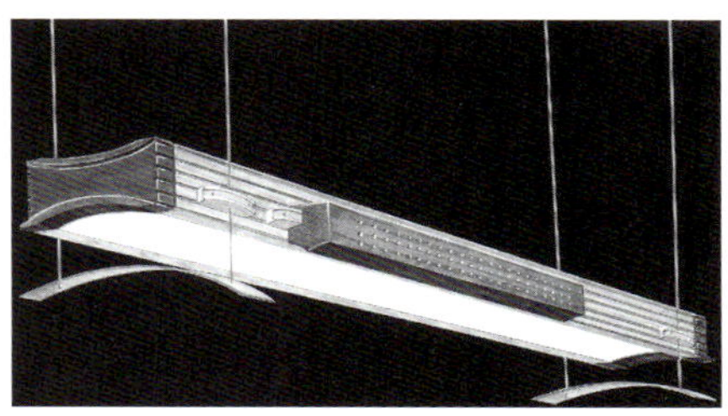

Figure 9.69.2
Direct/indirect light from luminaire diffusers for office applications with glass panel as information carrier
(Model: "Basic Line" with "Pikto", Thorn Licht GmbH, Arnsberg)

Figure 9.69.1
Modular micro-pyramidal lighting system, direct distribution "LIGHT FIELDS cluster version" model, Design by Sottsass Associati
(Image: Zumtobel Staff)

Figure 9.71
Office with combined indirect and direct lighting systems

9.7.1.2 Hallways

The illuminance selected for hallways should largely depend on the material reflection. As mentioned, a material with a reflectance of 15 % appears as bright under 200 lx as a material with a reflectance of 30 % under 100 lx. In hallways where detail recognition is important, the use of bright materials can translate into savings in power consumption.

When hallways are lit with direct systems (downlights), the floor is the most effective optical surface. On the other hand, if indirect lighting is used, the ceiling is the most effective surface. The lit surfaces define both the light ambience and the power consumption. In hallways, indirect lighting systems are more expensive in terms of consumption and investment than direct systems.

Long hallways do not necessarily require the same level of illuminance along the full length. If intersections or areas with information panels are brightly lit, other areas can be relatively dim without running the risk of creating a dreary ambience.

9.7.1.3 Lobbies / teller halls

The exterior design and location of a building are the public calling card of an enterprise. The teller hall of a bank is the place of interaction between bank staff and clients. The design, furnishing, material, colours and light in lobbies and teller halls are especially important for these reasons.

Figure 9.72
Lobby at Lamy Innovationswerkstatt
(ERCO, architects: Bertsch Friedrich Kalcher, Stuttgart)

Large teller halls are usually divided into a client service area, which serves as an orientation and waiting area and is frequently used for exhibitions as well. An understanding of how light and reflection interact is especially important in these areas, because they are generally not designed for visual tasks that require a high degree of concentration, with the exception, perhaps, of exhibitions. Brightness in these interiors is largely a matter of aesthetics. The incident illuminance is not directly perceived by the human eye and can therefore not be evaluated, as luminance is the measure for reflected light and stimulates the retina. Luminance is therefore the decisive parameter for brightness in a teller hall. **Figure 9.72** shows a good example of lobby lighting design, here at Lamy Innovationswerkstatt.

Light-coloured floors in atria – an example is shown in **Figure 9.73** – enhance the impression of brightness under daylighting and artificial lighting conditions.

Open client areas make special demands on lighting design with daylight and artificial light in banks. Overhead openings can frequently be used in atria to allow daylight to penetrate into the interior. Both the staff and the clients benefit from being able to experience the pulsating variations of daylight in the interior. The only conflicting factor is that incident daylight may result in excessive brightness, reducing visibility on screens as a result of reflections from surrounding surfaces and materials. Reducing the luminance with the help of daylight-deflecting elements is therefore as important as adjusting the relevant materials to the anticipated illuminance. The following images present several examples:

Figure 9.73 features a daylighting system in a savings bank, whose roof glazing consists of light-deflecting insulated glass. Direct solar radiation is transformed into diffuse solar radiation. The inner ceiling with specially structured glass is used to reduce luminance. Supplementary lighting is provided by means of fluorescent lamps installed between the two systems and formal lighting accents are added in the form of custom-designed chandeliers with halogen lamps.

Figure 9.73
Teller hall in savings bank, Bremen; lighting with chandeliers fitted with halogen lamps and OKALUX glazing

Figure 9.74
Teller hall in regional savings bank, Osnabrück; lighting with glass strut luminaires and halogen downlights

Figure 9.75
Savings bank in Bamberg, atrium with prismatic glazing and light from HQI pendant luminaires

In another bank building (**Figure 9.74**) incident daylight is restricted to the peripheral areas to avoid having to compensate for heat gain. The structural design ensures that excessive sky luminance is prevented from impacting the interior. The interior consists of a mirrored ceiling, which creates the illusion of extending the room height. Additional lighting is provided by means of fluorescent lamps mounted above glass struts in the ceiling. Formal lighting accents are introduced with integrated halogen downlights in combination with a large lens tip chandelier.

All systems obscure views through the roofs to the outside. The fluctuations in daylight are nevertheless experienced in the interior. Since views of the sky through glazed roofs are rarely feasible in teller halls with integrated workstations, several diffusing systems have been developed, based on the principle of sun-deflecting prisms (see **Figure 9.75**). Metallic deflecting systems between panes are another option for daylighting in teller halls; an example is shown in **Figure 9.76** with an OKA-Solar roof system.

9.7.1.4 Seminar and conference rooms

Seminar and conference rooms are used for formal events as well as professional meetings. They should therefore be equipped with lighting systems that can be adjusted to provide different types of ambience. These should include good working plane lighting of at least 300 lx at the conference table, options for accent lighting on walls or dimming for slide presentations and overhead projectors, as well as ambient lighting for dinners or festive events. This may mean that such rooms contain several different lighting systems, which are used separately for different events or in combination to create the desired effect.

Figure 9.77 shows a medium sized conference room; **Figure 9.78** shows a small conference room and **Figure 9.79** a large conference room.

Figure 9.77
Conference room, regional savings bank, Osnabrück

Figure 9.78
Small conference room at the Bayerische Landesbank, Munich

Figure 9.76
Atrium with OKASOLAR glazing

Figure 9.79
Large conference room, EPA, Munich

9.7.2 Factories and manufacturing sites

The tasks performed at manufacturing or production sites may be far more demanding from a visual performance perspective than those carried out in offices, meeting rooms and schools. Visual tasks in these environments can be very exacting with regard to detail, structure, form and colour.

Due to their massive size, production halls where difficult visual tasks are performed are rarely equipped with a lighting system that is equally intense across the entire space because the investment and operating costs would be astronomical. Instead, the standard approach is to design a general lighting scheme that is further complemented by localized lighting in specific zones. General lighting in large interiors of up to 6 m height is usually realized with luminaires for fluorescent lamps with reflectors, a solution that is efficient and economical. The luminaires are often mounted directly on the ceiling as strip lighting, or suspended (**Figure 9.80**).

Strip lighting should be arranged crosswise to work benches and machines to avoid undesired shadows and reflections.

The same vision laws outlined for office work apply to industrial tasks. The luminance of the luminaires should be as low as possible and reflector strips should be avoided, although these recommendations are sometimes ignored for reasons of cost efficiency. As mentioned, high luminance in the room (caused by open strip lighting) can lead to early onset of fatigue and errors. These errors can be far more expensive than the installation of good quality luminaires with diffusing grids.

Point light sources (metal halogen vapour lamps) in halogen reflecting luminaires are used (**Figure 9.81**) in hall interiors with extremely high ceilings. Hall luminaires with deep bowl reflectors, which can be very economically efficient, may produce reflection glare from glossy materials; it is important, therefore, to select the lighting system under consideration of the materials, which are processed or manufactured at the site and of the visual tasks.

Figure 9.80
Factory hall with suspended lighting strips for fluorescent lamps

Figure 9.81
Exhibition hall with halogen reflector luminaires

9.7.3 Hotels, congress centres and similar environments

In hotel lobbies and congress centres, the visitors' attention should be drawn to the reception area. The reception area can be emphasized by means of design and bright materials or by bright zone lighting. A different light ambience can be used for seating areas, for example, to provide a sense of visual zoning in large hotel lobbies.

Breakfast buffets and restaurants are lit to harmonize with the overall ambience of the restaurant. Spot lighting with incandescent or halogen bulbs can create an entirely different mood at night in an area that is partially daylit at lunch and breakfast times.

In hotels, corridors are a major factor for power consumption because they are almost always located in the core of the building and thus constantly illuminated with artificial lighting. Compact fluorescent tubes are the best choice for these areas because they are more efficient and have a longer lifespan than incandescent or halogen bulbs.

The lighting in hotel rooms and bathrooms is largely dependent on the interior design (furnishings) and the overall style of the hotel.
The minimum lighting features should include:
- Reading light by the bed
- General lighting in room with floor lamp
- Desk lamp
- Mirror light in bathroom
- General lighting in bathroom and hallway

9.7.4 Hospitals

The lighting in hospitals depends on the type of treatment provided. Intensive care units require bright lighting with good colour rendition to ensure good conditions for patient examination and treatment.

In rooms dedicated exclusively to patient examination, illuminance should be higher, providing an adjustable range from 500 to 1,000 lx.

Operating theatres are equipped with special lighting systems above the OR table, suitable for illuminance of 20,000 to 100,000 lx. The general lighting should be designed for 1,000 lx.

Rooms in which patients recover after surgery should have friendly and relaxing interiors where patients feel comfortable. The industry offers prefabricated systems for bed lighting, which are suitable for patient examination and treatment, although they are compromise solutions in terms of comfort.

Corridors and hallways occupy large areas in hospitals, and corridors in particular must be designed to be compatible with the treatment areas. This means they should have a very low illuminance at night and a higher illuminance during the day, which should also add brightness to the peripheral or transition zones into the rooms. Since patients are transported along these corridors and often "parked" in these areas, luminaires should be arranged to provide glare-free areas for patients on trolley beds who are looking at the ceiling (see **Figure 9.82.1**).

9.7.5 Athletics facilities

Athletics facilities include outdoor areas, as well as gyms and sports halls, skating, rollerblade or cycling arenas.

The lighting should create good visual conditions for the athletes, judges and spectators. Especially for sports with balls, pucks etc. such as table tennis or hockey, players, referees and spectators must be able to follow the rapid movement of the ball. This translates into high demands in terms of simultaneity and colour rendition. For training sessions, the illuminance can usually be reduced to half its full strength. For competition the necessary range of illuminance is 400 to 500 lx. Higher illuminance is required only when events are filmed for television broadcasts, e.g. in many multipurpose facilities. This higher illuminance is always beneficial for the athletes and the spectators at such events.

In interiors, fluorescent lamps with good colour rendition are used, while halogen metal vapour lamps with good colour rendition are employed for large halls (high ceilings) and outside facilities. **Figure 9.82.2** depicts the lighting installation in an outdoor facility.

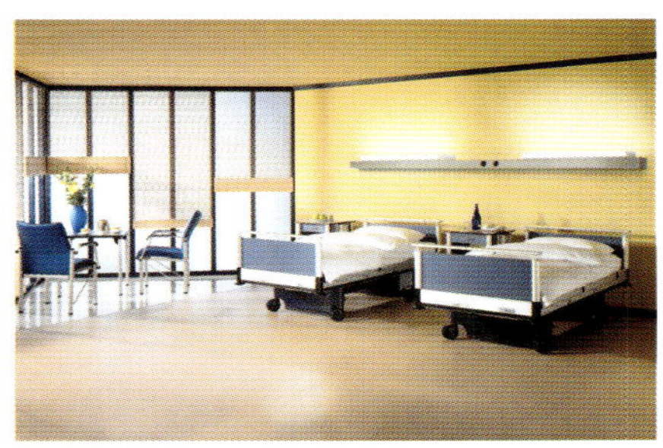

Figure 9.82.1
Lighting and medical supply system for the health and care sector "PURELINE" model, Design by Studio & Partners (Image: Zumtobel Staff)

Figure 9.82.2
Outdoor lighting at sports facility

9.7.6 Museums

In a museum, the exhibit (painting, sculpture etc.) must be the focus. The interior design or the lighting should never vie for attention with the exhibit. A general discussion on perception and stable perception in particular was presented in **Chapter 9.2.3**. It was noted that the object luminance should be higher than the surrounding luminance. In a museum, the object – for example, a painting – is characterized by various gradations of brightness. Old masters worked with luminous colours on dark backgrounds, while modern art favours light backgrounds, or rather, modern paintings are best displayed against a light background. The wall on which the painting is mounted should be understood as the immediate surroundings. It is easy to understand that the demand to keep the surrounding luminance lower than the object luminance evenly spread across the painting is difficult to fulfil for different exhibits. Moreover, conservation dictates that most paintings must be protected from too much light (heat exposure/fading or drying of colours). From a lighting perspective, there should be as little light as possible on the wall to enhance the effect of the paintings; but the wall should not be too dark in the interest of the overall ambience in the room. These parameters apply to diffuse reflection. If the wall covering is chosen to be partially diffusing and light is directed from above, the wall lighting can be fairly bright while still reducing the direct light incidence into the visual range of the observer and thus emphasizing the object luminance.

When a museum is lit with direct luminaires, it is more difficult to achieve this adjustment due to the reflection of the radiation on the paintings. Most museums are therefore lit with indirect artificial light or diffuse daylight. For energy efficiency, it would be preferable to work with direct luminaires (e.g. wall-washer luminaires) and direct daylight, especially since the plasticity of exhibits is increased under direct radiant light (see also **Chapter 10.3**).

9.7.7 Roads and town squares

There are standard regulations for street lighting, e.g. DIN 5044. Effective lighting is always the priority in street lighting, although it is important to use luminaires that do not produce glare, which would interfere with drivers' visual perception.

Town squares and other public places can be lit in a very individualized manner and planners have much more flexibility in terms of lighting design. The mean illuminance or mean luminance do not play a role in squares that are predominantly designed for pedestrians. Other tools are used in these areas to create an interesting environment, for example, floodlights on facades (see the office building in **Figure 9.83**). **Figure 9.84** shows the town hall square in Hamburg and **Figure 9.85** shows the Brandenburger Tor in Berlin.

Zone lighting can be very effective and above all energy efficient in arcades, as shown in the image of the arcade in the main railway station in Munich in **Figure 9.86**.

Parks and gardens are lit in an even more creative manner, for example the approach to a museum shown in **Figure 9.87**.

Figure 9.83
Facade lighting (floodlights) (Image: Sill)

Figure 9.84
Outdoor lighting on Rathausmarkt, Hamburg

9.7.8 Other situations

In other buildings, lighting must meet the demands for a wide variety of visual tasks, for example in
- Retail areas
- Shop windows
- Supermarkets
- Speciality workshops etc.

As described for the various rooms in the preceding sections, each room and each demand needs to be assessed individually.

From the perspective of energy consumption, lighting technology should always be viewed as a complement to daylighting. Lighting systems that respond to daylight should therefore be used whenever possible. The artificial light is controlled to provide a nominal level of light in response to the available daylight and user requirements. According to building type and use, energy savings may be 30 – 70 %. These solutions require adjustable luminaires with relevant accessories, preferably a bus system that meets EIB/LON or comparable standards.

Figure 9.86
Shopping concourse at Munich railway station

Figure 9.87
Entrance lighting, museum in Frankfurt on Main

Figure 9.85
Since October 3, 2002, the famous landmark of the formerly divided and now reunited city of Berlin once again looks out across the city with renewed vigour. (Image: Sill)

9.7.9 Emergency lighting

Emergency lighting is used for escape routes, safety lighting for high-risk work environments and back-up lighting. Emergency lighting must therefore also be capable of replacing general lighting in the event of power failure. It is also required in all areas where minimum illuminance along escape routes etc. is required in case of any interruption in the electrical power supply. Two types of emergency lighting are generally differentiated according to the purpose they fulfil:
- Safety lighting
- Back-up lighting

Emergency lighting is obligatory in all rooms for public gatherings, e.g. cinemas, theatres, exhibition rooms, dance halls, restaurants, hotels, hospitals, athletics facilities, high-rises, shopping centres, stairwells without natural light and electrical equipment rooms.

Emergency lighting systems must be connected to a power source that is independent of the grid in order to operate even in the event of total grid failure. The following components are used to supply emergency power for emergency lighting systems:
- Central battery plants to supply emergency power to many luminaires
- Accumulators with a charge device in the luminaires themselves
- Back-up systems for large building projects

The feed power for the lamps is provided by frequency transformers, which are connected to the batteries. Both can be integrated into the luminaires, provided there is sufficient space. It is important to ensure that the permissible ambient temperature is maintained to avoid reducing the lifespan and performance of the system. Operational tests should be undertaken every time an emergency lighting system has been in use.

The authorities (police, fire department, construction supervision etc.) generally require regular safety checks and supervision to ensure that emergency lighting systems are in working order. Fire and police department regulations specify that emergency lighting must be provided
- To identify escape routes and exits
- In underground car parks
- In high-rises
- In retail stores and shopping centres

Bus-monitored system luminaires are generally used to test and monitor safety and escape route lighting. The bus system monitors the working order of the entire system. The system sends out notification of any dysfunction, e.g. defective lamps; the notification is printed on a protocol printer or passed on to the central control station.

1. Building performance
2. Comfort
3. Integrated planning models
4. Heating systems
5. Sanitation and fire protection
6. Ventilation and air-conditioning technology
7. Refrigerating and cooling systems
8. Power plants
9. Lighting

10. Daylighting

11. Low-voltage systems
12. Transportation systems

10. Daylighting

10.1 Daylighting terms

(with Michael Schmidt, Munich)

Before we can begin to calculate internal daylighting conditions, it is important to understand celestial luminance and illuminance (sky factors).
These factors are defined for all sky conditions.

Overcast reference sky

The symmetrical, rotational distribution of luminance under an overcast sky is expressed as:

$$L(\varepsilon) = L_2 \frac{1 + \cos\varepsilon}{3}$$

or

$$L(\gamma) = L_Z \frac{1 + 2\sin\gamma}{3}$$

Where:

- $L(\varepsilon)$ Luminance of celestial points, distanced from the horizon at an angle ε
- $L(\gamma)$ Luminance of celestial points, distanced from the zenith at an angle γ
- ε Angle between observed celestial point and zenith
- γ Angle between observed celestial point and horizon (**Figure 10.1**, solar position angle reference)
- LC Luminance for clear sky

Clear reference sky

The relative distribution of luminance is defined according to CIE No. 22 (TC-4.2) for clear skies.
[CIE = Commission Internationale de l'Eclairage]
(CIE = daylight standard).

Mean reference sky

Overcast and clear sky, in the sense used in the preceding paragraphs, are idealized scenarios that are rare in reality. For long-term observation – e.g. efficiency calculations – it is necessary to factor in mean values resulting from a period spanning several years.

Duration of sunshine

is the sum of all time intervals within a given period (hour, day, month, year) when the radiation strength of direct solar radiation on a surface perpendicular to the direction of solar incidence is greater than or equal to 120 W/m² (roughly 11,000 lx). This radiation strength is recommended as the threshold value for bright sunshine by the WMO (World Meteorological Organization). Older data records are based on a threshold value of roughly 200 W/m².

Possible duration of sunshine

is the sum of time intervals within a given period when the sun is positioned above the true horizon, which may be obstructed by mountains, buildings, trees etc.

Relative duration of sunshine

is the ratio of the duration of sunshine to the possible duration of sunshine within the same given period.

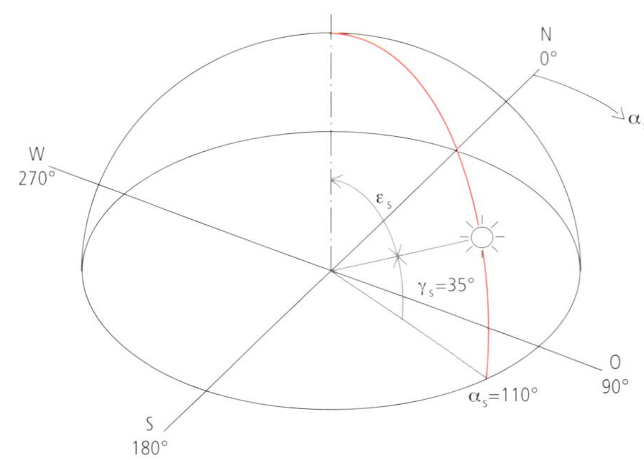

Figure 10.1
Solar position angle reference

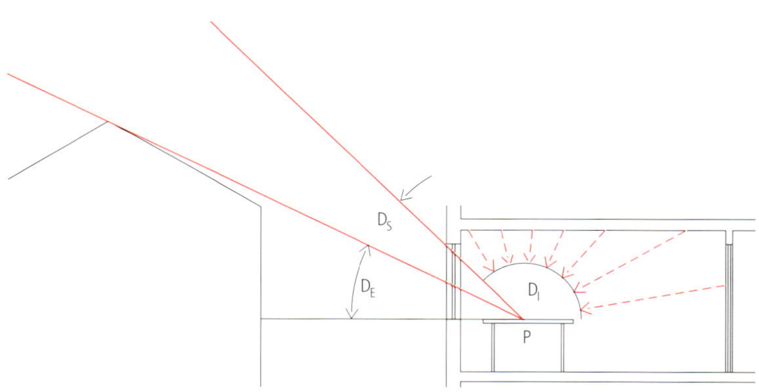

$E_i = D_S + D_E + D_I$

$D = E_i/E_{OU}$

- D_S = sky component
- D_E = external reflectance
- D_I = internal reflectance
- P = measuring point
- E_{OU} = horizontal outdoor illuminance unhindered by built environment
- E_i = illuminance at point P

Figure 10.2
Explanation of daylight factor components

Sunshine probability
is the long-term average of the instantaneous values of the relative duration of sunshine.

Solar constant E_o
expresses the radiation strength of extraterrestrial solar radiation on a surface perpendicular to the incident angle for average solar distance:
$E_o = 1.37\ kW/m^2$

Global radiation
is the sum of direct and diffuse solar radiation. Unless otherwise indicated, global radiation is always referenced to the horizontal plane. Diffuse solar radiation was formerly referred to as diffuse celestial radiation.

Turbidity factor T
is the ratio of the vertical optical density of a clouded atmosphere to the vertical optical density of the pure and dry atmosphere (Rayleigh atmosphere) in relation to the entire solar spectrum.

Built environment
indicates the spatial range that constantly fills one part of the horizon when seen from a particular point (houses, trees, mountains etc.).

Daylight factor D_f
indicates the ratio between internal and external illuminance.

$$D_f = \frac{E_i}{E_{ou}} \cdot 100 \text{ in } \%$$

E_i = illuminance at a point in the room produced by direct or indirect sky light for an assumed luminance distribution
E_{ou} = horizontal outdoor illuminance assuming the hemisphere is completely open (no built environment).

Direct sunlight is not taken into consideration when calculating daylight factors. However, it is essential to consider factors such as glazing, mullions and dirt on the windows, as well as the internal conditions of reflectance.

Hence, the daylight factor is the product of three components: the sky, and external and internal reflectance. The components of the daylight factor are illustrated in **Figure 10.2**.

The quality and quantity of daylight change according to time and place. A standard light D65 (similar to colour temperature 6,504 K) has therefore been set for calculations. The spectral composition of this standard light is illustrated in **Figure 10.3**. This standard light is only used for calculations since it cannot be technically reproduced.

The standard light type A shown in **Figure 10.3** was used formerly. It corresponds to the spectral range of an incandescent lamp. This spectral range features an increase in the red range and no longer corresponds to modern lamps such as discharge or fluorescent lamps.

Solar parameters

Luminous intensity of the sun	3 · 1027 cd
Mean luminance	1.93 · 108 cd/m²
Annual emission of energy	3.2 · 1027 kWh

Each m² of the earth's surface receives an avesage of 148 W solar ebergy per hour.

Solar altitude γ_s – solar azimuth α_s
Since the earth rotates around the sun and around its own axis, each location on the earth has a sun position dependent on hour and season. This sun position is defined by solar altitude and solar azimuth.

Solar altitude is the angle between the centre of the sun and the horizon, according to latitude, season and hour.

Solar azimuth is the horizontal angle between the reference direction North and the vertical circle through the centre of the sun (0° – 260°), again according to latitude, season and hour. The two angles are illustrated in **Figure 10.1**.

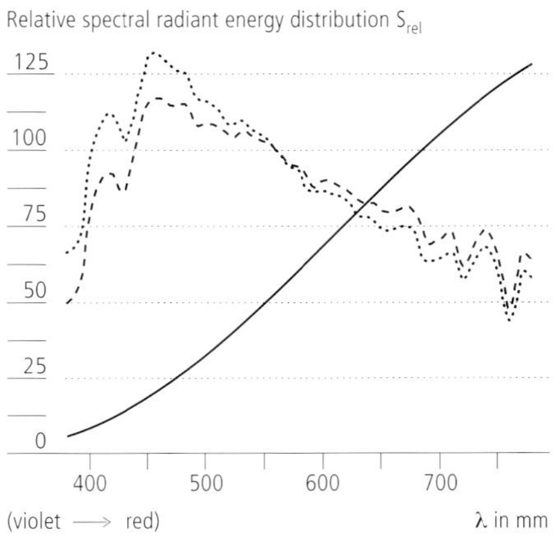

Figure 10.3
Relative spectral radiant energy distribution for standard light types A, D75 and D65

— A
- - - D65
······ D75

Solar declination δ

Solar declination is the angle between the centre of the sun and the celestial equator according to season. **Figure 10.4** plots the annual solar declination.

Illuminance/sky factor

The horizontal external illuminance depends on the solar altitude angle. This angle changes over the course of each day and the seasons at each specific location. **Figure 10.5** illustrates the horizontal illuminance E_{ou} in relation to hour and season. The illuminance in W/m² is also indicated. **Figure 10.6** also plots the illuminance under an overcast sky over the course of each day and the seasons in a manner that clearly identifies the diurnal illumination for each month.

The sky factor is expressed as a multiple of the zenith factor. When the sun is at its zenith, the sky factor between zenith and horizon fluctuates between 20 % and 50 % of zenith luminance (**Figure 10.7.1**).
In Central Europe, this is merely a theoretical model. **Figure 10.7.2** thus offers the same chart for a solar azimuth of 40°.

This graph illustrates that sky factors are up to five times higher when the sun is low on the horizon than when it is at its zenith.

Under an overcast sky, the relative distribution of the sky factor is symmetrical to the earth's rotation. The factor at the horizon is approximately one third of the factor at the sun's zenith (**Figure 10.8**).

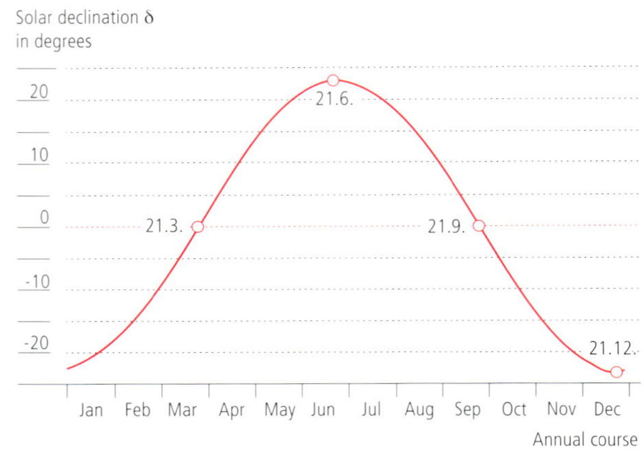

Figure 10.4
Annual solar declination δ

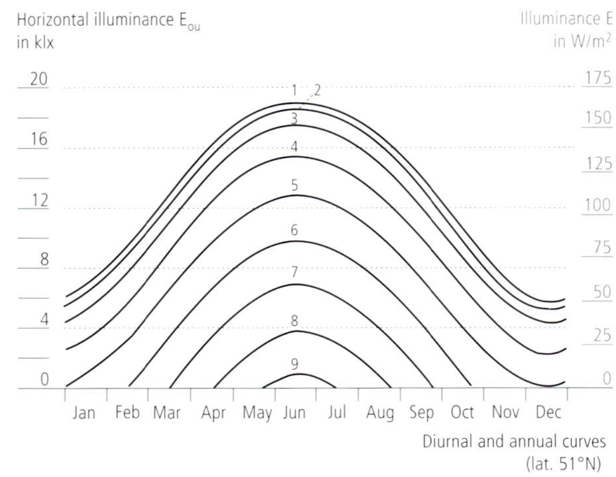

Figure 10.5
Horizontal illuminance E_{ou} in klx and illuminance E in W/m² in relation to hour and season (lat. 51° N)

1 12 o'clock
2 11 a.m. or 1 p.m.
3 10 a.m. or 2 p.m.
4 9 a.m. or 3 p.m.
5 8 a.m. or 4 p.m
6 7 a.m. or 5 p.m.
7 6 a.m. or 6 p.m.
8 5 a.m. or 7 p.m.
9 4 a.m. or 8 p.m.

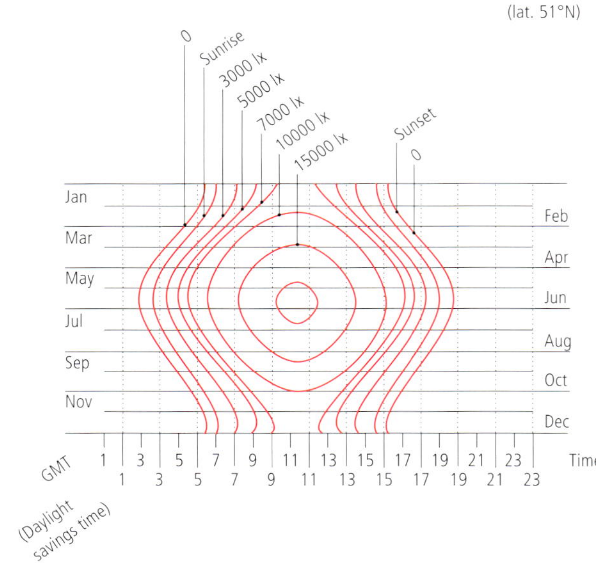

Figure 10.6
Diurnal and annual illumination factors for overcast skies

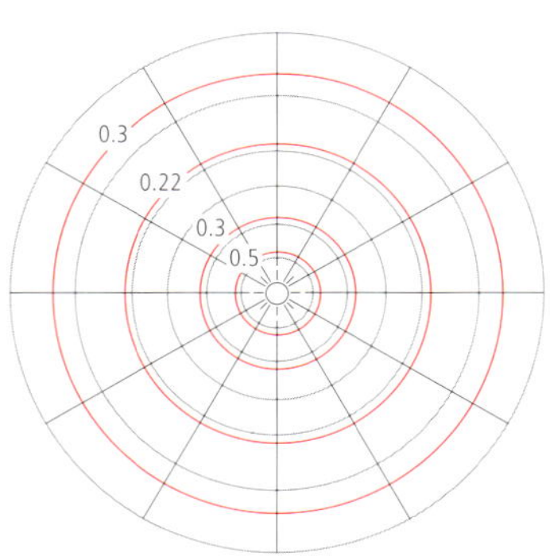

Figure 10.7.1
Luminance distribution under clear sky at solar azimuth of 90° expressed as a multiple of the zenith luminance

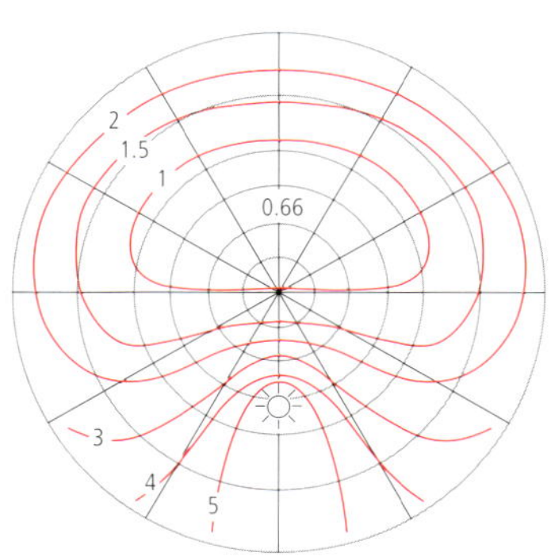

Figure 10.7.2
Luminance distribution at solar azimuth of 40°

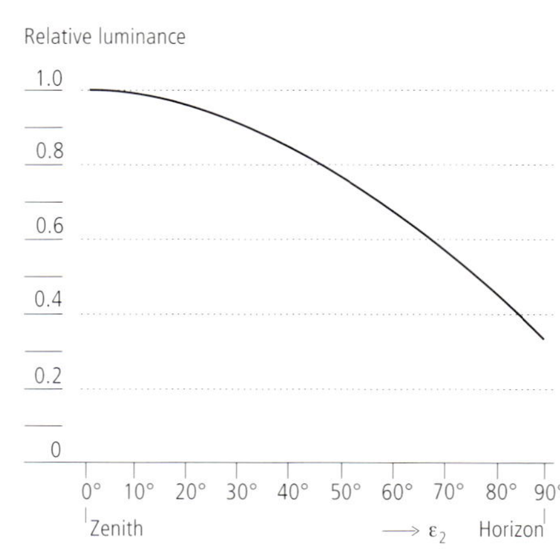

Figure 10.8
Relative luminance distribution under overcast sky

10.2 Daylight in buildings

10.2.1 Effectiveness criteria
10.2.1.1 General information

Like artificial lighting, daylighting should adhere to specific effectiveness criteria to provide optimum visual comfort for occupants, although it is important to remember that one essential function of a window is to offer visual contact with the outside. To feel comfortable, humans must be able to establish contact with their external environment through windows and have a sense of the light and weather conditions outside. Skylights should therefore at least provide a perception of the pulsation of daylight. This very important criterion relating to visual contact with the outside is not included in the standard lighting effectiveness criteria.

Again, as for artificial light, the distribution of illuminance in a room plays an important role in the sensory perception of the ambience. Daylighting at the work place should aim to provide even light at the working surface – a difficult feat with lateral daylight openings (cf. reduction of daylight factor in the depth of a room in **Figure 10.9**). Special glare protection and shading systems have been developed for the purpose of allowing daylight to penetrate deeper into the space. These systems will be discussed at a later point.

Daylighting of specific zones in halls and corridors may also be useful. In some cases, controlled direct incidence of sunlight may also be desirable to create a specific ambience.

10.2.1.2 Illuminance – daylight factor

The daylight factor provides information on the illuminance at a reference point in the room according to external illuminance. The daylight factor does not make a statement with regard to the total illuminance in the room, since the illuminance of daylight fluctuates with time and season, sun position and atmospheric clouding. The location and dimension of a window, its shading and glare protection features, as well as the built environment, all play a role in the daylight factor, as do the sky factor and the internal reflectance. **Figure 10.9** shows a typical daylight factor curve in a side-lit room; it also illustrates that rooms on the 1st floor of a house receive less daylight than rooms on the 4th floor.

The daylight factor is given for an overcast reference sky and the illuminance calculations are based on the same assumption. Side-lit rooms should feature the following minimum daylight factors:

- Living areas at mid depth and 1 m distance from side walls $D_{Fmin} = 0.9\ \%$
- Work areas at least favourable reference point $D_{Fmin} = 1\ \%$
- Work areas with light openings in more than one zone, predominantly side-lit $D_{Fmin} = 2\ \%$
- Work areas, predominantly top-lit $D_{Fmin} = 4\ \%$

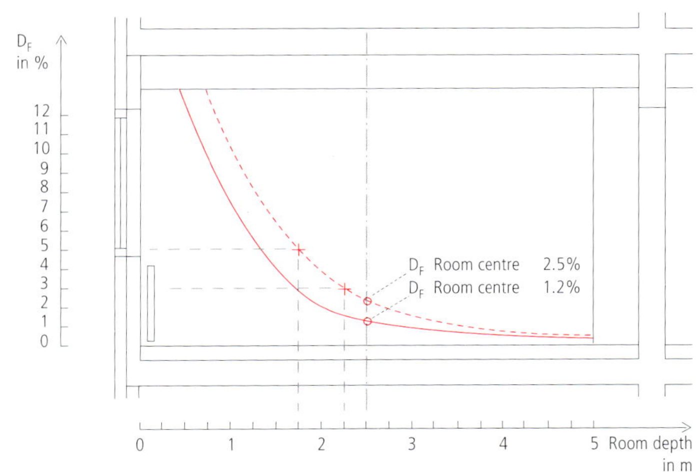

Figure 10.9
Daylight factor curve in a side-lit room

Window: clear glass
Transmittance 80%
Reflection 15%
Dirt factor 90%

—— Room 1st floor
- - - - Room 4th floor
–·– Room centre

10.2.1.3 Glare-free conditions

The effectiveness criterion of glare-free conditions demands a high degree of innovation on the part of all building experts to design an integrated window solution. The architect will be chiefly concerned with the aesthetic placement of the window and its effect on the facade. The heating and cooling engineer's principal goal is to achieve heat gain in winter and block unwanted heat gain in summer. The lighting engineer is faced with a conflict. On the one hand, the daylight factor should be as high as possible to minimize the need for artificial lighting; on the other, the illuminance even of an overcast sky may be so high that glare protection measures are necessary to reduce it, for example, to prevent glare on monitors and screens. Reducing the illuminance at the window and introducing shading or glare protection features will, however, inevitably reduce the amount of incident daylight and hence the daylight factor. Designers and clients are thus often faced with the difficult decision of having to choose whether more daylight will lead to less energy consumption for artificial lighting or to greater energy consumption (and costs) for cooling in air-conditioned buildings. In each case, the primary focus should be on providing the worker/occupant with a comfortable environment.

Shading and glare protection are therefore key components in any lighting strategy and to this end the following paragraphs present several specific examples.

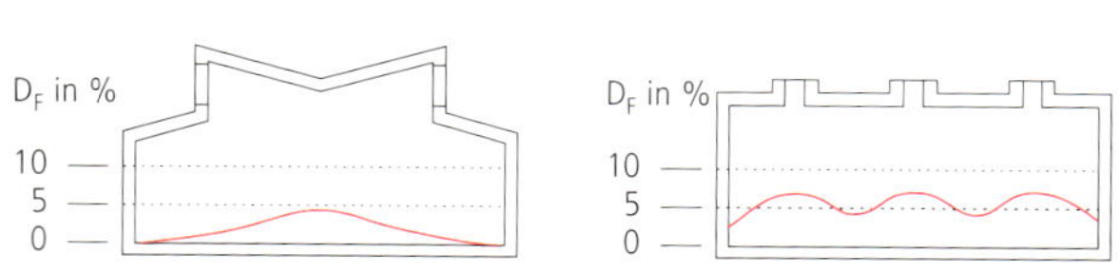

Figure 10.10
Daylight factors for different daylight openings of the same size in identical rooms

10.3 Daylight openings in buildings

10.3.1 Placement of daylight openings

The closer to the horizontal a daylight opening is in a building, the more daylight it will allow to penetrate into the interior. Skylights permit daylight to penetrate from the entire unobstructed hemisphere, while vertical windows on the sides admit only a percentage of the sky reflection. **Figure 10.10** indicates the daylight factors in identical rooms with identical window surfaces but different window placement.

10.3.2 Skylight designs

The following skylights are commonly used on flat roofs:
- Domes set right into the roof skin
- Domes with base (monitor skylight)
- Sawtooth overhead lights at various angles
- Northern light sheds

With the exception of the northern light shed, these skylights are installed without consideration for the building orientation. The required deflection of direct sunlight is achieved mainly by means of glazing with light-scattering properties.

The disadvantage of this type of glazing is, however, that the transformation of direct solar radiation into diffuse radiation may result in excessive luminance and interfere with the visibility of monitors and screens.

Various glass combinations are therefore used to reduce the luminance of skylights, despite the impact this has on the daylight factor and the amount of daylight that penetrates the room.

10.3.3 Shading and glare protection for skylights

The following shading and glare protection measures are frequently used for skylights:

10.3.3.1 Reflective glazing

Reflective glazing reduces the daylight factor by the visible transmittance of the pane and the diminution factors resulting from mullions and dirt. For a given roof size, the daylight factors, visible transmittance and dome size can be adapted to the specific requirements of the lighting strategy. In buildings and interiors where a certain amount of direct incident sunlight is permissible and where the energy reduction of reflective glazing is chiefly related to heating issues, this type of glazing is also acceptable from a lighting perspective.

This is especially true for internal streets or arcades. In entrance atria that include workstations, the reduction achieved by reflective glazing is generally insufficient, unless other shading features are added to ensure that direct sunlight cannot penetrate into the working area.

The image in **Figure 10.11** documents overhead reflective glazing above a shopping arcade in Hamburg.

Figure 10.12
Skylights with rigid external shading system

10.3.3.2 Clear glass skylights with external, rigid shading components

External, rigid shading components are a good solution if they shield against direct incidence of sunlight for all solar azimuth angles at the relevant location. On shed glazing (facing in a direction other than north) this effect is often achieved with horizontal louvres. However, since the incidence angle on south-oriented sheds is shallow as the sun describes its path from east to west, external shading is often a compromise solution. It protects mostly against the direct southern sun, whereas sunlight from all other directions (shallow incidence angles) continues to penetrate directly into the room with all the disadvantages this entails. **Figure 10.12** depicts an example of this type of rigid shading system.

"Intelligent" daylighting systems, described in section 10.3.3.6 as prismatic shading and in section 10.3.3.7 as insulated glazing with mirrored deflection panels, can overcome this disadvantage.

Figure 10.13
Awnings as moveable external shading features on a pagoda style skylight

Figure 10.11
Overhead reflective glazing, Hanse Viertel (shopping concourse), Hamburg

10.3.3.3 Skylights with moveable shading components

Moveable shading components on skylights almost fully eradicate the disadvantages associated with rigid shading components. Insulated glazing deemed suitable for an overcast reference sky in consideration of all relevant criteria is used for the skylight. Penetration of direct solar incidence is prevented by moveable shading systems using external louvres or internal screens. The daylight factor of such systems is quite good for both scenarios, since it corresponds to the calculated values for an overcast sky and reduces the disadvantages of shading when exposed to direct sunlight through the higher external brightness. However, these systems are elaborate in terms of design and construction, and expensive in terms of initial investment and maintenance. Internal shading by means of screens or fabric rarely achieves the desired reduction of luminance in terms of direct incidence.

Figure 10.13 depicts a skylight (pagoda style) with moveable external shading in the form of awnings. Naturally, moveable shading systems diminish or prevent visual contact with the outside when they are drawn.

10.3.3.4 Natural shading through plants

In some top-lit structures such as arcades or entrance atria, it is sufficient to shade the roofs by planting trees and shrubs, which provide protection against direct solar radiation.

10.3.3.5 Light-deflecting glass

The use of light-deflecting glass is a classic approach to reducing solar luminance and transforming direct solar incidence into diffuse light. The technical parameters of this type of glass can be predetermined and light-deflecting insulated glass is easily processed into skylights.

From a lighting perspective, the principal disadvantage of these systems is that the transformation of direct sunlight into diffuse light produces excessive luminance in some work areas, which may lead to reflections on monitors. Because diffuse light does not contain significant ratios of direct radiation, contours in the room may lack definition and be characterized by low plasticity (lack of shadowing). Another disadvantage lies in the fact that there is no direct visual contact with the outside. Skylights with light-deflecting glass should therefore only be used in spaces where vertical windows on walls overlooking a clear space (unobstructed exterior) provide the necessary contact with the exterior. **Figure 10.14** portrays a room with a skylight with light-deflecting glass.

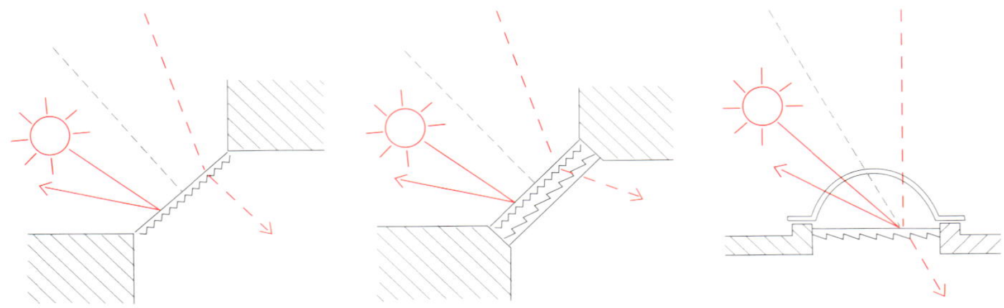

Figure 10.15
Prismatic shading system for skylights

Figure 10.16
Skylight with prismatic shading

10.3.3.6 Prismatic shading on skylights

Prismatic systems for skylights have been developed by the Siemens corporation in recent years. Prismatic panels are integrated into insulated glazing, manufactured especially for different orientations. As solar radiation falls onto the window at differing incident angles, the prisms deflect the radiation and reflect up to roughly 70 % away from the glass combination back towards the outside. **Figure 10.15** illustrates the functional principle of different systems with prismatic shading. This solution offers good shading, a low thermal transmittance (approx. 1.5 – 1.8 W/m^{2}K) and achieves satisfactory daylight factors. Light can be partially deflected into the room in a controlled manner, which is one of the great advantages. The sky luminance is sufficiently reduced to prevent visual interference at monitor workstations in the interior. Depending on the design of the prismatic shading components, visual contact to the outside may be more or less limited; complete transparency (visual contact to the outside) is not possible. **Figure 10.16** depicts a skylight with prismatic shading.

Figure 10.14
Skylight with light-deflecting glass
(Image: OKALUX)

Figure 10.18
Glazing over covered walkway with OKASOLAR system

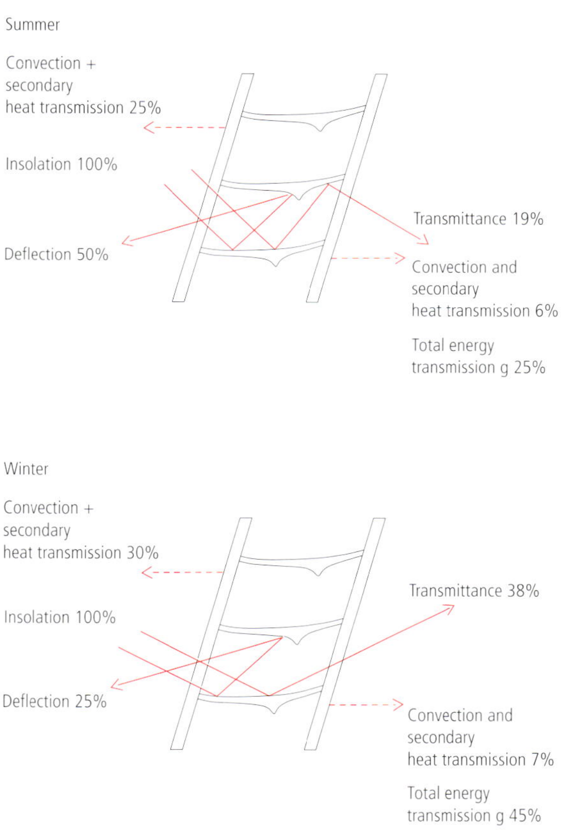

Figure 10.17
Thermopane system with deflecting units (OKASOLAR)

10.3.3.7 Insulated glass with mirror profiles

The firm OKALUX has developed a rigid shading system, in which specially formed mirror profiles are installed between two glass panes. The system is compatible with all standard glass types. The mirror profiles are formed to control the transmission of light independently of the altitude angle. In winter, when the sun is low on the horizon, much of the solar radiation is allowed to penetrate and in summer, when the sun is high, much of the radiation is deflected to the outside. Sunlight still penetrates into the room in summer, even when the sun is low on the horizon. **Figure 10.17** illustrates the functional principle of this shading system.

The technical data of this system depend on the solar altitude angle (total energy transmission g, daylight factor).

Visual contact with the outside is given at certain angles and completely blocked at others.

The detail in **Figure 10.18** depicts a covered walkway, whose glazing was combined with this OKA-Solar system.

10.3.3.8 Light-deflecting mirror systems in fixed skylights

The principle of deflecting sunlight independent of the incident angle can also be achieved with specially formed light-deflecting mirror systems instead of prismatic deflection systems. These systems are especially suitable for daylighting spaces with high ceilings, where daylight must fall onto a working plane across a great height and in applications where daylight needs to be deflected onto a wall.

The functional principle of these systems is explained in **Figure 10.19**.

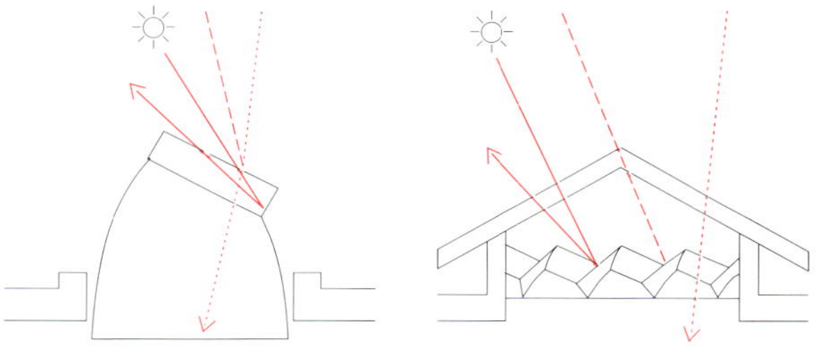

Figure 10.19
Deflecting system with fixed skylight

The example shown in **Figure 10.20**, the Museum of Natural History in Stuttgart, illustrates how light is deflected into a space from a great height. The image clearly shows how the skylight system reduces luminance and how the directed light promotes plasticity in the appearance of the exhibits.

Figure 10.21, portraying an interior and a roof view of the Kunstmuseum Bern, provides an example of a rigid light-deflecting system with aluminium profiles, directing light to the wall zones.

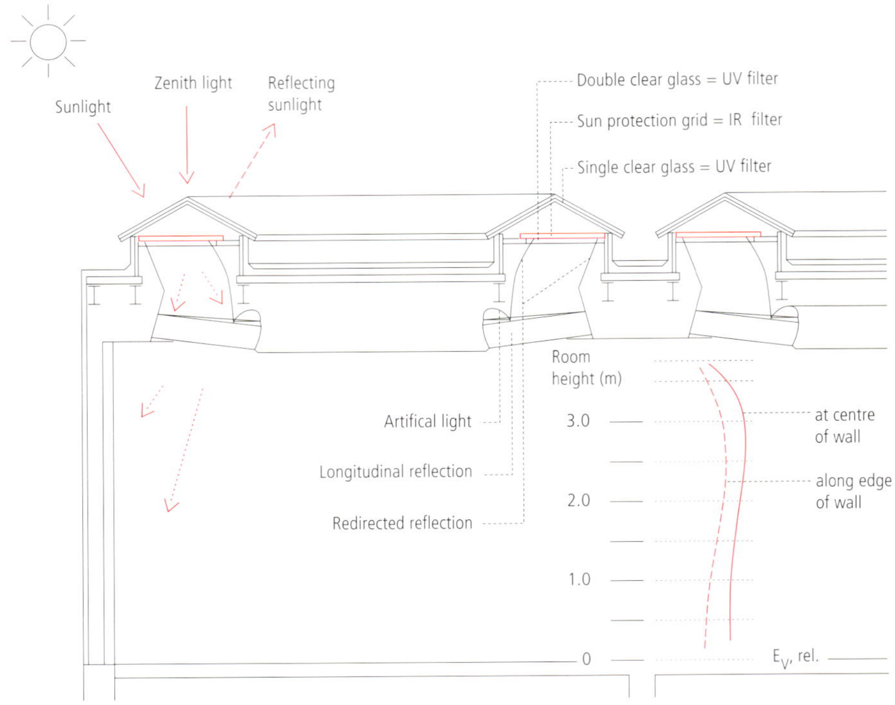

Figure 10.20
Aluminium skylight, Museum of Natural History, Stuttgart

Figure 10.21
Light cast onto walls in Bern

10.3.4 Shading and glare protection on vertical windows

10.3.4.1 General information

An unobstructed view is more important with vertical windows than with skylights. Shading and glare protection features should not result in a complete loss of visual contact with the outside. Reducing visual contact is tolerable only if the window is exposed to excessive direct sunlight, which interferes with perception and visibility at monitor workstations.

Shading has quite a different task from glare protection. Generally, both criteria cannot be solved by a single measure. Glare protection should reduce excessive luminances near windows, as they diminish perception in general and at monitors in particular, while shading should prevent direct sunlight from penetrating into the room. The latter is generally necessary for thermal control as well as for luminance control in the working plane.

The following paragraphs contain a brief overview of current, standard shading and glare protection measures on windows.

10.3.4.2 Shading and glare protection through window combinations with reduced light transmittance

A number of different window combinations make it possible to reduce luminance to any desired level (absorbing and reflecting panes). This can improve the criteria for glare protection. Group offices that are lit from several sides, and in which PCs and monitors are set up in various flexible locations, are characterized by excessively high luminance when clear glass is used, leading to poor visibility at visual display terminals and screens. For these types of interiors, window combinations can be designed so as to reduce the sky luminance by 80 %.

In open-plan spaces, typical of the 1970s and 1980s, such glare protection measures have yielded good results, since daylight penetration from windows does not play an important role in these rooms. On the contrary, by reducing the overall light transmittance, window areas are no longer the only sought-after locations for workstations in these interiors.

Shading systems rarely solve these problems, since glare can also occur on overcast days, and for this reason internal blinds and shades are also used. Due to the greatly reduced daylight transmittance, the days and hours when drawn blinds prevent visual contact with the outside can be limited to a considerable degree. Contact with the outside is provided by the fact that the illuminance of daylight is higher than the illuminance in the interior, which means that objects on the outside appear brighter than those on the inside. The construction of these types of rooms should be avoided whenever possible because the staff is forced to work almost exclusively under artificial light conditions.

In smaller rooms, greatly reduced light transmittance as a result of shading windows with reflective glazing is perceived as unnatural by most occupants, increasing the use of artificial light source. Windows in small rooms should therefore always feature clear glass or reflective glass with a transmittance of over 70 %, although this calls for additional glare protection.

10.3.4.3 External, moveable shading with external blinds

External blinds reduce direct solar incidence and heat gain in rooms (**Figures 10.22.1/2**). They are popular for this reason, although their use is limited in regions with strong prevailing winds or on high-rises. However, in combination with clear glass, the reduction of luminance near the window is not sufficient in all cases. Unless external blinds are carefully adjusted and operated, they interfere with visual contact with the outside (angle adjustment). It is therefore sensible to adjust the louvres to the azimuth position in order to provide a comfortable measure of contact with the surroundings. From a lighting perspective, external blinds should be used for smaller rooms only if the louvre positions can be individually adjusted to the azimuth, or in open-plan rooms if this adjustment is centrally controlled. Different colours on the top and bottom faces of the louvres can help to satisfy the need for more light in the room in combination with appropriate glare protection.

Figure 10.22.1
All-metal blinds as external shading system
(Image: Krülland)

Figure 10.22.2
Louvered blinds as external shading system
(Image: Krülland)

10.3.4.4 External, moveable shading with awnings

Special fabric panels in the form of awnings, roller blinds or screens can be used instead of external louvres. These features reduce or block direct solar incidence (**Figure 10.23**). It is important to ensure, however, that these fabric panels do not assume excessive luminance themselves when exposed to direct solar radiation, as this can also lead to glare and the associated difficulties for occupants (diminished legibility at monitors etc.).

Direct visual contact to the outside is generally prevented with these types of awnings etc., when they are operated in a vertical direction. Semitransparent screens used to provide both shading and glare protection while still allowing some degree of visual contact with the outside are a good solution. The best solution is to install external awnings etc. in a manner that structural projections create an open vista, thus maintaining visual contact with the surroundings.

Figure 10.23
Fabric blinds as external shading system
(Image: Krülland)

10.3.4.5 Internal, moveable shading with vertical louvered blinds, special fabrics or screens

From a lighting perspective, the issues arising with internal shading are the same as with external shading systems. The advantage of internal shading is that it is less elaborate in design and construction since it is not exposed to the elements (**Figure 10.24.1**). On the other hand, the disadvantage is that these systems are rarely ideal in terms of heat gain. Active cooling measures can be used to control undue heat loads in the room (**Figure 10.24.2**). Another approach is to employ a combination of reflective and absorbing glass or a combination of external and internal blinds, or quasi "external" blinds in double-skin facades.

Figure 10.24.1
Various interior shading/glare protection systems: vertical blind, louvered blind, roll-up blind
(Images: Hüppe)

Figure 10.24.2
A combination of internal shading features and water-filled cooling components serves to compensate for the external cooling load directly at the window.
(System Actiso, Kessler + Luch)

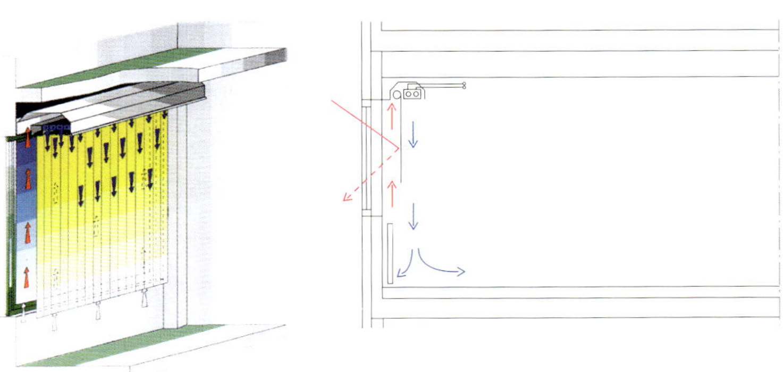

10.3.4.6 Shading with prisms

Prismatic shading systems have been developed for both skylights and windows. They operate by redirecting direct solar radiation in the required range of angles (approx. 20 – 60°), reflecting it back into the room and directing the zenith light into the interior. A section of the vertical window or the skylight is replaced by a prismatic element, while the remainder is realized with standard glazing, shading and glare protection features to ensure visual contact with the outside. However, daylight can be sufficiently redirected into greater depths of the room only if the ceiling is designed with special mirror elements and if light-deflecting louvres reflect the light from the ceiling. The effect of prismatic elements lies in their ability to reflect light in the azimuth range of possible sunshine hours (Central Europe, approx. 20 – 60°) and to redirect zenith light in a range of roughly 60 – 90°. While the sky luminance is very high at the zenith, the angle section is relatively small with side lighting. Light gain is therefore tolerable. If only one part of the window is equipped with prismatic elements, shading and glare protection must be provided for the remaining window areas. Flexible prismatic shading can be operated individually much like a standard blind. The light penetrating into the space is relatively high in this case.

The light direction differs, however, and a deflecting system must radiate the light entering into the room against the partly metallic reflecting ceiling in order to increase the daylight factor in the depth of the room. Light gain and glare protection are designed to reduce the need for artificial light on clear days drastically. However, contact with the surroundings is also reduced and the high-tech aesthetics of metallic light-deflecting louvres and metallic ceiling components are a matter of taste. While these system solutions offer improved responses to human visual demands, the occupants' comfort may nevertheless be diminished if visual contact with the surroundings is reduced. These systems may be an excellent solution for buildings where computer tasks are being performed all day long. However, in interiors where a variety of tasks are performed involving computer use only some of the time, system solutions of this kind are questionable.

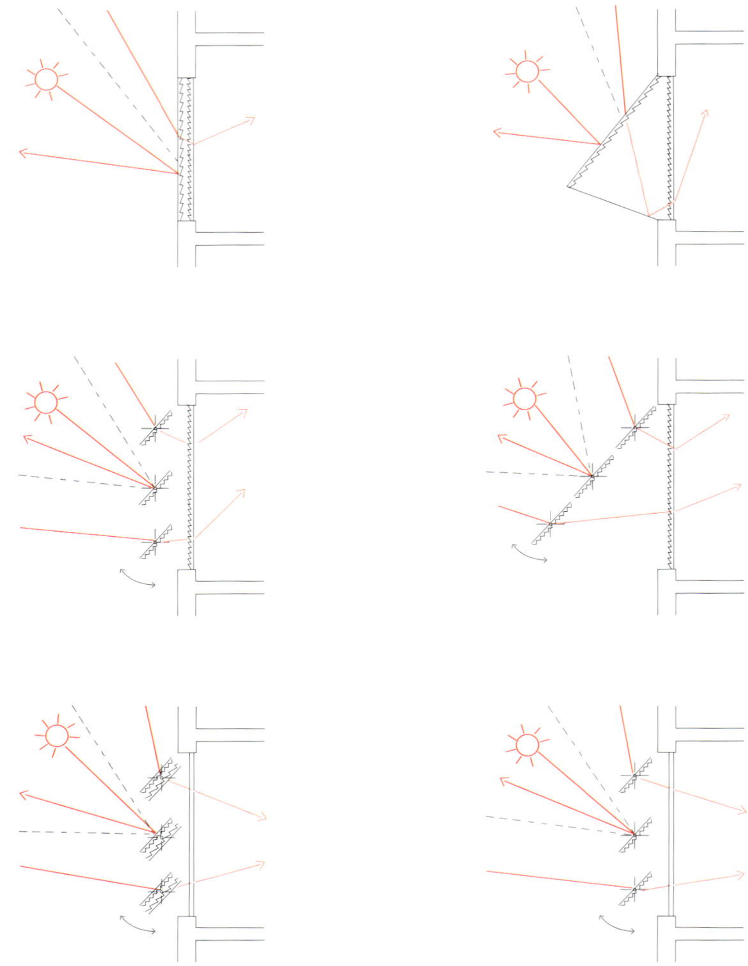

Figure 10.25
Principle of lateral sun protection with solar shading prisms

- - - max. solar altitude
—— direct solar incidence
—— diffuse daylight

non-coated, moveable solar shading prismatic panel

metal-coated, rigid solar shading prismatic panel

diffracting prismatic panel

Figure 10.26
Room with shading feature and daylight-deflection by means of a diffracting venetian blind (Image: Hüppe)

Figure 10.27
Room with lateral shading by means of mirror profiles

Figure 10.25 illustrates the functional principle of side-lit shading with prisms, **Figure 10.26** shows a room with light-deflecting blinds for daylight redirection. The initial investment costs for installing prismatic elements in buildings are often disproportionate to the savings achieved by reducing the burning hours of artificial light and the associated lower energy consumption and energy savings.

10.3.4.7 OKA-Solar louvres

In analogy to shading with prisms, the OKA-Solar louvre system is used to install laminated glass with specially formed mirror profiles that deflect light to the ceiling into part of a window. Shading with OKA-Solar systems allows part of the direct radiation to penetrate into the room when the azimuth is low, while deflecting direct radiation and much of the light to the outside when the azimuth is high.

Figure 10.27 shows a room with shading effected by means of mirror profiles. The seemingly excellent visual contact with the outside seen in this photograph is possible only from a nearly horizontal perspective. The sunlight-deflecting quality of this system is clearly illustrated by the fact that the window element with light-deflecting louvres does not permit any light to enter into the room; this is only possible through the square window section, which provides direct contact with the outside. **Table 10.1** contains an overview of the properties and developmental stage of different light-deflecting systems.

Type	System	Installation	Considerations for installation						
	No. description		Shading	Glare protection	Visual contact	Directing light into room depth	Evenly distributed light in room	Reducing need for artifical light	Application at night
1. Diffuse light deflection	1.1 Light strut	Window	M	M	Y	M	M	M	N
2. Shading with diffuse light transmittance	2.1 Prismatic panels	Window Skylight	Y	Y	M	M	M	M	Y/N
	2.2 Prisms and aluminium louvres	Window	Y	Y	N	Y	Y	Y	Y
	2.3 Shading mirror profiles	Skylight	Y	Y	N		Y	N	N
3. Sunlight deflection	3.1 Laser-cut panel	Window Skylight	N	N	Y	Y	Y	Y	N
	3.2 Light-deflecting glass	Window Skylight		Y	M	Y	Y		N
	3.3 Light strut for direct light deflection	Window Skylight	M	M	Y	Y	Y	Y	N
	3.4 Pivoting louvres	Window Skylight	Y	Y/M	M	Y	Y/M	Y	Y
	3.5 Venetian blinds for light deflection	Window	Y	Y	M	Y	Y	Y	Y
	3.6 Heliostat	Window				Y		Y	Y
4. Sunlight deflection	4.1 Laser-cut panel	Window Skylight	M	M	M	M	M	M	N

Table 10.1
Overview of the properties and developmental stage of different light-deflecting systems (after Müller and Kischkoweit-Lopin)

Y Yes
N No
M Marginal

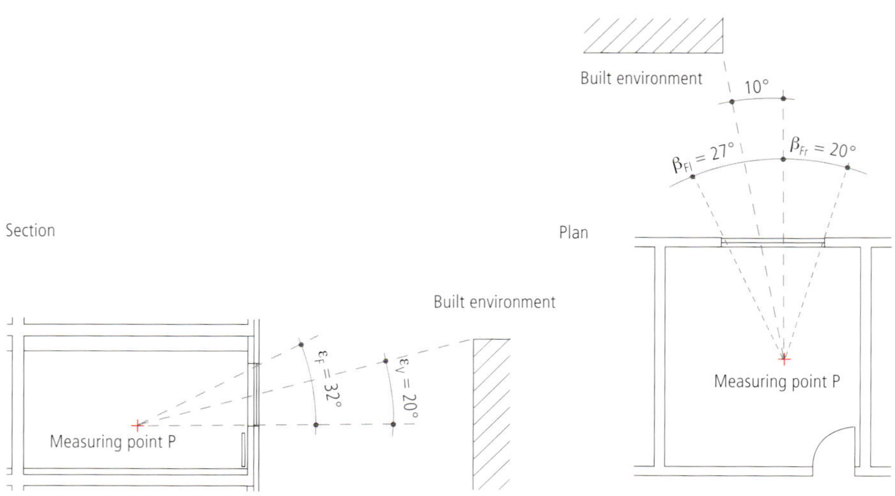

Figure 10.28
Daylight factor calculation, plan and section

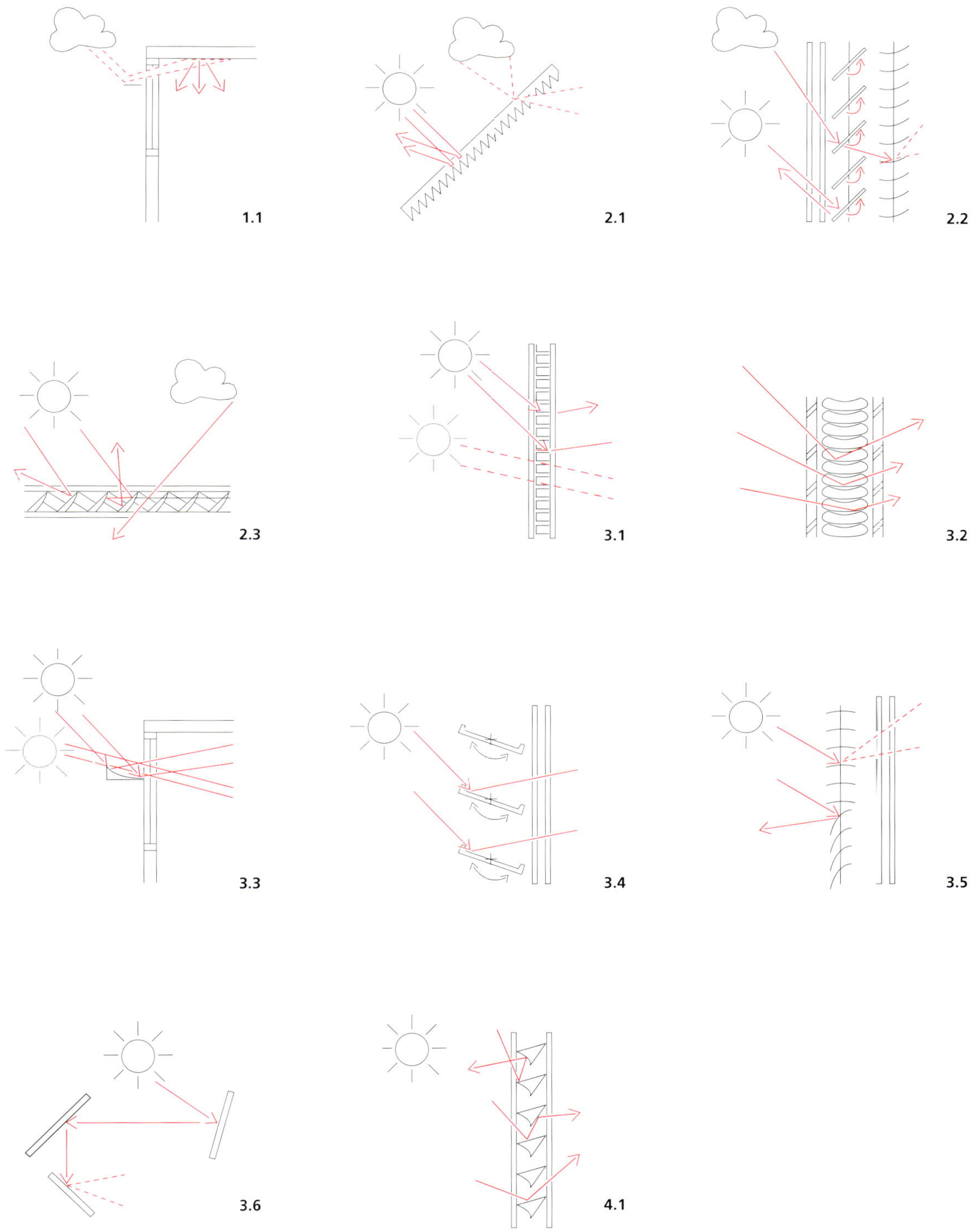

to Table 10.1
Schematic overview

10.4
Sample calculation of a daylight factor

Most daylight factor calculations for a specific point in a room and the course of daylight factors in that rooms are carried out on computers. However, DIN 5034 does allow for a manual calculation procedure. It can serve to determine the daylight factor at a point quickly, is relatively easy to carry out and yields practical results. An example is provided here.

As **Figure 10.28** (see page 442) shows, a point for which the daylight factor is to be determined is defined in the plan and section of a room. The next step is to include the angles for internal and external obstructions of the sky. In the diagram according to DIN 5034, the window is drawn in first, as shown in **Figure 10.29.1**, followed by the outline of the external obstructions (**Figure 10.29.2**). The remaining area, see **Figure 10.29.3**, represents the sky factor component. An additional external reflection component (reflection from surfaces in the built environment) is determined as shown in **Figure 10.29.3**.

The daylight factor is then calculated as follows: First, the number of fields exposed to direct skylight are counted in **Figure 10.29.3**. In this example, there are 21 fields.

Next, the fields representing the built environment are counted based on the lines in **Figure 10.29.4**. In this example, the result is 3.5 fields.

Each sky light field has a value of 0.1 % D_F. Each external reflection field has a value of 0.015 % DF. In the example shown, the resulting daylight factor is as follows, determined by external incident radiation (dependent on the number of fields):

21 · 0.1 %	=	2.10 %
3.5 · 0.015	=	0.05 %
Total		**2.15 %**

The following values must be subtracted from this sum

- Light transmission factor of the glass
- Dirt factor on glass
- Mullion factor

If the value is assumed to be 0.6, the daylight factor at this point would be 1.29 % due to the external built environment. The value is increased, however, as a result of daylight reflected from walls and windows. This increase, in turn, is strongly dependent on the reflectance of the materials in the surroundings and is therefore calculable only by means of a more elaborate process. In small, bright rooms, this increase may be as much as 100 % of the base value; in large, dark rooms, the indirect component of the daylight factor may be small enough to be negligible.

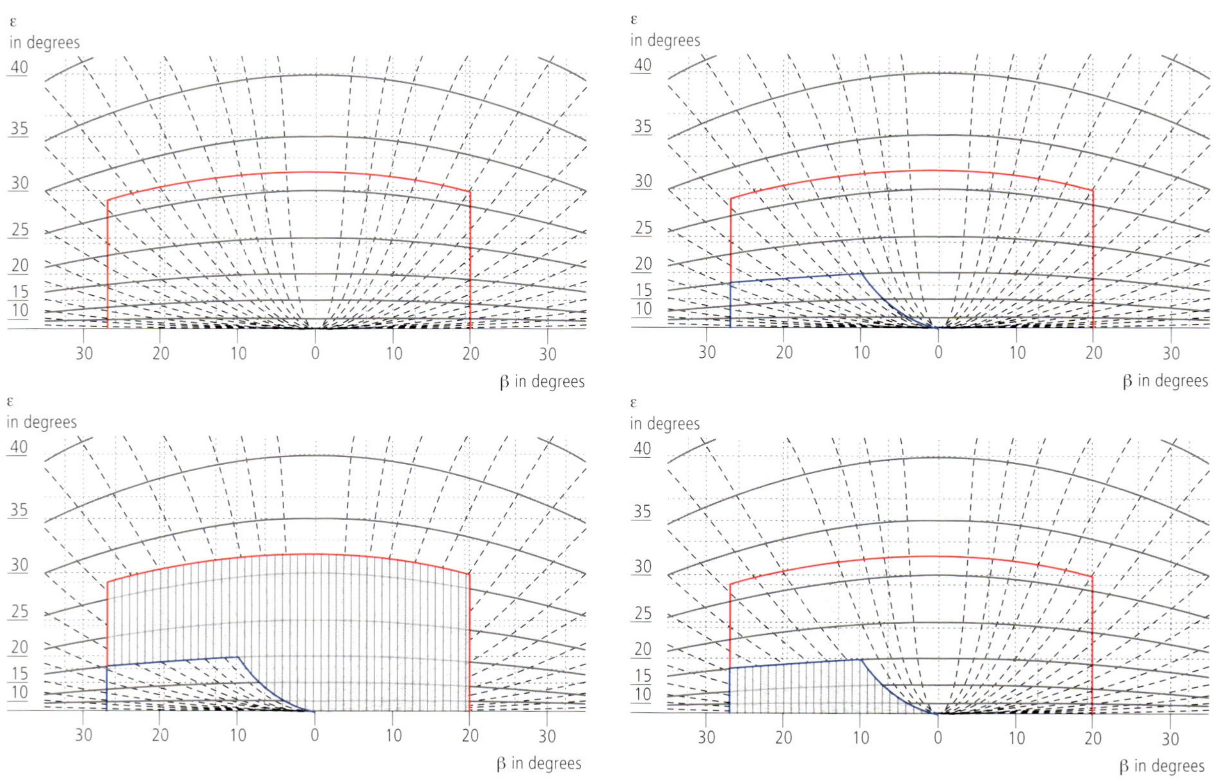

Figure 10.29
D_F calculation

Figure 10.29.1 (left)
Inputting the window dimensions

Figure 10.29.2 (right)
Inputting the outline of the built environment

Figure 10.29.3 (left)
Determining the sky factor component D_S

Figure 10.24.4 (right)
Determining the external reflection component

11. Low-voltage systems

(with Günther Faltin, Munich)

This chapter offers some brief notes on low-voltage systems.

Low-voltage systems include all those concerning signal, communication and control equipment. Communication systems for logging and channelling data of all kinds with speed and accuracy include:
- Telephone installations
- Data transmission networks
- Telex and fax systems
- Video conference systems
- Clocks and timers
- Intercom systems
- Paging installations (radio paging services etc.)
- Antenna arrays
- PA systems
- Wireless systems

Signal and surveillance systems include:
- Burglar alarms
- Fire alarms
- Security cameras (perimeter safety)
- Automatic parking systems (gate installations)
- Access surveillance systems
- Intrusion alarm devices
- Badge and flextime readers

These communication and signal systems are complemented by yet another system complex that safeguards the operation of all integrated building systems and minimizes operating costs. It is referred to as building automation or process control technology; in residential buildings, we speak of the EIB (European Installation Bus) system or BAS (building automation system).

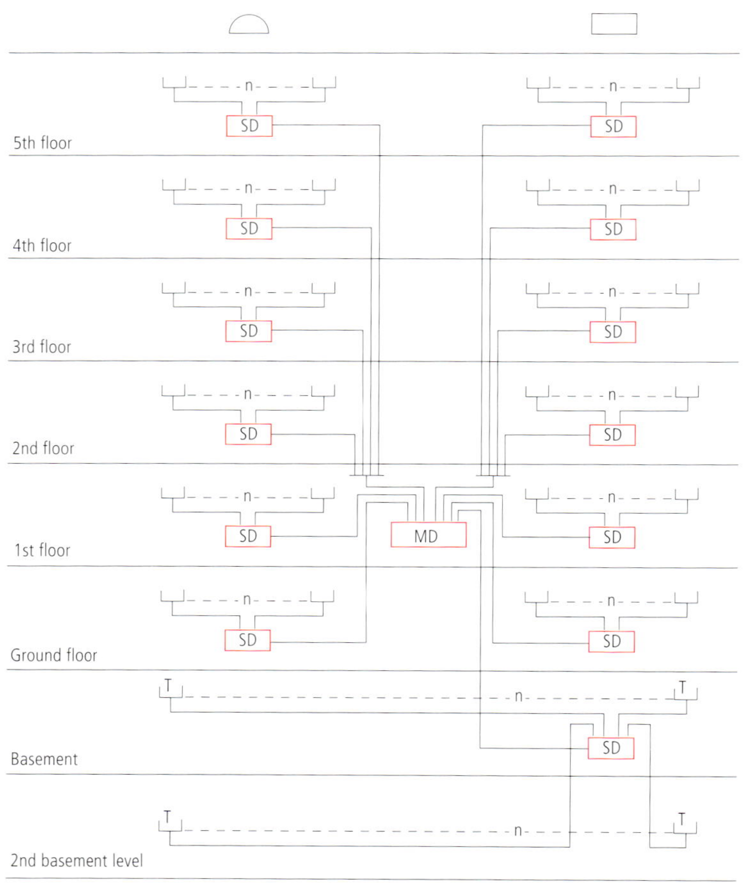

Figure 11.1.1
Overview of telephone and data transmission network

MD (Main distribution)
SD (Subdistribution)
T (Telephone)

11.1 Intercom systems

In large buildings, intercom systems are always configured as extension systems that direct dialling within the building, as well as a link to the public exchange. Extension or private branch exchange systems (P.B.X.) are classified as:

- Category 1: 1 (direct) exchange line, up to 10 extensions
- Category 2: up to 10 (direct) exchange lines and up to 100 extensions
- Category 3: more than 10 exchange lines and more than 50 extensions

The terminals for exchange lines are differentiated into connections for incoming, two-way and outgoing calls. Touch-tone telephones with a multitude of service characteristics are the standard today. The connection can be expanded one workstation at a time, providing configurations for visually and heating impaired users as well. All extensions of the exchange can be configured to operate individually within the system and not only through direct-dial numbers, as follows:

- Intraoffice extensions
- Intraoffice, partially restricted extensions
- Unrestricted extensions
- Partially restricted public exchange extensions (restricted authorization for trunk calls (GB))
- Unrestricted public exchange extensions (full authorization for trunk calls (GB))

Changes to the authorization levels for individual extensions are implemented at a central switchboard. With the growing trend toward digital telephony, most systems are being designed for ISDN capability (ISDN = Integrated Service Digital Network).

ISDN systems act as universal integrated, digital telecommunications networks. The architecture of the telephone network and individual extensions is shown in **Figure 11.1.1**.

11.2 Data line network

Each "intelligent" building should feature a data line network, to which existing and future data terminals can easily be connected. The capacity of the data terminals to be connected must be discussed with the user on a case-by-case basis in order to configure the entire system for current needs and future upgrades. It is essential to take the relevant cabling system of the major hardware manufacturers into consideration (copper or, increasingly, fibre optic cables; see **Figure 11.1.2**).

International standards allow for universal cabling installations in buildings for language, data and image transfer. The principal goal is to provide systems that are universally compatible (i.e. not dependent on products from any particular manufacturer). These cabling standards are divided into categories according to requirements and are being implemented more and more often.

Figure 11.1.2
Components for fibre optic networks
(Image: Dätwyler)

Optopatch Patchpanel 19", 1 HE for 24 connections. Here with ST ports.

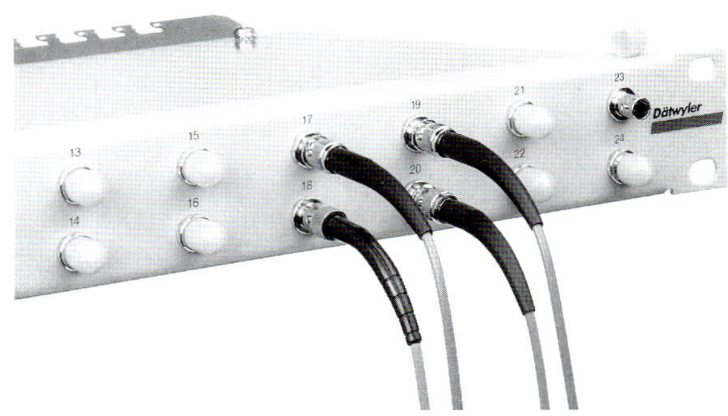

Optomod – ready-made fibre optic cable complete with insulating plastic tube and pulling ear, fully prepared for duct installation.

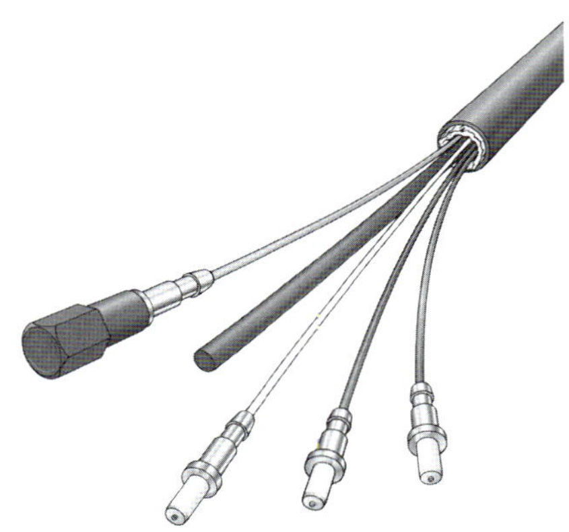

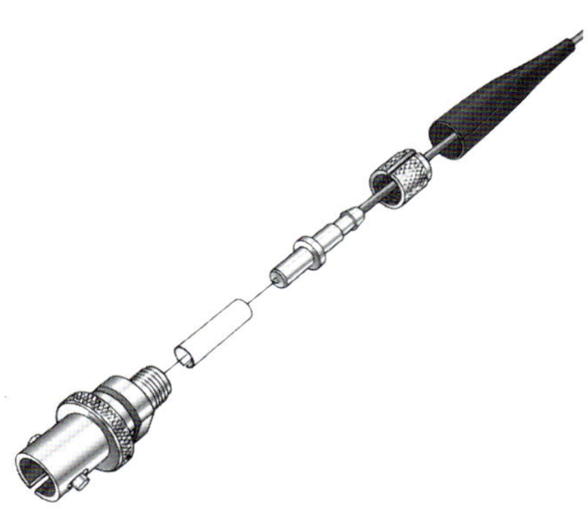

Principle of Optomod system
The ready-made ferrules are compatible with all standard plugs, e.g. ST, SC, SMA, Escon and FC/PC.
Since the plugs are the last pieces to be mounted, the diameter of the pullout tube is slightly wider than the cable diameter.

11.3
Clock systems

Many new buildings have a clock station comprising a crystal-controlled master clock and impulse-driven secondary or slave clocks to ensure synchronized time-keeping throughout the building or complex.

Oscillations generated by a quartz oscillator send an impulse signal per minute (pole change) via the master clock; the signal is then transferred to the connected clock devices via a pulse amplifier. Clock devices are secondary or slave clocks with analog or digital display. They are installed in all commonly used areas of a building, for example in classrooms, restaurant or cafeteria areas, telephone switchboards, building security stations and rooms, mail rooms (incoming and outgoing), porter's lodge etc.

Some digital clocks feature a date display in addition to the time. **Figure 11.2** illustrates the principal concept of a clock system.

Newer clock systems are based on the AFNOR (Association française de normalisation) code for time transmission and synchronization. The complete time information (time and date) is translated into binary code and transmitted from a master clock to the secondary or slave clocks by amplitude modulation with a carrier frequency of 1000 Hz, via a twin-core cable. This ensures that the entire time information is transferred once per second. The cable lengths may run to 40 km.

In addition to simple time display, the time signal may also record time or synchronize computers or automated equipment. The master clock, in turn, can be synchronized via a radio antenna, by means of either the radio time signal or the GPS time signal (Global Positioning System).

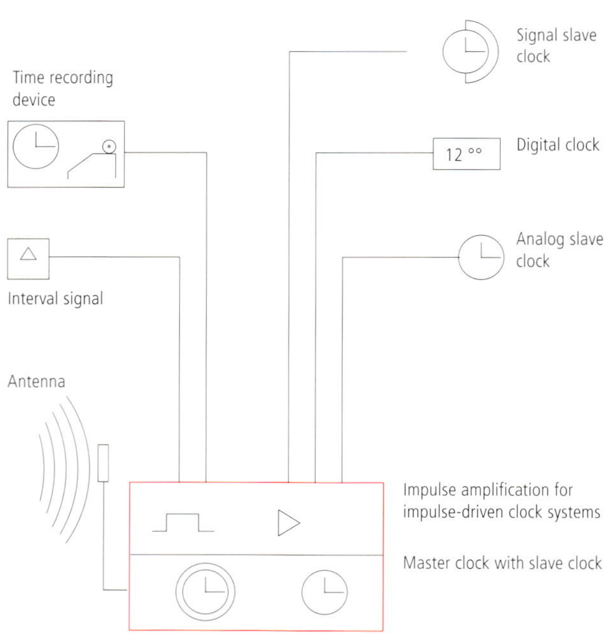

Figure 11.2
Plan of clock system

11.4
Intercom systems

Intercom systems should be installed to enable direct communication between external and internal areas; they should also be installed at access/entrance and security areas. Intercom systems service areas such as:
- Lodges/gates
- Garage entrance and exit points
- Main entrances
- Visitor parking
- Entrance/delivery
- Gates/locks
- Telephone switchboards
- Rentable areas

Intercom systems are generally configured as decentralized door intercoms; in entrance areas they are invariably installed where surveillance cameras are set up. The diagram in **Figure 11.3** illustrates an intercom system.

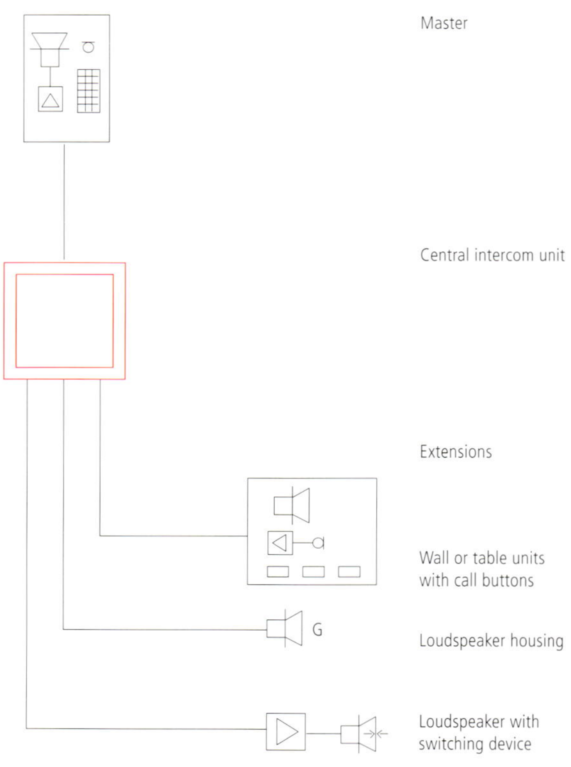

Figure 11.3
Plan of intercom system

11.5
Paging installations

Today, most paging installations are configured as wireless systems, functioning as:
– Radio paging systems (VHF or UHF)
– Paging systems operating on the induction loop principle

Radio paging systems are generally set up for contacting people in the grounds of a building or building complex. They aim to reach people who perform essential functions and must therefore be contactable at any time; some examples are:
– Doctors, factory physicians, paramedics
– Security personnel
– Maintenance and operating technicians for technical installations
– Important people within an organization

These people can be contacted individually or in predetermined groups by various means: audible signals (ringer, beeper), visual signals or numerical sequences on the pager itself. The entire paging system contains the components illustrated in **Figure 11.4**, and the call desk is generally located at a central telephone switchboard.

Wireless phone systems are increasingly replacing paging systems. The principal advantage is that they permit direct verbal communication.

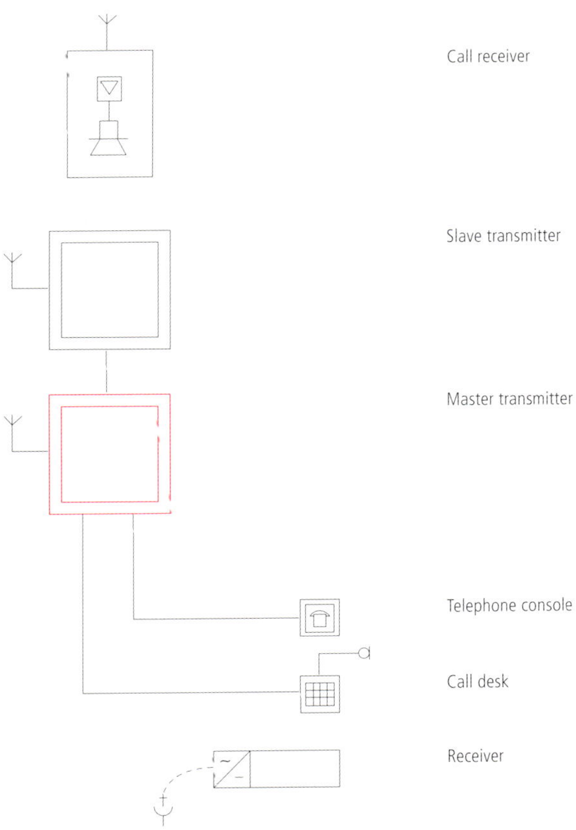

Figure 11.4
Plan of paging installation

11.6
Antenna arrays

Antenna arrays with rooftop installations and antenna amplifiers or cable connections are often set up on large building projects for public radio and television reception.

Antenna connector boxes should be installed in all rooms designated for meetings, as well as in executive, training and lounge areas. Coaxial or broadband cables are used for antenna arrays. The architecture of an antenna array is illustrated in **Figure 11.5**.

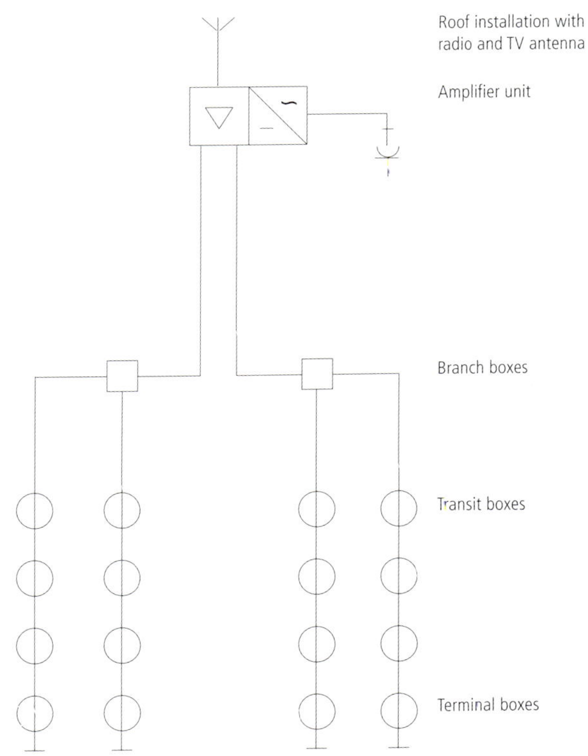

Figure 11.5
Plan of antenna array

11.7 Electroacoustic systems

Electroacoustic systems are used to amplify transmitted and recorded acoustic signals. The loudspeaker systems employed can be divided into the following groups according to the specific purpose they serve:
- Systems with a high quality of sound rendition (for language and music), also useable as public address (PA) systems in large areas
- Systems with a high quality of speech reproduction
- Systems with command quality

Systems with a high quality of speech reproduction are generally used for announcements addressed to specific employee groups within individual areas of a building.

Electroacoustic systems with command quality are recommended for alerting personnel in a building in the event of fire or other emergency.

Systems with a high quality of sound rendition are those where the loudspeakers themselves achieve high-quality rendition and are installed at close proximity to each other (in ceiling panels) to ensure that the human ear is unable to localize the sound source. Such electroacoustic systems comprise a central amplifier linked to a microphone station and the loudspeaker network for ceiling installation. In the event of fire or other emergency, a generator automatically triggers the sounding of an alarm upon receiving a signal from a fire alarm or ionization fire detector. When command-quality systems are used, loudspeakers are installed only in hall and circulation areas; in higher (sound) quality systems, loudspeakers are also positioned in the work areas. The loudspeakers are usually connected in series into a single speaker line per floor, allowing announcements to be made in selected areas within a building. The system should be designed to ensure that the signal is audible in all areas of a building with 65 % syllable intelligibility up to a distance of 15 m from the loudspeaker (voice announcements). The architecture of an electroacoustic system is illustrated in **Figure 11.6**.

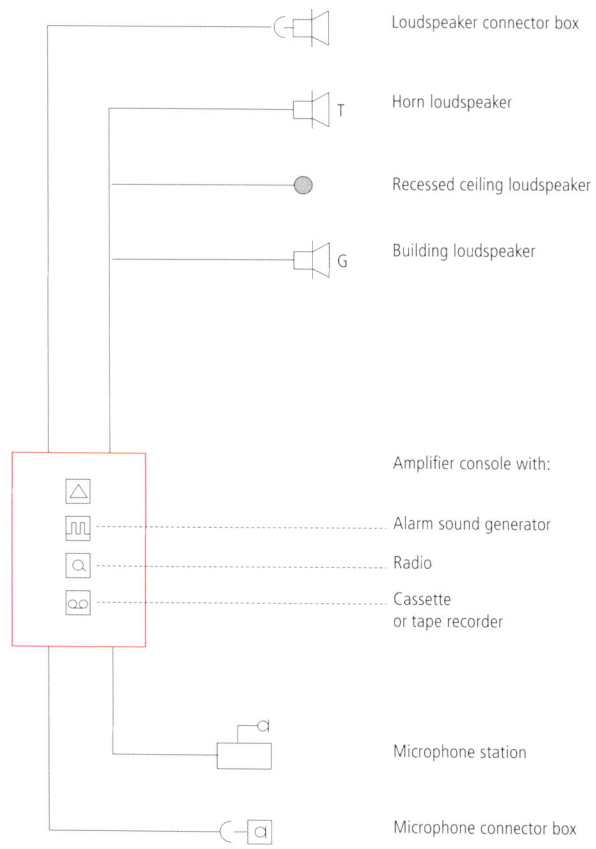

Figure 11.6
Plan of electroacoustic system

11.8 Burglar alarms and intrusion alarms

Burglar and intrusion alarms analyse and process different types of signal information produced by connected manual or automated accessory alarms. The following components are part of a burglar or intrusion alarm system:

- Master with plug-in for detector and sabotage (interference) lines for the entire building
- Master for special security areas (e.g. in banks, money processing, safes)
- Signalling device to relay alarms to the indicator panel or main alarm
- Manual distress indicators (hand- or foot-operated) for high-risk areas
- Monitoring contact for all doors, gates or grating leading to the outside
- Monitoring contact at communicating doors between high-security areas; broken glass detectors at all windows and glass doors still accessible from the outside (at least on the ground floor)
- Monitoring contact and impact sound detectors for high-security areas (e.g. safes)

At doors and windows, magnetic or bolt contacts are used to monitor opening and closing. For laminated glass, detection of shattering or breakage is effected by inlaid alarm wires; impact sound detectors are used for all other types of glass. Impact sound detectors are also installed on vulnerable wall areas to monitor shock/vibration or penetration.

Alarms triggered in high-security areas are relayed directly to police headquarters as "live" feeds via a main signal; alarms given by emergency and hand-operated alarms are also relayed directly.

In all other cases, alarms are usually relayed to a position that is manned 24-hours a day (concierge/gate keeper/central security desk), where they are indicated visually and/or acoustically. **Figure 11.7** illustrates the architecture of a burglar alarm system.

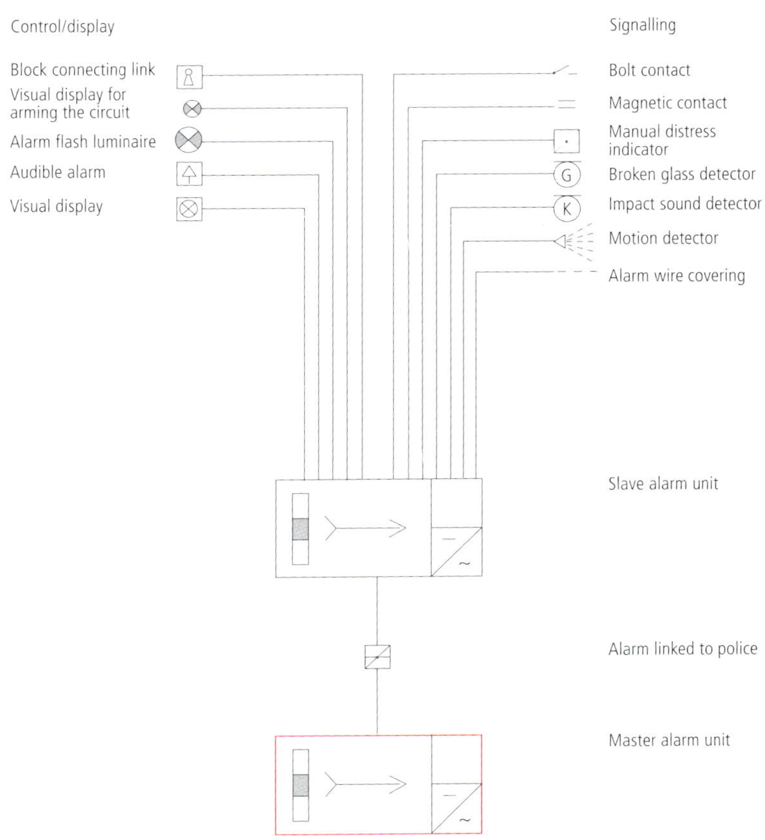

Figure 11.7
Plan of burglar alarm

11.9 Fire alarm systems

Automated fire alarm systems are installations that detect an outbreak of fire without human intervention and automatically signal the event to a central unit. This, in turn, will sound the alarm and initiate all necessary control functions without further assistance.

Manually operated alarms can also be integrated into an automated fire alarm system to good effect. While fire detection is no longer automatic in this case, all other operations following detection are still fully automated.

Automated fire alarm systems warn people of the danger of fire and enable them to instigate intervention as soon as they have been alerted to the risk, generally when the danger is still fairly slight. These systems serve to protect and save life, as well as goods of material, intellectual or sentimental value.

Automated fire alarm systems must therefore fulfil the highest standards in terms of dependability and readiness. Only professionally designed and operated fire alarm systems can meet these standards, see Figure 11.8.

The reliability and effectiveness of an automated fire alarm system also depends upon a chain of additional factors:
- Quality of system design
- Quality of assembly and installation
- Quality of commissioning (start-up)
- Quality of maintenance
- User training
- Alarm configuration

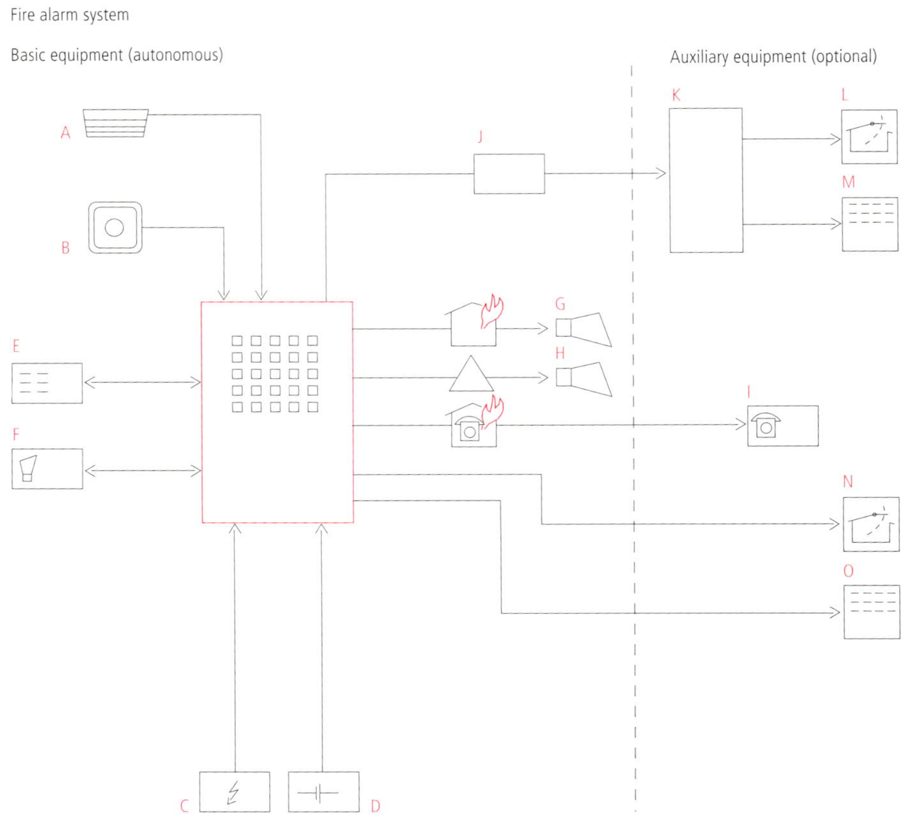

Figure 11.8
Plan of fire alarmsystem

A Automatic fire detectors
B Manual fire detectors
C Power supply from mains
D Emergency power supply
E Operating unit
F Alarm configuration
G Internal alarm
H Internal interference signal
I External alarm/interference signal
J Interface
K Building automation
L Fire spread control
M Distance signal
N Fire spread controls *
O Distance signal *

* Linked directly to the central fire alarm unit

11.10 Closed circuit television surveillance systems (CCTV)

CCTV systems relay information about events at obscured areas to one or more monitors, which are manned 24-hours a day. Video surveillance systems are the standard choice today; these comprise the following device configuration:

- CCTV centre with video crossbar
- Surveillance VDUs with selection keyboard set up at a central security desk or at the porter's lodge
- Surveillance VDUs in high-risk areas
- Surveillance cameras in high-risk areas
- Main entrance
- Visitor and short-term parking areas
- Entrance and exit areas for underground parking
- Emergency stairwells and exits that are not otherwise visible
- Stairs providing access to underground parking
- Underground parking areas

The CCTV system is generally operated on a 24-hour basis, recording the images captured by the various cameras on video (24-hour recording). **Figure 11.9** identifies the principal components of a CCTV system.

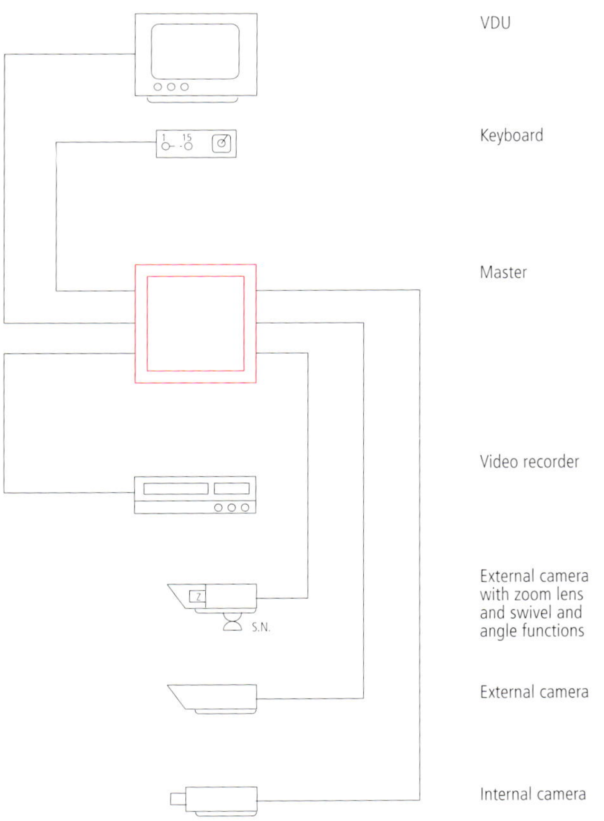

Figure 11.9
Plan of television surveillance system

11.11 Traffic systems

Traffic systems are often employed to direct vehicle entrances and exits, and to regulate the flow of traffic in parking areas (above ground and underground). These systems comprise the following components:

- Control centre with operating panel
- Entrance and exit gates with traffic detectors
- Entrance and exit control terminal
- "Free/Full" indicator panels
- Safety devices for rolling shutter doors (gates)

In public car parks, where individuals can freely select parking spaces, all vehicles entering and exiting should be counted by induction loops to transmit the appropriate message or notice automatically to the aforementioned indicator panel.
Figure 11.10 illustrates a traffic system.

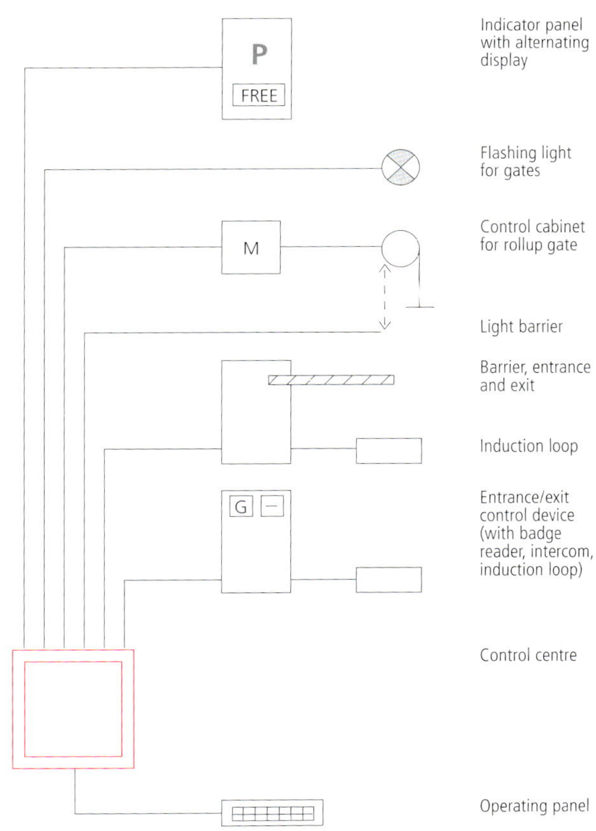

Figure 11.10
Plan of traffic system

11.12 Access control and flextime recording systems

Electronically controlled recording systems are installed to monitor access and to record work hours. Access control is used to prevent unauthorized access to a building, specific security areas or rooms.

Authorized employees are issued special ID cards that enable them to open a door controlled by a card reader, provided they have permission to enter. Program packages can be designed to allocate access authority by group or a variety of authorization levels (e.g. daytime/night, weekday or temporary authorization).

Magnetic contacts are used to control the door position (open/closed) in a time-dependent manner; if a door does not close within a pre-determined time or is opened without using the card reader, an alarm is transmitted to the central control unit. The required ID cards are encoded with prioritized security levels according to the security level of the supervised areas (rooms/open areas/building sections).

The same technology is used for time recording, which can be programmed to record individually for each employee. The basic information required is the standard hours of work, scheduled breaks, start and finish of work day, minimum hours per day or per week, all programmed per day or week for individual employees. Excess time can be accumulated as overtime and recorded per employee. A recording function for absenteeism, as well as for overtime and premium calculations with detailed breakdown per wage level, can be programmed into a matching data processing system. The dialogue generally executes an invoice period at any given time for individual employees, allowing for correction if required. **Figure 11.11** illustrates the configuration of an access control and time recording system.

General information

The central equipment of the systems described above is usually installed in a separate room or in the BAS equipment room. It must be protected against unauthorized tampering, since this equipment is an essential component of the overall information and security concept.

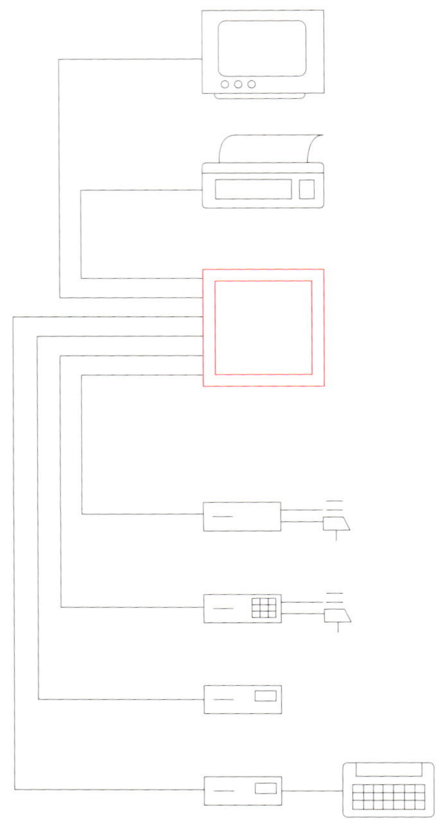

Figure 11.11
Plan of access control and flextime recording systems

11.13 Building automation systems (BAS)

Building automation systems have been developed to ensure simple, secure and efficient operation of building systems. They relieve maintenance personnel from having to carry out routine tasks by providing automatic monitoring or start-up/shut-down functions of system components. They also save energy costs, by optimizing system processes (load decrease programs).

Generally speaking, automated building systems monitor, control and regulate the following building systems:
- HVAC systems
- Refrigerating systems
- Heating systems
- Sanitation systems
- Fire-extinguishing systems
- High-voltage systems
- Lighting systems
- Elevator- and transportation systems
- Low-voltage systems

Systems that are subject to individual use, operation and hence supervision need not be included in the BAS; for example:
- EDP equipment
- Copiers, data terminals
- Kitchen appliances

Moreover, BAS should never be employed to assume safety or personnel data processing tasks such as:
- Fire alarms
- Burglar alarms
- Access control
- Flextime recording
- Canteen data recording etc.

This would impose special requirements both on the system itself and on the system operators, although relevant data protection issues can be excluded. These tasks can be fulfilled in a more cost-efficient manner using specialized information systems (**Figure 11.12**). The BAS simply monitors their operating status.

To execute its tasks, the BAS must override the individual building systems without interfering with their autonomy with regard to emergency or back-up operation. The interface conditions are defined in universal standards (e.g. VDI 3814).

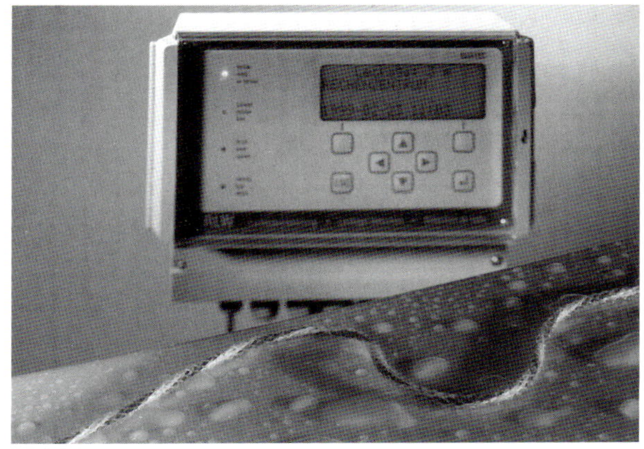

Figure 11.12
Water detection system, e.g. for double or false floors in EDPCs (electronic data processing centres)
(Image: Bartec)

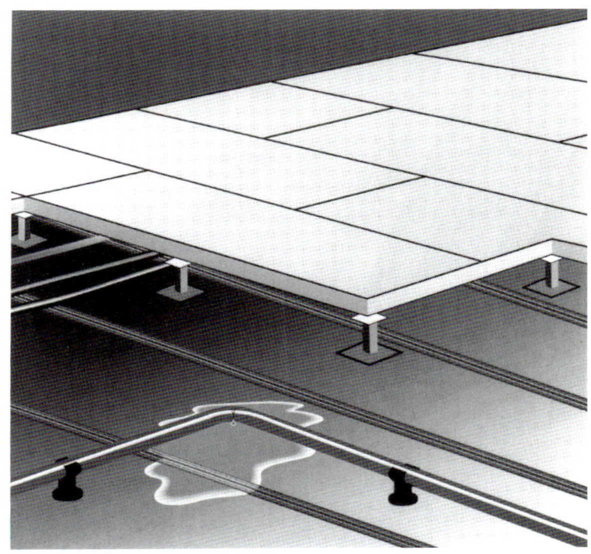

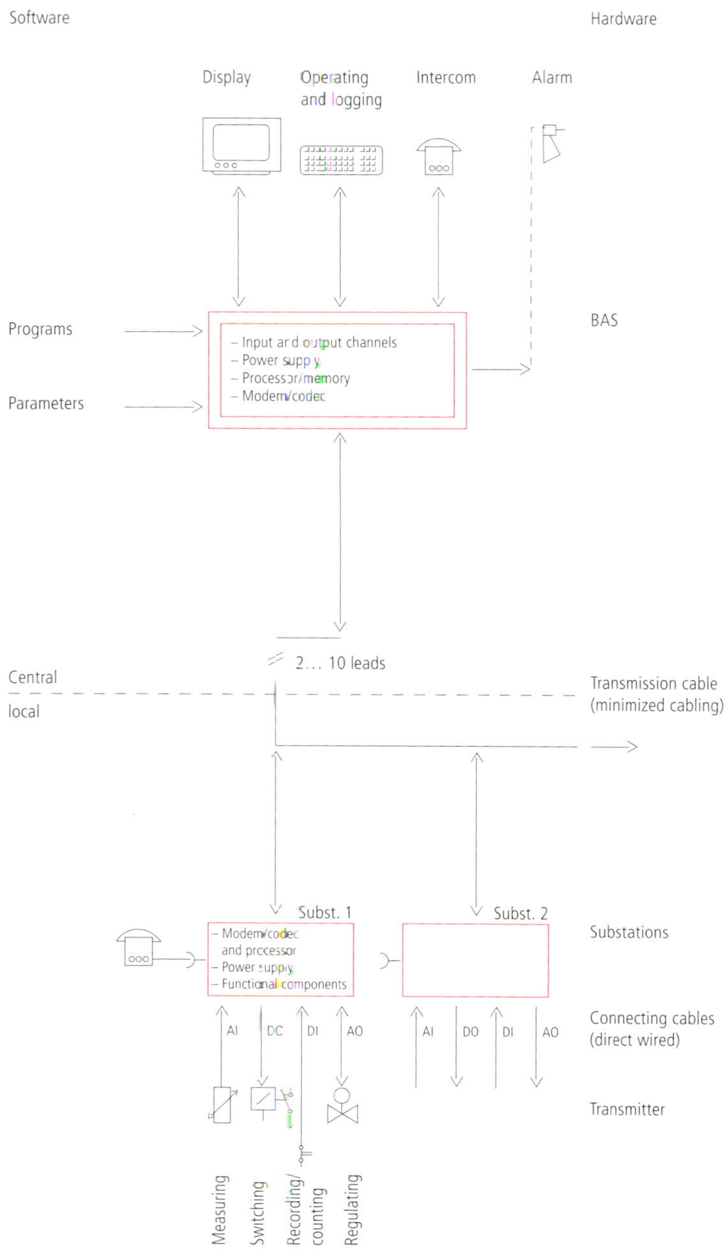

Figure 11.13.1
System architecture of a BAS (BAS = Building automation system)

System architecture

Figure 11.13.1/2 depicts the characteristic components of a BAS:

- Central BAS installation with operating devices
- Substations with functional components
- Efficient transmission network (i.e. minimized cabling) between substations and central BAS unit (individual and group supply)

The substations are housed together with or near switch cabinets, while the central BAS installation is located in a separate room (20 m^2) near the porter's lodge or security desk. In large buildings, the central BAS installation must be manned at all times to ensure that incoming operation notifications are recorded and processed according to priority. Since central BAS installations are 24-hour workstations designed for constant human occupation, they fall into the category of high-security office spaces and should be designed accordingly (**Figure 11.14**).

Figure 11.13.2
BAS components
(Image: Johnson Controls)

Figure 11.14
Building process control, surveillance and control centre, Neue Messe, Leipzig (Image: Honeywell)

BAS are not employed in smaller building projects or buildings with moderate technical building systems, where central monitoring systems (switchboard units) are used instead (see also Chapter 8.4.6, EIB systems). Designing a BAS is a highly complex task that should be executed by a trained systems engineer with a background in heating/cooling/HVAC, since these systems handle and process more data than any other. For large-scale building projects it is best to consult an experienced specialist who can document all operating processes and translate them into relevant BAS programs.

Larger building projects with sophisticated building technology can no longer operate without a BAS. The operational processes of many individual building systems are not designed to be controlled by one individual or even a small group of people. In addition to the energy-saving criterion, BAS also help to reduce personnel requirements, perform preventative maintenance and facility management tasks. The space requirements for the corresponding low-voltage systems and especially for a BAS range from 0.1 – 0.15 % of the gross floor area of a building, according to the scale and complexity of the technical installations (**Table 11.1**).

Space requirements for	**Area** in m^2	**Ventilation**	**Comments**
Minimum equipment (Operating unit, control cabinet)	10 – 20	Air-conditioning	Room with daylight incidence (e.g. office space) near workshop M
Additional other technical installations per BAS	2 – 3	Size according to estimated dissipated energy from technical installations and occupants	

Table 11.1
Building process control systems

12. Transportation systems

(with Renate Sprenger, Hamburg)

12.1 General information

Communication and information are fundamental to a functioning organization. Elevators and conveyer systems are indispensable technical installations that provide and ensure a continuous flow of information, passengers and goods. In combination with other communication systems such as public address and data transmission systems, transportation systems can greatly improve these fundamental requirements. In addition to the number and quality of the individual transportation means, optimum application is an essential factor in finding solutions for specific transportation problems.

New systems must comply with the EC standard pr EN 81 (later EN 81) and any relevant local or regional codes. Although other standards and codes also come into play, a detailed discussion would go beyond the scope of this chapter.

Transportation systems that fall within the scope of technical building systems include:
- Elevators
- Escalators and moving sidewalks
- Container conveyer systems
- Baggage conveyer systems
- Pneumatic post systems
- Facade access systems
- Hoisting equipment, gantry cranes and elevating platforms
- Cranes

Since the size and location of transportation systems can have a major impact on the architecture of a building, experts should be consulted at the very beginning of the preliminary design stage.

In buildings, vertical transportation of passengers and goods is provided by elevator installations.

The following elevator systems are differentiated according to use:
- Passenger elevators
- Freight elevators
- Simple cargo elevators
- Container elevators
- Hole-less elevators
- Service elevators

These are further differentiated according to operation type:
- Traction or cable-type elevators
- Hydraulic elevators

12.2 Passenger elevators

Passenger elevators should be located at the circulation core of a building and be grouped into banks when this is necessary or desirable.

The required number of elevators is determined by:
- Building type
- Building height
- Number of stops
- Floor use
- Passenger volume

The conveyance capacity is determined by the selected travel speed, load capacity and number of elevators, as well as by the individual elevator components.

12.2.1 Passenger elevator requirements

The elevator requirements are principally influenced by:
- Building use (e.g., office, hospital, hotel etc.)
- Location of building entrances (e.g., one or more entrances, underground parking access)
- Internal organization (e.g., canteen, data processing, central library)
- Work management (e.g., fixed work hours, flextime, dining or visiting hours)
- Special uses (e.g., emergency service in case of fire, VIP service, goods transportation etc.)

12.2.2 Calculating the required passenger capacity

To determine the total required passenger capacity, the number of people occupying the upper floors of the relevant building must first be identified. If these numbers are fixed, 50 % of the occupants on the floor above the entrance level can be deducted; another 10 % can be deducted to account for occupants who are on vacation, absent or on sick leave. These deductions apply only to buildings with a constant number of occupants.

Since this information is not always available when a traffic analysis is carried out for a building, the following equation is used to give an estimate (without deductions):

Gross floor area x 0.75 = principal floor space

Principal floor space:
approx. 12 m²/workstation = number of occupants

The number of occupants resulting from this calculation is the required passenger capacity on which all subsequent calculations are based.

12.2.3 Glossary of terms for conveyance calculation

Conveying capacity
Maximum achievable conveyance capacity within a 5-minute interval.
It is given as follows: passengers per interval as a percentage of building occupancy above loading level.

Car travel time
This is the time required by one elevator car to execute a full round trip. It includes all door opening and closing times, loading and unloading times, travel times between stops, as well as acceleration and delays.

Average car intervals
This is the time that passes between the departure of one elevator car and the arrival of another at the main stop.

Average waiting time
The waiting time for the passengers depends on the car intervals and conveying capacity. The conveying capacity during loading operation, for example, is sufficient if the maximum waiting time corresponds to the car intervals.

Car capacity
EN 81 stipulates specific load capacities per car floor area. This corresponds to the maximum load, which cannot be used as a basis for a traffic analysis. The maximum car capacity is taken to be 80 % of the square car floor area.

12.2.4 Conveyance recommendations

To ensure circulation within a building and comfortable elevator operation, universally recommended values have been established based on maximum building occupancy.

In office buildings with flexible work hours:
- Upward conveyance capacity of 12 – 14 % every 5 minutes
- Downward conveyance capacity of 2 % every 5 minutes
- Car intervals of 25 – 35 seconds

In office buildings with fixed morning hours and fixed public access hours:
- Upward conveyance capacity of 20 – 25 % every 5 minutes
- Downward conveyance capacity of 2 % every 5 minutes
- Car intervals of 25 – 35 seconds

12.2.5 Conveyance calculation

The following calculation model can serve to estimate elevator capacity; for complex buildings, experts should be consulted to design an optimum, object-specific solution.

The required conveyance (C_{requ}) is calculated as follows:

$$C_{requ} = \frac{\text{Number of passengers to be transported [p/5 min.]}}{\text{5 minutes}}$$

The numerical value of the required conveyance capacity is not, in itself, sufficient to determine the total number, size and speed of the necessary elevators. The following options can be employed for designing elevator systems:

– A few large elevators
Advantages: low investment and operating costs, efficient use of space
Disadvantages: long travel and waiting times, limited availability for maintenance and repair work, inconvenient for internal traffic outside of peak times

– Many small elevators
Advantages: short waiting and travel times, high availability for maintenance and repair work, convenient for internal traffic
Disadvantages: higher investment and operating costs, greater space requirements

12.2.6 Elevator car capacity

Over time, various car sizes with different load capacities have become the norm, and these are classified in universal standards (e.g. DIN 15306 and 15309).

The following capacities should be selected:

Office and administration buildings

Load	630 kg	800 kg	1000 kg	1250 kg	1600 kg
Passengers	6	10	13	16	21

12.2.7 Travel speed

Travel speed is a key factor in achieving efficient conveyance ratios. It influences the round-trip, waiting and travel times, and has a major impact on energy costs and operating type. However, travel speed cannot be randomly increased based on efficiency alone. Once a certain boundary value has been surpassed, transportation efficiency is in fact no longer improved. This is due to the fact that the time required for loading and unloading, approaching the stop, opening the doors, keeping them open and closing them is disproportionately high compared to the gain in travel time resulting from a higher travel speed.

Figure 12.1 illustrates the relationship between minimum travel distance and maximum achievable speed. This curve is based on standard values for an acceleration of 1 m/s^2 and a travel speed increase of 0.8 m/s^2.

The values in the accompanying table (**Table 12.1**) are not absolute; they also depend on the total conveyance height.

Time of winding cycle
Once the car capacities of an elevator system have been determined according to the aforementioned guidelines, the time required for one full winding cycle of that system can be calculated. The winding cycle T is calculated as follows:

$$T = (2\frac{h}{v}) + (c \cdot P) + S\,(t_1 + t_2).$$

where:

T = Time of winding cycle (s)
h = Mean hoisting height (m)
v = Operating velocity (m/s)
c = Loading and unloading time for one person (s/P)
P = Full capacity of elevator car (persons)
S = Number of probable stops
t_1 = Loss of time for acceleration and slow down (s)
t_2 = Time required for opening doors, keeping doors in open position and closing doors (s)

The mean hoisting height is a function of the total hoisting height used, the car capacity and the number of stops. The mean hoisting height can be established with the help of **Table 12.1**.

Number of floors above ground	630 kg	800 kg	1,000 kg	1,250 kg	1,600 kg
Maximum elevator car load (persons)	8	10	13	16	21
Actual elevator car load (persons)	6 = 75%	8 = 80%	10 = 77%	12 = 75%	15 = 71%
3	3.7	3.8	3.9	3.9	4.0
4	4.3	4.5	4.7	4.8	4.9
5	4.7	5.0	5.3	5.5	5.7
6	5.0	5.3	5.8	6.0	6.4
7	5.2	5.6	6.3	6.5	7.1
8	5.4	5.9	6.6	6.9	7.6
9	5.6	6.1	6.9	7.2	8.1
10	5.7	6.2	7.1	7.5	8.5
11	5.8	6.4	7.3	7.8	8.8
12	5.9	6.5	7.5	8.0	9.1
13	6.0	6.6	7.7	8.2	9.4
14	6.0	6.7	7.8	8.3	9.7
15	6.1	6.7	7.9	8.5	9.9
16	6.1	6.8	8.0	8.6	10.1
17	6.2	6.9	8.1	8.7	10.3
18	6.2	6.9	8.2	8.8	10.4
19	6.3	7.0	8.3	8.9	10.6
20	6.3	7.0	8.4	9.0	10.7
22	6.4	7.1	8.5	9.2	11.0
24	6.4	7.2	8.6	9.3	11.2
26	6.5	7.2	8.7	9.5	11.4
28	6.5	7.3	8.8	9.8	11.5
30	6.5	7.3	8.9	9.9	11.7
additional factors *	h = 0.90 x actual hoisting height		h = 0.95 x actual hoisting height		h = 1.00 x actual hoisting height

Table 12.1
Number of probable stops and average lifting height of passenger elevators servicing floors with similar number of occupants

* these factors are used to adjust the actual hoisting height

numbers in red indicate probable stops

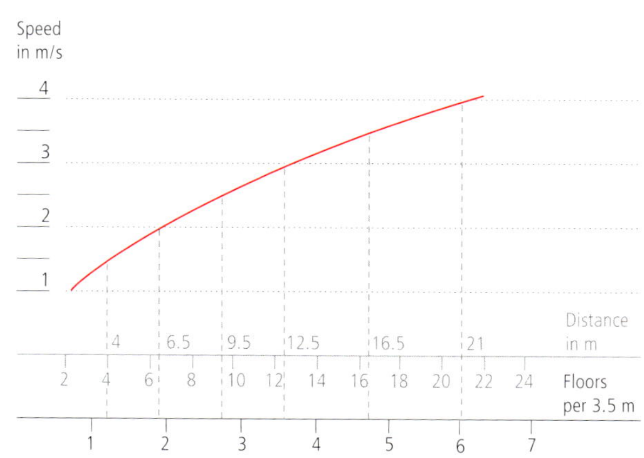

Figure 12.1
Achievable speed between floors

The following numerical values can be applied to loading and unloading times:

c = 2.0 to 2.5 seconds per person for elevators with doors of 900 mm clear width, and for all deep cars (car depth greater than car width).

c = 1.5 to 2.0 seconds per person for all elevators with doors of 1,100 mm clear width and more, provided the elevator car depth does not exceed its width.

Table 12.1 can also be used to determine the number of probable stops in relation to the total number of upper floors and the car capacity.

The travel or operating speed for passenger elevators for medium-high buildings (not high-rise) is listed in **Table 12.2**, while **Table 12.3** lists the loss of time for acceleration and deceleration. **Table 12.4** contains the times for opening and closing doors, and for maintaining doors in the open position for passenger elevators.

Once the time for a winding cycle has been established using the equation given above, the number of required elevators is calculated as follows:

$$C_{act.} = \frac{P \cdot 300\,s\ (5\ minutes)}{T}$$

$$C_{requ.} = \frac{\text{Number of passengers to be transported x 5 min.}}{\text{Time in which passengers must be transported}}$$

$$n = \frac{C_{requ.}}{C_{act.}}$$

$C_{act.}$ = Actual conveyance
$C_{requ.}$ = Required conveyance
P = Passenger capacity
T = Time of winding cycle
n = number of elevators

As a rule of thumb, the C_{requ} (required conveyance capacity) is approx. 25 % for buildings with fixed work hours, that is, all employees of a building can be transported within a period of 20 minutes.

The number of elevators can be adjusted by changing the car capacity.

Average car interval

$$K = \frac{T\ [seconds]}{n}$$

K = average car interval

Average waiting time

W = K · 0.5 [seconds]

Number of stops above reference stop (first reference stop, if applicable)	**Operating speed** in m/s
to 3	0.6* and 1
3 to 5	1.0
5 to 10	1.6
10 to 18	2.5
more than 18	4.0

* The operating speed of 0.6 m/s is predominantly used for hydraulic elevators.

Standard speeds according to DIN 15 306

Table 12.2
Operating speeds for passenger elevators

Operating speed in m/s	**Loss of time t_1** in s
0.63	2.0
1.0	2.5
For hydraulic drives with starting valve:	1 s more each
For hydraulic drives with starting valve and star-delta motor startup:	2 s more each
1.6	3.0
2.5	4.0
4.0	5.5

Table 12.3
Acceleration and deceleration times for passenger elevators

Door type and door width	**Time t_2** in s
Centre-opening doors, power-operated	
1. for standard doors, clear door width up to approx. 1,000 mm	6
2. for high-performance doors, clear door width up to approx. 900 mm	4.5
Doors opening to one side, power-operated	6

Table 12.4
Times required for doors to passenger elevators to open, stay open and close

12.2.8 Machine room dimensions and shaft pit locations (top and bottom)

The structural requirements are discussed in detail in subsequent sections. Tables 12.5.1/2 provide a preliminary overview of machine room dimensions, as well as dimensions for shaft pits at the top or bottom for cable-type elevators with standard heights.

Machine room for	Area	Clear room height
Passenger or freight elevators with traction (or cable) drive, machine room at top of elevator shaft	min. 10 m^2	2.1 to 2.5 m
Passenger elevators in high-rises Equipped with Ward-Leonard hoist	min. 10 m^2	2.1 to 2.5 m
Freight elevators for transporting hand carriages, with traction (or cable) drive, machine room at top of elevator shaft	Elevator pit area = machine room area	2.1 to 2.5 m
Elevators with hydraulic drive, machine room next to, in front of or behind first stop at bottom	min. 5.0 m^2	2.1 m

Table 12.5.1
Machine rooms for elevators

Placement of drive	Speed in m/s	Shaft head distance in m from upper edge of finished floor at top stop	Shaft pit distance in m from upper edge of finished floor at bottom stop
above shaft	0.63	approx. 3.80	approx. 1.40
	1.00	approx. 3.80	approx. 1.50
	1.60	approx. 4.00	approx. 1.70
	2.50	approx. 5.00	approx. 2.80
below and next to shaft (guide pulleys in shaft head)	0.63	approx. 4.20	approx. 1.65
	1.00	approx. 4.20	approx. 1.70
	1.60	approx. 4.40	approx. 1.80

Table 12.5.2
Shaft headroom and space below shaft for cable elevators with car heights up to 2.2 m

12.3 Freight elevators

Similarly to passenger elevators, conveyance calculations can also be carried out for freight elevators. This is only required, however, when freight elevators are part of production processes, or where specific freight conveyance tasks must be fulfilled.

Sample calculation:

$$T = 2\left[\frac{h}{v} + Lt + N(t_1 + t_2)\right]$$

h = Hoisting height
Lt = Loading and unloading time
N = Number of loading and unloading positions for each winding cycle
t_1 = loss of time for acceleration and deceleration (s)
t_2 = Time required for opening and closing doors (s)

Definition of terms:

Hoisting height
Since freight elevators generally stop at specific loading points, the hoisting height should be based on the stops most commonly used during peak hours.

Loading and unloading time
If goods are kept ready to load in rolling containers or similar transport containers at the elevator doors, 30 to 40 seconds can be taken as the time required to load each unit. In all other scenarios, it will be necessary to estimate the time. The location and other parameters for the loading areas in front of the elevator will play an important role. The time lost for acceleration and deceleration (t_1) is calculated as follows:

Operating speed	Loss of time (t_1)
0.25 m/s	approx. 1.0 s
0.40 m/s	approx. 1.5 s
0.63 m/s	approx. 2.0 s
1.00 m/s	approx. 2.5 s

Time for opening and closing doors t_2
– For single-hung, hand-operated doors: $t_2 = 7$ s
– For double-hung, hand-operated sliding doors: $t_2 = 10$ s.

Automatic doors:
– One-sided telescopic doors, 2 panels, 1.2 m wide $t_2 = 6.1$ s
– Centre-opening, 2 panels, 1.2 m wide $t_2 = 4.6$ s
– Centre-opening, 4 panels, 1.6 m wide $t_2 = 5.9$ s.

All freight elevators must also be equipped with cage doors.

The number of freight elevators required is the product of the ratio of winding cycle times to the available time for conveyance during peak hours, that is:

$$n = \frac{T}{t}$$

Fire department elevators and elevators used for gurneys are specialized applications. The necessity for these elevators is determined by building codes, unique technical requirements, specifications for erecting specific installations etc.

12.4 Technical and structural parameters for passenger and freight elevators

12.4.1 Location

Passenger elevators should be located at the focal point of a building and be conveniently accessible from all main entrances. Elevator cores provide an obvious answer to these requirements.

When passenger elevators are grouped, it is important to note that no more than 3 elevator systems can be housed in a common shaft according to national building codes. For groups of 4 or more elevators, the elevators should be located in facing pairs.

The waiting or loading area in front of a bank of elevators or between two facing groups of elevators should be 3 m or 2 – 2.5 times the depth of the elevator car.

It is important to ensure that the elevator lobby is not required to be a circulation area as well. If necessary, elevators can be recessed in niches.

Freight elevators should not be located in the same area as passenger elevators. Many buildings are equipped with separate loading zones for loading and unloading delivery trucks. Freight elevators should be allocated to these loading zones. The loading space in front of these types of elevators should be at least the full depth of the elevator cage plus an area corresponding to the turning radius of rolling containers (for goods delivered in such containers) and space for handling the containers. Where goods are manually shifted by personnel (that is, no rolling containers), an additional 0.5 m of space is sufficient.

12.4.2 Motive power

The choice of motive power for elevators has a major impact on the structure of a building as a result of the equipment required for the different types of drives. The three most common types are:
- Traction (or cable) type
- Hydraulic
- Hydraulic/cable type (indirect hydraulic drive)

Traction or cable type drive
The most common drive is the traction or cable type drive.

In this system, the cage is suspended from cables, which pass over a traction sheave on a winch located above the cage and balanced with a counterweight at the end.

The counterweight generally corresponds to the weight of the cage and half its load. This means that for any given load, only half of the load at most and the traction must be overcome. The drive system is located above the shaft head (**Figure 12.2**).

Arrangements next to the shaft at the top or at the bottom of the shaft are also possible (**Figure 12.3**).

This arrangement is fairly common today, even though it puts a greater amount of strain on the cables.

Traction elevators that travel up to 2.5 m/s are generally equipped with drives consisting of a three-phase alternating current motor, gearbox and driving wheel. Elevators with operating velocities greater than 2.5 m/s are operated with gearless elevator equipment, which does not include a gearbox between the drive motor and the driving wheel. Instead, the driving wheel sits directly on the drive shaft, and the hoisting machines are direct current motors (most recently also available with alternating current motors).

Cable-type elevators are differentiated according to the following two drive control systems:
- Voltage-controlled three-phase drive
- Frequency-controlled drive

Today, voltage-controlled three-phase drives are predominantly used for elevator speeds up to a maximum of 1.0 m/s. They require far less electrical power than pole-changing drives and provide a more comfortable ride.

Frequency-controlled drives are used for of roughly 1.0 m/s and greater. Since this type of control always activates the drive motor at its optimal operating point, this solution offers additional energy savings over voltage-controlled drives due to the low start-up current and the high degree of efficiency while still ensuring optimal comfort.

Open-loop elevators with pole-changing drives are rarely used today, chiefly due to their high power consumption and limited comfort. They no longer correspond to modern standards of technology.

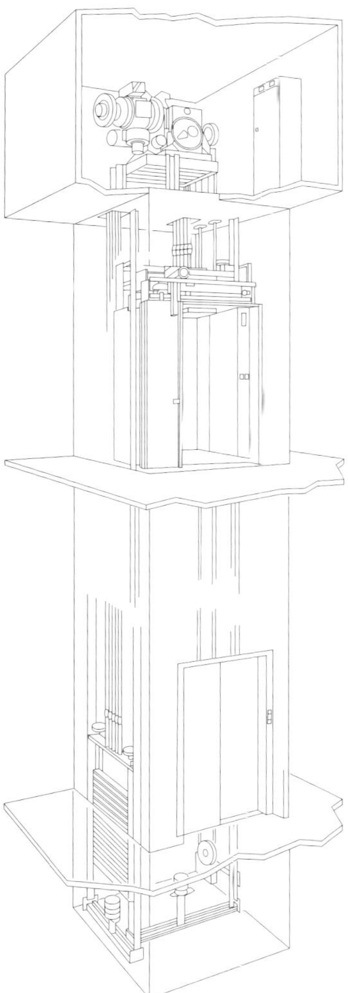

Figure 12.2
Traction drive, winch above shaft

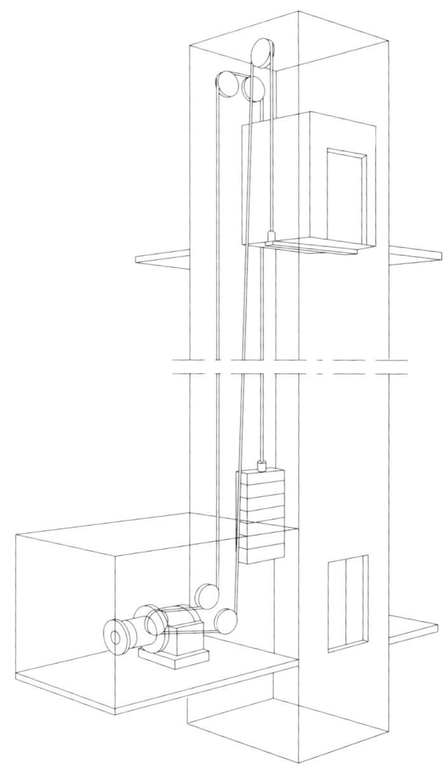

Figure 12.3
Traction drive, winch at bottom next to shaft

Hydraulic drive

An oil-hydraulic drive consists of an electric motor, an oil pump, tank and filter, hydraulic pistons and cylinders. Since the drive unit is linked to the hydraulic cylinder via an oil-pressure pipe, the machine room can be located up to 15 m (true length) from the elevator shaft.

The travelling curve of a hydraulic elevator is influenced by controlled or electronically regulated valves.

The number of required upward rides, i.e. motor-starts, is an important factor in the use of hydraulic elevators. A maximum of 120 rides per hour can be achieved with oil cooling. This benchmark must be strictly adhered to for morning peak operation in office buildings .

Hydraulic elevators are employed for elevating heights of up to 20 m. The recommended travel speed of hydraulic installations is 0.63 m/s. Codes restrict the maximum speed to 1.0 m/s. A differentiation is made between directly and indirectly operated hydraulic installations and furthermore between pressure and thrust piston designs.

Direct hydraulic drive
The car is moved by one or two hydraulic pistons in the shaft (**Figure 12.4**).

Depending on the hoisting height and the bearing capacity, the hydraulic pressure piston can be centred, i.e. placed directly beneath the car, or located to the side of the car (the most common hydraulic drive solution).

The "classic" hydraulic drive with a centred ram directly beneath the car calls for a boring directly beneath the shaft pit into which the pressure piston is lowered inside a protective pipe. It should be noted that local water utilities or other authorities may not grant permission for the deep borings required. The thrust piston offers another alternative for direct hydraulic drive; in this design, the car is pulled upwards by means of a thrust piston.

Indirect hydraulic drive
(Suspension 1:2)
In indirectly driven hydraulic elevators, the pistons are connected to the car via steel cables (**Figure 12.5**). A guide block across which the steel cables are run is located at each piston head. These hydraulic drives are currently designed with one or two lateral hydraulic cylinders.

The following types are differentiated:
– Indirect cable-hydraulic pressure piston elevator (**Figure 12.5**)
– Indirect cable-hydraulic thrust piston elevator (**Figure 12.6**).

Figure 12.4
Direct hydraulic drive, ram beneath car

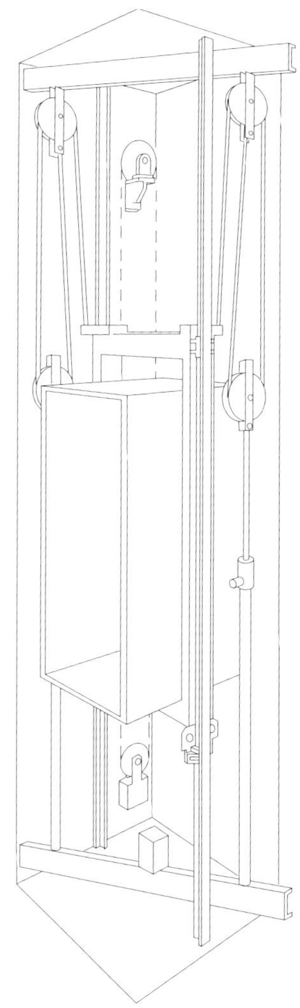

Figure 12.5
Hydraulic thrust piston elevator, indirect drive, two lateral cylinders

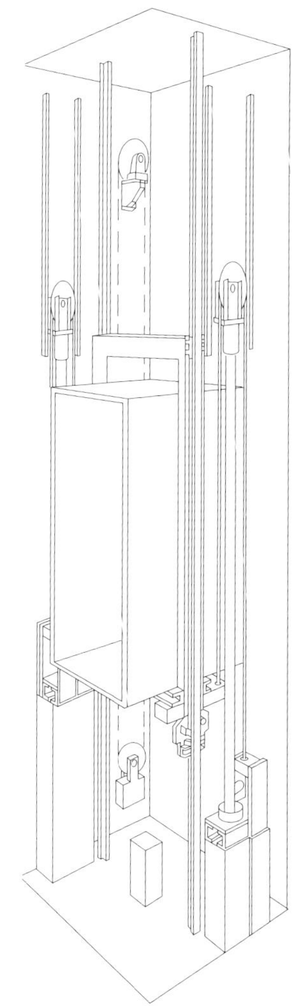

Figure 12.6
Indirect hydraulic pressure piston elevator, ram next to car

12.4.3 Construction

Elevator shaft

The shafts contain building components necessary for the operation of the elevator only. Any lines or components not related to elevator operation may not be included in the shafts.

The shaft dimensions largely depend on the elevator model (car measurements with and without load), the type of drive and the door design (**Figure 12.7**). The requirements stipulated in the elevator codes also determine the measurements of the shaft, especially the headroom at the top and the pit at the bottom. Elevator shafts must have ventilation and smoke extraction openings. The cross section of these openings is generally 2.5 % of the shaft floor area, with the minimum cross section stipulated as 0.1 m^2. Details are defined in the national building law.

All elevator shafts must be constructed in compliance with pr EN 81, later EN 81. This code differentiates between fully protected shafts in buildings classified for protection against the spread of fire (DIN 4102) and partially protected shafts for building areas where this type of protection is not required, e.g. "scenic" or observation elevators in atria.

Shaft pit (SP)

The bottom end of the shaft is commonly called the pit. The depth of the pit is measured from the top edge of the finished floor at the lowest elevator stop to the top edge of the finished floor of the pit floor.

The minimum depth of the pit is determined by:
– Space required for construction
– Over-run and safety space stipulated by regulations

The pit generally sits directly on the building foundation (unspoilt land). If the space beneath the pit is accessible, additional measures, clearly defined in the regulations, must be taken.

Shaft pits that are 1 to 2.5 m deep must be equipped with a removable access ladder. Pits with a depth greater than 2.5 m must have a secure (lockable) access door to a building floor (underground level).

Shaft head (SH)

The shaft head is the upper section of the shaft, measured from the top edge of the finished floor at the uppermost stop to the bottom edge of the shaft ceiling. The height of the shaft head is derived from the sum of the measurements for
– Construction space requirements
– Over-run and safety space stipulated in codes and regulations

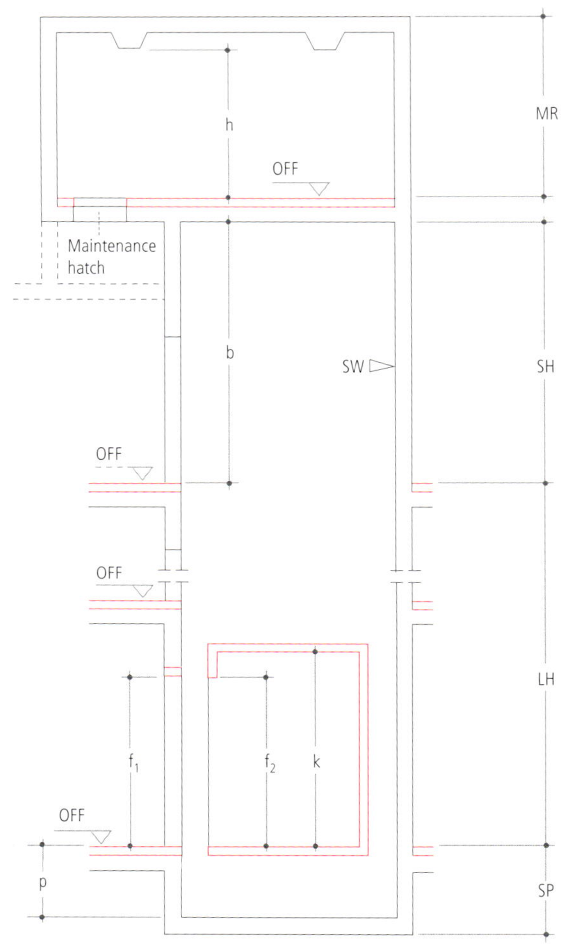

SW Shaft wall
SP Shaft pit
LH Lifting height
SH Shaft head
MR Machine room

h clear height, machine room
b clear height, elevator car, top stop to bottom edge of machine room
f_1 clear height of external door frame
f_2 clear height of telescoping door (head room)
c clear height of elevator car/cabin
p between upper edge of shaft pit and upper edge of bottom stop

Figure 12.7
Elevator shaft and machine room

Shaft access

The size of the shaft access points is determined by the door design, while their location is defined by the shaft symmetry. A basic distinction is made between mounted and built-in doors.

Mounted doors are sliding doors mounted on the interior shaft wall, onto angle clamps or concrete lips. Door models are differentiated into centre-opening or side-opening, telescopic or non-telescopic types. The shaft doors must be absolutely plumb one above the other, since they are moved by the nosings of the outer car door. Side-opening, telescopic sliding doors require the least shaft width. Individual door leafs should be no less than 0.3 m in width to achieve the highest degree of synchronization of the door leaf action.

Built-in doors are automatic sliding doors as well as side-hung doors, where the door leaf closest to the shaft interior is flush with the inside shaft wall at the access side of the elevator car.

Once the door type has been selected, the corresponding dimension of the opening in an unfinished state can be determined as follows:

For automatic sliding doors, add roughly 15 cm to the door size in order to arrive at the dimension of the opening in an unfinished state. The dimensional guidelines provided by the manufacturer should always be taken into consideration.

Reserve openings for external indicators and call signals

The operating buttons and indicators of elevators that are operated as individual systems are usually integrated into the doorframe (case). However, when elevators are grouped (elevator bank), the operating features are located between two sets of doors.
A reserve opening 15 cm-wide and 20 cm-high is recommended in these cases. The bottom edge of this reserve opening should be 1 m above the top edge of the finished floor. If out-of-service indicators or up and down arrows are also planned for these external operating panels, the reserve openings must be correspondingly larger.

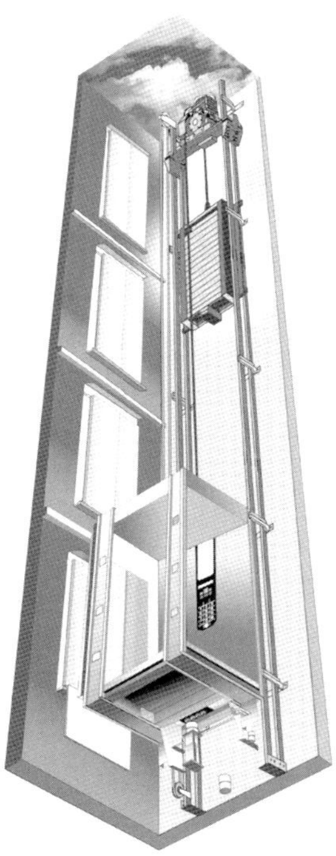

Figure 12.8.1
Construction of elevator without machine room, with drive located directly in the elevator shaft; seen from below

Machine room

Depending on the type of drive, the machine room is located either at the top above the shaft, or at the bottom next to it. Today, there are a number of elevator systems with different types of drive on the market, which can be operated without the need for a machine room. These systems offer limited speed, transportation height, load and number of stops. They are available from different manufacturers with hoisting machines, with or without gearbox, and also as hydraulic models. The drive is installed directly in the elevator shaft (**Figure 12.8.1**) or on the elevator car (**Figure 12.8.2**).

Machine room for hydraulic elevators

The machine room and drive unit of all hydraulic elevators are generally located directly next to the shaft, level with the lowest stop. However, the machine room can also be placed at several metres' distance to the shaft. It is important, in the latter case, to ensure that the hydraulic lines are installed in a manner that renders them accessible for inspection at all times and along the full length of the lines. Local authorities may also insist upon additional fire protection measures.

Figure 12.8.2
Elevator without machine room with car-attached drive

Machine room for cable-type elevators

Machine rooms for traction (cable-type) elevators should preferably be located above the shaft.

When designing the machine room it is vital to make allowances for installing the drive and the control panel. The elevator contractor determines the necessary reserve openings between machine room and shaft since these depend on the specific elevator components used. To facilitate drive installation and replacement in case of repairs, an installation girder or similar feature should be included below the ceiling above the drive (**Figure 12.8.3**).

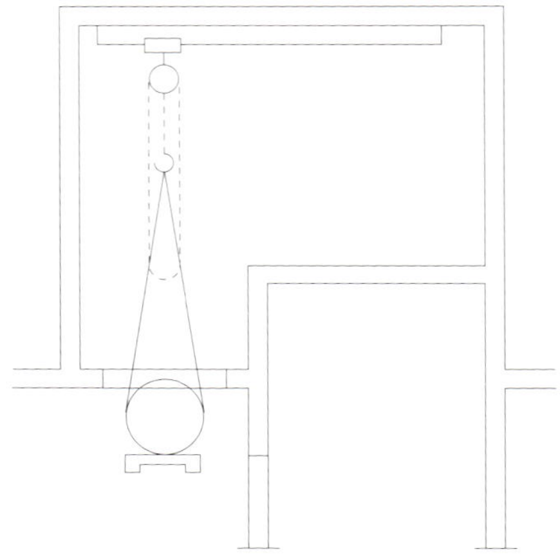

Figure 12.8.3
Machine room for traction elevator with maintenance hatch

A fastening device for safe hoisting and moving of heavy loads must be mounted on the machine room ceiling.
A transport opening must be provided in the floor of the machine room itself or in immediate proximity to the machine room.

Access

All access features (doors, hatches etc.) to the machine rooms must be man-sized to allow maintenance workers access at all times (see also the relevant regulations). A clear height of 2 m and an approximate width of 0.9 m are generally considered to be sufficient.

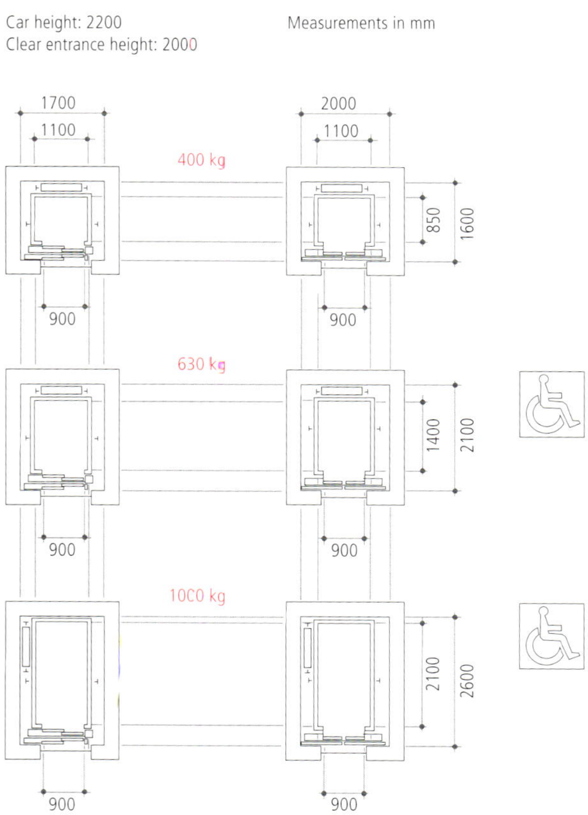

Figure 12.9.1
Elevators for residential buildings (Measurements in mm)

12.4.4 Elevator cars

In addition to the doors, the size of the elevator shaft is also largely determined by the dimensions of the elevator car. Some of the most common car dimensions are presented in **Figures 12.9.1 and 12.9.2**.

All elevator cars must be lit; they are also equipped with grid-independent safety lights, which are battery operated. The battery is installed in the control cabinet of the elevator system. Lighting design can be more individual for passenger elevators; freight elevators, on the other hand, should generally be lit from recessed lights in the car ceiling to prevent the fixtures being damaged during loading and unloading; they should also be designed with greater dust protection.

Passenger and freight elevator cars must be ventilated. Air intake and exhaust openings must be placed to ensure sufficient diagonal and vertical cross-ventilation; integrated small ventilators in the cars are recommended. If required during use or construction, an escape hatch should be built into the car ceiling as an escape route for persons stuck in an elevator. If goods are moved into and out of freight elevators on trolleys, the cars should be equipped with protective rails. All trolleys used for transporting goods in freight elevators must have parking brakes.

Special features for elevators designed to transport disabled and wheelchair users

Full compliance with the regulations specified in DIN 18024/025 as well as the relevant regional or local codes is required for elevators designed to transport disabled and wheelchair users on a regular basis.

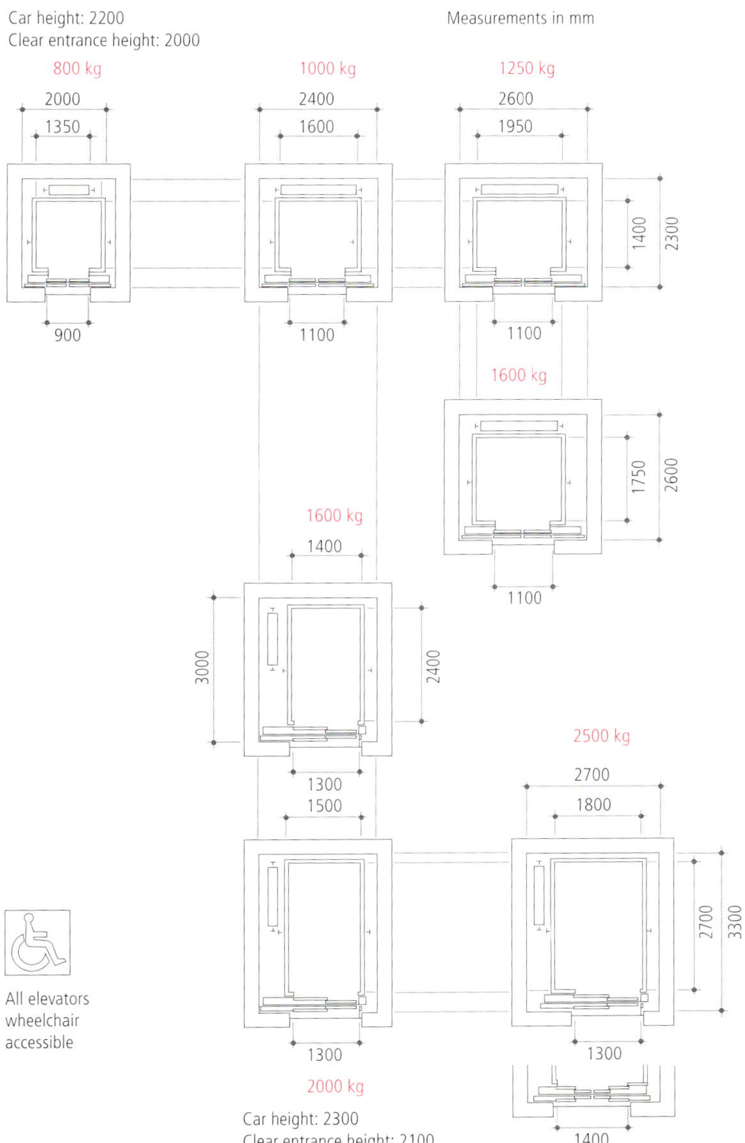

Figure 12.9.2
Elevators for office buildings and hotels
(Measurements in mm)

Figure 12.10
Users enter the destination floor into this microprocessor control unit; the system selects a specific car, which then drives directly to the destination floor
(System Micionic 10®, Schindler)

12.4.5 Elevator controls

The type of controls considerably effects the capacity of elevator installations. The following controls are recommended for passenger elevators:

Directional elevator control
All internal and external control commands are stored. The external control panels at each stop feature two control buttons, one for each direction "Up" or "Down". The elevator stops only if the direction in which the car is travelling corresponds to the selected direction. This results in more efficient traffic management, which has proven especially advantageous in installations with numerous stops between the ground and top floor.

Group elevator control
When several passenger elevators are gathered into a group, the basic module of the directional elevator control is used.

The microprocessor controls used today respond to each incoming call by allocating the car best placed to respond to the call in terms of waiting time and efficiency. This decision is checked several times per second to ensure that it is indeed the best choice (**Figure 12.10**).

The capacity can also be improved with the help of an "occupied" feature. If elevator cars are frequently filled to capacity in a particular system, stops caused by external control calls can lead to unnecessary delays. An "occupied" feature prevents reception of an external call until the individual car is able to take on new passengers (i.e., after passengers have exited). The "occupied" feature can be programmed to a range of 60 to 80 % of the total capacity.

Overload features are generally installed to prevent cars loaded beyond the defined capacity from taking off.

12.5 Scenic elevators

Scenic elevators, often simply called glass elevators, are increasingly popular architectural features in the design of distinguished buildings. They loosen the rigour of the architecture and give passengers a visually stimulating ride between floors. Passengers can participate actively in the ride and perceive the space through which they are moving. In a hotel setting, they can enjoy the atmosphere, and in office buildings, they can relax between meetings, while glass elevators in shopping centres offer a quick overview and facilitate orientation.

A superficial enquiry into glass elevators may give the impression that this type of elevator is exclusively suited for luxurious buildings. Often, however, simple models can also contribute to designing bright and welcoming lobbies or stairwells by virtue of their transparency.

In the area of public transportation, glass elevators abolish the anonymity of closed elevator cars, and increase the passenger's sense of security (**Figures 12.11.1/11.2**). Since the interior of scenic elevator cars is also visible from the outside while the elevator is in motion, passengers are aware that they can be observed and incidents of vandalism are generally reduced.

Figure 12.11.1
Car for scenic elevator at the ICE railway station in the airport in Frankfurt/Main
(Architects: Bothe Richter Teherani, Hamburg, manufacturer: Otis)

The architectural design must be determined in advance, before detailed plans for a scenic elevator are even drafted. The next question is whether the technical components of the elevator (machine etc.) are to be exposed as a visible element of the architecture or hidden in a shaft.

If the technical components are to be hidden, the scenic elevator itself consists of an entrance area and a viewing area. The entrance area is surrounded by an enclosed shaft that contains the necessary elevator technology (counterweight, car mechanics, shaft installation). The car itself is enclosed in this area.

The car walls must be constructed with laminated glass in compliance with EN 81, appendix J; otherwise pendulum impact tests must be carried out. Laminated glass is, however, used in all cases.

Any glazed sections in elevator shafts within the transportation area must also be constructed with laminated glass.

The safety and practicability of all other glass constructions deviating from the EN 81 standard must be proven on the basis of risk analyses provided by the manufacturer.

Depending on the architecture, opaque sheet metal doors can be replaced with translucent glass doors in scenic elevators. Glass doors can, however, only be employed if the elevator does not travel across fire lobbies. Glass doors are not suitable for fireproof shafts under DIN 4102.

When glass doors are a suitable option, they must correspond to the safety guidelines for laminated glass doors. These guidelines specify what type and thickness of glass may be employed. The closing edges of all glass doors should be framed with a profile to render the movement of the door visible and to prevent the edges being damaged. When glass doors are used in elevators subject to specific performance targets, it should be remembered that the added weight of glass doors considerably increases the travel time of these elevators in comparison to models with standard steel doors. The longer floor-to-floor times that result hamper the overall transportation performance of the elevator.

Like the observation area itself, glass doors also invite passengers to look out at the surroundings. Consequently, passengers often lean against the car walls or the doors. When the doors open, there is therefore a risk that a passenger's hands may be caught as they adhere to the glass surface. Children particulary are at risk; the regulations specify relevant safety measures.

In Central Europe and similar climate zones, external elevators on building facades are practical only to a limited degree, since even those designed for all weather conditions may have to be shut down in extreme conditions. In other words, these elevators should only be employed to a limited degree for specific transportation requirements. When designing exposed external elevators it is also important to remember that the shaft doors cannot provide a sufficient seal at the stops; gated antechambers at each stop are therefore necessary. The best recommendation for glass elevators in these climate zones is to enclose external elevators on the facade in a weatherproof glass shaft.

Figure 12.11.2
Scenic elevators in the SPD Headquarters, Willy Brandt House, Berlin
(Image: Vestner)

12.6 Simplified service, container, hole-less and small cargo elevators

Service elevators are employed as an efficient and economical means of transporting goods vertically in factories, workshops and warehouses, and for transporting large waste containers in hotels and administration buildings.

These elevators are limited to a maximum of three stops and are not designed to carry passengers. Different types are distinguished according to use.

Simplified service elevators are systems designed specifically to transport goods between no more than three levels, with a maximum load capacity of 2,000 kg and a maximum car floor area of 2.5 m^2. The operating speed is set at a maximum of 0.3 m/s.

Container elevators are designed to transport goods inside specially designed containers between a maximum of three stops, with a maximum load capacity of 1,000 kg, a car floor area of 2.0 m^2 and an operating speed not exceeding 0.3 m/s.

Hole-less elevators are simplified service or container elevators whose shaft ends flush with the floor level of the uppermost terminal; in other words, they do not require shaft space above the top terminal.

Small cargo elevators are used in offices, hotels, department stores, hospitals and all other types of buildings where small loads totalling no more than 300 kg must be transported frequently.

12.7 Technical and constructional measures for service, container, hole-less and small cargo elevators

Simplified service, container, hole-less and small cargo elevators are generally powered by hydraulics (**Figures 12.12.1/12.12.2**).

The drive and controls are housed in a lockable cabinet for which no separate machine room is required. Hole-less elevators should be covered by a canopy. **Figure 12.13** illustrates different types of canopy.

The car is generally lifted and lowered by direct hydraulics. Indirect hydraulic models with chain beams and two pulleys for a cable-suspended travelling or hoisting carriage are also conceivable.

Small cargo elevators

Small cargo elevators are usually erected inside shaft towers (**Figure 12.14**). The shaft tower absorbs all forces originating from the driving mechanism and diverts them to the building fabric. Small cargo elevators are powered by three-phase alternating current motors. The entire drive equipment is mounted on a frame and located in the machine room above the shaft tower. The machine room is separated from the shaft tower by a simple sheet steel structure with openings for cables and installation.

The cages consist of a supporting frame constructed with sectional steel, while the individual cars are manufactured from sheet steel. The openings at each stop are covered with sliding doors or single-leaf, side-hung doors; the sliding doors are usually vertical sliders. The clear door dimensions correspond to the clear car dimensions. Special designs for automatic container loading and unloading into small goods elevators are also available. The load capacity is generally in the range of 50, 100 or 300 kg for standard speeds between 0.2 and 0.63 m/s.

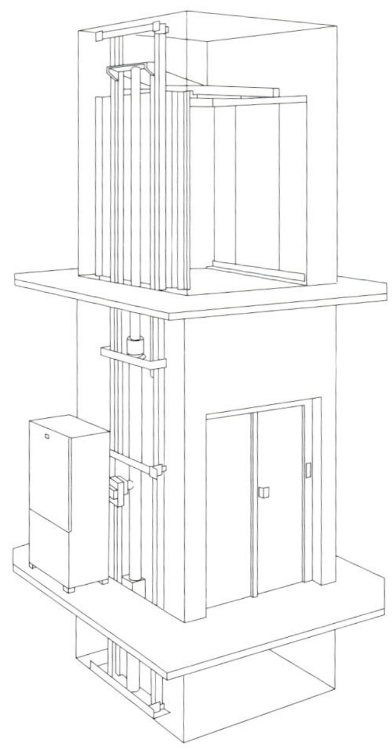

Figure 12.12.1
Simplified freight elevator

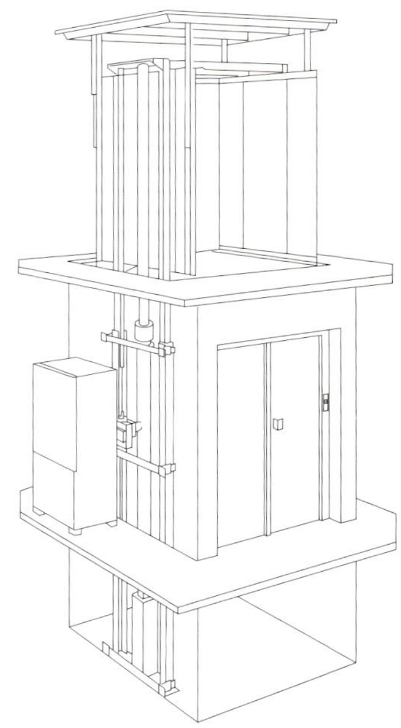

Figure 12.12.2
Hole-less elevator

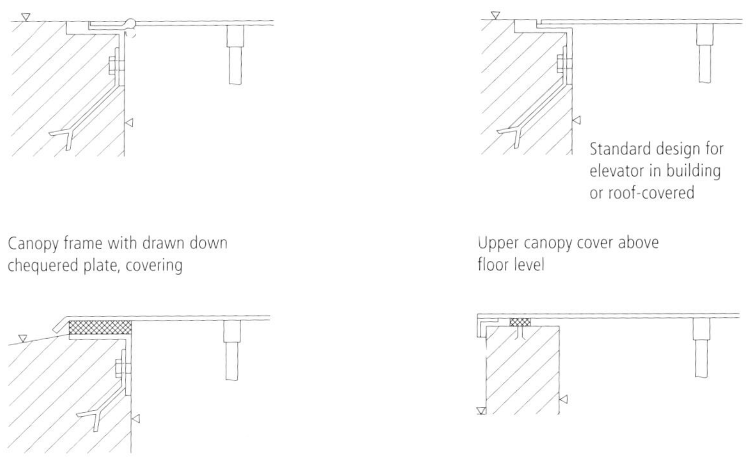

Figure 12.13
Canopy coverings for hole-less elevators

▽ Floor level of uppermost stop
▽ Elevator shaft wall, inside

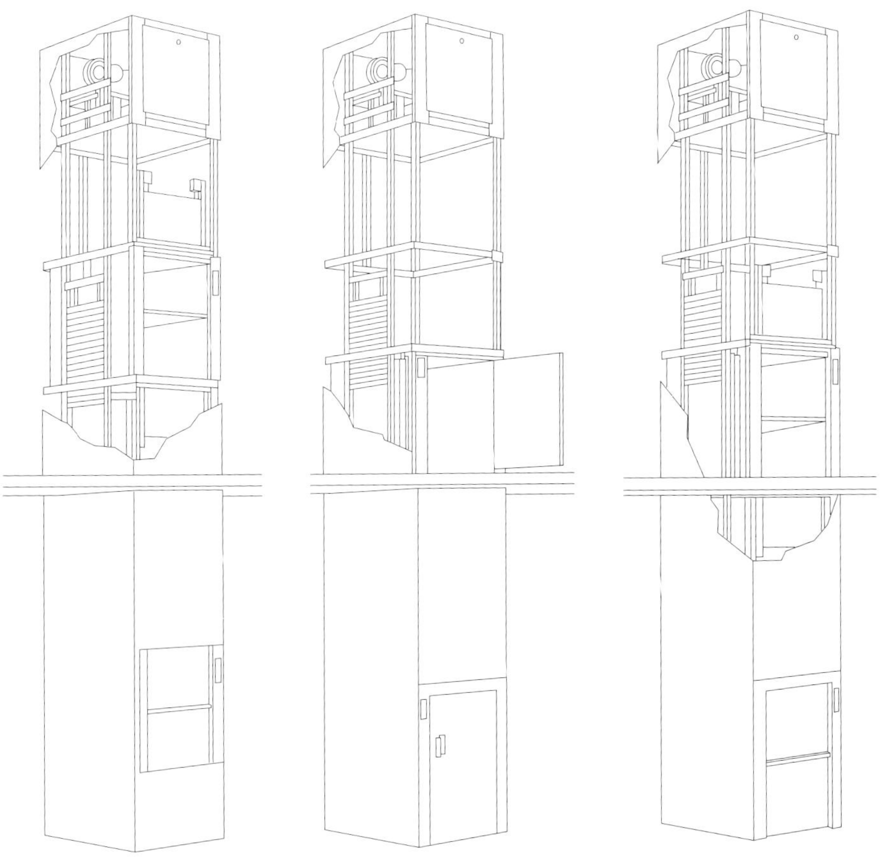

Figure 12.14
Small cargo elevator

12.8
Escalators and moving sidewalks

12.8.1 Escalators

Escalators are reliable means of continuous transportation for persons. They have high capacities and are suitable for a wide variety of applications.

12.8.1.1 Applications

The most common application for escalators is in department stores and shopping malls, where the main objective is to transport shoppers effortlessly between retail levels. The central location of the escalators also gives customers visual information about the different levels as they are moving up or down. In administration buildings with high rates of public traffic and equally high levels of traffic between floors, escalators offer the fastest and most comfortable means of transportation. In multi-storey exhibition halls, escalators ensure that visitors can move between the various exhibition levels without waiting times even during peak hours.

In airports, railway and subway stations and over- or underpasses, escalators are high-capacity and efficient means of public transport. Here, too, they keep the flow of traffic moving even during peak times.

A differentiation is made between commercial and traffic escalators.

In public areas, e.g. railway stations, traffic escalators are designed to be especially hard-wearing, while the demands in terms of frequency and running times tend to be somewhat less for commercial escalators in department stores, shopping malls and office buildings.

Incline and transportation height
Escalators are generally designed on an incline of 27.3, 30 and 35 degrees.

The 35-degree escalator is the most efficient solution since it requires the least amount of space. In compliance with regulations, this incline is applicable up to a total transportation height of 6 m. If the height between the bottom and top of the escalator exceeds 6 m, the incline should be 27.3 or 30 degrees. For heights up to 6 m, the escalator must be designed with two horizontally retracting steps, while three horizontal steps are required for heights greater than 6 m.

The different installation lengths for identical floor heights and varying inclines are illustrated in **Figure 12.15. Figure 12.16** depicts various installation options.

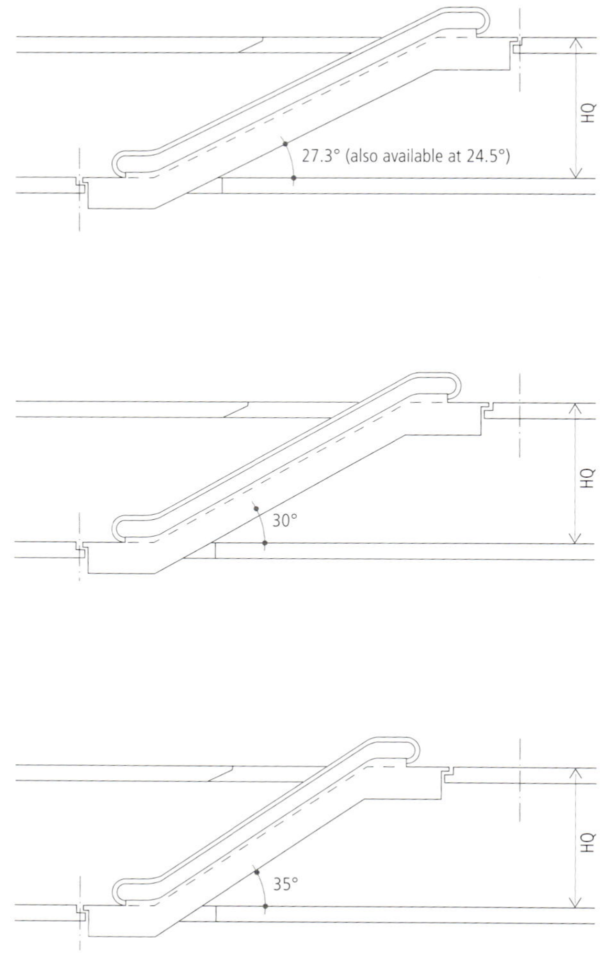

Figure 12.15
Escalators, inclines and heights

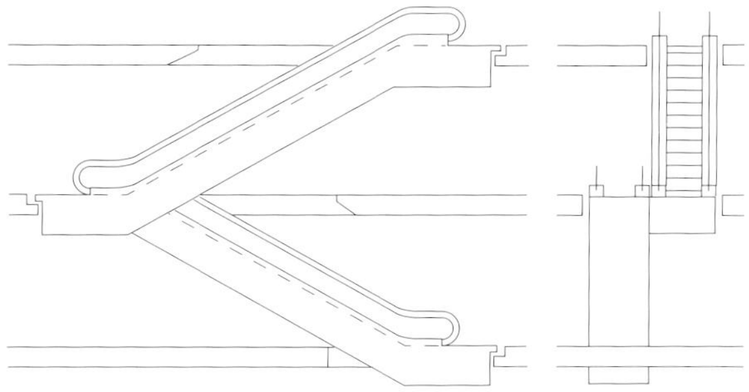

Escalators operating in one direction – continuous

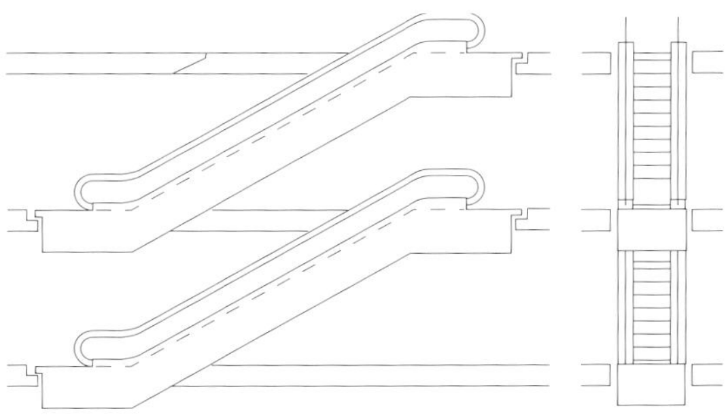

Escalators operating in one direction – with interruption

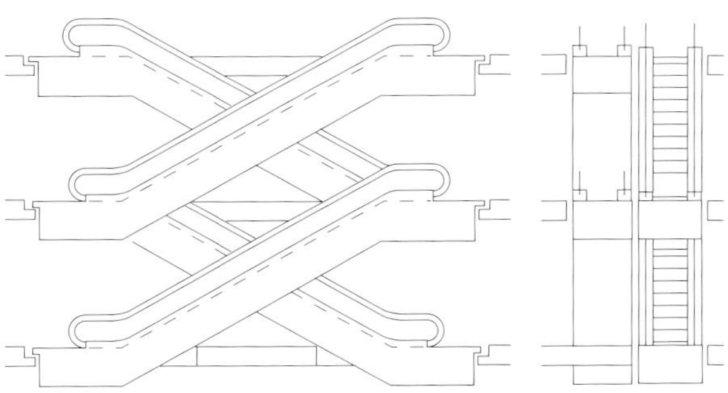

Escalators operating in both directions – criss-cross arrangement

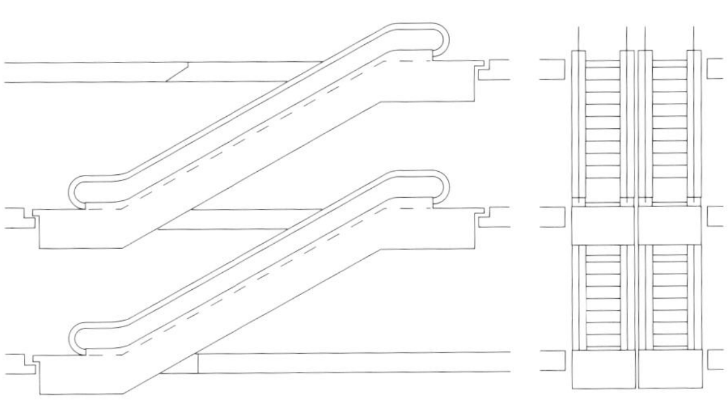

Escalators operating in both directions – parallel arrangement

12.8.1.2 Escalator capacity

Most escalators are designed with 1,000-mm-wide steps, which allow persons to move onto it comfortably even when carrying luggage or shopping bags. 600-mm-wide steps (rarely requested) and 800-mm-wide steps are also available, although they are generally only used in low-traffic areas or in situations where space constraints preclude the use of wider steps. When several escalators are used in a building, all steps should be of equal width to prevent dust accumulating.

The standard transportation speeds range between 0.5 and 0.65 m/s.

For a speed of 0.5 m/s, the theoretical capacity is:

- 600 mm step width = 4,500 persons/hour
- 800 mm step width = 6,750 persons/hour
- 1,000 mm step width = 9,000 persons/hour

The practice has shown, however, that these theoretical capacities are never achieved because people hesitate as they step onto the escalator, deliberately maintain a greater distance to others on the escalator, or do not fully use the option of walking past others on moving escalators.

Increasing the travelling speed does not increase capacity proportionally because people hesitate even longer before stepping on and off the escalator.

Figure 12.16
Escalators, installation options

Whenever possible, it is best to install two or more parallel sets of escalators. This increases capacity and ensures backup in case one escalator should fail. Each escalator is designed for transportation in both directions (up and down). The desired direction is determined with a key switch. It is important to ensure that the landings at the bottom and top of each escalator are sufficiently large.

The operating options are as follows:

Continuous operation

The escalator continues to run without stopping until it is switched off.

Intermittent operation

The escalator is triggered into operation by means of a contact floor covering or a light barrier at the landing; it stops running after a safe interval once the person has disembarked.

Alternatively, the escalator continues to operate at a low speed of roughly 0.2 m/s when not in use (frequency transformer), with the speed accelerating if someone steps past the light barrier at the landing.

There is a wide variety of options with regard to design for escalators:
- Glass balustrades (e.g. with etched glass)
- Strip lighting on the steps
- Handrail lighting
- Handrails without visible balustrade covers
- Handrails in a variety of colours
- Glass-enclosed escalator body with visible mechanism, steps etc. and lit from the inside

and many other options.

Escalator balustrades are generally constructed with laminated glass, or as steel structures covered in sheet metal on the inside and the outside. There are few limitations to the design of these surfaces. **Figure 12.17** depicts examples of escalators.

Figure 12.17
Examples of escalator installations
Karstadt department store, Düsseldorf
Stadttor Düsseldorf
(Images: Otis)

12.8.2 Moving sidewalks
12.8.2.1 Applications

Moving sidewalks have proved to be comfortable, safe and efficient means of transportation across long horizontal or slightly inclined routes. They are generally employed where long distances must be covered in comfort.

The international standards for inclines of moving sidewalks range between 0 and 12 degrees. **Figure 12.18/19** illustrate two types of moving sidewalk (horizontal and slight incline).

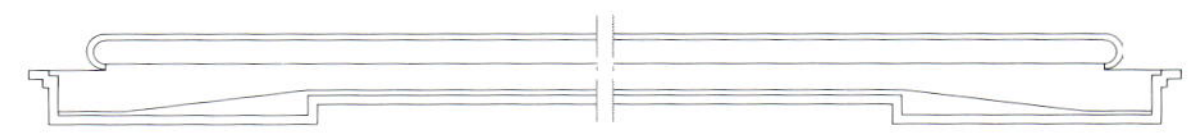

Figure 12.18
Horizontal moving sidewalk

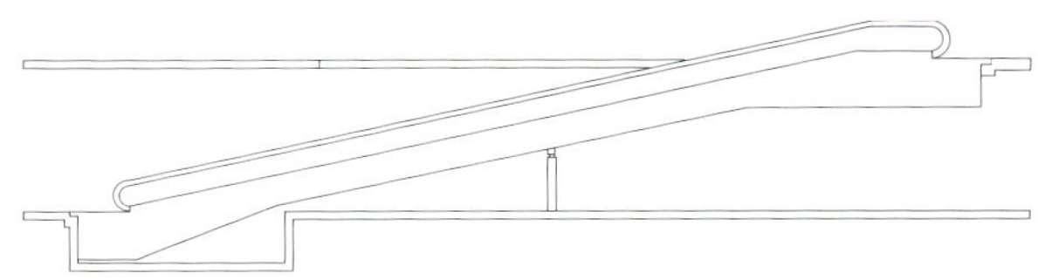

Figure 12.19
Inclined moving sidewalk

12.8.2.2 Capacity of moving sidewalks

The following theoretical capacities can be calculated for fully occupied moving sidewalks for three different travelling speeds. Once again, these capacities are rarely achieved in practice as users tend naturally to keep a greater distance to others on the moving sidewalk.

Speed	0.5 m/s	0.65 m/s
Pallet width		
800 mm	6,750 Pers/h	8,800 Pers./h
1,000 mm	9,000 Pers/h	11,700 Pers./h
1,200 mm	10,800 Pers/h	14,040 Pers./h
1,400 mm	12,600 Pers/h	16,380 Pers./h

12.8.2.3 Installation of moving sidewalks

The most common moving sidewalks thus far are installed with 1,000-mm-wide pallets, although pallet widths of 1.2 and 1.4 m are also available. These pallet widths are suitable for shopping centres, airports and railway stations, where shopping and luggage trolleys are used. The travel speed of inclined moving sidewalks is 0.5 m/s. Horizontal moving sidewalks can be operated at speeds ranging from 0.65 to a maximum of 0.75 m/s depending on total length.

Like escalators, all moving sidewalks can be programmed for operation in both directions. Balustrades, too, are similar to those used for escalators. Loading and unloading zones must be generously designed to ensure safe use.

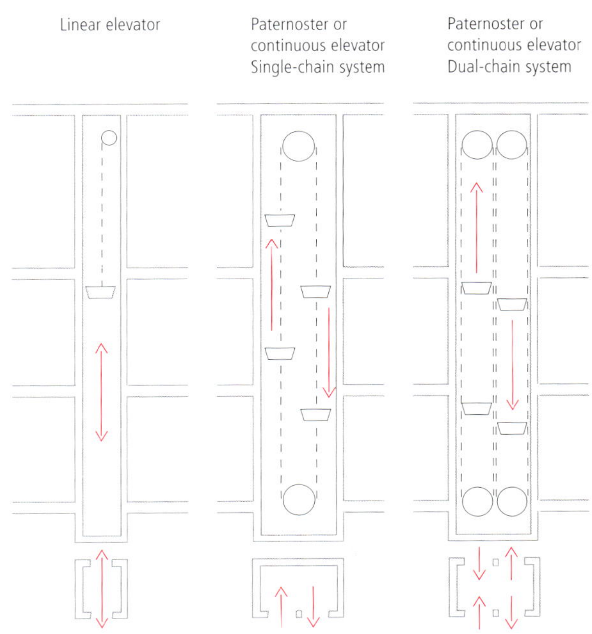

Figure 12.21
Vertical transportation for track-mounted container conveyer systems

12.9 Container conveyer systems

Smooth transportation of documents and goods is often essential for efficiency in operation. Specialized conveyer systems are used for these applications. Container conveyer systems provide fully automated transportation of small cargo from the loading station to the desired destination.

These systems are used, for example, in hospitals for transporting medication, patient files, blood supplies or similar materials, in insurance headquarters for original claims files, and on large manufacturing sites for automated conveyance of components and materials etc.

12.9.1 Track-mounted container conveyer systems

In track-mounted container conveyer systems, the cargo is loaded into a container that is programmed for transport to a specific destination. The container is then transferred onto a track and transported to the destination. The maximum load of the containers is 70 kg. Vertical conveyers are essential components of these conveyer systems. In all installations reaching across several floors for container transport, they move the container to the desired destination floor. The illustration in **Figure 12.20** shows a model of such an installation.

Different types of conveyers are employed for vertical transportation according to conveyer capacity (**Figure 12.21**).

Linear elevators and container elevators for a capacity of 120 containers per hour. As soon as the cage has reached the level where a loaded container is ready, a shaft door opens automatically. The container enters the cage, the door closes and the cage continues automatically towards the programmed floor (and destination). Upon arrival, the container rolls through the automatically opening shaft door onto the conveyer track or belt and continues its path horizontally to the destination.

For conveyance capacities of up to 240 containers/h, fixed tray conveyers are used, which operate on the principle of an endless chain or bucket system. When all loading surfaces are stacked one directly above the other, a single-chain system is sufficient. If operations require that containers are loaded from two sides, a two-chain system is necessary. Horizontal transport is provided on flat conveyer belts, drive-operated travolators and/or gravity drive travolators.

Flat conveyer belts carry containers across long horizontal distances and up inclines up to 20°. Drive-operated travolators and gravity drive travolators complement the transportation route at specific positions, for example at loading and unloading positions, storage areas and curves. Drive-operated travolators can accelerate, reroute and steadily transport containers. As their name suggests, gravity drive travolators make use of gravity to transport the container goods.

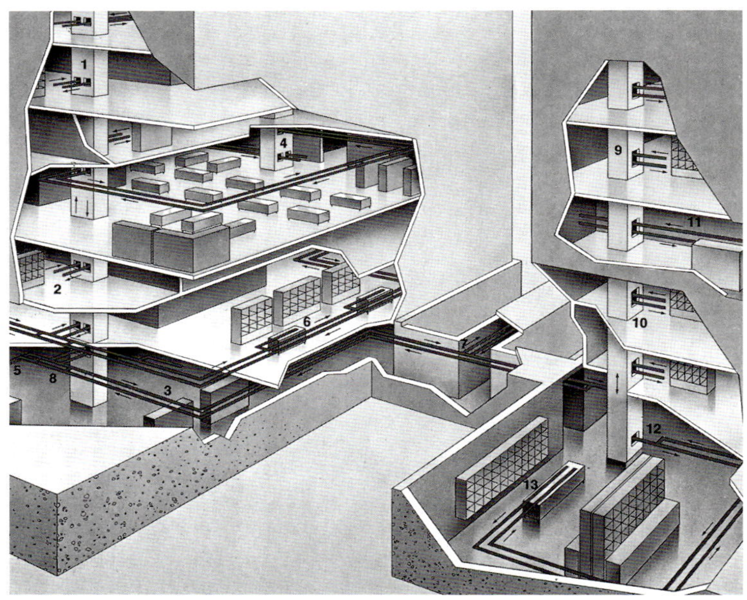

1 Fixed tray conveyer with automatic loading and unloading during ride
2 Dispatch and receiving stations in container conveyer system
3 Double-tracked horizontal transportation routes in ceiling area, parallel arrangement
4 Altitude adjuster to compensate for level changes within same floor
5 Angled conveyer for an incline of up to 20°
6 Receiving station in horizontal route section
7 90° turns with automatic control
8 Switch point for 180° turns
9 Linear container conveyer with automatic loading and unloading
10 Dispatch and receiving station at container elevator
11 Double-tracked horizontal transportation routes, stacked arrangement
12 Combined loading station for at container elevator
13 Double-tracked horizontal transportation route at waist level

Figure 12.20
Track-mounted container conveyer system

12.9.2 Self-propelling container conveyer systems

The track network of self-propelling container conveyer systems (**Figure 12.22**) consists of horizontal and vertical C-shaped aluminium rails with integrated brass alloy conductors. Switches for a wide variety of functions create connections to all possible destinations. The individual containers are designed to carry a load of up to 10 kg. The container is attached to a running gear with a motor designed for operation on horizontal and vertical routes under full load capacity. Power transmission from running gear to track is effected by a friction wheel on horizontal routes and a toothed wheel or gearwheel on vertical routes.

Contemporary systems are largely microprocessor controlled, with sensors in the track system guiding the coded containers through the system. Automated loading and unloading stations (**Figures 12.23**) optimize the system and can also be used to pre-sort the transported goods.

Stations are incorporated where containers need to be loaded and unloaded. The track network is comprised of straight tracks, bends and switches, in the horizontal and vertical direction. Depending on the size of the installation, one or more empty container depots can be built into the transportation route. Calling keys are used at the stations to request delivery of an empty container.

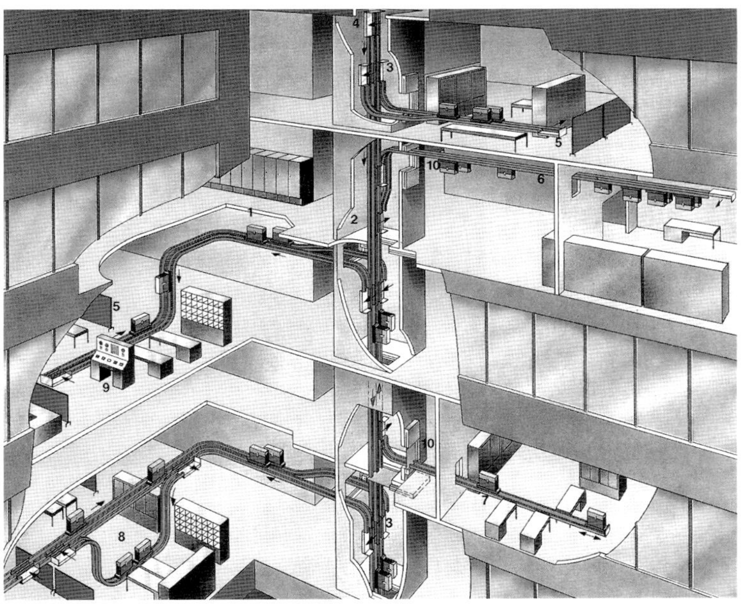

Figure 12.22
Self-propelling container conveyer systems

1 Drive with containers
2 Vertical transportation route, double-tracked for up and down traffic
3 Transfer switch points, link between stations and continuing horizontal transportation routes
4 Switch point, container continues in opposite direction
5 Transit stations with switch point
6 Empty container storage, containers suspended upside down from ceiling
7 Return transfer station: single track
8 Transit station without switch points
9 Monitoring panel
10 Fire-proof doors, horizontal and vertical installation, closed or open door

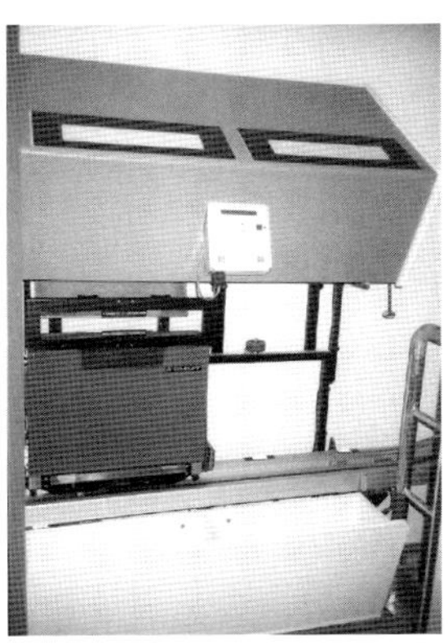

Figure 12.23.1
ADAL station
Automatic loading and unloading
(Image: Telelift)

Figure 12.23.2
Postal service station
(Image: Telelift)

12.10
Facade access systems

Facade access systems (or facade hoists) are used where facades require external cleaning or maintenance services.

The principal components of a facade access system are:
- The carriage with the travelling gear, accommodating the hoisting gear with cable drum and controls
- The boom, connected to the carriage
- The car, which travels up and down on the facade

Operators of facade hoists or access systems generally step into the car on the roof; the controls are operated either from the carriage or directly by the operators in the car. Facade access systems are differentiated into trackless and rail-mounted types.

12.10.1 Trackless systems

The trackless facade hoist is by far the most economical system. It is used primarily on buildings that already have a load-bearing and passable roof covering for other applications. Appropriate routes must be installed along the length of the parapet. The carriages of facade hoists are generally powered by electricity, making them dependent on power supply (cabling). The design of the carriage, which is equipped with wide, travelling wheels, allows free movement within the areas defined for accessibility.

The advantages of trackless systems are their mobility and economic viability.

Figure 12.24 illustrates the system design of the facade access system with carriage, boom and car.

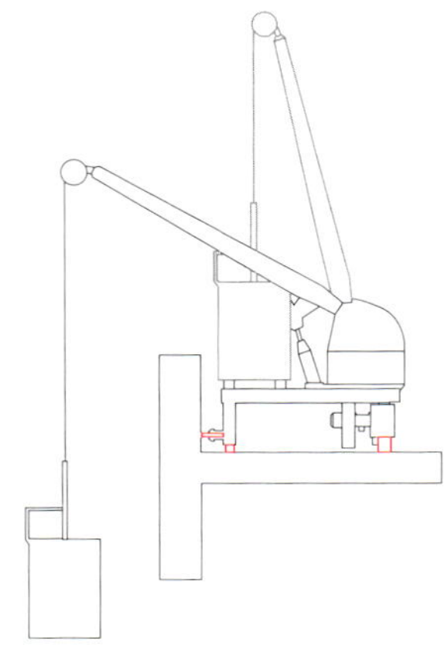

Figure 12.24
Layout of a facade access system with drive, boom and cage

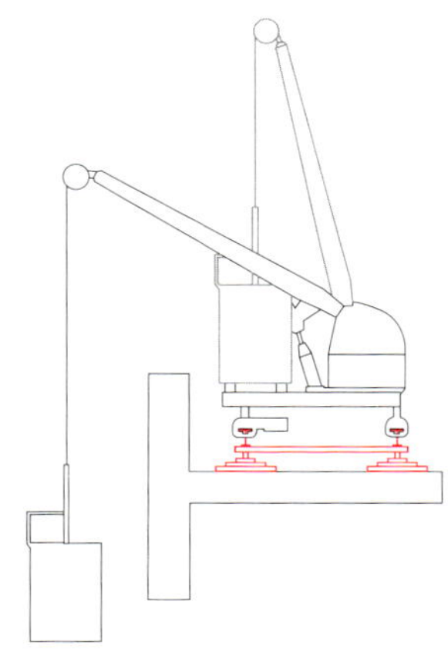

Figure 12.25
Facade access system – horizontal track installation

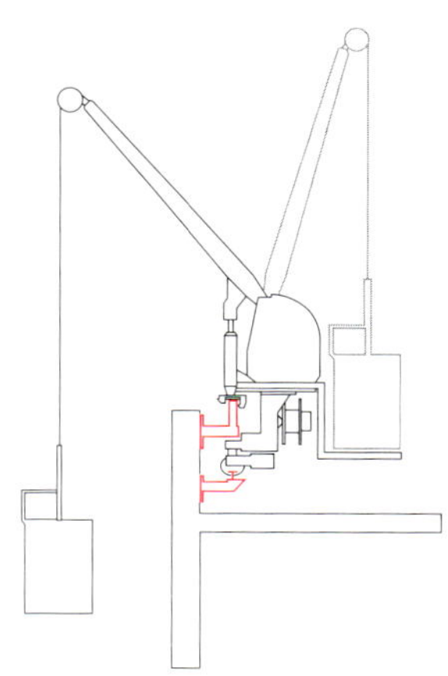

Figure 12.26
Facade access system – vertical track installation

12.10.2 Rail-mounted facade hoists

In rail-mounted facade hoists, the carriage is guided across a network of rails or tracks installed on the roof. The route of the facade hoist is therefore precisely defined.

The tracks are mounted either horizontally on the rooftop or vertically on the inside walls of the parapet.

In horizontal track arrangements (**Figure 12.25**) the tracks are either anchored directly in the roof structure or mounted freely on top of the structure. The latter are generally preferable since they do not puncture the roof skin. However, they can be employed only where there are no extreme conditions – e.g. narrow tracks, large booms or increased loads – that require anchored track supports. As **Figure 12.25** shows, this facade system consists of the same principal components as described for the trackless system.

Vertical track mounting (**Figure 12.26**) is possible only on taller parapets with very stable and load-bearing construction. The advantage of this arrangement is that it provides very narrow travelling routes. This is, however, coupled with the disadvantage that the parapet must be correspondingly high, which may have an adverse effect on the visual appearance of the roof area.

12.10.3 Load-carrying member

The car, which is generally made of light metal to achieve the lowest possible self-weight, serves as the load-carrying member. The car is usually 1 m high and equipped with a railing on the rear side to minimize the risk of falling. It is standard to secure car occupants with safety belts, which are tied off at the anticipated landing points. The water required for facade cleaning is transported in containers inside the car.

The cars are equipped with rubber buffers on the side facing the facade to avoid damaging it.

Regulations stipulate that guides are used for cars on buildings exposed to above-average wind loads and wind velocities, as well as on buildings more than 30 m high. The guide generally consists of vertical sections covering the full height of the facade; these must be adapted to the facade architecture and integrated into its structure.

Figure 12.27 illustrates different guide sections, which can accommodate the necessary guide wheels for the car. Where the distance of the sections (building grid) is greater than the length of the car, the guide wheel units are linked to the load-carrying member by a roller crossbar to allow for lateral movement across the entire width of that section of the facade. If guide sections cannot be installed because of a particular architectural design, or because the facade access system has been retrofitted, the load-carrying member can be controlled with guy ropes on the building or on the facade hoist itself. It should, however, be ensured that any ropes attached to the building are anchored at both parapet and ground level.

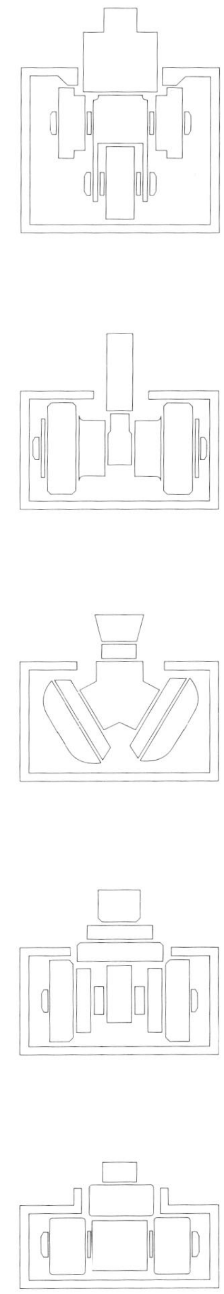

Figure 12.27
Different guide sections to accommodate the necessary guide rollers of the cage

Illustration credits

1.4	von Gerkan, Marg & Partner, Hamburg
1.5	PFP – Planung Fahr Partner, Munich
3.5, 3.6.1, 3.6.2	Auer + Weber, Munich
4.10	after: Energie & Management Jahrbuch 2000
4.14, 4.15, 4.17, 4.17.1	Cetetherm Wärmetauscher GmbH, Düsseldorf
4.20	Deutsche Hoval GmbH, Rottenburg Neckar
4.21, 4.22, 4.25, 4.26, 4.28	Viessmann Werke, Allendorf
4.30, 4.31, 4.32	August Brötje GmbH, Rastede
4.43	Solar-Energie-Technik, Altlußheim
4.45.1	Thermo-Solar, Regensburg
4.45.2	Reflex Winkelmann + Pannhoff GmbH, Ahlen
4.47	Viessmann Werke, Allendorf
4.55.1	Studer + Partner AG, Zurich, CH
4.59	Fotoatelier Bernhard GmbH, Hardheim
4.61	after: ASUE, Blockheizkraft-werke
4.68.2	F. W. Oventrop KG, Olsberg
4.68.3	Reflex Winkelmann + Pannhoff GmbH, Ahlen
4.69, 4.70, 4.71, 4.72	after: Grünzweig + Hartmann AG, Ladenburg
4.78	Baufa-Werke GmbH, Menden
4.79, 4.81, 4.83, 4.84	Kermi GmbH, Plattling
4.82	August Brötje GmbH, Rastede
4.85, 4.87.1	Baufa-Werke GmbH, Menden
4.87.2	Purmo AG, Hanover
4.92	after: EMCO, Lingen
4.95, 4.95.1	Purmo AG, Hanover
4.95.2, 4.96.1, 4.98.2	D.F. Liedelt Velta, Norderstedt
4.96.2, 4.99	Polytherm GmbH, Ochtrup
4.97.2	Cuprotherm, Ulm
4.100.1, 4.100.2	Schütz eht, Bad Honnef
4.104	Zent-Frenger, Heppenheim
4.106.1, 4.106.2	after: Gartner & Co, Gundelfingen
4.107.1 (1), 4.107.1 (2), 4.107.3	Theodor Heimeier Metallwerke GmbH, Erwitte
4.107.1 (3), 4.107.1 (4), 4.107.4	F. W. Oventrop KG, Olsberg
4.108	Kermi GmbH, Plattling
4.115, 4.116	Grundfos GmbH, Wahlstedt
5.3, 5.16	Wilo GmbH, Dortmund
5.23.1	August Brötje GmbH, Rastede
5.23.2	Viessmann Werke, Allendorf
5.24.1	Cetetherm Wärmetauscher GmbH, Düsseldorf
5.38.2, 5.41.1, 5.41.2, 5.41.3, 5.42	after: Geberit International GmbH, Rappersswill, CH
5.50.1, 5.50.2	KSB AG, Pegnitz
5.58.2	Mallbeton, Pfohren
5.66.1, 5.66.9 – 12	Freidrich Grohe AG, Hemer
5.66.2 – 5.66.8	Hansa Metallwerke AG, Stuttgart
5.71, 5.72	Krantz TKT, Bergisch-Gladbach
5.75.3	DAL-Georg Rost & Söhne, Porta Westfalica
6.10.1, 6.10.2	ROM Rud. Otto Meyer GmbH, Hamburg
6.12.1, 6.12.2, 6.13	von Gerkan, Marg & Partner, Hamburg
6.25 – 28	Kohn Pedersen Fox Associates New York, London
6.53, 6.54.1, 6.54.2, 6.57.1	Delbag, Berlin
6.58, 6.59	Gebr. Trox GmbH, Neukirchen Vluyn
6.61	Kessler Tech GmbH, Gießen
6.63	ABB, Butzbach
6.72, 6.75.2	Axair Deutschland GmbH, Garching near Munich
6.77	Kessler Tech GmbH, Gießen
6.82, 6.84.1, 6.90 D, 6.93 E,F	LTG Aktiengesellschaft, Stuttgart
6.84.2, 6.85, 6.87, 6.90 A–C, E, 6.93 A-D, 6.98.2 A	Gebr. Trox GmbH, Neukirchen Vluyn
6.98.2 B	Krantz TKT, Bergisch-Gladbach
6.98.2 C	Heinrich Nickel F+E Komponenten, Betzdorf
6.100 B, D, E, 6.109.3	LTG Aktiengesellschaft, Stuttgart
6.102	ABB, Butzbach
6.103, 6.104	Menerga Apparatebau GmbH, Mülheim a. d. Ruhr
6.106, 6.107	Liebert-Hiross, Kirchheim
6.105.2, 6.109.2	Gebr. Trox GmbH, Neukirchen Vluyn
6.114, 6.116, 6.118, 6.120	IKL, Gießen
6.121.1, 6.121.2, 6.121.3	Kessler + Luch GmbH, Gießen
6.137, 6.138.2 B, 6.139 C	Krantz TKT, Bergisch-Gladbach
6.138.1, 6.139 B	Zent-Frenger, Heppenheim
6.138.2 A	Götz GmbH, Würzburg
6.139 A, 6.142.2	Stulz, Hamburg
6.139 D	Clestra Hausermann GmbH, Dreieich
6.142.1	Kessler + Luch GmbH, Gießen
6.142.2	LTG Aktiengesellschaft, Stuttgart
6.143	PS Company, Tokyo
6.146.1	Philipp Holzmann AG, Frankfurt a. M.
6.157.1, 6.157.2	M+W Zander Facility Engineering GmbH, Stuttgart/Nürnberg
7.19, 7.23, 7.24, 7.26.1, 7.28.2	Trane, Gilching
7.22, 7.26.2, 7.27, 7.28.1, 7.28.3	YORK International GmbH, Mannheim
7.29, 7.29.1	after: ILK Dresden
7.30	ILK Dresden
7.31, 7.32	after: W.J.Eugster et.al.
7.34.1 – 5	after: K.-J. Albers
7.35.1	ROM Rud. Otto Meyer GmbH, Hamburg
7.36.1, 7.36.2, 7.37	after: ROM Rud. Otto Meyer GmbH, Hamburg
7.46	after: ILK Dresden
7.47	Carrier, Munich
7.48, 7.54	Trane, Gilching
7.50, 7.52.1	E. W. Gohl GmbH, Singen
7.52.2, 7.53.2, 7.53.3, 7.53.4, 7.63.2	Baltimore Aircoil International N.V., Heist-op-ten-berg, B
7.55 below	Yäggi – Güntner, Bern
8.7	after: Deutsche Verbundgesellschaft, Heidelberg
8.15, 8.32, 8.33, 8.34, 8.40, 8.41, 8.45, 8.46, 8.47	after: RWE Energie Bau-Handbuch
8.23	GEAFOL
8.35	after: RICO Rieth & Co, Kirchheim-Teck
8.49, 8.51	after: Siemens, Munich
8.52	Busch-Jaeger Elektro-GmbH, Lüdenscheid
8.68	Waldner Laboreinrichtungen GmbH, Wangen i.A.
8.76 B, F, G, 8.77 A	Busch-Jaeger Elektro-GmbH, Lüdenscheid
8.76 C	Siemens AG, Munich
8.79	Piller, Osterode
8.82, 8.86	AGREBA, Dreieich
8.85	mtu, Friedrichshafen
8.93	Schüco International, Bielefeld
8.94, 8.95.1, 8.95.2	after: Schüco International, Bielefeld
9.15, 9.24, 9.25	Regent, Basel, CH
9.16	ERCO Leuchten GmbH, Lüdenscheid
9.29, 9.30	Zumtobel Staff, Dornbirn, A
9.32	ERCO Leuchten GmbH, Lüdenscheid

9.33	Adelmann Lichtsysteme, Munich
9.34	Spectral Ges. für Lichttechnik GmbH, Freiburg
9.35	BEGA Gantenbrink, Menden
9.37.2	WILA Leuchten, Iserlohn
9.39	Franz Sill GmbH, Berlin
9.43, 9.44, 9.45, 9.46, 9.47, 9.48, 9.49.1	Siemens AG, Munich
9.49.2	ERCO Leuchten GmbH, Lüdenscheid
9.50	Thorn Licht GmbH, Dortmund
9.59	WILA Leuchten, Iserlohn
9.65	Krantz-TKT, Bergisch-Gladbach
9.68.2, 9.69.1	Zumtobel Staff, Dornbirn, A
9.68.3, 9.69.2	Thorn Licht GmbH, Dortmund
9.72	ERCO Leuchten GmbH, Lüdenscheid
9.73, 9.76	Okalux Kapillarglas GmbH, Marktheidenfeld
9.83, 9.98	Franz Sill GmbH, Berlin
10.14, 10.17, 10.18	Okalux Kapillarglas GmbH, Marktheidenfeld
10.22.1, 10.22.2, 10.23	Krülland Sonnenschutz, Neuss
10.24.1, 10.26	Hüppe Form, Oldenburg
10.24.1	Kessler + Luch GmbH, Gießen
11.1.2	Dätwyler Kabel + Systeme, Neufahrn
11.12	Bartec, Bad Mergentheim
11.13.2	JCI Regelungstechnik GmbH, Essen
11.14	Honeywell AG, Offenbach
12.8.1, 12.8.2, 12.10	Schindler Aufzügefabrik GmbH, Berlin
12.11.1, 12.17	Otis GmbH, Berlin
12.11.2	Vestner Aufzüge GmbH, Garching near Munich
12.20, 12.22, 12.23	Swisslog Telelift GmbH, Munich
1.1 – 1.4, 1.6 – 1.11, 3.4, 3.6.1, 3.6.2, 3.7, 4.29, 4.42, 4.46, 4.48, 4.50, 4.51.1, 4.51.2, 4.52, 4.62, 5.5, 6.9.2, 6.11, 6.31, 6.36, 6.37.1, 6.37.2, 6.38.1, 6.38.2, 6.39.1, 6.39.2, 6.40, 6.41.1, 6.41.2, 6.42, 6.43, 6.44, 6.45, 6.46, 6.146.2, 7.33, 7.35, 7.38, 7.42.1, 7.55 above, 7.56, 9.55 below ,10.6	after: The Technology of Ecological Building, W. Schwaiger, A. Alber, Ch. Albrecht
2.2, 2.7, 2.11, 2.12, 4.1 - 4.4, 4.68.1, 7.44	after: Recknagel, Sprenger, Schramek, Oldenbourg Verlag
Other diagrams, technical illustrations	HL-Technik AG, Munich, Riemer Design, Munich
Tables	
2.1, 2.2, 2.5, 4.1, 4.2	after: Recknagel, Sprenger, Schramek, Oldenbourg Verlag
5.11, 5.16, 5.17.1, 5.17.2, 5.18, 5.19.1, 5.19.2	after: Geberit International GmbH, Rapperswill, CH
7.1, 7.2	after: ILK Dresden
7.4	Trane, Gilching
9.7	Regent, Basel, CH
10.1	after: M. Kischkoweit-Lopin anc H. Müller
Cover Illustration	Competition for the new headquarters for the CCTV-Central China Television, Peking Solar panel/solar protection elements Dominique Perrault Architecte, Paris © Perrault Projets/Image: Yves Moreau

Index

Bibliography

Index

Bibliography

Bradshaw, Vaughn;
Kenneth E. Miller:
Building Control Systems,
New York: John Wiley and Sons
1993

Daniels, Klaus:
Low-Tech – Light-Tech –
High-Tech,
Building in the Information Age,
Basel, Boston, Berlin: Birkhäuser
– Publishers for Architecture
1998

Daniels, Klaus:
The Technology of Ecological
Building,
Basel, Boston, Berlin: Birkhäuser
– Publishers for Architecture
1997

Gauzin-Müller, Dominique:
Sustainable Architecture and
Urbanism, Concepts,
Technologies,
Basel, Boston, Berlin: Birkhäuser
– Publishers for Architecture
2002

Kreider, Jan F.; Curtiss, Peter:
Heating and Cooling of
Buildings, Design for Efficiency,
Columbus: Mc Graw Hill
1994

Lechner, Norbert:
Heating, Cooling, Lighting,
Design Methods for Architects,
New York: Wiley and Sons
2000

Stein, Benjamin;
John F. Reynolds:
Mechanical and Electrical
Equipment for Buildings,
New York: John Wiley and Sons
1999

Tao, William K.Y.; Janis,
Richard R.:
Mechanical and Electrical
Systems in Buildings,
Upper Saddle River: Prentice
Hall
2001

Company profiles

AUGUST BRÖTJE GmbH
Postfach 1354
D-26171 Rastede
(Germany)
Phone: +49 (0) 44 02 / 80-0
Fax: +49 (0) 440 2 / 80-5 83
www.broetje.de

August Brötje GmbH:
The heating division, responsible for the marketing of heating appliances (boilers, units, wallhung gas boilers, condensing boilers, burners, storage water tanks, controllers, solar heating systems, radiators) to specialist stockists.

In matters of heating Brötje puts its trust in high tech.

High tech heating, energy saving and environmental protection are the three concepts associated with the August Brötje GmbH.

The company, which was founded in 1919 and has its headquarters in Rastede in the Oldenburg region, can justifiably claim that in the field of heating and environmental engineering it has had a significant share in many important innovations producing optimal energy use and maximum heating luxury.

Brötje is a member of the European BAXI GROUP Ltd., based in Derby, in England. In the quest for more environmentally friendly use of energy resources Brötje is ahead of the field in the heating market. Many Brötje products carry the "Blue Angel" environmental emblem. In fact, the emission levels of the condensing wallhung gas boiler Ecotherm Plus are almost too low to be detected.

Brötje high tech products are marketed through a wide network of wholesalers based throughout the Federal Republic of Germany.

ERCO Leuchten GmbH
Postfach 24 60
58505 Lüdenscheid
Brockhauser Weg 80-82
58507 Lüdenscheid
(Germany)
Phone.: +49 (0) 2351 / 551-0
Fax: +49 (0) 2351 / 551-300
info@erco.com
www.erco.com

ERCO – Company objective: light
ERCO Leuchten GmbH, Lüdenscheid, is one of the leading companies in the European luminaire industry. The family business founded in 1934 now operates world-wide with over 60 subsidiaries, branches and agencies. Working together with internationally renowned designers, lighting engineers and architects, ERCO develops product ranges for all areas of architectural lighting.

In perfect harmony with the corporate philosophy of "Light not luminaires," luminaires are offered as instruments of lighting design. The firm's strategy is based on the use of modern technology for optimising lighting design and lighting control. Design and implementation of lighting projects in museums and in architectonically challenging structures are placed to the fore. ERCO can boast an impressive list of reference projects in this field. These include the Guggenheim Museum in Bilbao, the Grand Louvre in Paris and the Berlin Reichstag Building, all of which are illuminated by ERCO products.

The transformation into the Light Factory: Light not luminaires
ERCO first presented its corporate concept of "Light not luminaires" at the Hanover fair over thirty years ago and the philosophy has been successful ever since. This was the idea with which Managing Director Klaus Jürgen Maack set the company on a new course. Behind this little phrase is the notion of understanding light to be the fourth dimension of architecture: light interprets rooms and helps us to perceive and experience their atmosphere. The first product ranges of the new type appeared at the end of the nineteen-sixties and the lighting branch had to take notice. With spotlights and mounting tracks plus flush-fitting, ceiling-mounted downlights, ERCO conquered the growing market for architectural lighting and the lamp factory was transformed into the Light Factory.

Shaping the future of professional lighting
By making targeted use of visualisation technology and the Internet, ERCO is shaping new standards for architectural workflows. ERCO's website Light Scout at www.erco.com represents a hub for information logistics, where clients and interested parties - ranging from architects, design engineers and fitters through to new building technology service providers such as system integrators – can quickly and conveniently access the information relevant for them. The Light Scout is designed around the three cornerstones: Projects, Products and Guide, whereby numerous links between the three are featured in the manner that characterizes this media.

The market position
Today, ERCO is lighting up museums, universities, shop windows, churches, airports, hotels, chain stores, trade fair stands, administrative buildings and private houses. In 2000, ERCO's approximately 1050 strong world-wide workforce turned over a consolidated sales of EUR 135 Millions. The share of the turnover due to foreign trade, currently at 64 %, is it to be expanded to 75 % in the coming years.

Fläkt Woods Group
Affolternstrasse 40
CH-8050 Zurich
(Switzerland)
Phone: +41/43/288 38 00
Fax: +41/43/288 38 10
info@flaktwoods.com
www.flaktwoods.com

Fläkt Woods was created by bringing together two of the worlds leading players in the air management business, Fläkt and Woods Air Movement. Fläkt brought a wide technology base covering air climate systems and industrial air movement and cooling applications. Woods is the world leader for axial fan technology. As a combined business, the Fläkt Woods Group has the capability to provide its customers with application specific solutions and unrivalled levels of service worldwide.

The Group has operations in over 30 countries and a distributor network in all other major markets. Fläkt Woods offers products and services for building ventilation, infrastructure and industrial processes.

Creative Products
The Group has a unique combination of proven technical expertise and forward thinking, as well as a portfolio of market leading products that are well known for value, quality, reliability, innovation and customer service.

- The Group's comprehensive range of Fans has a global rep utation for creating and meeting demanding specifications for performance, efficiency and safety in the widest range of applications.

- Our Air Handling Units have a worldwide reputation for providing innovative, market leading solutions from standardised designs to customised configurations that suit all key requirements.

- We also produce a range of Air Terminal Devices which lead the world in technical superiority, aesthetic excellence and value.

- Our wide range of Fan Coils, Ducts and Heat Exchangers complete the portfolio.

Creative Solutions
Fläkt Woods offers a wide range of creative air climate solutions for buildings and air movement solutions for industrial and infrastructure applications:

Indoor Air Quality
Hygienic Environments
(such as for hospitals & clean rooms)
Safer Tunnels and Metros
Fans and coils for industrial cooling
Heavy Duty Fans and After Market Services
(for sectors such as Power, Cement, Steel & Water Treatment)

The Group's Creative Vision is to play not only a leading role in our industry's evolution, but to push the boundaries of global air management to the very limits of new materials and scientific endeavour, whilst at the same time delivering the highest levels of value and professionalism to our customers. With demand for improved air quality and safer air management growing, the Fläkt Woods Group is ideally placed to offer the best 'Air Management Systems and Solutions' in the world.

E. W. GOHL GmbH
Im Haselbusch 12
D-78224 Singen
(Germany)
Phone: +49 (0) 77 31/88 06 15
Fax: +49 (0) 77 31/650 51
info@gohl.de
www.gohl.de

E.W.Gohl GmbH was founded in 1933. 150 persons are currently employed. The company has her own manufactory building in Singen.

Over 40 years ago, E.W. Gohl has specialized in development and manufacture of the modern **"blow-thru"** design in cooling towers. The range of our products includes:
- cooling towers for open circuit operation
- evaporative cooler for closed circuit operation
- air-cooled water cooler
- hybrid water cooler
- evaporative condenser
- air-cooled condenser

E.W.Gohl GmbH is one of the leading manufactures of cooling towers in Europe. The firm has an excellent reputation on the market.

In our constant search for improvements in design and performance, E.W.Gohl GmbH has manufactured cooling towers und one or more West-German patents.

The benefits of this research and experience are build into every **GOHL cooling tower**.

We have foreign representatives in Switzerland, Austria, Belgium, Denmark and England.

Friedrich Grohe AG & Co. KG
Postfach 1361
D-58675 Hemer
(Germany)
Phone: +49 (0) 2372 93-0
Fax: +49 (0) 2372 93-1322
info@grohe.com
www.grohe.com

Based in Hemer (Germany), Grohe is a leading supplier of complete water technology solutions. Grohe solutions are specifically designed to enable people to use this valuable resource with care and responsibility. The brand, Grohe Water Technology, positions the company as a single-source supplier of top-quality sanitary products and systems. Targeting both residential and contract applications, Grohe Water Technology combines supreme functionality with path-breaking design.

"Articulation" is the name of the company's new premium range which comprises selected designer faucets, bathroom accessories, shower systems and sanitary systems. All "Articulation" products meet the very highest standards in terms of engineering and design. In addition, the company continues to market its proven ranges of high-quality bathroom faucets, versatile hand showers and shower systems, convenient kitchen faucets as well as sanitary systems which facilitate creative bathroom design.

After having won the German Marketing Prize in 1996, Grohe received the EVA Event Award in 2002 and the 2003 red dot design award of the Design Zentrum Nordrhein-Westfalen.

Grohe is the undisputed market leader in Germany and Europe. Continued further development in the fields of R & D, production, finance, marketing and corporate communication has been instrumental to the fact that Grohe sales have more than doubled to EUR 898 million in the past ten years. About 75 percent of sales are achieved outside Germany. Some 6,000 employees in 140 countries are behind the company's success, explaining the global stature of the Grohe brand.

Grohe has 19 sales companies, 11 branch offices and nine production facilities including five in Germany and one each in Thailand, Portugal, Canada and China. All international Grohe factories work to the same strict standards of quality as their German counterparts.

HL-Technik AG
beratende Ingenieure
Wolfratshauser Str. 54
D-81379 Munich
(Germany)
Phone: +49 (0) 724 06 0
Fax +49 (0) 724 06 139
info@hl-technik.de
www.hl-technik.de

HL-Technik develops techniques with its origins out of the urban structure up to a succesful operation af a building based on environmentally sensible solutions.

Our goals are:

Progress through innovation, close contact to our clients, integrated design solutions and environmentally sound technologies – for the client's benefit.

Our field of activities includes:

- Planning of infrasrtucture
- Thermal physics of buildings
- Building services
- Daylight and artificial lighting
- Service management

The HL-Technik team consists of:

HL-Technik AG Munich, with central design-consulting departments:
- Lighting design
- Information technologies
- Thermal physics

and regional offices in
- Düsseldorf
- Frankfurt
- Hamburg
- Munich

HL-Technik AG, Zurich

HL-Technic
Engineering Consultancy, Dubai

HL-Service Management GmbH, Siegen, Frankfurt a. Main

interferenz® LICHTsysteme

Interferenz LICHT*systeme* GmbH
Lenenweg 27,
D-47918 Tönisvorst
(Germany)
Postfach 1264, 47908 Tönisvorst
Phone: +49 (0) 21 51 / 979 89-0
Fax: +49 (0) 21 51 / 979 89-99
info@interferenz.de
www.interferenz.de

Throughout the ages LIGHT has always been an element of architectural design.

The firm of INTERFERENZ has made it its business to use daylight and artificial light in light architecture that is out of the ordinary.

The "INTERFERENZ" production and sales programme has grown up over a period of 30 years, in often close cooperation with well-known architects, specialist planners and engineers.

The product range of this specialised lighting manufacturer includes lighting systems both for general purposes and for public halls, foyers, hotels, concert halls and floodlighting. The equipment employed includes elegant uplighters, mirror deflectors, modern chandeliers and sophisticated secondary and downlighting technology.

The utilisation of daylight as a renewable energy source is achieved by, for example, the employment of heliostats.

Uniquely attractive design and construction, value for money, flexibility and prompt delivery are the hallmarks of the company.

"INTERFERENZ" sees itself as a source of new ideas and a standard-bearer for innovation in the lighting industry. It aims to give an even higher profile to lighting and light systems with the registered brand name of "INTERFERENZ", renowned throughout Europe.

OKALUX GmbH
Am Jöspersthecklein 1
97828 Marktheidenfeld
(Germany)
Phone: +49 (0) 9391 / 9000
Fax: +49 (0) 9391 / 9001 00
info@okalux.de
www.okalux.de

In the beginning....

In the early 1960's, the long-established family-owned textile company Heinrich Otto Company began investigating the possibility of producing hollow man-made fibres for improved permeability and wearability of garments. By the mid-1960's, coinciding with their 150th anniversary, the company's research division had developed a pilot production plant. However, euphoric dreams were soon quashed – the hollow fibre that could be produced could not subsequently be processed to yarn, hosiery or woven fabrics due to insurmountable problems caused by knotting.

The technology to produce a new product existed, but there seemed to be no clear market or application for such a product. Further development was undertaken into possible applications including a new form of artificial turf, technical micro filters to isolate plankton and wall linings. Finally, a procedure was developed which enabled the hollow fibres to be made into a slab where the capillaries were aligned parallel to each other and perpendicular to the faces of the slab; the light diffusing and insulating capillary panel.

Initially, it was intended that this capillary slab would be used in conjunction with clear plastic facing materials. However, whilst the European insulating glass industry was still in its infancy at that time, it was beginning to develop at the expense of single glazing, and it seemed logical to combine the capillary slab with glass. Entry into the glass industry was now inevitable.

The textile spinning division of Heinrich Otto KG, Fils-Textil GmbH, had established Otto-Kapillar as a new division to undertake the hollow fibre development project. Otto-Kapillar became a legally independent company in 1971, taking its name from the initial letters of the original division and combining them with the Latin term for light. OKALUX was born.

In these early days, OKALUX light diffusing and insulating glass was mainly used for roof glazing in industrial buildings and sports halls to permit evenly diffused, glare-free natural illumination rather than for heat insulation.
However, the energy crisis of the early 1970's changed this radically. OKALUX presented their innovative capillary honeycomb glazing products for the first time at the first ever environmental trade fair, "Reinhaltung der Luft" in Dusseldorf. An overwhelming response was received, and this led to an ever-increasing demand for OKALUX glass, which has resulted in continuous growth.

The product range was steadily growing too, and the current range of innovative architectural glass products is outlined on the following pages. More detailed information is included in literature for each specific product, although the nature of our business is such that every product is produced to the individual customer's specifications.

This special position within the insulating glass industry resulted in many export opportunities from an early stage, and a remarkable list of reference projects exists, not only throughout Europe, but also in the Middle East, Far East, United States and Australasia.

Continuous research and development

As a result of its origins, OKALUX has maintained continuous investment in product research and development as well as in the application of its products. This has proved invaluable over the years, and is proved by the continued use of wired glass for roof glazing by OKALUX.

The use of insulating glass for roof glazing was rather unusual when OKALUX was formed, with most applications being simply single glazed with wired glass.

Like many other insulating glass products, OKALUX insulating glass was produced using rough cast glass for the external leaf and wired glass for the internal leaf to ensure that the glass would remain in position in the event of breakage, minimising the risk of injury to occupants from falling glass.

In these days, insulating glass units were installed in the same way as single glazing, but many designers and installers did not realise that the thermal stresses applied to the internal wired glass leaf were significantly higher with insulating glass than with single glass.

Extensive areas of roof glazing suffered dramatic failures, with some cases involving breakage of every unit.

OKALUX also suffered from this industry-wide problem, but following intense technical analysis of the environmental conditions and installation detailing, OKALUX were able to develop a detailed understanding of the nature of the problem. Together with specialist installation contractors, OKALUX have developed a range of recommended glazing methods and details which reduce the risk of thermal stress fracture.

Today, OKALUX are one of only a few insulating glass producers prepared to offer a standard warranty which includes thermal stress fracture, subject to approval of construction drawings.

This is just one example of the persistence and determination exhibited by OKALUX in solving problems. In this way, OKALUX actively seek to ensure that their products are developed and produced to the complete satisfaction of the design team, specialist installation contractor, general contractor and client or end user.

The active pursuit of this philosophy, backed with an extensive range of specialist apparatus and ongoing liaison with other specialist research institutions and engineering organisations working within the architectural glass industry leads to further improvements and the development of new products and techniques.

OKALUX can provide indicative heating and cooling load calculations at scheme design/development stage, as well as recommended glass thicknesses for specific load/dimension conditions.

Experience, expertise and quality assurance

OKALUX has more than 35 years experience in the manufacture and application of specialist insulating glass products in a variety of locations throughout the world, but recognise that our expertise is not in the field of glass manufacture or processing.

OKALUX does not manufacture glass, nor do we process glass by toughening, laminating, bending, drilling, coating etc. Our expertise lies in sourcing such processed products from the most appropriate specialist supplier, then incorporating these products into an insulating glass unit. As an independent company, we have extensive contacts and ongoing relationships with all of the main glass processors throughout Europe, and accordingly, are in a position to satisfy even the most demanding specification requirements on competitive terms.

Over the last few years, the demand for more complex glass products within ever-tighter tolerances and specifications has increased dramatically, with the requirement for corresponding levels of quality control to be implemented.

Okalux has an externally monitored quality control programme in operation which is in accordance with industry standards. The specialised nature of much of our production enables further specific project-related internal quality control programmes to be implemented as required.

OKALUX recognise that with more intensely processed glass products now being used, careful organisation of procurement and co-ordination of production, transportation and delivery of materials is essential to ensure product quality and avoid unnecessary delay.

Today, OKALUX GmbH remains an independent and internationally successful company owned by the Otto family and run by a management and staff of around 100 people with an annual turnover in excess of 30 million DM.

F.W.Oventrop GmbH & Co. KG
Paul-Oventrop-Straße 1
D-59939 Olsberg
(Germany)
Phone: +49 (0) 29 62 / 82-0
Fax: +49 (0) 29 62 / 82-400
mail@oventrop.de
www.oventrop.de

– Valves, Controls and Systems for the Building Services Industry –

F.W. Oventrop GmbH & Co. KG, Olsberg, is one of the leading manufacturers of valves, controls and systems for the Building Services Industry. In the areas of heating, sanitary and industry the company offers a comprehensive range with over 4000 products, which are marketed in the majority of the world's countries. The company is an expert partner for stockists, contractors, consultants and industry. Its aim is to offer its partners a long-term beneficial relationship.

Factories

Oventrop was founded in 1851. Ever since, the firm has been independent and family owned. The main works of the company are located in Olsberg. There is another factory in Brilon. To take care of foreign sales, Oventrop has more than seven subsidiaries in Europe and agencies in all other important countries.
Oventrop has a workforce of highly qualified and committed employees, including a high proportion of trainees.
The company is an important factor in the economies of the regions where it is located. Job security and environmental protection are given high priority.

Innovation, Quality and Design

It is the task of an innovative enterprise to recognise the demands of the future. Knowledge gained leads to new and improved products and services.
Existing markets are developed and new ones opened up. Thinking has undergone a transformation: Away from isolated solutions and towards the bigger picture. Oventrop offers systems in which individual components form a meaningful part of a harmonious whole.
For Oventrop quality is of prime importance. It goes without saying that quality management is certified according to DIN – EN – ISO 9001.
Good design is also extremely important for Oventrop. Already the firm has received many national and international awards. Design and construction are carried on in house. Furthermore Oventrop cooperates with leading outside designers.

Software, Training, Internet

Oventrop supports its partners with comprehensive IT programmes, technical literature and catalogues. Intensive training, instruction and seminars are held in Olsberg and locally. Worldwide contacts are made through the Internet, where the firm has a very lively and successful presence. Information is provided in twelve different languages.

Guarantees and Certification

Since 1998 Oventrop has had a liability agreement with the ZVSHK and a warranty agreement with the BHKS. In this way the company is able to support individual contractors in matters of contractual agreement.

Handwerkermarke (Master Craftsman Brand)

Since 2000 Oventrop has joined eleven top brand manufacturers of heating and sanitary equipment granted the right to use the ZVSHK title of *Handwerkermarke*.

ZUMTOBEL STAFF
DAS LICHT ®

Zumtobel Staff GmbH
Head office
Schweizer Str. 30
Postfach 72
A-6851 Dornbirn (Austria)
Phone: +43 (0) 5572 / 390-0
Fax: +43 (0) 5572 / 22826
info@zumtobelstaff.co.at
www.zumtobelstaff.com

ZUMTOBEL STAFF:
"Creating worlds of experience using light ..."

Zumtobel Staff: one of the leading suppliers of lighting solutions, with five production plants on three continents, with customer-friendly sales organisations and consulting centres in more than 60 countries worldwide – in brief: one of the few global players operating in all areas of application of professional interior lighting, be it office and education, industry and engineering, arts and culture, presentation and retail, hotels and catering, leisure and sports, health and care, transit areas and car parks, or orientation and safety. During the 2000/01 financial year, the company realised total sales of EUR 485.9 million. The export share amounted to 82.7 percent (Austrian location: 77.4 percent), exports to Eastern Europe were 2.1 percent (Austrian location: 3.4 percent).

Lighting solution: comprises a holistic, dynamic concept that does not focus on luminaires exclusively but centres on individuals and their needs, that includes space and architecture, seeking to achieve perfection in terms of design, constantly striving for technological innovation and – in collaboration with the avant-garde of European designers and architects – awareness of form, in order to enable customers to experience light in all its diversity; a concept that wants to create worlds of experience using light. Lighting solution – that involves the areas of research and development, manufacturing and distribution, and customer advice and service. The complete package from one single supplier, conclusive in itself, perfectly coordinated.

Research and development: always up to date, ahead of the time, no matter whether it is about LED technology for fascinating interplays of light, the innovative micro-prism optic for glare-free task area lighting, the lighting chamber principle, Balanced Lighting for mellow, virtually natural light, or the revolutionary CLIX system which makes installation safe, easy and extremely fast. A maximum of ingenuity and creativity, profound technological expertise, subtle sophistication and inventiveness: Zumtobel Staff is perfectly ready for finding individual, pioneering solutions to challenges in terms of technology and design – our solutions are quick, reliable, close to the customer, tailor-made, high-tech, and they also go beyond the product catalogue, if necessary.

Manufacture: ultra-modern production plants with unique special tools and equipment especially developed and built by the company's own tool-making division, in order to optimise the production of luminaires and accessory components. For series production as well as non-standard manufacture of special products.

Sales and advice: decentralised organisation close to the customer, including Light Forums, Lighting Centres and Consulting Centres. Professional, friendly staff - lighting solution consultants rather than salespeople – are available for information and project consulting also on site. It is possible to test lighting experiences. At the Light Forums and Lighting Centres, against a background of attractive architectural settings by Hans Hollein, Bolles Wilson, Sauerbruch & Hutton, to name a few, and varied room situations, it is possible to experience what lighting solutions look like and which effects they create, and how Active Light reproduces the transition of light according to the time of day and season, with a specific purpose in mind. Thanks to innovative VR technology, ideas and designs can be visualised three-dimensionally and close to reality, already long before they are implemented – in their original size, with correct photometric data, and with the option of control by the viewer.

Service: comprehensive services for customers in all matters concerning light, lighting technology and lighting design. Up-to-date knowledge is provided in seminars and workshops; the light dictionary, the "light manual for practitioners" and technical lighting software indirectly provide practical assistance for design and implementation. The Sales Division provides direct design support. And completion of a project does by no means imply that customer service is finished: the maintenance, servicing and configuration of computer-backed Luxmate lighting management systems, for instance, is also part of the range of services offered by Zumtobel Staff.

List of advertisers

A 1 Axima Engineering
Contracting GmbH
Heßbrühlstr. 51
D-70565 Stuttgart/Germany

Axima
Management Services AG
Zürcherstrasse 12
CH-8401 Winterthur/
Switzerland

A 2 Durlum-Leuchten GmbH
An der Wiese 5
D-79650 Schopfheim/Germany

A 3 Ferdinand Schad KG
D-78600 Kolbingen/Germany

A 4 AL-KO THERM GMBH
Maschinenfabrik
Hauptstraße 248-250
D-89343 Jettingen-Scheppach/
Germany

A 4 KOREX Apparatebau GmbH
Mühlfelder Straße 58
D-82211 Herrsching/Ammersee/
Germany

A 5 Zumtobel Staff GmbH
Head office
Schweizer Str. 30
Postfach 72
A-6851 Dornbirn/Austria

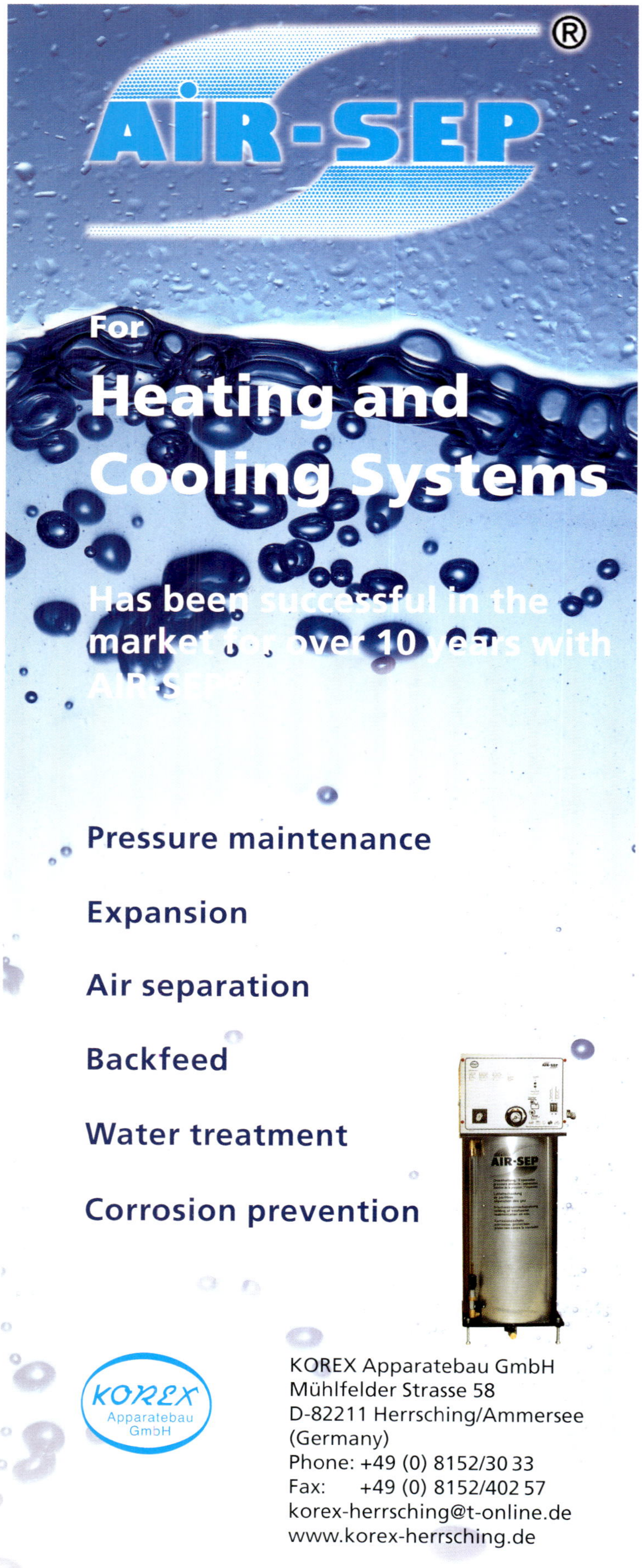
AIR-SEP®
For
Heating and
Cooling Systems
Has been successful in the market for over 10 years with
AIR-SEP
Pressure maintenance
Expansion
Air separation
Backfeed
Water treatment
Corrosion prevention
KOREX Apparatebau GmbH
KOREX Apparatebau GmbH
Mühlfelder Strasse 58
D-82211 Herrsching/Ammersee
(Germany)
Phone: +49 (0) 8152/30 33
Fax: +49 (0) 8152/402 57
korex-herrsching@t-online.de
www.korex-herrsching.de

AL-KO
Lufttechnik
Professionelle Luftbehandlung
RLT-Geräte für die Gebäudeklimatisierung
Klimatechnik für Komfort- und Spezialapplikationen
AT4 RLT-Geräte
Standard · RAL · Hygiene (VDI 6022, DIN 1946 T3) · integrierte Kälte · wetterfest
AT2 Flachgeräte
für den Einbau in Zwischendecken
AT4 Klimaschränke
für die Klimatisierung thermisch hochbelasteter Räume
AT58 RLT-Groß- und Sondergeräte
für Volumenströme ab 50.000 m³/h
www.al-ko.de

Translation into English:
Elizabeth Schwaiger, Toronto

With thanks to Ralph E. Hammann, Dr. Eng. Sc., Associate Professor, College of Architecture, Planning, and Landscape Architecture, University of Arizona, Tucson, USA

Managing editor of the original German edition:
Dr. Andreas Colli, Munich

Design and typesetting:
Riemer Design, Munich

The original German edition "Klaus Daniels, Gebäudetechnik, Ein Leitfaden für Architekten und Ingenieure" has been published by:
Oldenbourg Industrieverlag GmbH, Munich

A CIP catalogue record for this book is available from the Library of Congress, Washington D.C., USA

Bibliographic information published by
Die Deutsche Bibliothek
Die Deutsche Bibliothek lists this publication in the Deutsche Nationalbibliografie;
Detailed bibliographic data is available on the Internet at http://dnb.ddb.de.

Printed on acid-free paper produced from chlorine-free pulp. TCF ∞

Printed in Germany
ISBN 3-7643-6723-7

9 8 7 6 5 4 3 2 1